AF788006

Ecological Studies

Analysis and Synthesis

Edited by

W. D. Billings, Durham (USA) F. Golley, Athens (USA)

O. L. Lange, Würzburg (FRG) J. S. Olson, Oak Ridge (USA)

H. Remmert, Marburg (FRG)

Volume 43

Mediterranean-Type Ecosystems

The Role of Nutrients

Edited by

F.J. Kruger D.T. Mitchell J.U.M. Jarvis

With 143 Figures

Springer-Verlag
Berlin Heidelberg New York Tokyo 1983

Frederick J. Kruger
South African Forestry Research Institute
Department of Environment Affairs and Fisheries
Pretoria 0001, South Africa

Professor Derek T. Mitchell
Department of Botany
University of Cape Town
Rondebosch, South Africa

Professor Jennifer U.M. Jarvis
Department of Zoology
University of Cape Town
Rondebosch, South Africa

ISBN-13: 978-3-642-68937-6 e-ISBN-13: 978-3-642-68935-2
DOI: 10.1007/978-3-642-68935-2

Library of Congress Cataloging in Publication Data. Main entry under title: Mediterranean-type ecosystems. (Ecological studies; v. 43). Includes part of the proceedings of a symposium held at the University of Stellenbosch, South Africa, Sept. 22–23, 1980. Bibliography: p. Includes indexes. 1. Biogeochemical cycles – Congresses. 2. Mediterranean climate – Congresses. 3. Ecology – Congresses. I. Kruger, F. J. (Frederick J.), 1944– . II. Mitchell, D. T. (Derek T.), 1942– . III. Jarvis, J. U. M. (Jennifer U. M.), 1939 –. IV. Series. QH344.M43. 1983. 574.5′222. 83-363.

This work is subject to copyright. All rights are reserved, whether the whole or part of the material is concerned, specifically those of translation, reprinting, re-use of illustrations, broadcasting, reproduction by photocopying machine or similar means, and storage in data banks.

Under § 54 of the German Copyright Law where copies are made for other than private use, a fee is payable to "Verwertungsgesellschaft Wort" Munich.

© by Springer-Verlag Berlin · Heidelberg 1983
Softcover reprint of the hardcover 1st edition 1983

The use of registered names, trademarks, etc. in this publication does not imply, even in the absence of a specific statement, that such names are exempt from the relevant protective laws and regulations and therefore free for general use.

Printing and binding: Brühlsche Universitätsdruckerei, Giessen
2131/3130–543210

Preface

The theory of ecological convergence underlies the biogeographers' maps of world biome-types. It also determines the degree to which ecological principles, derived from research on particular populations, communities or ecosystems, are generally valid, and hence also to what extent resource management principles are general. To quote Di Castri and Mooney (1973): "In effect, in order to assess the transfer of technology, it is essential to know to what extent information acquired from studying one particular ecosystem is applicable to another ecosystem of the same type but situated in a different location."

The five relatively small, isolated, mediterranean-climate zones of the earth, each with its distinct fauna and flora, have provided the ideal testing grounds for this theory. A heritage of precisely focused ecosystems research has resulted, beginning with the international comparative analyses conducted by Specht (1969a, b) but with antecedents in earlier studies in South Australia (Specht and Rayson 1957, Specht 1973). Cody and Mooney (1978) reviewed the information available at the time for the four zones excepting Australia and concluded that the arrays of strategy-types to be found among the different biotas were so similar that they could be explained only in terms of the convergence hypothesis; nevertheless, evident differences in community organization and dynamics, especially phenology, required closer study of resource availability and resource-use patterns to better explain relations between form and function overall, and to assess the degree of convergence at higher levels of organization than the population.

Furthermore, the biogeographers' concept of a mediterranean-climate biome-type, established by Grisebach (1872) and Schimper (1903), has been challenged by Specht (1979) and found untenable on the grounds that major sclerophyllous shrublands, the Australian heathlands and South African fynbos, are determined by nutrients, not by climate. Both of these vegetation formations occur within mediterranean-climate zones but extend well beyond the boundaries of the mediterranean-type climates as normally defined and in Australia occur extensively within the tropics; their distribution correlates rather with soils exceptionally poor in nutrients. Specht proposed that the floras of this vegetation had ancient origins and evolved in edaphically determined oligotrophic habitats in the humid tropical to sub-tropical environments of the Cretaceous and early Tertiary: the physiological drought of the intermittently waterlogged soils preadapted these plants to the seasonally arid climates that arose in the Tertiary, and the later inception of mediterranean climates.

This book incorporates part of the proceedings of a conference called to assess the role of nutrients in species and ecosystem convergence in mediterranean-climate ecosystems. Specifically, the first aim was to draw the contrast between the nutrient-poor shrublands of Australia and South Africa and the physiognomically similar vegetations of richer soils. This comparison was to indicate how the environmental contrasts are expressed in the adaptive responses of the biota as well as in key ecosystem processes. The second question addressed, had two parts: why is it that the biotas of these regions, especially the plants, are unusually rich in species, and how is this richness maintained in resource-poor environments?

The symposium was held at the University of Stellenbosch, South Africa, from 22 to 23 September 1980 and was followed by a four-day workshop at Hermanus. It was the third in a series of international conferences, beginning with the symposium at the Universidad Austral de Chile, Valdivia, in March 1971 (see Di Castri and Mooney 1973), and followed some time later by the meeting at Stanford University in 1977 (Mooney and Conrad 1977). This series began with a critical overview of information relevant to studies of convergence, with the emphasis essentially on the Mediterranean Basin and the New World. The Stanford meeting focused on applied ecological questions, principally the role of fire in resource management. The Stellenbosch meeting saw a return to the question of convergence, though in a wider context. These trends reflect the vigorous and encouraging interplay between fundamental and applied research that has arisen in these zones.

These conferences have to a large degree been part of a continuous process of development of integrated ecological research programmes involving comparative studies of mediterranean-type ecosystems recently reviewed by Mooney (1981a, b).

The papers from the Stellenbosch symposium included here are of two kinds: invited review papers, and contributed papers that met the requirements of the editorial panel; all papers considered for publication were subject to peer review and revised accordingly, some extensively. The workshop proceedings are to be published separately (Day 1983). Although at the outset the intention was to provide a review of all major topics relevant to the conference theme, this proved too ambitious. We found that certain topics were easily covered, and that Australian heathlands and South African fynbos in particular seem inevitably to get the greater attention, perhaps justly so. We therefore do not claim that this book is comprehensive; nevertheless it introduces a new dimension and balance to the subject and raises many novel questions. The international studies of vegetation function in the mediterranean-type ecosystems of Chile and California (Mooney 1977, Miller 1981) provided the first truly synthetic comparative accounts of ecosystem function. The Stellenbosch meeting proved a major step toward further synthesis for several continents and these proceedings will hopefully help to maintain the momentum for intercontinental comparative studies.

The text of 27 chapters is organized in six sections. The first, on the evolution and characters of mediterranean-type ecosystems, proceeds from a review of the evolutionary history of the systems, through major features of vegetation-environment patterns, to the question of detecting patterns in the geographic distribution of plant forms. The second, on plant form and function, is aimed at assessing our understanding of how environment is functionally linked to plant forms. In the third, nutrient cycling is considered at the ecosystem level, while plant nutrition is considered in the fourth. The fifth section contains eight chapters that deal with community pattern and diversity, specifically how communities are structured in these resource-poor environments, and how their dynamics contribute to maintenance of richness. The final section, on plant-animal interactions, has two papers that concentrate on herbivory.

The symposium and workshop from which this volume proceeds was convened by the Fynbos Biome Project, one of several large interdisciplinary cooperative research projects of the National Programme for Environmental Sciences. The support of the Council for Scientific and Industrial Research and the Department of Environment Affairs and Fisheries is gratefully acknowledged.

Credit is owing to many for assistance in seeing this volume to the press. We would especially like to acknowledge Suzanne Winkelman, then of the CSIR, for assistance in editing copy during the early and crucial phases of the work. Minja Martin contributed unstintingly in the latter stages. Margaret Orton undertook the entire typography. For the support of these and other members of the section for Cooperative Scientific Programmes of the Council for Scientific and Industrial Research we are extremely grateful.

Pretoria and Cape Town, April 1983

FRED KRUGER
DEREK MITCHELL
JENNY JARVIS

References

Cody ML, Mooney HA (1978) Convergence versus nonconvergence in mediterranean-climate ecosystems. Annual Review of Ecology and Systematics 9:265–321

Day J (ed) (1983) Mineral nutrients in mediterranean ecosystems. Results of a workshop in Hermanus, South Africa, September 1980 (in press)

Di Castri F, Mooney HA (eds) (1973) Mediterranean-type ecosystems: origin and structure. Springer, Berlin Heidelberg New York, 405 pp

Grisebach AHR (1872) Die Vegetation der Erde nach ihrer klimatischen Anordnung. W Engelmann, Leipzig

Miller PC (ed) (1981) Resource use by chaparral and matorral: a comparison of vegetation function in two mediterranean-type ecosystems. Springer, Berlin Heidelberg New York, 455 pp

Mooney HA (ed) (1977) Convergent evolution in Chile and California mediterranean climate ecosystems. Dowden, Hutchinson and Ross, Stroudsberg, Pennsylvania, 224 pp

Mooney HA (1982) Mediterranean-type ecosystems – research progress and oppurtunities. South African Journal of Science 78:5–7

Mooney HA, Conrad CE (technical coordinators) (1977) Proceedings of the symposium on the environmental consequences of fire and fuel management in mediterranean ecosystems. USDA Forest Service General Technical Report WO-3, 498 pp

Schimper AFW (1903) Plant-geography upon a physiological basis. Clarendon Press, Oxford, 824 pp

Specht RL (1969a) A comparison of sclerophyllous vegetation characteristic of mediterranean-type climates in France, California and southern Australia. I. Structure, morphology and succession. Australian Journal of Botany 17:277–292

Specht RL (1969b) A comparison of the sclerophyllous vegetation characteristic of mediterranean-type climates in France, California, and southern Australia. II. Dry matter, energy, and nutrient accumulation. Australian Journal of Botany 17:293–308

Specht RL (1973) Structure and functional response of ecosystems in the mediterrranean climate of Australia. In: Di Castri F, Mooney HA (eds) Mediterranean-type ecosystems: origin and structure. Springer-Verlag, Berlin, pp 113–120
Specht RL (1979) Heathlands and related shrublands of the world. In: Specht RL (ed) Ecosystems of the world, vol 9B. Heathlands and related shrublands. Part A: Descriptive studies. Elsevier, Amsterdam, pp 1–18
Specht RL, Rayson P (1957) Dark Island heath (Ninety-Mile Plain, South Australia). I. Definition of the ecosystem. Australian Journal of Botany 5:52–85

Contents

Section IV. Plant Nutrition

Section V. Community Patterns and Diversity

Section VI. Plant-Animal Interactions

Contributors

BEARD, JOHN S.	6 Fraser Road, Applecross, Western Australia 6153, Australia
BOND, WILLIAM	Saasveld Forestry Research Station, Department of Environment Affairs, Private Bag X6531, George 6530, South Africa
CATLING, PETER C.	CSIRO Division of Wildlife Research, P.O. Box 84, Lyneham, ACT 2602, Australia
CODY, MARTIN L.	Department of Biology, University of California, Los Angeles, CA 90024, USA
CROWE, TIM M.	Percy FitzPatrick Institute, University of Cape Town, Private Bag, Rondebosch 7700, South Africa
DEACON, HILARY J.	Department of Archaeology, University of Stellenbosch, Stellenbosch 7600, South Africa
ETCHEGARAY, JULIA	Instituto de Ciencias Biologicas, Pontificia Universidad Catolica de Chile, Casilla 114-D, Santiago, Chile
FOX, BARRY J.	School of Zoology, University of New South Wales, P.O. Box 1, Kensington 2033, Australia
FUENTES, EDUARDO R.	Instituto de Ciencias Biologicas, Pontificia Universidad Catolica de Chile, Casilla 114-D, Santiago, Chile
GROVES, RICHARD H.	CSIRO Division of Plant Industry, P.O. Box 1600, Canberra City, ACT 2601, Australia
GULMON, SHERRY L.	Department of Biological Sciences, Stanford University, Stanford, CA 94305, USA
HASTINGS, STEVEN J.	Systems Ecology Research Group, San Diego State University, San Diego, CA 92182, USA
KRUGER, FREDERICK J.	South African Forestry Research Institute, P.O. Box 727, Pretoria 0001
KUMMEROW, JOCHEN	Department of Botany, College of Sciences, San Diego State University, San Diego, CA 92182, USA

LAMONT, BYRON B. School of Biology, Western Australian Institute of Technology, Kent Street, Bentley, Western Australia 6102, Australia

MILLER, PHILIP C. Department of Biology, College of Sciences, San Diego State University, San Diego, CA 92182, USA

MITCHELL, DEREK T. Department of Botany, University of Cape Town, Private Bag, Rondebosch 7700, South Africa

MOLL, EUGENE J. Department of Botany, University of Cape Town, Rondebosch 7700, South Africa

MOONEY, HAROLD A. Department of Biological Sciences, Stanford University, Stanford, CA 94305, USA

MORROW, PATRICE A. Department of Ecology and Behavioral Biology, University of Minnesota, 318 Church Street SE, Minneapolis, MN 55455, USA

NEWSOME, ALAN E. CSIRO Division of Wildlife Research, P.O. Box 84, Lyneham, ACT 2602, Australia

OECHEL, WALTER C. Systems Ecology Research Group, San Diego State University, San Diego, CA 92182, USA

ORSHAN, GIDEON Department of Botany, The Hebrew University of Jerusalem, Jerusalem, Israel

PRESSINGER, FRANCES Botanical Research Institute, c/o Department of Botany, University of Cape Town, Rondebosch 7700, South Africa

RABINOVITCH-VIN, AVIVA Nature Reserves Authority, 78 Yirmeyahu Street, Jerusalem 94467, Israel

READ, DAVID J. Department of Botany, University of Sheffield, Sheffield S10 2TN, U.K.

RUNDEL, PHILIP W. Department of Population and Environmental Biology, University of California, Irvine, CA 92664, USA

SHAVER, GAIUS R. The Ecosystems Center, Marine Biological Laboratory, Wood's Hole, MA 02543, USA

SIEGFRIED, W. ROY Percy FitzPatrick Institute, University of Cape Town, Rondebosch 7700, South Africa

SOMMERVILLE, JAQUI Department of Botany, University of Cape Town, Rondebosch 7700, South Africa

SPECHT, RAY L. Department of Botany, University of Queensland, St. Lucia, Queensland 4067, Australia

TILMAN, DAVID Department of Ecology and Behavioral Biology, University of Minnesota, 318 Church Street SE, Minneapolis, MN 55455, USA

TRABAUD, LOUIS Centre d'Études Phytosociologiques et Ecologiques, Centre National de la Recherche Scientifique, Route de Mende, BP 5051, 34033 Montpellier, Cedex, France

WESTMAN, WALTER E. Department of Geography, University of California, 1255 Bunche Hall, Los Angeles, CA 90024, USA

Section I. Evolution and Character of Mediterranean-Type Ecosystems

This section includes papers that present the evolutionary and biographic framework for the sections that follow.

The evolutionary histories of mediterranean-type ecosystems are of special importance to current research and resource management. Climates that characterize the present-day environments arose only some three million years ago, while the ancestral forms of modern biotas came to prominence over the preceding 25 million years or so, not necessarily in the present-day mediterranean zones. Much environmental fluctuation is evident during this 30-million-year time-span. During the recent past, over a quarter of a million years or more in the case of the Mediterranean Basin and southern Africa, man has played a greater or lesser but largely conjectural role in moulding landscapes and communities. Thus the processes of assembly of modern communities have been complex and the adaptations of constituent species relatively diverse, while ecosystem stability in the face of environmental change is a matter for surmise. DEACON presents a detailed synthesis of our knowledge of the palaeoenvironments from the Mesozoic to the present, incorporating the wealth of information that has become available recently from deep-sea drilling programmes and other novel sources. This review clarifies the picture of a progressive transition from the Oligocene onward from weakly zoned, humid tropical to sub-tropical environments to the strongly patterned, largely seasonally arid patterns that obtain today. The present-day environments "... relate to an interglacial mode that will not continue indefinitely".

Axelrod (1975) and Specht (1979) present essentially similar models for the evolution of sclerophyllous shrublands: the sclerophyllous taxa evolved locally within a forest environment on seasonally waterlogged, nutrient-poor sites or on ridges and other patches of shallow soil during the late Cretaceous and early Cenozoic, and then spread progressively as forest or woodland understoreys as climates deteriorated; modern communities developed as the vegetation opened in response to ice-age aridity and the activities of man. DEACON's review shows that the present-day mediterranean environments of Australia and South Africa were indeed forested or largely so until the late Miocene, with sclerophyllous taxa increasingly dominant thereafter, but that fossil data from present-day adjacent arid zones are lacking and that the scenarios for the evolution of the sclerophyllous taxa cannot yet be tested. Various detailed reconstructions of past vegetation change within the South African and Australian regions are assessed and compared. These hypothetical models need to be expressed with greater precision, at least for the Southern Hemisphere, to allow proper tests of their validity.

This introduction was prepared by F J Kruger

If Specht (1979) has shown that the concept of a climatically determined mediterranean-type ecosystem is not valid, what biogeographic framework can replace it? SPECHT and MOLL deal with this in the next chapter of the section. They review the bioclimatic relationships as now understood and present general, predictable relations between the cover and stature of the evergreen overstorey that characterizes all mediterranean-type ecosystems, and parameters of the water balance of the site. No such relationships hold for the understoreys. Here, soil factors override climate. SPECHT and MOLL propose three major formation categories: on moderately leached soils, seasonal savanna, with or without trees or tall shrubs; on strongly leached soils, evergreen heathlands, with or without stunted trees or tall shrubs; and on calcium-rich/hig', pH soils seasonal savanna, with or without stunted trees or shrubs. They note the ecophysiological convergences to be found among the biotas of each type, but emphasise also the different adaptations evident, so that uniform application of general management principles to the different formation types without cognisance of these divergences would be hazardous at best.

The two succeeding chapters present detailed examples of biogeographic patterns. BEARD summarizes vegetation patterns in mediterranean Western Australia and shows how in these very old landscapes with little differentiation in soil fertility, regular suites of vegetation, ranging from tall-forest to heath, are ordered on a rainfall gradient or by physical properties of the soil as they determine moisture relations. RABINOVITCH-VIN, by contrast, presents the vegetation patterns within a relatively uniform rainfall zone of northern Israel, and shows marked correlations between different vegetation formations and particular features of soil nutrition.

Finally, ORSHAN addresses the problem of discriminating plant growth forms characteristic of different mediterranean-climate environments as a step toward meaningful characterization of vegetation type and function by growth-form composition. He shows also that different vegetation units are each characterized by a typical growth-form spectrum, rather than a single dominant form.

References

AXELROD DI (1975) Evolution and biogeography of Madro-Tethyan sclerophyll vegetation. Annals of the Missouri Botanic Garden 62: 280-334

SPECHT RL (1979) Heathlands and related shrublands of the world. In: SPECHT RL (ed) Ecosystems of the world, vol 9B. Heathlands and related shrublands. Descriptive studies. Elsevier, Amsterdam, pp 1-18

1. The Comparative Evolution of Mediterranean-Type Ecosystems: A Southern Perspective

H. J. DEACON

1.1 Introduction

The mediterranean-type climate areas of the world (Aschmann 1973:13) show considerable variation in geographical extent, physiography, lithology and structure as well as differences in the biotas. This can be explained at a general level by the remoteness of these areas from each other, differences in earth history of each and their long isolation. Historical studies give a perspective on the scale of change in environmental parameters and biotas over longer periods than can be observed directly from contemporary experiments and further they promote a dynamic rather than a static viewpoint. The mediterranean areas, no less than the tropical regions (Simpson and Haffer 1978), carry an imprint of past changes in climate and related environmental effects, and evolution, dispersal and extinction through time have determined the modern biotas. The purpose of this paper is to discuss some aspects of climatic and vegetation change with specific emphasis on southern Australia and southern Africa to illustrate the kind of information currently available for comparing the selective pressures that have contributed to the evolution of the mediterranean-type ecosystems.

Thrower and Bradbury (1973) have outlined the general geological history of the mediterranean-type areas. Possibly the most striking comparison is that the areas on two continents, Australia and Africa, are essentially rifted or sheared margins, whereas three areas, Chile, California and the European Mediterranean margin, are associated with plate subduction and mountain building that has resulted in the elevation of the coastal ranges during the Cenozoic. The contrasting geological histories of the Cape and the Mediterranean illustrate the differences between these kinds of continental margins. The continental margin at the Cape has formed in concordance with the structural trends of two earlier phases of deformation through folding, the younger representing the present Cape Folded Mountains which were folded in the late Palaeozoic. The margin was outlined at the end of the Jurassic 140 million years ago (m.y.) by rifting along the west and by shearing along the south through the separation of Africa and South America. Outward flexuring of the continental margin and epeirogenic uplift have contributed to the relief, but the area has been relatively stable. The Mediterranean area on the other hand represents the borders of an inland relict of the former Tethys Sea. This Mesozoic sea-way was disrupted by structural developments associated with the Alpine orogeny which were initiated in the mid- and late Cretaceous. The main phase of mountain building was completed in the Miocene although the area has suffered further uplift since that time. Mountain building isolated the Mediterranean from the eastern European Paratethys extension (Hsü et al. 1977) in the Miocene but the present southeastern basin was preserved as an older remnant of the Tethys (Read and Watson 1975:205). The geological history of the Mediterranean is thus complex relative to that of the continental margin of the Cape and not only are there contrasts in the age and stability of these margins but also differences in the nature of the associated sediment accumulations with quartzose and sand-derived substrates found in the Cape and more calcareous substrates in the Mediterranean area.

It is purely fortuitous that the present-day landscapes of mediterranean-type ecosystems have become positioned between latitudes 30° and 40° on the globe; any similarities in geological history and substrate are also fortuitous. Similarities in climate, however, are not due to chance, but to the symmetry of the atmospheric circulation. Mediterranean-type climates are governed by the positioning and seasonal movement of the subtropical anticyclone belt on the equatorial side and the positioning of areas of cyclogenesis in the belt of the mid-latitude westerlies on the poleward side. The present strongly zonal atmospheric circulation reflects the effects of the pole-to-equator temperature gradient which is enhanced by the polar ice caps; therefore the earth's glacial history is pertinent to the discussion of the evolution of mediterranean-type climates.

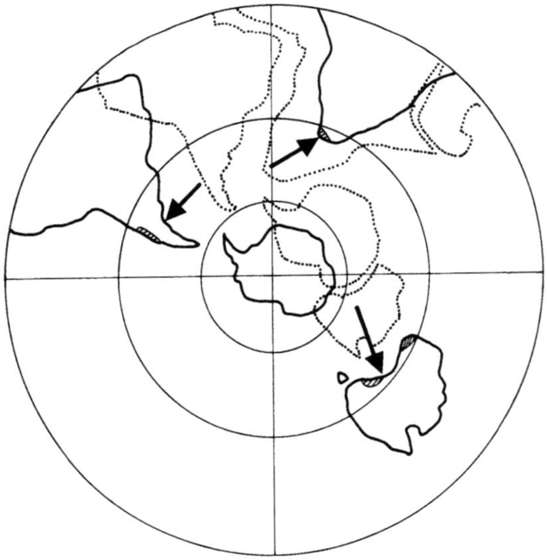

Fig. 1. South polar stereographic projection showing the relative positions of the present-day mediterranean-type areas now (solid line) and in the early Cretaceous 120 m.y. ago (dotted line). After Smith and Briden (1977).

The nuclei of the biotas found in these areas antedate the Late Cenozoic Ice Age when modern climates developed. Distributions of some groups of lower vertebrates, invertebrates and plants reflect patterns that were established prior to the Mesozoic break-up of the continents. Angiosperms diversified at a time when the rupture of Gondwanaland had reached an advanced stage. Disjunct distributions of the floras of the southern continents mirror this as well as subsequent long-distance dispersal incidents (Fig. 1). Mammals diversified in the early Cenozoic and because of greater separation of the land masses from that time, the faunas evolved in relative isolation. As a result the mammals of Africa, Australia and South America differ fundamentally (Keast 1972). Land connections existed between Eurasia and North America in the Atlantic area until the Eocene (Nilsen 1978) and through Beringia in later times and between Africa and Eurasia in the Palaeocene and early and late Miocene (Cooke 1972), so that these are periods marked by faunal interchange between Africa and the northern continents.

As Johnson (1973) has pointed out, ecosystems evolve not as units but through the addition and subtraction of organisms and changes in the attributes of component species. Climate and vegetation are emphasized here although it is appreciated that, for example, edaphic factors play a role in ecosystem evolution. Cenozoic climate and vegetation history with particular reference to Australia and southern Africa are considered for two broad time periods, the first pre-dating and the second post-dating the inception of summer-dry climates. For the latter time range the last glacial maximum is used to illustrate the complexity of even recent climatic change as one of the prime factors contributing to ecosystem evolution. From a palaeoecological perspective, the observations of present-day processes operating to maintain mediterranean-type ecosystems relate to an interglacial climatic mode which will not continue indefinitely.

1.2 Pre-Cenozoic background

Numbers of recent reviews have discussed the implications of continental drift for the biological sciences and for particular groups of organisms (Keast et al. 1972; Tarling and Runcorn 1973; Raven and Axelrod 1974).

The delayed radiation of the mammals until the Cenozoic is a response to the end-Cretaceous extinction of the dinosaurs, but the mid-Cretaceous radiation of the angiosperms is less easily explained (Hughes 1976). Evidence from both sides of the Tethys Sea indicates the appearance in the Barremian (Fig. 2) of simple but undoubted angiosperms, monosulcate pollen grains, and in the succeeding Aptian in north and west Africa and Brazil, tricolporate grains as found in the non-magnoliid dicots (Doyle 1978:368). The primary diversification of the angiosperms took place then at the time when Africa and South America were separated and the separation sequence is now reasonably well known and dated (Simpson 1977).

Fossils of closely related species of a notosuchian terrestrial crocodilian found in both Niger and Brazil in deposits of Aptian age, together with the other evidence, suggest that West Africa and Brazil were connected by land at this time (Buffetaut and Taquet 1979). In the succeeding Albian (Förster 1978) the ammonite faunas indicate the first connection between the northern and southern proto-Atlantic and by this time the pollen record shows further diversification of morphological types. Although it is debatable whether west Gondwanaland was the centre of evolution of the angiosperms, current evidence suggests that from here

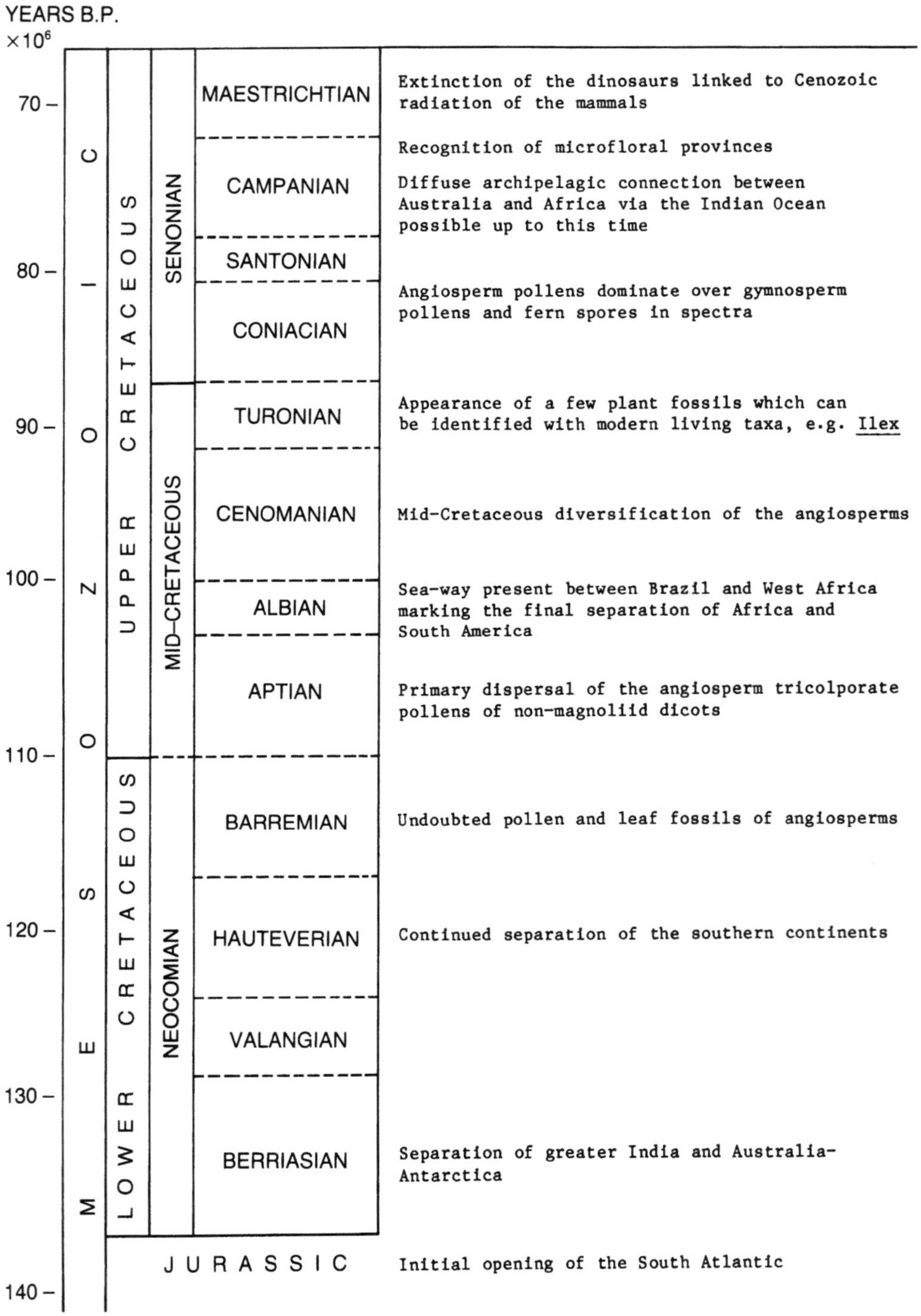

Fig. 2. Geological time scale for the Mesozoic based on data from Muller (1970), Simpson (1977) and Doyle (1978), showing the dating of major events relevant to the development of modern biotas.

they dispersed over the southern continents. Raven and Axelrod (1974:635) propose that the original flora of Australia, for example, derived from west Gondwanaland via the Indian Ocean links in or prior to the mid-Cretaceous.

In the early Cretaceous (130 m.y.) Australia and greater India and the fragments making up the Mascarene-Madagascar continent separated (Veevers and McElhinny 1976); diffuse connections across the Indian Ocean may have persisted into the late Cretaceous (75 m.y.) (Schuster 1976). The earliest angiosperm pollen found in Australia dates from the upper Albian. Dettman (1973) reports that not only were particular morphological types of pollen grains introduced later in Australia, but also that assemblages are less diverse than the mid-Cretaceous assemblages from West Africa and the Americas. She concludes that there probably was migration into Australia rather than evolution within that continent.

This evidence does not suggest that heathland taxa dispersed world wide from a primary centre in Antarctica (Specht 1979:10); however, it is not implausible that some of the families are derived from mid-Cretaceous ancestors that were dispersed among the southern continents. There are problems in identifying modern families prior to the Eocene or later (A R H Martin 1973; Hughes 1976) but a number of living families have a history extending back to the Senonian and the Turonian (Muller 1970). An example is the recent report of Ilex from Turonian deposits in Australia (Martin 1977a). Numerous researchers (Levyns 1962, 1964; Goldblatt 1978; Taylor 1978) have commented on the appearance of austral taxa in southern African flora, but given the time Australia and southern Africa have been separated, and allowing for evolution and extinction, one would expect relationships at the generic level to be few. Goldblatt (1978) in particular has emphasized the tropical origin of the flora of southern Africa. It is not possible or perhaps even meaningful to factor out an austral element versus a tropical element in the southern African flora because the elements involved are ultimately derived from the same source and the older taxa shared with other southern continents have had separate evolutionary histories on these continents for a long time. Nevertheless, the parallels in the floras between the mediterranean-climate regions of the Cape and southern Australia are striking, the more so when the relationship of sister families such as the Ericaceae and the Epacridaceae is taken into account. As remarked by Livingstone (1975), the biota of the Cape is a bizarre relict and the same could be said of that of Western Australia. This implies long isolation.

By contrast, the floras of California and the Mediterranean region do not show the same degree of isolation from one another. This is reflected in the number of common genera of woody plants (Raven 1973:215). Somewhat conflicting explanations have been offered for the taxonomic similarities in the respective floras. Raven (1973:215), Axelrod (1975) and Raven and Axelrod (1978) have postulated that some genera derived from ancestral groups were continuously distributed throughout the area occupied by an Arcto-Tertiary Geoflora in the Palaeogene or earlier and that others were transported across the Atlantic between mid-Eocene sub-humid climatic belts. Wolfe (1971, 1975) in a more general review of the plant geography of the northern hemisphere in the late Cretaceous and Tertiary has criticized the concept of an Arcto-Tertiary Geoflora and has argued against the postulate of a dry or sub-humid belt stretching across North America with an equivalent extension of the Mediterranean region. Instead, he has stressed that by the Eocene the floras of western Europe and western North America were highly similar with numerous genera in common and that a close relationship existed prior to the Oligocene climatic deterioration. Wolfe

(1975) contends that the Neogene floras (Miocene-Pleistocene) derived primarily from the Palaeogene (Palaeocene-Oligocene) lineages which had existed on the respective continents.

Evidence that the climate at the time of the North Atlantic link was not temperate is provided by a 30-m-thick lateritic palaeosol intersected during deep-sea drilling on the Iceland-Faeroe Ridge. Lateritic weathering, at present restricted to lower latitudes, takes place under warm, humid conditions (Frakes and Kemp 1973). The palaeosol, resting on a basalt dated to some 42 m.y., is overlain by Eocene-Oligocene marine sediments which represent the later subsidence of the ridge below sea level (Nilsen 1978). As the Iceland-Faeroe Ridge in the Eocene was a land bridge between North America and western Europe, its subsequent climatic deterioration and disruption may have contributed to the divergence apparent in the biotas of Eurasia and North America in the Eocene.

Solbrig et al. (1977:17) have suggested that the mediterranean-type flora of Chile developed from a cool-temperate flora primarily derived from taxa of warm neotropical origin. Elements of the cool-temperate flora like Nothofagus evidence, in their present or past distribution between South America, Antarctica, Australia and New Zealand, what has been described in Croizat's terminology as a track (Cracraft 1975), implying a pathway for the dispersal of individual taxa. Such tracks apart, South America has been as isolated as the other southern continents since the Mesozoic, but at present has a richer temperate flora. This continent has not changed its latitudinal position since the Mesozoic separation, in contrast to Africa and Australia which both drifted northwards. The substantial elevation of the Andean ranges in the Cenozoic is another contrast with other southern continents. The Panamanian land bridge,established some 3 m.y. ago (Marshall et al. 1979), linked South and North America after separation since the earlier Mesozoic, but was more important for the interchange of terrestrial faunas than it has been for the migration of plant taxa.

1.3 The Cenozoic: 65-3 million years ago

For the sake of convenience, the climate and vegetation history of the mediterranean-type ecosystems are discussed in two informally defined periods, the earlier covering the greater part of the Cenozoic from the Palaeocene through to the mid-Pliocene (65-3 m.y.), the latter covering the latest Cenozoic from the mid-Pliocene to the present (3-0 m.y.). Mediterranean-type ecosystems are defined on the basis of climate and the evolution of climate is therefore of direct interest. Raven (1973) mentions that mediterranean climates are geologically recent and younger than the biotas found in these regions, which makes it pertinent to survey what is known of the climates of prior geological times (Fig. 3).

1.3.1 PHYSICAL EVIDENCE FOR PALAEOCLIMATES

Cores from deep-sea drilling programmes, which have provided more continuous stratified sequences for study than have been available from terrestrial environments, have shown that while upland glaciation may date from the pre-Miocene in Antarctica, global climates under essentially ice-free conditions in the Palaeocene and Eocene were warm and humid even in the high latitudes. A more pronounced cooling trend beginning in the Oligocene culminated in the late mid-Miocene growth of the Antarctic ice

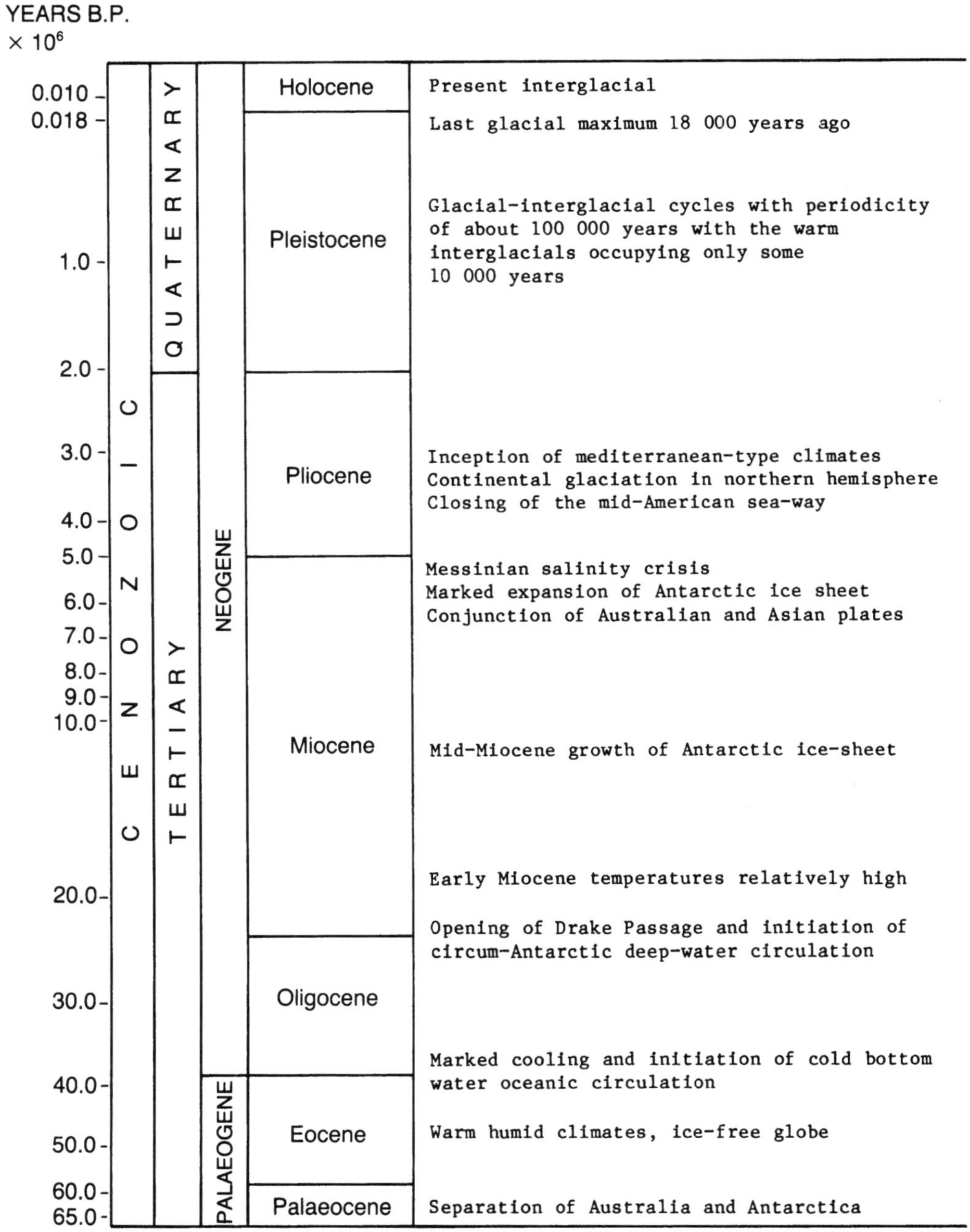

Fig. 3. Geological time scale for the Cenozoic with dating of main events relevant to the development of mediterranean-type climate regions.

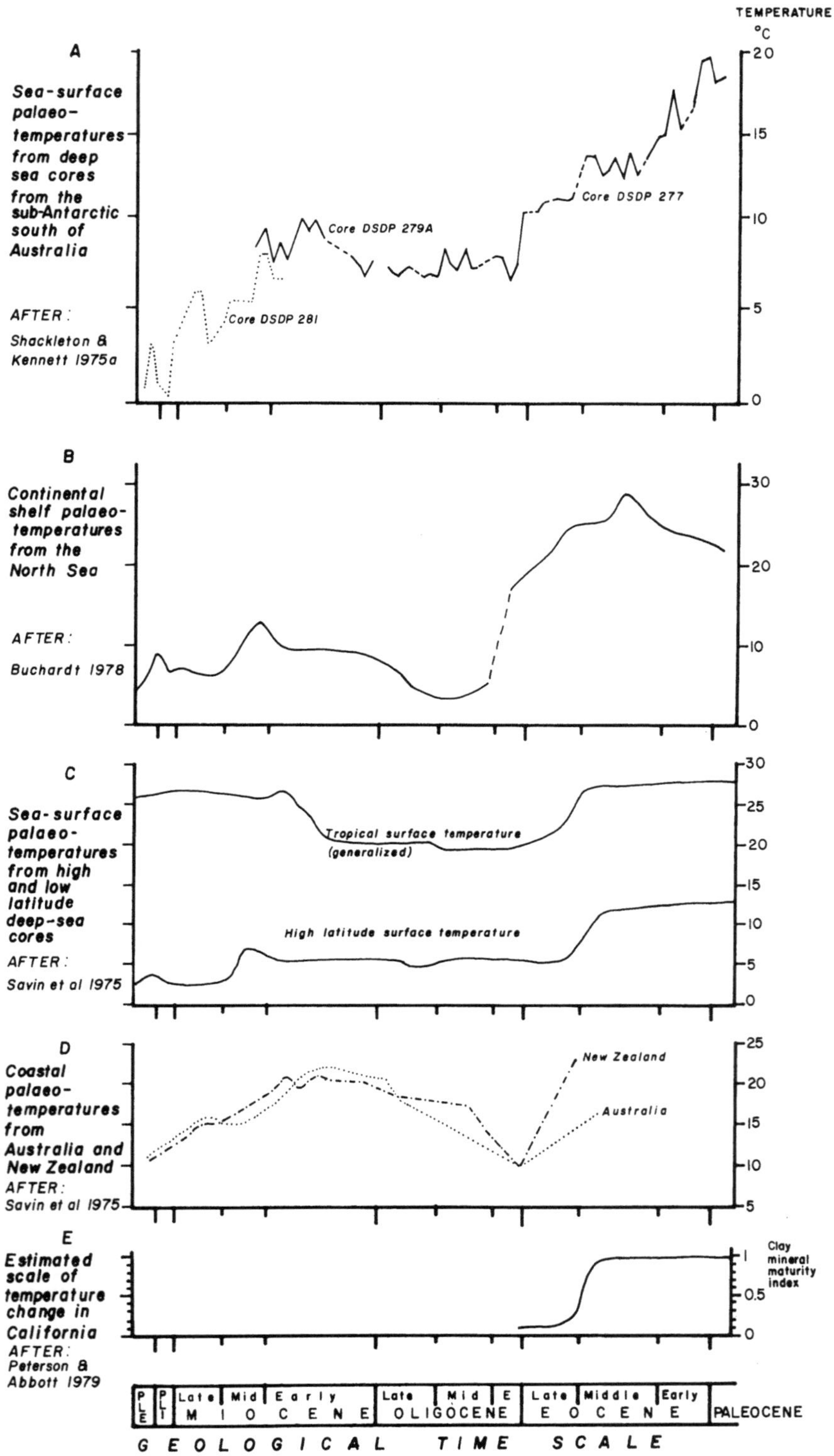

Fig. 4. Palaeotemperature records for the Cenozoic.

sheet (10-14 m.y.) and the later development of glaciation on other continents.

As Wilson and Hendy (1971) point out, the predicted temperatures for the equatorial regions in the early Cenozoic under ice-free conditions are only slightly higher than the present, whereas those for the high latitudes are significantly higher than the present, thereby reducing the temperature gradient from the poles to the equator with the consequence that wind strengths would have been lower. Estimates of temperatures for surface and bottom waters of the oceans at high and low latitudes are possible using oxygen isotope studies of planktonic and benthic foraminifera from deep-sea cores for the whole Cenozoic and are supported by isotopic measurement analyses on shell material from coastal shelf areas (Fig. 4).

Shackleton and Kennett (1975a) have reported a sharp cooling of surface waters from the Palaeocene to the Eocene in the sub-Antarctic at the equivalent of a palaeolatitude of 60°S. Particularly noteworthy is the abrupt decline of some 5°C in both surface and bottom waters in the earliest Oligocene, 38 m.y. ago, within a period of between 75,000 and 100,000 years. Kennett and Shackleton (1976) relate it to the onset of Antarctic bottom water formation and surface temperatures near freezing in the ocean adjacent to Antarctica. This cooling has world-wide correlates and is, for example, strongly registered in the oxygen isotope composition of shell material from the coastal shelves of the North Sea area (Buchardt 1978; Fig. 4). Imprecise dating hampers correlation with terrestrial records but the clear indication that climate was strongly modified implies far-reaching biological changes.

The evidence for mid-Eocene climatic change in south-west California and northwestern Baja California can be cited as an example of the problems that arise in correlation. A widespread palaeosol of Palaeocene and early Eocene age indicates a tropical palaeoclimate similar to the modern equatorial belt, and an overlying late Eocene sedimentary succession indicates a semi-arid climate (Peterson and Abbott 1979). In this area the modern mean annual temperature is 16°C and the winter rainfall is some 250 mm yr^{-1}. It is estimated that although the late Eocene temperatures and precipitation were higher than 18-20°C and less than 630 mm yr^{-1} respectively, they were significantly lower than the 20-25°C mean annual temperature and the 1250 mm yr^{-1} precipitation estimated for the Palaeocene-early Eocene. The authors have correlated this change with the ocean surface temperatures curves of Savin et al. (1975) which show an abrupt ocean temperature drop in the late middle to early late Eocene. In turn, however, Shackleton and Kennett (1975a) consider the cooling in the earliest Oligocene to be isochronous with that recorded by Savin et al. This suggests that the different time scales used mask the wider significance of what is a single rather than several events. As the palaeolatitude of the southern Californian area in the Eocene was similar to that of the present (Peterson and Abbott 1979), this example gives some indication of the scale of climatic change in these mid-latitudes (Fig. 4).

Frakes and Kemp (1973), who modelled the contrast in oceanic circulation patterns between the Eocene and Oligocene, suggest that the longer residence time of Pacific surface waters and deflection of the Pacific equatorial currents to the high latitudes as well as circulation through the Caribbean and heat stored in the Tethys would have contributed to warm and humid climates in the Palaeogene and Eocene. They see the Oligocene cooling as related to the reduced efficiency of meridional heat transfer and the initial step towards the evolution of present climates.

Van Zinderen Bakker (1975) has suggested that the formation of Antarctic bottom waters in the Oligocene could be linked to the initiation of a proto-Benguela cold-water current and consequent aridity along the Namibian coast, bordering on the present winter-rainfall area of the Cape. Although by the early Oligocene, the stage-by-stage opening of the Atlantic had progressed to a point where the south Atlantic was open to Antarctic bottom waters (Sclater et al. 1977) and a cold-water coastal current may have existed, data from deep-sea cores on the Walvis Ridge (Siesser 1978, 1980; Diester-Haas and Schrader 1979) show the first signs of the modern Benguela current and upwelling system in the late Miocene. The Namib later became desert (Tankard and Rogers 1978) and although aridity is reinforced by cold-water current patterns the prime cause of hyperaridity is the type of atmospheric circulation (Lancaster 1979b, see below) that developed in the latest Cenozoic. The history of upwelling that can be documented from the study of deep-sea cores is an important record not only of evolution of ocean currents but also of the development of the trade wind system (Diester-Haas and Schrader 1979:48). Although few Oligocene deposits are known, one could expect to see the effects of the Oligocene cooling in the terrestrial fossil record in southern Africa. It is in this time range however that the tropical flora, which reached its furthest extension south in Chile in the Palaeocene and Eocene, was replaced by a temperate *Nothofagus*-dominated flora, suggesting major vegetation adjustments (Mercer 1978).

The sub-Antarctic oxygen isotope record shows some increase in ocean surface temperatures relative to the Oligocene in the early and late Miocene, and a significant decrease in the latest Miocene-Pliocene (Shackleton and Kennett 1975a). Changes in isotope composition of the foraminifera reflect the growth of the East Antarctic ice sheet in the late mid-Miocene (10-14 m.y.). The cooling of surface waters in the latest Miocene-early Pliocene correlated with the Kapitean Stage in the New Zealand terminology is related to a markedly increased ice storage in Antarctica relative to the present (Shackleton and Kennett 1975b), a development which had world-wide consequences. One of these is the estimated eustatic lowering of sea level by 40-70 m at about 5.5 m.y. (Loutit and Kennett 1979). This lowering of sea level has been linked by a number of authors (for example Ryan 1973) to the isolation of the Mediterranean basin and the deposition of evaporites that characterize the youngest part of the Messinian Stage, the so-called Messinian salinity crisis, when the Mediterranean Basin became a deep depression with saline lakes. Hsü et al. (1977:402) have suggested that one result of this was an increased selection for annuals in the herbaceous flora of the Mediterranean region.

1.3.2 MODELLING EARLY CENOZOIC CLIMATES

In the early Cenozoic under ice-free conditions, with more efficient meridional circulation and lower wind strengths, synoptic climates differed greatly from the present. Flöhn (1978:8) has suggested that the subtropical anticyclones would be located between 50° and 60° latitude with a seasonally shifting tropical Hadley circulation bringing convective summer rainfall to over 80% of the earth's surface. Under this model a weak Ferrel-type polar circulation is proposed with polar frontal disturbances extending in winter to between latitudes 40° and 50°. Lamb (1972:119), however, has pointed out that such a positioning of the high pressure belt is unlikely in terms of the mechanics of global circulation and suggests that under ice-free conditions, weak and chaotic circulation on a relatively small scale with variable wind directions, rather than trade winds, would have characterized the global surface equatorwards of the westerlies located at some 60-80° latitude. The effect of changes in oceanic circulation and associated cooling in the Oligocene would have been to increase the temperature gradient from the poles to the Equator resulting in increased wind speeds in the westerlies and in their movement equatorward. This would have intensified with the thermal isolation of Antarctica and the development of a circum-Antarctic current in the latest Oligocene at 23.5 m.y. (Kennett 1978b) and with the subsequent growth of the Antarctic ice sheets. From the Oligocene with a single glaciated polar continent, as Flöhn (1978) has pointed out, there would have been an asymmetry of global climates, a balance that was not restored until present-day Arctic sea ice was developed in the Pliocene some 3 m.y. ago.

As the westerlies moved equatorward they focused atmospheric subsidence into a belt of high-pressure cells. The stability of these descending air masses is the prime cause of the subtropical zones of aridity. The preferred position of the high pressure cells and seasonal changes in their positioning is an important control of the mediterranean-type climates at present.

1.3.3 CLIMATIC EVOLUTION AND VEGETATION HISTORY: 65-3 M.Y.

The mediterranean-type climate areas have not occupied their present geographic positions throughout the Cenozoic. In particular, the South West Botanical Province of Australia has shifted from its location at some 50°S latitude in the Eocene as a result of the northward drift of the continent from the Palaeocene to the late Miocene. The northward movement of the African continent since its position in the Campanian (80 m.y.; Simpson 1977) through some 10° of latitude was completed by the beginning of the Miocene and is thus a smaller and earlier northward displacement. The estimated Eocene palaeo-latitude of Baja California, within 4° of latitude of the present (Peterson and Abbott 1979), has suffered relatively less displacement. Not only have the mediterranean-type climate regions moved in their relative geographical positions, but atmospheric and ocean circulation patterns have been evolving over the same time range.

Australia. The recent review of available data on the vegetation history of Australia by Kemp (1978) is noteworthy because it tests data on vegetation history, primarily from palynological studies, against reconstructions of palaeoclimates. Her attempt to apply Lamb's suggestions on global atmospheric circulation under ice-free conditions when reconstructing the evolution of Cenozoic climates is a step forward from Beard's (1977) palaeoclimatic model, which assumes that Cenozoic climatic zones were the same as at present. Beard's assumption means the northward drift of the Australian continent through bioclimatic zones can be used as a basis for inferring that desert xerophytes may date from the Eocene and that the first mediterranean climatic zone, and therefore mediterranean-type xerophytes, appeared in the Oligocene in Australia. There is at present no direct evidence to support the prediction that areas of desert in any modern sense existed in the north and north-west from the Eocene-Oligocene onward but this does not preclude the existence of drier areas important in the evolution of xerophytes from such times. Beard contends that Australian plants developed xeromorphy or sclerophylly in response to early climatic changes and that features of Australian soils have contributed to the special forms of xeromorphy exhibited.

Kemp assumes that the westerlies in the Eocene were confined to between 60° and 80°S latitude with the zone of the westerlies south of the continent, while Beard's model suggests that the greater part of the continent lay in the path of the westerlies. Kemp interprets the Oligocene cooling in terms of an increased meridional temperature gradient, intensified circulation and a northward movement of the westerlies, with the southern part of the continent then coming under the influence of the westerlies and variable or irregular circulation patterns being maintained over the rest of the continent. With the development of the circum-Antarctic Current and the initiation of the Antarctic Convergence by the early Miocene and its northward progression through this period, Kemp suggests that the palaeocirculation approached modern intensities only by the mid- to late Miocene. The dramatic rapid northward movement by some 300 km of the Antarctic Convergence, which is marked by a belt of siliceous biogenic sediments, is correlated with the Kapitean Stage and the Messinian Event and marks the end of the Miocene (Kemp 1978:203; Fig. 3). Kemp argues against zonation of vegetation under uniformly warm, humid conditions of the Palaeocene and Eocene although at least by the Miocene her data are consistent with the development of relatively arid areas in the north and west of the continent. Presumably desertification, however, is more recent and would not have developed until the irregular or chaotic surface winds were transformed into a pattern of persistent subtropical high-pressure cells and trade winds.

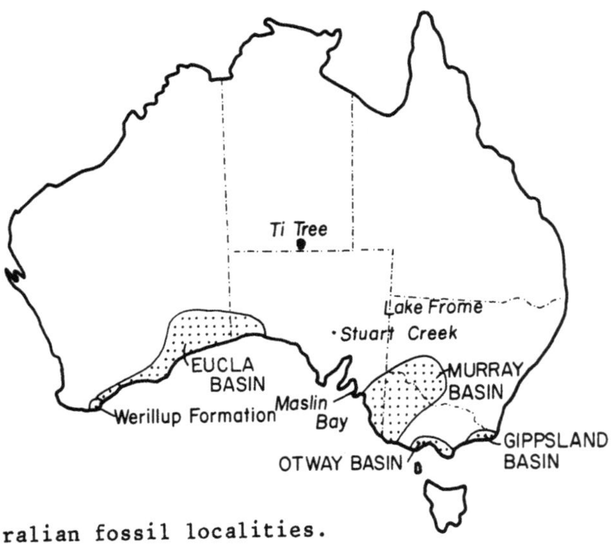

Fig. 5. Australian fossil localities.

The fossil record from Australia (Fig. 5) is mainly from epi-continental basins along the southern margin but, despite this limitation, is a much more complete record of the Cenozoic vegetation history than is available for other southern continents. The record for the southern margin is primarily one of wet humid forest until the mid-Miocene or later. For example, from Albany in Western Australia, Hoss (1975) has described an upper Eocene palynological assemblage dominated by Nothofagus pollen primarily of the brassi group, a type now restricted to New Guinea and New

Caledonia. Included as sub-dominants are over 30 proteaceous species comprising about a quarter of the total microflora, as well as a form species that principally represents Casuarina. Podocarps, pteridophytes and pollen types related to taxa with tropical affinities complete the assemblage. This spectrum is taken to represent a subtropical forest in this part of the continent which was then situated at a palaeolatitude of some 50°S and had a mangrove-fringed coast (Churchill 1973:84). This area now has a mediterranean climate and the sclerophyllous vegetation of the South West Botanical Province. An important point noted by Kemp (1978:189) in relation to these Eocene rain forests with their apparent admixture of taxa indicative of both warm-tropical and cool-temperate climatic regimes is that these forests probably have no modern analogue. The evidence from the continent as a whole in the Palaeocene-Eocene is for rain forest (Kemp 1978).

As the Oligocene record is not well represented, it is difficult to gauge the effects of marked climatic change in this time range. However, the pollen record from the Gippsland Basin for the transitional Eocene-Oligocene Upper Nothofagus asperius Zone indicates some reduction in diversity (Stover and Partridge 1973:243), and Nipa palm pollen disappears from the record at the end of the Eocene (Kemp 1978:193), indicating perhaps both sea-level regression and cooler waters on the southern coast. The sea-level regression, estimated at about 250 m, was thought to have been on a scale sufficient to expose the continental shelf (Quilty 1977). The most relevant record is that from the Murray Basin where the accumulation of 150 m of sediments with carbonaceous clays and lignites, interpreted as forming under swamp or lake conditions from the early Oligocene to the mid-Miocene, suggests continued humid climates (Martin 1977b). Palynological evidence of Nothofagus rain forest supports this (Martin 1977b). On Kemp's (1978:195) reconstruction of the palaeocirculation, the southern margin of Australia would have come under the influence of the westerlies from Oligocene times and in this sector rainfall is unlikely to have been seasonal.

There is evidence for a more open vegetation mosaic and seasonal precipitation inland of the southern coast from the mid-Miocene onward. Callen and Tedford (1976:139) reconstructed an environment in the Lake Frome area with Nothofagus and Podocarpus gallery forest inhabited by arboreal marsupials along water courses among extensive areas of grassland. While this suggests seasonal climate the gallery forest nevertheless indicates relatively high humidity in what is now an arid zone inland of the mediterranean climate area. There is also evidence of up to 10% grass pollens in a Nothofagus-dominated spectrum of similar age from Ti Tree in central Australia (Kemp 1978:199) which, while again indicating more open Miocene environments, still contrasts with the present vegetation and climate of the Simpson Desert.

The vegetation changes at the end of the Miocene and the Plio-Pleistocene are at present best understood in terms of changes in the vegetation mosaic in the Murray Basin area. Martin (1973, 1977b, 1979) has recognized a succession of phases with two Myrtaceae-Casuarina phases separated by Nothofagus and gymnosperm phases, the latter marking an expansion of rain forest in the Pliocene. These are succeeded in the Plio-Pleistocene by vegetation that shows an increase in Asteraceae and Poaceae in the pollen spectrum which she equates with the development of a herbaceous ground cover and possibly savanna as found in that area today (Martin 1973:47). At present the dating of these successive phases lacks precision, as does the definition of the Plio-Pleistocene boundary (Martin 1979:75). The evidence from the Murray Basin underscores that important changes in vegetation took place along the southern margin of the continent in the end-Miocene and the more modern vegetation associations date from this time.

The paucity of evidence from the interior of Australia in the now arid areas precludes discussion of the evolution of vegetation under more seasonal climates from Oligocene times onward. Among taxa well represented in Western Australia is Acacia. Although Acacia pollen is recorded from the base of the Miocene (Kemp 1978:196), this genus can be suggested to have had a long history on the continent on the basis of its seed chemistry which indicates a primitive pattern shared only with A. heterophylla, native to the Mascarene Islands (Bell and Evans 1978). Eucalyptus is another prominent Australian genus with a poor fossil record. Eucalyptus-type pollen is recorded from the Murray Basin in the Pliocene, but the only authentic macrofossil record is of fruits from a silcrete locality north of Woomera, Stuart Creek, that dates to the Miocene or Eocene-Oligocene (Ambrose et al. 1979). Lange (1980) has failed to authenticate the occurrence of Eucalyptus in a study of Eocene-Oligocene samples of fossil leaves from the southern coastal margin deposits and thus this genus does not appear to have been part of the southern rain forest associations.

South Africa. The few Cenozoic occurrences (Fig. 6) which have yielded botanical or palaeontological evidence from the mediterranean-type climate area of southern Africa are for the most part either understudied or still being studied. Dating possibly to the late Eocene-Oligocene is the infill of sediments in the Arnot kimberlite pipe on the farm Banke in Namaqualand which has yielded a sample of 70 fragmentary dicotyledonous leaves representing 12 taxa (Rennie 1931). Most common are 20 leaves with dentate margins referred to as being of Myrica-type. Kirchheimer (1934) described and illustrated a limited range of pollens from the carbonaceous sediments

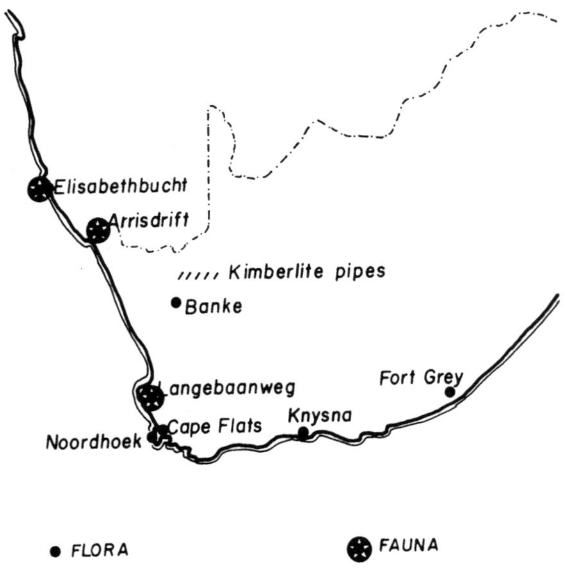

Fig. 6. South African fossil localities.

including two types of podocarp and Adamson (1931) identified opalized wood from the upper horizons as Ficus. Axelrod and Raven (1978) interpret the evidence from this site as indicating that the temperate forest was depleted in the late Eocene and had an admixture of more xeric elements ancestral to those found in present fynbos communities. A relatively dry climate and summer rainfall have been posited for this situation some 40 m.y. ago (Tankard and Rogers 1978) but it seems premature to interpret the Banke leaf and pollen flora in specific terms of past climatic and vegetation patterns without the benefit of a more complete investigation.

Along the southern coast at Fort Grey near East London (Adamson 1934) macroscopic remains including Podocarpus and possibly Widdringtonia, fruits referred to Curtisia, as well as the remains of other dicots and monocots have been found. Du Toit (1939:417) suggests that these are Cretaceous to Eocene and perhaps older than the Knysna lignites, which are a sequence of cross-bedded sands with intercalated lignites associated in places with macroscopic plant remains. Rogers (1908) concluded that the deposits post-dated the cutting of the coastal platform. The deposits appear to occupy channels and to be fluviatile and terrestrial in origin. Phillips (1927) has reported Widdringtonia, Podocarpus, Gonioma and Curtisia in the macroscopic remains. Thiergart et al. (1966) note high frequencies of pollens of Restionaceae and fern spores in addition to Podocarpaceae, Proteaceae and Myrtaceae, taxa which together with other suggested identifications are found in the present forest flora (Axelrod and Raven 1978). Although the data are consistent with the area having been forested, forest was not necessarily continuous. This is supported by preliminary examination of a pollen spectrum from near the base of the sequence which shows low frequencies of Podocarpaceae and high frequencies of Proteaceae and in quite dissimilar spectra from the modern Knysna forest (Deacon and Scholtz, unpublished). Levyns (1964) has commented that the absence of Erica from such a wet site is anomalous in terms of the modern vegetation but this could be a function of the age of the deposits. A point of interest is that Thiergart et al. (1966) record transported carbonized material in the samples they examined and this suggests a fire-prone vegetation in the catchment area.

Coetzee (1978) has reported on the microflora from a number of buried channels in the southwestern Cape which preserve a discontinuous record possibly from the Oligocene through to the Pleistocene. An alternating succession of pollen assemblages has been documented primarily from the Noordhoek core (Coetzee 1978:116) that suggests to her marked changes between palm-dominated vegetation and forest with Podocarpaceae and other forest taxa. She considers the first signs of Asteraceae to indicate the base of the Miocene, but it is only at the end of the Miocene that there is evidence for pollen spectra which can be related to fynbos. Interpretation of this evidence depends on more precise dating of the older deposits, but it may indicate altitudinal zonation of the vegetation, with forest in the upland area and more subtropical vegetation in the lowlands of the Cape Peninsula, persisting to the end of the Miocene.

Fossil pollen assemblages that include typical fynbos taxa (Tankard and Rogers 1978:324) have been dated from the associated mammal fauna in the Quartzose Sand Member of the Langebaanweg site near Saldanha. The stratigraphy and fauna have been described in a number of papers by Hendey and his associates (Hendey 1974, 1976, 1981; Dingle et al. 1979). Correlation with radiometrically dated fauna from East Africa provides a narrow range centred on 5.0 m.y. for the age of the Langebaanweg fauna. A range between 4.2 and 5.2 m.y. is supported by reference to the local history of marine transgressions and regressions (Hendey 1981). The site records the shifting course of the proto-Berg River and the mammalian taxa include many extinct genera with both browsers like giraffids, indicating woodland, and grazers present. These data from the early Pliocene show that neither the faunal communities nor the vegetation mosaic were wholly modern. The subsequent Plio-Pleistocene period, which would shed light on the evolution of the fynbos ecosystem, is poorly documented in the fossil record at the Cape, but Tankard and Rogers (1978:325) assume progressive desiccation with more xerophyllous vegetation replacing grassland from the Pliocene into the Pleistocene.

Comparisons between Australia and southern Africa. When the palaeoecological evidence for the mediterranean-type climate areas of the Cape and Australia are compared parallels emerge. The early Cenozoic vegetation recorded so far is forest, dominated by gymnosperms in the Cape and by *Nothofagus* in Australia, but with Proteaceae important in both these associations. There is little direct evidence that Ericaceae was represented in any vegetation association prior to the late Miocene. In Australia the fossil record of the sister family Epacridaceae is similarly poor, but Specht (1979) and Westman (1978) suggest an association of heath families including Epacridaceae and proteaceous species was present from the early Cenozoic. Forest persisted in the coastal areas that were later to come under the influence of mediterranean-type climates until the end-Miocene with evidence from more northerly inland areas of more open environments existing in the Miocene. Lake Frome and Ti Tree have been noted in this connection in Australia and in southern Africa the early and mid-Miocene fauna in the Namib and on the Orange River indicate a woodland mosaic but with more open habitats developing (Corvinus and Hendey 1978). In Australia the disappearance of *Nothofagus* pollen of the *brassi* type from the coastal forest spectra has been suggested by Kemp (1978) to mark the terminal Miocene and possibly to correlate with the abrupt cooling and northward extension of the Antarctic Convergence documented at this time. This is in turn correlated with the Messinian event. Hendey (1981) has noted possible correlations with eustatic changes in sea level in the Langebaanweg sequence with the Messinian and in both the Cape and Australia it would seem to be an important stage in the evolution of modern ecosystems. In the Pliocene in the Murray Basin an extension of forest is recorded that as yet has no documented parallel from the Cape or adjacent regions. What does seem important is that on neither the south coast of Australia nor that of the Cape is there evidence for summer-dry climate or the dominance of modern shrub taxa prior to the end of the Miocene or possibly into the Pliocene.

Comparisons with other mediterranean-type areas. Axelrod has commented extensively on the history of sclerophyllous vegetation and on the evolution of mediterranean-type ecosystems in California (Axelrod 1973; Raven and Axelrod 1978) and the Mediterranean (Axelrod 1975). Axelrod and Raven (1978) have proposed a model for the evolution of fynbos vegetation which can be conveniently discussed in the following section.

Sclerophylly in Axelrod's view is an ancient and possibly primitive character that long predates the evolution of summer-dry climates. He has stressed that in both North America and Eurasia sclerophyllous vegetation occurs in summer- and winter-rainfall areas and that its once more continuous and greater geographic distribution has been fragmented by zones of arid climate (Axelrod 1975). Axelrod (1973:63) traces sclerophyllous vegetation in North America-Eurasia back to the Madro-Tertiary and

Mediterrano-Tertiary Geofloras respectively which had their origins in the southwestern portion of continents in the late Cretaceous. In the late Cretaceous and early Tertiary subtropical and tropical laurophyllous forests covered the low-middle latitudes of Eurasia and North America; Axelrod (1973) suggests that the sclerophyllous tree and shrub taxa derived therefrom occupied drier sites with thin, poor soils. The fossil evidence is from the somewhat later mid-Eocene when some woody taxa, whose descendants at present contribute to the sclerophyllous woodland chaparral and woodland-maquis alliances, are recorded in assemblages of macroscopic plant remains from the central and southern Rocky Mountains in North America and sites in western Asia respectively. Using this evidence Axelrod (1975) suggests broad-leaved evergreen sclerophyllous taxa occupied a sub-humid belt stretching across North America and from the Mediterranean region eastwards into central Asia by the mid-Eocene. This sub-humid or dry belt he associates with a zone of descending dry air, the Palaeogene equivalent of the subtropical high pressure circulation of today (Axelrod 1975:310). Axelrod and Raven (1978:93) expand on the assumption in stating that there have always been high pressure cells on the edge of the tropics, but that the intensity of the circulation associated with these cells has increased with time. This is a different concept of early Cenozoic atmospheric circulation from that proposed by Lamb (1972) and would seem to need substantiation.

Following the Eocene, spreading dry climates favoured increased distribution of sclerophyllous vegetation (oak-laurel forest, oak woodland-savanna, chaparral, mixed evergreen forest) which assumed dominance in southeastern California by the Miocene and had a continuous distribution across North America in the Miocene and Pliocene. In the Mediterranean region it is suggested that the lowlands were covered by sclerophyllous woodland (oaks, laurels, olives and palms) with numbers of sclerophyllous shrubs from the Oligocene through the Miocene (Axelrod 1975). An expansion of forest is recorded in the late Pliocene in California similar to that recorded in the Murray Basin in Australia. In the mediterranean regions, as summer-dry climates evolved after the first glacial, decreasing precipitation and decreasing temperatures led to the segregation of what were, relative to the Miocene and Pliocene, somewhat impoverished woodland and chaparral or maquis communities.

History of the sclerophyllous vegetation in southern Africa. When Axelrod and Raven (1978) attempted to reconstruct the late Cretaceous and Cenozoic vegetation history of the African continent they assumed a symmetrical distribution of climate and vegetation relative to the palaeo-equator through this time range. The main trend deduced is the progressive narrowing of the moist tropical belt and an increase in the areas of savanna, thorn scrub, grassland and more arid environments in the Cenozoic. Although factors such as the northward drift of the continents and epeirogenic uplift have contributed to this trend, the main factor postulated was climatic deterioration, increasing selection for taxa that could withstand drought.

By analogy with the northern hemisphere data, Axelrod and Raven (1978:93) infer that sclerophyllous woodland and scrub occupied the drier part of the subtropical savanna belt in southwestern Africa by the late Eocene, being derived from taxa that had evolved xeromorphous adaptations at local dry sites during a prior time period (Fig. 7). The sclerophyllous vegetation is considered to have occupied areas now desert, semi-desert and grassland in the Miocene and Pliocene, with semi-desert environments spreading from the end of the Miocene (7-5 m.y.). These authors hold that this relatively richer sclerophyllous vegetation persisted in the now more arid areas into the Pleistocene, thereafter invading the present fynbos area and incorporating the isolated local pockets of sclerophyllous vegetation there. In response to pluvial conditions and the expanding temperate forest, fynbos migrated into now drier areas during the Pleistocene. The fynbos was largely swept into its present areas as a result of drier climates at the end of the last pluvial 10,000 years ago.

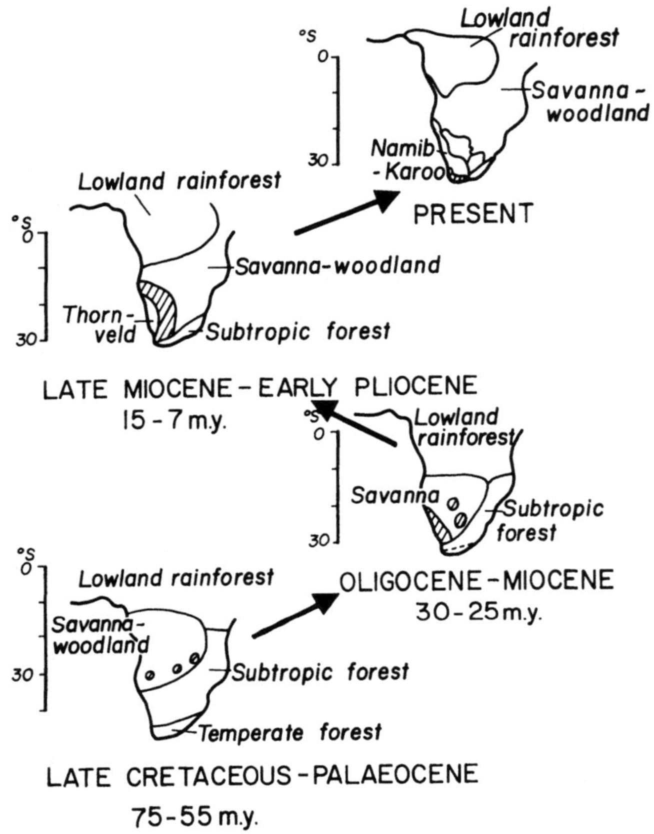

Fig. 7. Model of the history of southern African sclerophyllous vegetation. Regions with sclerophyllous vegetation are shaded. After Axelrod and Raven (1978).

The climatic change in the earliest Oligocene, given imprecision in dating, might be considered to substantiate the appearance of pockets of vegetation with sclerophyllous taxa prominent in southwestern Africa. Banke or similar occurrences may serve to test this postulate, as pointed out by Axelrod and Raven. Other than faunal evidence for a mosaic of forest-woodland-savanna (Corvinus and Hendey 1978) in the early Miocene, however, there is no direct information on the vegetation in present-day arid areas in the southwestern region in the Miocene and Pliocene. The proposed later migration of the fynbos vegetation poses some problems. Taylor (1980) has commented that a vegetation category like the fynbos sensu stricto, which is associated with particular substrates, cannot be considered to migrate as a unit and that the response to climatic changes is towards adjustment in the range of some individual taxa and not mass migrations. That Axelrod and Raven (1978) equate the term fynbos with a broader category of sclerophyllous vegetation than is current among local researchers contributes in part to the difference in opinion. There is currently no accepted sequence of pluvials for southern Africa and it is too simplistic to equate glacial periods with pluvials or prolonged periods of increased rainfall. Axelrod and Raven (1978:94) recognize four pluvial cycles in the last 2.7 m.y. with the last pluvial ending 10,000 years ago, but as no definition can be given to pluvials, the arguments on the timing and scale of responses to climate changes in the Plio-Pleistocene must be considered tentative.

1.4 The latest Cenozoic: 3.0-0 m.y.

The evidence for continental glaciation outside of Antarctica is from 3.5 m.y. ago. Among the earliest evidence is that from Patagonia at 3.5 m.y. (Mercer 1976) and from Bolivia at greater than 3.27 m.y. (Clapperton 1979). Continental glaciation in the northern hemisphere dates from some 3.2 m.y. The fact that ice-sheet formation in the northern hemisphere was retarded by some 10 m.y. relative to Antarctica where it was established 14 m.y. ago (Kennett 1978a) is as yet unexplained. While the Arctic sea-ice cover and the Greenland ice sheet are permanent features, as is the Antarctic ice sheet, the Plio-Pleistocene has been marked by repeated growth and decay of continental ice sheets in Eurasia and North America that reflect globally synchronous climatic changes. The deep-sea core records and continental loess deposits preserve a clear picture of these cycles in the Pleistocene, with a periodicity of some 100,000 years (Fig. 8). Kukla (1977) notes evidence for 17 glacials or longer periods of cool and cold climate, and interglacials or short periods of warm climate, in the last 1.7 m.y., which is close to the duration of the Pleistocene. All the glacial cycles in the period extending back to 3.5 m.y. were not of the same intensity, however, and many of the glacial stages prior to 1.0 m.y. are less pronounced than the most recent glacial maximum (Van Donk 1976). A further point in relation to glacial and interglacial cycles is that significant climatic changes can be relatively abrupt (Woillard 1979). As repeated episodes of climatic change, the sequence of Pliocene and Pleistocene glaciations are biologically important in that they have had large-scale effects on animal and plant distributions and community composition.

The last glacial cycle is the best known and gives an example of the scale of changes involved in a 100,000-year period. It falls within the Late Pleistocene, defined by Shackleton (1977) as covering the time between 125,000 B.P. and 70,000 B.P. (before the present) which equates with oxygen isotope stages 5a-e, and the last glacial which lasted from 70,000 to 10,000 years ago through oxygen isotope stages 4-2 (Kukla 1977:322). The last maximum advance of the global ice sheets falls within stage 2 (32,000-10,000 B.P.) and is marked by the maximum eustatic lowering of sea level at 85 m to 130 m below present sea level. The last glacial (stages 4-2) was a complex succession of cold and cool intervals with shorter warm interstadials, but only in the last interglacial, stage 5e, were conditions as warm or warmer than the present (Fig. 8). As the greatest contrast was between the interglacial and glacial maxima, considerable effort has gone into the detailed study of climatic conditions at the last glacial maximum

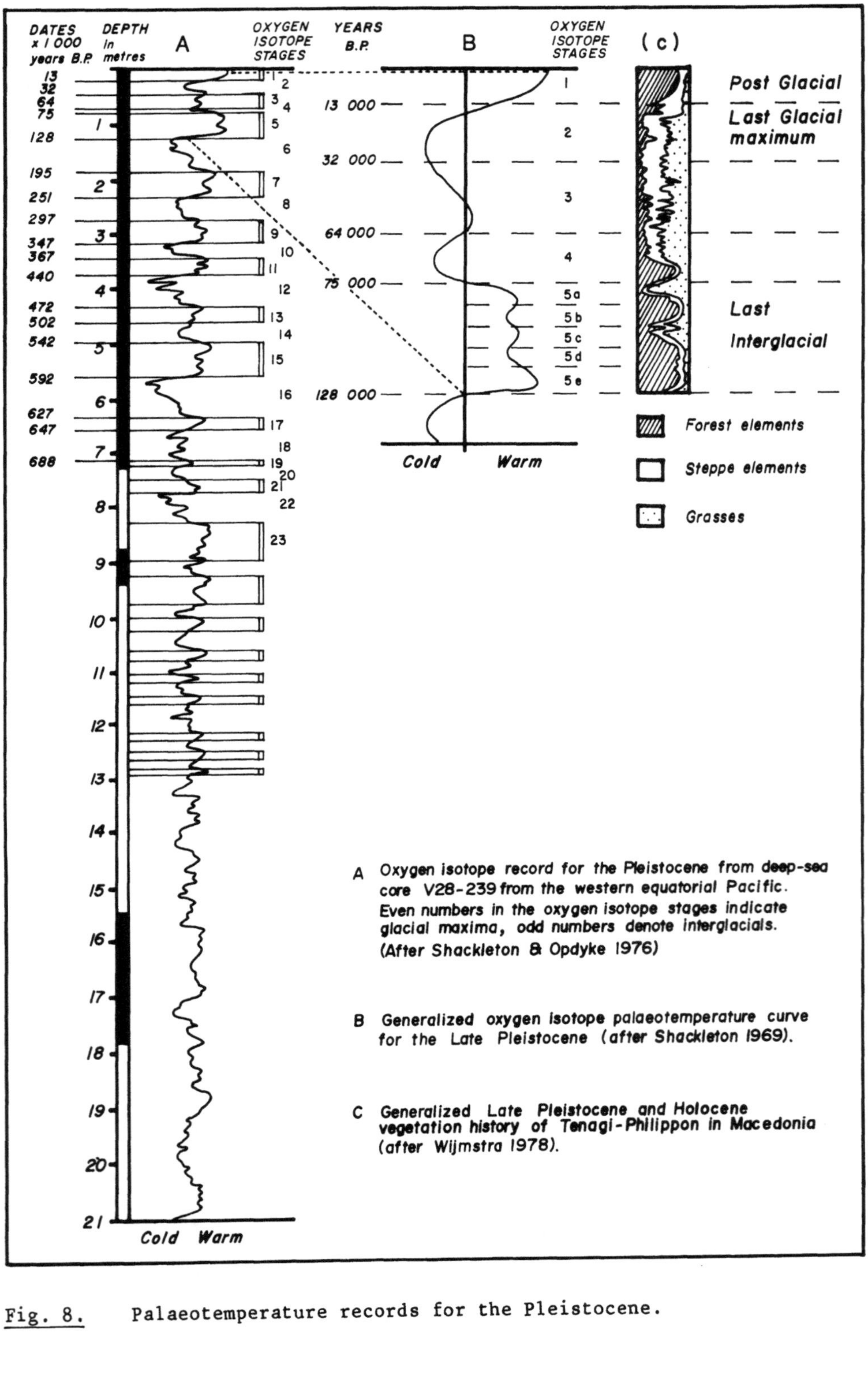

Fig. 8. Palaeotemperature records for the Pleistocene.

18,000 years ago in contrast to the present representing the Holocene interglacial (stage 1; 10,000-0 B.P.).

1.4.1 THE DATING OF THE INCEPTION OF MEDITERRANEAN-TYPE CLIMATES

Seasonal precipitation patterns are difficult to demonstrate directly although it should be possible using rhythmically banded sediments. An example from a Lower Pleistocene lake in northeastern Spain shows a spring peak in sediment accumulation inferred from the seasonal differences in pollen rain (De Deckker et al. 1979), although in this instance the peak is probably due to spring melt waters carrying a higher sediment load via the river that fed the lake and thus does not directly reflect local seasonal precipitation. The question, then, is how to recognize winter rainfall in mediterranean-climate regions in prior times, and how to demonstrate that winter rainfall regimes have their inception in the Plio-Pleistocene period of continental glaciation. One line of reasoning concerns evolution of modern circulation patterns. It suggests that although essentially modern circulation patterns in the southern hemisphere date from the end of the Miocene (5.2 m.y.), the stable high-pressure systems that cause a summer deficit in rainfall in the mediterranean-climate regions would not have been established prior to development of permanent sea-ice in the Arctic and glaciation of the northern hemisphere continents. This reasoning can be tested against independent lines of evidence. Among the most persuasive biological evidence that mediterranean climates are recent is the observation that some plant taxa have not adapted their growth rhythm to summer-dry climate (Levyns 1964). Other biological evidence that can be cited is that communities or individual taxa indicative of different climatic requirements were formerly found in mediterranean-type climate areas. An example is of the progressive extinction of the Taxodiaceae in western Europe that Michaux et al. (1979) have documented. That this non-synchronous pattern of extinction was initiated along the western Mediterranean area is significant and is explicable in that it marks the appearance there of summer-dry climate. The later extinction elsewhere indicates persistence of adequate year-round precipitation until the late Pliocene when marked cooling is apparent. The mid-Pliocene dating for the inception of summer-dry climate in Spain and southeastern France is made in reference to the local biostratigraphy, but it may correlate with a dating of 3.2 m.y. for the initiation of continental glaciation in the northern hemisphere, an event also reflected in an abrupt 3-4°C drop in surface temperatures in the Mediterranean Sea (Thunell 1979b).

1.4.2 CLIMATES OF THE LAST GLACIAL MAXIMUM AT 18,000 YEARS AGO

Information on global climates during the last glacial maximum come from terrestrial records, deep-sea cores and mathematical models of global circulation. In particular, the CLIMAP Project (1976) has had as one of its goals the mapping of sea-surface temperatures of the oceans at 18,000 B.P. These data and others on ice distribution, land elevation (taking into account the regression of sea-level) and the earth's albedo have been used to specify boundary conditions in several versions of mathematical global circulation models (Gates 1976; Manabe and Hahn 1977; Heath 1979) developed to hindcast climatic conditions of the last glacial maximum. All show global climates markedly cooler at the glacial maximum than they are at present.

An example of the precise data used as input for such models is that available on sea-surface temperatures for the Mediterranean (Thiede 1978). The winter and summer sea-surface temperatures have been reconstructed from data on fossil planktonic foraminifera populations in a series of dated cores from sample points throughout the basin. These show that the glacial maximum summer and winter temperatures of 13°C and 7°C along the Spanish and French coasts (as against modern summer sea-surface temperatures of 22°C) were considerably lower than off the Levantine coast where the summer and winter temperatures were 25°C and 19°C respectively and were more comparable to the present. Under a glacial climate the differences between the western and the eastern Mediterranean were intensified. Thunell (1979a) notes that the marked temperature gradient across the Mediterranean Basin characterizes only the last 1 m.y. of more intense glacial and interglacial fluctuations; previously temperatures were more uniform and the annual range was lower. The models are capable of generating not only temperature patterns on the continents, but also patterns of precipitation, evaporation, cloud cover and wind strength. This approach permits a more complete reconstruction not possible using isolated sets of observations of the incomplete terrestrial geological record.

As a complement to the CLIMAP Project, Peterson et al. (1979) have begun compiling maps on a world-wide basis with evidence from palynological and palaeontological studies as well as studies of past lake levels to be able to estimate temperature and precipitation conditions from terrestrial environments at 18,000 B.P. This corpus of data will help specify boundary conditions in global circulation models and is useful here in pointing out some anomalies between the data from the mediterranean-climate areas at 18,000 B.P.

For the Mediterranean Basin there are some 18 sites from Spain to Iran listed by Peterson et al. (1979) that provide consistent information that the climate along the northern Mediterranean shore was cooler and drier than at present. Estimates of the reduction in mean annual temperature are about 3-6°C. Noteworthy also is the palynological evidence that steppe vegetation was widespread. One of the sites, the Tenagi-Philippon site in Macedonia, preserves a long record of Pleistocene glacial and interglacial cycles in a deep peat deposit and can be correlated with the deep-sea core record (Fig. 8). The evidence from this site shows a grass-*Artemisia* steppe vegetation containing *Pinus* at the last glacial maximum. As climate improved at the end of the Pleistocene a series of vegetation changes culminated in the appearance of evergreen oaks and maquis vegetation some 8000 years ago (Bottema 1978; Wijmstra 1978). Apart from the evidence for cooler, drier conditions at 18,000 B.P., inferred from the requirements of taxa like *Eurotia* present in the pollen spectrum, the data demonstrate that the ecology of the area was markedly different from the present, with numbers of woody taxa restricted to upland refugia. There is the suggestion that the southern border of the Mediterranean, the Atlas Mountains of the Maghreb (Rognon and Williams 1977:310) and Israel in the east (Van Zinderen Bakker and Maley 1979) may have been out of phase with the rest of the basin in receiving higher rainfall. For the Atlas Mountains the evidence for higher precipitation is inferred from fluviatile deposits but supporting biological evidence is needed.

California is a mediterranean-climate region for which Peterson et al. (1979:254) have proposed moister hydrological conditions in the last glacial maximum. Temperature estimates, similar to those for the Mediterranean Basin, range from 2°C to 6°C lower than the present mean annual temperatures. Increased precipitation is also indicated for the adjacent arid Southwest on the basis of the dating of high levels in the desert lakes. A contrary view, however, is that of Brakenridge (1978) who holds that higher lake levels in the Southwest can be accounted for by lower temperatures alone. Van Devender and Spaulding (1979) have concluded from a detailed palaeobotanical study of a well-dated series of fossil packrat middens in Nevada, Arizona and California that during the last glacial maximum and the later part of the Pleistocene the zone of winter rainfall extended further eastwards into the present summer rainfall area of the arid Southwest. They account for the plant distributions observed in the middens in the late Pleistocene by suggesting increased winter rainfall, relatively mild winter temperatures and cooler summers. Under these conditions in the Southwest there was an extension of the range of woodland, chaparral and cold desert plants in California, Arizona and Nevada and the authors note that modern analogues for the late Pleistocene associations are not readily observed today because plant species have responded individually and not as a community to climatic change.

Johnson (1977a) maintains that throughout the last glacial from 70,000 to 10,000 years ago precipitation and temperature changes in California have been modest; thus the area constitutes an ice-age refugium. Evidence for more extensive coastal coniferous forests with some taxa represented more than 100 km south of their modern range Johnson explains in terms of increased fogginess and cooler though still mild temperatures. Consistent with this view of the low amplitude of late Pleistocene climatic changes, Johnson (1977b) ascribes the extinction of the Rancholabrean fauna at the end of the Pleistocene to human impact on the environment. A study of the pollen sequence from a 12,000-year record in offshore rhythmites in the Santa Barbara Basin (Heusser 1978), however, provides good evidence for dynamic changes in the vegetation in southern California extending into the Holocene. Coniferous communities are prominent in the spectrum from 12,000 B.P. until 7800 B.P., when they were succeeded in importance by lowland and cismontane taxa with *Quercus* and Asteraceae. Chaparral and sage scrub associations become increasingly important after 5700 B.P., particularly after 2300 B.P. Heusser considers these changes to be responses to climatic factors with the latest change possibly reinforced by human activities.

It is not possible to evaluate whether there is real conflict in the evidence from pollen studies such as that of Heusser and the inferences drawn from primarily macroscopic plant remains by Johnson, but clearly more palynological studies are required. The geographic situation of California and the Southwest in relation to a major ice sheet, the Cordilleran, may have contributed to features of atmospheric circulation that are not mirrored in other mediterranean-climate regions. In particular, the effect of the ice sheet in deepening the Aleutian low, the site of cyclogenesis, is given importance in the explanations of the climate patterns of the last glacial by Van Devender and Spaulding (1979) and Johnson (1977a). The weight of the available evidence then suggests that the Southwest had an increased winter rainfall at the last glacial maximum that continued into the early Holocene and that precipitation also increased in California (Peterson et al. 1979), or at least there was no moisture deficit.

In the Chilean lake district (40°S) Heusser (1974), who estimates that the temperatures at the last glacial, the Llanquihue Glaciation, were 8°C cooler in summer, suggests decreased precipitation too. The palynological data suggest a treeless or park-type vegetation in this section of the Chilean central valley with forest established only after 10,000 B.P.

For Australia there is good field evidence for cool climates at 18,000 B.P. in the extensive glaciation in Tasmania and the more limited glaciers in the Snowy Mountains in Australia reaching their maximum extent (Bowler et al. 1976; MacPhail 1979). These cold climates were associated with general aridity in southern Australia (Bowler 1976; Rognon and Williams 1977) and information from coastal south Australia (Dodson 1977) suggests that the climate was drier than the present from 50,000 B.P. to 10,000 B.P. with the driest period from 26,000 B.P. to 11,000 B.P. In the semi-arid areas of southern Australia prior to 25,000 B.P. lake levels were high and the movement of dunes in the arid desert was limited (Bowler 1976, 1978), providing evidence for a wet phase that preceded the peak of aridity at 18,000 years ago. In the southern coastal situations, Martin and Peterson (1978) have stressed that eustatic sea-level changes markedly influenced environmental gradients. Thus in the Eucla area the increased aridity between 20,000 B.P. and 10,000 B.P. and the rise in rainfall to near present-day values at 5000 B.P. are reflected in the vegetation changes and are explained by the changing distance from the coast rather than any absolute reduction in precipitation at 18,000 B.P. Eustatic sea-level changes would also have influenced the environmental gradient in this time range in the mediterranean-climate areas in western and southern Australia and elsewhere because of their coastal situations.

The Subtropical Convergence, which in the southern hemisphere marks the northerly extent of the West Wind Drift, was displaced some 4° northwards at 18,000 B.P. in the eastern Indian Ocean. This displacement suggests that the West Wind Drift would have been partly deflected by the Australian landmass thereby strengthening the West Australian Current (Prell et al. 1979) and increasing the aridity of the west and north-west coasts. The corresponding northward shift of the westerlies and the accompanying cyclonic rainfall belt at 18,000 B.P. would be expected to have resulted in increased precipitation in southern Australia and the interior, but almost all data suggest rainfall was much lower than today (Walker 1978:92) and fossil dune orientations suggest a different and stronger wind pattern (Bowler 1976). Webster and Streten (1978) have attempted to explain this potential anomaly. They show that present-day dry conditions in northern and southern Australia occur when there is a southern and western displacement of the high-pressure cells. They therefore suggest that during the last glacial maximum the configuration of the Antarctic ice sheet and the exposure of the land bridge between Australia and Tasmania encouraged the formation of a semi-permanent ridge pattern in the upper westerlies over the eastern part of Australia and intensified troughs (i.e. high-pressure cells) in the Indian Ocean, western Pacific and Tasman Sea regions. With this situation and with a northerly displacement and strengthening of the westerlies due to the increased temperature gradient between the South Pole and the Subtropical Convergence, together with cooler oceans restricting evaporation and atmospheric disturbance, southern and southeastern Australia would have had a predominantly northwesterly circulation with most of the rain-bearing disturbances being deflected to the south of Tasmania. Thus most of the stronger winds from the west would not have brought rain and the interior of Australia would have been much drier than at present.

In pioneering attempts to model the glacial climates of southern Africa Van Zinderen Bakker (1976) has assumed a more northerly shift of almost 10° of latitude of the zone of westerlies and the penetration of cyclonic winter rains to the interior of the subcontinent to a latitude of some 24°S during glacial episodes. Under this model the present area of winter rainfall would have been somewhat reduced by an extension of the area of biseasonal rainfall. Other authors have followed Van Zinderen Bakker in proposing greater year-round precipitation with a winter maximum in the southern parts of South Africa at the last glacial maximum and Tankard and

Rogers (1978), for example, link this to an invasion of temperate forest displacing winter rainfall sclerophyllous vegetation from the fynbos area into the arid areas of the western Karoo and Namib. Axelrod and Raven (1978) have proposed a similar reconstruction of vegetation history. The Subtropical Convergence was displaced about 1° northwards at the glacial maximum in this section of the South Atlantic (Morley and Hays 1979) and this does not support the significant latitudinal shift of the westerlies demanded by the model.

More recent information for the interior of southern Africa from pan deposits suggests that evidence for increased precipitation, dating primarily between 16,000 B.P. and 15,000 B.P., was linked to greater summer rainfall as a consequence of a more southerly position of the Inter-Tropical Convergence Zone (Lancaster 1979a). The dating of observations is not precise enough to establish whether these more humid conditions spanned the whole of the last glacial maximum, or whether they were initiated later. As the last glacial maximum is widely correlated with expanding dune systems, Sarnthein (1978) has suggested that Kalahari dunes at present covered with vegetation were active in the recent past. This is in potential conflict with evidence for humid conditions in the present arid interior of southern Africa at 18,000 B.P. Sarnthein (1978) has pointed out that the reduction in deserts and thus in the extent of active dune systems preceded the onset of full de-glaciation and changes from desert to vegetated land are rapid. It may be that the increase in summer rainfall is related to atmospheric circulation conditions developed immediately after the last glacial maximum.

While evidence that winter cyclonic rains penetrated further into the interior of southern Africa is largely lacking at 18,000 B.P., evidence from biseasonal rainfall areas in the southern Cape shows that precipitation was reduced. The clearest evidence comes from a cave deposit in the foothills of the Swartberg (Deacon 1979) that preserves microflora and micromammals from the last glacial maximum. Avery (1979) found relatively high proportions of *Otomys unisulcatus*, a karoid-habitat taxon indicating that the last glacial maximum was the driest part of the Late Pleistocene. Cold indicators are also present and diversity is reduced. The occurrence of Alcelaphine antelopes and other grazers in the large mammal faunas from the same Swartberg site and others in the fynbos area (Klein 1980) in the Late Pleistocene, after the last glacial maximum, is further evidence of cooler, drier conditions and probably reflects the expansion of C_3 grasses in the vegetation mosaic of the intermontane valleys and coastal forelands. Under last glacial maximum conditions, at least in the intermontane valleys, temperature inversion and ground frosts would have been a strong selective factor in the distribution of individual taxa. From the winter rainfall area proper the occurrence of well-developed block streams and block fields in some mountain areas (Goede, personal communication) indicates freeze-thaw wedging of blocks from krans outcrops and significantly lower temperatures.

Data from Schalke's (1973:38) study of the Rietvlei sequence can be taken to indicate that effective precipitation was reduced in the Upper Pleniglacial or glacial maximum. Much has been made of the evidence Schalke presented for expansion of forest at several periods in the Late Pleistocene prior to the glacial maximum (Axelrod and Raven 1978; Tankard and Rogers 1978). The relative frequencies of *Podocarpus* pollen, sometimes but not invariably associated with *Ilex mitis* and *Curtisia dentata* (Schalke 1973:27), have been used to define biostratigraphic zones and the presence of these taxa in particular zones has been interpreted as indicating mixed *Podocarpus* forest on the Cape Flats, or the central part of it, and a climate similar to that of present-day Knysna which has a rainfall of 860 mm yr^{-1}, distributed biseasonally (Schalke 1973:38). This interpretation is not completely accepted (Van Zinderen Bakker 1976:182) because claims of forest vegetation growing on a site in the past cannot be based solely on the frequencies of Podocarpus grains. However, the presence of more than 20% Podocarpus in spectra from the Cape Flats borehole dated to some 40,000 B.P. (GrN-5550) is still impressive evidence of some forest expansion in the Late Pleistocene. As the dating in this time range could be a minimum rather than a finite age determination, it requires support from further observations. The temporary linking up of new, isolated forest patches in kloof refuges in the fynbos area at repeated times in the Pleistocene is an acceptable postulate, but large-scale displacement of fynbos by podocarp forest at 18,000 years or earlier in the Late Pleistocene is at present unsupported. In the southern Cape periods of local forest expansion can be shown in the Holocene and dry conditions which prevailed prior to 8000 B.P. (Martin 1968) presumably extended back to the last glacial maximum.

1.4.3 PLEISTOCENE CLIMATES AND BIOGEOGRAPHY

Most modes of speciation involve a geographical component of some kind (White 1978) and as climatic fluctuations are a possible source of geographic change, the study of Pleistocene climates has relevance for the discussion of speciation in general and, in particular here, the species richness of the floras and degree of endemism associated with mediterranean-type ecosystems. Not all the mediterranean-climate regions show the same level of species diversity; the Cape region, for example, is notable for its anomalously high delta diversity (Kruger and Taylor 1979). The regions differ in terrain diversity, fertility of substrates, geographical extent, as well as the nature and age of the biotas and these factors contribute to interregional differences. As a rule, however, these regions are richer in species than adjoining non-mediterranean-climate regions.

Stebbins and Major (1965) make the point that the presence of old species, essentially Tertiary relicts, and/or active speciation in recent times can contribute to the diversity of a flora. In the Californian context the older relict species appear to be associated with geographically distinct mesic refuges and more recent speciation, during which neo-endemics appeared, seems associated with ecotonal or border areas of high ecological diversity (topography, soil and climate combined) between mesophytic forest and xerophytic chaparral and scrub formations. It is their hypothesis that border areas are particularly sensitive to climate changes such as took place in the Pleistocene and that climatic change caused migration of resident species or evolution and migration. In conjunction with edaphic diversity these processes would most actively promote evolution and differentiation of new species. It is noteworthy that their data show no correlation between species richness and increasing altitude in California.

The increased effect of Pleistocene climatic fluctuations in areas of transitional climate, defined as occurring between those regions with permanently arid and permanently humid climate, again in conjunction with edaphic factors, are similarly held by Hopper and Maslin (1978) to have been important in promoting speciation of Acacia in Western Australia. In a more recent paper Hopper (1979) has restated the concept that the semi-arid transitional climatic zones, equated with an annual rainfall of

300-800 mm in Western Australia, are particularly favourable sites for speciation whereas permanently humid and permanently arid zones favour evolutionary stability.

The concept that border or ecotonal centres in the sense of Stebbins and Major (1965) and areas of transitional climates in the sense of Hopper (1979) are areas of high stress may be questionable since "permanently" arid and "permanently" humid areas also show marked changes in vegetation in response to Pleistocene climatic regimes. A good example is the scale of Pleistocene changes in equatorial forest biomes (Hamilton 1976; Simpson and Haffer 1978) where many populations of humid forest organisms were restricted to isolated refugia under cooler, drier Pleistocene climates. An alternate hypothesis is that ecotonal centres of species richness in the mediterranean areas are lower rather than higher stress environments. The association of these centres with intermediate altitudes as in California may reflect the more pronounced effect of cooler Pleistocene climates on plant growth at higher altitudes and, conversely, the effect of drier Pleistocene climates on more xerophytic plant formations at lower altitudes. Lower extinction rates could have occurred in these centres with the scale of climatic change still adequate to allow for adjustments in geographic range of taxa and allopatric speciation.

The question of the role of Pleistocene climates and climatic stress in promoting speciation is also pertinent to discussion of the species richness and the high degree of endemism in the mountain fynbos of the Cape. The vegetation occupies what are in relative terms intermediate elevations in a terrain of high diversity and an area of nutrient-poor soils (Kruger and Taylor 1979). The occurrence of numbers of monotypic genera and endemic taxa suggests relict species contribute significantly to the anomalously high level of species diversity. Taylor (1980) has pointed out that the greatest concentrations of taxa and endemics are found in the southwestern Cape, equivalent to the Caledon centre of Nordenstam (1969), which has the highest rainfall and the highest terrain diversity. Although the mountain fynbos occupies an essentially transitional climate zone it may not have been a zone of more marked stress during Pleistocene climatic fluctuations. Under a cold, dry climate such as that of the last glacial maximum, reduced precipitation is likely to have been less critical in the mountains where orographic effects are marked. Because of the low tolerance of many fynbos taxa to frost (Levyns 1962), the effects of climatic changes on the mountain peaks and in the intermontane valleys and foothills would have been more pronounced than at intermediate elevations, zones which occupy the greatest area. The very high delta diversity of the mountain fynbos suggests that the vegetation was fragmented into large landscape units, but Kruger and Taylor (1979) argue that climatic fluctuations would not have disrupted the distribution of fynbos along the mountain chains. They do, however, allow that Pleistocene climatic changes may have created isolated subpopulations in some cases, but do not state whether they believe that this promoted speciation, emphasizing instead possible changes in community composition. There are at present no explicit testable models to account for species richness and endemism in the fynbos and Goldblatt (1978:409), for example, considers that a unique combination of factors including topography, climates of different periods, soils and the presence of an old sclerophyllous vegetation are pertinent. Progress in understanding the role of these different factors may ultimately lie in gaining a better knowledge of the principal modes of speciation. This requires more information on pollination and seed dispersal mechanisms and the relevant genetic systems than is currently available.

One of the more impressive biogeographic patterns that emerges from the mapping of the numbers of species per unit area associated with mediterranean and other ecosystems is phytogeographic foci or centres of species concentration. Centres can be defined for single taxa as has been done, for example, for Aloe (Holland 1978), Euryops (Nordenstam 1969) and Aspalathus (Dahlgren 1963) in southern Africa and Acacia, Eucalyptus, Proteaceae and Rutaceae (Hopper and Maslin 1978; Hopper 1979) in Western Australia. Overlapping phytogeographic centres of several taxa characterize the southwestern Cape, for instance. In southern Africa, centres of species richness are associated with upland areas which suggests that it is ultimately the reliability of precipitation which promotes continued species richness. This could be true in the Cape where the source of precipitation is cyclonic rains. Under Pleistocene climates even with reduced precipitation due to the influence of colder ocean surface waters, extinction rates may have been lower in favourable terrain because the frequency of cyclones would have been maintained if not increased. Present biogeographic patterns represent a stage in a trend to isolate centres and diminish floristic interchange between them as a consequence of drying Cenozoic climates. Kemp (1978) believes that in the South West Botanical Province of Australia this process began at the end-Miocene and others (Marchant 1973; Hopper 1979) possibly earlier. Changes in the Cape centre may also date from that time. Pleistocene aridity would have maintained the isolation of centres.

There is virtually no information available on the rates of speciation of either woody or herbaceous taxa for mediterranean-type ecosystems, although they are characterized as regions of species richness. Hopper and Maslin (1978) estimate 30,000 years for some level of genetic divergence in Acacia in Western Australia and Holland (1978) has estimated the rate of speciation in Aloe to be more than 100,000 years. Rourke (1972) has explained the disjunct distribution of two subpopulations of a subspecies of Leucospermum between points on the Cape coast 160 km apart as due to the existence of a suitable pathway around the present rocky mountain barrier at a time of lowered sea level, which must have been 10,000 years ago. Separation of the subpopulations over this time range has apparently not led to greater genetic divergence. This example shows that Pleistocene climatic events are repetitive and may not be of great value in calibrating rates of speciation. If it is assumed that a continuous suitable sandy substrate habitat only existed at a sea level close to that of the last glacial maximum (120 m below the present), then the separation of these Leucospermum subpopulations may date from 18,000 years ago, or again from 145,000, 240,000, 530,000 or 750,000 years ago. These are the times in the last million years when glacial stages occurred with maxima comparable to that of the last glacial and with correspondingly low sea levels (Van Donk 1976). The geographical separation of palaeoendemics with a known long geological history and neoendemics in California (Stebbins and Major 1965) suggests that species classified as neoendemics are of Pleistocene age. Again, it is not established whether speciation rates are gradual or whether bursts of speciation (punctuated equilibria) occurred for example in response to epeirogenic uplift (Axelrod and Raven 1978) or abrupt climatic changes.

1.5 Concluding remarks

The differences in the evolutionary potentials of taxa of northern and southern hemisphere mediterranean-type ecosystems are only explained by the fact that the nuclei of these biotas had their origins on different sides of the Tethys. The biotas of the southern continents are more distantly related to each other than those of the north because of continental separation since the Cretaceous. The simple north-south dichotomy does not include comparisons of the continental margins on which mediterranean-type ecosystems are found. Those that have evolved on margins characterized by Cretaceous-Cenozoic mountain building (California, Chile and the larger part of the Mediterranean Basin) contrast with those evolved in essentially older landscapes on more stable margins in southern Africa and Australia. It is in the latter regions that base-poor substrates are important. All continental margins have been subjected to eustatic changes of sea level during the Cenozoic with only those of the later Cenozoic clearly related to changes in glacial ice volumes (Vail and Hardenbol 1979). On the stable margin of Western Australia the record is relatively well preserved (Quilty 1977) and the late Eocene transgression (Hopper 1979) and the Oligocene regression (Kemp 1978) for example seem correlated with important biological changes. A similar sequence of transgressive and regressive cycles is recorded for the Cape coast (Siesser and Dingle 1981). The glacio-eustatic changes of the late Miocene through the Pleistocene have potential biogeographic significance.

Climate has a primary influence on vegetation and, local effects aside, globally synchronous climatic changes have been documented through the Cenozoic so that broadly similar climatic histories can be posited for the mediterranean-type regions. The trend in the Cenozoic has been towards cooler, drier climates culminating in cyclic glacial and interglacial episodes in the Pleistocene. The biological implications are that the larger-scale changes such as those registered in the earliest Oligocene at 38 m.y. represented periods of relatively intense selection for new adaptations. These are indicated in the changeover from a tropical weathering regime to a strongly seasonal if not semi-arid sedimentary regime in southern California (Fig. 4; Peterson and Abbott 1979) that is suggested as being correlated with the Oligocene event. Given uncertainties in correlation, it may be significant that in North America and western Asia sclerophyllous taxa first enter the fossil record at the Eocene-Oligocene boundary or apparently just prior to it (Axelrod 1975).

Under globally ice-free conditions the Palaeocene and Eocene climates between 65 and 38 m.y. ago were warm and humid even in the high latitudes. The association of taxa linked to present-day tropical and temperate environments in the example of the microflora from the Werillup Formation of Western Australia (Hoss 1975) suggests weak vegetation zonation in the earlier Palaeogene. Associations of taxa in the vegetation differs from the present so that the relationship to modern ecosystems is very tenuous. However, some older taxa like the Proteaceae in the Werillup Formation microflora, for example, are well represented and are highly diversified even at this early time. One expects that progressively more modern associations of taxa should be discernible in the fossil record from the Oligocene onwards, although the record is limited from areas other than California and the Mediterranean Basin until Miocene times.

Johnson (1973) has posited that the evolution of dominants is the most important aspect of ecosystem history. Evidence from the mediterranean-climate regions of southern Africa and southern Australia indicates that vegetation dominated by forest taxa persisted until the later Miocene with sclerophyllous shrub vegetation increasingly dominant thereafter. The paucity of palaeo-botanical evidence from the present-day adjacent arid areas precludes adequate discussion of the history of these sclerophyllous vegetations. In the comparable North American context where the fossil record is better known, a large part of the history of the sclerophyllous vegetation (Raven and Axelrod 1978) was played out under the selective pressures of increasing aridity in the now bordering drier areas. Specht (1979) and Axelrod (1975) have stressed the importance of the pre-adaptation of ancestral sclerophyllous taxa at local dry or waterlogged sites with thin or infertile soil in the late Cretaceous and earlier Cenozoic and their subsequent wider distribution under developing seasonal climates as forest understorey or component of sclerophyllous woodland. When the vegetation was further thinned in response to ice-age aridity and man disturbed the environment, modern communities resulted. This broad scenario of vegetation history has yet to be adequately tested in the fossil record for the southern continents.

Considering fynbos, our evolutionary history must account for features like the high delta diversity and the unwieldy speciation in a few prominent taxa. This is best done by postulating that the present vegetation evolved from a restricted biota in a highly diverse terrain over an extended period. The fynbos landscape has served as a refugium for palaeo-endemics but recruitment probably occurred as it also includes taxa common to biogeographic centres elsewhere in Africa. However, establishment of fynbos can be viewed as a gradual in situ process rather than a migration event. It is noteworthy that several more widely distributed taxa are represented by more numerous species with morphologically advanced characters in the fynbos. This was the Levyns's (1962, 1964) basis for holding that speciation in the fynbos is secondary, meaning simply that it took place after the primary dispersal of the relevant ancestral taxa. There are several centres of species richness for southern Africa as a whole which have been suggested as both refugia and sites of active speciation (Nordenstam 1969). The difference between the fynbos and other centres, more specifically the Caledon centre of Nordenstam, is in degree and not in kind of species richness. The common feature is that they are associated with areas of high terrain with concomitant edaphic diversity and contain many varied habitats. It seems that allopatric speciation has been favoured at these loci and that it is the extremely rugged and old landscape of the southwestern Cape which renders it a super centre in terms of species richness. The same pattern of biogeographic centres has been noted for Western Australia (Hopper 1979) where relief is more subdued but where erosion has created varied landscape units. The contrasts between the Cape and the Australian mediterranean-type regions and those elsewhere suggest that the age of the landscapes and the base status of the substrate are important in determining species richness. Studies of speciation processes fall outside the scope of direct palaeoecological observation but they are relevant to the better understanding of the constraints, including climatic change and the development of landscape, under which the processes have been operative. Palaeoecological studies in turn provide a measure of the results of these evolutionary processes in terms of the scale and rate of change in the biotas.

The pattern of extinction of the Taxodiaceae in Spain and south-east France exemplifies the kind of evidence available for dating the inception of mediterranean-type climates. Comparable extinction leading to impoverishment of the richer sclerophyllous vegetation that flourished in California and Nevada in the Miocene and Pliocene have been recorded for North America (Axelrod and Raven 1978). Coetzee (1978) has noted that the end-Miocene marks the disappearance of older Tertiary taxa in the southwestern Cape and as in Australia extinctions in this time range may be linked to cooling associated with the expanding Antarctic ice sheet and increasing dryness. Dating and duration and scale of change in the Pleistocene are much better understood now. It is now generally appreciated that plants have responded on an individualistic basis (Livingstone 1975) and migration of whole communities as units was not involved. There remains however the task of assessing more fully the biological implications of Pleistocene climatic change since the glacial maximum 18,000 years ago in all the mediterranean-type regions. It is only for the northern Mediterranean littoral that substantial data are available.

Acknowledgements

Financial support for the preparation of this paper was received from the CSIR, Pretoria. It also reports some results from research in the southern Cape funded by the HSRC, Pretoria. I am grateful to A Scholtz for assistance and Q B Hendey for providing samples for study. Comments on an earlier draft of the manuscript have been received from D I Axelrod and F J Kruger and these are gratefully acknowledged. The illustrations were prepared by J Deacon.

References

ADAMSON RS (1931) Notes on some petrified wood from Banke, Namaqualand. Transactions of the Royal Society of South Africa 19: 255-258

ADAMSON RS (1934) Fossil plants from Fort Grey near East London. Annals of the South African Museum 31: 67-96

AMBROSE GJ, CALLEN RA, FLINT RB, LANGE RT (1979) Eucalyptus fruits in stratigraphic context in Australia. Nature 280: 387-389

ASCHMANN H (1973) Distribution and peculiarity of mediterranean ecosystems. In: DI CASTRI F, MOONEY HA (eds) Mediterranean type ecosystems: origin and structure. Springer-Verlag, Berlin, pp 11-19

AVERY DM (1979) Upper Pleistocene and Holocene palaeoenvironments in the southern Cape: the micromammalian evidence from archaeological sites. PhD Thesis. University of Stellenbosch, Stellenbosch

AXELROD DI (1973) History of the mediterranean ecosystem in California. In: DI CASTRI F, MOONEY H (eds) Mediterranean type ecosystems: origin and structure. Springer-Verlag, Berlin, pp 225-277

AXELROD DI (1975) Evolution and biogeography of Madrean-Tethyan sclerophyll vegetation. Annals of the Missouri Botanic Garden 62: 280-334

AXELROD DI, RAVEN PH (1978) Late Cretaceous and Tertiary vegetation history of Africa. In: WERGER MJA (ed) Biogeography and ecology of southern Africa. Junk, The Hague, pp 77-130

BEARD JS (1977) Tertiary evolution of the Australian flora in the light of latitudinal movements of the continent. Journal of Biogeography 4: 111-118

BELL EA, EVANS CS (1978) Biochemical evidence of a former link between Australia and the Mascarene Islands. Nature 273: 295-296

BOTTEMA S (1978) The late glacial in the eastern Mediterranean and Near East. In: BRICE WC (ed) The environmental history of the Near and Middle East since the last ice age. Academic Press, London, pp 15-28

BOWLER JM (1976) Aridity in Australia: age, origins and expressions in aeolian landforms and sediments. Earth Science Reviews 12: 279-310

WLER JM (1978) Glacial age aeolian events at high and low latitudes: a southern hemisphere perspective. In: VAN ZINDEREN BAKKER EM (ed) Antarctic glacial history and world palaeonenvironments. Balkema, Rotterdam, pp 149-172

WLER JM, HOPE GS, JENNINGS GN, SINGH G, WALKER D (1976) Late Quaternary climates of Australia and New Guinea. Quaternary Research 6: 359-394

AKENRIDGE GP (1978) Evidence for a cold dry full glacial climate in the American Southwest. Quarternary Research 9: 22-40

CHARDT B (1978) Oxygen isotope palaeotemperatures from the Tertiary Period in the North Sea area. Nature 275: 121-123

FFETAUT E, TAQUET P (1979) An early Cretaceous terrestrial crocodilian and the opening of the South Atlantic. Nature 280: 486-487

LLEN RA, TEDFORD RH (1976) New Late Cainozoic rock units and depositional environments, Lake Frome area, South Australia. Transactions of the Royal Society of South Australia 100: 125-167

URCHILL DM (1973) The ecological significance of tropical mangroves in the early Tertiary floras of Southern Australia. Special Publications of the Geological Society of Australia 4: 79-86

APPERTON CM (1979) Glaciation in Bolivia before 3.27 M.yr. Nature 277: 375-377

IMAP MEMBERS (1976) The surface of the ice age earth. Science 191: 1131-1137

ETZEE JA (1978) Late Cainozoic palaeoenvironments of southern Africa. In: VAN ZINDEREN BAKKER EM (ed) Antarctic glacial history and world palaeoenvironments. Balkema, Rotterdam, pp 115-127

OKE HBS (1972) The fossil mammal fauna of Africa. In: KEAST A, ERK FC, GLASS B (eds) Evolution, mammals and southern continents. State University of New York Press, Albany, pp 89-140

RVINUS G, HENDEY QB (1978) A new Miocene vertebrate locality at Arrisdrift in Namibia (South West Africa). Neues Jahrbuch für Geologie und Paläontologie, Monatshefte 4: 193-205

ACRAFT J (1975) Historical biogeography and earth history: perspectives for a future synthesis. Annals of the Missouri Botanical Gardens 62: 227-250

HLGREN R (1963) Studies on Aspalathus: phytogeographical aspects. Botaniska Notiser 116: 431-472

ACON HJ (1979) Excavations at Boomplaas Cave: a sequence through the Upper Pleistocene and Holocene in South Africa. World Archaeology 10: 241-257

DECKKER P, GEURTS M-A, JULIA R (1979) Seasonal rhythmites from a Lower Pleistocene lake in north-eastern Spain. Palaeogeography, Palaeoclimatology, Palaeoecology 26: 43-71

TTMANN ME (1973) Angiospermous pollen from Albian to Turonian sediments of eastern Australia. Special Publications of the Geological Society of Australia 4: 3-34

ESTER-HAAS L, SCHRADER H-J (1979) Neogene coastal upwelling history off northwest and southwest Africa. Marine Geology 29: 39-53

NGLE RV, LORD AR, HENDEY QB (1979) New sections in the Varswater Formation (Neogene) of Langebaan Road, south-western Cape, South Africa. Annals of the South African Museum 78: 81-92

DSON J (1977) Late Quaternary palaeoecology of Wyrie Swamp, south-eastern South Australia. Quaternary Research 8: 97-114

YLE JA (1978) Origin of angiosperms. Annual Review of Ecology and Systematics 9: 365-392

TOIT AL (1939) Geology of South Africa. Oliver and Boyd, Edinburgh.

FLÖHN H (1978) Comparison of Antarctic and Arctic climate and its relevance to climatic evolution. In: VAN ZINDEREN BAKKER EM (ed) Antarctic glacial history and world palaeoenvironments. Balkema, Rotterdam, pp 3-13

FÖRSTER R (1978) Evidence for an open seaway between northern and southern proto-Atlantic in Albian times. Nature 272: 158-159

FRAKES LA, KEMP EM (1973) Palaeogene continental positions and evolution of climate. In: TARLING DG, RUNCORN SK (eds) Implications of continental drift to the earth sciences. Academic Press, London, pp 539-558

GATES WL (1976) The numerical simulation of ice-age climate with a global general circulation model. Journal of the Atmospheric Sciences 33: 1844-1873

GOLDBLATT P (1978) An analysis of the flora of southern Africa: its characteristics, relationships and origins. Annals of the Missouri Botanical Gardens 65: 369-436

HAMILTON A (1976) The significance of patterns of distribution shown by forest plants and animals in tropical Africa for the reconstruction of Upper Pleistocene palaeoenvironments: a review. Palaeoecology of Africa 9: 63-97

HEATH GR (1979) Simulations of a glacial palaeoclimate by three different atmospheric simulation models. Palaeogeography, Palaeoclimatology, Palaeoecology 26: 291-303

HENDEY QB (1974) The Late Cainozoic carnivora of the south-western Cape Province. Annals of the South African Museum 69: 1-369

HENDEY QB (1976) The Pliocene fossil occurrences in 'E' Quarry, Langebaanweg, South Africa. Annals of the South African Museum 69: 215-247

HENDEY QB (1981) Geological succession at Langebaanweg, Cape Province, and global events of the Late Tertiary. South African Journal of Science 77: 33-38

HEUSSER CJ (1974) Vegetation and climate of the southern Chilean lake district during and since the last interglacial. Quarternary Research 4: 290-315

HEUSSER L (1978) Pollen in Santa Barbara Basin, California: a 12,000 yr record. Geological Society of America Bulletin 89: 673-678

HOLLAND PG (1978) An evolutionary biogeography of the genus Aloe. Journal of Biogeography 5: 213-216

HOPPER SD (1979) Biogeographical aspects of speciation in the southwest Australian flora. Annual Review of Ecology and Systematics 10: 399-422

HOPPER SD, MASLIN BR (1978) Phytogeography of Acacia in Western Australia. Australian Journal of Botany 26: 63-78

HOSS D (1975) Preliminary investigation of the palynology of the Upper Eocene Werillup Formation, Western Australia. Journal of the Royal Society of Western Australia 58: 1-13

HSÜ KJ, MONTADERT L, BERNOULLI D, CITA MB, ERIKSON A, GARRISON RE, KIDD RB, MÉLIÈRES F, MÜLLER C, WRIGHT R (1977) History of the Mediterranean salinity crisis. Nature 267: 399-403

HUGHES NF (1976) Palaeobiology of angiosperm origins. Cambridge University Press, Cambridge

JOHNSON AW (1973) Historical view of the concept of ecosystem convergence. In: DI CASTRI F, MOONEY HA (eds) Mediterranean type ecosystems: origin and structure. Springer-Verlag, Berlin, pp 3-7

JOHNSON DL (1977a) The late Quarternary climate of coastal California: evidence for an ice-age refugium. Quarternary Research 8: 154-179

JOHNSON DL (1977b) The California ice-age refugium and the Rancholabrean extinction problem. Quarternary Research 8: 149-153

KEAST A (1972) Comparisons of contemporary mammal faunas on southern continents. In: KEAST A, ERK FC, GLASS B (eds) Evolution, mammals and southern continents. State University of New York Press, Albany, pp 19-87

KEAST A, ERK FC, GLASS B (eds) (1972) Evolution, mammals and southern continents. State University of New York Press, Albany

KEMP EM (1978) Tertiary climatic evolution and vegetation history in the south-east Indian Ocean region. Palaeogeography, Palaeoclimatology, Palaeoecology 24: 169-208

KENNETT JP (1978a) Cainozoic evolution of circumantarctic palaeooceanography. In: VAN ZINDEREN BAKKER EM (ed) Antarctic glacial history and world palaeoenvironments. Balkema, Rotterdam, pp 41-56

KENNETT JP (1978b) The development of planktonic biogeography in the southern ocean during the Cenozoic. Marine Micropalaeontology 3: 301-345

KENNETT JP, SHACKLETON NJ (1976) Oxygen isotopic evidence for the development of the psychrosphere 38 m.y. ago. Nature 260: 513-515

KIRCHHEIMER F (1934) On pollen from the Upper Cretaceous dysodil of Banque, Namaqualand (South Africa). Transactions of the Royal Society of South Africa 21: 41-50

KLEIN RG (1980) Environmental and ecological implications of large mammals from Upper Pleistocene and Holocene sites in southern Africa. Annals of the South African Museum 81: 223-283

KRUGER FJ, TAYLOR HC (1979) Plant species diversity in Cape fynbos: gamma and delta diversity. Vegetatio 41: 85-93

KUKLA GJ (1977) Pleistocene land-sea correlations 1: Europe. Earth Science Reviews 13: 307-374

LAMB HH (1972) Climate: present, past and future, vol 1: Fundamentals and climate now. Methuen, London

LANCASTER IN (1979a) Evidence for a widespread late Pleistocene humid period in the Kalahari. Nature 279: 145-146

LANCASTER IN (1979b) Quarternary environments in the arid zone of southern Africa. Occasional Paper of the Department of Geography and Environmental Studies, University of the Witwatersrand 22: 1-75

LANGE RT (1980) Evidence of lid-cells and host specific microfungi in the search for Tertiary Eucalyptus. Review of Palaeobotany and Palynology 29: 29-33

LEVYNS MR (1962) Past plant migrations in South Africa. Annals of the Cape Provincial Museums 2: 7-10

LEVYNS MR (1964) Migrations and origin of the Cape Flora. Transactions of the Royal Society of South Africa 37: 85-107

LIVINGSTONE DA (1975) Late Quarternary climatic change in Africa. Annual Review of Ecology and Systematics 6: 249-280

LOUTIT TS, KENNETT JP (1979) Application of carbon isotope stratigraphy to Late Miocene shallow marine sediments, New Zealand. Science 204: 1196-1199

MACPHAIL MK (1979) Vegetation and climates in southern Tasmania since the last glaciation. Quarternary Research 11: 306-341

MANABE S, HAHN DG (1977) Simulation of the tropical climate of an ice age. Journal of Geophysical Research 82: 3889-3911

MARCHANT NG (1973) Species diversity in the south-western flora. Journal of the Royal Society of Western Australia 56: 23-30

MARSHALL LG, BUTLER RF, DRAKE RE, CURTIS GH, TEDFORD RH (1979) Calibration of the great American interchange. Science 204: 272-279

MARTIN ARH (1968) Pollen analysis of Groenvlei Lake sediments, Knysna (South Africa). Review of Palaeobotany and Palynology 7: 107-144

MARTIN ARH (1973) Re-appraisal of some palynomorphs of supposed proteaceous affinity. Special Publications of the Geological Society of Australia 4: 73-78

MARTIN HA (1973) Upper Tertiary palynology in southern New South Wales. Special Publications of the Geological Society of Australia 4: 35-54

MARTIN HA (1977a) The history of Ilex (Aquifoliaceae) with special reference to Australia: evidence from pollen. Australian Journal of Botany 25: 655-673

MARTIN HA (1977b) The Tertiary stratigraphic palynology of the Murray Basin in New South Wales. 1. The Hay-Balranald-Wakool Districts. Journal and Proceedings of the Royal Society of New South Wales 110: 41-47

MARTIN HA (1979) Stratigraphic palynology of the Mooki Valley, New South Wales. Journal and Proceedings of the Royal Society of New South Wales 112: 71-78

MARTIN HA, PETERSON JA (1978) Eustatic sea-level changes and environmental gradients. In: PITTOCK AB, FRAKES LA, JENSSEN D, PETERSON JA, ZILLMAN JW (eds) Climatic change and variability: a southern perspective. Cambridge University Press, Cambridge, pp 1-8

MERCER JH (1976) Glacial history of southernmost South America. Quarternary Research 6: 125-166

MERCER JH (1978) Glacial development and temperature trends in the Antarctic and in South America. In: VAN ZINDEREN BAKKER EM (ed) Antarctic glacial history and world palaeoenvironments. Balkema, Rotterdam, pp 73-93

MICHAUX J, SUE J-P, VERNET J-L (1979) Climatic inferences from the history of the Taxodiaceae during the Pliocene and the early Pleistocene in western Europe. Review of Palaeobotany and Palynology 27: 185-191

MORLEY JJ, HAYS JD (1979) Comparison of glacial and interglacial oceanographic conditions in the South Atlantic from variations in calcium carbonate and radiolarian distributions. Quarternary Research 12: 396-408

MULLER J (1970) Palynological evidence on early differentiation of angiosperms. Biological Review 45: 417-450

NILSEN TH (1978) Lower Tertiary laterite on the Iceland-Faeroe Ridge and the Thulean land bridge. Nature 274: 786-788

NORDENSTAM B (1969) Phytogeography of the genus Euryops (Compositae): a contribution to the phytogeography of southern Africa. Opera Botanica 23: 1-77.

PETERSON GL, ABBOTT PL (1979) Mid-Eocene climatic change, south-western California and north-western Baja California. Palaeogeography, Palaeoclimatology, Palaeoecology 26: 73-87

PETERSON GM, WEBB T, KUTZBACH JE, VAN DER HAMMEN T, WIJMSTRA TA, STREET FA (1979) The continental record of environmental conditions at 18,000 yr B.P.: an initial evaluation. Quaternary Research 12: 47-82

PHILLIPS JFV (1927) Fossil Widdringtonia in lignite of the Knysna Series, with a note on fossil leaves of several other species. South African Journal of Science 14: 188-197

PRELL WL, HUTSON WH, WILLIAMS DF (1979) The Subtropical Convergence and late Quaternary circulation in the southern Indian Ocean. Marine Micropalaeontology 4: 225-234

QUILTY PC (1977) Cenozoic sedimentation cycles in Western Australia. Geology 5: 336-340

RAVEN PH (1973) The evolution of mediterranean floras. In: DI CASTRI F, MOONEY HA (eds) Mediterranean type ecosystems: origin and structure. Springer-Verlag, Berlin, pp 213-224

RAVEN PH, AXELROD DI (1974) Angiosperm biogeography and past continental movements. Annals of the Missouri Botanical Gardens 61: 539-673

RAVEN PH, AXELROD DI (1978) Origin and relationships of the California flora. University of California Publications in Botany 72: 1-134
READ HH, WATSON J (1975) Introduction to geology, vol 2. Earth history: part II: Later stages of earth history. Macmillan, London
RENNIE JVL (1931) Note on fossil leaves from the Banke clays. Transactions of the Royal Society of South Africa 19: 251-253
ROGERS AW (1908) Notes on a journey to Knysna. Annual Report of the Geological Commission of the Cape of Good Hope 13: 129,134
ROGNON P, WILLIAMS MAJ (1977) Late Quaternary climatic changes in Australia and North Africa: a preliminary interpretation. Palaeogeography, Palaeoclimatology, Palaeoecology 21: 285-327
ROURKE JP (1972) Taxonomic studies on Leucospermum R. Br. Supplement to the Journal of South African Botany 8: 1-194
RYAN WBF (1973) Geodynamic implications of the Messinian crisis of salinity. In: DROOGER CW (ed) Messinian events in the Mediterranean. North Holland, Amsterdam, pp 26-38
SARNTHEIN M (1978) Sand deserts during glacial maximum and climatic optimum. Nature 272: 43-46
SAVIN SM, DOUGLAS RG, STEHLI FG (1975) Tertiary marine palaeotemperatures. Geological Society of America Bulletin 86: 1499-1510
SCHALKE HJWG (1973) The upper Quaternary of the Cape Flats area (Cape Province, South Africa). Scripta Geologica 25: 1-57
SCHUSTER RM (1976) Plate tectonics and its bearing on the geographical origin and dispersal of Angiosperms. In: BECK CB (ed) Origin and early evolution of Angiosperms. Columbia University Press, New York, pp 48-138
SCLATER JC, HELLINGER S, TAPSCOTT C (1977) The palaeobathymetry of the Atlantic Ocean from the Jurassic to the present. Journal of Geology 85: 509-552
SHACKLETON NJ (1969) The last interglacial in the marine and terrestrial records. Proceedings of the Royal Society of London B 174: 135-154
SHACKLETON NJ (1977) The oxygen isotope stratigraphic record of the Late Pleistocene. Philosophical Transactions of the Royal Society of London 280: 169-182
SHACKLETON NJ, KENNETT JP (1975a) Palaeotemperature history of the Cenozoic and the initiation of Antarctic glaciation: oxygen and carbon isotope analyses in DSDP sites 277, 279 and 281. Initial Reports of the Deep Sea Drilling Project 29: 743-755
SHACKLETON NJ, KENNETT JP (1975b) Late Cenozoic oxygen and carbon isotope changes at DSDP site 284: implications for glacial history of the northern hemisphere and Antarctica. Initial Reports of the Deep Sea Drilling Project 29: 801-807
SHACKLETON NJ, OPDYKE ND (1976) Oxygen-isotope and palaeomagnetic stratigraphy of Pacific core V28-239: late Pliocene to latest Pleistocene. In: CLINE RM, HAYS JD (eds) Investigation of late Quaternary palaeooceanography and palaeoclimatology. Geological Society of America Memoir 145: 449-464
SIESSER WG (1978) Aridification of the Namib Desert: evidence from oceanic cores. In: VAN ZINDEREN BAKKER EM (ed) Antarctic glacial history and world palaeoenvironments. Balkema, Rotterdam, pp 105-113
SIESSER WG (1980) Late Miocene origin of the Benguela upwelling system off northern Namibia. Science 208: 283-285
SIESSER WG, DINGLE RV (1981) Tertiary sea-level movements around southern Africa. Journal of Geology 89: 89-96
SIMPSON BB, HAFFER J (1978) Speciation patterns in the Amazon forest biota. Annual Review of Ecology and Systematics 9: 497-518

SIMPSON ESW (1977) Evolution of the south Atlantic. The Geological Society of South Africa. Annexure to vol 80: Alex L du Toit Memorial Lectures 15: 1-15
SMITH AG, BRIDEN JC (1977) Mesozoic and Cenozoic palaeocontinental maps. Cambridge University Press, Cambridge
SOLBRIG OT, CODY ML, FUENTES ER, GLANZ W, HUNT JH, MOLDENKE AR (1977) The origin of the biota. In: MOONEY HA (ed) Convergent evolution in Chile and California: Mediterranean climate ecosystems. Dowden, Hutchinson and Ross, Stroudsberg, Pennsylvania, pp 13-26
SPECHT RL (1979) Heathlands and related shrublands of the world. In: SPECHT RL (ed) Ecosystems of the world, vol 9A. Heathlands and related shrublands. Descriptive studies. Elsevier, Amsterdam, pp 1-18
STEBBINS GL, MAJOR J (1965) Endemism and speciation in the California flora. Ecological Monographs 35: 1-35
STOVER LE, PARTRIDGE AG (1973) Tertiary and Late Cretaceous spores and pollens from the Gippsland Basin, south-eastern Australia. Proceedings of the Royal Society of Victoria 85: 237-286
TANKARD AJ, ROGERS J (1978) Late Cenozic palaeoenvironments on the west coast of southern Africa. Journal of Biogeography 5: 319-337
TARLING DH, RUNCORN SK (1973) Implications of continental drift to the earth sciences, vol 1. Academic Press, London
TAYLOR HC (1978) Capensis. In: WERGER MJA (ed) Biogeography and ecology of southern Africa. Junk, The Hague, pp 171-229
TAYLOR HC (1980) Phytogeography of fynbos. Bothalia 13: 231-235
THIEDE J (1978) A glacial Mediterranean. Nature 276: 680-683
THIERGART F, FRANTZ R, RAUKOPF K (1966) Summary of the pollen flora of Tertiary deposits near Knysna (Cape Province). Palaeoecology of Africa 1: 100-102
THROWER NJW, BRADBURY DE (1973) The physiography of the mediterranean lands with special emphasis on California and Chile. In: DI CASTRI F, MOONEY HA (eds) Mediterranean type ecosystems: origin and structure. Springer-Verlag, Berlin, pp 37-52
THUNELL RC (1979a) Eastern Mediterranean Sea during the last glacial maximum: at 18,000 year B.P. reconstruction. Quaternary Research 11: 353-372
THUNELL RC (1979b) Climatic evolution of the Mediterranean Sea during the last 5.0 million years. Sedimentary Geology 23: 67-79
VAIL PR, HARDENBOL J (1979) Sea-level changes during the Tertiary. Oceanus 22: 71-79
VAN DEVENDER TR, SPAULDING WG (1979) Development of vegetation and climate in the south-western United States. Science 204: 701-710
VAN DONK J (1976) O^{18} Record of the Atlantic Ocean for the entire Pleistocene Epoch. In: CLINE RM, HAYS JD (eds) Investigation of late Quaternary palaeooceanography and palaeoclimatology. Geological Society of America Memoir 145: 147-163
VAN ZINDEREN BAKKER EM (1975) The origin and palaeoenvironment of the Namib desert biome. Journal of Biogeography 2: 65-74
VAN ZINDEREN BAKKER EM (1976) The evolution of Late Quaternary palaeoclimates of southern Africa. Palaeoecology of Africa 9: 160-202
VAN ZINDEREN BAKKER EM, MALEY J (1979) Late Quaternary palaeoenvironments of the Sahara region. Palaeoecology of Africa 11: 83-104
VEEVERS JJ, McELHINNY MW (1976) The separation of Australia from other continents. Earth Science Reviews 12: 139-159
WALKER D (1978) Quaternary climates of the Australian region. In: PITTOCK AB, FRAKES LA, JENSSEN D, PETERSON JA, ZILLMAN JW (eds) Climatic change and variability: a southern perspective. Cambridge University Press, Cambridge, pp 82-97

WEBSTER PJ, STRETEN NA (1978) Late Quaternary ice age climates of tropical Australasia: interpretations and reconstructions. Quaternary Research 10: 279-309

WESTMAN WE (1978) Evidence for the distinct evolutionary histories of canopy and understorey in the Eucalyptus forest-heath alliance of Australia. Journal of Biogeography 5: 365-376

WHITE MJD (1978) Modes of speciation. Freeman, Reading

WILSON AT, HENDY CH (1971) Past wind strength from isotope studies. Nature 234: 344-345

WIJMSTRA TA (1978) Palaeobotany and climatic change. In: GRIBBEN J (ed) Climatic change. Cambridge University Press, Cambridge, pp 25-45

WOLFE JA (1971) Tertiary climatic fluctuations and methods of analysis of Tertiary floras. Palaeogeography, Palaeoclimatology, Palaeoecology 9: 27-57

WOLFE JA (1975) Some aspects of plant geography of the northern hemisphere during the late Cretaceous and Tertiary. Annals of the Missouri Botanical Gardens 62: 264-279

WOILLARD G (1979) Abrupt end of the last interglacial s.s. in north-east France. Nature 281: 558-563

2. Mediterranean-Type Heathlands and Sclerophyllous Shrublands of the World: An Overview

R. L. SPECHT and E. J. MOLL

2.1 Introduction

The mediterranean regions of the world are defined by the distinctive annual climatic sequence in which a hot, dry summer alternates with a cool to cold, humid period lasting for a period of 5-10 months from autumn through winter to spring. This characteristic climate is found on all continents.

Evergreen, sclerophyllous shrubs or trees have dominated the five mediterranean landscapes. Sclerophyllous forests and sclerophyllous shrublands are thus typical of mediterranean areas. In any particular locality the structure of the overstorey of these plant formations is developed primarily in response to the water balance of the ecosystem, although the structure is also moulded by the nutrient status of the soil and the impact of man. The understorey also responds strikingly to changes in nutrient status and human disturbance.

A wide range of soil types are common in most regions of the world, and the mediterranean regions are no exception. On soils particularly low in plant nutrients, with a toxic level of certain nutrients, with an imbalance of cations, or a high or low pH, specialized plant communities have evolved. Pedologists rarely consider soil fertility in the classification of soils, except indirectly by the degree of leaching. For the ecophysiologist, a scale of soil fertility, based on maximum response of an indicator species when limiting plant nutrients are applied to the soil, is necessary (Specht 1981e). In this paper, soil fertility levels are assessed on the total soil nitrogen and total soil phosphorus recorded in the major rooting zone, about 300 mm deep. It must be stressed that these values of total phosphorus and nitrogen overstate the availability of these nutrients. For southern Australian soils, a wide range of values have been observed (Fig. 1). Distinct fertility categories can be distinguished, i.e. strongly leached soils (or soils derived from nutrient-poor substrate), moderately leached soils, weakly leached soils, and nutrient-rich soils.

A marked discontinuity in the structure and floristic composition of the plant communities is apparent at the transition between strongly leached and moderately leached soils (Fig. 1). In contrast, the more fertile categories appear to be intergraded.

This generalized series of soil fertility classes (Fig. 1) may not include soils where nutrient imbalance, nutrient toxicity or extreme pH are problems. For example, serpentine rocks produce magnesium-rich/calcium-poor soils; saline soils show toxic levels of sodium; calcium-rich/high-pH soils reduce the availability of phosphorus, iron, copper and cobalt.

In the mediterranean regions three major soil suites may be distinguished, each of which develops a distinctive sequence of vegetation structure from humid to semi-arid regions:

(a) On moderately leached soils: an overstorey of evergreen, sclerophyllous trees or shrubs, with an understorey of seasonal annual and perennial grasses and herbs. This ground stratum may be termed the savanna understorey.

(b) On strongly leached soils (termed oligotrophic in Specht (1979a,b)): stunted overstorey of evergreen, sclerophyllous trees or shrubs, with an understorey of evergreen nanophanerophytes, chamaephytes and hemicryptophytes (also evergreen and sclerophyllous). This complex of sclerophyllous plants may be termed the heathland understorey (Specht 1979a).

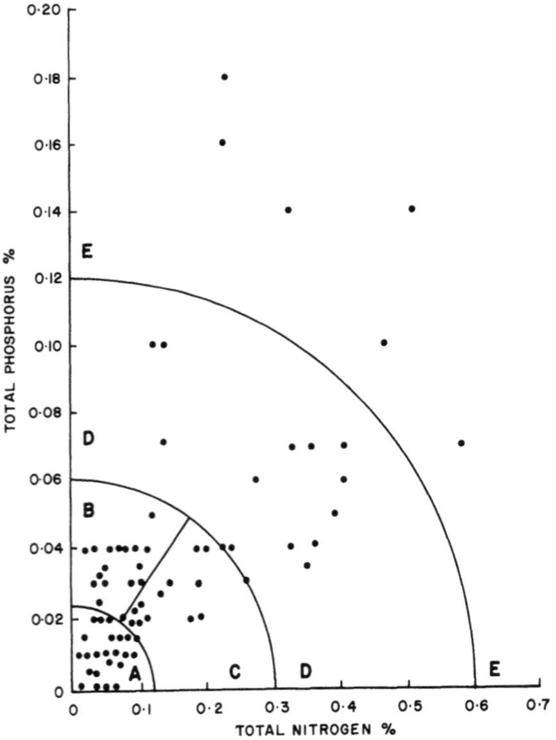

Fig. 1. Relative nutrient levels of surface soils in mediterranean-type ecosystems in southern Australia, based on total soil nitrogen (H_2SO_4 extraction) and total soil phosphorus (HCl extraction) - after Specht (1981d). A: strongly leached soils. B: moderately leached soils (semi-arid zone). C: moderately leached soils (humid, sub-humid zone). D: weakly leached soils. E: nutrient-rich soils.

(c) On calcium-rich/high-pH soils: a stunted overstorey of evergreen, sclerophyllous trees or shrubs with a ground stratum graded from seasonal grasses and herbs to evergreen, sclerophyllous chamaephytes, hemicryptophytes and succulents (South Africa).

These three major soil suites may be recognized in southern and southwestern Australia (Specht 1969, 1972b, 1973, 1979b, 1981a; George et al. 1979), around the Mediterranean Sea (Quézel 1981), and in South Africa (Kruger 1979; Boucher and Moll 1981). The first two soil suites have been recognized in California (Whittaker 1979). Apparently only weakly leached soils are characteristic of central Chile (Rundel 1981).

It is the purpose of this review to define the climatic sequence (from humid to semi-arid regions) of the mediterranean ecosystems which may be developed according to the three nutritional categories outlined above. The mediterranean ecosystems of southern Australia and South Africa will be investigated in depth. Nutritional aspects of the other three mediterranean regions will then be discussed.

2.2 The mediterranean climate

Köppen (1923) distinguished the mediterranean climatic region as a Cs type - a warm, climate (C) with a mild, humid winter (at least one month below 18°C) and a dry summer season (s). Within this climatic region Köppen distinguished a gradation in the intensity of the dry summer season from a mild summer (Csb climate, hottest month below 22°C and at least four months above 10°C) to a hot summer (Csa climate, hottest month above 22°C).

Within these types, adequate soil water is available for growth of the mediterranean vegetation from autumn through to spring. In the humid zone the potential growing season (when soil water in the surface horizon is not limiting) may extend for 9-10 months of the year with only 2-3 months of summer drought (Prescott and Thomas 1949; Prescott et al. 1952). At the drier limit of the Csa/Csb climates the potential growing season is shorter, only 6-7 months of the year. However, cultivated pastures and crops can be produced in even drier areas with a growing season as short as five months during which surface soil water is in adequate supply. The mediterranean climatic zone may thus be regarded as extending into Köppen's semi-arid (BS) climatic type with a mean annual temperature below 18°C (Köppen BSk), though still with a pronounced drought period during summer.

The extent of these mediterranean climatic regions in Australia and South Africa are shown in Figs 2 and 3 (calculated according to the steps outlined in Dick 1975).

Although Köppen stressed the severity of summer drought in defining the mediterranean climate, inland and more elevated localities within the mediterranean climatic zone may experience severe winter frosts which inhibit growth during this period (Trumble 1939). K Mitrakos (personal communication 1979) believes that plant life throughout the Mediterranean is determined by the independent stresses induced by both summer drought and winter cold. The influence of these two periods of stress on plant growth has been explored by Specht (1981b), using a simple growth model.

2.3 Availability of water

In any area subjected to seasonal drought the seasonal and annual water balance of the habitat is of paramount importance. The water balance

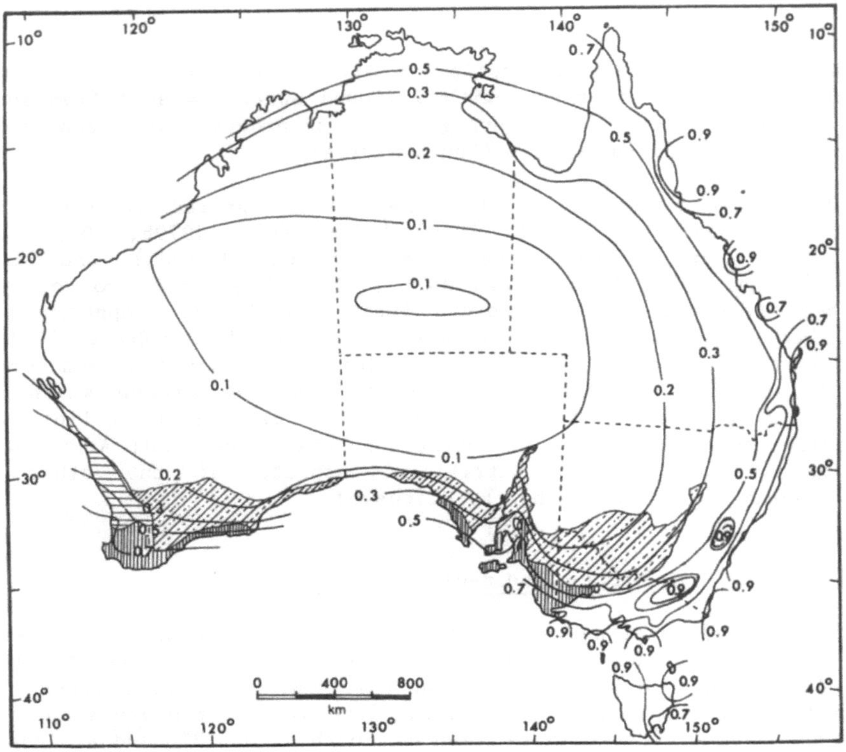

Fig. 2. Distribution of mediterranean climates (Köppen types Csa: horizontal hatching, Csb: vertical hatching and BSk: stippled) in Australia (after Dick 1975), superimposed with isolines of annual moisture index (Ea/Eo).

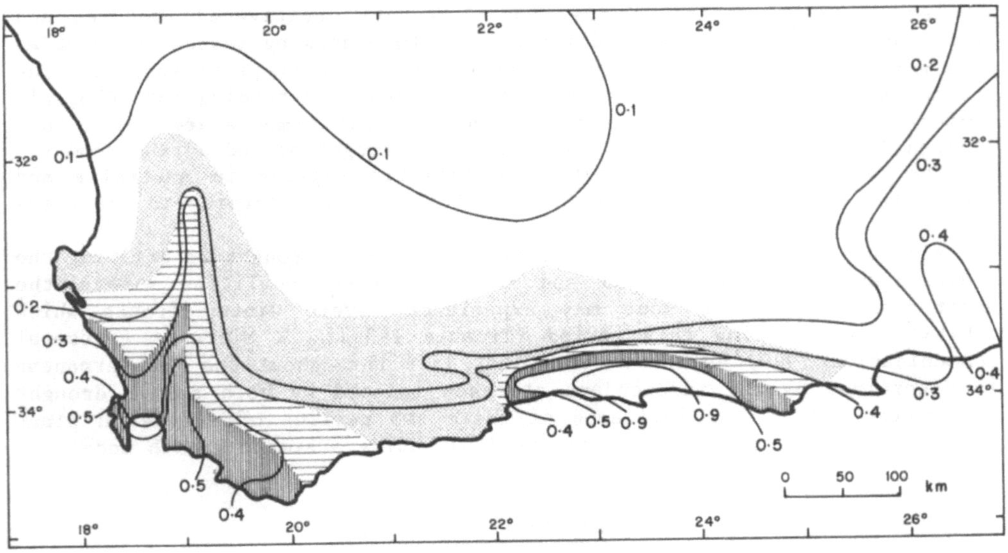

Fig. 3. Distribution of mediterranean climates (Köppen types Csa: horizontal hatching, Csb: vertical hatching and BSk: stippled, calculated according to Dick 1975) in South Africa, superimposed with isolines of annual moisture index.

equation, over a stated time interval (t_1 to t_2), may be expressed as follows:

$$\text{Input (Precipitation + Runon + Seepage + Extractable Soil Water}_{t_1} + \text{Biotic Water}_{t_1})$$

$$= \text{Output (Evapotranspiration + Runoff + Drainage + Extractable Soil Water}_{t_2} + \text{Biotic Water}_{t_2})$$

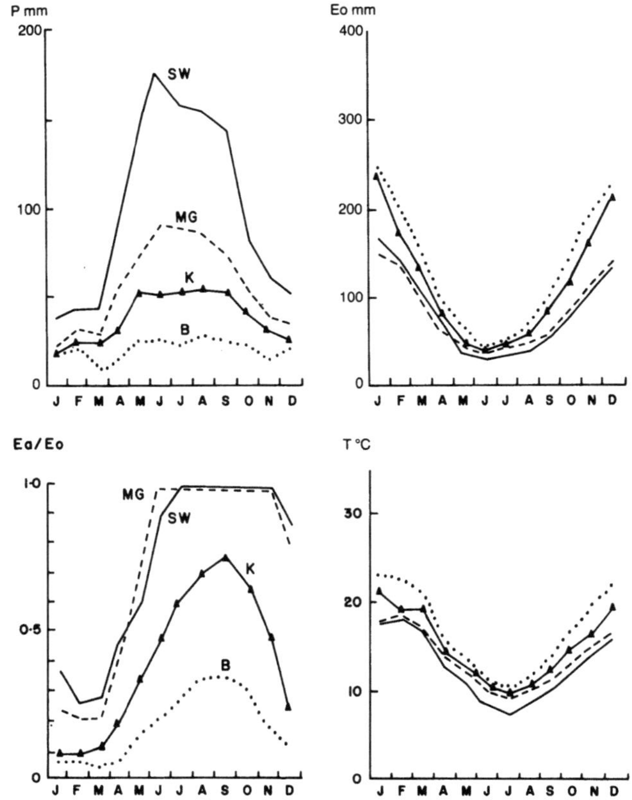

Fig. 4. Monthly climatic data of four representative localities in the mediterranean-climate region of South Australia.

Locality	Köppen	Annual climatic means P(mm)	Eo(mm)	T_{max}(°C)	T_{min}(°C)	Ea/Eo
Stirling West (SW)	Csb	1183	1021	17.8	8.1	0.72
Mount Gambier (MG)	Csb	682	1010	18.7	9.1	0.71
Keith (K)	Csb	457	1395	21.1	9.4	0.39
Berri (B)	BSk	250	1559	23.4	10.2	0.19

The amount of evapotranspiration is firstly dependent on the evaporative power of the atmosphere, which can be estimated by measuring the loss of water from a standard evaporimeter (the international Class A pan evaporimeter). But evapotranspiration is markedly affected by the amount of available water and the water potentials thus produced in the soil-plant-atmosphere continuum within the ecosystem. The ratio (Ea/Eo) of water lost from the ecosystem by actual evapotranspiration (Ea) to water lost from a standard pan evaporimeter (Eo) (the Moisture Index, Fitzpatrick and Nix 1970) can provide a valuable index of the degree of stomatal control exerted, during daylight hours on the transpiration from the leaf canopy.

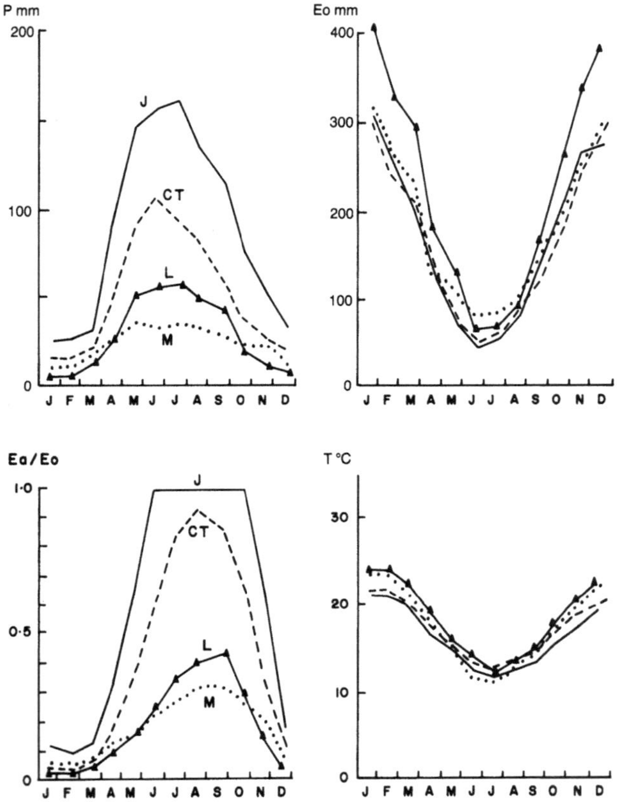

Fig. 5. Monthly climatic data of four representative localities in the mediterranean-climate region of Cape Province, South Africa.

Locality	Köppen	Annual climatic means				
		P(mm)	Eo(mm)	T_{max}(°C)	T_{min}(°C)	Ea/Eo
Jonkershoek (J)	Csb	1068	2065	22.4	12.3	0.56
Cape Town (CT)	Csb	627	2049	22.2	12.5	0.43
Langgewens (L)	Csa	368	2763	24.0	12.0	0.19
Montagu (M)	BSk	312	2226	25.1	10.1	0.18

Values of the Moisture Index have been estimated for evergreen woody vegetation in both Australia and South Africa, using the techniques outlined in Specht (1972a) based on the equation:

Monthly Moisture Index (Ea/Eo) = k (Available water in cm per month) (1)

where k is a constant (the evaporative coefficient) for the particular plant community.

Monthly values for four representative sites from the humid to semi-arid sections of the mediterranean climatic zone within Australia and South Africa are shown in Figs 4 and 5 respectively. Values of the Moisture Index greater than 0.2 would indicate the months when sufficient moisture is stored in the surface soil to ensure rapid growth of seasonal understorey vegetation. Within the mediterranean climatic zone values of the annual Moisture Index range from 0.9 to 0.2 in southern and southwestern Australia, and from 0.6 to 0.15 in South Africa (Figs 2 and 3).

The climates of the two regions reveal an important difference. Although monthly precipitation, solar radiation and temperature may show essentially the same sequence in the two continents, values of pan evaporation are consistently larger in South Africa (Figs 4 and 5). The differences in evaporation appear to be related to higher wind speeds experienced in South Africa (Table 1), particularly during the dry summer months.

The high wind speeds lead to increased evapotranspiration as well as to increased pan evaporation, and thus a decreased Moisture Index and growth potential.

Table 1. Mean wind speeds (km hr^{-1}) recorded in centres within the mediterranean regions of Australia and South Africa.

Centre	January	July	Year
Perth, Western Australia	17.5	14.2	15.6
Adelaide, South Australia	12.6	11.6	12.2
Melbourne, Victoria	13.5	13.0	12.7
Cape Town, South Africa	22.4	15.4	17.4
Port Elizabeth, South Africa	19.5	13.8	17.1

2.4 Community structure

2.4.1 FOLIAGE PROJECTIVE COVER OF OVERSTOREY

In spite of the seasonal variation in rainfall, virtually all the dominant woody vegetation of the mediterranean region is evergreen. It is only towards the driest limits of the climatic range that the dominant woody plants may become semi-deciduous, seasonally dimorphic, semi-succulent or even succulent.

The evergreen canopy remains relatively constant in foliage projective cover (the proportion of the landscape covered by photosynthetic tissue, projected vertically) throughout the year because seasonal shoot growth and leaf abscission tend to occur almost simultaneously (see Chapter 7, this

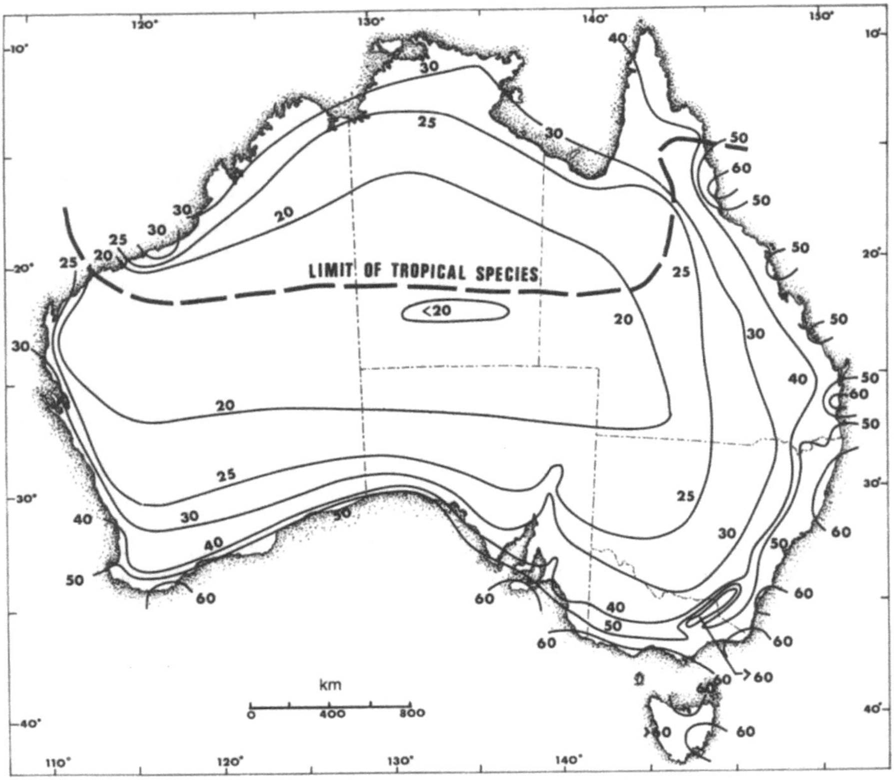

Fig. 6. Isolines of foliage projective cover (%) of the evergreen climax overstorey stratum in Australia.

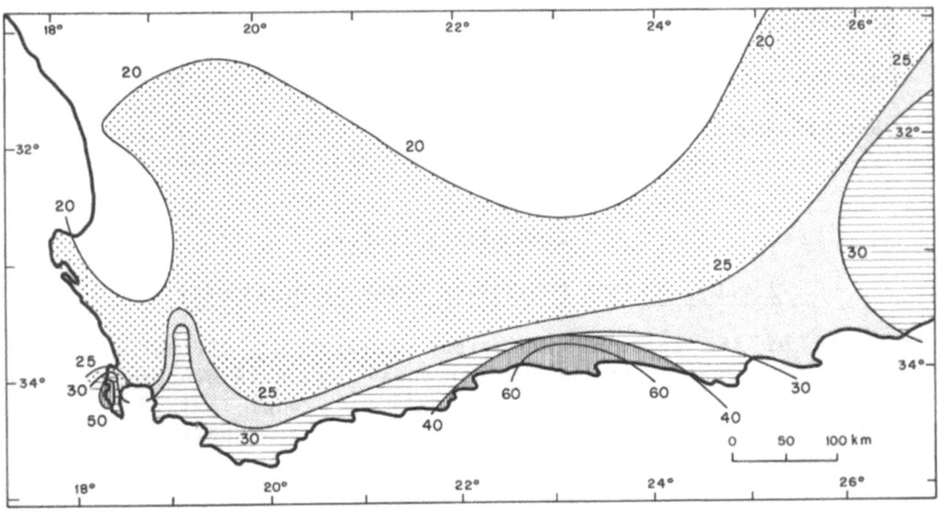

Fig. 7. Isolines of foliage projective cover (%) of the evergreen climax overstorey stratum in Cape Province, South Africa.

volume). Studies on water balance of evergreen plant communities (Specht 1972a) showed that, in any particular area, the foliage projective cover (FPC) of the overstorey tends to reach a maximum ("climax") value in equilibrium with the climate of the area. This can be expressed as follows (Specht 1972a, 1981c):

$$\text{FPC (\%) of climax overstorey} = 896.5k - 6.4 \qquad (2)$$

where k indicates the evaporative coefficient of the particular plant community as shown in Equation (1). Thus, the foliage projective cover of the climax overstorey can be estimated for a particular area, following the computation of the seasonal changes in the water balance of that area. The estimates of foliage projective cover (of climax overstorey) made for the mediterranean areas of Australia and South Africa are shown in Figs 6 and 7. Within these regions the values range from over 70% in the humid area to below 25% towards the dry limits.

2.4.2 STRUCTURE OF OVERSTOREY

(a) Moderately leached soils. The growth potential of the photosynthetic canopy of a plant community can be estimated using the growth index approach (Fitzpatrick and Nix 1970; Specht 1972a, 1981c,e):

Net Photosynthetic Index of photosynthetic canopy = Foliage Projective Cover x Moisture Index x Light Index x Thermal Index x Soil Fertility Index (3)

Within any particular mediterranean region the water balance of the ecosystem exerts the major controlling influence on the development of the plant community.

In broad terms, the plant formations developed on moderately leached soils in the mediterranean region of Australia follow the gradation with decreasing annual moisture index shown in Table 2. The potential formation grades in structure from closed forest/tall open forest over 30 m tall, to open forest, and woodland 20 m tall, and to open woodland 10 m tall as the moisture balance becomes more unfavourable from the humid to semi-arid limits of the mediterranean climate.

(b) On strongly leached soils and calcium-rich/high-pH soils. In at least four of the five mediterranean regions of the world, considerable areas of either strongly leached soils or calcium-rich/high-pH soil may be intermixed with moderately leached soils (Boucher and Moll 1981; Specht 1981a) from which the plant community can extract an adequate and reasonably well-balanced quota of nutrients essential for plant growth.

A deficit of essential nutrients (produced by a low concentration or an imbalance in the soil) appears not to affect the foliage projective cover of the plant community, but does have a marked effect on the potential growth of the community. Hence, in the mediterranean zone of southern Australia, the optimum community structure developed on moderately leached soils is markedly depressed on strongly leached soils (Table 2).

Considerable stunting of the overstorey results, the tree/shrub species often producing several stems arising from a large lignotuber. Such a community is termed open-scrub in the Australian nomenclature (Specht 1970, 1972b), and may be regarded as equivalent to other mediterranean-type shrublands (Di Castri et al. 1981). In certain very deficient sandy areas

Table 2. Structure of plant formations characteristic of moderately leached soils, contrasted with formations on strongly leached soils and on calcium-rich/high-pH soils, from the humid to semi-arid limits of the mediterranean region of southern Australia (after Specht 1972b, 1973, 1981a).

Annual Moisture Index (Ea/Eo)	Overstorey foliage projective cover (%)	Community structure[a]		
		Moderately leached soils	Strongly leached soils	Calcium-rich/ high-pH soils
1.0	>80	closed-forest	-	-
0.9-0.8	60	tall open-forest/open-forest	open-forest	-
0.8-0.7	55	open-forest	woodland/open-scrub	open-scrub
0.7-0.6	50	open-forest/ woodland	woodland/open-scrub	open-scrub
0.6-0.5	45	woodland	open-scrub/ dry-heathland	open-scrub
0.5-0.4	40	woodland	open-scrub/ dry-heathland	open-scrub
0.4-0.3	35	woodland/ open-woodland	open-scrub	open-scrub
0.3-0.2	30	open-woodland	open-scrub	open-scrub

[a] Structural classification and nomenclature of plant formations follows Specht (1970, 1972b) except that the term "woodland" (mean canopy height of 10-30 m and FPC of 50-30%) is used to distinguish the more xeric end of the formation formerly included under "open-forest" (with FPC 70-30%).

the tree/shrub species fail to survive, leaving the understorey of heath species to constitute a "dry-heathland" formation.

2.4.3 NATURE OF THE UNDERSTOREY

The nutrition of the soil types on which the community grows markedly affects the nature of the understorey in southern Australia (Table 3). A savanna (grass/herb) understorey develops on the more fertile soils, a heathland understorey on the infertile soils. On the more fertile soils, semi-succulent chenopodiaceous plants tend to increase as the arid zone is approached.

Such a sequence tends to be characteristic of other mediterranean regions in the world, though semi-deciduous rather than semi-succulent plants appear to be more characteristic of the more arid mediterranean areas around the Mediterranean Sea and in California.

It would appear that a simple inverse linear relationship may be expected between the foliage projective covers of the overstorey plants and the understorey plants (Specht and Morgan 1981). In any particular area, the climax community reaches a maximum value for the FPC of the overstorey trees/shrubs (see Equation (2) above). The FPC of the understorey is in equilibrium with that of the climax overstorey. However, due to small changes in the microenvironment or to manipulation by man, the maximum FPC of the overstorey may be reduced; FPC of the understorey increases in compensation.

Table 3. Nature of the understorey characteristic of moderately leached soils, strongly leached soils and calcium-rich/high-pH soils in the humid to semi-arid climate of the mediterranean region of southern Australia

Annual Moisture Index (Ea/Eo)	Understorey		
	Moderately leached soils	Strongly leached soils	Calcium-rich/ high-pH soils
1.0	-	-	-
0.9-0.8	grass/herb	heath	-
0.8-0.7	grass/herb	heath	sclerophyllous herb
0.7-0.6	grass/herb	heath	sclerophyllous herb
0.6-0.5	grass/herb	heath	grass/herb
0.5-0.4	grass/herb	heath	grass/herb
0.4-0.3	grass/herb + chenopods	heath	grass/herb + chenopods
0.3-0.2	grass/herb + chenopods	*Triodia*/heath	grass/herb + chenopods

2.4.4 COMMUNITY STRUCTURE IN THE MEDITERRANEAN ZONE OF SOUTH AFRICA

Most of the mediterranean climatic region of South Africa is relatively dry. Apart from a small humid area around Knysna, the majority of the area experiences a climate with an annual moisture index of less than 0.6 (Fig. 3).

By comparison with the ecosystems of southern and southwestern Australia (less disturbed by man, fire and grazing animals), trees would appear to have been an insignificant part of the mediterranean landscape in South Africa (Kruger 1979) though there is some evidence to suggest that at one time trees or large shrubs were a more significant feature of the landscape (Moll et al. 1980; Boucher and Moll 1981). *Widdringtonia*

Table 4. Structure of plant formations (veld types) characteristic of moderately leached soils, strongly leached soils and calcium-rich/high-pH soils in the mediterranean region of South Africa (after Acocks 1953, Kruger 1979, Boucher and Moll 1981).

Annual moisture index (Ea/Eo)	Overstorey foliage projective cover (%)	Moderately leached soils	Strongly leached soils	Calcium-rich/high-pH soils
0.6-0.5	45	open-scrub[a] coastal renosterveld[b]	mountain fynbos[c] (± small trees)	-
0.5-0.4	40	open-scrub coastal renosterveld	mountain fynbos (± small trees)	-
0.4-0.3	35	open-scrub coastal renosterveld	coastal fynbos mountain fynbos	calcareous coastal fynbos[d]
0.3-0.2	30	coastal renosterveld	coastal fynbos mountain fynbos	calcareous coastal fynbos
<0.2	25	karoid broken veld/ succulent karoo	arid fynbos	strandveld

a Open-scrub formation, dominated by Olea europaea subsp. africana, Euclea racemosa and Sideroxylon inerme, was probably the original vegetation of the moderately leached soils.
b "Renosterveld" dominated by the woody composite, Elytropappus rhinocerotis, appears to be a disclimax community following the destruction of the original open-scrub formation.
c "Fynbos" is a plant formation equivalent to "heathland".
d "Calcareous coastal fynbos" (on the southern coast between Cape Agulhas and Mossel Bay) is developed on coastal dune sands leached to form a calcarenite core.

cedarbergensis and W. schwarzii occasionally form low open-forests or woodlands, but populations are normally very sparse and restricted to special habitats on cliffs, in boulder fields and occasionally along watercourses. Leucadendron argenteum forms a low woodland with heathland understorey on the cool and moist eastern, shady, south-facing slopes of the Cape Peninsula and Stellenbosch area. Under favourable conditions, Maytenus oleoides, Protea nitida and P. laurifolia may reach 6 m, while Olea europaea subsp africana and Euclea racemosa may become trees if protected from fire.

In the absence of trees most of the plant communities of the Cape fall into the formation categories of Olea-Sideroxylon open-scrub, disclimax Elytropappus shrubland (renosterveld), Protea-Erica heathland (fynbos) and coastal variants of heathland/shrubland (strandveld).

The relationships of these formations are shown in Table 4. It is only (a) in the more humid areas near Knysna, (b) in sheltered areas of the Cape Peninsula, and on the Hottentots-Holland and Langeberg mountain ranges which receive extra seepage water, and (c) on the Cape Flats where the water table is not far below the surface, that water is in optimal supply throughout the year; on these sites, the open-structured communities change sharply to closed-scrubs and closed-forests (as shown for Australian vegetation by Specht (1972a, 1981c)).

The mediterranean shrublands (with an open-scrub overstorey of evergreen sclerophyllous shrubs over grass-herb understorey) on moderately leached soils must have been originally clearly distinct from the heathland formations (with evergreen sclerophyllous strata of Proteaceae, Ericaceae and Restionaceae) on Table Mountain Sandstone.

2.5 Mineral nutrition of mediterranean ecosystems

The moderately leached soils of mediterranean ecosystems are reasonably well-balanced nutritionally, although some nutrient elements (nitrogen, phosphorus) may not be in optimal supply. The overstorey developed on these soils shows a structural sequence from tall open-forest to open woodland (or even shrubland) as the climate changes from humid to semi-arid (see Table 2 for Australian ecosystems), all with a savanna understorey. Terms such as savanna open-forest, savanna woodland and savanna shrubland can be used to describe the formations.

In contrast to the moderately leached soils many of the mineral nutrients in the strongly leached soils are at a very low level or unavailable to plants. In particular nitrogen and phosphorus tend to be in short supply, but one or more nutrient elements (potassium, sulphur, copper, zinc, molybdenum) may also be limiting. The plant communities have heathland understoreys (Specht 1979a) though on well-drained xeric sites and on seasonally waterlogged sites, trees and tall shrubs fail to establish (see Specht 1979a) and the community is left as a true "dry-heathland" or "wet-heathland" formation. However, away from these extreme habitats evergreen sclerophyllous trees and shrubs survive, growing to produce an open-forest, woodland or shrubland depending on the water balance of the site. These trees and shrubs overtop the heathland understorey to produce a heathy open-forest, heathy woodland or heathy shrubland. The stature (height and volume) of these formations on strongly leached soils is usually markedly less than that observed in adjacent plant formations on moderately leached soils.

Calcium-rich/high-pH soils, although apparently containing a reasonable balance of mineral nutrients, develop a high pH at which a number of elements may be precipitated in a form unavailable to plants. In particular the trace elements (iron, copper, manganese, cobalt) may be rendered unavailable, and even PO_4^{-3} may be precipitated as insoluble calcium phosphate. The vegetation developed on these soils is thus subjected to a nutrient stress often as severe as that experienced by vegetation on strongly leached soils. Stunted evergreen sclerophyllous shrublands are typical of these calcium-rich/high-pH soils even at the more humid end of the mediterranean-climatic zone (where open-forests and woodlands are normally developed on moderately leached soils). In the humid mediterranean zone of southern Australia, these communities are developed on shallow red and grey calcareous soils developed on eroded calcarenite. In some areas the eroded calcarenite may be covered with a veneer of leached sand, but provided the calcarenite is within reach of the roots of the sclerophyllous dominants, the high pH of the subsoil limestone exerts an unfavourable nutritional influence on the overstorey stratum; an open-scrub/tall shrubland, rather than an open-forest/woodland, results.

The ground stratum of the formation developed on the calcium-rich/ high-pH soils depends on the pH and fertility of the surface soil. If the surface soil is neutral and reasonably fertile a savanna understorey develops. A neutral, infertile surface soil supports a heathland understorey. A surface soil high in free calcium carbonate with a high pH tends to support a sparse layer of evergreen sclerophyllous hemicryptophytes and chamaephytes. All intergradations may be observed.

The three major soil suites discussed above may be recognized in southern Australia (Specht 1969, 1972b, 1973, 1979b, 1981a) and in South Africa (Kruger 1979; Boucher and Moll 1981). Chemical analyses of surface soils typical of the three major soil systems are shown in Table 5 for southern Australia and Table 6 for South Africa. The strongly leached soils are nutritionally less favourable than the moderately leached soils which grade into weakly leached soils with well-balanced levels of available nutrients (Fig. 1). Calcium-rich/high-pH soils show a high level of free calcium carbonate, higher pH, a lower level of available phosphorus (and probably of available iron, copper and cobalt, which were not measured).

These soil suites also occur around the Mediterranean Sea (Quézel 1981). Many of the calcium-rich/high-pH soils have probably resulted from the destructive impact of man on the original ecosystems over the last 2000 to 3000 years. In southern France, taller vegetation (such as the *Quercus ilex* forest on moderately leached soils developed over limestone) has been destroyed and the soils eroded almost to the bedrock. A calcium-rich/ high-pH soil has resulted, which has favoured the development of a low (less than 2 m), dense, sclerophyllous shrubland (garrigue) dominated by *Quercus coccifera*.

In the mediterranean region of California two soil suites may be recognized - moderately leached and strongly leached soils (Whittaker 1979), while apparently only weakly leached soils are found in central Chile (Rundel 1981).

It is often very difficult to relate the chemical analysis of the soil (see Tables 5 and 6) with the availability of plant nutrients. Foliar analyses (Figs 8 to 11) of dominant plants from mediterranean ecosystems clearly segregate the nutritional status of these communities. Further information on the nutrient levels in the leaves of dominant species in other mediterranean ecosystems is collated in Table 7.

The three major nutritional groups of ecosystems may be defined as given below.

Table 5. Chemical analyses (mean and standard deviation) of surface soils typical of the three major soil systems in the mediterranean-climate region of South Australia (after French 1958; Stace et al. 1968; Specht 1972b, 1979b, 1981a).

CHEMICAL PROPERTY	Moderately leached soils	Strongly leached soils	Calcium-rich/high-pH soils	
MAJOR SOIL GROUPS	Red-brown earths Alluvial soils Grey-brown podzolics	Podzolics (± laterite) Solodized solonetz	Solonized brown soils	Terra rossa Shallow red and grey calcareous soils
MAJOR PLANT FORMATIONS	Savanna woodland	Heathy open forest/woodland/mallee/heathland	*E. socialis* Mallee open scrub	*E. diversifolia* Mallee open scrub
pH	6.47 ± 0.11	5.70 ± 0.07	7.98 ± 0.05	7.7-9.0
$CaCO_3$ (%)	0	0	6.43	0.39 to much free lime
%N	0.127 ± 0.012	0.074 ± 0.010	0.061 ± 0.004	0.045 to 0.110
%P[a]	0.026 ± 0.004	0.007 ± 0.001	0.009 ± 0.001	0.02
Available P ($\mu g\ g^{-1}$)	20-35	1-6	2-9	2-9
%K[a]	0.29 ± 0.04	0.15 ± 0.07	0.38 ± 0.03	0.67
Total soluble salts (%)	0.022 ± 0.004	0.016 ± 0.003	0.14	0.04-0.08
Chlorides (% NaCl)	0.008 ± 0.002	0.006 ± 0.002	0.026	0.01
Exchangeable cations (meq 100 g^{-1})	10.24	4.89	6-16	9.7-25.3
Exchangeable Ca (%)	72	63	63	70
Exchangeable Mg (%)	18	28	25	20
Exchangeable K (%)	7	6	7	8
Exchangeable Na (%)	3	3	5	2

[a] Total P and K following HCl digestion

Table 6. Chemical analyses of surface soils typical of two of the major soil systems in the mediterranean-climate region of South Africa.

CHEMICAL PROPERTY	Moderately leached soils	Strongly leached soils	
MAJOR SOIL GROUPS	Ferralitic soils Illuvial soils Planosolic soils	Sandstone lithosols	Arenosols
MAJOR PLANT FORMATIONS	Renosterveld	Mountain fynbos	Coastal fynbos
pH	4-7	3.5-5	5-8
%N	0.04-0.50	0.04-0.10	0.005
%P	0.015-0.0075	0.0027-0.0015	0.06
Exchangeable cations (meq 100 g^{-1})	3.22	4.30	0.48
Exchangeable Ca	1.35	0.69	0.15
Exchangeable Mg	0.56	0.27	0.15
Exchangeable K	0.30	0.87	0.02
Exchangeable Na	0.11	0.29	0.02

2.5.1 STRONGLY LEACHED SOILS

(a) Very strongly leached soils, characteristic of the Australian heathland and South African fynbos, produce foliar-nutrient polygons of minimal dimensions, showing a lack of most major nutrients (Figs 8 and 9). Magnesium is apparently reasonaly available even in nutrient-poor soils.

(b) Strongly leached soils appear to show low levels of one or more of the major nutrients. In Australia foliar samples from the mallee-broombush ecosystem are as low in phosphorus and nitrogen as in the heathland ecosystem, whereas the other major nutrients (calcium, magnesium, potassium) all show reasonable levels (Fig. 9).

2.5.2 MODERATELY/WEAKLY LEACHED SOILS

(a) Moderately leached soils reveal reasonable levels of all of the major nutrients in the leaves of dominant species (see the foliar-nutrient polygon for *Eucalyptus odorata*-etc. (savanna woodland) alliance in Fig. 9).

(b) Weakly leached soils lead to non-limiting, possibly luxury uptake of all major nutrients (see the foliar-nutrient polygon for montane matorral, Chile, in Fig. 8).

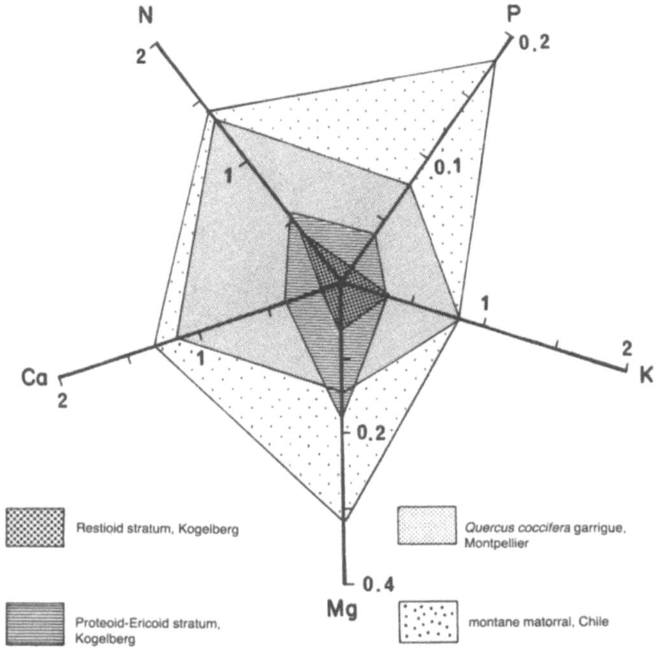

Fig. 8. Foliar analysis (% dry mass) of four representative mediterranean shrublands. Weakly leached soils - montane matorral, Chile (P W Rundel, personal communication 1979). Calcium-rich/high-pH soil - *Quercus coccifera* garrigue, Montpellier, France (Rapp and Lossaint 1981). Very strongly leached soil - Proteoid-Ericoid stratum and Restioid stratum, Kogelberg, South Africa (P W Rundel and F J Kruger, personal communication 1979).

2.5.3 CALCIUM-RICH/HIGH-pH SOILS

These soils produce a foliar-nutrient polygon characteristic of the moderately leached soils (see the polygon for *Quercus coccifera* garrigue in Fig. 8 and the figures for the *Eucalyptus diversifolia* alliance in Table 7). However, the high pH appears to reduce the availability of minor elements, iron and manganese, to almost one half of the typical moderately leached ecosystem (Table 7).

2.6 Foliar analyses of understorey species

The savanna understorey of moderately leached soils has a distinctly seasonal foliage. Agronomic studies indicate that the nutrient levels of the leaves are high, much higher than in the sclerophyllous leaves of the overstorey species. Mean values of nitrogen (1.81%, range 0.78% to 2.76%), phosphorus (0.21%), potassium (1.44%) and calcium (1.50%) were observed in 29 samples of the grass/herb ground-stratum ("natural pastures") of mediterranean savanna woodlands near Adelaide, South Australia (Davies and Sim 1931).

The heathland understorey is evergreen, most leaves being held on the plants for two to three years. These communities are often burned, but

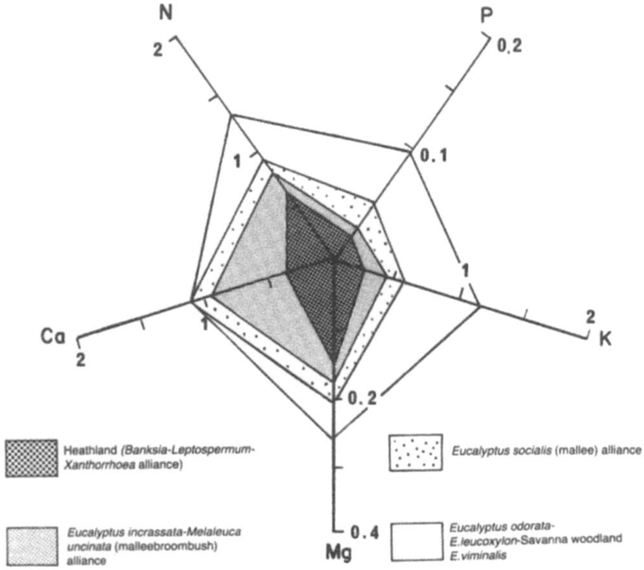

Fig. 9. Foliar analyses (% dry mass) of four representative mediterranean communities in South Australia. Very strongly leached soils - heathland on deep sand: Banksia-Leptospermum-Xanthorrhoea alliance. Strongly leached soils: Eucalyptus incrassata-Melaleuca uncinata alliance on sandy solodized solonetz. Transitional calcium-rich/high-pH soil - Eucalyptus socialis alliance on brown solonized soils. Moderately leached soil - Eucalyptus odorata - E. leucoxylon - E. viminalis alliance on grey-brown podzolics.

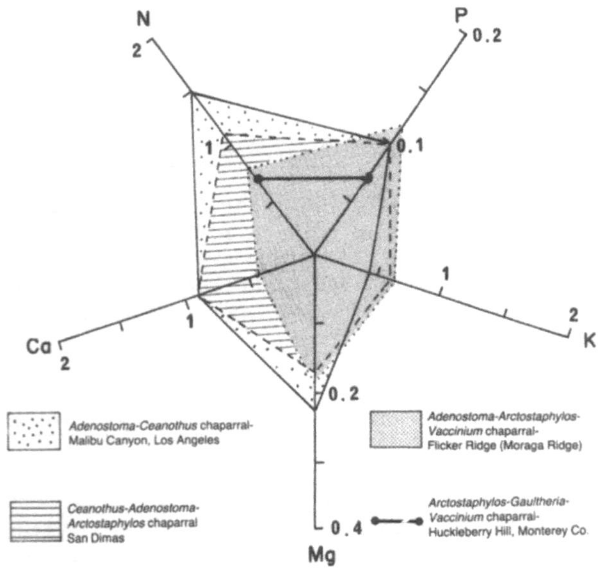

Fig. 10. Foliar analyses (% dry mass) of four representative mediterranean shrublands in California: Adenostoma-Ceanothus chaparral - Malibu Canyon, Los Angeles; Ceanothus-Adenostoma-Arctostaphylos chaparral - San Dimas; Adenostoma-Arctostaphylos-Vaccinium chaparral - Flicker Ridge (Moraga Ridge), near Berkeley; Arctostaphylos-Gaultheria-Vaccinium chaparral - Huckleberry Hill, Monterey County.

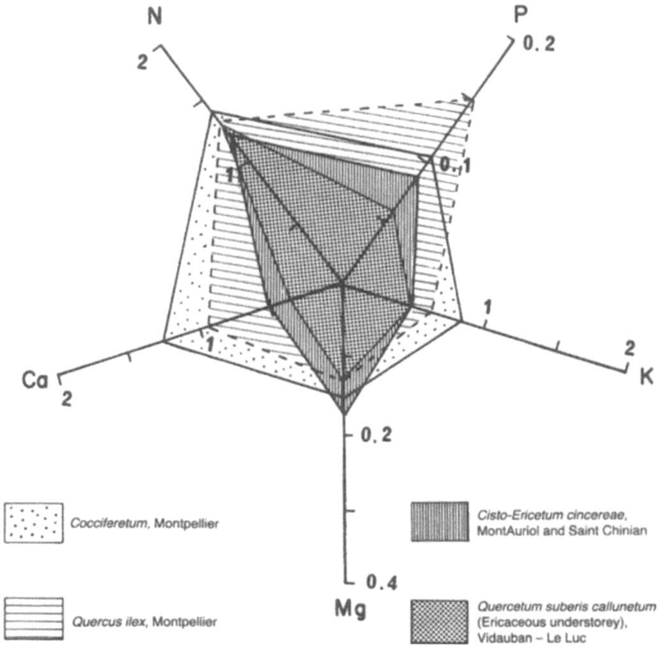

Fig. 11. Foliar analyses (% dry mass) of four representative mediterranean communities in southern France: Cocciferetum, Montpellier; Quercus ilex, Montpellier; Cisto-Ericetum cinereae, Mont Auriol and Saint Chinian; Quercetum suberis callunetum (ericaceous understorey), Vidauban-Le Luc.

most component species regenerate (from root stocks or seed) shortly after the fire and as the community develops, species are gradually lost (see Chapter 7, this volume).

In general the long-lived overstorey species show low levels of foliar nutrients; the shorter-lived species of both the mid-stratum and ground stratum show higher levels of foliar nutrients; long-lived, ground-stratum species (sclerophyllous members of the Cyperaceae, Restionaceae, Xanthorrhoeaceae and some Poaceae such as the genus Triodia) may have exceedingly low levels of foliar nutrients (Table 7). It would appear that low foliar nutrient levels may be associated with increased longevity of the species.

The semi-succulent chenopods (Atriplex and Maireana), which are found as understorey species in the driest part of mediterranean Australia, show high levels of the foliar nutrients (especially nitrogen, phosphorus and potassium - see Table 7) - values comparable to the high levels observed in the leaves of deciduous shrubs in California (Salvia) and Chile. These foliar analyses of deciduous and semi-succulent species are similar to those observed above in the seasonal foliage of savanna understoreys in mediterranean South Australia.

2.7 Conclusions

In the mediterranean regions of the world, the availability of soil nutrients appears to be the major environmental factor affecting the nature and distribution of vegetation.

Table 7. Foliar analysis (expressed on a dry mass basis) of mature leaves of species characteristic of forest, woodland, shrubland and heathland ecosystems in the five mediterranean regions of the world.

Locality and plant community	No. of spp.	N (%)	P (%)	K (%)	Ca (%)	Mg (%)	Fe ($\mu g\ g^{-1}$)	Mn ($\mu g\ g^{-1}$)	Reference
AUSTRALIA									
Savanna woodland (Green Hill, S Aust)									
Eucalyptus odorata-*E. leucoxylon*-*E. viminalis*	3	1.34	0.10	1.15	1.07	0.26	68	294	Specht, unpublished
Heathy open forest (Mt Lofty, S Aust)									
Eucalyptus obliqua-*E. baxteri*	2	0.88	0.040	0.40	0.78	0.24	66	209	Specht, unpublished
Understorey shrubs	2	0.97	0.040	0.47	0.69	0.24	98	222	
Heathland									
(Dark Island, S Aust)									
Xanthorrhoea	2	0.55	0.024	0.86	0.49	0.21	40	17	Specht, unpublished
Many spp.	37	-	0.033	-	-	-	-	-	
(Sydney, NSW)	8	1.02	0.044	0.65	0.54	0.18	-	-	Maggs and Pearson 1977
(Beerwah, Qld)									
Xanthorrhoea	1	0.56	0.010	0.53	0.24	0.10	104	14	Specht and Bolton, unpublished
Shrub spp.	11	0.74	0.023	0.24	0.34	0.17	95	59	
"Sedge" spp.	2	0.46	0.010	0.28	0.12	0.11	165	142	
Mallee open-scrub (NSW - S Aust - Vic)									
Eucalyptus behriana	2	1.19	0.080	0.51	0.78	0.22	108	116	Specht, unpublished
Understorey *Melaleuca*	1	1.30	0.060	0.60	0.66	0.42	180	14	"
Eucalyptus socialis	5	0.96	0.048	0.60	1.08	0.20	134	81	"
Understorey *Melaleuca*	1	1.25	0.060	0.42	1.00	0.40	182	8	"
Understorey *Atriplex*-*Maireana*	3	2.57	0.110	2.43	0.81	0.48	287	93	"

Eucalyptus incrassata	4	0.74	0.033	0.46	1.04	0.17	101	106	"
Understorey *Melaleuca*	3	1.19	0.050	0.58	0.83	0.24	101	35	"
Understorey *Triodia*	1	0.56	0.020	0.23	0.28	0.04	204	26	"
Eucalyptus diversifolia	2	0.64	0.030	0.39	0.72	0.17	70	52	"
Understorey *Lasiopetalum*	1	1.02	0.050	0.90	1.56	0.14	306	14	"
Understorey *Gahnia*	2	0.42	0.015	0.33	0.48	0.05	228	60	"
CHILE									
Matorral									
Evergreen	5	1.40	0.17	0.75	1.43	0.33	-	-	Rundel 1981
Deciduous	2	2.23	0.23	0.92	2.55	0.30	-	-	Rundel 1981
Montane matorral	3	1.37	0.18	0.82	1.29	0.32	-	-	P W Rundel, pers comm 1979
FRANCE									Braun-Blaunquet et al. 1952; Loisel, 1976
Quercetum suberis callunetum (Vidauban-Le Luc)			See Fig. 11						R Loisel and L Trabaud, pers comm 1981
Quercus suber, *Arbutus unedo*	2	1.39	0.08	0.71	1.00	0.20	-	-	"
Understorey shrubs									
Cistus spp.	3	1.44	0.19	0.63	1.34	0.18	-	-	"
Erica, *Calluna*	3	1.31	0.06	0.51	0.35	0.13	-	-	"
Cisto-Ericetum cinereae									
(*Mont Auriol*, *Beziers*)	6	1.11	0.07	0.56	0.55	0.17	300	200	L Trabaud, pers comm 1980
(Saint Chinian, Herault)	7	1.29	0.09	0.64	0.53	0.20	300	400	L Trabaud, pers comm 1980
Cocciferetum (Montpellier)									
Quercus coccifera	1	1.35	0.08	0.81	1.15	0.14	60	260	Rapp and Lossaint 1981
Garrigue spp.	8	1.47	0.13	0.84	1.39	0.18	-	-	H A Mooney, pers comm 1979
Quercetem ilicis gallo-provinciale (Montpellier)									
Quercus ilex	2	1.30	0.154	0.61	0.92	0.13	180	370	Lossaint and Rapp 1978
Understorey shrubs	7	2.25	0.076	1.17	2.47	0.23	180	150	Lossaint and Rapp 1978

Table 7. (Continued)

Locality and plant community	No. of spp.	N (%)	P (%)	K (%)	Ca (%)	Mg (%)	Fe ($\mu g\ g^{-1}$)	Mn ($\mu g\ g^{-1}$)	Reference
CALIFORNIA									
(Malibu Canyon, Los Angeles)									
Salvia mellifera coastal sage	1	2.18	0.28	2.90	1.30	0.14	450	54	Specht, unpublished
Mixed chaparral	3	1.50	0.10	0.49	0.92	0.23	169	77	"
(San Joaquin Hills)									
Adenostoma chaparral	1	1.04	0.06	0.39	0.82	0.22	240	184	"
San Dimas									
Mixed chaparral	7	1.07	0.10	0.57	0.90	0.17	141	192	"
(Yosemite National Park)									
Arctostaphylos chaparral	1	0.80	0.11	0.63	1.16	0.14	150	62	"
(Flicker Ridge near Berkeley)									
Adenostoma-Arctostaphylos-Vaccinium chaparral	3	0.78	0.12	0.62	0.44	0.19	80	78	"
(Huckleberry Hill, Monterey Co)									
Arctostaphylos-Gaultheria-Vaccinium chaparral	4	0.79	0.07	-	-	-	-	-	P W Rundel, pers comm 1979
(Ash Mountain, Sequoia N.P.)									
Quercus douglasii (deciduous) savanna woodland	1	2.19	0.41	0.98	0.86	0.08	-	-	P W Rundel, pers comm 1981
Quercus wislizenii (evergreen) savanna woodland	1	1.54	0.10	-	-	-	-	-	P W Rundel, pers comm 1981
SOUTH AFRICA									
Fynbos									
(Kogelberg)									
Proteoid-Ericoid	2	0.54	0.04	0.28	0.43	0.19	-	-	P W Rundel and F J Kruger
Restioid	2	0.39	0.01	0.30	0.01	0.07	-	-	P W Rundel and F J Kruger
Proteaceae	7	0.48	0.024	-	-	-	-	-	H A Mooney and F J Kruger
Evergreen shrubland (*Cassine*, *Rhus*, *Pterocelastrus*)	3	0.72	0.047	-	-	-	-	-	H A Mooney and F J Kruger pers comm 1979

(a) Strongly leached soils are associated with plant communities in which both the overstorey and understorey are sclerophyllous.

(b) Moderately leached soils support vegetation with a sclerophyllous overstorey and a seasonal (annual and perennial) savanna ground stratum.

(c) Calcium-rich/high-pH soils produce a stunted sclerophyllous overstorey with an understorey varying from savanna to sclerophyllous depending on the degree of leaching of the surface soil.

A distinctive sequence of plant formations (from the humid to semi-arid zones) has developed on each soil suite in response to both the water balance and the nutritional status of the soil.

The mediterranean moderately leached and calcium-rich/high-pH soils have been exploited by man over many centuries (or millennia, around the Mediterranean Sea). In comparison, the vegetation on strongly leached soils has been affected only slightly by man. In recent years, the pressures for alternative land use of the strongly leached ecosystems have increased rapidly.

The nutritional subdivisions of the mediterranean vegetation have many ecophysiological attributes in common, but others are markedly divergent. Management and conservation programmes must be aware of these divergences.

References

ACOCKS JPH (1953) Veld types of South Africa. Memoirs of the Botanical Survey of South Africa 28: 1-192

BOUCHER C, MOLL EJ (1981) South African mediterranean shrublands. In: DI CASTRI F, GOODALL DW, SPECHT RL (eds) Ecosystems of the world, vol 11. Mediterranean-type shrublands. Elsevier, Amsterdam, pp 233-248

BRAUN-BLANQUET J, ROUSSINE N, NÈGRE R (1952) Les groupements végétaux de la France Méditerranéenne. Centre National de la Recherche Scientifique, Montpellier, 297 pp

DAVIES JG, SIM AH (1931) The influence of frequency of cutting on the productivity, botanical and chemical composition, and the nutritive value of "natural" pastures in southern Australia. Council for Scientific and Industrial Research, Australia, Pamphlet No 18, 28 pp

DI CASTRI F (1981) Mediterranean-type shrublands of the world. In: DI CASTRI F, GOODALL DW, SPECHT RL (eds) Ecosystems of the world, vol 11. Mediterranean-type shrublands. Elsevier, Amsterdam, pp 1-52

DICK RS (1975) A map of the climates of Australia: according to Köppen's principles of definition. Queensland Geographical Journal. Ser 3,3: 33-69 and map

FITZPATRICK EA, NIX HA (1970) The climatic factor in Australian grassland ecology. In: MOORE RM (ed) Australian grasslands. Australian National University Press, Canberra, pp 3-26

FRENCH RJ (1958) Soils of Eyre Peninsula. South Australian Department of Agriculture Bulletin No 457: 1-53

GEORGE AS, HOPKINS AJM, MARCHANT NG (1979) The heathlands of Western Australia. In: SPECHT RL (ed) Ecosystems of the world, vol 9A. Heathlands and related shrublands. Descriptive studies. Elsevier, Amsterdam, pp 211-230

KÖPPEN W (1923) Der Klimate der Erde. Bornträger, Berlin, 369 pp

KRUGER FJ (1979) South African heathlands. In: SPECHT RL (ed) Ecosystems of the world, vol 9A. Heathlands and related shrublands. Descriptive studies. Elsevier, Amsterdam, pp 19-80

LOISEL R (1976) La végétation de l'Étage Méditerranéen dans le sud-est continental Francais. DSc Thesis, l'Université d'Aix, Marseille

LOSSAINT P, RAPP M (1978) La forêt méditerranéenne de chênes verts (Quercus ilex L). In: LAMOTTE M, BOURLIÈRE F (eds) Problèmes d'écologie. Structure et fonctionnement des écosystèmes terrestres. Masson, Paris, pp 129-185

MAGGS J, PEARSON CJ (1977) Minerals and dry matter in coastal scrub and grassland at Sydney, Australia. Oecologia 31: 227-237

MOLL EJ, MCKENZIE B, MCLACHLAN D (1980) A possible explanation for the lack of trees in the Fynbos, Cape Province, South Africa. Biological Conservation 17: 221-228

PRESCOTT JA, THOMAS JA (1949) The length of the growing season in Australia as determined by the effectiveness of the rainfall. Proceedings of the Royal Geographical Society of Australia, South Australia Branch 50: 42-46

PRESCOTT JA, COLLINS JA, SHIRPURKAR GR (1952) The comparative climatology of Australia and Argentina. Geographical Review 42: 118-133

QUÉZEL P (1981) Floristic composition and phytosociological structure of sclerophyllous matorral around the Mediterranean. In: DI CASTRI F, GOODALL DW, SPECHT RL (eds) Ecosystems of the world, vol 11. Mediterranean-type shrublands. Elsevier, Amsterdam, pp 107-121

RAPP M, LOSSAINT P (1981) Some aspects of mineral cycling in the garrigue of southern France. In: DI CASTRI F, GOODALL DW, SPECHT RL (eds) Ecosystems of the world, vol 11. Mediterranean-type shrublands. Elsevier, Amsterdam, pp 289-301

RUNDEL PW (1981) The matorral zone of central Chile. In: DI CASTRI F, GOODALL DW, SPECHT RL (eds) Ecosystems of the world, vol 11. Mediterranean-type shrublands. Elsevier, Amsterdam, pp 175-201

SPECHT RL (1969) A comparison of the sclerophyllous vegetation characteristic of mediterranean-type climates in France, California and southern Australia. I. Structure, morphology and succession. II. Dry matter, energy and nutrient accumulation. Australian Journal of Botany 17: 277-292, 293-308

SPECHT RL (1970) Vegetation. In: LEEPER GW (ed) The Australian environment, 4th ed. CSIRO, Melbourne University Press, Melbourne, pp 44-67

SPECHT RL (1972a) Water use by perennial evergreen plant communities in Australia and Papua New Guinea. Australian Journal of Botany 20: 273-299

SPECHT RL (1972b) The vegetation of South Australia, 2nd ed. Government Printer, Adelaide, 328 pp

SPECHT RL (1973) Structure and functional response of ecosystems in the mediterranean climate of Australia. In: DI CASTRI F, MOONEY HA (eds) Mediterranean-type ecosystems: origin and structure. Springer-Verlag, Berlin, pp 113-120

SPECHT RL (1979a) Heathlands and related shrublands of the world. In: SPECHT RL (ed) Ecosystems of the world, vol 9A. Heathlands and related shrublands. Descriptive studies. Elsevier, Amsterdam, pp 1-18

SPECHT RL (1979b) The sclerophyllous (heath) vegetation of Australia: The eastern and central states. In: SPECHT RL (ed) Ecosystems of the world, vol 9A. Heathlands and related shrublands. Descriptive studies. Elsevier, Amsterdam, pp 125-210

SPECHT RL (1981a) Mallee ecosystems in southern Australia. In: DI CASTRI F, GOODALL DW, SPECHT RL (eds) Ecosystems of the world, vol 11. Mediterranean-type shrublands. Elsevier, Amsterdam, pp 203-231
SPECHT RL (1981b) Primary production in mediterranean-climate ecosystems regenerating after fire. In: DI CASTRI F, GOODALL DW, SPECHT RL (eds) Ecosystems of the world, vol 11. Mediterranean-type shrublands. Elsevier, Amsterdam, pp 257-267
SPECHT RL (1981c) Ecophysiological principles determining the biogeography of major vegetation formations in Australia. In: KEAST A (ed) Ecological biogeography of Australia. Junk, The Hague, pp 299-332
SPECHT RL (1981d) Evolution of the Australian flora: some generalizations. In: KEAST A (ed) Ecological biogeography of Australia. Junk, The Hague, pp 783-806
SPECHT RL (1981e) Growth indices - their rôle in understanding the growth, structure and distribution of Australian vegetation. Oecologia 50: 347-356
SPECHT RL, MORGAN DG (1981) The balance between the foliage projective covers of overstorey and understorey strata in Australian vegetation. Australian Journal of Ecology 6: 193-202
STACE HCT, HUBBLE GD, BREWER R, NORTHCOTE KH, SLEEMAN JR, MULCAHY MJ, HALLSWORTH EG (1968) A handbook of Australian soils. Rellim Technical Publication, Glenside, South Australia, 435 pp
TRUMBLE HC (1939) Climatic factors in relation to the agricultural regions of southern Australia. Transactions of the Royal Society of South Australia 63: 36-43
WHITTAKER RH (1979) Appalachian balds and other northern American heathlands. In: SPECHT RL (ed) Ecosystems of the world, vol 9A. Heathlands and related shrublands. Descriptive studies. Elsevier, Amsterdam, pp 427-440

3. Ecological Control of the Vegetation of Southwestern Australia: Moisture versus Nutrients

J. S. BEARD

3.1 Introduction

Extensive maps of the vegetation of southwestern Australia, recently completed, allow analysis of vegetation in relation to environment on a broad scale. The facts given here are extracted from the Explanatory Notes to the 1:1,000,000 vegetation map "Swan" (Beard 1981).

Fig. 1 shows the mediterranean region of southwestern Australia, contained within a line curving from Shark Bay to the Great Australian Bight (Beard 1979) and known as the South West Botanical Province (Diels 1906). The Eremaean Province of the interior is a part of the Australian Arid Zone. Between the two there is a transitional area known as the Southwestern Interzone (Burbidge 1960). The area of the South West Province is estimated at 306,000 km^2 and that of the Interzone at 128,000 km^2 (Beard, unpublished).

Rainfall is highest on the southwestern coast where it exceeds 1500 mm annually, and declines in a northeasterly direction. The boundary of the South West Province as determined by vegetation mapping coincides with the 250 mm isohyet in the north, shifting to the 300 mm isohyet in the south. Winter rainfall predominates. Within the Southwestern Interzone it changes from this winter or mediterranean distribution to non-seasonal in the east. Climatic classification based on ombrothermic diagrams (Bagnouls and Gaussen 1957) shows three different regimes within the South West Province. A Moderate Mediterranean regime with less than four dry months and rainfall exceeding 750 mm annually, mostly over 1000 mm, occupies a belt along the southwestern coast from Cape Naturaliste to Bremer Bay. Vegetation is tall forest, forest and swamps; there is little agriculture and forestry is the chief occupation. A Dry Mediterranean regime with five or six dry months and 500-1000 mm annual rainfall extends northwards up the west coast and corresponds to the remainder of the Darling Botanical District. This belt is still largely forest covered; occupations are forestry, dairying and fruit growing. The remainder of the South West Province has an Extra-Dry Mediterranean regime with seven or eight dry months and 300-500 mm annual rainfall. Nearly all natural vegetation has been cleared as this is the Wheat Belt of Western Australia, devoted to

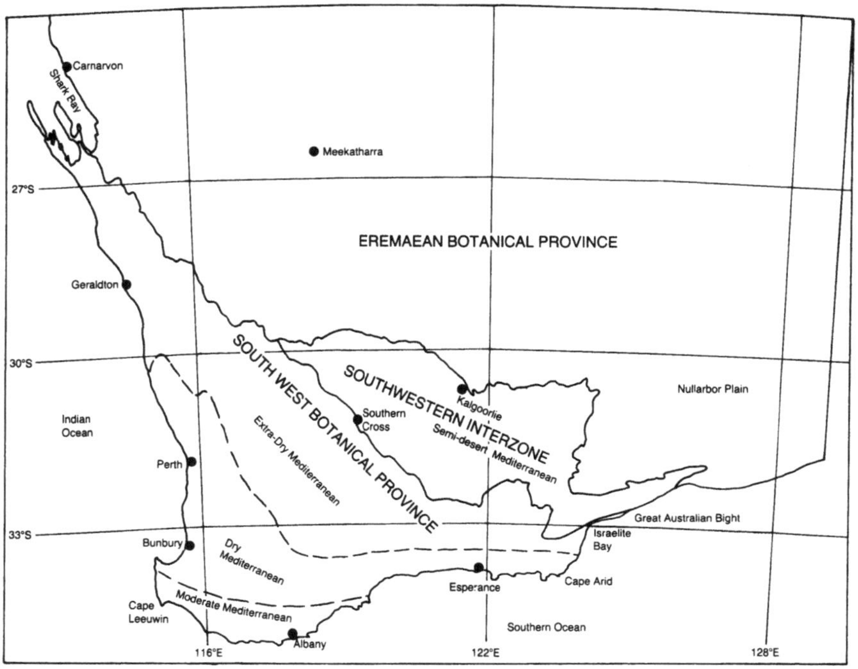

Fig. 1. Southwestern Australia showing Botanical Provinces and their division into climatic zones.

wheat and sheep. The Interzone has a Semi-desert Mediterranean climate with 9-10 dry months, 300 mm rain grading eastward to Desert Nonseasonal with no regular growing season. Natural vegetation is largely intact. There is no agriculture or forestry, and very limited pastoral use.

Geologically southwestern Australia is dominated by the Darling Fault which arose as the eastern edge of a rift valley formed during the breakup of Gondwanaland. The fault roughly parallels the western coast from 25 km to 125 km inland from it. The rift valley - the Perth Basin - became filled with Palaeozoic and Mesozoic sediments. Jurassic and Cretaceous members, predominantly arenaceous, now form the bulk of the outcrop, superficially mantled by Pleistocene deposits which are of calcarenite in a coastal belt up to 20 km wide, otherwise sands. East of the Darling Fault and underlying the greater proportion of southwestern Australia is the Western Shield, an ancient rigid area of Precambrian rocks which has been tectonically inert throughout Phanerozoic time. The rocks are predominantly granite or granitoid about 2800 million years old, with some belts of Archaean metasediments and metaigneous rocks.

On the Western Shield important ecological factors are the great age of the landscape and its lack of physiographic diversity. Erosion after the Permian glaciation brought the Shield more or less to base level by the end of the Cretaceous. Surface rocks then weathered under a warm, humid climate to depths of 60 m. On granite and gneiss the weathering gave rise to a white, kaolinitic "pallid zone" which has lost iron, aluminium, combined silica and bases. It may be capped by a ferruginous or bauxitic duricrust.

A general uplift of the Shield during the Tertiary raised the surface to its present level of about 250 m above sea level, rising gently inland by another 100 m some 500 km from the coast. The uplift caused rivers to deepen their beds but partly due to later onset of aridity this process has not proceeded very far. Rejuvenation has not reached more than 100 km inland from the west or south coasts. In the interior the landscape has not been substantially modified in the past 50 million years.

Therefore southwestern Australia today consists predominantly of a very gently undulating plateau 250-350 m above sea level, without marked relief, a monotonous landscape broken only by occasional granite domes and tors, covered by the products of its own weathering. Extensive sandplains mantle the uplands, usually mixed with or overlying ironstone gravel which may crop out as massive duricrust. Red loamy soils occur in the broad valleys. At the lowest points drainage is ponded into salt lakes. Thus, virtually all soils in southwestern Australia are nutrient deficient. The red loams can be made agriculturally productive by adding only superphosphate but the sandplains require also nitrogen, potassium, copper zinc and sometimes manganese, molybdenum and cobalt.

3.2 Vegetation in relation to climate and substrate

Mapping of the vegetation of the south west (Beard 1981) has distinguished 18 structural formations, eight of which consist of forest and woodland of varying height and density, seven of shrublands, one of sedgeland, one of hummock grassland and one of chenopodiaceous halophytes. All tall-forests and forests are eucalypt dominated. Only in the case of low forest and low woodland may other genera, e.g. *Banksia*, *Callitris*, *Casuarina*, *Melaleuca* and *Agonis*, become co-dominant or replace *Eucalyptus*. These trees are all evergreen sclerophylls. The communities which they dominate have understoreys of scleromorphic shrubs and were classified by Diels (1906) as "Sklerophyllen-Wälder" in contradistinction to "Savannen-Wälder" frequently encountered in the north and east of Australia which have grassy understoreys.

Among the shrublands, *Eucalyptus* is dominant in the mallee formation and conspicuous in mallee heath. All shrublands are scleromorphic, with ground layers of low "heath" shrubs and taller plants of scleromorphic *Acacia*, *Casuarina* and *Melaleuca*. The sedgeland formation may be regarded as scleromorphic, consisting of mixed species of Cyperaceae and Restionaceae with scattered sclerophyllous shrubs. Hummock grassland is likewise scleromorphic, consisting of a special growth form of sclerophyllous grass (*Triodia* spp.) unique to Australia (Beard 1976b). Only the chenopodiaceous halophytes are not scleromorphic. These succulent and semi-succulent sub-shrubs include *Halosarcia*, *Atriplex*, *Maireana* and numerous other genera. In the Arid Zone of Australia such chenopod-dominated communities occur on base-rich saline and alkaline soils whereas scleromorphic communities of hummock grasses and shrubs occur on deficient siliceous soils (Beard 1976b).

It is now widely accepted in Australia and elsewhere that many scleromorphic plant forms have developed in response to nutrient deficiencies and imbalances, rather than in response to aridity as was formerly supposed. Walter (1979) has given us the useful term "peinomorphosis" for this. An environment dominated by nutrient deficiency is a "peinobiome". Southwestern Australia is almost entirely a peinobiome, virtually the only exception being the vegetation of saline areas. Otherwise peinomorphosis is only relaxed to some extent in the understoreys of *Eucalyptus diversicolor* (karri) tall forest and *E. loxophleba* (York gum) woodland where these occupy young soils on bedrock.

Within the southwestern peinobiome there is a general climatic control of vegetation but no simple climatic zonation, owing to the very strong influence of soils. Rather, in each climatic zone there are particular sequences of vegetation types determined by soil. Owing to their great age, the landscapes upon the Western Shield are covered by highly evolved soils arranged in catenary sequences, of which a typical model is given in Fig. 2. The close relationship between vegetation and soil results in catenary sequences of vegetation. The catena illustrated in Fig. 2 occurs in the Extra-Dry Mediterranean climatic zone and is generally typical of it, with local modifications. The profound influence of soil under uniform climate is graphically illustrated, since height of the vegetation ranges from 25 m to less than 5 m. In areas of higher rainfall the components of the vegetation catena change but the basic sequence of soil and land-form types is the same.

Some typical catenary sequences for the principal climatic zones are as follows, taken from the lowest to the highest point in the landscape:

(a) Moderate mediterranean: Sedgeland and *Melaleuca* low forest in swamps; *Banksia-Eucalyptus marginata* low woodland on bleached sands; *E. diversicolor* tall forest on red loam of lower and middle slope; *E. marginata-E. calophylla* forest on upper slopes and laterite cappings, *E. marginata-Casuarina* low woodland on ill-drained plateau sites.

(b) Dry mediterranean: *Eucalyptus rudis* and *Melaleuca* spp. lining creek banks; *E. calophylla-E. wandoo* woodland on red loams on slopes, *E. marginata-E. calophylla* forest on laterite-capped uplands.

(c) Extra-dry mediterranean: Salt lake fringed by halophytes; *E. salmonophloia* woodland on heavy loam on lower slopes; *E. loxophleba-E. wandoo* woodland on light loam on upper slopes; *Casuarina campestris* thicket on laterite; heterogeneous scrub-heath on sand.

(d) Semi-desert mediterranean: Salt lake fringed by halophytes; *E. salmonophloia-E. salubris* woodland on heavy loam on lower slopes; *E. transcontinentalis-E. flocktoniae* woodland on light loam of upper slopes; mallee on duplex soil below the breakaway; *Casuarina* spp. thicket on laterite; *Acacia resinomarginea* thickets on sand.

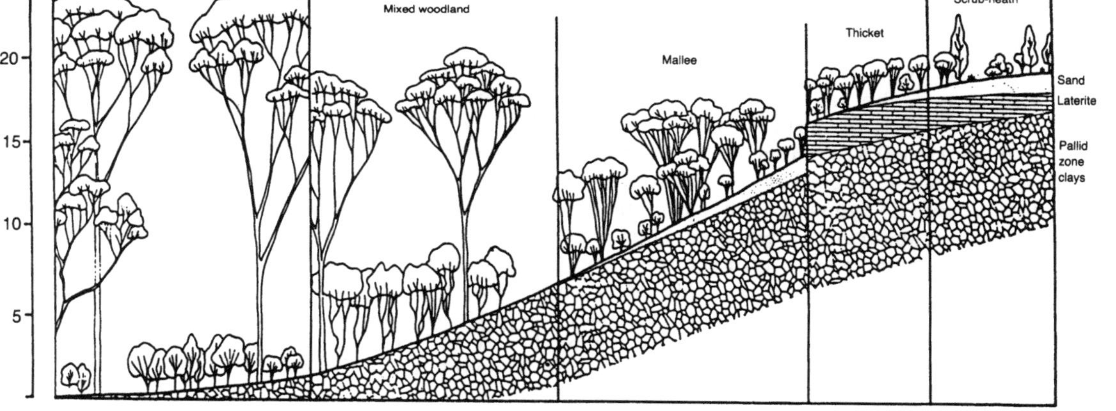

Fig. 2. Illustration of the catena of the Boorabbin vegetation system, Coolgardie Botanical District. The formations are shown for convenience as occupying equal portions of the catena which they do not do in the field (after Beard 1981).

Shrublands which occur on sandplains or on associated laterites go by the collective name of "kwongan" (Beard 1976a), an Aboriginal word equivalent to maquis, macchia, chaparral and fynbos. In any given catenary sequence of which kwongan is a component it might be concluded that its structural reduction is a response to nutrient deficiency since kwongan soils are known to be more deficient than the red loam soils of the valleys. On the other hand kwongan soils also have poorer moisture relations.

This question can be clarified by studying the various plant formations in relation to climate and soil by arrangement into vegetation series. Applying this concept to tropical America, Beard (1944, 1955) postulated a single climatic optimum formation, the tropical rain forest, from which five series of formations were conceived to radiate from optimum to pessimum, each showing progressive reduction of structure with increasing adversity of the habitat. Each series comprised a single "essential habitat". One series was on well-drained deep soils, a second on well-drained shallow soils, both responding to increasing dryness of climate. Two other series were on ill-drained soils, the one permanently wet, the other seasonally waterlogged: these tended to be dependent upon depth of inundation rather than climate. The fifth series was montane and controlled by the effects of increasing altitude. Although the concept cannot be so usefully applied in Western Australia, Perry (1970) showed that in the vegetation of the West Kimberley there were three different suites of plant communities varying with the underlying rocks and soils through three climatic zones. In the south-west five major substrate types can most usefully be illustrated. There are others such as coastal limestone whose occurrence is more limited, but the major types (after Northcote 1965) are massive ironstone duricrust soils; sands containing or overlying ironstone gravel; deep sands; pallid zone soils; young soils on bedrock.

The vegetation upon each of these under varying rainfall is illustrated in Fig. 3. It has to be expected that there is some interaction between rainfall and substrate, so that for example the deep sands become less and less leached as rainfall decreases, and the pallid zone soils change from yellow to red and from neutral to alkaline.

All substrates are forested at maximum rainfall, either with karri (*E. diversicolor*) tall-forest or with jarrah-marri (*E. marginata*-*E. calophylla*) tall-forest and forest. At some point of rainfall varying with the substrate, forest is reduced to woodland, low woodland, scrub heath and thicket. In the massive ironstone duricrust series woodlands of varying composition persist down to 450 mm rainfall, since tree roots penetrate the apparently inhospitable duricrust into the deep moisture storage of the pallid zone beneath (Doley 1967; Kimber 1974). At the 450 mm threshold there is a sudden reduction to heath dominated by *Dryandra* spp., which occupies a relatively narrow rainfall band down to 375 mm, below which types of thicket appear. It is interesting that biomass in these is apparently greater than in the *Dryandra* heath.

On sands with ironstone gravel jarrah-marri forest is reduced to *Banksia* low woodland at 750 mm of rainfall and this to scrub heath of a heterogeneous composition at 625 mm. *Acacia*-*Casuarina* thickets, again with apparently greater biomass, supervene below 400 mm. Deep sands, because of their poor moisture retention, cannot support tall forest or forest below 1000 mm, nor even eucalypt woodland below 900 mm where *Banksia* low woodland takes over. Scrub heath appears at 625 mm and *Acacia* thicket at 325 mm. Pallid zone soils behave very differently. There is no transition to kwongan anywhere in this series which continues to support eucalypt woodland throughout. Biomass in the salmon gum-gimlet (*E. salmonophloia*-*E. salubris*) woodland appears as in the previous series to rise at the dry

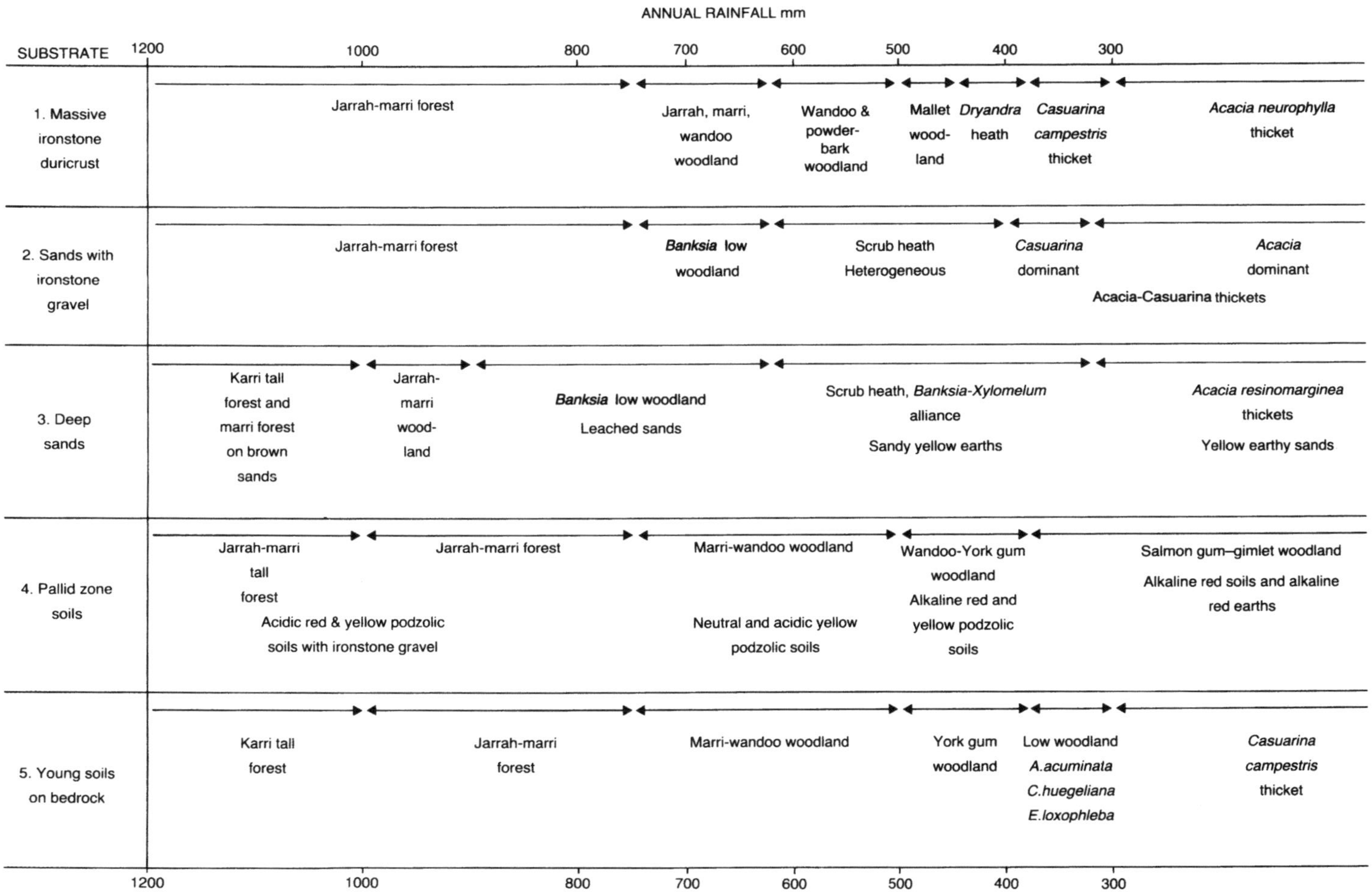

Fig 3. Vegetation changes along rainfall gradients on five major substrate types of the South West Botanical Province.

end of the rainfall spectrum. Young soils on bedrock behave in the same way as pallid zone soils down to the 375 mm rainfall level, below which the shallowness of the soil influences moisture-retaining capacity adversely, so that there is a decline to low woodland and Casuarina thicket.

3.3 Conclusion

The conclusion drawn from Fig. 3 is that moisture relations rather than nutrient availability control gross physiognomy in Western Australian vegetation. Owing to the ancient deep-weathering of the soil mantle under prior climates, the inherent capability of the substrate to supply nutrients can be assumed to remain approximately constant within each vegetation series except for the young soils on bedrock where parent material may be of some importance. Differences in efficiency of recycling may occur between vegetation types but that is a secondary consideration. A nutrient threshold cannot be supposed at the transition from woodland to kwongan, whereas it is quite conceivable that with the declining rainfall a limiting moisture threshold is reached. The variations between forest, woodland, heath and thicket within and between the vegetation series are therefore controlled by differences in moisture relations, not fertility. Nutrient availability has influenced selection of species whose scleromorphic growth forms characterize the overall physiognomy, but not to the extent of determining gross structural variation. The concept of moisture relations includes moisture-holding capacity, which will be more important in drier areas, and permeability, drainage and aeration, important chiefly in the higher rainfall areas. The properties are determined by the physical characteristics of the soil.

Since the sandy and lateritic kwongan soils are known to be more nutrient deficient for agricultural purposes than the loamy pallid zone soils, it might be assumed in the context of individual vegetation catenas that kwongan vegetation is depauperate for that reason. However, the fact that equivalent sandy and lateritic soils carry woodland and forest under higher rainfall show this to be false. As previously shown, all substrates support forest at maximum rainfall. This general conclusion is supported by the fact that after agricultural development the crops and pastures consume less moisture than the natural vegetation they replaced so that water tables rise and salt is brought to the surface. It is clear that the natural vegetation, whether woodland or heath, was highly moisture efficient. This could not have been possible if biomass had been reduced by nutrient starvation.

Each vegetation series is a continuum in which structural types and species dominance determine readily recognizable segments. Plants having typical heath growth forms occur throughout, forming the understorey in forest and woodland, becoming the dominant layer in heath and again taking an understorey role in the dry Acacia-Casuarina thickets. Component species vary with rainfall but the physiognomy is remarkably constant.

George et al. (1979) listed species richness as number of species per 500 m^2 sample as follows:

Western Australian heath and scrub heath:	37 to 95, mean 59.5
Eastern Australian heath and scrub heath:	30 to 56
Mallet (Eucalyptus accedens-E. wandoo) woodland:	28
Wandoo (E. wandoo) woodland:	35
Jarrah-marri (E. marginata-E. calophylla) forest:	43
Eastern Australian forests and woodlands (with a heathy understorey):	26-96, mean 40

Most of the species richness in the forests and woodlands cited is accounted for by the understorey. Other woodlands in both eastern and Western Australia on less sandy soils which do not have heathy understoreys are poorer in species. Greatest species richness in Western Australian heathlands occurs on sandy soils containing ironstone gravel and the least richness on deep sands and heavy laterites.

The almost universal scleromorphic shrub layer in southwestern Australia is a response to the widespread nutrient deficiency, since as has been discussed earlier the whole region constitutes a peinobiome. Within the constraints of the peinobiome vegetation is determined by climate and soil. Physical characteristics of soil, especially those which affect moisture relations, are of paramount importance.

References

BAGNOULS F, GAUSSEN H (1957) Les climats écologiques et leur classification. Annals of Geography 66: 193-220

BEARD JS (1944) Climax vegetation in tropical America. Ecology 25: 127-158

BEARD JS (1955) The classification of tropical American vegetation types. Ecology 36: 89-99

BEARD JS (1976a) An indigenous term for the Western Australian sandplain and its vegetation. Journal of the Royal Society of Western Australia 59: 55-7

BEARD JS (1976b) The evolution of Australian desert plants. In: GOODALL W (ed) Evolution of desert biota. University of Texas Press, Austin

BEARD JS (1979) Phytogeographic regions. In: GENTILLI J (ed) Western landscapes. University of Western Australia Press, Nedlands.

BEARD JS (1981) Vegetation survey of Western Australia, 1:1,000,000 series. Explanatory notes to Sheet 7, "Swan". University of Western Australia Press, Nedlands

BURBIDGE NT (1960) The phytogeography of the Australian region. Australian Journal of Botany 8: 75-211

DIELS L (1906) Die Vegetation der Erde, 7. Die Pflanzenwelt von West-Australien südlich des Wendekreises. Engelmann, Leipzig

DOLEY D (1967) Water relations of Eucalyptus marginata Sm. under natural conditions. Journal of Ecology 55: 597-614

GEORGE AS, HOPKINS AJM, MARCHANT NG (1979) The heathlands of Western Australia. In: SPECHT RL (ed) Ecosystems of the world, vol 9A. Heathlands and related shrublands. Descriptive studies. Elsevier, Amsterdam

KIMBER P (1974) The root system of jarrah (Eucalyptus marginata). Bulletin of the Forests Department of Western Australia 10

NORTHCOTE KH (1965) A factual key for the recognition of Australian soils. Division of Soils Divisional Report 2/65, CSIRO, Perth

PERRY R (1970) Vegetation, in Stewart et al. Lands of the Ord-Victoria area, Western Australia and Northern Territory. Division of Land Research Technical Paper Series 28, CSIRO, Perth

WALTER H (1979) Vegetation of the earth (2nd English edn). Springer-Verlag, Berlin

4. Influence of Nutrients on the Composition and Distribution of Plant Communities in Mediterranean-Type Ecosystems of Israel

A. RABINOVITCH-VIN

4.1 Introduction

A vegetation survey of the Upper Galilee of Israel showed that different climax plant communities grow in the same altitudinal belt under similar climatic conditions (Rabinovitch-Vin 1979). This contradicts the generally accepted opinion that climate is the main factor determining the distribution and development of the climax plant communities in the Middle East under mediterranean-type climates (Zohary 1960, 1973). It was found also that where annual rainfall exceeds 600 mm, rock type is the main factor determining the nature of plant communities. Consequently the composition of rock and soil types were examined and compared to the composition of the leaves of the dominant plants.

4.2 The area under investigation - environment

The study area is the Upper Galilee in northern Israel (Fig. 1). Its altitude ranges from 300 m to 1200 m above sea level. It is dissected by valleys up to 140-200 m deep with steep slopes and has a mediterranean-type climate. Average annual rainfall increases with altitude from 550 mm to 1100 mm in the higher places. Average annual temperatures vary from 15°C to 19°C, and humidity is higher in the western part of the area. Marine Lower Cretaceous to Eocenian sedimentary rocks underlie much of the study area. Some of the plateaux are covered by basalt flows of Miocene to Pleistocene age.

In hard, karstic rocks, water penetrates rapidly through pores, voids and joints. Since such rocks do not retain water, they are almost dry and the reddish-brown terra rossa soil formed from them is extremely leached, oxidized and permeable to water, some of which is available to plants. There are, however, several types of terra rossa, containing different clay minerals. Soils containing mainly montmorillonite are developed on the Cenomanian-Turonian rocks of the Judaea Group forming most of the Galilee ridges. Terra rossa soils containing kaolinite, on the other hand, are developed on the karstic rocks of the Eocenian series in the environs of

Table 1. Description of substrate types and available soil water (% of mass) for the plant alliances in the Galilee.

Plant alliance	*Quercus boissieri*	*Arbutus andrachne*	*Quercus calliprinos*	*Laurus nobilis*	*Sarcopoterium spinosum*	*Hordeum bulbosum* herbland
Geological formation	Deir Hanna, Ce_2	Deir Hanna, Ce_2 and Yirka, T	Sakhnin, Ce_3	Bina, T	Chalk Group-Senonian	Bar Kokhba Middle E_2
Rock type	Alternate hard and soft dolomite	Marl	Karstic dolomite	Karstic limestone	Chalk	Karstic limestone
Soil type	Brown soil	Marly rendzina	Montmorillonitic terra rossa	Montmorillonitic terra rossa	Grey rendzina	Kaolinitic terra rossa
Percentage of available water	14.5-29.0	12.6-18.4	10.1-13.0	12.4-18.7	13.9-19.2	7.1-9.8
Sample size	23	14	16	14	9	20

Ce_2 = Cenomanian, Deir Hanna formation; Ce_3 = Cenomanian, Sakhnin formation; T = Turonian Bina and Yirka formation; E_2 = Eocene, Bar-Kokhba formation

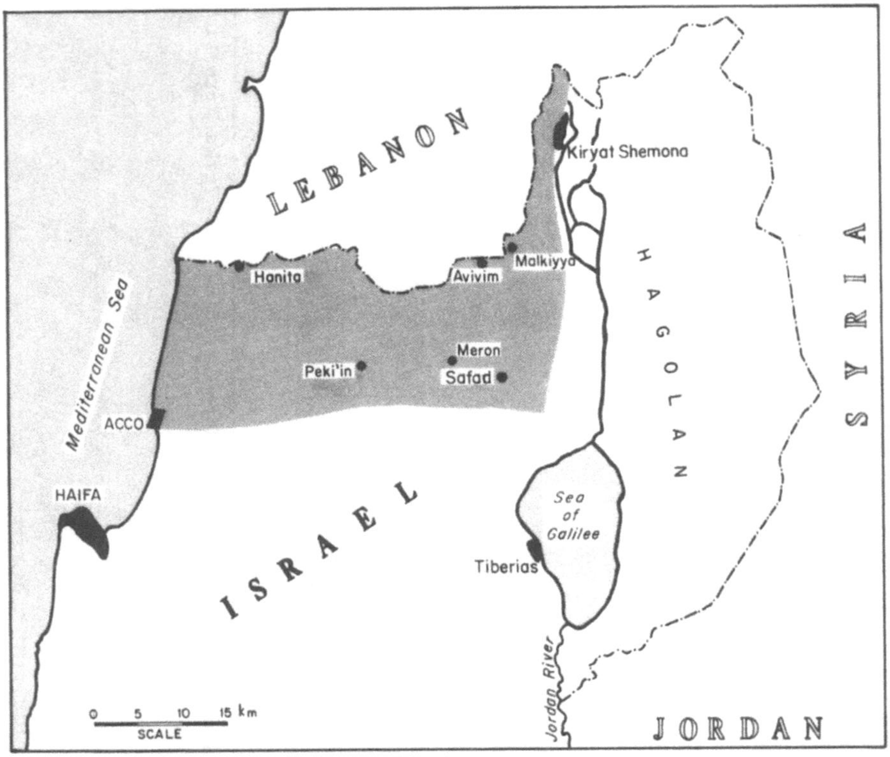

Fig. 1. Map of the study area, showing localities mentioned in the text.

Malkiyya, Avivim and Ramot Razim (east of Safed). Soil water content varies and so does, to a great extent, the amount of water available to plants (Table 1). Alternating hard and soft rocks yield partly leached brown soil, which has the highest water retention potential. Being water-absorbent, chalky rocks are leached less than hard limestones or hard dolomites. Unlike soils which develop on hard, karstic rocks and which comprise mainly insoluble residue, much of the light to dark grey rendzina soil that develops on chalks contains many small particles of unchanged chalky rock. Water penetrates the chalks through joints and voids, and aeration in such rock is quite good during the greater part of the year.

Marls are rocks consisting of limestone or dolomite and clay. The clay part in the marl absorbs the water, swells and fills all free space. The marly rocks are porous but not permeable and aeration is very poor. Although they contain relatively large amounts of water the poor aeration inhibits oxidation. The soils which develop on the marls are similar in their properties to the rock itself and have been defined in this study as "marly rendzinas".

4.3 Vegetation

Six plant alliances have been defined for the study area (Rabinovitch-Vin 1979), as follows.

(a) The Quercus calliprinos alliance. This is the most widespread alliance in the Galilee. Q. calliprinos is a shrub 3-5 m high but under favourable conditions may become a tree. As a shrub it forms dense maquis communities rich in annual species although the cover of these latter is small in closed stands. These communities occupy hard rugged limestones and dolomites mainly of the Cenomanian-Turonian Judaea Group but also on Eocenic Bar-Kokhba limestones.

(b) Pinus halepensis-Arbutus andrachne alliance. P. halepensis is a tree up to 10 m high while A. andrachne is a low tree or more often a shrub up to 4 m high. Here, few annual species are found, and their cover is low. Several dwarf- and half-shrubs characterize the alliance.

(c) Laurus nobilis alliance. This occurs in the western Galilee at altitudes exceeding 300 m. Soil developed on these rocks is a reddish brown montmorillonitic terra rossa. Laurus is an evergreen shrub 3-5 m high. This plant community is rich in annuals.

(d) Quercus boissieri alliance. Q. boissieri is a small winter-deciduous oak up to 5-8 m high. These plant comunities are generally rich in companion evergreen shrubs and annuals.

(e) Ceratonia siliqua-Pistacia lentiscus alliance. This alliance is found on altitudes lower generally than 300 m above sea level in the Western Upper Galilee.

(f) Pistacia atlantica alliance. This occupies the upper eastern slopes of the Upper Galilee facing the Jordan Valley. P. atlantica is a winter-deciduous tree which forms an open forest with a rich herbaceous layer as well as shrub layer.

The Ceratonia siliqua-Pistacia lentiscus and the P. atlantica alliances grow in the low parts of the study area, where annual rainfall is less than 600 mm and therefore are not considered further in this account.

(g) Sarcopoterium spinosum alliance. Of several secondary half- and dwarf-shrub communities in the study area this is the most important and widespread. It occupies open stands with well-aerated soils, mainly montmorillonitic terra rossa developed on hard limestones and dolomites and grey rendzina developed on chalks. On the latter it persists for a relatively longer time.

4.4 Methods

The results here are based on an earlier comprehensive study (Rabinovitch-Vin 1979). The vegetation at sample sites was recorded by the Braun-Blanquet method. Soils were sampled subjectively only in the centre of areas dominated by each of the plant communities investigated to avoid transitional areas. Samples were taken on hilltops to avoid the effects of runoff and soil erosion.

Apart from my own samples, soil samples taken by pedologists (Dan 1965; Gal 1965; Koyumdjisky 1968, 1972) were used if taken in places which fitted the above rules.

The analytical methods were the same for all the soils examined. Most analyses were carried out in the same laboratory. For carbonates the volumetric method was used (Schlichting and Blume 1966). Exchangeable cations were measured according to Bower (1955) and Bower et al. (1952). Available ammonium and nitrate were determined by extraction with KCl (Bremner 1965) and available phosphorus by extraction with sodium bicarbonate (Olsen et al. 1954) and subsequent colorimetry (Watanabe and Olsen 1965).

4.5 Soil nutrients as related to the distribution and composition of plant communities

4.5.1 AMMONIUM:NITRATE RATIO

In soils of low pH and at low temperatures, ammonium is formed and accumulated in the soil; nitrification is favoured by increasing pH and rising temperatures (Russell 1961). Plants often take up nitrogen in the form of nitrate, and where nitrogen is available only in the form of ammonium, growth of certain species is inhibited to the point of death. Several studies, however, have shown that not all plants conform to this last rule, and that some of them prefer ammonium. In fact, such plants showed lack of vitality in cases where nitrate is the source of nitrogen, with reactions ranging from delayed to completely arrested development. Chlorosis is symptomatic of such damage to the plants (mostly Ericaceae). Thus the ratios of ammonium to nitrate in soils can profoundly affect vegetation composition.

Previous studies (Pardo 1935; Russell 1961; van Tuil 1965) have been carried out under culture conditions and in pots. In this paper I present field evidence.

(a) Pinus halepensis-Arbutus andrachne alliance

Arbutus andrachne. Soils developed on Cretaceous marl form a suitable habitat for *A. andrachne*, except for places where Nari-type caliche develops, with improved aeration and water penetration. Characteristically, these rocks are soft, porous, rich in clay and lacking joints. They retain water well and are poorly aerated. There is a gradual transition from rock to soil. A variety of soils develop on these rocks as far as colour and ion composition are concerned, but the presence of ammonium as a source of nitrogen for plants is common to all of them (Table 2).

It was found that all the ammonium-containing soils examined are basic with pH from 7 to 8 (Table 2). Such soils have also been found in the Carmel, Judaea and Samaria areas. Hence we conclude in this case that the pH of the soil has nothing to do with ammonium formation; formation of ammonium is due to anaerobic conditions.

Hitherto in Israel, soils were defined mainly by colour, and by this criterion soils developing on marl or marly chalk were classified as either grey rendzina or terra rossa. Unlike these types, however, the soils under discussion do not contain nitrate and hence a new type of soil has been defined here, i.e. a "marly rendzina", characterized by poor aeration and the presence of ammonium, which forms a habitat preferred by ammonophilic plants, the *Arbutus andrachne-Pinus halepensis* alliance.

A. andrachne is the only representative of Ericaceae in this area. Communities dominated by ericaceous plants occur usually in high, cool, wet areas, on acidic soils containing ammonium-nitrogen. The only existing Israeli representative of this family also grows in a habitat in which nitrogen is supplied as ammonium, in spite of the fact that the soil

Table 2. Soil contents of ammonium- and nitrate-nitrogen, soil pH values, and Ca:Mg ratios for representative locations in the Galilee. Abbreviations as for Table 1.

Location	Plant community	Soil type	Rock type	Ca:Mg	pH	NH_4^+	NO_3^-
Mt Hazon	Arbutus andrachne	Marly rendzina	LC (Lower cretaceous)	2.0	7.50	4.06	0.10
Mt Meron	A. andrachne	Marly rendzina	Clay and soft dolomite, Ce_2	2.3	7.85	1.54	0.00
Mt Meron	A. andrachne	" "	" "	2.4	7.90	3.36	0.00
Mt Matat	A. andrachne	" "	" "	2.1	7.90	5.60	0.00
Yirka	Arbutus and Pinus	" "	Marl, T	26.2	8.10	2.66	0.00
Yirka	Pinus and Arbutus	" "	Marl, T	13.0	7.97	2.52	0.00
Mt Sulam Sour	Pinus halepensis	" "	Marly chalk Ce_2	15.3	8.04	4.48	2.94
Kelakh	P. halepensis	" "	" "	14.0	?	11.20	3.08
Kelakh	P. halepensis	" "	" "	15.8	?	9.80	2.80
Rihaniya	Sarcopoterium spinosum and Quercus calliprinos	Grey rendzina	Chalk, S	13.1	7.90	6.70	6.02
Alona	Q. calliprinos	" "	Chalk, S	14.5	7.90	2.63	4.48
Bar'am 1	Q. calliprinos	Montmorillonitic terra rossa	Hard dolomitic limestone, Ce_3	9.1	7.20	0.70	3.64
Bar'am 2	Q. calliprinos	" "	" "	?	7.20	3.24	2.92
Mt Avivim 1	Hordeum bulbosum	Kaolinitic terra rossa	Hard limestone, E_2	15.1	6.90	2.38	8.68
Mt Avivim 2	Q. calliprinos	" "	" " "	9.2	7.0	1.68	8.40

is basic. This suggests that the procurement of nitrogen from ammonium is of greater importance to plants of this family than the acidity of the soil. There even seem to be genetic factors causing a single species of this family to obtain its nitrogen from ammonium compounds. Such ecological adaptation certainly deserves further investigation.

Pinus halepensis. Like A. andrachne, P. halepensis grows on soft formations of the Judaea Group. Many plants, particularly those belonging to the batha and herbaceous communities, are common to both associations but whereas A. andrachne grows on marl and on fairly clayey soil, P. halepensis prefers marly chalk which has somewhat better aeration which is in accordance with McFee and Stone 1968.

Though ammonium is the primary source of soil nitrogen in this habitat there is some nitrate present and hence the greater numbers of annual herbs than in the A. andrachne habitat.

(b) The Quercus calliprinos alliance

Plant associations of the Q. calliprinos alliance are found in the Galilee in most cases on hard, rugged limestones and dolomites of the Cenomanian-Turonian Judaea Group, though less on Eocenic Bar-Kokhba Limestones from which terra rossa soil is developed. This plant association also occurs on the rendzina soils formed on Senonian and certain Eocenian chalky rocks. These rocks and soils have good aeration with nitrates readily formed in them (Table 2).

4.5.2 CALCIUM:MAGNESIUM RATIO

The distribution pattern of Laurus nobilis is difficult to explain from the results of this study. The main habitat of this species in Israel is the western Galilee though it also occurs rarely in the Meron area with the exception of the Peki'in ridge. A few specimens grow in wadis east of the divide and in the northernmost part of the Galilee. There is no apparent reason for its limited distribution in areas with adequate amounts of rainfall outside the western Galilee. The results of independent studies of rock and soil composition in the Galilee suggest that the distribution of this Mediterranean mesophilic plant is related to the ratio of exchangeable calcium and magnesium cations. This led me to examine the results of four other independent studies in the Galilee dealing with magnesium in rock and soil (Dan 1965; Gal 1965; Kafri 1972; Koyumdjisky 1972) and to analyse soil samples at 40 sites. All results were similar and are presented in Fig. 2 which illustrates the relationships between the ratio of calcium and magnesium ions and soil and rock types.

The Cenomanian layers (Kamon, Deir Hanna and Sakhnin formations) are mainly dolomitic, supplying relatively high quantities of magnesium to the soil. The Deir Hanna and the Turonian Bina formations in the western Galilee consist mainly of chalks and limestones respectively which supply more exchangeable calcium to the soil. A high calcium:magnesium ratio alone, however, does not determine the distribution of L. nobilis, since, like other mesophilic plants, it does not grow on the leached, calcium-rich Eocenic terra rossa, due to limited water in the karstic Eocenic limestones and kaolinitic terra rossa soil.

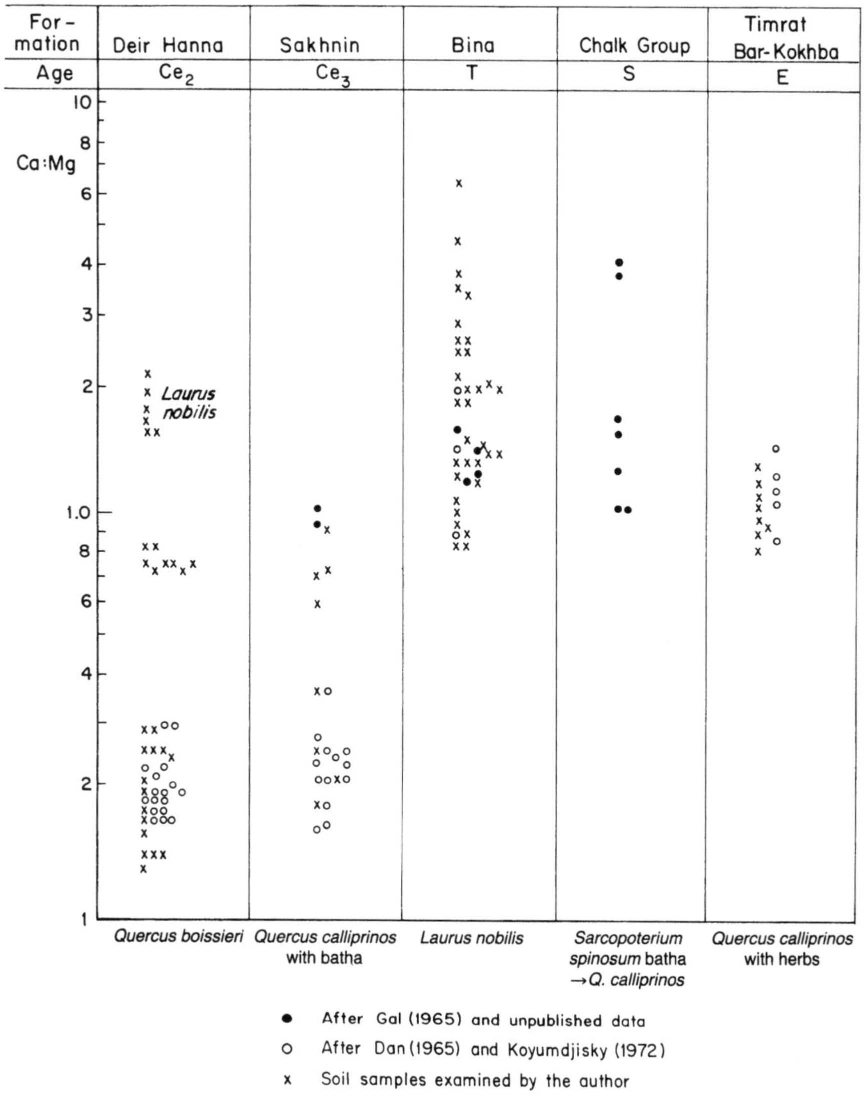

Fig. 2. Distribution of calcium:magnesium ratio in soils formed on different rock formations in the Galilee. Dots are for data from Gal (1965) and unpublished, circles from Dan (1965) and Koyumdjisky (1972), and crosses for soils sampled in this study. Symbols for ages as in Table 1.

4.5.3 PHOSPHORUS

The sedimentary rocks of the Galilee vary in their phosphorus content. Thus, Cenomenian-Turonian rocks of the Judaea Group and the soils formed from them are low in phosphorus, whereas the younger Senonian and Eocenian rocks and their soils contain higher amounts.

The Pliocene basalt rocks and their soils in the Galilee are also rich in phosphorus (Koyumdjisky 1968). In certain locations where maquis is well developed there is a high phosphorus content in the uppermost layers of the soil under the leaf litter, even in low phosphorus soil. This seems due to the fact that the tree roots collect and concentrate phosphorus from a large volume of rock and soil.

Many studies throughout the world have shown that a high phosphorus content in soil is essential for herbaceous plants, whereas low phosphorus soils are dominated by ligneous plants (bushes and trees). This phenomenon has also been observed in the present study. Herbaceous associations are developed in kaolinitic terra rossa overlying hard Eocenian limestones and also on basaltic soils.

Batha communities are missing from these habitats, and tree development is also very slow. On the other hand, soils developed on rocks of the Judaea Group are dominated by ligneous plants of batha and maquis.

4.5.4 SARCOPOTERIUM SPINOSUM BATHA AND HIGHLY CALCAREOUS SOILS

The *Sarcopoterium spinosum* alliance, a type of batha, occurs on grey rendzina soils developed from Senonian chalks. This calcium-rich, aerated soil contains nitrate and phosphorus. This combination makes the *S. spinosum* batha very rich in a variety of annual plants, in contrast to the bathas of Cistaceae and Lamiaceae, which grow on marly rendzinas and are almost completely deficient in annual plants.

The strong calciphilic character of *S. spinosum* has been experimentally demonstrated. It has been grown in kaolinitic terra rossa both in the field and in pots but the plants failed to grow normally. They remained in rosette form, did not produce spiny dolichoblasts, and later showed nutritional deficiency. Adding calcium resulted in immediate improvement and normal growth (Rabinovitch-Vin and Orshan 1974).

S. spinosum does grow on montmorillonitic terra rossa, which is less subject to leaching than kaolinitic terra rossa and contains nitrogen mainly as nitrate (Table 2).

4.5.5 CLAY MINERALS AND SOIL MOISTURE AVAILABILITY

It is known that in moderately leached soil the common clay mineral is montmorillonite; in well-leached soil it is kaolinite, the product of montmorillonite weathering. Nonetheless, the soil in the rainy areas of Mt Meron and the western Galilee (annual precipitation 800 mm to 1100 mm) is mainly montmorillonitic, whereas that in the drier areas of Malkiyya-Avivim, Rosh Pina, Arbel and Gilboa (annual precipitation 500 mm to 700 mm) soils are kaolinitic.

Apparently, the type of clay in the Galilee depends on the silica content in the rock on which the soil is developed rather than on the amount of rain. Thus, in leached soil where silica is not available the montmorillonite disintegrates into kaolinite. Indeed, the Turonian strata in the Galilee are rich in silica and montmorillonitic soils develop on them whereas Eocenian strata lack silica and yield kaolinitic soil. This conclusion is in good agreement with Jackson (1965), who claims that silica accumulation is necessary for montmorillonite reformation. Thus the chemical composition of the rock determines the make-up of clay minerals in the soil which, in turn, determines the amount of water available for plants.

The nitrophilic *Quercus boissieri* alliance occurs on brown soils formed from alternating hard and soft layers of the Judaea group, which has relatively large amounts of available water and exchangeable cations. Such soils exist in the high and central Galilee in the dolomitic part of the Cenomanian Deir Hanna formation. Similar soils, but with a higher calcium:magnesium ratio (10:1 as opposed to 2:1 in the previous case), which are developed in the western Galilee on limestones of the same formation, support the Laurus nobilis-Acer syriacum association.

4.6 Conclusions

This study is concerned with relatively cool, rainy areas with precipitation exceeding 600 mm. Since this sample was from climax communities, the conclusion is that the principal factor determining the nature and distribution of plant communities in the upper Galilee is the type of rock on which they develop and its chemical and physical properties. Since the rock characteristics are so important, classifying them according to age (Cenomanian, Senonian, etc.) only is insufficient. Insufficient, too, is mere classification by rock type (limestone, dolomite, chalk etc.) because there are for example substantial differences between karstic limestones or dolomites, in which air circulation and oxidation are good, and soft limestone and dolomite, which lack good aeration. Similarly, there is a difference between air circulation in Senonian chalky rocks, which are vertically jointed and hence maintain good air circulation, and Cenomanian chalks, which are non-jointed and therefore poorly aerated. Nitrate-containing soils (terra rossas and grey rendzinas) have developed from hard rocks (dolomite, limestone) as well as from jointed chalk. Ammonium-containing, poorly aerated marly rendzinas have developed from soft, clay-rich rocks. Types of clay minerals in the soil (montmorillonite or kaolinite) are determined by the mineral composition of the underlying rock. These, in turn, determine the amount of available water in the soil. The amounts of magnesium, calcium and phosphate in the rock determine the availability of these minerals in the overlying soil.

In summary, the properties of the parent rock and its chemical composition determine the properties of the overlying soil, and hence the nature and composition of the plant communities:

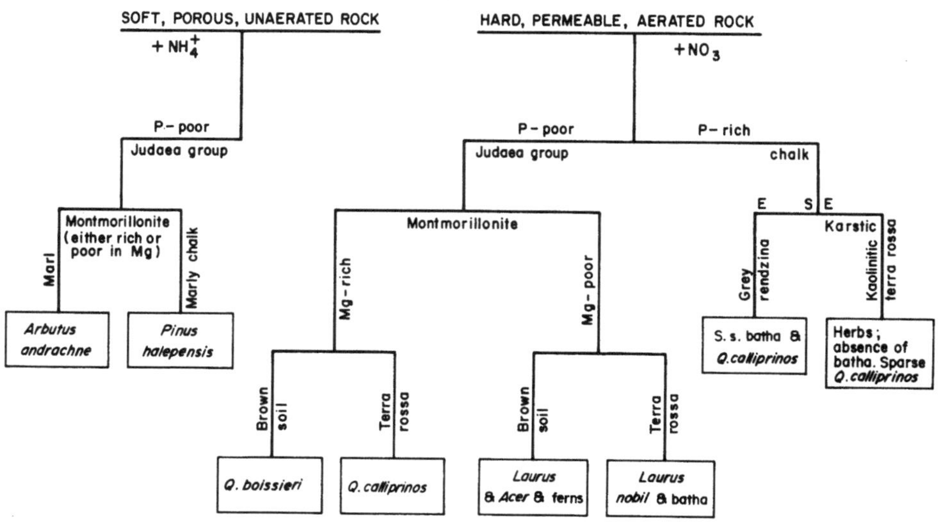

Fig. 3. Rocks which determine soil properties and plant communities receiving amounts of rainfall greater than 600 mm.

(a) Well-aerated, nitrate-containing soils support plant associations defined here as nitrophilic. They include maquis and batha alliances of *Quercus calliprinos*, *Q. boissieri*, *Laurus nobilis*, and *Sarcopoterium spinosum*. *Q. calliprinos* is dominant on soils that are developed on dolomitic rocks where magnesium is abundant. *Q. boissieri* characteristically grows on brown soils with a high cation exchange capacity and a large amount of available water (Table 1). *L. nobilis* competes with other forest trees where magnesium content in soil is low and water retention good. *S. spinosum* grows on aerated soil after the destruction of the maquis, but preferentially on rendzinas which contain free $CaCO_3$. Herbaceous associations are competitive with ligneous ones in phosphorus-rich soils.

(b) Plant communities growing on poorly aerated, ammonium-containing soils such as those of the *Arbutus andrachne*-*Pinus halepensis* alliance, are well defined. Annuals are virtually absent from such biotopes in marked contrast with the nitrophilic alliances, all of which contain annual herbs. These relationships are summarized in Fig. 3.

References

BOWER CA, REITEMEYER RF, FIREMAN H (1952) Exchangeable cation analysis of saline and alkaline soils. Soil Science 73: 251-261

BOWER CA (1955) Determination of exchangeable magnesium in soils containing dolomite. Proceedings. Soil Science Society of America 19: 40-42

BREMNER JM (1965) Inorganic forms of nitrogen. In: BLACK CA et al (eds) Agronomy, vol 9. Methods of soil analysis. American Society of Agronomy, Madison, Wisconsin, pp 1195-1201

DAN J (1965) Effect of relief on the formation of soils in Israel and their distribution. PhD Thesis. Hebrew University of Jerusalem

GAL M (1965) Genesis, properties and distribution of soils formed on limestones in the Upper Galilee. PhD Thesis, The Hebrew University of Jerusalem

JACKSON ML (1965) Clay transformations in soil genesis during the quaternary. Soil Science 99: 15-22

KAFRI U (1972) Lithostratigraphy and environment of deposition. Judea Group, Western and Central Galilee. Geological Survey Bulletin No 54

KOYUMDJISKY H (1968) Nutrient availability as related to genetic characteristics of Israel soils. Report September 1968, p 118

KOYUMDJISKY H (1972) Behaviour of magnesium during weathering of carbonate and silicate rocks and its influence on soil formation in Israel. PhD Thesis. Hebrew University of Jerusalem

MCFEE WW, STONE EL Jr (1968) Ammonium and nitrate as nitrogen sources for *Pinus radiata* and *Picea glauca*. Proceedings. Soil Science Society of America 32: 879-884

OLSEN SR, COLE CM, WATANABE FS, DEAN LA (1954) Estimation of available phosphorus in soil by extraction with sodium bicarbonate. Circular. US Department of Agriculture 939

PARDO JH (1935) Ammonium in the nutrition of higher green plants. The Quarterly Review of Biology 10(1)

RABINOVITCH-VIN A (1979) Influence of parent rock on soil properties and composition of vegetation in the Galilee. PhD Thesis. Hebrew University of Jerusalem

RABINOVITCH-VIN A, ORSHAN G (1974) Ecological studies on the vegetation of the upper Galilee, Israel. II. Factors determining the absence of batha and garrigue components on middle Eocenian strata. Israel Journal of Botany 23: 111-119

RUSSELL EW (1961) Soil conditions and plant growth, 9th eds. Longmans, New York, p 688

SCHLICHTING E, BLUME HP (1966) Bodenkundliches Praktikum. Paul Parey, Hamburg, pp 107-108

VAN TUIL HDW (1965) Organic salts in plants in relation to nutrition and growth. Centre for Agricultural Publication and Documentation, Wageningen. Agricultural Research Report No 657

WATANABE FS, OLSEN RS (1965) Test of an ascorbic acid method for determining phosphorus in water and $NaHCO_3$ extracts from soil. Proceedings of the Soil Science Society of America 29, 677-678

ZOHARY M (1960) The maquis of *Quercus calliprinos* in Israel and Jordan. Bulletin of the Research Council of Israel 9b(2): 51-72

ZOHARY M (1973) Geobotanical foundations of the Middle East, vols 1 and 2. Fischer, Stuttgart

5. Approaches to the Definition of Mediterranean Growth Forms

G. ORSHAN

5.1 Introduction

Mediterranean-type ecosystems, characterized by evergreen sclerophyllous small trees or shrubs, are fairly similar in their physiognomy. There have been relatively extensive comparisons among mediterranean-type ecosystems of Chile and California from the evolutionary, morphological and physiological points of view during the last two decades. This work, which continues, was summarized by Di Castri and Mooney (1973), Mooney (1977) and Cody and Mooney (1978). The mediterranean-type ecosystems of California have also been compared to those of the Mediterranean Basin and Australia (Naveh 1967; Specht 1968a,b).

The mediterranean-type plant growth forms and vegetation formations are thought to be the result of convergent evolution, the nature of which regretfully does not reveal much about the morphogenetic processes which resulted in these growth forms, the related physiological properties which are important as adaptations to mediterranean-type environments, or the nature of the processes involved in natural selection by these environments.

There is more than one growth form for plants growing in mediterranean conditions. Trees, shrubs, half-shrubs and perennial and annual herbs grow side by side in the same stand in the typical mediterranean area. Growth forms are clearly adapted in various ways to the niches occupied by the plants. Thus any mediterranean ecosystem comprises a complex or a spectrum of growth forms, a fact which had been realized and dealt with by Mooney et al. (1974), Mooney (1974) and Parsons and Moldenke (1975), who analysed changes in growth forms and physiological properties of plants along ecological gradients. The fact that differences in growth forms found in adjacent stands along gradients within Chile or California were larger than between climatically matched stands from both countries suggests that each gradient examined contained several mediterranean-type environments with distinct edaphic complexes and microclimates. Hence one should define each mediterranean-type environment in order to compare more accurately its growth form complexes to those of other mediterranean-type environments or environmental gradients.

5.2 Definition of the mediterranean-type climate

Daget and Michel-Villag (1975) consider that there are several mediterranean-type climates differing from each other in many respects.

It is generally accepted that the mediterranean-type climates are characterized by biseasonality, i.e. a hot, dry summer and a mild, rainy winter. There are, however, different opinions about the ranges of annual amounts of rainfall and the extent of summer dryness characterizing them.

Aschmann (1973) delimits the mediterranean-type climate by amounts of rainfall ranging between 275 and 900 mm of which at least 65% falls in winter. Köppen (1936) defines it in a somewhat less restrictive way whereas Specht (1979) and others think that rainfall might be either high or low. In the latter case semi-arid conditions prevail.

Daget and Michel-Villag (1975), who try to delimit the mediterranean-type climate of the Mediterranean Basin on the basis of precipitation regime alone, also suggest that seasonality and not the total amount of rainfall is the most important parameter. In the map they present the northern Sahara, Libya and Egypt as well as the greater parts of Arabia, Iran and even Afghanistan are included within the mediterranean climatic region. Thus according to them, there are not only the mesic mediterranean-type climates but also mediterranean-type semideserts, deserts and even extreme deserts.

From the biological point of view the duration and number of seasons with optimal conditions for growth and development on the one hand, and of the unfavourable ones with one or more limiting environmental factors on the other, should be considered.

Throughout the mediterranean region water is a limiting factor either seasonally under more humid conditions or nearly throughout the year under more arid ones. The rainfall values suggested by Aschmann (1973) as delimiting the mediterranean-type climate agree well with conditions prevailing in the Middle East. Here an annual rainfall exceeding 300 to 350 mm causes thorough leaching of the entire rock and soil profile every year with a full recharge of water used by plants. These conditions allow normal growth of phanerophytes which are consequently the dominants of undisturbed vegetation. Under lower amounts of rainfall the dominants are generally chamaephytes - dwarf-shrubs or half-shrubs.

According to Aschmann, the high value of rainfall delimiting the mediterranean-type climate also seems to be indirectly related to water as a limiting factor. A Rabinovitch-Vin (personal communication 1976 and 1979) suggests that for the Galilee of Israel where rainfall is lower than 650 mm, water is the main environmental factor which determines the distribution of plant communities, while in higher rainfall areas the mineral composition of the soil and bedrock are decisive. Water may still be a limiting factor up to a rainfall of about 900 mm.

Throughout the areas with mediterranean-type climates defined by narrow ranges of rainfall, temperature, especially low temperature, is seldom a major limiting factor. Within wider ranges, however, low temperatures seem to be limiting especially at higher altitudes, and high temperatures sometimes in the more arid regions.

Where temperatures are not limiting, there is only one season of stress - summer, when organisms are subjected to water stress. Wherever low temperatures are also limiting there is a temperature-stress season - winter - in addition to the summer one. In both cases, however, spring is the season optimal for growth and development.

It is thus evident that the growth forms of mediterranean ecosystems can only be precisely defined for limited, well-defined mediterranean climatic variants.

5.3 Criteria for determination of growth forms

Various systems of growth forms have been developed since the days of Theophrastos and Von Humboldt (1806). They have been reviewed in detail by Du Rietz (1931). Most of them are based on various combinations of morphological, anatomical and phenological characters. Raunkiaer's system (Raunkiaer 1934), however, is unique in being based upon one single character - the location of the renewal buds. This character was supposed to be the best expression of the adaptations of plants to withstand the unfavourable season. It is therefore no wonder that Raunkiaer's classification of phytoclimates based on life form spectra was recognized and used by others (Cain 1950).

Even Raunkiaer himself subdivided his main life forms on the basis of several characters other than the location of the renewal buds. These subdivisions were revised by his followers (e.g. Braun-Blanquet 1964).

One of the more recent comprehensive systems of growth forms is that proposed by Ellenberg and Mueller-Dombois (1967b). For the self-supporting autotrophic cormophytes it is based on Raunkiaer's life forms. Each life form is subdivided on the basis of many criteria. For the phanerophytes, for example, the following characters were used, in this sequence: (a) shape of stems, (b) number of stems, (c) size of the whole plant, (d) seasonality, (e) bud protection, (f) leaf structure, (g) crown shape, (h) leaf shape, (i) root morphology, (j) bark features, (k) thorniness and (l) position of inflorescences. Some of these and other characters were used for subdividing other life forms.

This rather complex system aptly describes the physiognomy of plants and no doubt helps to describe different vegetation types (Ellenberg and Mueller-Dombois 1967). The question remains, however, whether it adequately expresses adaptations of plants or plant communities to their environment.

Sometimes we hardly know the adaptive value of each of the characters used (Mooney 1975) or the extent to which they result from the pressures of selective environmental factors (Gigon 1978). Another problem is the relation between different adaptive characters, and how each of them is related to the sum of the plant's adaptations to its environment. A possible solution to this problem is to first look for correlation between environmental factors or environmental complexes on the one hand, and single growth form characters or growth form character complexes on the other, and then try to look for causes and effects.

Parsons (1976) used multivariate analysis in comparing 24 morphological, phenological and to a certain extent physiological growth form character criteria of mediterranean-type ecosystems in California and Chile and obtained as a result 9 character groups. It is doubtful, however, whether the combinations of characters represent character complexes that are adapted to mediterranean-type environments in general or more specifically to those of Chile and California. If one considers for example two groups of Parsons's growth forms - leafless green shrubs and spiny drought-deciduous green-stemmed shrubs - it seems that green stems do not necessarily characterize shrubs only or are correlated with spininess or with absence of leaves. In fact there are evergreen green-stemmed geophytes (*Asparagus* spp. which are spiny), hemicryptophytes (*Foeniculum* spp. which are not spiny and bear leaves), chamaephytes (*Ochradenus baccatus* which is not summer deciduous and lacks spines) and leafless articulate plants (*Anabasis* spp., *Hammada* spp. and others) as well as phanerophytes (*Retama raetam* which is not spiny and practically leafless) and *Calycotome villosa* (which is spiny and summer deciduous).

Sclerophylly, on the other hand, seems to be linked with evergreenness. Malacophyllous mediterranean plants are generally deciduous although not necessarily summer deciduous (e.g. Cercis siliquastrum). It is, admittedly, not easy to clearly define a sclerophyll and distinguish it from other leaf types (A Fahn, personal communication 1980) although characteristics like density may be used. Went (1971), however, does not consider sclerophylly to be an adaptive character, although it has been shown that under drought stress sclerophyllous plants reduce their transpiration and assimilation rates (Mooney et al. 1968, 1975). The question remains open whether this is due to the plant's anatomical structure or to other aspects of their soil-plant-air water fluxes. One should point out however that under favourable water conditions both assimilation rates and water output are higher for malacophyllous plants than for sclerophyllous ones (Mooney et al. 1968).

Given our present state of knowledge examination of correlations between single growth form characters and environmental factors or environmental complexes as carried out, for example, by Parsons and Moldenke (1975) seem promising. In order to examine such correlations thoroughly one should prepare a phenomorphological "plant passport" in which as many morphological and phenological characters as possible will be listed for every species. Equipped with such a passport, one could analyse the distribution of characters listed in different plant communities or geographical territories whether small or large.

Preparing such a passport for the entire flora of a region or a country is a formidable task because various characters like the nature of the plant parts which are seasonally shed, life duration of leaves and stems or even location of renewal buds may be unknown for many species. Therefore one should first choose the characters that are allegedly of adaptive value.

As stated above, seasonality seems to be the outstanding feature characterizing all mediterranean-type climates. Therefore first plants persisting throughout the year and withstanding summer drought, with above-ground living organs, and ephemeral and ephemeroid ones disappearing during the stress seasons, should be distinguished and taken into account when analysing other characters.

Figs 1 to 5 present an attempt to analyse the distribution of some growth forms defined in terms of a single character in a few plant communities of the more humid upper Galilee and the Negev desert of Israel. The analysis is based on phytosociological tables summing vegetation records made in 500 m^2 quadrats by Rabinovitch-Vin (1979) in the Galilee and by Danin (1970) in 100 m^2 quadrats in the Negev (see also Danin et al. 1975). The annual amount of rainfall in the Galilee is 750 to 1000 mm and in the Negev 80 to 150 mm.

The Galilee plant communities analysed are climax associations dominated by evergreen trees or shrubs with a total cover generally exceeding 80%.

They are:

(a) Quercus calliprinos - Pistacia palaestina association (QP). Q. calliprinos is an evergreen shrub or small tree generally 4 to 5 m high. Its dense stands can hardly be penetrated. The cover as well as number of species of herbaceous plants is rather small in mature and dense stands. In more open ones however, the number of species recorded in 100 m^2 quadrats generally exceeds 100 sometimes reaching 150 or more. This association generally occupies terra rossa soil developed from hard limestones or dolomites (14 records; 206 species).

(b) Pinus halepensis - Juniperus oxycedrus association (PJ). Both dominants are evergreen trees 5 to 10 m high, though Juniperus often has a shrub habit. The total cover of the upper tree layer seldom exceeds 40%. There is a distinct dwarf-shrub layer and also an annual layer of medium density. This plant community grows on a moderately aerated rendzina soil developed on chalks or marly chalks (7 records; 73 species).

(c) Arbutus andrachne - Helianthemum syriacum association (AH). A. andrachne is a small evergreen tree whereas H. syriacum is a dwarf-shrub. Arbutus and tall shrubs do not form a continuous layer of high cover (cover is no higher than 30 to 40%). The cover of the dwarf-shrubs and half-shrubs is relatively high. Annuals on the other hand are scarce. This plant community is characterized by geophytes, especially Orchidaceae. It grows on a compact marly soil developed on lower Cretaceous rocks (19 records; 138 species).

The Negev plant communities differ from the Galilee ones in being small-shrub climax communities. Their physiognomy is similar. The persisting plants are the dwarf-shrubs or half-shrubs which are 0.25 to 0.50 m high. Their cover is 15 to 20% of the area. They shed the greater part of their leaf or assimilating branch cover or both during the summer but remain evergreen. Ephemeral annuals and perennial herbs form a higher cover during the spring but disappear in early summer. The three Negev plant communities examined are:

(a) Artemisia herba alba - Reaumuria negevensis association (AR). This is an Irano-Touranian dwarf-shrub steppe community occupying the Negev brown lithosols developed on hard chalks (12 records; 71 species).

(b) Hammada scoparia association (HS) in the Negev occupies the loessial plains in the transition zone between the Irano-Touranian steppes and the Saharo-Arabian deserts. The soil is a loessial sierozem (22 records; 106 species).

(c) Zygophyllum dumosum - Reaumuria hirtella association (ZR), a typical Saharo-Arabian desert community on brown lithosols (21 records; 154 species).

The distribution of each of the following characters was analysed for each of the above plant communities and presented in Figs 1 to 5 as percentages of total numbers of species: (a) seasonality - the seasons in which the plant bears green organs, (b) location of renewal buds - Raunkiaer's life forms, (c) the nature of plant organs seasonally shed, (d) lignification - parts of plants which are permanently lignified and (e) life duration of transpiring and assimilating organs.

5.3.1 SEASONALITY

The four seasonality types considered were: evergreens (E), winter active (W), biseasonals (B) - plants active during parts of winter and summer - and summer active (S).

Fig. 1 presents the distribution of the seasonal types in each of the plant communities examined. All winter-active plants are ephemerals and ephemeroids and their percentages are higher in the Negev than in the Galilee. There are no persisting summer-deciduous plants. The summer-active winter-deciduous plants are altogether absent in the Negev. Thus all persisting plants in the Negev are evergreens which are partial drought evaders, shedding parts of their transpiring organs in summer (Orshan 1964).

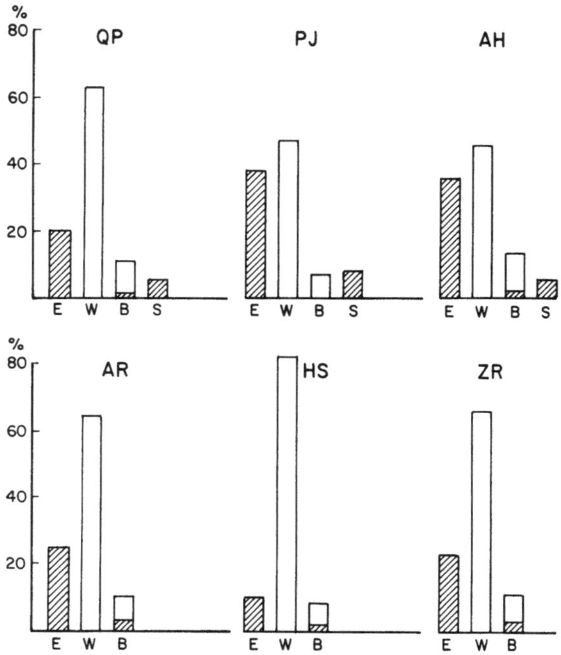

Fig. 1. Distribution of seasonality types in plant communities of the Galilee and Negev expressed as percentage of total number of species. Hatched columns represent percentage of species that are persistent. E: evergreen plants, W: plants green in winter, B: Biseasonals, S: summergreen plants. Plant communities identified as in text.

5.3.2 LOCATION OF RENEWAL BUDS

Raunkiaer's five main life forms (i.e. therophytes, cryptophytes, hemicryptophytes, chamaephytes and phanerophytes) are supposed to express plant adaptation to unfavourable or stress seasons - not necessarily water stress ones. The results of the analysis show clearly (Fig. 2) that percentages of growth forms less resistant to stress periods tend to decrease with increasing aridity, with phanerophytes nearly absent in the Negev communities. The percentage of therophytes, however, is high in both the Negev and the Quercus-Pistacia association in the Galilee and relatively low in the PJ and AH communities, a fact which is probably related to specific edaphic conditions of more compact and less aerated soils which differ in their nutrient composition (Litav 1963; Rabinovitch-Vin 1979). One should also note that there are persisting plants among the cryptophytes and hemicryptophytes.

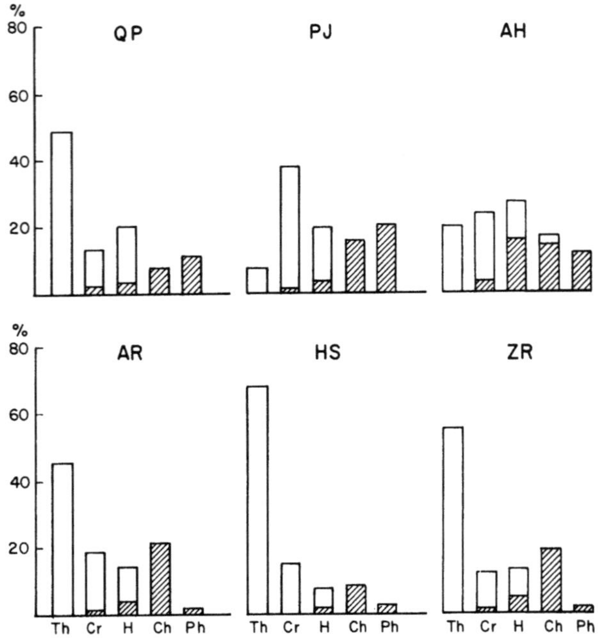

Fig. 2. Distribution of Raunkiaer life forms in plant communities of the Galilee and Negev expressed as percentage of total number of species. Hatched columns represent percentage of species that are persistent. Th: Therophytes, Cr: Cryptophytes, H: Hemicryptophytes, Ch: Chamaephytes, Ph = Phanerophytes. Plant communities identified as in text.

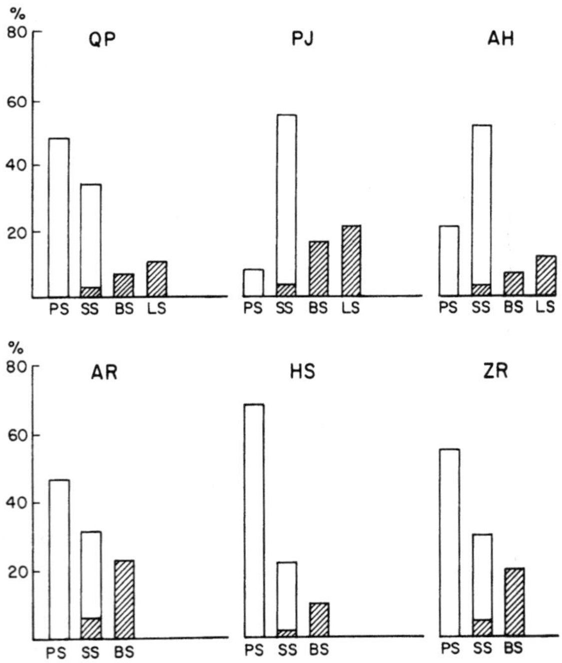

Fig. 3. Distribution of shedding types in plant communities of the Galilee and Negev expressed as percentage of total number of species. Hatched columns represent percentage of species that are persistent. PS: whole plant shedders, SS: shoot shedders, BS: branch shedders, LS: leaf shedders. Plant communities identified as in text.

5.3.3 THE NATURE OF SEASONALLY SHED PLANT ORGANS

The nature of plant organs which are seasonally renewed or shed was suggested as a suitable criterion of a plant's adaptation to seasonal drought stress (Orshan 1953). The following four types were recognized: (a) whole plant shedders - plants which are seasonally renewed from seeds, i.e. annuals (PS), (b) shoot shedders - plants seasonally renewing their whole shoot (SS), (c) branch shedders - plants seasonally shedding and renewing the upper parts of their branches (BS) and (d) leaf shedders - plants shedding only their leaves and inflorescences (LS).

The distribution of these types, presented in Fig. 3, shows that there are no leaf shedders in the Negev communities; the few desert phanerophytes recorded (Fig. 2) are branch shedders. The percentages of ephemeral and ephemeroid shoot shedders are clearly higher in the Galilee compared to the Negev. There is no clear-cut difference in the percentages of the persisting shoot shedders.

5.3.4 LIGNIFICATION

The degree of permanent lignification in shoots of plants is widely considered to be an important criterion for determination of growth forms (Du Rietz 1931). This criterion, while related to persistence, does not necessarily overlap with it. A plant like Phoeniculum vulgare, for example, is a persistent but not lignified evergreen. The following three types have been distinguished following Du Rietz and others: (a) holoxyles

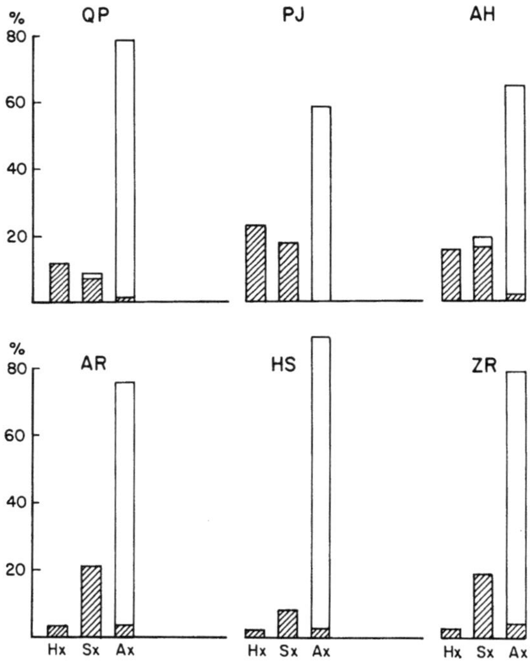

Fig. 4. Distribution of lignification types in plant communities of the Galilee and Negev expressed as percentage of total number of species. Hatched columns represent percentage of species that are persistent. Hx: Holoxyles, Sx: Semixyles, Ax: Axyles. Plant communities identified as in text.

- the whole plant is lignified (HX), (b) semixyles - plants with lower parts of branches lignified and upper parts herbaceous (SX) and (c) axyles - herbaceous plants (AX).

Fig. 4, which presents the distribution of lignification types, shows quite clearly that percentages of holoxyles are markedly higher in the Galilee than in the Negev and the percentages of axyles are lower except for the QP community, a fact related to the high percentage of annuals in this community (compare to Figs 2 and 3).

5.3.5 LIFE DURATION OF TRANSPIRING AND ASSIMILATING ORGANS

This character seems to represent an optimized balance between extra water-use during stress periods by long-lasting organs and extra energy input in frequent replacements of short-lived ones.

For the sake of simplicity leaves and green stems were considered jointly and only the following three longevity classes were considered: (a) the life duration of the assimilating organs is less than 6 months, (b) 6 to 12 months and c) greater than 12 months. The analysis in Fig. 5 clearly shows that the percentage of persistent plant with short-lived organs is nil or markedly lower in the Negev but this category is of considerable importance in the Galilee. On the other hand there is no marked difference in the distribution of the two upper longevity classes between the Galilee and the Negev.

The analysis of the distribution of the monocharacter growth forms shown above is based on presence of species only. Tables 1 and 2 present the distribution of the persisting species in the Galilee and the Negev respectively based on their relative total cover. In the Galilee, tree

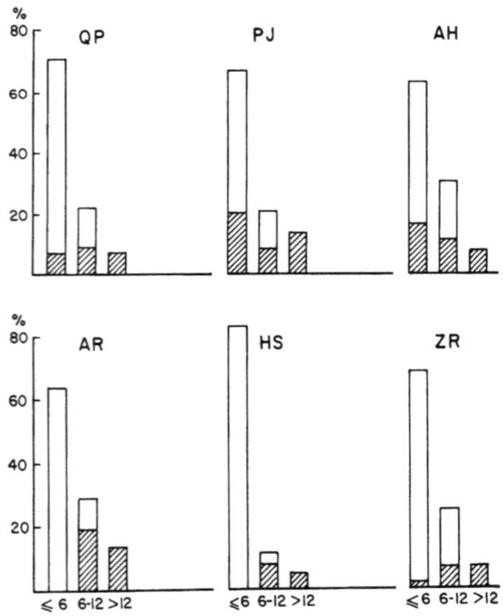

Fig. 5. Distribution of types according to life span of green organs in plant communities of the Galilee and Negev expressed as percentage of total number of species. Hatched columns represent percentage of species that are persistent. Figures under columns denote life span in months. Plant communities identified as in text.

and shrub layers and dwarf-shrub and semi-shrub layers were analysed separately for plant communities. With regard to seasonality it is evident that the evergreen plant community consists mainly of the trees and shrubs in the Galilee although there are some summergreen plants especially in the QP association. All the dwarf-shrubs and semishrubs in the Galilee and all the persisting plants in the Negev Desert are evergreen.

All the dominant trees and shrubs in the Galilee are phanerophyte leaf shedders. Most of the dwarf-shrubs and half-shrubs (or all in the AH association) are chamaephyte and branch shedders. The percentage of phanerophytes and leaf shedders in this layer decreases as soil compactness increases and nitrates among soil nitrogen compounds decrease (Rabinovich-Vin 1979). All the dominant trees and shrubs are holoxyles while the major part of the dwarf-shrubs and semishrubs are semixyles (Table 1).

Table 1. Distribution of growth-form types of the dominant persisting species in (a) the tree and shrub layer and (b) the dwarf-shrub and half-shrub layer in Galilee plant communities.

Growth form types	Estimated percentage of total cover of layer					
	Quercus-Pistacia association		Pinus-Juniperus association		Arbutus-Helianthemum association	
	a	b	a	b	a	b
SEASONALITY TYPES						
Evergreen	76	100	95	100	95	100
Summergreen	24	-	5	-	5	-
Biseasonal	-	-	-	-	-	-
RAUNKIAER LIFE FORMS						
Phanerophytes	100	33	100	8	100	-
Chamaephytes	-	65	-	87	-	100
Hemicryptophytes	-	2	-	5	-	-
SHEDDING TYPES						
Leaf shedders	100	33	100	7	100	11
Branch shedders	-	66	-	87	-	89
Shoot shedders	-	1	-	6	-	-
LIGNIFICATION TYPES						
Holoxyles	100	36	100	36	100	11
Semixyles	-	64	-	64	-	89
LIFE SPAN OF GREEN ORGANS						
6 months	8	64	4	93	2	89
6-12 months	17	-	28	-	52	-
12 months	75	36	68	7	46	11

The life span of green organs of most of the Galilee trees and shrubs exceeds 12 months in the QP and PJ communities with smaller percentages of shorter-lived green parts, while the percentages of plants with green parts living 6 to 12 months and more than 12 months in the AH community are more or less equal (Table 1).

All the dominant persisting plants of the Negev communities are branch shedders and all of the AR and HS communites are chamaephytes and semixyles. The ZR community differs in that the majority of the persisting dominants are phanerophytes and holoxyles (Table 2).

The life of green organs of most persisting dominants of the AR and ZR communities is 6-12 months while that of the HS exceeds 12 months (Table 2). This is no doubt because Hammada scoparia is an articulate plant.

5.4 The definition of a mediterranean growth form

5.4.1 PREREQUISITIES FOR CHARACTERIZING A REGION BY ITS GROWTH FORMS

In order to characterize a plant community or a vegetational region by its growth form the following steps are proposed:

Table 2. Distribution of growth form types of dominant persisting species in the Negev plant communities.

Growth form types	Estimated percentage of total cover		
	Artemisia-Reaumuria association	Hammada scoparia association	Zygophyllum-Reaumuria association
SEASONALITY TYPES			
Evergreen	100	100	100
RAUNKIAER LIFE FORMS			
Phanerophytes	-	-	71
Chamaephytes	100	100	39
SHEDDING TYPES			
Branch shedders	100	100	100
LIGNIFICATION TYPES			
Holoxyles	-	-	71
Semixyles	100	100	37
Axyles	-	-	2
LIFE SPAN OF GREEN ORGANS			
6 months	9	-	-
6-12 months	91	2	95
12 months	-	98	5

(a) Clearly define and delimit the community or the region to be characterized; describe its diversity in composition and environment. Wide-ranging units may be subdivided into more homogeneous ones and each characterized separately. In many cases it is advantageous to subdivide the unit under consideration not only geographically but also morphologically, i.e. to different synusiae or layers, or temporally to successional stages which usually have different characteristic growth forms.

(b) Decide whether a vegetation unit should be characterized by the growth forms of its dominant species or of all its species. Obviously it is easier to analyse a relatively smaller number of dominant species which seem best adapted to their environment and are therefore the best competitors with other species. On the other hand the presence of every species in a vegetation unit indicates that neither physical nor biotic factors prevent it from completing its life cycle. Obviously, therefore, analysis of the dominant species of a vegetation unit reflects to a certain extent its optimal environmental conditions while analysis of all its species seems to more accurately reflect the limiting factors. One should also point out that dominance may change with the season, as is the case in the more arid territories of the Mediterranean Basin where annuals attain dominance in spring. Another aspect of the same problem is whether to take into account only the presence of species whether dominants or others, or also to consider their relative biomass or cover.

(c) Decide upon growth-form characters to be used for the analysis. There seem to be three alternative methods which admittedly are not mutually exclusive:

- Either use an accepted growth-form system or devise a new one. Generally the accepted systems are based on several characters.

- Use as many monocharacter growth-form types as possible and analyse each character independently from the others as demonstrated in this chapter and other papers (e.g. Parsons and Moldenke 1975). Then look for the character or character combinations which best typify the vegetation unit or vegetation gradients under consideration. The disadvantage of this method is that it is extremely laborious even when computers are used.

- Determine selected growth characters for the species examined and use multivariate analysis or other mathematical methods to group them into synthetic multicharacter growth form types. The drawbacks of this method are that it sometimes causes loss of information and, more seriously, that in many cases it is not easy to extrapolate from the results.

5.4.2 IS THERE A DEFINITE MEDITERRANEAN GROWTH FORM?

The answer to this question might be positive or negative depending on the assumptions made and methods of analysis used.

The data presented above show that each vegetation unit is characterized by a growth-form spectrum rather than by one growth form if all the species are taken into consideration. This seems to be true irrespective of the character or character combination used. If on the other hand dominance is chosen as the criterion, one may in most cases define one growth form which characterizes the vegetation unit provided that the climatic ranges of its environment are not too wide.

It is thus possible to make the general statement that mediterranean territories as defined climatically by Aschmann (1973) are characterized by climax communities dominated by evergreen sclerophyllous trees or shrubs. With a decrease in rainfall other growth forms take over, such as seasonally dimorphic evergreen dwarf-shrubs and half-shrubs in the Mediterranean Basin, or summer-deciduous plants and succulents in California and Chile (Mooney et al. 1974). With decreasing winter temperatures, winter-deciduous (spring and summer active) trees and shrubs play an increasing role as dominants of the climax vegetation.

Two major problems remain outside the scope of the present paper, viz. the means by which the above-mentioned growth forms are adapted to the mediterranean environments, and the manner in which they were selected by it. It is worth pointing out, however, that the evergreen sclerophyllous plants which practically do not change their transpiring parts throughout the dry summer are in full disharmony with the mediterranean climatic rhythm. The dwarf-shrubs and half-shrubs dominating the semi-arid and sometimes arid territories of the Mediterranean Basin, which markedly reduce their transpiring parts as well as their transpiration rates in summer, seem to be in greater harmony with mediterranean seasonality.

Acknowledgements

I would like to acknowledge the contribution of the late Professor C Sauvage towards clarifying various phenomorphological problems through long and stimulating discussions. Thanks are due also to M Zohary and A Fahn for discussions and criticism, to A Danin and A Rabinovitch-Vin and A Shmida for supplying information about growth form parameters of various species, and to P Grossman for drawing the graphs.

References

ASCHMANN H (1973) Distribution and peculiarity of Mediterranean ecosystems. In: DI CASTRI F, MOONEY HA (eds) Mediterranean type ecosystems: origin and structure. Springer-Verlag, Berlin, pp 11-19

BRAUN-BLANQUET J (1964) Pflanzensoziologie, Grundzuge der Vegetationskunde, 3rd edn. Springer-Verlag, Wien. 865 pp

CAIN SA (1950) Life forms and phytoclimates. Botanical Review 16: 1-32

CODY ML, MOONEY HA (1978) Convergence versus nonconvergence in mediterranean climate ecosystems. Annual Review of Ecology and Systematics 9: 265-321

DAGET PH, MICHEL-VILLAG JP (1975) Délimitation de la région méditerranéenne selon les régimes des précipitations. In: Proceedings of Symposium Israel-France; ecological research and development of arid zones (mediterranean deserts) with winter precipitation. Special publication 39 Volcani Centre Bet-Dagan, Israel. pp 3-13

DANIN A (1970) A phytosociological ecological study of the northern Negev of Israel. PhD Thesis, The Hebrew University of Jerusalem (Hebrew with English summary)

DANIN A, ORSHAN G, ZOHARY M (1975) The vegetation of the northern Negev and the Judaean Desert of Israel. Israel Journal of Botany 24: 118-172

DI CASTRI F, MOONEY HA (eds) (1973) Mediterranean type ecosystems: origin and structure. Springer-Verlag, New York, 405 pp

DU RIETZ GE (1931) Life forms of terrestrial flowering plants. Acta Phytogeographica Suecica 3: 1-95

ELLENBERG H, MUELLER-DOMBOIS D (1967a) Tentative physiognomic ecological classification of plant formations of the earth. Bericht uber das Geobotanische Forschungsinstitut Rübel in Zürich 37: 21-56

ELLENBERG H, MUELLER-DOMBOIS D (1967b) A key to Raunkiaer's plant life form with revised subdivision. Bericht über das Geobotanische Forschungsinstitut Rübel in Zürich 45: 64-133

LITAV M (1963) Effect of soil types and competition on the occurrence of Avena sterilis L. in the Judaean Hills (Israel). Israel Journal of Botany 14: 74-89

MOONEY AH (1974) Plant form in relation to environment. In: STRAIN BR, BILLINGS WD (eds) Vegetation and environment. Handbook of vegetation science, Part IV. Junk, The Hague

MOONEY HA (1975) Plant physiological ecology - a synthetic view. In: VERNBERG FJ (ed) Physiological adaptation to environment. Proceedings of a symposium held at the 1973 meeting of the American Institute of Ecological Sciences, pp 19-36

MOONEY HA (1977) Convergent evolution in Chile and California: mediterranean climate ecosystems. Dowden, Hutchinson and Ross, Stroudsberg, Pennsylvania, 224 pp

MOONEY HA, BRAYTON R, WEST M (1968) Transpiration intensity as related to vegetation zonation in the White Mountains of California. American Midland Naturalist 80: 407-412

MOONEY HA, DUNN EL, SHROPSHIRE F, SONG L (1972) Land use history of California and Chile as related to the structure of the sclerophyll scrub vegetation. Madrono 21: 305-319

MOONEY HA, GULMON SL, PARSONS DJ, HARRISON AT (1974) Morphological changes within the chaparral vegetation types as related to elevational gradients. Madrono 22 (6): 281-285

MOONEY HA, HARRISON AT, MORROW PA (1975) Environmental limitations of photosynthesis on a California evergreen shrub. Oecologia 19: 293-301

MOONEY HA, KUMMERSON J, JOHNSON AW, PARSONS DJ, KEELEY S, HOFFMANN A, HAYS RI, GILIBERTO J, CHU C (1977) The producers, their resources and adaptive responses. In: MOONEY HA (ed) Convergent evolution in Chile and California: mediterranean climate ecosystems. Dowden, Hutchinson and Ross, Stroudsberg, Pennsylvania, pp 85-143

NAVEH Z (1967) Mediterranean ecosystems and vegetation types in California and Israel. Ecology 48: 445-459

ORSHAN G (1953) Notes on the application of Raunkiaer's life forms in arid regions. Palestine Journal of Botany, Jerusalem Series 6: 120-122

ORSHAN G (1964) Seasonal dimorphism of desert and mediterranean chamaephytes and their significance as a factor in their water economy. In: RUTTER AJ, WHITEHEAD FH (eds) Water in relation to plants. Blackwell, Oxford, pp 206-222

PARSONS JD (1976) Vegetation structure in the mediterranean scrub communities of California and Chile. Journal of Ecology 64: 435-447

PARSONS DJ, MOLDENKE AR (1975) Convergence in vegetation structure along analogous climatic gradients in California and Chile. Ecology 56: 950-957

RABINOVITCH-VIN A (1979) Influence of parent rock on soil properties and composition of vegetation in the Galilee. PhD Thesis. The Hebrew University of Jerusalem (Hebrew with English summary)

RAUNKIAER C (1934) Life forms of plants and statistical plant geography. Clarendon Press, Oxford, 632 pp
SPECHT RL (1968a) A comparison of the sclerophyllous vegetation characteristic of mediterranean type climates in France, California and southern Australia. I. Structure, morphology and succession. Australian Journal of Botany 17: 277-292
SPECHT RL (1968b) A comparison of the sclerophyllous vegetation characteristic of mediterranean type climates in France, California and southern Australia. II. Dry matter, energy and nutrient accumulation. Australian Journal of Botany 17: 293-308
SPECHT RL (1979) Heathlands and related shrublands of the world. In: SPECHT RL (ed) Ecosystems of the World, vol 9A. Heathlands and related shrublands. Descriptive studies. Elsevier, Amsterdam, pp 1-18
VON HUMBOLDT A (1806) Ideen zu einer Physiognomik der Gewachse. Goth, Stuttgart, 28 pp
WENT FN (1971) Parallel evolution. Taxon 20: 197-226

Section II. Plant Form and Function

Although considerable progress has been made in recent years on studies of plant form and function in Californian and Chilean mediterranean-climate plants, little information is available on similar plants of South Africa and Australia.

In this section, SPECHT and co-authors have distinguished mediterranean shrublands from mediterranean heathlands by their situation on either moderately or strongly leached soils, shoot growth strategies and foliage projective covers (FPC) of the overstorey and understorey strata throughout the year. Applications of inorganic nutrients which are of limited supply in the soil to heathlands of southern Australia produced growth responses in the shorter-lived (understorey) plants whereas negligible changes occurred in the long-lived overstorey species (Specht 1963; Heddle and Specht 1975). The overstorey/understorey foliage relationships appear to be determined by the annual water balance of the ecosystem.

Mediterranean-type plants growing during spring are mainly shallow-rooted and are dependent upon soil moisture whereas the late-growing species are deep-rooted and have access to water in the deeper soil layers (see KUMMEROW, this volume). The timing of shoot growth, flowering and cambial growth appears to be temperature-dependent whereas soil moisture may be the most important factor controlling root growth and litter fall.

KUMMEROW (Chapter 17) has attempted to predict the different phenophases of chaparral plants based upon threshold values of 0.1 g of moisture m^{-1} of soil in the preceding autumn and the accumulation of warming hours (= hours 15°C).

MILLER and his co-workers are developing simulation models to provide a better understanding of the interactions between the environment and vegetation in the mediterranean regions of the world. Incoming solar irradiance is absorbed by the canopy and ground and then lost as infrared radiation, convection, and evapotranspiration. Evaporative loss is 10-20% of the annual total energy exchanged and is low during the dry summer period. Many mediterranean-type plants elevate their leaves to a cooler position off the soil surface, reducing heat and water stress during the summer. Leaf area index appears to be directly proportional to photosynthesis and transpiration but this will depend upon the shock tolerance of the leaf and maximum leaf conductance to water loss (i.e. 2 mm s^{-1}).

As well as an annual pattern of temperature and water availability, length of active plant growth and duration of drought, another feature controlling plant form in mediterranean-type ecosystems is the time required for leaves to accumulate carbon to be used in various structural

---

This introduction was prepared by D T Mitchell

components. Carbon accumulation patterns between Californian and Chilean shrubs appear to be convergent whereas the allocation to various structural components is not (Oechel and Lawrence 1981).

In California and Chile evergreen shrubs have been shown to differ from drought-deciduous shrubs in their capacity to gain carbon from photosynthesis (see Mooney, this volume). Evergreen sclerophylls are characterized by relatively low CO_2 exchange rates, broad photosynthetic temperature responses and a capacity to photosynthesize at relatively low water potentials. However, leaves of evergreen species have low nitrogen contents and are less efficient at utilizing nitrogen than those of drought-deciduous species. Mooney has shown that the photosynthetic rates of many Californian shrubs are directly proportional to leaf nitrogen content. Diplacus aurantiacus, a Californian chaparral shrub, was studied in detail by GULMON and produces mesophytic drought-deciduous leaves of high nitrogen content and smaller drought-resistant leaves of low nitrogen content. The nitrogen efficiencies of the leaves of D. aurantiacus produced at different times were compared.

Clearly, more research on other plant growth forms in mediterranean-type ecosystems is required to test the general validity of the above statements.

References

HEDDLE EM, SPECHT RL (1975) Dark Island heath (Ninety-Mile Plain, South Australia). VIII. The effect of fertilizers on composition and growth, 1950-1972. Australian Journal of Botany 23: 151-164

OECHEL W C, LAWRENCE W (1981) Energy and carbon acquisition. In: MILLER PC (ed) Resource use by chaparral and matorral. Springer-Verlag, Berlin, pp 185-235

SPECHT RL (1963) Dark Island heath (Ninety-Mile Plain, South Australia). II. The effect of fertilizers on composition and growth, 1950-1960. Australian Journal of Botany 11: 67-94

6. Carbon-Gaining Capacity and Allocation Patterns of Mediterranean-Climate Plants

H. A. MOONEY

6.1 Introduction

There have been a number of recent reviews on the photosynthetic characteristics and productive structure of mediterranean-climate plants of the world (Mooney 1977, 1981; Mooney et al. 1977; Cody and Mooney 1978; Oechel et al. 1981; Ehleringer and Mooney 1982). Here I will summarize the principal findings of these reviews, adding what limited new data are available. As will become evident, data are particularly lacking from South Africa and Australia. This is unfortunate, since it is from these nutrient-deficient regions that we might expect exceptions from the general trends that are emerging from other mediterranean-climate areas. Hopefully, research in the next few years will provide the data necessary to make a meaningful comparison of the productive characteristics of plants of all the mediterranean-climate regions of the world.

Even though we do have enough information on the physiological characteristics of plants from the Mediterranean region and mediterranean-climate Chile and California to discern general trends, it should be noted that the data available are far from definitive. This is because, as our measuring techniques become more accurate, and simply as more measurements are being made, it is becoming clear just how variable measurement results can be on the same species. Plant factors such as tissue and plant age, reproductive stage, and genetically based carbon fixation potential (e.g., carboxylating enzyme content, diffusive pathway) can all affect observed photosynthetic rates (Fig. 1). Further, the particular conditions of the growth environment such as light, temperature, humidity and nutrients can also be a strong determinant of rate capacity. Thus, for example, plants growing in dissimilar or variable environments or sampled at different times can yield very different results. This effect can be so strong as to mask completely any assessment of variability in genetic potential between genotypes within a population. Variability in rates between plants may reflect differences in their environmental rather than genetic history.

Further, as our capacity to control the environment of a plant during measurement has increased, certain parameters that were often unmeasured or uncontrolled in early studies (such as humidity) prove to have a strong effect on the photosynthetic rate (Fig. 1). Of equal importance to controlling and measuring environmental factors that influence rate capacity are the duration, direction and rate of change of these factors during analysis. Quite often leaves subjected to rapid change in magnitude of a given parameter will respond by closing their stomata, thus giving a lowered capacity to exchange CO_2.

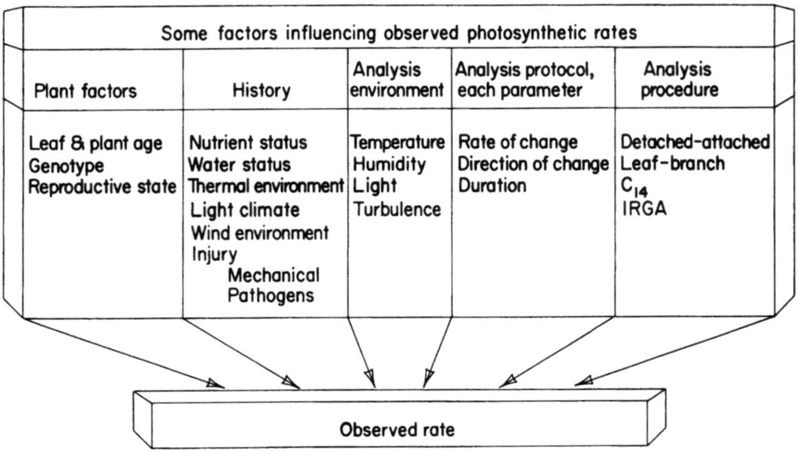

Fig. 1. Factors and procedures influencing observed photosynthetic capacity.

Finally, the type of tissue which is used, whether attached or detached branches, single leaves or whole branches, is important in determining the rate that is observed. Many of the early studies of the photosynthetic capacity of mediterranean-climate plants utilized branches detached from field plants and then transported to the laboratory (e.g., Larcher 1961; Harrison et al. 1971). Now, measurements are made on tissue in situ (Oechel and Lawrence 1979; Mooney et al. 1976). Although the general trends in rate capacity that are discernible from the literature between different ecological groupings of mediterranean-climate plants are no doubt valid, attempts to make precise comparisons may not be justified. The best comparisons to be made are often those published by a single research group working among different plant types using the same experimental protocol rather than among results of different research groups, each working on a different plant type.

6.2 Rate comparisons

One of the most fundamental differences between the trees and shrubs of mediterranean-climate regions is whether they are evergreen or deciduous. Although the principal adaptive mode in all of the mediterranean-climate regions is the evergreen shrub, drought-deciduous shrubs can also be important, particularly in disturbed or drier than normal habitats (Harrison et al. 1971). All of the comparative data gathered so far indicate that evergreen trees and shrubs have similar photosynthetic

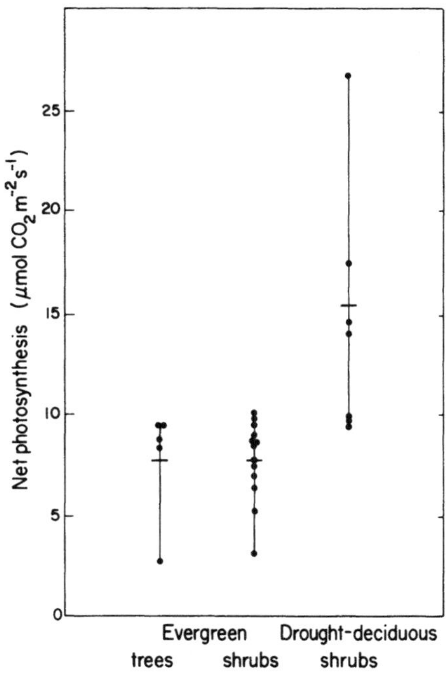

Fig. 2. Comparative photosynthetic rates of mediterranean-climate and drought-deciduous shrubs as measured principally on intact plants in California, Chile, and the Mediterranean region (see Mooney 1981, for original sources). Data are for the evergreen shrubs Prunus ilicifolia, Rhus ovata, Heteromeles arbutifolia, Lithraea caustica, Kageneckia oblonga, Colliguaya odorifera, Pistacia lentiscus, Rhamnus alaternus, and Phillyrea media; drought-deciduous shrubs Encelia californica, Salvia apiana, Salvia leucophylla, Salvia mellifera, Podanthus mitiqui; evergreen trees Arbutus menziesii, Quillaja saponaria, Quercus ilex, and Olea europaea.

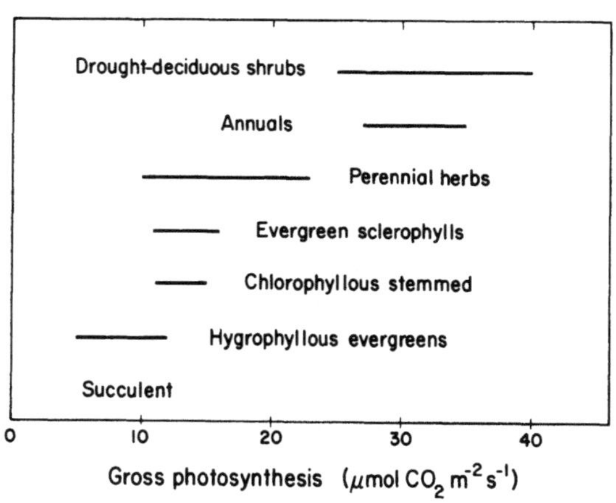

Fig. 3. Comparative photosynthetic capacity of diverse Chilean mediterranean-climate growth forms (from P Martinez, unpublished). Gross photosynthesis data based on carbon-14 technique.

capacities which, however, are considerably lower than those that characterize drought-deciduous shrubs (Fig. 2).

Data are now available for a larger array of mediterranean-climate growth forms, including not only leafy shrubs but also annuals and perennial herbs, and chlorophyllous-stemmed shrubs, as well as a succulent (P Martinez, unpublished) (Fig. 3). These measurements of Chilean plants indicate that drought-deciduous species have a photosynthetic capacity in the same high range as that of drought-deciduous shrubs. Chlorophyllous-stemmed shrubs, on the other hand, have low rates comparable to evergreen sclerophyllous shrubs. The only succulent studied had the lowest photosynthetic rate measured of any growth form.

What are the ultimate and proximate factors that lead to these differences? First, it appears that evergreen leaves generally have considerably lower amounts of nitrogen on a dry-mass basis than do deciduous species (Fig. 4). It has been shown for a number of species that photosynthetic capacity is linearly related to leaf nitrogen content (Fig. 5A). These results were obtained both by varying nitrogen availability to the plants and (in the case of Diplacus and Eucalyptus species) by contrasting rates of old vs. young leaves (Lepechinia). Thus evidence is available to predict that the drought-deciduous species generally have higher photosynthetic rates on a mass basis (Fig. 5B).

Based on results from California species it appears that the drought-deciduous species are generally more efficient in utilizing their nitrogen than evergreen species (Fig. 5C). This probably results from the fact that at very low nitrogen contents the ratio of carboxylating enzyme to non-photosynthetic nitrogen is low. In fact, the intercept of photosynthetic rate vs. nitrogen content probably is an indication of leaf contents of non-photosynthetic nitrogen (see Fig. 5A).

The low photosynthetic efficiency per unit nitrogen of evergreen leaves results in their having low photosynthetic rates even on an area basis. However, because of the high leaf-specific masses of evergreens their nitrogen contents per unit area may equal or exceed deciduous species (Fig. 4).

The question, then, becomes "why is the nitrogen content lower in evergreen leaves than in leaves of drought-deciduous species?" This answer probably relates in part to leaf turnover rates and the magnitude and periodicity of resource availability. Mooney and Gulmon (1982) have described general patterns where nitrogen is translocated to the position on the plant of greatest potential return in carbon-gaining capacity, that is, to the newest leaves at the top of the canopy. The faster the growth rate of a plant in a closed community, the faster the movement of the nitrogen from old to new leaves and the shorter the life of this leaf. Where nitrogen, and hence growth rate, is limited, leaf turnover is reduced and, in fact, there is selection toward increased leaf duration. The longest leaf durations generally occur on plants occupying the most nutrient-deficient habitats (Fig. 6).

Thus, it would appear that leaf duration is an index of habitat nutrient availability, the habitats on which drought-deciduous species are found are often disturbed or quite xeric. They are evidently characterized by flushes of nutrient availability when the rains commence. The plants in these habitats produce low-specific-mass leaves of a high photosynthetic capacity, which turn over rapidly as they are overtopped or as the drought ensues.

Evergreen plants in mediterranean-type ecosystems tend to be found in habitats where soils have a low nutrient status and nutrients are more slowly recycled than in habitats where drought-deciduous species are

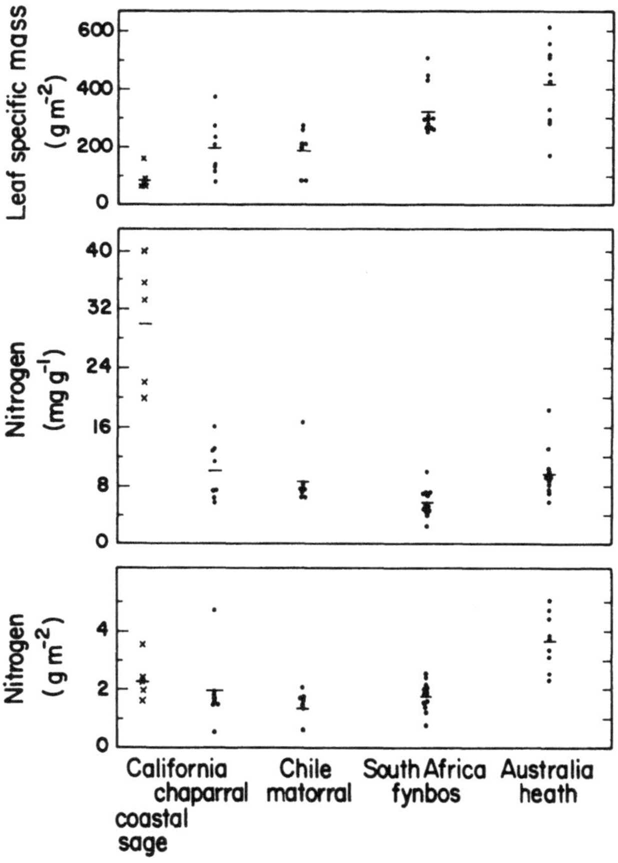

Fig. 4. Leaf specific mass and nitrogen content (dry-mass basis) of leaves of evergreen trees and shrubs of four mediterranean-climate regions. For comparison, values are also given for drought-deciduous species of the Californian coastal sage community. Values for the California chaparral and Chilean matorral are from Mooney et al. (1977). California chaparral: Rhus ovata, Ceanothus leucodermis, Heteromeles arbutifolia, Arctostaphylos glauca, Adenostoma fasciculatum, Quercus agrifolia, Quercus dumosa, Ceanothus greggii; Chilean matorral: Lithraea caustica, Trevoa trinervis, Kageneckia oblonga, Colliguaya odorifera, Satureja gilliesii, Cryptocarya alba, Quillaja saponaria. South African fynbos (H A Mooney and C Chu, unpublished; values for one year-old leaves collected in February): Leucadendron nervosum, L. conocarpodendron, L. gandogeri, Protea cynaroides, P. obtusifolia, P. amplexicaulis, P. nitida, P. lorifolia, P. repens, Rhus glabra, Pterocelastrus tricuspidatus, Cassine barbara; Australian heath (H A Mooney and C Chu, unpublished): Acacia myrtifolia, Banksia marginata, B. ornata, B. prostrata, Eucalyptus diversifolia, E. endesmioides, E. tetragona, Hakea baxteri, H. crassifolia, H. cucullata; Californian coastal sage (H A Mooney, C Chu and C Field, unpublished): Artemisia californica, Encelia californica, Salvia mellifera, Lepechinia calycina, Diplacus aurantiacus.

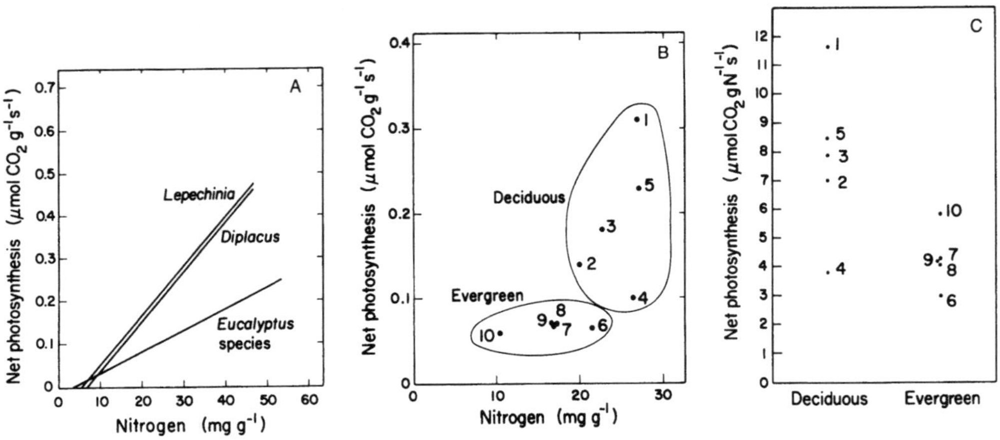

Fig. 5. Light-saturated photosynthetic capacities in relation to leaf nitrogen content (dry-mass basis). A. Correlation between photosynthetic rate and leaf N content within species: data for two Californian drought-deciduous species *Lepechinia calycina* (C Field, unpublished) and *Diplacus aurantiacus* (S L Gulmon and C Chu, unpublished) and for various species of *Eucalyptus* from diverse climatic regions (Mooney et al. 1978). The *Diplacus* and *Eucalyptus* data are from controlled environment-grown plants and those of *Lepechinia* from leaves produced in the summer and measured in situ in their native habitat. B. Differentiation between Californian drought- deciduous, (1) *Diplacus aurantiacus* (2) *Salvia mellifera*, (3) *Ceanothus thrysiflorus* (partially drought-deciduous), (4) *Quercus kelloggii* (winter deciduous), (5) *Ribes sanguineum*; and evergreen species (6) *Heteromeles arbutifolia*, (7) *Quercus agrifolia*, (8) *Ribes viburnifolium*, (9) *Ceanothus maritimus*, (10) *Arctostaphylos densiflora* (from Winner 1981; data from greenhouse-grown plants.) C. N-use efficiency in Californian evergreen and drought-deciduous species. Numbers refer to same species indicated in Fig. B legend (from Winner 1981).

found. Obviously we need to gather data on the dynamics of soil nutrient concentrations and plant-uptake patterns to confirm the relationships that appear from leaf analyses alone. Further, we need methods to assess the ratio of carboxylating enzyme to nonphotosynthetic protein. Such analyses are generally not possible because of the high phenolic contents of evergreen leaves in general.

As leaves age the nutrient content becomes diluted as a result of an accumulation of carbon (Mooney and Gulmon 1982). Leaves that live for long periods are expected to have low ratios of nitrogen to carbon, and generally have high specific masses. Leaves of plants that grow on nutrient-deficient habitats will, in addition, have a low nutrient to carbon ratio at an age comparable to that of plants growing in sites with moderate nutrients.

Unfortunately, we have limited data on leaf turnover in the evergreen species of the various mediterranean-climate regions. Kruger (1981) notes that in the genus *Protea* some species have leaves that live for only a year, whereas leaves in other species live for as long as six years. Comparable within-genus variation in leaf duration may be found in the *Ceanothus* spp. of California.

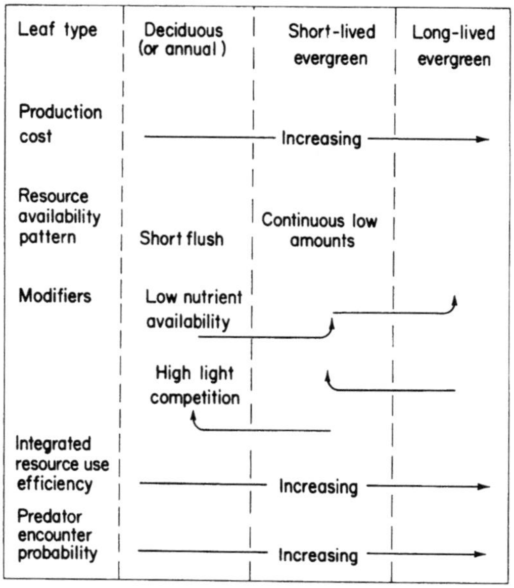

Fig. 6. Relationship between leaf longevity and resource availability (from Mooney and Gulmon, 1982).

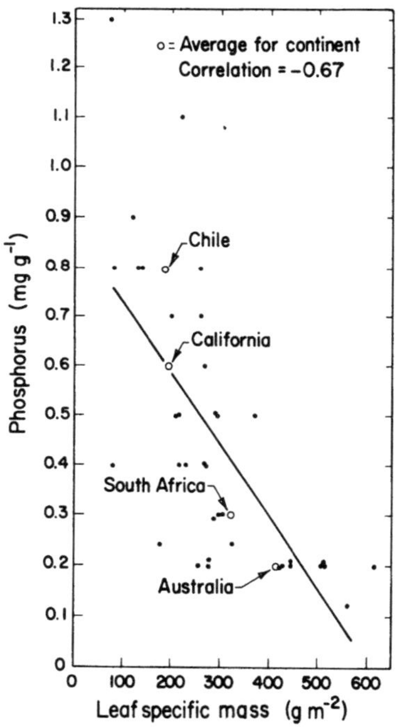

Fig. 7. Relationship of leaf specific mass with leaf phosphorus content (dry-mass basis) of evergreens of four mediterranean-climate regions. Values are for the same species and source as given in Fig. 4.

It is known that leaf specific mass varies with water supply (Chapter 8, this volume) and nutrient content, particularly phosphorus (Fig. 7). Comparable relationships are found with nitrogen (Gulmon and Chu 1981). This nutrient content versus leaf specific mass trend results from the age and nutrient availability relationships described above. That is, leaf duration increases with decreasing nutrient availability.

In addition to the more or less passive relationships leading to high leaf specific mass, there are, of course, selective factors, such as drought stress and herbivory patterns. We need to study systems where we are able to separate these influences so that we can understand why, for example, leaves of evergreens of South Africa and Australia generally have higher specific masses (Fig. 7) than their Californian and Chilean counterparts.

6.3 Photosynthetic capacity and biomass accumulation

Photosynthetic capacity is one of the principal biological determinants of the primary productivity of a plant community. Thus one would predict that in areas of comparable climate, if the photosynthetic capacities of the dominants were similar, so would be the productivities. As noted above, the photosynthetic capacities of evergreen sclerophylls are generally low and similar for plants from the different mediterranean-climate regions. In general, above-ground annual biomass accumulation is between 100 and 200 g m^{-2} yr^{-1}, as determined by harvests at different post-fire time intervals (Fig. 8). Some of the Australian heath communities are somewhat below this range, whereas at least one fynbos sample is above the range. It would thus appear that, although it is likely that the photosynthetic capacities of plants of the different evergreen communities vary because of differing nutrient relationships, other factors related to productivity apparently compensate for these limitations - such as leaf longevity or root allocation patterns. Although leaves of the fynbos plants have low nitrogen contents on a mass basis,

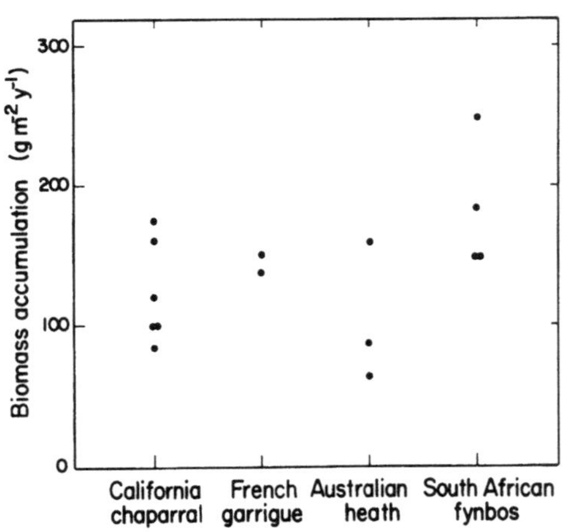

Fig. 8. Above-ground biomass accumulation of evergreen scrub communities of various mediterranean-climate regions. Data for California from Specht (1969), Mooney et al. (1977), Sampson (1944), Schlesinger and Gill (1980), DeBano and Conrad (1978); for France from Specht (1969), Lossaint (1973); for Australia, Specht et al. (1958), Specht 1969; for South Africa from Kruger (1977a).

they have a high specific mass and thus they maintain a high nitrogen content on an area basis (Fig. 4).

6.4 Environmental limitations on photosynthetic capacity and photosynthetic rates

As indicated above, nitrogen availability, and hence leaf nitrogen content, is one of the prime determinants of the intrinsic photosynthetic capacity of leaves. There are a number of other environmental factors that, of course, also affect photosynthetic capacity as well as the rate of photosynthesis at any given moment.

6.4.1 LIGHT

An example of the interaction of factors that limit photosynthesis is leaf nitrogen content and photon flux density. As indicated earlier, the higher the leaf nitrogen content, the greater the intrinsic photosynthetic capacity of a leaf at light saturation. At less than light-saturated levels photosynthetic rates are much less nitrogen-dependent (see Fig. 1 in Chapter 9, this volume).

Since most mediterranean-climate evergreen shrubs have low leaf nitrogen contents, they should have relatively low light-saturated photosynthetic rates. On the other hand, drought-deciduous species from these same regions have, as noted above, high leaf nitrogen contents, and thus they should have a high light requirement for photosynthesis. This prediction is realized in a comparison of a Californian evergreen sclerophyll (Heteromeles arbutifolia) and the drought-deciduous species, Encelia californica both of which have the C_3 pathway (Fig. 9) (Mooney 1981). Thus plants that can acquire the most nitrogen will also have the capacity to harvest the most light energy.

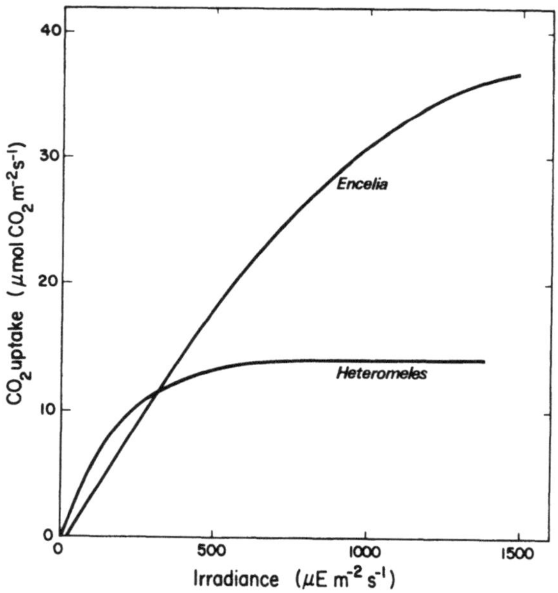

Fig. 9. Photosynthetic light response of drought-deciduous species, Encelia californica and an evergreen species, Heteromeles arbutifolia (from Mooney 1981).

6.4.2 TEMPERATURE

The temperature of optimum photosynthesis of a range of Californian mediterranean-climate growth forms falls within the range between 16°C and 29°C (Fig. 10). This range includes annuals, drought-deciduous as well as evergreen shrubs. The optimum temperature of photosynthesis of most mediterranean-climate evergreens is near 20°C (Mooney 1981). These plants are characterized by little acclimation potential (Harrison 1971) and very broad temperature response curves (Oechel et al. 1981). These responses may be contrasted with those of Californian desert plants, which have a wide range of temperature optima, generally temperature-sensitive photosynthetic responses (Fig. 10), as well as many examples of very large shifts in temperature response dependent on growth environment (Mooney 1980b).

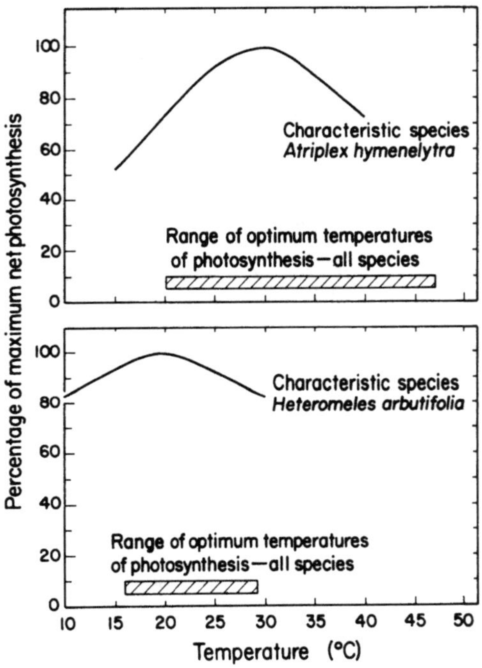

Fig. 10. Range of optimum temperatures of photosynthesis for a number of Californian mediterranean-climate plants (*Lepechinia calycina*, C Field, unpublished; *Plantago erecta*, *Clarkia rubicunda*, and *Hemizonia luzulaefolia*, N Chiariello and H Mooney, unpublished; *Ceanothus greggii*, *Arctostaphylos glauca*, *Rhus ovata*, *Adenostoma fasciculatum*, Oechel et al. 1981; *Heteromeles arbutifolia*, Mooney et al. (1975) (bottom). Optimum temperatures for photosynthesis of California desert plants of Death Valley, *Camissonia claviformis*, *Larrea divaricata*, *Atriplex hymenelytra* and *Tidestromia oblongifolia*, Mooney et al. (1976) (top). Temperature response of photosynthesis of an evergreen shrub of Death Valley, *Atriplex hymenelytra* (Mooney et al. 1976) (top) and of a characteristic evergreen shrub of the California chaparral, *Heteromeles arbutifolia* (Harrison 1971).

6.4.3 WATER

There is very little information available on the direct effects of water stress on the photosynthetic capacity of mediterranean-climate plants (Rundel 1977). Oechel et al. (1981) have recently shown that Californian drought-deciduous species have their carbon-gaining capacities affected more by drought stress than do evergreen species. By extrapolation they predicted that drought-decidous species of Salvia would have no net carbon gain when dawn water potentials were less than -3.5 MPa, whereas evergreen species such as Arctostaphylos glauca and Adenostoma fasciculatum would have positive net carbon gain during days with dawn water potentials above -5.0 MPa and -5.5 MPa, respectively.

Miller and Poole (1979) provide values of leaf conductance versus xylem water potential for a series of Californian evergreen chaparral plants (Fig. 11). Their data indicated considerable between-species variation in stomatal sensitivity to drought stress and, by extrapolation, photosynthetic sensitivity to water stress. More data are needed on the photosynthetic drought response of all mediterranean-climate growth forms. We do not have sufficient descriptive information even to ask meaningful questions about the kinds of mechanisms that evergreen plants have evolved to maintain positive carbon gain while under considerable drought stress or the mechanisms they have to avoid stress. Studies of the comparative water relations of South African evergreens would be particularly valuable, since they may provide the basis for understanding the asynchrony of growth characteristics of fynbos communities (Kruger 1981).

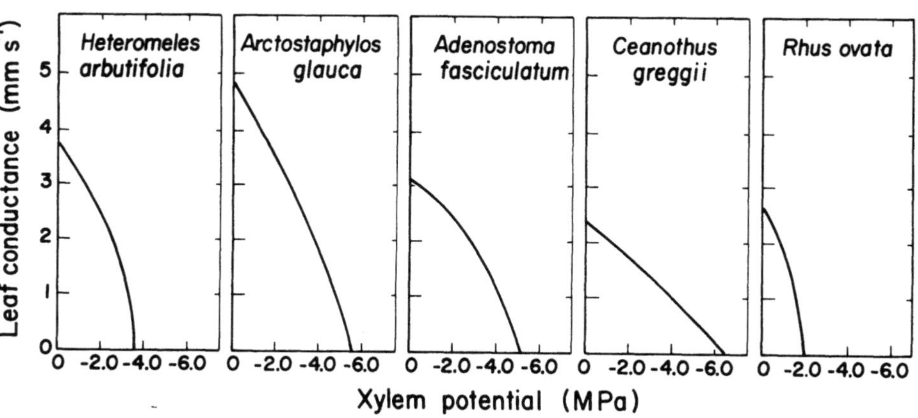

Fig. 11. Relationship between leaf conductance and xylem water potential of a series of evergreen chaparral plants growing at Echo Valley, California (from Miller and Poole 1979).

6.5 Allocation patterns

6.5.1 GENERAL

Considerable progress has been made in recent years in quantifying carbon allocation patterns in woody Californian and Chilean mediterranean-climate plants. Oechel and Lawrence (1981) have found that 30-40% of the carbon gained by these plants is used to build and maintain their root systems. Further, a large fraction of carbon is devoted to maintaining the stem system. Oechel and Lawrence (1981) found that, although carbon acquisition patterns between Californian and Chilean shrubs are convergent, allocation patterns to various structural components are not. They note in particular that the Chilean shrubs may have larger amounts of secondary chemicals than Californian counterparts, although no direct evidence is presented. An understanding of the significance of the various chemical component allocation patterns in mediterranean-climate plants will come only as we learn the adaptive roles these components play. A clue to the significance of various allocation patterns comes from the seasonality of these patterns in relation to the seasonality of potential physical and biotic selective agents.

6.5.2 RESOURCE AVAILABILITY AS A CONTROL ON ALLOCATION

The growth of plant tissue in a sense represents a commitment either of resources that are acquired by the plant simultaneously with growth or those drawn totally or partially from stored reserves. These resources include carbon as well as nutrients. Plants generally do not grow all of their parts simultaneously; rather, the growth of various parts is separately phased (e.g., roots, stem, etc.). The argument has been made (Mooney et al. 1977) that this time separation is in part due to the limitation of the building materials and that priority is given at any time to the most critical growth phase. For example, it was argued that canopy growth during the early spring takes precedence over allocation to protection of the leaves from herbivores which takes place primarily during periods of no canopy growth (Mooney and Chu 1974). However, the time separation of growth events may also be related to different periodicities of selective factors (which is not exclusive of the above argument). The most direct example is between canopy growth and flowering. The factors selecting for the timing of these events may be totally independent. In one case, there appears to be a synchronization of canopy growth between dominant species, such as the woody plants of the chaparral, whereas there may be separation between species of this same community in flowering time. Competition for pollinators may have resulted in different flowering times, since there is a year-round pool of different pollinators available. However, if the same resources are utilized by all species, such as the water and nutrients of a given rooting zone, there cannot be a time separation in use, since the resource would be depleted by the first user. In the California warm deserts, species of the same growth form co-occur, yet they have different growth synchronies (Mooney 1980). This is due to the fact that in this light-unlimited habitat rain may come at any season and species may be adapted to photosynthesize and grow efficiently at different temperatures (Fig. 10). Thus a species may deplete the moisture available at a given prevailing temperature, but not at another.

In the chaparral there are growth forms other than the dominant evergreen trees and shrubs. It has been shown that these different growth forms, such as the annuals and perennial herbs and the drought-deciduous shrubs, have different activity patterns which relate to the resources available to them (Cody and Mooney 1978). The shallow-rooted herbs start growing as soon as the winter rains commence, followed by the somewhat deeper rooted drought-decidous shrubs, and finally the deep-rooted evergreen shrubs and trees (Fig. 12). These growth forms also tend to be separated spatially. Very few herbs grow in the understorey of the evergreen scrub; rather, they are found mainly in openings caused by disturbance or edaphic factors. The same is true of the drought-deciduous shrubs characteristic of disturbed or xeric sites.

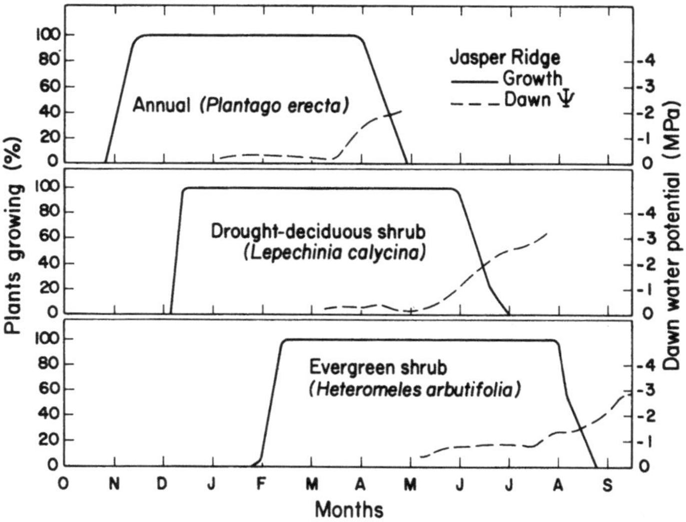

Fig. 12. Growth cycles of three different growth forms of California mediterranean-climate plants (Jasper Ridge) as related to dawn water potential. The implications of these three patterns is that the end of the growth period is similarly controlled by water availability but that each is utilizing different water resource (Mooney, unpublished).

The herbs and the drought-deciduous species are generally synchronized in their respective habitats in their growth patterns since they all share common water and nutrient resources. However, since they have a limited seasonal carbon-gaining period, they do not show pronounced asynchrony of flowering. The flower types of the various species are generally quite diverse in colour and form and attract diverse pollinators even though flowering time is synchronous.

The point of the above discussion is to show how, by understanding the pattern of resource availabilities to different plant types and the mechanisms possible for resource sharing, one can understand the phenological patterns of various members of a community.

Unfortunately, however, this kind of information, particularly on resource availability, is scarce. For example, it has been a puzzle why drought-deciduous trees of the tropical savannas initiate new leaf growth before the end of the dry season (Daubenmire 1972) or why phreatophytes such as Prosopis become deciduous. Apparently, in both cases, water that is depleted while the trees are in leaf is replenished from ground water while they are leafless. Thus, they can produce new leaves prior to any input of precipitation, since their limiting resource has been replenished (Mooney et al. 1979).

We do not have sufficient information available to understand the basis of the growth patterns that has been described for the heath and fynbos which differs from that of the mediterranean-climate evergreen scrub communities of Europe and North Africa, California and Chile. In the former case, there is asynchrony in the growth of the dominants, with some growth even during the dry season (Kruger 1981; Specht 1969), whereas in the latter communities growth is restricted principally to spring. The greater variability of growth forms in the Australian and South African communities certainly indicates a greater variability of resource-gathering strategies present. This variability is no doubt in part possible because of the low nutrient status of these regions. Data are needed particularly on the water and nutrient-use patterns of the principal growth forms of these regions in order to explain this interesting divergence between mediterranean-climate regions.

6.5.3 ALLOCATION TO REPRODUCTION

There have been a number of comparisons of the reproductive modes of various mediterraean-climate plants, particularly in reference to fire regime (Gill 1977; Keeley 1977; Kruger 1977b). All of the mediterranean-climate regions have plants that possess one or more features that adapt them to particular fire regimes. Certain features, however, are particularly prevalent in the plants of only a couple of these areas and, since these features involve considerable carbon allocation, they deserve more detailed study.

One such feature is woody fruits in which seeds may be stored on plants for a large number of years. This feature is commonly found in western Australia (Cupressaceae, Casuarinaceae, Proteaceae, and Myrtaceae) (Gill 1977) and South Africa (Kruger 1977b). Kruger (1979) indicates that nearly a quarter of the above-ground biomass of certain proteaceous South African shrubs can be reproductive organs while in South Australia Specht et al. (1958) report proportions that exceed this. Although persistent woody fruits are found on plants in the Mediterranean and California, they are on conifers only and of importance only in local areas. A V Milewski (unpublished) has pointed to the differences in the fauna of various mediterranean-climatic regions as the possible evolutionary cause for dissimilarities in fruiting patterns among the plants of these regions (Weins and Rourke 1978).

Thus, in order to understand the patterns of allocation in plants, we must consider not only the physical resource base but also the availability and habits of the fauna. Since the physical resource base is more similar among mediterranean-climatic regions than are the fauna, one might expect the greatest divergences in plant allocation patterns to be in those features subject to coevolution.

Acknowledgements

Appreciation is expressed to S L Gulmon and B Lilley for their help in this manuscript and to the National Science Foundation for their support of my studies covered in this review (DEB 78-02067).

References

CODY ML, MOONEY HA (1978) Convergence versus nonconvergence in mediterranean-climate ecosystems. Annual Review of Ecology and Systematics 9: 265-321

DAUBENMIRE R (1972) Phenology and other characteristics of tropical semideciduous forest in northwestern Costa Rica. Journal of Ecology 60: 147-170

DEBANO LF, CONRAD CE (1978) The effects of fire on nutrients in a chaparral ecosystem. Ecology 59: 489-497

EHLERINGER J, MOONEY HA (1982) Photosynthesis and productivity of desert and mediterranean-climate plants. In: LANGE OL (ed) Encyclopedia of plant physiology. Springer-Verlag, Berlin. In press

GILL AM (1977) Plant traits adaptive to fires in mediterranean land ecosystems. In: MOONEY HA, CONRAD CE (eds) Symposium on the environmental consequences of fire and fuel management in mediterranean-climate ecosystems. USDA Forest Service General Technical Report WO-3, pp 17-26

GULMON SL, CHU C (1981) The effects of light and nitrogen on photosynthesis, leaf characteristics, and dry matter allocation in the chaparral shrub, Diplacus aurantiacus. Oecologia 49: 207-212

HARRISON AT (1971) Temperature-related effects on photosynthesis in Heteromeles arbutifolia M. Roem. PhD Thesis, Stanford University, California

HARRISON AT, SMALL E, MOONEY HA (1971) Drought relationships and distribution of two mediterranean-climate California plant communities. Ecology 52: 869-875

KEELEY JE (1977) Fire-dependent reproductive strategies in Arctostaphylos and Ceanothus. In: In: MOONEY HA, CONRAD CE (eds) Symposium on the environmental consequences of fire and fuel management in mediterranean-climate ecosystems. USDA Forest Service General Technical Report WO-3, pp 391-396

KRUGER F (1977a) A preliminary account of aerial plant biomass in fynbos communities of the mediterranean-type climate zone of the Cape Province. Bothalia 12: 299-305

KRUGER FJ (1977b) Ecology of Cape fynbos in relation to fire. In: MOONEY HA, CONRAD CE (eds) Symposium on the environmental consequences of fire and fuel management in mediterranean-climate ecosystems. USDA Forest Service General Technical Report WO-3, pp 230-244

KRUGER FJ (1979) Plant ecology. In: DAY JE, SIEGFRIED W, LOUW E, JARMAN M (eds) Fynbos ecology: a preliminary synthesis. South African National Scientific Programmes Report 40, Council for Scientific and Industrial Research, Pretoria, pp 88-126

KRUGER FJ (1981) Seasonal growth and flowering rhythms: South African heathlands. In: SPECHT RL (ed) Ecosystems of the world, vol 9B. Heathlands and related shrublands. Analytical studies. Elsevier, Amsterdam, pp 1-4

LARCHER W (1961) Jahresgang des Assimilations- und Respirationvermögens von Olea europea L. ssp. sativa Hoff. et Link., Quercus ilex L. und Quercus pubescens Willd. aus dem nördlichen Gardaseegebiet. Planta 56: 575-606

LOSSAINT P (1973) Soil vegetation relationships in mediterranean ecosystems of southern France. In: DI CASTRI F, MOONEY HA (eds) Mediterranean type ecosystems: origin and structure. Springer-Verlag, Berlin, pp 199-210

MILLER PC, POOLE D (1979) Patterns of water use by shrubs in southern California. Forest Science 25: 84-98

MOONEY H (1977) The carbon cycle in mediterranean-climate evergreen scrub communities. In: MOONEY HA, CONRAD CE (eds) Symposium on the environmental consequences of fire and fuel management in mediterranean-climate ecosystems. USDA Forest Service General Technical Report WO-3, pp 107-115

MOONEY HA (1980) Seasonality and gradients in the study of stress adaptation. In: TURNER N, KRAMER P (eds) Stress physiology of plants. Wiley, New York, pp 279-294

MOONEY HA (1981) Primary production in mediterranean-climate regions. In: DI CASTRI F, GOODALL DW, SPECHT RL (eds) Ecosystems of the world, vol 11. Mediterranean-type shrublands. Elsevier, Amsterdam, pp 249-255

MOONEY HA, BJÖRKMAN O, EHLERINGER J, BERRY J (1976) Photosynthetic capacity of in situ Death Valley plants. Carnegie Institution Year Book 75: 410-413

MOONEY HA, CHU C (1974) Seasonal carbon allocation in Heteromeles arbutifolia, a California evergreen shrub. Oecologia 14: 295-306

MOONEY HA, FERRAR PJ, SLATYER RO (1978) Photosynthetic capacity and carbon allocation patterns in diverse growth forms of Eucalyptus. Oecologia 36: 103-111

MOONEY HA, GULMON SL (1982) Constraints on leaf structure and function. BioScience 32: 198-206

MOONEY HA, HARRISON AT, MORROW P (1975) Environmental limitations of photosynthesis on a California evergreen shrub. Oecologia 19: 203-301

MOONEY HA, KUMMEROW J, JOHNSON AW, PARSONS DJ, KEELEY S, HOFFMANN A, HAYS RI, GILIBERTO J, CHU C (1977) The producers - their resources and adaptive responses. In: In: MOONEY HA (ed) Convergent evolution in Chile and California: mediterranean climate ecosystems. Dowden, Hutchinson and Ross, Stroudsburg, Pennsylvania, pp 1-12

OECHEL WC, LAWRENCE WT (1979) Energy utilization and carbon metabolism in mediterranean scrub vegetation of Chile and California. I. Methods: a transportable cuvette field photosynthesis and data acquisition system and representative results for Ceanothus greggii. Oecologia 39: 321-335

OECHEL WC, LAWRENCE W (1981) Carbon allocation and utilization. In: MILLER PC (ed) Resource use by chaparral and matorral. Springer-Verlag, Berlin, pp 185-235

OECHEL WC, LAWRENCE WT, MUSTAFA J, MARTINEZ J (1981) Energy and carbon acquisition. In: MILLER PC (ed) Resource use by chaparral and matorral. Springer-Verlag, Berlin, pp 151-183

RUNDEL DW (1977) Water balance in mediterranean-climate ecosystems. In: MOONEY HA, CONRAD CE (eds) Symposium on the environmental consequences of fire and fuel management in mediterranean-climate ecosystems. USDA Forest Service General Technical Report WO-3, pp 95-106

SAMPSON AW (1944) Plant succession on burned chaparral lands in northern California. University of California Agricultural Experiment Station Bulletin 685. 144 pp

SCHLESINGER WH, GILL DS (1980) Biomass, production, and changes in the availability of light, water and nutrients during the development of pure stands of the chaparral shrub, Ceanothus megacarpus, after fire. Ecology 61: 781-789

SPECHT RL (1969) A comparison of the sclerophyllous vegetation characteristic of mediterranean-type climates in France, California, and southern Australia. Australian Journal of Botany 17: 277-292

SPECHT RL, RAYSON P, JACKMAN M (1958) Dark Island heath (Ninety-mile Plain, South Australia). VI. Pyric succession: changes in composition, coverage, dry weight, and mineral nutrient status. Australian Journal of Botany 6: 59-88

WIENS D, ROURKE JP (1978) Rodent pollination in southern African _Protea_ spp. Nature 276: 71-73

WINNER WE (1981) The effect of SO_2 on photosynthesis and stomatal behaviour of mediterranean-climate shrubs. In: MARGARIS N, MOONEY H (eds) Photosynthesis, primary production and biomass utilization in mediterranean-type ecosystems. Junk, The Hague, pp 91-103

7. Moisture Regime and Nutrient Control of Seasonal Growth in Mediterranean Ecosystems

R. L. SPECHT, E. J. MOLL, F. PRESSINGER, and J. SOMMERVILLE

7.1 Mediterranean sclerophyllous shrublands and heathlands

In Chapter 2 of this volume Specht and Moll showed that soil nutrient status plays a major role in the development of the structure of the plant community in the mediterranean regions of southern Australia and South Africa. Three major suites of soils were distinguished:

(a) moderately leached soils, with an adequate balance of essential plant nutrients;

(b) strongly leached soils, containing low levels of available nutrients; and

(c) calcium-rich/high-pH soils, which may induce various nutrient deficiencies.

In the subhumid to semi-arid zone characteristic of the mediterranean climate of much of southern Australia and South Africa, these soils support sclerophyllous shrublands (= open-scrub) and heathlands (= fynbos) with or without sclerophyllous trees. In a few fertile habitats, scattered, evergreen trees may form a savanna woodland formation.

On moderately leached soils, the mediterranean shrublands are dominated by evergreen, sclerophyllous, multistemmed microphanerophytes with a seasonal, herbaceous ground stratum. In South Africa much of the mediterranean shrublands have been converted into farmlands. A few areas have been revegetated by a disclimax community of Elytropappus rhinocerotis, the renosterveld (Boucher and Moll 1981).

Heathlands (fynbos) develop on strongly leached soils and in contrast to the mediterranean shrublands all strata in the heathlands possess evergreen, sclerophyllous leaves.

The two plant formations - mediterranean shrublands and heathlands - are therefore segregated nutritionally on markedly different soils. The plant communities developing on these contrasting soils are clearly defined both by the structure and by the floristic composition of the dominants within each stratum:

(a) Mediterranean shrublands

Overstorey	- evergreen, sclerophyllous tall shrubs
	- *Euclea*, *Olea*, *Pterocelastrus*, *Rhus*, *Sideroxylon* (in South Africa)
	- *Eucalyptus socialis*, *E. dumosa* (in southern Australia)
Understorey	- seasonal grasses and herbs (South Africa and southern Australia)

(b) Heathlands

Overstorey	- evergreen, sclerophyllous shrubs (Proteaceae, etc.)
Mid-storey	- evergreen, sclerophyllous sub-shrubs (Ericaceae, Epacridaceae, etc.)
Ground stratum	- evergreen hemicryptophytes (Restionaceae, etc.) and evergreen chamaephytes.

7.2 Pyric succession

Fire is an integral part of mediterranean ecosystems and the vegetation shows a remarkable capacity to regenerate following a burn which may destroy all living tissue above ground.

After a fire, heathland species regenerate rapidly from rootstocks and seed. In the Dark Island heathland in South Australia, 38 species of flowering plants were recorded soon after fire (Specht et al. 1958). Many of these species germinated and flourished on the mineral nutrients returned to the soil in the ash. But the additional supply of nutrients was soon exhausted and the species recorded shortly after fire soon began to disappear. By 25 years only 20 remained, while only 10 species were found after 50 years (Specht et al. 1958).

Mediterranean shrublands are also razed by fire. The ground stratum of seasonal grasses and herbs may respond for a few years to the additional nutrients released by the fire, but this nutritional increment is relatively small compared to the nutrients available in the moderately leached soils. The overstorey species also appear to be well buffered nutritionally and do not show the same decrease in species composition after fire as observed in the heathland communities (Specht et al. 1958).

7.3 Overstorey-understorey foliage projective cover

The long-lived overstorey species gradually dominate the understorey species in the pyric succession, illustrated for the Dark Island heathland (Specht et al. 1958). Estimates of the Foliage Projective Cover (FPC - the proportion of the landscape covered by photosynthetic tissue, recorded vertically) of species found in the heathland at various ages after fire show that the total foliage cover of the community reaches a maximum within a few years after the fire and then remains relatively constant for a long period of time (Fig. 1). Early in the pyric succession short-lived understorey species regenerate rapidly, while the long-lived overstorey species regenerate slowly. FPC of the short-lived understorey species peaks rapidly, but the slowly regenerating overstorey species soon cause the understorey species to decline (Fig. 1A). During this competitive phase, there appears to be an inverse linear relationship between the FPC of the overstorey and understorey strata (Fig. 1B).

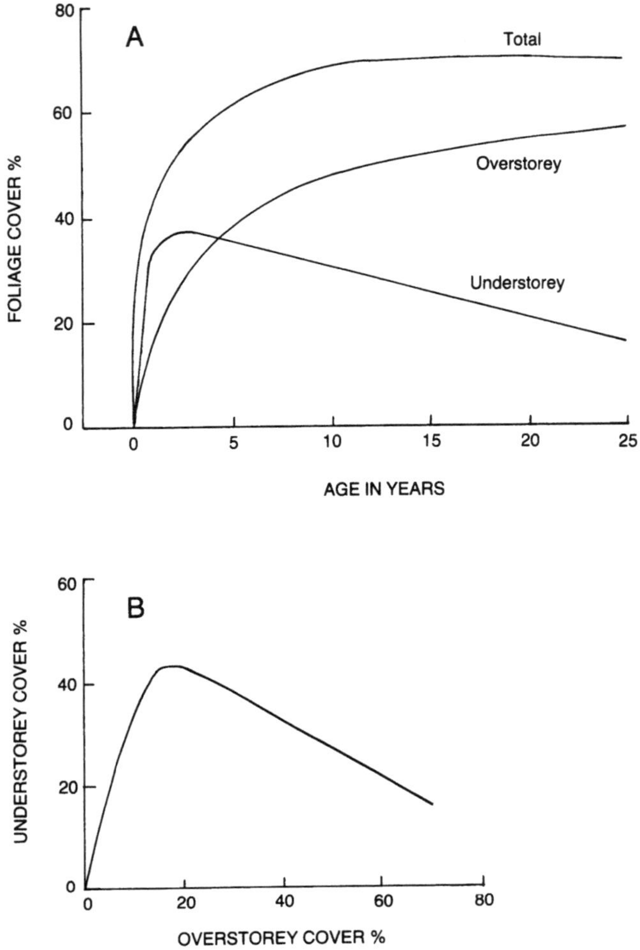

Fig. 1. Foliage projective cover of the overstorey and understorey species in the Dark Island heathland, South Australia, are plotted against age after fire in Fig. 1A and against each other in Fig. 1B (after Specht et al. 1958; Specht and Morgan 1981).

This gradual suppression of short-lived understorey by long-lived overstorey is also seen in FPCs of three regenerating stands of South African fynbos (Fig. 2). A few years after the fire the regenerating fynbos reaches a maximum FPC value which remains relatively constant for many years. The level of this "steady-state" FPC value appears to be controlled by the water balance of the ecosystem (See Specht and Moll, Chapter 2). A value of 68% has been recorded by the methods of Specht and Morgan (1981), at the humid environment of Bain's Kloof, while a value of 55% was observed at the semi-arid environment of Algeria, Cedarberg, Cape.

Within the fynbos community, the slowly regenerating overstorey gradually suppresses the short-lived understorey. The FPC crossover of these two strata may be reached within 9 to 10 years at the humid sites where growth is rapid, but in the more arid environments competitive equilibrium may not occur for 35 to 40 years. Eventually, FPC of the overstorey will reach a maximum value in equilibrium with the water balance of the ecosystem (see Fig. 3 in Chapter 2, this volume). In this mature ("climax") community, FPC of overstorey and understorey will be in balance (according to the inverse linear relationship shown in Fig. 1). Unfortunately heathlands rarely escape fire for more than 10 years and the distribution of FPC within degenerating stands is in the realms of speculation.

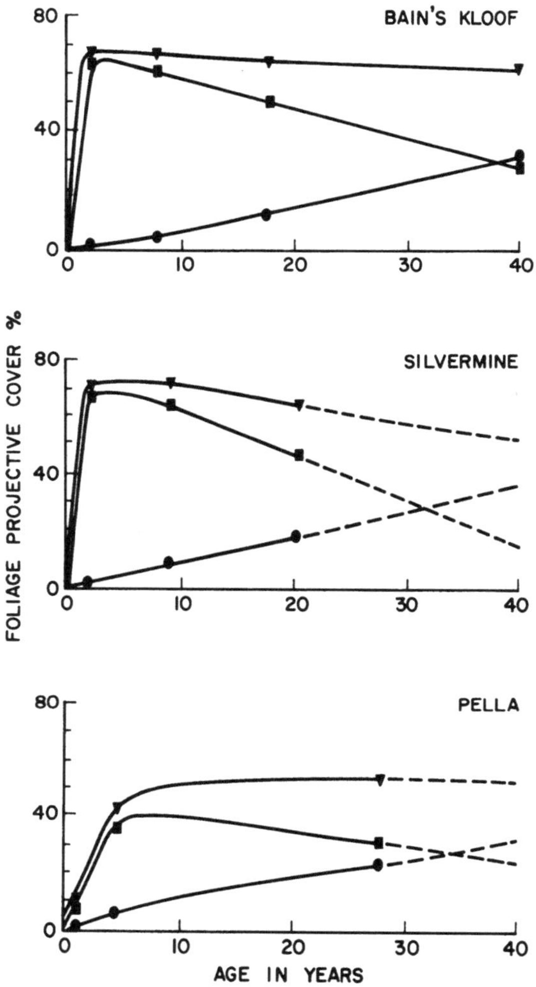

Fig. 2. Foliage projective cover of the overstorey (dot), understorey (square) and total (triangle) species of fynbos regenerating after fire are plotted against age in years for South African localities with subhumid to semi-arid mediterranean climates: Bain's Kloof (annual precipitation ± 800 mm); Silvermine (annual precipitation ± 900 mm), and Pella (annual precipitation ± 550 mm) (Specht and Moll, unpublished).

7.4 Seasonality of foliage

Specht and Rayson (1957a) have shown that long-lived *Banksia* spp. (Proteaceae) in Dark Island heathland produce new foliage shoots during summer (November to February), the driest season of the mediterranean climate. In the last decade, much more information on the seasonality of foliage (including leaf fall) has been assembled for other mediterranean shrublands and heathlands (Table 1) and for South Africa (Table 2). Although many more studies are needed, some conclusions can be drawn from present knowledge.

Table 1. Seasonal shoot growth (S) and leaf fall (L) of characteristic mediterranean shrubland-heathland species in southern Australia, California and southern France. (Lossaint and Rapp 1978; Mooney and Kummerow 1981; Rapp and Lossaint 1981; Specht et al. 1981).

Species	Winter			Spring			Summer			Autumn		
Southern Hemisphere	(June-Aug)			(Sept-Nov)			(Dec-Feb)			(March-May)		
Northern Hemisphere	(Dec-Feb)			(March-May)			(June-Aug)			(Sept-Nov)		
DARK ISLAND HEATHLAND, S. AUSTRALIA (strongly leached soils)												
Banksia ornata						S L	S L	S L	S L	L		
Casuarina pusilla					S L	S L	S L	S L	L	L		
Hakea spp.				S	S L	S L	S L	L				
TUTANNING HEATHLAND, W. AUSTRALIA (strongly leached soils)												
Microphanerophytes					L	S L	S L	S L	L	L		
Nanophanerophytes					S L	S L	S L	L	L			
CHAPARRAL, CALIFORNIA (moderately leached soils)												
Adenostoma fasciculatum			S	S	S	S	S	L	L	L	L	L
Heteromeles arbutifolia			S	S	S	S	S	L	L	L	L	L
Rhamnus crocea			S	S	S	S	S	L	L	L	L	L
SCLEROPHYLLOUS FOREST, SHRUBLANDS MONTPELLIER, FRANCE												
Pinus pinea forest					S	S	S	S L	L			
Quercus ilex forest (moderately leached soils)				S	S	L	L					
Quercus coccifera garrigue (calcium-rich/high-pH soils)					S L	S L						

Table 2. Seasonal vegetative shoot growth (S), flowering shoot growth (F) and leaf fall (L) of characteristic fynbos species at Jonkershoek (Kruger 1981) and Bain's Kloof (Moll, unpublished), South Africa.

Species	Winter (June-Aug)			Spring (Sept-Nov)			Summer (Dec-Feb)			Autumn (March-May)		
DOMINANT/SUBDOMINANT SPP.												
Cliffortia ruscifolia				F	F	S	S L	S L	S			
Erica hispidula						S	S L	S L				
Erica pinea						S	S L	L				
Erica sphaeroidea						S	S	S L	L			
Leucadendron salignum								S	S L	S L		
Leucospermum tottum								S	S	S		
Protea nitida					S	S		L	(S) L	(S)		
Protea laurifolia				S	S	S	S L	S L	S L			
Protea neriifolia						S	S					
Protea repens					S	S L	S L	S L	S L			
UNDERSTOREY RESTIOIDS												
Chondropetalum paniculatum			S	S	S							
Restio cuspidatus			S	S	S			L	L			
SHORT-LIVED FYNBOS SPP.												
Asparagus thunbergianus (drought-deciduous)	S	(S)						L	L	L		S
Helichrysum rutilans	S	S	S	S		L	L	L	L			
Psoralea obliqua			S	S	S					L	L	L
SHRUBLAND SPP. OF STREAMS AND ROCKY OUTCROPS												
Olea capensis	S					L	L	L	S	S	-	S
Metrosideros angustifolia					S	S	S L	L	L			
Brabejum stellatifolium			S	S	S	S						
Maytenus acuminata						L	L	L				

(a) Heathlands on strongly leached soils

Overstorey species. Shoot growth is initiated in late spring and continues through the mediterranean summer. Leaf fall and shoot growth tend to occur simultaneously, though abscission may increase markedly towards the end of the dry summer period.

Understorey species. Shoot growth of the sclerophyllous mid and ground strata tends to occur in autumn or spring, the most favourable periods with respect to available soil and water temperature. Leaf fall is delayed until the dry summer season.

(b) Mediterranean shrublands on moderately leached soils

Overstorey species. Shoot growth normally occurs in spring, sometimes in autumn, while leaf fall follows shoot growth as soil drought increases during late spring and summer.

Understorey species. The seasonal grasses and herbs which constitute the ground stratum of the mediterranean shrublands sprout from seed, rhizome, bulb or tuber with autumn rains, and complete their annual cycle by the end of spring.

(c) Mediterranean shrublands on calcium-rich/high-pH soils

Overstorey species. The only example of this ecosystem studied in detail is the *Quercus coccifera* garrigue in Montpellier, France by Rapp and Lossaint (1981) who suggest that shoot growth and leaf fall coincide - as noted for overstorey species in the strongly leached heathland vegetation. More information from communities in calcium-rich/high-pH soils is needed to establish the general pattern.

In the drier regions of the Mediterranean Basin (e.g. Greece and Israel) another phenological strategy has been demonstrated in *Sarcopoterium spinosum*, the dominant species of the low drought-deciduous mediterranean shrubland termed batha. This species is seasonally dimorphic, losing its "normal" leaves towards the end of spring and sprouting modified leaves during the summer period (see Chapter 5, this volume).

In summary, mediterranean shrublands on moderately leached soils and short-lived species on strongly leached soils (surviving on nutrients released from the aerial biomass burnt during a fire) appear to show autumn and/or spring shoot growth, while leaf fall is delayed until the dry season of the year (in late spring to summer). Thus during the active growing season of spring, these communities (or strata) carry almost double the amount of photosynthetic tissue in new and old leaves. It may be assumed that the FPC of these species will increase during that growing season. The herbaceous understorey also shows marked seasonality in foliage cover.

In contrast, the long-lived overstorey species of the heathlands show late spring to summer shoot growth, with leaf fall occurring essentially at the same time. The FPC of this vegetation thus remains relatively constant throughout the year. The seasonal behaviour of maquis vegetation on strongly leached soils around the Mediterranean Sea and in California needs investigation to ascertain whether the sequence observed in the Southern Hemisphere applies also to the Northern Hemisphere.

The seasonality of mediterranean shrublands on calcium-rich/high-pH soils indicates that the nutrient stresses induced by high pH favour the development of communities where shoot growth and leaf fall coincide. In southern France this phenophase occurs during spring; from superficial observations it occurs during late spring-summer in southern Australia where summer drought may not be as severe as it is around the Mediterranean Sea and in California. The marked difference between the phenological phases observed in mediterranean heathlands and shrublands of South Africa are illustrated in Fig. 3.

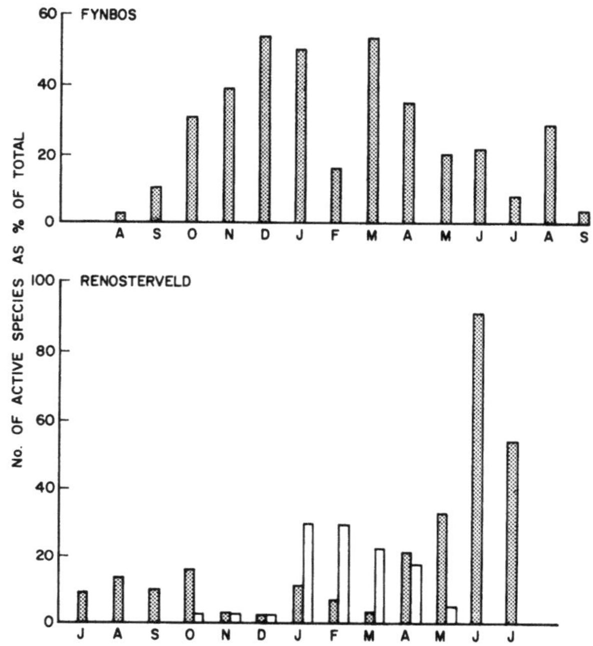

Fig. 3. Seasonal shoot growth and leaf fall in coastal fynbos (on strongly leached soils) and renosterveld (on moderately leached soils) in the mediterranean region of South Africa (data from J Sommerville, unpublished). Open bars: percentage of species losing leaves, shaded bars: percentage of species growing.

7.5 Addition of nutrients to nutrient-poor heathlands

If the sequence of phenophases of the mediterranean heathlands is a response to soils of a low nutrient status it should be possible to influence the sequence by adding the limiting nutrients to the soil. In Dark Island heathland, South Australia (Specht 1963; Heddle and Specht 1975), the addition of phosphatic fertilizer produced a differential growth response in the component species. Short-lived understorey species (Leptospermum and Casuarina) showed the greatest response to added nutrients (up to 100 g m^{-2} superphosphate), while the long-lived overstorey species (Banksia ornata and Xanthorrhoea australis) showed only a small growth response, which was not increased by adding superphosphate in excess of 20 g m^{-2}. This suggests that understorey species possess physiological strategies which enable them to respond to the plant nutrients released in the ashes of a bush fire while the overstorey species have evolved strategies which favour nutrient conservation. One nutrient conservation strategy may be the unusual late spring-summer shoot-growth phenophase which occurs simultaneously with leaf fall. Short-lived heath species, species which are able to fix atmospheric nitrogen, and seasonal members of the mediterranean shrublands readily invade areas which are fertilized with phosphate. The resultant vegetation will show a series of seasonal phenophases characteristic of mediterranean shrublands, rather than those formerly present in the unfertilized heathland.

The success of many exotic plants and weeds in southern Australia appears to be due to their different phenological strategies compared to the native species (Specht 1975).

7.6 Water balance

Studies on the water balance of evergreen heathland and sclerophyllous forest communities have been made in the mediterranean zone of southern Australia (summarized in Specht 1972). The essential point to emerge from these studies is that water is conserved during the wet winter-spring season to be used during the summer season. Transpiration, albeit low, does occur from the leaves of deep-rooted overstorey species even in the driest months of summer (Specht 1957b). Only the shallow-rooted species may experience periods of severe drought, during which time the evergreen leaves of these species may reach exceedingly low leaf water potentials and may even be lost by defoliation.

Community structure (with respect to foliage area, distribution and leaf resistances) may conserve water during the humid period of the year for use (albeit at a low level) during the dry summer period. The delicately balanced pattern of conservation and usage is governed by the following relationship:

$$Ea/Eo = k(P - R - D + Sext)$$

where Ea = actual evapotranspiration, Eo = pan evaporation, P = precipitation, R = runoff, D = drainage, Sext = extractable soil moisture at the beginning of the month (all measured in cm), and k = a constant. For the Dark Island heathland, South Australia, moisture index (Ea/Eo) is correlated with available soil moisture (Fig. 4A) while Fig. 4B indicates how precariously close the community comes to running out of water during the dry mediterranean summer period.

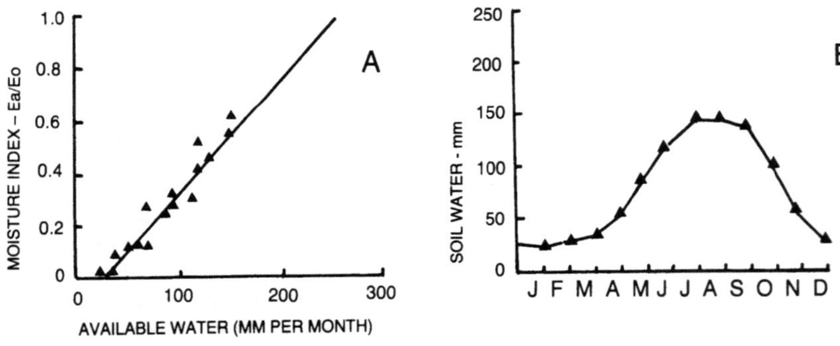

Fig. 4A. Relationship between monthly values of the moisture index (Evapotranspiration Ea/Pan Evaporation Eo) and available water (cm) in Dark Island heathland, South Australia. B: Calculated monthly values of available soil water (mm) in Dark Island heathland.

7.7 Conclusions

It would appear that different shoot growth strategies have been developed in mediterranean heathlands on strongly leached soils and in mediterranean shrublands on moderately leached soils. The strategies are summarized in Fig. 5. The timing of shoot growth and leaf fall appear to confer decided ecophysiological advantages:

(a) In mediterranean shrublands on moderately leached soils, where shoot growth is followed by leaf fall, the extra foliage during the favourable spring season would boost photosynthetic production.

(b) In mediterranean heathlands on strongly leached soils, shoot growth in late spring-summer follows the release of nutrients from decomposing litter during spring when most of the nutrients are trapped (physically and chemically) by fine rootlets which interlace the decomposing litter (Coleman and Specht 1981; Specht 1981). Nutrient cycling, externally via decomposing leaf litter and internally from abscising leaves, is thus maximized by both shoot growth and leaf abscission coinciding after nutrient release from decomposing litter.

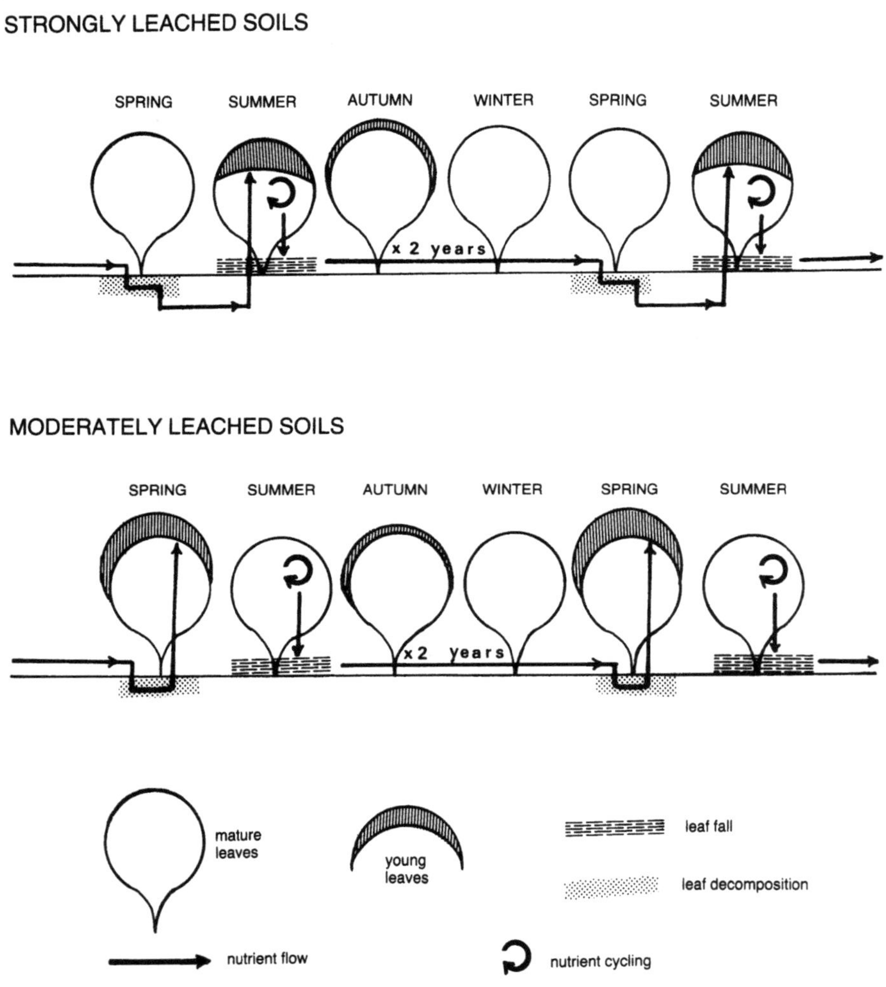

Fig. 5. Summary diagram showing (a) seasonal phenophases of foliage of overstorey species in heathlands (on strongly leached soils) and in mediterranean shrublands (on moderately leached soils), and (b) seasonal nutrient flow through these two ecosystems.

Table 3. Ecophysiological characteristics of overstorey and understorey species, regenerating after fire in Dark Island heathland, South Australia.

Ecophysiological characteristic	Understorey	Overstorey	References
Leaf size	narrow	broad	Specht and Rayson (1957a)
Short-wave reflectivity	14%	20%	Yates (1981)
Rainfall interception - stem flow	minimal	maximal	Specht (1957a)
Shoot growth	spring (temperature range 10-20°C)	late spring-summer (temperature range 15-25°C)	Specht and Rayson (1957a) Specht et al. (1981)
Rooting depth	shallow	deep	Specht and Rayson (1957b)
Growth response (to superphosphate fertilizer)	response up to 1000 kg ha^{-1}	growth response up to 200 kg ha^{-1}	Specht (1963), Heddle and Specht (1975)
Phosphate conservation	minimal(?)	high	-
Longevity	5-20 years	20-50+ years	Specht et al. (1958)

Either strategy induces problems in the annual water balance. The extra foliage held by the mediterranean shrublands increases evapotranspiration during spring, thus reducing water conserved for drought survival during summer. In the mediterranean heathlands water conserved during spring is rapidly used by the new shoots produced during late spring-summer.

Some ecophysiological attributes which may be necessary for plant survival under both nutrition- and water-stressed conditions (Table 3) are broad, highly reflective leaves, rainfall interception and stem flow to concentrate rainfall, and a deep root system to tap subsoil moisture.

More research on seasonality, in relation to water balance and nutrient flow, is necessary throughout the mediterranean region to elucidate the above points.

7.8 References

BOUCHER C, MOLL EJ (1981) South African mediterranean shrublands. In: DI CASTRI F, GOODALL DW, SPECHT RL (eds) Ecosystems of the world, vol 11. Mediterranean-type shrublands. Elsevier, Amsterdam, pp 233-248

COLEMAN RG, SPECHT RL (1981) Mineral nutrition of heathlands: the possible role of polyphosphate in the phosphorus economy of heathland species. In: SPECHT RL (ed) Ecosystems of the world, vol 9B. Heathlands and related shrublands. Analytical studies. Elsevier, Amsterdam, pp 197-207

HEDDLE EM, SPECHT RL (1975) Dark Island heath (Ninety-Mile Plain, South Australia). VIII. The effect of fertilizers on composition and growth, 1950-1972. Australian Journal of Botany 23: 151-164

KRUGER FJ (1981) Seasonal growth and flowering rhythms: South African heathlands. In: SPECHT RL (ed) Ecosystems of the world, vol 9B. Heathlands and related shrublands. Analytical studies. Elsevier, Amsterdam, pp 1-4

LOSSAINT P, RAPP M (1978) La forêt méditerranéenne de chênes verts. In: LAMOTTE M, BOURLIERE F (eds) Problèmes d'ecologie. Ecosystèmes terrestres. Masson, Paris, pp 129-185

MOONEY HA, KUMMEROW J (1981) Phenological development of plants in mediterranean-climate regions. In: DI CASTRI F, GOODALL DW, SPECHT RL (eds) Ecosystems of the world, vol 11. Mediterranean-type shrublands. Elsevier, Amsterdam, pp 303-307

RAPP M, LOSSAINT P (1981) Some aspects of mineral cycling in the garrigue of southern France. In: DI CASTRI F, GOODALL DW, SPECHT RL (eds) Ecosystems of the world, vol 11. Mediterranean-type shrublands. Elsevier, Amsterdam, pp 289-301

SPECHT RL (1957a) Dark Island heath (Ninety-Mile Plain, South Australia). IV. Soil moisture patterns produced by rainfall interception and stem-flow. Australian Journal of Botany 5: 137-150

SPECHT RL (1957b) Dark Island heath (Ninety-Mile Plain, South Australia). V. The water relationships in heath vegetation and pastures on the Makin Sand. Australian Journal of Botany 5: 151-172

SPECHT RL (1963) Dark Island heath (Ninety-Mile Plain, South Australia). VII. The effect of fertilizers on composition and growth, 1950-1960. Australian Journal of Botany 11: 67-94

SPECHT RL (1972) Water use by perennial evergreen plant communities in Australia and Papua New Guinea. Australian Journal of Botany 20: 273-299

SPECHT RL (1975) A heritage inverted: our flora endangered. Search 6: 472-477

SPECHT RL (1981) Nutrient release from decomposing leaf litter of Banksia ornata, Dark Island heathland, South Australia. Australian Journal of Ecology 6: 59-63

SPECHT RL, MORGAN DG (1981) The balance between the foliage projective covers of overstorey and understorey strata in Australian vegetation. Australian Journal of Ecology 6: 193-202

SPECHT RL, RAYSON P (1957a) Dark Island heath (Ninety-Mile Plain, South Australia). I. Definition of the ecosystem. Australian Journal of Botany 5: 52-85

SPECHT RL, RAYSON P (1957b) Dark Island heath (Ninety-Mile Plain, South Australia). III. The root systems. Australian Journal of Botany 5: 103-114

SPECHT RL, RAYSON P, JACKMAN ME (1958) Dark Island heath (Ninety-Mile Plain, South Australia). VI. Pyric succession: changes in composition, coverage, dry weight and mineral nutrient status. Australian Journal of Botany 6: 59-88

SPECHT RL, ROGERS RW, HOPKINS AJM (1981) Seasonal growth and flowering rhythms. In: SPECHT RL (ed) Ecosystems of the world, vol 9B. Heathlands and related shrublands. Analytical studies. Elsevier, Amsterdam, pp 5-13

YATES DJ (1981) Optical properties of heathlands. In: SPECHT RL (ed) Ecosystems of the world, vol 9B. Heathlands and related shrublands. Analytical studies. Elsevier, Amsterdam, pp 91-97

8. Canopy Structure of Mediterranean-Type Shrubs in Relation to Heat and Moisture

P. C. MILLER

8.1 Introduction

8.1.1 GENERAL BACKGROUND

The similar vegetative characteristics in regions with similar mediterranean-type climates imply that these characteristics should be important for plant survival and optimal resource use within the constraints of irradiance, temperature, annual precipitation and soil moisture. These similar vegetative characteristics include shrub form, moderate leaf area indices and hard, small, evergreen leaves. Usually, where the annual precipitation decreases, the vegetation includes smaller shrubs, lower leaf area indices and drought-deciduous, soft leaves. The links between the vegetative morphological characteristics and soil characteristics involve the physiological characteristics of the species. The vegetation has developed similar forms despite separate phylogenetic histories and different intensities of various selective forces, including climatic factors, low nutrient conditions and frequent fire (Walter 1973; Specht 1979). Regardless of the vegetation histories, differences in distribution and diverse previous selective forces within the regions, the broad similarities between vegetation form and climate in the present mediterranean regions imply current functional-physiological inter-relations.

The objectives of this paper are to analyse some of the interactions between vegetation and climate characteristics in terms of the physical processes of heat and water exchange in order to isolate plant-climate interactions from the total suite of interacting plant, climate and soil factors; to quantify some of the conditions necessary for plant survival in mediterranean-type climates, and to evaluate some of the plant structural features in terms of the concepts of survival and optimization of resource use. The interrelations between these characteristics and resource use were analysed by means of simulation experiments and interpretation of data.

The mediterranean-type climate is characterized by cool, wet winters with low solar irradiance and hot, dry summers with high solar irradiance. Soil moisture is low during the summer (Kramer 1952; Specht 1957a; Pillsbury et al. 1963; Shachori et al. 1967; Mooney and Dunn 1970a,b; Mooney and Parsons 1973; Ng 1974; Miller and Poole 1979; Ng and Miller 1980). Annual mean temperatures are 12-20°C (Miller et al. 1977). The interplay between temperature and water with respect to plant form has been emphasized in many papers (Schimper 1898; Cooper 1922; Mooney and Dunn 1970a,b; Miller 1979) which have pointed out the inhibition of growth and photosynthesis in winter by cool temperatures and the necessity of avoiding both excessive plant temperatures and excessive water loss during the summer drought.

Much of the discussion in this paper will be based on research in mediterranean regions of southern California and central Chile (Mooney 1977a; Thrower and Bradbury 1977; Miller 1981). In these regions precipitation is 250-350 mm yr^{-1} and annual mean temperature 14-17°C along the coast in the coastal sage vegetation where drought-deciduous shrubs occur, 350-650 mm yr^{-1} and 12-14°C where evergreen shrubs predominate and 650-800 mm yr^{-1} and 10-12°C at the top of the coastal mountains where forests occur (Miller et al. 1977). In the evergreen shrub region two chaparral types can be distinguished, chamise chaparral dominated by *Adenostoma fasciculatum*, and mixed chaparral dominated by *Arctostaphylos* spp., *Ceanothus* spp., *Quercus dumosa* and *A. fasciculatum*. The same general pattern of annual precipitation and vegetation forms also seems to be present in mediterranean regions of South Africa and Israel (Cooke 1965; Shachori and Michaeli 1965). In South Africa and Australia shrublands often occur with relatively high precipitation instead of forests because of low nutrient availability or low soil water-holding capacity (Kruger 1979; Specht 1979).

In the mediterranean-type climate the year can be divided into three periods: (a) winter when precipitation occurs and soil moisture levels are high; (b) spring when little precipitation occurs but soil moisture levels are relatively high; and (c) summer when little precipitation occurs and little soil moisture is present (Wallén and de Brichambaut 1962; Cochemé and Franquini 1967; Cochemé 1968; Wallén 1968). In southern California and southern Europe about 17% of the annual precipitation is received in spring and in Chile, south-west Australia, and south-west Africa about 9% of the annual precipitation is received in spring (Gentilli 1971; Schulze 1972; Miller et al. 1977). This spring season precipitation is readily available for photosynthesis and growth and may affect the length of the season of active growth. In addition to the spring precipitation, precipitation frequently occurs during the summer months at higher elevations in southern California. In central Chile most of the precipitation occurs during the winter months at all elevations. In southern France the development of anti-cyclonic conditions increases the precipitation in fall (Linés Escardo 1970).

8.1.2 ENERGY EXCHANGE PROCESSES

Climate and vegetation structure interact through the physical processes of energy exchange, which include radiation, convection, evaporation and conduction. The plant absorbs solar and far infrared radiation and loses this absorbed energy by radiating in the far infrared, by evaporation and by convection. If air temperatures are above plant temperature, the plant can gain energy by convection. The processes of energy exchange between the environment and the individual leaf or stem are affected by several structural and physiological properties of the plant part, including its absorptance, emittance, width or diameter, inclination to the sun and conductance to water vapour diffusion. Within the plant canopy the energy exchange is affected additionally by the total leaf and stem area indices, clustering of leaves and stems, vertical profiles of leaf and stem areas and of leaf and stem inclinations, and density of the foliage with respect to air movement within the canopy. The canopy will affect the exchange of energy at the soil surface by shading the surface from the direct solar beam and by radiating infrared to the surface. In return the canopy receives reflected solar radiation and far infrared radiation from the soil surface. Energy processes of the whole plant are also affected by the amount of absorbing root because evaporation from the canopy cannot continue without water absorption.

The processes of energy exchange affect plant survival directly and indirectly by determining equilibrium plant temperatures and plant water contents, which may affect its survival, and, through the effect of moisture stress on photosynthesis, affect the carbohydrate available for leaf growth, root growth and reproduction. Reduced reproduction may reduce the species population and cause species replacement.

It is clear that the processes of energy and water exchange affect and are affected by plant activities on several scales of time - from minutes to generations. The effect of the energy exchange processes is cumulative through the life of the plant so that small effects on a minute-to-minute basis develop cumulatively into large effects on a life cycle basis. Also, the effects of the processes of energy and water exchange differ through the life cycle of the plant. Seedlings and root crown sprouts which live close to the soil surface must endure a more severe thermal environment than do mature plants which are well above the soil surface. Seedlings, root crown sprouts and mature plants also differ in their access to soil moisture.

8.1.3 GEOGRAPHIC PATTERNS OF CANOPY CHARACTERISTICS

Changes in canopy characteristics along climatic gradients have been related to processes of energy exchange. Mooney et al. (1974) noted that with increasing aridity leaves in chaparral became narrower, more steeply inclined and less absorptive. Shaver (1978) noted that leaves of *Arctostaphylos* spp. became more steeply inclined and less absorptive with increased aridity. Werger and Ellenbroek (1978) showed that along a gradient of annual precipitation in South Africa, leaves of the riparian forest become narrow with decreasing precipitation. Parsons and Moldenke (1975) and Steward and Webber (1981) noted that narrow-leaved shrub forms are more common in the drier chaparral than in mesic chaparral and are more common in chaparral than in matorral, which is more mesic. These canopy characteristics affect the processes of energy exchange and reduce daytime leaf temperatures in drier climates. The reduction in temperatures reduces the chance of high, lethal temperatures, decreases the chance of low, lethal water contents and promotes longer periods of photosynthesis.

8.2 Processes of energy exchange in shrub canopies

The processes of energy exchange in canopies of mediterranean shrubs may be illustrated by measurements and simulations for chaparral and matorral.

8.2.1 INTERCEPTION AND ABSORPTION OF SOLAR IRRADIANCE

The interception of solar irradiance in the canopy affects plant temperature, transpiration and photosynthesis. The importance of interception has led to numerous studies in agriculture and forestry but to few studies in scrub communities (Specht 1972b). Lowe and Hinds (1971) and Dirmhirn (1972) measured solar irradiance beneath desert shrubs. A simulation model for the interception of solar irradiance by shrub canopies (Miller 1975; Miller et al. 1976) has been evaluated in stands of *A. fasciculatum* and *Ceanothus greggii* in the chaparral of southern California and on *Baccharis linearis* and *Colliguaya odorifera* in the matorral in central Chile (Lawrence 1975). A model calculating irradiances on leaves in different parts of widely separated shrubs has been evaluated in the Chilean shrubs *Lithraea caustica* and *C. odorifera* (Roberts and Miller 1977). The penetration of solar irradiance through the canopy, i.e. the complement of interception, was expressed as follows:

$$I = I_o \exp(-k_\ell c_\ell A_\ell - k_s c_s A_s) \quad (1)$$

where I_o is the irradiance above the canopy, I is the irradiance below leaf and stem area indices A_ℓ and A_s, respectively, k_ℓ and k_s are the appropriate extinction coefficients, and c_ℓ and c_s are the appropriate clustering coefficients (<1 when leaves or stems are clustered, >1 when regularly dispersed). Shrubs of chaparral and matorral tend to have clustered leaves. For example, the clustering coefficients for *A. fasciculatum*, *C. greggii*, *C. odorifera* and *B. linearis* were 0.3-0.7 with an average of 0.5 (Lawrence 1975), though Roberts and Miller (1977) estimated clustering coefficients of 1.0-1.3 with an average of 1.1 for *L. caustica* and *C. odorifera*; the different estimates for *C. odorifera* may be related to the scale of the pattern of clustering (Greig-Smith 1964) and different scales of measurements. Where leaves in these shrub canopies are clustered this results in almost 100% more solar irradiance penetration than where leaves are randomly dispersed.

Leaf and stem area indices are not easily measured in the shrub canopies which contain many small, often adpressed leaves and many small stems, but available data indicate low leaf area indices in mediterranean shrublands. Stem area indices are about half the leaf area indices, but species differ widely (Mooney 1977a). Leaf area indices in the southern Californian drought-deciduous community are similar to those on the equator-facing slope in chaparral (Mooney 1977b).

Interception of irradiance is also governed by leaf orientation in relation to the solar path. The higher the sun, the shorter the path of the direct beam through the canopy and the less the interception (de Wit 1965; Duncan et al. 1967; Miller 1981). Interception is decreased by more steeply inclined leaves when the sun is high but increased when the sun is relatively low. A canopy with horizontally inclined leaves should intercept a constant fraction of the incoming direct beam from spring to fall. Interception in a canopy with steeply inclined leaves should vary more throughout the year than interception in a canopy with more nearly horizontal leaves. In winter a canopy with steeply inclined leaves intercepts a greater fraction of incoming direct beam than a canopy with nearly horizontal leaves. In the Chilean matorral *C. odorifera* has steeply inclined leaves while *L. caustica* has horizontally inclined leaves. The pattern of interception of these two shrubs (Roberts and Miller 1977) is consistent with the shallower roots, cessation of physiological activities and higher leaf drop in summer of *C. odorifera*

(Montenegro and Riveros de la Puente 1977; Thrower and Bradbury 1977) and with the deeper roots and the year-long physiological activity of L. caustica (Avila et al. 1975; Giliberto and Estay 1978).

At Echo Valley, California, mixed chaparral on the pole-facing slope absorbed about 69% of the total incoming solar irradiance (300-3000 nm) in the canopy (Miller and Poole 1980). This percentage varied little through the season (68-70%). About 18% was absorbed by the soil. In chamise chaparral about 54% was absorbed in the canopy varying between about 48% in summer and 60% in winter. About 32% was absorbed by the soil. The reduced seasonal variation of absorbed irradiance in mixed chaparral compared to the variation in chamise chaparral was related to the higher leaf area index in mixed chaparral. The absorbed irradiance on a daily basis was satisfactorily calculated by using only the leaf area index and an extinction coefficient of 0.7 in the negative exponential equation. At Fundo Santa Laura, Chile, 75% of the incoming solar irradiance on the pole-facing slope was absorbed in the matorral canopy and 12% was absorbed at the soil surface (Miller, unpublished). On the ridgetop 39% was absorbed in the canopy and 48% in the soil. An individual shrub on the ridgetop absorbed slightly more solar irradiance than on the pole-facing slope, but the vegetative cover was less and the reflection of solar irradiance from the soil surface was greater on the ridgetop.

The albedo at Echo Valley and Fundo Santa Laura was 12-14%, increasing about 2% as the soil dried. At Fundo Santa Laura the albedo of the bare soil on the ridgetop was about 30%, while the albedo of the soil and litter on the pole-facing slope was about 11%. At Echo Valley the albedo of the bare soil on the ridgetop was about 24%, while the albedo of the soil and litter on the pole-facing slope was 21%. The albedo on the pole-facing slope was 1-2% less than that on the equator-facing slope because of the greater foliage area index on the pole-facing slope and the greater trapping of reflected irradiance within the canopy. With open canopies ($A_\ell < 1$ m^2 m^{-2}) a high proportion of incoming solar irradiance reaches the soil surface and is reflected back. Thus, the heat load on leaves in open canopies can be higher than on leaves in closed canopies (Cooper 1922). The white surface of the podzolic soils in South Africa and in South and Western Australia may create high heat loads on the leaves.

The fraction of the intercepted solar irradiance which is absorbed by leaves depends on leaf chlorophyll content, thickness and inclination. Shaver (1978) concluded that leaf inclination was more related to geographic distribution than leaf absorptance for several Arctostaphylos spp. Mooney et al. (1974) found that the incidence of pubescence, which is often negatively correlated with absorptance, decreased from drier to mesic areas in the chaparral zone.

8.2.2. INFRARED IRRADIANCE, CONVECTION AND CONDUCTION

Infrared irradiance from the sky was usually 24.1-27.2 MJ m^{-2} day^{-1} throughout the year at Echo Valley and Fundo Santa Laura with the higher values occurring in summer (Miller 1981). The net infrared over the canopy was more variable at Echo Valley than at Fundo Santa Laura because of the higher air temperatures at Echo Valley. Infrared loss from the canopy increased and infrared loss from the soil decreased as leaf area indices increased. During Santa Anas at Echo Valley, canopy temperatures were warmer than the soil surface temperatures and the soil gained infrared energy. On the ridgetops slightly more than 40% of the absorbed solar irradiance was lost by infrared irradiance, while on the pole-facing slopes 50-60% of the absorbed solar was lost by infrared irradiance (Miller and Poole 1980).

Convection accounted for about 46% of the solar irradiance absorbed by the canopies on the ridgetops and for about 29% of the solar irradiance absorbed by the canopies on the pole-facing slopes at Echo Valley and Fundo Santa Laura (Miller 1981; Fig. 1). The open canopies on the equator-facing slopes allow freer air movement through the canopy, greater convectional loss of heat, reduced temperatures and decreased infrared radiation loss. Smaller leaves further increase convectional loss. About 80-90% of the convectional heat loss was from stems. Stimulations of

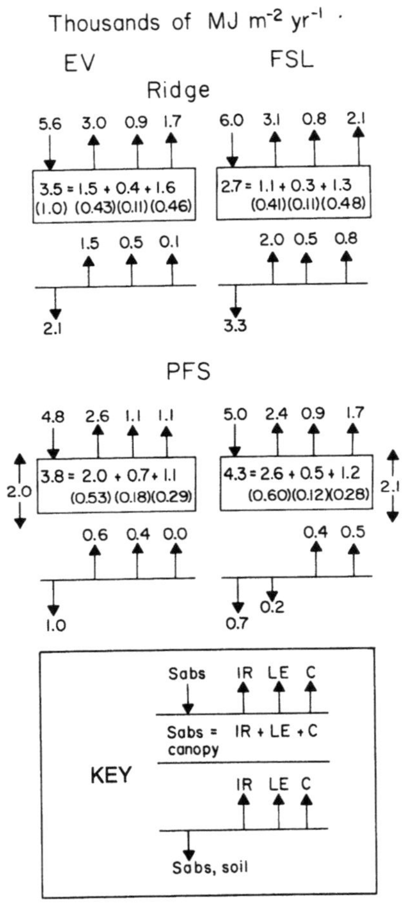

Fig. 1. Partitioning of solar irradiance absorbed in the canopy and soil into infrared irradiance, evaporation and convection. Units are thousands of megajoules per square metre per year. Fraction of solar absorbed is given in parentheses. IR: infrared radiation, LE: latent heat of evaporation and C: convection exchange. (From Miller et al. 1981)

convectional loss in the canopy were supported by measurements. In simulations of Ceanothus greggii in May, leaf temperatures were 0.6° C below air temperature. Measured leaf temperatures of Arctostaphylos glauca in May were 0-3°C below air temperature (D K Poole and P C Miller, unpublished). The low leaf temperatures relative to air temperatures resulted from the steep leaf inclination and the glaucous coating in A. glauca. Convectional loss from the soil was small on all slopes.

Heat conduction into the soil was generally positive in the spring and negative in the fall at Echo Valley and Fundo Santa Laura (D K Poole and P C Miller, unpublished). However, in the late spring even though the temperatures in the soil profile increased, the heat content of the profile decreased because of the loss of water from the upper soil layers. Annually soil heat conduction was near zero, indicating no general warming or cooling of the soil.

8.2.3 EVAPORATION AND PLANT WATER STRESS

About 11% of the solar irradiance absorbed in the canopy was lost by transpiration on the ridgetop sites in both chaparral and matorral, and 12-18% was lost on the pole-facing slopes. At Echo Valley the pole-facing slope with mixed chaparral lost about 0.3 MJ m^{-2} day^{-1} (annual average) through interception losses, 2.0 by transpiration, and 0.7 by evaporation from the soil surface. On the equator-facing slope the chamise chaparral lost about 0.3 MJ m^{-2} day^{-1} by interception losses, 1.0 by transpiration and 1.2 by soil evaporation. In the matorral at Fundo Santa Laura, interception losses accounted for about 0.4 MJ m^{-2} day^{-1}, transpiration for 1.3 and soil evaporation for 0.9.

The annual pattern of soil drought and air temperature along an elevational transect from 0 m to 2000 m in southern California was calculated from the annual courses of precipitation and potential evapotranspiration (Miller 1979). Such a pattern was postulated by Mooney et al. (1972) and Mooney and Dunn (1970a,b). Xylem pressure potentials, which were measured to check the calculations (Miller and Poole 1980), decreased earlier in the summer at the coastal and desert edges of the chaparral and later toward the centre of the chaparral. Poole and Miller (1975) also showed that xylem pressure potentials of Rhus spp. and Heteromeles arbutifolia decreased earlier at the coast than in the centre of chaparral. The pattern of drying was affected by the vegetation cover (Miller and Poole 1980). At the edges of the distribution, water use was about 200 mm yr^{-1} per unit leaf area index ($A\ell$), which occurred with precipitation of about 400 mm yr^{-1}. Soil evaporation and interception losses account for the remainder. With higher precipitation the leaf area index increases until the resulting transpiration rate decreases to about 200 mm yr^{-1} $A\ell$. Without disturbance chaparral vegetation should develop until the water use per unit leaf area and the duration of the summer soil drought are similar throughout the distribution of the chaparral. The resulting period of summer drought should be about three months long. This hypothesis is consistent with the patterns of water and soil drought measured under the vegetation of pole- and equator-facing slopes at Echo Valley (Miller and Poole 1979; Ng and Miller 1980) and with the pattern of several leaf properties (Miller and Poole 1980).

In California the diurnal course of leaf conductance and transpirational water loss varied seasonally by slope and by species (Poole and Miller 1975; Miller and Poole 1979; Roberts et al. 1981;

Roberts et al., unpublished). High conductances occurred when soil moisture was available; low conductances occurred during the summer drought. The seasonal variation was greatest in Arctostaphylos glauca, less in C. greggii, and still less in Adenostoma fasciculatum, where it was more pronounced on the pole- than on the equator-facing slope. Rhus ovata, R. integrifolia and R. laurina showed the lowest leaf conductances. R. integrifolia and R. laurina have stem growth throughout the summer because of their deep roots (Watkins and de Forest 1941) and conservative water use (Shapiro and de Forest 1932).

The seasonal progressions of predawn and midday xylem pressure potentials, measured in coastal sage and chaparral in California and in matorral in Chile on both pole- and equator-facing slopes, indicated species and slope differences in water stress (Poole and Miller 1975; Giliberto and Estay 1978). The deep-rooted Quillaja saponaria, L. caustica, and Rhus spp. showed relatively high midday potentials (-2.0 MPa) throughout the year, which were consistent with their leaf conductance-pressure potential relationships. Pressure potentials of soft-leaves shrubs, Artemisia californica, Satureja gilliesii and Trevoa trinervis, decreased markedly in spring. When potentials of A. californica were below -6.0 MPa, the plant was leafless. The other species maintained leaves during these periods of low pressure potentials. Predawn potentials were below -6.0 MPa for 2.5 months in A. californica; about two months in A. glauca, C. odorifera and Cryptocarya alba; about one month in C. greggii and T. trinervis; and less than one month in A. fasciculatum. Predawn xylem pressure potentials, which are commonly equated with soil water potentials, were 3.0-5.0 MPa lower than the potentials estimated from the bulk soil moisture data. At Echo Valley predawn and midday potentials of A. glauca, A. fasciculatum and C. greggii were lower during the summer drought on the pole- than on the equator-facing slope. Potentials of shallow-rooted species rose more rapidly than those of deep-rooted species after the first fall rains but potentials of the Rhus spp. near the coast and inland at Echo Valley rose before the fall rains, probably because of coastal fog and increased air humidity. The recovery at the coast and inland was generally synchronous because the winter rains were widespread. During the winter xylem pressure potentials of all species were high.

In California chaparral transpiration in the evergreen shrubs varied seasonally (Miller and Poole 1979). Although leaf conductances were high in winter, transpiration was low because net radiation was low and vapour density deficits were small. In the spring transpiration increased because of increased daylength, since leaf conductances and vapour density deficits were nearly constant. In June and July transpiration decreased because leaf conductances decreased, even though vapour density deficits increased. A. glauca ceased transpiration for almost 2.5 months. R. ovata, H. arbutifolia and C. greggii showed no transpiration for about one month. Similar annual patterns were shown for heath in South Australia (Specht 1957a). In mature shrubs transpiration during spring was 0.7-2.0 mm day^{-1} m^{-2} of leaf area. Annual water use on a square metre of leaf basis was 150-190 mm yr^{-1} for Adenostoma fasciculatum, 145-195 from C. greggii, 280-290 for Arctostaphylos glauca and 127 for R. ovata (Miller and Poole 1979).

Several sets of measurements and calculations indicate that transpiration of mature mixed chaparral is higher than that of mature chamise chaparral (Miller and Poole 1979; Miller and Stoner 1979; Ng and Miller 1980). The estimates indicate that mature chamise chaparral, grasslands and bare soil lose water similarly while mature mixed chaparral loses more water (Shachori and Michaeli 1965; Miller and Poole 1979; J H Patric, unpublished).

8.3 Influence of canopy form on heat exchange, water loss, length of drought and photosynthesis

8.3.1 SHRUB FORM

The shrub form affects light interception, especially during the winter, and affects convectional cooling during the summer. With simulations Miller and Mooney (1976) showed that, in winter, photosynthesis was light-limited and photosynthetic rates were highest at the top of the canopy. In early spring photosynthesis was limited by low temperatures and rates were highest at the bottom of the canopy because of the higher soil temperatures there. In late spring photosynthetic rates were uniform throughout the canopy. In summer photosynthesis was limited by lack of water and by high temperature, and rates were highest at the top of the canopy where temperatures were lower. These simulations used a relatively low leaf area index and data for H. arbutifolia (Harrison 1971) which indicated a relatively large temperature dependence of photosynthesis. More recent data (Oechel et al. 1981) for other species from Echo Valley and Fundo Santa Laura indicated less temperature dependence. Using data of Oechel et al., the simulated rates of photosynthesis and transpiration and the photosynthesis:transpiration ratio were highest at the top of the canopy throughout the year. Thus, the shrub form has an advantage over shorter vegetation forms because of its higher efficiencies of resource use and higher photosynthetic rates at the top of the shrub canopy. Growth should be greater at the top of the shrub canopy because carbohydrate reserves are greater there.

The abundance of the shrub form relative to the tree form may relate to the more rapid regrowth of shrubs following disturbance. The rapidity of regrowth of lucerne after cutting was related to the number of shoots coming from the root crown, rather than to the storage carbohydrate content (Leach 1968). Regrowth potential can be analysed in terms of plant shoot population dynamics (White 1979). Biomass increase by an individual shoot is limited, yet biomass increase by the plant is important to increase photosynthesis and nitrogen uptake while competing with other individuals. Increasing the number of shoots is an expedient way to increase biomass rapidly. The population of dormant buds which can potentially form active shoots is increased each year by the number of axillary buds produced each year by each active shoot. Thus, regrowth potential is higher in the sprouting shrub form, with a higher initial population of axillary buds, in contrast to the tree form.

The shrub form also maintains lower temperatures during the warm summer drought. With moderate annual precipitation the water can support only a relatively small leaf area index, and a period of the year with high temperature and little evaporative cooling occurs. With a small leaf area and an open canopy, a large fraction of the incoming solar irradiance is absorbed at the soil surface. This energy is dissipated, when no water is available for evaporation, by increasing long-wave irradiance and convection. Because the absorptance of the plant to infrared irradiance is greater than its absorptance to solar irradiance, the total amount of radiation absorbed by the plant can increase under these circumstances. An open canopy of scattered shrubs with a low leaf area index, rather than a continuous canopy, decreases the absorption of the outgoing infrared irradiance and is more effective in dissipating the absorbed radiation by convection. Increased shading also decreases the absorption of solar irradiance. In simulations with *C. greggii* and *L. caustica* using a constant foliage area index, changing the vertical distribution of the foliage area changed plant temperature and transpiration. As the canopy became shorter, leaf and soil surface temperatures increased 0.2-0.8°C at Echo Valley and decreased 0.1-2.1°C at Fundo Santa Laura. Transpiration was less with leaves concentrated at the top of the canopy than with leaves concentrated at the bottom or with leaves distributed in the canopy as they actually occur. Changing the leaf area from that which naturally occurred resulted in higher leaf temperature by 0.2-1.0°C and higher soil surface temperature by 0.2-1.9°C.

Stems absorb solar irradiance and convect part of this energy to the air. Although the interception and dissipation of irradiance by stems was indicated in simulations to be beneficial during the late spring and summer when canopy temperatures were above the optimum for photosynthesis, during the early spring interception of solar irradiance by stems decreased photosynthesis by 40%, transpiration by 15% and the photosynthesis:transpiration ratio by 40%.

8.3.2 LEAF AREA INDEX

The leaf area indices of the vegetation in mediterranean regions are commonly 1-2 m_{ℓ}^{2} m_{g}^{-2}, increasing to 3-4 as precipitation increases or the site water balance becomes more favourable (Cody and Mooney 1978). In simulations canopy photosynthesis increased with increasing leaf area index and reached a plateau for Californian and Chilean evergreen shrubs except *Adenostoma fasciculatum* (Miller and Stoner 1979). The highest canopy photosynthetic rates occurred in *Arctostaphylos glauca* with a leaf area index of 3-4. The highest rates occurred in *C. greggii* and *R. ovata* with leaf area indices of 2.0-2.5. But using a canopy of *A. fasciculatum*, the highest canopy photosynthetic rates occurred at a leaf area index of about 1.0. The leaf area indices measured for these species at Echo Valley were near the values associated with maximal photosynthetic rates. The overall leaf area index for Fundo Santa Laura (Mooney 1977a) was near the leaf area index for high production in a mixed stand of the three Chilean species simulated, *L. caustica*, *C. odorifera* and *S. gilliesii*, but the leaf area indices of the individual species were slightly below optimal. The simulations indicated that *A. fasciculatum* can survive in open canopies where the leaf area index is < 1 but cannot survive in closed canopies where the leaf area index is > 1 because it is intolerant of shading. *Arctostaphylos glauca* or *C. greggii* could occur in closed canopies. These conclusions are consistent with field observations of community structure. The three Chilean shrubs *L. caustica*, *C. odorifera* and *S. gilliesii* should be able to coexist and do at Fundo Santa Laura. The simulations indicate that the light-photosynthesis relations controlling shade tolerance are consistent with the leaf area index. Higher leaf area indices should be associated with increased shade tolerance. These results would be especially interesting to check in the mediterranean-type ecosystems of South Africa and Australia.

In these simulations transpirational water loss increased with foliage area. Soil evaporation decreased with foliage area and was 200-300 mm yr^{-1} in southern California and about 200 mm yr^{-1} in central Chile. Total site evapotranspiration increased with foliage area index in species which had high transpiration rates but remained constant in species which had moderate or low transpiration rates. In these latter species increased transpirational losses were compensated by decreased soil evaporation. Nitrogen release decreased as foliage area increased, because the foliage area increased transpiration which decreased the length of the decomposition season.

8.3.3 LEAF SIZE AND ORIENTATION

Leaves in mediterranean regions are characteristically small and steeply inclined (Walter 1973; Cody and Mooney 1978). Leaf width affects the potential for convectional and evaporational exchange of energy by affecting the resistance of the leaf air boundary layer to heat and water exchange (Gates 1962, 1965, 1968; Taylor 1975). The effect of increasing leaf widths on transpirational loss depends on the relative magnitude of the leaf resistance to water loss and the leaf air boundary layer resistance and on the humidity of the air. As leaf widths increase, leaf temperature may increase because of the increased resistance to heat and water vapour exchange. The increasing leaf temperature increases the vapour density gradient from leaf to air, tending to increase transpiration. Because increasing leaf width increases the resistance to the diffusion of water vapour from leaf to air, transpiration can decrease as leaf width increases under certain environmental conditions. The environmental conditions in the mediterranean regions are such that transpirational loss usually will increase if leaf widths increase. The increase in temperature and transpiration may influence photosynthesis directly, depending on the temperature and the temperature response curve, and indirectly via the effect on leaf water balance and stomatal closure. Mooney et al. (1974) pointed out that wide leaves might be beneficial in winter and narrow leaves might be beneficial in summer, depending on the prevailing temperatures and on the temperature optima for photosynthesis. Simulations indicated increased production and water-use efficiency with wider leaves in winter and with narrower leaves in summer (Miller and Mooney 1976). The overall effect of wider leaves was to decrease annual production and water-use efficiency. The simulations indicated that plants with narrow leaves should be more common in southern California than in central Chile (Miller and Mooney 1976); measurements of leaf size supported these predictions (Parsons 1973; Mooney et al. 1977). The narrower leaves in southern California compensate for the higher annual mean temperature, so the leaf temperatures in southern California and in central Chile should be more similar than the air temperatures. The predominance of broad-leaved species on the pole-facing slope in contrast to the narrow-leaved species found on the equator-facing slopes and on the ridgetops may result partly from the need for more effective convectional exchange on these exposures because of higher soil surface temperatures.

The potential effects of leaf width on transpiration, photosynthesis and water-use efficiency were simulated. Using parameters for *A. fasciculatum*, as leaf width decreased, photosynthesis increased and water-use efficiency increased. The advantage appeared to be with narrow leaves. With *C. greggii* as leaf width decreased transpiration increased; photosynthesis increased in February and May but not in September; the water-use efficiency decreased in May and September. The advantage of narrow leaves was not as marked as for *A. fasciculatum*. With increased leaf width in *C. odorifera*, transpiration decreased, photosynthesis was increased in July and November; and the water-use efficiency increased.

Narrow leaves appear advantageous for C. odorifera. As leaf width decreased in L. caustica transpiration decreased in March but increased in July and November; photosynthesis increased in November, and water-use efficiency increased in March and November and decreased in July. Wider leaves were more beneficial to L. caustica than to the other species. The results are consistent with the actual leaf sizes of the species.

Both leaf width and leaf inclination had less effect on production and water-use efficiency than did vertical position within the canopy, location with respect to the hot soil surface, and leaf duration-carbon balance considerations. Leaf temperatures were usually not critical, so modifications of width and inclination only constitute fine-tuning of a plant system which already survives the extremes of the physical environment. The effect of increased leaf clustering was similar to that of increased stem area and resulted in decreased photosynthesis and decreased transpiration, but had a variable effect on the water-use efficiency.

8.4 Influences of environment on canopy form

8.4.1 SEASONAL PATTERN OF TEMPERATURE AND WATER AVAILABILITY

The geographic distribution of evergreen and deciduous plant growth forms is related to the annual patterns of water availability and temperature via their influence on the plant carbon balance. Because growth requires higher temperatures than does photosynthesis, the winter temperatures of the mediterranean climate suppress growth more than photosynthesis (Fig. 2). At mid-elevations in southern California, leaves must grow in the spring and survive the summer drought to be available for the winter photosynthetic season (Miller 1979). This region has an annual mean temperature of 13.4°C, a range from 2-30°C and an annual precipitation of 475 mm. During the drought photosynthesis is reduced because of stomatal closure (Mooney and Dunn 1970a,b; Harrison 1971; Harrison et al. 1971; Dunn 1975; Poole and Miller 1975) and growth ceases

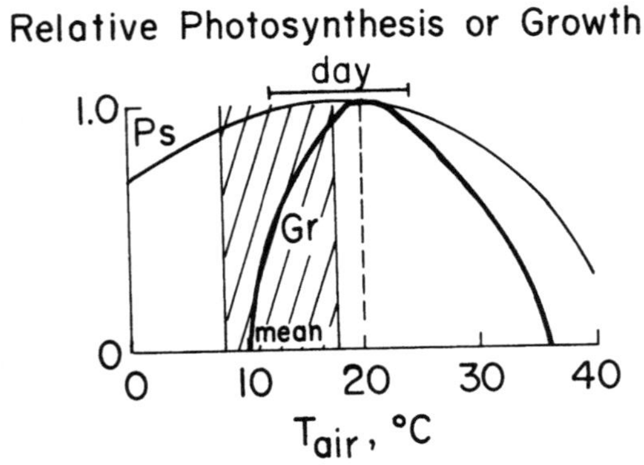

Fig. 2. Relative rates of photosynthesis (Ps) and growth (Gr) with respect to temperature. The annual range of daily mean temperature is indicated by the shaded area. The annual range of daytime temperature is indicated by the bar.

(Kummerow and Fishbeck 1977; Mooney et al. 1977). Leaf survival requires strengthening with structural tissue, small cell size, a thick cuticle and other xeromorphic features. At lower elevations along the coast, temperatures during the winter are sufficiently high to allow growth as well as photosynthesis. Summer-deciduous plants are possible and are favoured by the longer drought. With increased precipitation along the coast, evergreen forms are more favoured than drought-deciduous forms because of the decreased length of drought. At higher elevations the low winter temperatures reduce photosynthesis; the summer drought is shorter and growth is possible during the summer. Winter-deciduous forms are possible and advantageous. A similar pattern was observed in Israel (D Zohary, personal communication). The calculated elevational patterns of the annual production of evergreen and deciduous forms followed the measured elevational distributions of these forms (Parsons and Moldenke 1975; Fig. 3). Production by evergreens increased with elevation, while production by drought-deciduous shrubs decreased at mid-elevations because low temperatures suppressed leaf growth during the winter. The length of the active growing season was bounded by low temperature and soil drought and decreased from coast to mid-elevations. At all elevations growth of low plants is greater during the winter because the soil surface temperatures are higher than the air temperatures.

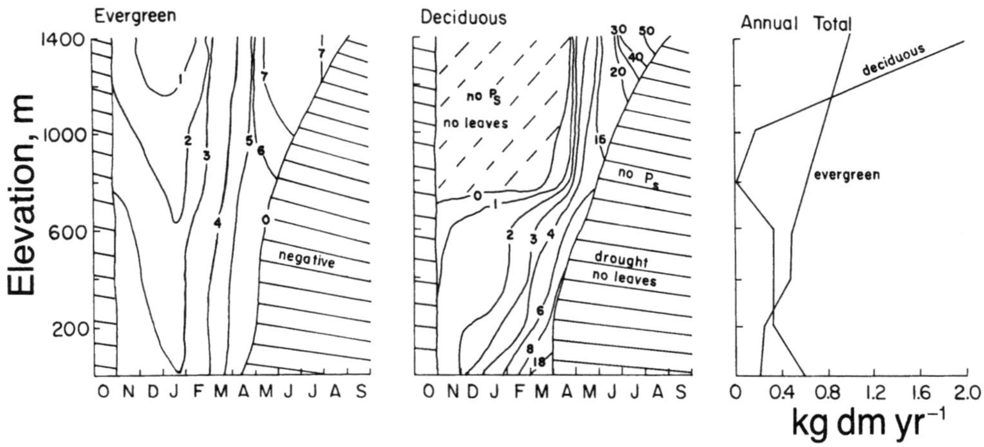

Fig. 3. Simulated dry-mass gain of evergreen and deciduous leaf species along an elevational gradient in southern California. Values given in the seasonal progression are grams dry mass per square metre of leaf per day for evergreen species and grams dry mass per plant per day for deciduous species. Values for the annual totals are kilograms dry mass per year. (From Miller et al. 1981)

The growing season would be lengthened if the temperature requirement for growth were decreased because growth would be possible during the winter. Although many species can grow at temperatures below 10°C (Billings and Mooney 1968; Savile 1972), apparently most species in the mediterranean-type climates cannot. The predominant species in chaparral grow when daily mean temperatures are above 10°C. Australian shrubs and species such as R. ovata require temperatures above 18.5°C for growth (Jacobson et al. 1981; Specht 1972a). Species in the matorral grow at temperatures below 10°C. The physiological adjustments required to increase growth at temperatures below 10°C may be incompatible with the drought and high temperatures of summer that are common in mediterranean-type climates.

The distribution of leaf forms is also related to the length of the summer drought. In southern California evergreen shrubs occur where the carbon cost of leaf maintenance through the summer drought is less than the carbon cost of new leaf growth. Deciduous shrubs occur where the carbon cost of leaf maintenance is greater than the cost of leaf recreation (Miller and Mooney 1976). The costs of maintenance and new leaf growth were equal with a drought of about 100 days. Evergreen forms should occur with a drought shorter than 100 days and deciduous forms with a drought longer than 100 days, which agrees with the distribution of plant forms and lengths of drought in the chaparral and coastal sage zones.

The leaf must also function until the carbon costs of its construction are recovered. The period required to recover the costs depends on the photosynthetic rates, specific leaf density, biochemical composition of the leaf, the amount of supportive stems and roots associated with the leaf and the time of year when leaf growth takes place. In simulations based on the data for these factors, A. fasciculatum required the shortest time to recover construction costs; R. ovata and L. caustica the longest (Miller and Stoner 1979; Fig. 4). A. fasciculatum should be favoured under conditions of periodic catastrophic mortality such as fire because of its relatively low costs of leaf construction. Leaves of A. fasciculatum, which were initiated between October and early April, recovered their construction costs by September. Leaves initiated in May and June did not recover their costs until the following spring. Of the species at Fundo Santa Laura, S. gilliesii had the shortest recovery time, which is appropriate for a drought-deciduous species. However, leaves initiated in December, if they remained active, should not recover their growth costs until April after the summer drought. Leaves initiated later in the spring become brown and dry in summer but persist until the following fall (Montenegro et al. 1979). The longest required leaf duration was in R. ovata. Leaves initiated in spring or summer did not recover their construction costs until almost a year later. The calculated long carbon recovery time is supported by the observed low frequency of branches which produce new growth each year (Kummerow and Fishbeck 1977), because a branch cannot grow each year wihout developing a carbon deficit.

Because photosynthetic rates decrease with shading, the required length of the recovery period increases with foliage area index. Thus, leaf area index, stem area index and leaf duration are interrelated. To coexist C. greggii, A. fasciculatum and R. ovata require leaf durations of at least 2.5-3.0 years to recover the costs of construction, but Arctostaphylos glauca requires only one year. Leaf durations of A. fasciculatum by itself should be at least 1.0-1.5 years. In matorral a leaf area index of 2.0 should be associated with leaf durations of one year in C. odorifera, 1.5 in L. caustica and 1.0 in S. gilliesii. Leaf durations calculated from Thrower and Bradbury (1977) were 1.8 years in A. glauca, 2.5 in C. greggii, 2.3 in R. ovata, 2.2 in C. odorifera, 0.4 in S. gilliesii and 1.8 in L. caustica.

Deciduous shrubs can support higher leaf area indices than evergreen shrubs because of the lower cost of producing leaves, about 1.3 g glucose g^{-1} dry mass for deciduous leaves versus 1.7 g glucose g^{-1} for evergreen leaves, and because of the higher photosynthetic rates in deciduous shrubs (Miller 1979). However, the deciduous form requires at least an 80-day period with temperatures and moisture favourable for

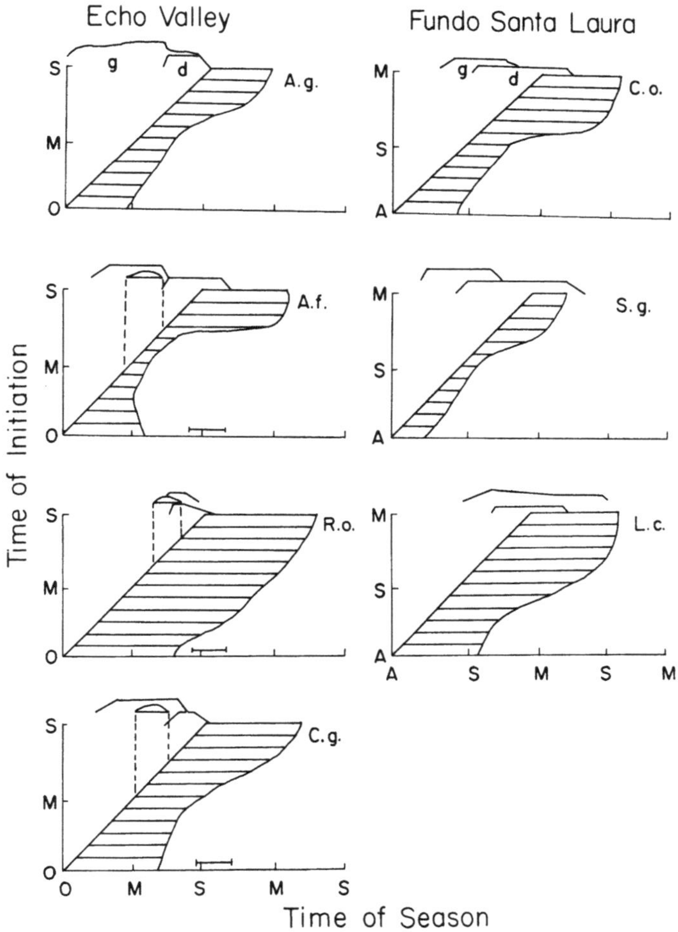

Fig. 4. Length of time (horizontal lines) required to regain the cost of constructing a new leaf and associated stem and root biomass when initiated at different times of the year. Months (October, March, September) are indicated. The observed period of leaf growth (g) and leaf death (d) are given for each species. The period of greatest growth is indicated by the dashed lines. The period of most frequent fires for California is indicated by the heavy horizontal line. Species at Echo Valley are Arctostaphylos glauca (A.g.), Adenostoma fasciculatum (A.f.), Rhus ovata (R.o.) and Ceanothus greggii (C.g.); and at Fundo Santa Laura Colliguaya odorifera (C.o.), Satureja gilliesii (S.g.) and Lithraea caustica (L.c.). (From Miller and Stoner 1979; Miller 1981)

photosynthesis to regain the carbon costs of leaf growth; such a period is usually not available in the mid-elevations of the mediterranean shrub region in southern California.

The long leaf duration affects the availability of nitrogen. Evergreen leaves require strengthening by ligneous compounds and synthesis of herbivore defence compounds. Both kinds of compounds can reduce the mineralization rate and the quality of the substrate for decomposition.

Selection of evergreen leaves by climate favours leaves of about one-year duration but does not necessarily favour leaves with more than one-year duration. The length of the period needed to recover the carbon costs of leaf production depends on climatic and plant factors and may require leaves with longer than one-year duration. The evergreen leaves should have structural modifications to maintain rigidity under water stress and reduce the intensity of water stress (Beadle 1966; Cutler et al. 1977). According to heat transfer theory, such leaves should be rigid, narrow, steeply inclined and light coloured to reduce cellular distortion, plant temperatures and transpiration rates. Small cell size is advantageous (Cutler et al. 1977). Leaf longevities may increase with nitrogen limitation (Monk 1966). Although evergreenness caused by nitrogen limitation does not require the adaptations to survive drought (Beadle 1966), the general biochemical composition of nitrogen-limited leaves may be similar to that of water-limited leaves, resulting in similar sclerophyll indices. The sclerophyll index is based on lignin, cellulose and nitrogen content (Loveless 1961). Water limitation creates a direct need for sclerophyllous tissue, while the nitrogen or phosphorus limitation may reduce protein formation due to low nutrient availability, which may increase cellulose and lignin formation because of excess carbohydrate (Loveless 1961; Beadle 1966; Steubing and Alberdi 1973).

Of the five mediterranean regions of the world, central Chile is relatively nutrient-rich and its vegetation is climatically selected; leaf turnover rates are relatively high. In southern California and southern Europe soils are less nutrient-rich, but climate is still a strong agent; leaf turnover rates are lower than in central Chile but are still 1-2 years. In the mediterranean regions of Australia and South Africa where old, nutrient-poor soils occur and leaf turnover is lower than in California (Kruger 1979; Specht 1979; Miller et al., unpublished) the climatic effect is reinforced by nutrient limitations.

8.4.2 AIR AND SOIL TEMPERATURES

In southern California the mean annual air temperatures were 15.9°C in the coastal sage, 13.4°C in the chaparral and 11.3°C in the montane forest. In matorral the mean annual temperature was 12.1°C. During the year daily mean temperatures were 8-20°C at Echo Valley and 7.5-17°C at Fundo Santa Laura. The mean annual soil temperatures at 0.02 m depth on the ridgetops were 18.8°C at Echo Valley and 19.8°C at Fundo Santa Laura but reached 55°C from April to August during a dry year and 38°C during a wet year at Echo Valley and 44°C at Fundo Santa Laura. Temperatures at 0.02 m were usually above freezing at both sites and were usually above 10°C during the day throughout the year. The mean annual soil temperatures were higher than the mean annual air temperatures. At 0.32 m depth the mean annual soil temperature was greater than air temperature by 3.4°C at Echo Valley and 3.8°C at Fundo Santa Laura. The difference was even greater in the surface soil (Fig. 5) and decreased with vegetative cover.

Heat sums accumulated as degree days above 10°C after 1 October were higher at the soil surface than in the air. Heat sums of 1000 degree days occurred in early August in the montane forest, in early July in the chaparral, between February and July in the coastal sage and in the equivalent of mid-September at Fundo Santa Laura. The year-to-year variability in the coastal sage vegetation zone in southern California was large because the winter temperatures were normally close to 10°C. Soil surface heat sums reached 1000 degree days two months earlier than air temperature heat sums at the bottom of the canopy at Echo Valley, three months earlier than air temperature heat sums at Echo Valley and 3-4.5 months earlier at Fundo Santa Laura. Thus, the heat environment near the soil surface in chaparral is similar to that of the air at the coast. Leaf growth is possible during the winter only near the soil surface in chaparral but is possible higher in the canopy at the coast. In the mid-elevation chaparral an evergreen overstorey is necessary because of low air temperature, but a deciduous understorey is possible because of higher temperatures near the soil. This combination may only be possible in sites without nutrient constraints. Specht (1979) has pointed out that in the mediterranean regions of the world the overstorey trees were evergreen because of water stress, while the ground stratum may be evergreen or seasonally herbaceous depending on the fertility of the soil.

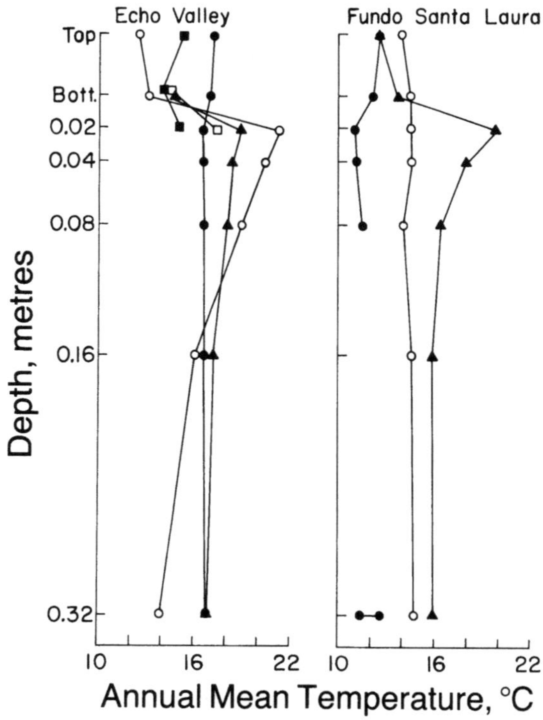

Fig. 5. Vertical profiles of annual mean air temperature in 1972 at the top and bottom of the canopy and soil temperatures at five soil depths at Echo Valley and Fundo Santa Laura on the pole-facing slope (●), equator-facing slope (○) and ridgetop (△). Data for 1979 (D K Poole, unpublished) is shown as pole-facing slope (■) and equator-facing slope (□). (From Miller et al. 1981)

8.4.3 SOIL DEPTH AND WATER AVAILABILITY

Since the period of high transpiration, photosynthesis and growth is limited by the decline of soil moisture in early summer, its duration depends on the time required for evapotranspiration to remove all available water from the soil once temperatures become favourable. The length of time depends on the amount of water held in the soil at the beginning of the growing season, the precipitation during the period, and the rate of evapotranspiration, which is related to the potential evapotranspiration. Thus, the length of the growing season (λ) can be expressed as

$$\lambda = [\Theta - \Theta_{min})z(1 - f) + P]/[(PET)(\phi)] \tag{2}$$

where Θ is soil water content at the beginning of the period (g $m\ell^{-1}$), z is soil depth, f is the fraction of rocks in the profile, P is precipitation in this period, PET is potential evapotranspiration and ϕ is the ratio of actual water removal to potential evapotranspiration. With 17% of the annual precipitation occurring on the average in the growing season in southern California and about 9% occurring in central Chile, the growing season should lengthen about 9 days per 100 mm increase in annual precipitation in southern California and about 4.5 days per 100 mm increase in central Chile. The period of active plant growth is more clearly defined at Echo Valley than at Fundo Santa Laura because of the greater diversity in the patterns of growth of the Chilean shrubs. In simulations the growth rate of the evergreen shrubs was nearly constant during the growing season, about 4 g dry mass m^{-2} day^{-1} at Echo Valley and 1.5-2.5 g dry mass m^{-2} day^{-1} at Fundo Santa Laura. Because only a small fraction of the annual precipitation occurs during the growing season, annual production is not strongly correlated with annual precipitation. The correlation appears lower at Echo Valley than at Fundo Santa Laura. According to the simulations and to the interrelations between precipitation and growing season length, growth should increase about 34 g dry mass m^{-2} yr^{-1} 100 mm^{-1} in annual precipitation at Echo Valley and about 9 g dry mass m^{-2} yr^{-1} 100 mm^{-1} at Fundo Santa Laura. Similar relations are expected in the other mediterranean-type ecosystems of the Northern and Southern Hemispheres. A particular year may deviate markedly from this pattern because most of the annual precipitation is received in a few storms which can occur at any time during the winter, including during the growing season. The length of the growing season should decrease about 12 days per 10% increase in soil rockiness and lengthen about 10 days per 0.1 m increase in soil depth. The length of the soil drought should change conversely to the length of the growing season. In general the soil drought should occur at nearly the same time of year regardless of annual precipitation, vegetative cover and topographic position of the vegetation, a generalization which is supported by data from Echo Valley in southern California (Poole and Miller 1975; Poole and Miller 1981; Fig. 6).

Because precipitation increases with elevation, interrelations between elevation, precipitation, soil depth, length of the drought and plant growth form can be approximated (Fig. 7). The interrelations indicate that with increasing precipitation or elevation evergreen shrubs can be expected on shallower soils. With steep precipitation-elevation relations such as occur in South Australia in the Mount Lofty Range and in South Africa, evergreen shrubs can be expected on a wide range of soil depths and types (Schwerdtfeger 1976; Kruger 1979). The lower spring

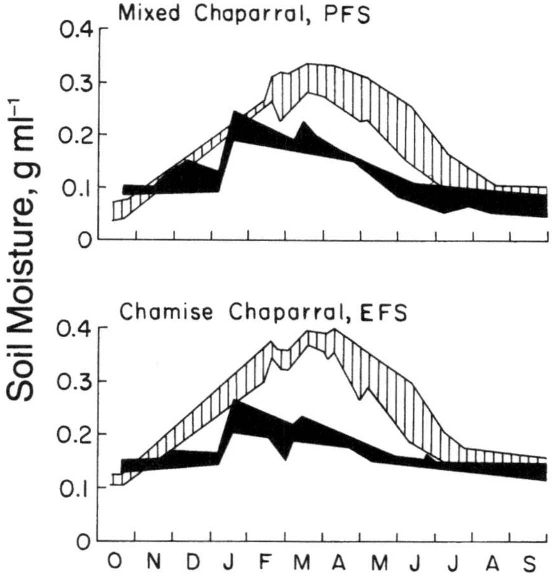

Fig. 6. Annual progression of soil moisture between 0.3 m and 1.0 m depths in chamise and mixed chaparral in two years, one with 666 mm yr^{-1} precipitation (hatched) and one with 331 mm yr^{-1} precipitation (solid). Soil moisture contents ($m\ell\ m\ell^{-1}$) was measured in 150 mm depth intervals. (Miller and Poole, unpublished)

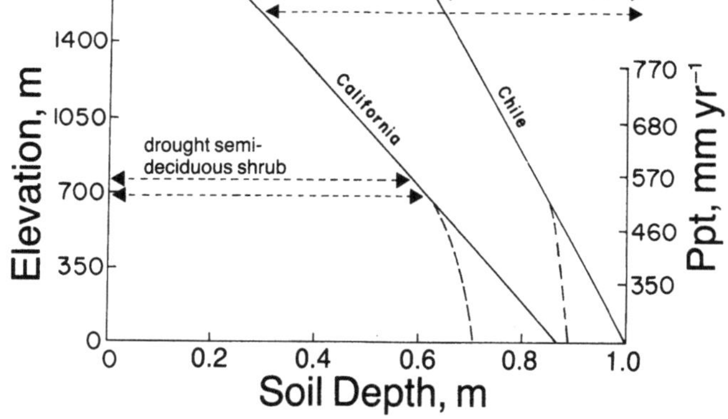

Fig. 7. Hypothetical relations among precipitation, elevation, soil depth and on the distribution of evergreen and drought semi-deciduous shrubs, via their influence of the length of the drought.

precipitation in the Southern Hemisphere should increase the distribution of drought semi-deciduous shrubs, which is apparent in central Chile compared to southern California. Specht (1972b) presented a similar equation for evaluating the water held in the soil and the possible evapotranspiration from the community.

8.4.4 WATER AVAILABILITY AND LEAF AREA

The transpiration efficiency (calculated as transpiration divided by precipitation) in mediterranean climates is directly related to the foliage area index and shading of the soil. With increasing foliage area indices, shading decreases soil evaporation while transpiration and interception losses increase. Because interception losses are less than soil evaporation losses, the vegetation becomes more efficient at using water as foliage area index increases from about 0 to 2 m_ℓ^2 m_g^{-2}. The foliage area index reaches a maximum at which photosynthesis just sustains tissue maintenance and replacement costs. At these high foliage areas transpiration reaches a plateau even though precipitation increases, and in this range of precipitation transpiration efficiency decreases. The maximum foliage area which can be maintained is constrained by the amount of photosynthate fixed, which is related to the water transpired, the ratio of photosynthesis to transpiration, the costs of replacing leaves and stems, the turnover rates of leaves and stems and the maintenance costs. The steady state biomass (B) can be expressed in terms of the transpiration efficiency, the water-use efficiency, the respiratory costs and longevity by

$$B = (Ppt)\ (Tr/Ppt)\ (P_s/Tr)/(r_m + r_g\tau) \qquad (3)$$

where Ppt is annual precipitation, Tr is transpiration, P_s is photosynthesis, r_m is maintenance respiration rate, r_g is growth respiration and τ is the turnover rate.

With relatively low precipitation (< 400 mm yr^{-1}) only low foliage areas can be maintained (< 2.0 m^2 m^{-2}), soil evaporation is high and transpiration efficiency is low (Fig. 8). In simulations as the foliage area index increased from 0 to 2 m^2 mg^{-2}, soil evaporation decreased to two-thirds. A similar relation between transpiration efficiencies and annual precipitation has been noted for grasslands and other semi-arid lands (Fischer and Turner 1978; Sims and Singh 1978). Based on field observation in southern California and theory, an annual precipitation of 400-450 mm yr^{-1} is required to maintain a foliage area index of about two (Poole and Miller 1981). Thus, with precipitation below about 400 mm yr^{-1} transpiration efficiency can be expected to decrease rapidly with decreasing precipitation. Above 450 mm yr^{-1} precipitation the foliage area which can be maintained is limited by the photosynthesis-light responses of the species and leaf turnover during the drought. Transpiration is limited by foliage area, and transpiration efficiency decreases as precipitation increases above 450 mm yr^{-1}. The lower transpiration efficiency is accompanied by increased drainage. The low water-capture efficiency of chamise chaparral at low elevations and on drier sites and the higher efficiency of mixed chaparral at mid-elevations and on more mesic sites is consistent with the precipitation constraints on transpiration efficiency. Fundo Santa Laura receives enough precipitation to place it above the high transpiration efficiencies. Thus, the matorral stands at Fundo Santa Laura have relatively low

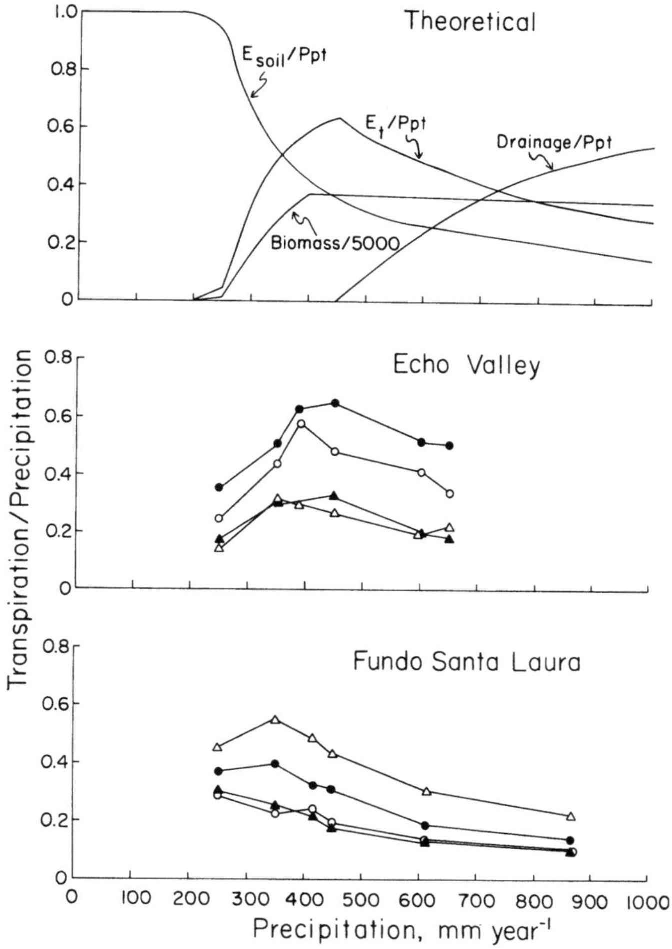

Fig. 8. Relations between transpiration efficiency and annual precipitation for Californian and Chilean shrubs at their steady-state foliage area indices for different precipitation amounts. Theoretical curves are calculated from the general relationships described in the text. E_{soil}/Ppt is the ratio of annual soil evaporation to annual precipitation, E_t/Ppt is transpiration to precipitation, Drainage/Ppt is the water drained from the soil in relation to precipitation. Biomass is given in g m^{-2} divided by 5000. Species at Echo Valley are *Adenostoma fasciculatum* (▲), *Arctostaphylos glauca* (●), *Ceanothus greggii* (○) and *Rhus ovata* (△); and at Fundo Santa Laura *Colliguaya odorifera* (●), *Lithraea caustica* (△), *Satureja gilliesii* (▲) and *Trevoa trinervis* (○) (From Miller 1981).

transpiration efficiencies because of relatively high precipitation and low net radiation in winter. Chamise chaparral has a relatively low foliage area index, high soil evaporation and relatively low transpiration efficiency because of its shade intolerance. The low transpiration efficiency of matorral is caused by the low foliage area. Similar constraints are expected to reduce transpiration efficiences in other

Southern Hemisphere mediterranean climate regions compared to Northern Hemisphere regions. In contrast to the evergreen shrubs, the transpiration efficiency of the drought semi-deciduous shrubs decreases with increasing precipitation. With respect to elevation, the maximum transpiration efficiency occurs at elevations below those at which the maximum abundance of the species occurred, supporting the generality for mediterranean-type ecosystems that water is the limiting factor at the lower elevational limit of a species.

Year-to-year variations in precipitation affect transpiration efficiencies because leaf growth occurs after the period of receipt of the precipitation and varies less than precipitation. The simulated response of production to precipitation was flatter than the usual production-precipitation relations in arid regions (Chang 1968; Whittaker 1975) and was flatter at Fundo Santa Laura than at Echo Valley. The flatness of the response was due to the use of constant initial conditions for foliage area index and to the low fraction of annual precipitation received during the growing season.

The maximum leaf area index which could be maintained with different annual precipitation amounts was estimated by simulation. With a given annual precipitation the initial leaf area index was increased until the final leaf area index after a simulated year was the same as the initial leaf area index. These steady-state leaf area indices increased as precipitation increased to about 400 mm yr^{-1} with the Californian shrubs and to 350 mm yr^{-1} with the Chilean shrubs. The calculated steady-state leaf area indices were similar to those measured in the field (Miller 1981).

Annual precipitation and nitrogen taken up by the vegetation should be interrelated; such a relationship can be derived for the steady state situation. The amount of nitrogen taken up by the vegetation limits the maximum biomass attainable. In the steady state nitrogen taken up must equal that lost as plant parts are shed.

$$B = (N_{up})/[(N_d)\tau] \qquad (4)$$

where B is the biomass, N_{up} is the nitrogen taken up, and N_d is the nitrogen content of plant parts at death. The equations relating the maximum biomass attainable to precipitation (Equation 3) and to nitrogen uptake (Equation 4) can be combined to calculate the nitrogen uptake expected with different annual precipitation amounts when the biomass is at a maximum, given the plant turnover rates and the nitrogen contents of the shed parts (Fig. 9). The equations indicate that the steady-state biomass should increase by 400-500 g m^{-2} for every 100 mm yr^{-1} increase in annual precipitation and by about 400 g m^{-2} for every additional g N m^{-2} yr^{-1} taken up. Increasing leaf longevities more than two years has smaller effects on the steady-state biomass than increasing leaf longevities from one to two years. The nitrogen-uptake demands follow the precipitation and water efficiency relationships. Within a precipitation regime leaf longevity should increase in accordance with nitrogen limitations because turnover rate is more closely related to nitrogen uptake than to precipitation.

Other plant characteristics correlate with the precipitation gradient. With low precipitation (< 400 mm yr^{-1}) and low foliage area index, rapid use of water by plants is advantageous because otherwise the water evaporates. Transpiration and transpiration efficiency increase with high leaf conductances, steeply inclined leaves, narrow leaves, high leaf reflectances and high leaf area:dry mass ratios, which increase soil

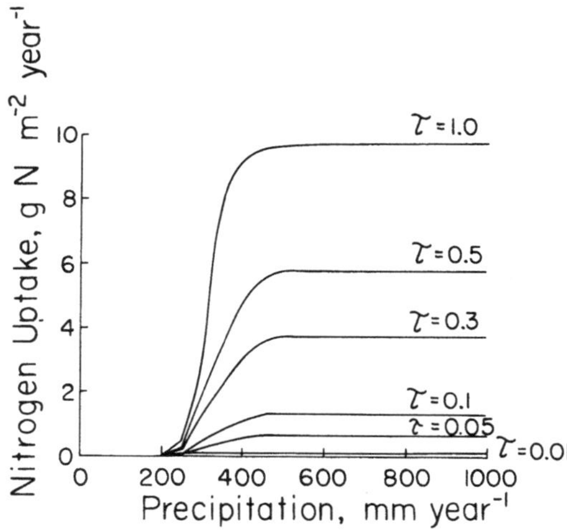

Fig. 9. Relations between annual precipitation and nitrogen uptake for vegetation with different turnover rates of leaves and stems (τ, g g^{-1} yr^{-1}). The lines indicate combinations giving full use of both nitrogen and precipitation.

shading. Because the soil drought is lengthened by the soil evaporation and high transpiration rate, drought-deciduous species are favoured. The advantage of steep leaf inclination is balanced against the advantage of shading the soil surface and reducing soil evaporation. With extremely low annual precipitation (< 250 mm), the evaporation loss is reduced by storing the water in the plant rather than in the soil and by using the water slowly, as occurs in succulents. At higher precipitation levels (> 450 mm yr^{-1}) water is lost by drainage; transpiration is increased with higher leaf conductances; the length of the drought is short. Leaf width, inclination and colour can be more variable without affecting water loss. At intermediate precipitation levels (400-550 mm yr^{-1}) the composition of the vegetation changes the duration of the soil drought by changes in the abundance of species with different leaf conductances and different leaf area indices. The transpiration efficiency is controlled by the vegetation composition and can be relatively high. Mixed chaparral usually occurs in the intermediate precipitation range and has a high water-capture efficiency and moderate leaf conductances.

The expected trends in water relations occur in chaparral and matorral species (Table 1). Maximum leaf conductances of Californian species arranged from highest (about 10.0 mm s^{-1}) to lowest (about 2.0 mm s^{-1}) were: *Artemisia californica*, *Salvia apiana*, *Arctostaphylos glauca*, *C. greggii*, *Adenostoma fasciculatum*, *Quercus dumosa* and *R. ovata*. The sequence was similar for the lowest xylem pressure potentials observed in a species, from a low of < -6.5 MPa to a high of -2.0 MPa, and for the duration of low xylem pressure potentials for a species. The sequence of species was one of increasing hydrostability, i.e., decreasing variability in pressure potentials through the year (Walter 1973). Osmotic concentrations at turgidity were similarly moderate, ranging from -2.0 MPa to -3.0 MPa. The depth of rooting probably also increases in the same

Table 1. Summary of water use characteristics of different chaparral and matorral shrubs. (From Miller et al. 1981)

Life history type	Drought semi-deciduous			ECHO VALLEY Evergreen				
	Artemisia californica	*Salvia apiana*	*Eriogonum fasciculatum*	*Arctostaphylos glauca*	*Ceanothus greggii*	*Adenostoma fasciculatum*	*Quercus dumosa*	*Rhus ovata*
Maximum leaf conductance (mm s^{-1})	10.0	10.0	13.0	4.5	5.0	4.0	3.0	2.5
Minimum xylem potential during year (MPa)	-6.5	-4.0	-6.5	-6.5	-6.5	-6.0	-5.0	-2.5
Xylem potential at zero conductance (MPa)	6.5	ND	ND	-6.0	-6.0	-5.0	ND	-2.0
Osmotic potential at turgidity (MPa)	ND	ND	ND	-3.0	-2.8	ND	ND	-3.0
Annual transpiration (mm yr^{-1} m^{-2} leaf)	ND	ND	ND	280-290	145-195	150-190	ND	127
Rooting depth				SHALLOW			DEEP	

Life history type	Drought semi-deciduous		FUNDO SANTA LAURA Evergreen			
	Satureja gilliesii	*Trevoa trinervis*	*Colliguaya odorifera*	*Cryptocarya alba*	*Quillaja saponaria*	*Lithraea caustica*
Maximum leaf conductance (mm s^{-1})	5.0	4.0	3.0	2.5	3.0	1.5
Minimum xylem potential during year (MPa)	-4.5	-6.0	-6.0	-6.0	-3.0	-3.5
Xylem potential at minimum conductance (MPa)	ND	ND	-4.5	-3.5	-3.5	-3.0
Osmotic potential turgidity (MPa)	ND	ND	-3.0	-3.5	-2.5	-2.5 at
Annual transpiration (mm yr^{-1} m^{-2} leaf)	60	60	130	ND	ND	200
Rooting depth		SHALLOW			DEEP	

sequence. A. californica, S. apiana, C. greggii and A. glauca are recognized as shallow-rooted, while Q. dumosa and R. ovata are thought to be deep-rooted (Hellmers et al. 1955; Kummerow et al. 1981). The sequences of maximum conductance, annual variations in conductance, hydrostability and rooting depths also segregate obligate seeders and sprouting shrubs. The sequence of species occurs along a gradient of increasing annual precipitation. Different patterns of reproduction may reinforce the patterns of water-use characteristics. Seedlings and resprouts have different water environments when young. Schlesinger and Gill (1980) showed lower water potentials during the summer in seedlings than in mature individuals of Ceanothus megacarpus. Radosevich et al. (1977) indicated that resprouts of A. fasciculatum generally had higher water potentials than the unburnt control vegetation. The higher soil moisture under Banksia ornata (Specht 1957b) and decreased water use with reduced stem densities of Acacia aneura (Pressland 1976) appear to be consistent with the expected interaction between leaf area, leaf conductance and water loss.

The Chilean species, arranged by maximum leaf conductance from highest (about 5.0 mm s^{-1}) to lowest (about 1.5 mm s^{-1}), are S. gilliesii, T. trinervis, C. odorifera, Q. saponaria, C. alba and L. caustica. The sequence was similar for the lowest xylem pressure potentials measured but was not as consistent as with the Californian species. Root depth may be similarly arrayed (Avila et al. 1975; Giliberto and Estay 1978) in the Chilean species. The sequence is not as clearly related to reproductive patterns as with the Californian shrubs.

Specht (1972a,b) suggested that evergreen species in a mediterranean climate have a pattern of moderate water use throughout the year so soil moisture is not depleted until just prior to the onset of the wet season following the summer drought. A pattern of moderate water use would allow sufficient water throughout the drought period for photosynthesis to offset maintenance respiration and allow for growth. Specht's suggestion is consistent with the pattern of water use on the equator-facing slope at Echo Valley (Ng and Miller 1980) which was dominated by a single species, A. fasciculatum, and with the pattern of water use in Q. dumosa and R. ovata and in the Chilean shrubs. R. ovata also has a high temperature requirement for growth similar to Australian shrubs (Specht 1969a,b). The Chilean shrubs appear to have low temperature requirements for growth (Jacobson et al. 1981). But Specht's suggestion is not consistent with the pattern of soil moisture on the pole-facing slope which is dominated by several species including A. glauca and C. greggii, which are also evergreen shrubs. Thus, the patterns of water use must be viewed in the context of various physiological characteristics and the floral history of the area. The Australian mediterranean vegetation has no species from cool, temperate regions, but such species are present in the chaparral and matorral because of the migrations of the flora during the Pleistocene (Axelrod and Raven 1978; Solbrig et al. 1977). The need to conserve water is more important for species with high temperature requirements for growth.

The photosynthetic efficiency, calculated as photosynthesis:solar irradiance, will be small relative to the transpiration efficiency and nitrogen-uptake efficiency because much of the incoming solar irradiance will be lost as reflection, infrared irradiance and convection, processes over which the plant has little control. Transpiration occurs concomitantly with CO_2 diffusion into the plant. The water-use efficiency will be small because of the low concentration of CO_2 relative to the amount of water vapour in the air and because of the higher resistances in the path of carbon diffusion into the leaf than in the path of water vapour diffusion out of the leaf. Consequently, the highest rates of photosynthetic efficiency are about 5%, and in most natural communities are 2%.

Because southern California and central Chile differed in the seasonal progression of solar irradiance even though the annual totals were similar (Miller et al. 1977; Miller and Hajek 1981), the effects of changing solar irradiance were simulated (Miller 1981). Increasing irradiance increased net photosynthesis, growth and carbon balance. The photosynthetic efficiency decreased with increasing solar irradiance because the foliage area index was constrained by the constant initial foliage area indices and by the availability of water.

Nitrogen availability is expected to increase with the community turnover rate (Stoner et al. 1978; Miller and Stoner 1979). Because plant turnover rates and mineralization are interdependent, the nitrogen-use efficiency may be similar over a range of nitrogen availabilities. If the nitrogen-use efficiency is not high, unabsorbed nitrogen can be lost from the soil by leaching. The ecosystem can develop a relatively high nitrogen-use efficiency both by the vegetation developing efficient nitrogen-capturing mechanisms and by the loss of nitrogen through leaching. Because the nitrogen-use efficiency depends mainly on the behaviour of the vegetation-soil system, it contrasts with the transpiration efficiency which varies with the physical environment. Nitrogen additions should always increase growth regardless of the type of vegetation they are applied to, unless some other factor is strongly limiting. Partial evidence for this is the fact that fertilization increased growth in semi-arid grasslands where water is generally limiting (Specht 1963, 1972a; van Keulen 1975; F W T Penning de Vries, personal communication 1977).

The availability of inorganic nutrients can affect water use. When Specht (1963) added phosphorus to heath in Australia, the increased growth increased transpiration, but eventually the Banksia in the heath was killed by drought which could have been made more severe by the increased growth. Water availability and nitrogen interactions were simulated. A gram of nitrogen taken up and incorporated into above-ground tissue yields an increment of about 166 g dry mass g^{-1} N. Leaf area is increased by about 0.004 m^2 g^{-1} dry mass, although some of the new biomass is contained in stems. This increment of leaf area results in an increase in transpiration of about 670 g day^{-1} g^{-1} N. Allowing for the dry mass and water costs involved in the growth of leaf area, the net increase in transpiration is about 330 g day^{-1} g^{-1} N. The increased transpiration decreases the length of the season in which soil moisture is available. The length of the growing season changes with transpiration according to the amount of water available divided by the square of the daily evapotranspiration rate. Thus, the reduction in the growing season because of added nitrogen would be about 0.33 x 38 or 12 days per gram nitrogen taken up by the plants in southern California and by about 10 days per gram nitrogen taken up in central Chile. This shortening of the growing season should reduce growth by about 48 g dry mass m^{-2} yr^{-1} in southern California and by 20 g dry mass m^{-2} yr^{-1} in central Chile, while the increased nitrogen should increase growth by about 166 g dry mass. Thus, net growth should increase about 118 g dry mass g^{-1} N taken up in California and 146 g dry mass g^{-1} N taken up in Chile, even though the growth period is limited by temperature and water. Transpiration efficiency should increase with fertilization because increasing the leaf area growth rate should increase shading and decrease evaporation from the soil.

8.5 Conclusions regarding the relative importance of climatic and plant factors in affecting canopy form

The factors recognized here as important in affecting the use of water, light, energy and nitrogen indicate that the transpiration: precipitation ratio is largely controlled by the physical environment, although several plant properties are also important, while photosynthetic efficiency and nitrogen-uptake efficiency are largely controlled by plant properties. The plant properties most influential in affecting resource use and resource-use efficiency are expected to be most strongly selected for, if the use of resources is important in species survival.

The transpiration:precipitation ratio is mainly determined by the annual precipitation, seasonal distribution of precipitation, solar or net total irradiance in winter and soil depth. This efficiency will be affected to a lesser degree by the air temperature, vapour pressure and wind, which affect potential and actual evapotranspiration and the water-use efficiency. Transpiration efficiency is also influenced by the successional stage of the vegetation, the rapidity of shrub regrowth and by leaf conductance. Populations of species with higher maximum leaf conductances increase as soil evaporation becomes increasingly important and accelerates the loss of soil moisture, shortens the growing period and increases the length of drought. Decreasing the minimum water potential which can be endured and increasing survival during drought increases the period during which stomates can be open and photosynthesis is possible. Survival during drought can be increased by smaller cell size and cell wall strengthening but at the cost of increased resistance to carbon dioxide diffusion. Concentrating roots near the soil surface should increase the transpiration efficiency, especially if soil evaporation is important, but the rapid, sporadic drying of the surface soil may cause high root turnover rates resulting in high carbon and nitrogen costs to the plant. When biomasses are low, increasing biomass can increase transpiration efficiency. Biomass can increase with decreased turnover rates of leaves, stems and roots, increased shade tolerance, increased stem area and increased leaf area. Increased shade tolerance increases the maximal attainable leaf area indices and increases the length of the drought.

Photosynthetic efficiency is mainly affected by height in the canopy, length of the photosynthetic season, leaf and stem area indices, water use efficiency and maximum photosynthetic rate. The water-use efficiency and maximum photosynthetic rates are affected by the cellular and wall resistances to CO_2 diffusion.

Nitrogen-uptake efficiency is determined by both nitrogen availability and nitrogen loss. Nitrogen availability is affected by soil temperature, soil water and by substrate quality which decreases in the case of evergreen leaves. Both soil temperature and soil water are affected by the vegetation cover. Nitrogen uptake is affected by the total root length or surface area and by the distribution of roots through the soil profile. Root length is affected by the root extension per unit biomass and by the root mass. The root mass is influenced by the carbon available for root growth.

Ecological generalities are difficult to test. A set of simulation experiments (Miller, unpublished) indicates that the range of natural variation in an ecosystem may make it impossible to completely disprove reasonable generalities. An implicit assumption often made in community theory is that competition and niche specialization among the species in a community results in optimal or maximal resource use by the community (Odum and Pinkerton 1955; Lewontin 1969; Cohen 1970; MacArthur 1970, 1972; Rosenzweig 1974; Orians 1975). The assumption of optimal resource use asserts that the natural mix of species in a community produces a higher resource use than any other mix of the same species. In simulation experiments for chaparral and matorral using root distributions measured at Echo Valley and Fundo Santa Laura, the resource use by mixed species communities was always less than resource use by single species communities (Miller et al. 1978). However, in earlier simulations using the distribution of roots in the soil given by Hellmers et al. (1955) for San Dimas, resource use in the mixed-species community was greater. Factors other than root distributions also allow alternate conclusions. Simulations using older measurements of root extension per unit dry mass indicated that Arctostaphylos glauca would use resources more fully than Adenostoma fasciculatum or than the two together. Later simulations, which used newer data on root extension, indicated that A. fasciculatum could use resources more fully than A. glauca or than a mixture of the two. The range of variation encountered in the natural system indicates caution in accepting or rejecting ecological theories.

The existence of many plant properties which increase resource use and resource-use efficiency indicates that the species are not merely surviving. Species show climatic adaptations which increase resource use, but it cannot be said that the mature mediterranean shrub vegetation uses all resources fully. This ecological generality may not be true in mature mediterranean shrub vegetation because the generality is false or because resource use is more important in earlier successional stages or during the total successional sequence after fire.

The resource use and resource-use efficiency in mediterranean climates is environmentally constrained by interactions and feedbacks between several environmental and vegetational properties. Low irradiance in winter rather than plant or soil properties limits the use of water and light in central Chile. Even with climatic limitations, the vegetation appears nutrient-limited because the turnover rates of the vegetation are adjusted to the rates at which nutrients become available. Leaf strengthening is necessary if the leaves are evergreen and if they must survive the summer drought. The relatively high lignin and cellulose contents of evergreen leaves is related to the low nitrogen content of the leaves and may not be due to low nitrogen availability but to dilution of the cytoplasm by supporting materials. The low nitrogen content of leaves increases the carbon:nitrogen ratio of the litter and decreases the rates of nitrogen release by litter decomposition. This low rate of release in turn favours evergreen leaves and low nitrogen contents. Infertile parent materials and leaching losses may also contribute to low nitrogen availability in the soil. Thus, the sclerophyllous evergreen form is reinforced by low available nutrients. The summer drought also may lead to narrow, steeply inclined leaves that are closely appressed or are clustered in order to reduce leaf temperatures in spring and summer and decrease transpiration. The decreased water loss rate prolongs the period of available soil moisture, the growth period and the photosynthetic period. Growth at the top of the canopy leads to increased ratios of photosynthesis: transpiration and photosynthesis:solar irradiance which creates a more favourable carbon balance for a plant with leaves at the top of the canopy. Physiological and morphological compensations result in the general similarity of vegetation form in spite of dissimilar phylogenetic histories.

Acknowledgements

This paper is largely based on research sponsored by the National Science Foundation on grants DEB 75-19491 and DEB 77-13944. The preparation of the paper was also supported on these grants. I thank Ms Marti Poole and Ms Patsy Miller for their editorial criticism and improvements and Ms Beth Sigren for typing the several revisions of the manuscript. Many others have helped with the field work.

References

AVILA G, ALJARO M, ARAYA S, MONTENEGRO G, KUMMEROW J (1975) The seasonal cambium activity of Chilean and Californian shrubs. American Journal of Botany 62: 473-478

AXELROD DI, RAVEN PH (1978) Late Cretaceous and Tertiary vegetation history of Africa. In: WERGER MJA (ed) Biogeography and ecology of southern Africa. Junk, The Hague, pp 77-130

BEADLE NCW (1966) Soil phosphate and its role in molding segments of the Australian flora and vegetation, with species reference to xeromorphy and sclerophylly. Ecology 47: 992-1007

BILLINGS WD, MOONEY HA (1968) The ecology of arctic and alpine plants. Biological Reviews (Cambridge) 43: 481-529

CHANG J (1968) Climate and agriculture, an ecological survey. Aldine Publishing Co, Chicago

COCHEMé J (1968) FAO/UNESCO/WHO agroclimatology survey of a semi-arid area of West Africa south of the Sahara. In: Agroclimatological methods. UNESCO, Paris, pp 235-248

COCHEMé J, FRANQUINI P (1967) An agroclimatology survey of a semi-arid area in Africa south of the Sahara. World Meteorological Organization Notes 86 and 210: 1-109

CODY ML, MOONEY HA (1978) Convergence versus nonconvergence in mediterranean-climate ecosystems. Annual Review of Ecology and Systematics 9: 265-321

COHEN JE (1970) A Markov contingency table mode for replicated Lotka-Volterra systems near equilibrium. American Naturalist 104: 547-559

COOKE HBS (1965) The Pleistocene environment in southern Africa. In: DAVIS DHS (ed) Ecological studies in southern Africa. Junk, The Hague, pp 1-23

COOPER WS (1922) The broad-sclerophyll vegetation of California: an ecological study of the chaparral and its related communities. Carnegie Institution of Washington Publication 319

CUTLER JM, RAINS DW, LOOMIS RS (1977) The importance of cell size in the water relations of plants. Physiologia Plantarum 40: 255-260

DE WIT CT (1965) Photosynthesis of leaf canopies. Agricultural Research Reports, Wageningen 663

DIRMHIRN I (1972) The radiative environment and the surface temperature on a microscale in a sagebrush biome. US International Biological Program Desert Biome Research Memorandum RM 72-36: 1-13

DUNCAN WC, LOOMIS RS, WILLIAMS WA, HANAU R (1967) A model for simulating photosynthesis in plant communities. Hilgardia 38:181-205

DUNN EL (1975) Environmental stresses and inherent limitation affecting CO_2 exchange in evergreen sclerophylls in mediterranean climates. In: GATES DM, SCHMERL RB (eds) Perspectives of biophysical ecology. Springer-Verlag, New York, pp 259-181

FISCHER RA, TURNER NC (1978) Plant productivity in the arid and semiarid zones. Annual Review of Plant Physiology 29: 277-317
GATES DM (1962) Energy exchange in the biosphere. Harper and Row, New York, 151 pp
GATES DM (1965) Energy, plants and ecology. Ecology 46: 1-13
GATES DM (1968) Transpiration and leaf temperature. Annual Review of Plant Physiology 19: 211-238
GENTILLI J (1971) Climates of Australia and New Zealand. World survey of climatology, vol 13. Elsevier, Amsterdam
GILIBERTO J, ESTAY H (1978) Seasonal water stress in selected shrub species in the mediterranean region of central Chile. Botanical Gazette 139: 236-240
GREIG-SMITH P (1964) Quantitative plant ecology, 2nd ed. Butterworths, London, 256 pp
HARRISON AT (1971) Temperature related effects on photosynthesis in Heteromeles arbutifolia M. Roem. PhD Thesis, Stanford University, Stanford
HARRISON AT, SMALL E, MOONEY HA (1971) Drought relationships and distribution of two mediterranean climate California plant communities. Ecology 52: 869-875
HELLMERS H, HORTON JS, JUHREN G, O'KEEFE J (1955) Root systems of some chaparral plants in southern California. Ecology 36: 667-678
JACOBSON MB, STONER WA, RICHARDS SP 1981. Models of plant and soil processes. In: MILLER PC (ed) Resource use by chaparral and matorral: A comparison of vegetation function in two mediterranean type ecosystems. Springer-Verlag, Berlin, pp 287-368
KRAMER PJ (1952) Plant and soil water relations on the watershed. Journal of Forestry 50: 92-95
KRUGER FJ (1979) South African heathlands. In: SPECHT RL (ed) Ecosystems of the world, vol 9A. Heathlands and related shrublands. Descriptive studies. Elsevier, Amsterdam, pp 19-80
KUMMEROW J, FISHBECK K (1977) Part 2. In : THROWER NJW, BRADBURY DE (eds) Chile-California mediterranean scrub atlas: A comparative analysis. Dowden, Hutchinson and Ross, Stroudsburg, Pennsylvania, pp 78-237
LAWRENCE WT (1975) A radiation model for chaparral canopies. MS Thesis, San Diego State University, San Diego
LEACH GJ (1968) The growth of the lucern plant after cutting: The effects of cutting at different stages of maturitiy and at different intensities. Australian Journal of Agricultural Research 19: 517-530
LEWONTIN RC (1969) The meaning of stability. Brookhaven Symposium of Biological Diversity and Stability in Ecological Systems No 22
LINéS ESCARDO A (1970) The climate of the Iberian Peninsula. In: WALLéN CC (ed) Climates of Northern and Western Europe. World survey of climatology, vol 5. Elsevier, Amsterdam, pp 295-240
LOVELESS AR (1961) A nutritional interpretation of sclerophylly based on differences in the chemical composition of sclerophyllous and mesophytic leaves. Annals of Botany 25: 168-184
LOWE CH, HINDS DS (1971) Effects of Paloverde (Cericidium) trees on the radiation flux at ground level in the Sonoran Desert in winter. Ecology 52: 916-922
MACARTHUR RH (1970) Species packing and competitive equilibrium for many species. Theoretical Population Biology 1: 1-11
MACARTHUR RH (1972) Geographical ecology. Harper and Row, New York, 269 pp

MILLER PC (1975) A comparison of short-term effects of thermal addition on photosynthesis and plant-water stress in three ecosystems. In: Environmental effects of cooling systems at nuclear power plants. International Atomic Energy Agency, Vienna, pp 623-636

MILLER PC (1979) Quantitative plant ecology. In: HORN D, STAIRS GR, MITCHEL RD (eds) Analysis of ecosystems. Ohio State University Press, Columbus, pp 179-232

MILLER PC (1981) Similarities and limitations of resource utilization in mediterranean type ecosystems. In: MILLER PC (ed) Resource use by chaparral and matorral: a comparison of vegetation function in two mediterranean type ecosystems. Springer-Verlag, Berlin, pp 369-407

MILLER PC, HAJEK E (1981) Resource availability and environmental characteristics of mediterranean type ecosystems. In: MILLER PC (ed) Resource use by chaparral and matorral: a comparison of vegetation function in two mediterranean type ecosystems. Springer-Verlag, Berlin, pp 17-42

MILLER PC, MOONEY HA (1976) The origin and structure of American arid-zone ecosystems. The producers: interactions between environment form and function. In: DE WIT CT, ARNOLD GW (eds) Critical evaluation of systems analysis in ecosystems research and management. Centre for Agricultural Publishing and Documentation, Wageningen, pp 38-56

MILLER PC, POOLE DK (1979) Patterns of water use by shrubs in southern California. Forest Science 25: 84-98

MILLER PC, POOLE DK (1980) Partitioning of solar and net irradiance in mixed and chamise chaparral in southern California. Oecologia 47: 328-332

MILLER PC, STONER WA (1979) Canopy structure and environmental interactions. In: SOLBRIG O, JAIN S, JOHNSON GB, RAVEN PH (eds) Plant population biology. Columbia University Press, New York, pp 163-173

MILLER PC, STONER WA, RICHARDS SP (1978) MEDECS, a simulator for mediterranean ecosystems. Simulation 30: 173-190

MILLER PC, STONER WA, TIESZEN LL (1976) A model of stand photosynthesis for the wet meadow tundra at Barrow, Alaska. Ecology 57: 411-430

MILLER PC, BRADBURY DE, HAJEK E, LAMARCHE V, THROWER NJW (1977) Past and present environment. In: MOONEY HA (ed) Convergent evolution in Chile and California, mediterranean climate ecosystems. Dowden, Hutchinson and Ross, Stroudsburg, Pennsylvania pp 27-72

MILLER PC, HAJEK E, POOLE DK, ROBERTS SW (1981) Microclimate and energy exchange. In: MILLER PC (ed) Resource use by chaparral and matorral: a comparison of vegetation function in two mediterranean type ecosystems. Springer-Verlag, Berlin, pp 97-122

MONK CD (1966) An ecological significance of evergreenness. Ecology 47: 504-505

MONTENEGRO G, HOFFMANN AJ, ALJARO ME, HOFFMANN AE (1979) Satureja gilliesii, a poikilohydric shrub from the Chilean mediterranean vegetation. Canadian Journal of Botany 57: 1206-1213

MONTENEGRO G, RIVEROS DE LA PUENTE F (1977) Comparison of differential environmental responses of Colliguaya odorifera. Flora 166: 125-135

MOONEY HA (ed) (1977a) Convergent evolution in Chile and California, mediterranean climate ecosystems. Dowden, Hutchinson and Ross, Stroudsburg, Pennsylvania, 224 pp

MOONEY HA (1977b) Southern coastal scrub. In: BARBOUR MG, MAJOR J (eds) Terrestrial vegetation of California. John Wiley, New York, pp 471-489

MOONEY HA, DUNN EL (1970a) Photosynthetic systems of mediterranean-climate shrubs and trees of California and Chile. American Naturalist 104: 447-453
MOONEY HA, DUNN EL (1970b) Convergent evolution of mediterranean-climate evergreen sclerophyll shrubs. Evolution 24: 292-303
MOONEY HA, DUNN EL, SHROPSHIRE F, SONG L Jr (1972) Land-use history of California and Chile as related to the structure of the sclerophyll scrub vegetations. Madrono 21 305-319
MOONEY HA, PARSONS DJ (1973) Structure and function of the California chaparral - an example from San Dimas. In: DI CASTRI F, MOONEY HA (eds) Mediterranean type ecosystems: origin and structure. Springer-Verlag, Berlin, pp 83-112
MOONEY HA, TROUGHTON JH, BERRY JA (1974) Arid climates and photosynthetic systems. Carnegie Institution Yearbook 73: 793-805
MOONEY HA, TROUGHTON JH, BERRY JA (1977) Carbon isotope ratio measurements of succulent plants in southern Africa. Oecologia 30: 295-305
NG E (1974) Soil moisture relations in chaparral. MS Thesis, San Diego State University, San Diego
NG E, MILLER PC (1980) Soil moisture relations in the southern California chaparral. Ecology 61: 98-107
ODUM HT, PINKERTON RC (1955) Time's speed regulator. American Scientist 43: 331-343
OECHEL WC, LAWRENCE W, MUSTAFA J, MARTINEZ J (1981) Energy and carbon acquisition. In: MILLER PC (ed) Resource use by chaparral and matorral: A comparison of vegetation function in two mediterranean type ecosystems. Springer-Verlag, Berlin, pp 151-184
ORIANS GH (1975) Diversity, stability and maturity in natural ecosystems. In: VAN DOBBEN WH, LOWE-MCCONNELL RH (eds) Unifying concepts in ecology. Junk, The Hague, pp 139-150
PARSONS DJ (1973) A comparative study of vegetation structure in the mediterranean scrub communities of California and Chile. PhD Thesis, Stanford University, Stanford
PARSONS DJ, MOLDENKE AR (1975) Convergence in vegetation structure along analogous climatic gradients in the mediterranean climate ecosystems of California and Chile. Ecology 56: 950-957
PILLSBURY AF, OSBORN JF, PELESHEK RE (1963) Residual soil moisture below the root zone in southern California watersheds. Journal of Geophysical Research 68: 1089-1091
POOLE DK, MILLER PC (1975) Water relations of selected species of chaparral and coastal sage communities. Ecology 56: 1118-1128
POOLE DK, MILLER PC (1981) The distribution of plant water stress and vegetation characteristics in southern Californian chaparral. American Midland Naturalist 105: 32-43
PRESSLAND AJ (1976) Effect of stand density on water use of mulga (Acacia aneura F. Muell.) woodlands in south-western Queensland. Australian Journal of Botany 24: 177-191
RADOSEVICH SR, CONRAD SG, ADAMS DR (1977) Regrowth responses of chamise following fire. In: MOONEY HA, CONRAD CE (eds) Proceedings of the symposium on the environmental consequences of fire and fuel management in mediterranean ecosystems. US Department of Agriculture, Washington, DC, pp 378-382
ROBERTS SW, MILLER PC (1977) Interception of solar radiation as affected by canopy organization on two mediterranean shrubs. Oecologia Plantarum 12: 273-290

ROBERTS SW, MILLER PC VALAMANESH V (1981) Comparative field water relations of four co-occurring chaparral shrub species. Oecologia 48:360-363

ROSENZWEIG ML (1974) On the evolution of habitat selection. In: Proceedings of the first international congress of ecology. Pudoc, Wageningen, pp 405-409

SAVILE DBO (1972) Arctic adaptations in plants. Canada Department of Agriculture Research Branch Monograph No 6:1-81

SCHIMPER AFW (1898) Pflanzen-geographie auf physiologischer Grundlage. G Fischer, Jena

SCHLESINGER WH, GILL DS (1980) Biomass, production, and changes in the availability of light, water, and nutrients during the development of pure stands of the chaparral shrub, Ceanothus megacarpus, after fire. Ecology 61: 781-789

SCHULZE BR (1972) South Africa. In: GRIFFITHS JF (ed) Climates of Africa. Elsevier, Amsterdam, pp 501-586

SCHWERDTFEGER P (1976) Climate. In: TWODALE CR, TYLER MJ, WEBB BP (eds) Natural history of the Adelaide region. Royal Society of South Australia, Northfield, pp 75-86

SHACHORI AY, MICHAELI A (1965) Water yields of forest, maquis, and grass covers in semi-arid regions. A literature review. In: ECKARDT FD (ed) Methodology of plant ecophysiology. UNESCO, Paris, pp 467-477

SHACHORI A, ROSENZWEIG D, POLJAKOFF-MAYBER A (1967) Effect of mediterranean vegetation on the moisture regime. In: SOPPER WE, LULL HW (eds) International symposium on forest hydrology. Pergamon, Oxford, pp 291-311

SHAPIRO AA, DE FOREST H (1932) A comparison of transpiration rates in chaparral. Ecology 13: 290-295

SHAVER GR (1978) Leaf angle and light absorptance of Arctostaphylos species (Ericaceae) along environmental gradients. Madrono 25: 133-138

SIMS PL, SINGH JS (1978) The structure and function of ten western North American grasslands. III. Net primary production, turnover and efficiencies of energy capture and water use. Journal of Ecology 66: 573-597

SOLBRIG OT, CODY ML, FUENTES ER, GLANZ W, HUNT JH, MOLDENKE AR (1977) The origin of the biota. In: MOONEY HA (ed) Convergent evolution in Chile and California, mediterranean climate ecosystems. Dowden, Hutchinson and Ross, Stroudsburg, Pennsylvania, pp 13-26

SPECHT RL (1957a) Dark Island heath (Ninety-Mile Plain, South Australia). V. The water relationships in heat vegetation and pastures on the Makin Sand. Australian Journal of Botany 5: 151-172

SPECHT RL (1957b) Dark Island heath (Ninety-Mile Plain, South Australia). IV. Soil moisture patterns produced by rainfall interception and stem-flow. Australian Journal of Botany 5: 137-150

SPECHT RL (1963) Dark Island heath (Ninety-Mile Plain, South Australia). VII. The effect of fertilizers on composition and growth 1950-1960. Australian Journal of Botany 11: 67-94

SPECHT RL (1969a) A comparison of the sclerophyllous vegetation characteristic of mediterranean type climates in France, California, and southern Australia. I. Structure, morphology, and succession. Australian Journal of Botany 17: 277-292

SPECHT RL (1969b) A comparison of the sclerophyllous vegetation characteristic of mediterranean type climates in France, California, and southern Australia. II. Dry matter, energy, and nutrient accumulation. Australian Journal of Botany 17: 293-308

SPECHT RL (1972a) The vegetation of South Australia. AB Jones, Adelaide, 32 pp
SPECHT RL (1972b) Water use by perennial evergreen plant communities in Australia and Papua New Guinea. Australian Journal of Botany 20: 273-299
SPECHT RL (ed) (1979) Ecosystems of the world, vol 9A. Heathlands and related shrublands. Descriptive studies. Elsevier, Amsterdam, 498 pp
STEUBING L, ALBERDI M (1973) The influence of phosphorus deficiency on the sclerophylly. Oecologia Plantarum 8: 211-218
STEWARD D, WEBBER PJ. 1981 The plant communities and their environments. In: MILLER PC (ed) Resource use by chaparral and matorral: A comparison of vegetation function in two mediterranean type ecosystems. Springer-Verlag, Berlin, pp 43-68
STONER WA, MILLER PC, RICHARDS SP, BARKLEY SA (1978) Internal nutrient cycling as related to plant life-form: A simulation approach. In: ADRIANO DC, BRISBIN IL, JR (eds) Environmental chemistry and cycling processes. US Department of Energy, Washington, DC, pp 165-181
TAYLOR SE (1975) Optimal leaf form. In: GATES DM, SCHMERL RB (eds) Perspectives of biophysical ecology. Springer-Verlag, Berlin pp 73-86
THROWER NJW, BRADBURY DE (eds) (1977) Chile-California mediterranean scrub atlas: A comparative analysis. Dowden, Hutchinson and Ross, Stroudsburg, Pennsylvania, 237 pp
VAN KEULEN H (1975) Simulation of water use and herbage growth in arid regions. Centre for Agricultural Publishing and Documentation, Wageningen, 176 pp
WALLéN CC (1968) Agroclimatological studies in the Levant. In: Agroclimatological methods. UNESCO, Paris, pp 225-233
WALLéN CC, DE BRICHAMBAUT GP (1962) A study of agroclimatology in semi-arid and arid zones of the Near East. FAO, Rome, 185 pp
WALTER H (1973) Vegetation of the earth in relation to climate and the eco-physiological conditions. (Translated from 2nd German edn by WIESER J). Springer-Verlag, Berlin, 273 pp
WATKINS VM, DE FOREST H (1941) Growth in some chaparral shrubs of California. Ecology 22: 79-83
WERGER MJA, ELLENBROEK GA (1978) Leaf size and leaf consistence of a riverine forest formation along a climatic gradient. Oecologia (Berlin) 34: 297-308
WHITE J (1979) The plant as a metapopulation. Annual review of ecology and systematics 10: 109-145
WHITTAKER RH (1975) Communities and ecosystems, 2nd edn. MacMillan, New York

9. Carbon and Nitrogen Economy of *Diplacus aurantiacus*, a Californian Mediterranean-Climate Drought-Deciduous Shrub

S. L. GULMON

9.1 Introduction

Accumulating evidence suggests a fairly direct link between leaf nitrogen content and photosynthetic rate per unit leaf mass (Takeda 1961; Natr 1972; Yoshida and Coronel 1976; Brown 1978; Mooney et al. 1978). In addition to specific relationships obtained on individual species (as cited above), Winner (1980) has observed an overall correlation of 0.64 ($P < 0.001$) between nitrogen content and photosynthetic rate among nine different chaparral species in which the measurements of photosynthesis were not controlled for leaf age or stomatal conductance.

The photosynthetic rate, total leaf area and leaf duration are primary components of plant productivity and, for plants considered in aggregate, of community productivity. Thus, the way in which plant nitrogen is allocated among individual leaves will bear significantly on the relationship between soil nitrogen available to the system and community productivity.

In this study we examined the allocation of nitrogen to individual leaves and its effect on leaf carbon gain throughout the growing season in Diplacus aurantiacus. Diplacus is a successional, semi-drought-deciduous shrub common in chaparral vegetation throughout central California. New leaf production starts with the onset of autumn rains and ceases in midspring. Reproduction occurs from early spring to midsummer. The first leaves produced in the autumn age and die by midspring, and most of the leaves are dead by early summer. However, the last leaves, produced in the spring at the branch terminus and in the axils of earlier leaves, survive through the summer until the following winter. These leaves do not expand fully until autumn, and never become as large as the leaves that are produced during the early and middle parts of the rainy season.

The photosynthetic rate in Diplacus is a linear function of the leaf nitrogen content (Fig. 1). This proved to be generally true, even when photosynthesis was not light-saturated. The nitrogen content of individual leaves declines steadily with age (Table 1), and newly initiated leaves have lower nitrogen contents as the season progresses (Mooney et al. 1980). In this paper, the nitrogen contents of specific leaves throughout their lifespans and the photosynthesis to nitrogen-content relationship were used to compute the cumulative return in fixed carbon for leaves produced at different times during the growing season. In particular, the efficacy of the oversummering terminal leaves and its possible effect on overall nitrogen-use efficiency were considered.

Table 1. Nitrogen content per unit leaf dry mass (N $mass^{-1}$), specific leaf mass (SLM) and area per leaf of *Diplacus aurantiacus* grown at three rates of nitrogen supply (from Gulmon and Chu 1981).

NO_3^- supply ($mmol\ day^{-1}$)	N $mass^{-1}$ ($mg\ g^{-1}$)[a]			SLM ($g\ m^{-2}$)[b]			Area per leaf pair ($m^2 \times 10^4$)		
	4.0	1.6	0.64	4.0	1.6	0.64	4.0	1.6	0.64
Leaf position									
1-3 (youngest)	51	42	29						
4	49	39	25	56	63	78	15.6	13.4	9.3
5	45	34	22	49	63	76	26.4	20.9	12.4
6	41	31	21	47	63	73	27.5	22.7	16.8
7	39	30	20	44	56	65	20.8	18.8	12.2
8	35	30	14	41	53		17.6	13.7	
9 +	20	20							

[a] Treatment, leaf position effects significant to $P < 0.001$ by two-way analysis of variance.
[b] Treatment, leaf position effects sigificant to $P < 0.001$, and $P < 0.05$, respectively, by two-way analysis of variance.

9.2 Growth pattern and leaf nitrogen content

Stem growth and leaf production were measured by marking stems at the start of the growing season and recording stem length, node count and leaf condition at biweekly intervals. Leaf nitrogen contents were determined by marking randomly selected stems and harvesting them at intervals throughout the season. Leaves were stripped, divided into age classes, and total nitrogen was determined on them by the Kjeldahl method.

Leaf initiation at the primary branch terminus occurred at two-week intervals from November to mid-May (Fig. 2B). Primary stem length increased exponentially during the same time period (Fig. 2A). The difference in growth rate of the two components occurred because the leaf internodes continued to elongate after the leaves matured. As a consequence, a relatively higher proportion of the carbon budget is allocated to leaves early in the season and to stem during the reproductive period.

The distribution of total nitrogen among leaves and reproductive tissue within a single stem, including its side shoots, is shown for one growing season in Fig. 2C. At the start of the season, the leaves that had over-summered contained about 10 mg of nitrogen. The maximum amount of nitrogen accumulated in leaves was 32.5 mg, just before the start of reproduction in April. Thus, in relation to the overall seasonal nitrogen

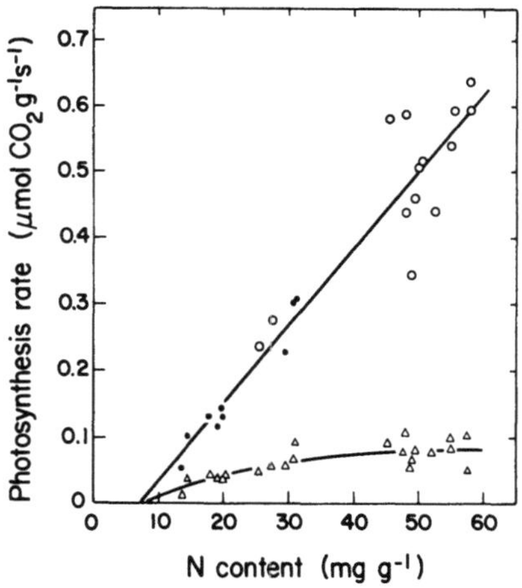

Fig. 1. Photosynthesis rate versus leaf nitrogen content in *Diplacus aurantiacus*. Upper curve is photosynthesis at light saturation: $Y = 0.0117X - 0.082$ ($r = 0.96$, $P < 0.001$). Lower curve shows photosynthetic rates of the same leaves at 130 $\mu E\ m^{-2}\ S^{-1}$. The curve is a quadratic fit, but linear correlation analysis was statistically significant ($r = 0.76$, $P < 0.001$). From Gulmon and Chu 1981.

budget, the over-summering leaves constituted a substantial input to the system at the start of the growing season.

Figs 3A, B and C show the seasonal course of nitrogen content (% of dry mass) from leaf initiation until death for the 14 pairs of leaves on the main stem. Leaves on the lateral branches, which were mainly produced during the latter half of the growth season, were equivalent to terminal leaves produced at the same time. The first four leaves had the highest nitrogen contents throughout their life spans. The second four were similar to the first group at leaf initiation, but their nitrogen contents declined relatively more rapidly as the leaves aged. The final six leaves had much lower nitrogen contents at initiation and remained low throughout their life spans. At the start of the autumn rains, the nitrogen content of these oversummering leaves increased, suggesting that they can serve as immediate sinks for nitrogen which is taken up as soon as the soil becomes moist.

9.3 Computation of cumulative carbon gain

The lifetime cumulative yield in mol CO_2 fixed per gram of final leaf mass was estimated for representative leaves from each of the three age categories depicted in Fig. 3, using the data and assumptions outlined below.

Photosynthesis at light saturation was assumed to be proportional to nitrogen content as shown in Fig. 1. The photosynthetic rates in Fig. 1 were measured with an infrared gas analyser in a flow-through system which is described in detail by Winner and Mooney (1980).

Total weekly photon flux had been measured throughout the year for two years at a site near the sample stand of *Diplacus*. Light intensity was estimated by assuming that the total daily photon flux was distributed over the daylight period as a sine wave.

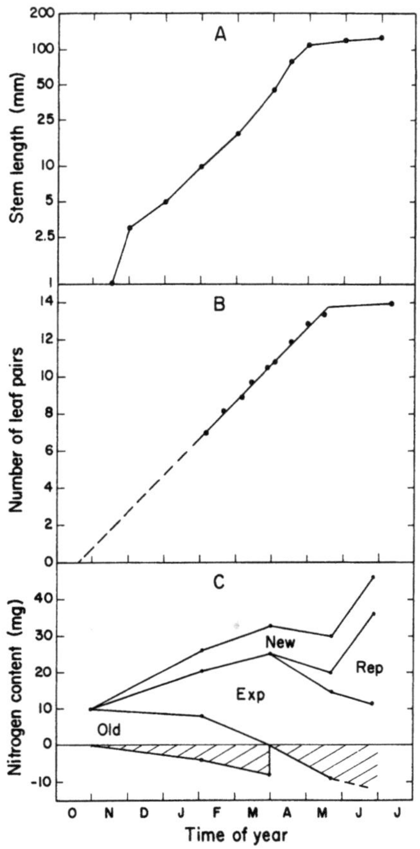

Fig. 2A. Length of a primary stem during one growing season. B. Number of leaf pairs on the stem. C. Allocation of nitrogen among the previous year's (oversummering) leaves (Old), fully expanded leaves (Exp), unexpanded leaves (New), and reproductive tissue (Rep). Shading indicates loss of nitrogen through leaf death. C is based on the entire stem, including the side branches.

Leaf temperature was not taken into account, but measurements on other chaparral plants indicate that temperature response curves are generally broad. Leaf temperature is not normally a significant limiting factor to carbon gain (Mooney et al. 1976; Oechel et al. 1981), an assumption also substantiated by the growth data on Diplacus. Night respiration was also not taken into account. Since night temperatures are generally quite cool compared to day temperatures, especially during winter, when the dark period is longest, this factor would constitute an error of less than 5% (Oechel and Lawrence 1981).

Photosynthesis was computed for 18 intervals within each day and summed over the life of the leaf. Nitrogen content was adjusted at weekly intervals, using the curves in Fig.3. Leaf expansion took eight weeks from the time a pair of leaves first became visible. Carbon gain for each week during this period was multiplied by the fraction of final leaf mass that was obtained at the middle of the week.

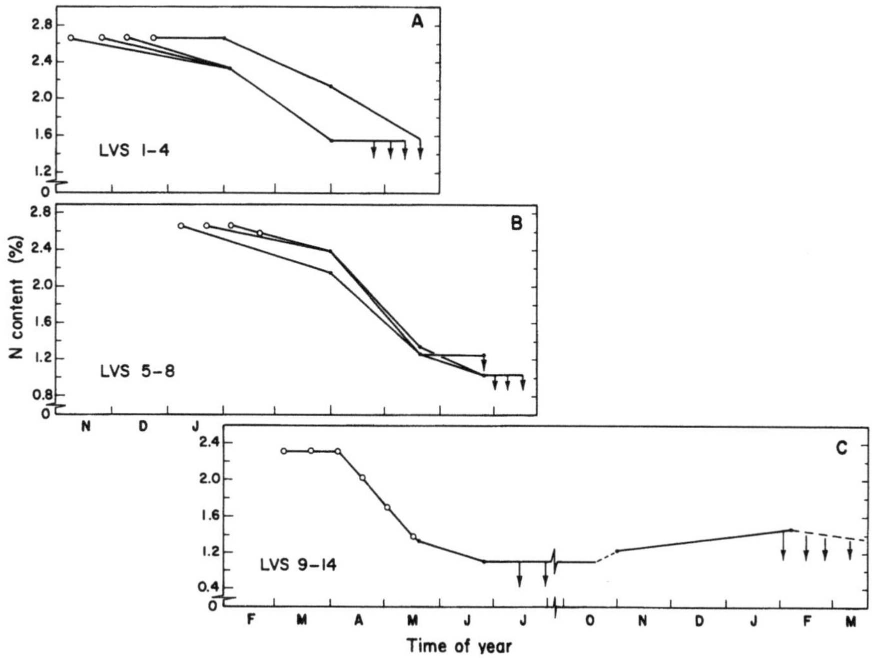

Fig. 3. Nitrogen content through the leaf life span of the 14 pairs of leaves on a main stem. A. Leaves 1-4; B. leaves 5-8; C. leaves 9-14. A circle indicates the date of leaf initiation. An arrow indicates complete senescence.

The lifetime cumulative carbon gain computed for leaves 1,5 and 12 is shown in Fig. 4. The lag period at the start corresponds to the period of leaf expansion. During this interval most of the fixed carbon is used in situ to build the leaf. The slopes of cumulative carbon gain are nearly linear despite the decline in leaf nitrogen concentration because the reduced photosynthetic rate is balanced by a longer photosynthetic period and higher light intensity as day length increases.

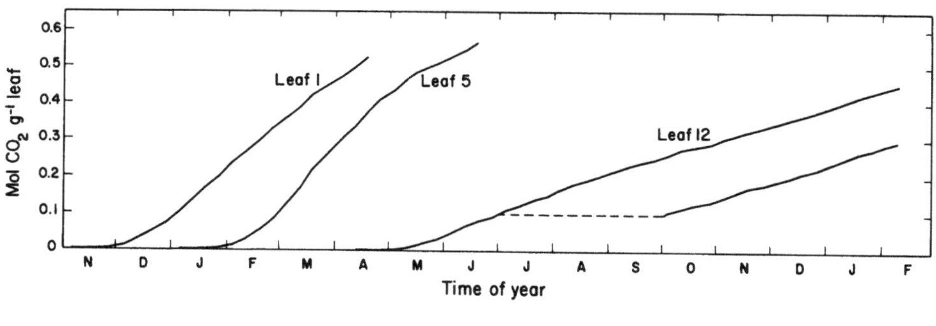

Fig. 4. Cumulative carbon gain of representative leaves per gram of final leaf mass. The dotted line for leaf 12 represents the assumption that no net carbon gain occurs during the summer drought period. The solid line for leaf 12 is carbon gain not limited by drought stress.

Leaves produced early in the season, which had relatively high nitrogen contents, gained an average 0.55 mol CO_2. This is equivalent to 15.68 g dry matter assuming a CO_2 to polysaccharide conversion ratio of 0.648 (Kvet et al. 1971). If a cost ratio of 1.3 g glucose g^{-1} leaf is assumed for the leaves alone (Miller and Stoner 1979), and a cost ratio of 2.52 g glucose g^{-1} resin (Lincoln 1980) is added for the external leaf resin which comprises about 15% of the leaf dry mass (Mooney et al. 1980), then about 1.48 g of glucose would be needed to produce 1 g of leaf.

At the end of the growing season (May 21), the stem-to-leaf mass ratio was 1.06 ± 0.39. Thus the stem cost may be approximated by assuming that the investment in leaves is matched by an equal investment in stem growth and respiration (see Oechel and Lawrence (1980) for discussion of stem respiration in chaparral species). If it is further assumed that the cost of root growth, turnover and respiration is also equal to the leaf cost (Kummerow et al. 1977; Gulmon and Chu 1981), then the total root, stem and construction cost per g of leaf would be 4.44 g of glucose. If the cumulative carbon gain of a leaf is 15.68 g of glucose, then the potential seasonal biomass increase, or growth ratio, would be 15.68/4.44 = 3.53 (Fig. 5).

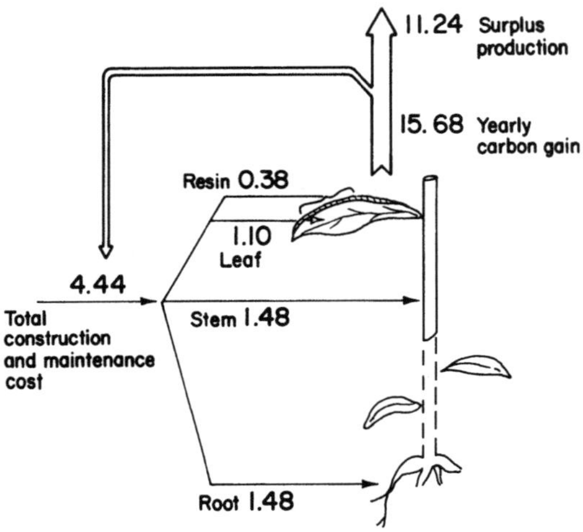

Fig. 5. Total carbon gain over the life of the leaf compared to the estimated leaf costs for predator protection, leaf tissue, and stem and root growth and respiration, all in g glucose g^{-1} of leaf.

At the start of the season, mean total branch mass was 2.24 ± 0.67 g. The mean mass of branches on May 21 was 5.0 ± 2.7 g, and the reproductive biomass on June 26 was 2.76 ± 0.75 g. The total growth ratio was therefore (5.0 + 2.76)/2.24 = 3.46. The coincidence of this figure and the growth ratio based on cumulative leaf carbon gain suggests that the computations of carbon gain represent a fairly realistic picture of the productivity of individual leaves.

The oversummering leaves (represented by leaf 12 in Fig. 4) produced about 0.3 mol CO_2, or 8.55 g glucose g^{-1} of leaf, which despite their longer life was substantially less than the earlier leaves. Nevertheless, with a payback cost of 4.44 g, this would still result in substantial net carbon gain over the life of the leaf, even if it is assumed that there is no carbon gain during the summer period of drought stress. In fact, measurements made in August at midday showed that photosynthetic rates were reduced to one-half the expected rates based on nitrogen concentration.

Nitrogen-use efficiency can be defined as the total mass of carbon fixed by a leaf divided by the mass of nitrogen which is lost when the leaf is shed. Using this definition, the nitrogen-use efficiency of leaves 1, 5, and 12 was 404, 452 and 360 g C g^{-1} N, respectively.

9.4 Discussion

The nitrogen-use efficiency of *Diplacus* leaves can be compared with values obtained for Alaskan tundra vegetation by Chapin et al. (1980). For tundra leaves with a one-year life span, nitrogen-use efficiency varied from 340 to 1370. *Diplacus* thus falls on the low end of this range. However, day-length for the tundra vegetation was twice as long as it was for *Diplacus*, and Chapin et al. (1980) assumed maximum photosynthetic rates throughout the leaf life-span. Nevertheless, *Diplacus* seems to have a relatively moderate nitrogen-use efficiency; this is to be expected, since it is a successional shrub which is not characteristic of extremely low nitrogen habitats.

Several authors have proposed that long-lived (evergreen) leaves are an adaptation to severely nutrient-limited habitats because they permit more total carbon gain per unit of nutrient invested and reduce the rate of nutrient loss to the environment (Small 1972; Schlesinger and Chabot 1977; Moore 1980). There are two components to nitrogen-use efficiency as previously defined; the amount of nitrogen remaining in discarded leaves, which is lost to the plant, and the total lifetime photosynthetic gain of the leaf. Measurements of the nitrogen content of fresh litter (Bocock 1964; Rapp 1969; Anderson 1973; Gosz et al. 1973; Maggs and Pearson 1977; Lousier and Parkinson 1978; MacLean and Wein 1978; Schlesinger and Hasey 1981) range from 0.4 to 4.44% nitrogen for different species from different ecosystems, but it appears that, for any given species, a maximum of about 60%, and often much less, of the leaf nitrogen is actually translocated out before abscission or complete senescence (eg, Small 1972; Grubb 1977; H A Mooney, unpublished). Thus, photosynthetic capacity and leaf duration would appear to be the primary components determining nitrogen-use efficiency in plants. Evergreen leaves normally have substantially lower rates of carbon gain than do deciduous leaves (Larcher 1975; Oechel et al. 1980). Therefore, although long-lived leaves should be selected in habitats with low nutrient availability, they must live, and function, substantially longer than deciduous leaves in order to attain a higher nitrogen-use efficiency.

In the case of *Diplacus*, the oversummering leaves actually had a somewhat lower nitrogen-use efficiency than the shorter-lived leaves produced earlier in the season. However, their function must be viewed in light of the alternatives. If the leaves were dropped at the start of drought stress in early summer, then their nitrogen-use efficiency would be extremely low and, more important, their total carbon gain would be less than the cost of their construction. If these late leaves were not produced at all, the nitrogen they contain could either be diverted to reproduction, stored elsewhere in the plant, or not taken up in the first place. Since the oversummering leaves more than pay the costs of construction, even with a conservative estimate of total carbon gain, it seems most adaptive to retain the nitrogen in this way. These leaves also provide the additional advantages of a "sink" for nitrogen uptake and capacity for immediate carbon gain at the onset of the rainy season.

A final consideration is why the oversummering leaves are much lower in nitrogen content, even at leaf initiation, than the leaves produced earlier. One possible reason is simply that nitrogen is diverted to reproduction at the expense of these leaves, but the possibility that the low nitrogen content is, in itself, adaptive, should be explored. A high nitrogen content would result in a higher respiration rate, which in turn might result in a net loss of carbon during the summer drought when photosynthesis is curtailed. There is also some indication that low-nitrogen leaves are more tolerant to drought stress (Radin and Parker 1979a,b).

An alternative, possibly ancillary, explanation may lie in the plant-herbivore interaction between *Diplacus* and its specialized, and most important, herbivore, *Euphydryas chalcedona*. Prediapause larvae hatch in mid-June and feed briefly before going into summer diapause. The population of post-diapause larvae that emerges in late winter to feed on the winter and spring foliage is reduced in number by a factor of ten (Mooney et al. 1980). The low-nitrogen leaves that are available to the newly hatched prediapause larvae are an extremely marginal food source (Lincoln et al. 1982), and may thus be a factor in controlling the population size of the chief herbivore of *Diplacus*.

In conclusion, the production of two classes of leaves by *Diplacus*, highly productive short-lived leaves and less productive longer-lived ones, may be viewed as enhancing total carbon gain and simultaneously increasing the capacity of the plant to take up nitrogen when it is available and use it more efficiently. *Diplacus*, primarily found in disturbed ares in the chaparral, is highly productive compared to the evergreen sclerophylls in the adjacent vegetation. The short-lived, rather mesophytic leaves conform to this role. However, the oversummering leaves may serve to make *Diplacus* more competitive over a slightly longer time span within the predominantly evergreen vegetation.

Acknowledgements

N Chiariello, C Field, H Mooney and J Roy provided helpful discussion and suggestions. Technical assistance was provided by C Chu, B Lilley, and M Roantree. This research was supported by NSF Grant DEB78 02067.

References

ANDERSON JM (1973) The breakdown and decomposition of sweet chestnut (*Castanea sativa* Mill.) and beech (*Fagus sylvatica* L.) leaf litter in two deciduous woodland soils. Oecologia 12: 275-288

BOCOCK KL (1964) Changes in the amounts of dry matter, nitrogen, carbon, and energy in decomposing woodland leaf litter in relation to the activities of the soil fauna. Journal of Ecology 52: 273-284

BROWN KL (1964) Changes in the amounts of dry matter, nitrogen, carbon, and energy in decomposing woodland leaf litter in relation to the activities of the soil fauna. Journal of Ecology 52: 273-284

BROWN RH (1978) A difference in N use efficiency in C_3 and C_4 plants and its implications in adaptation and evolution. Crop Science 18: 93-98

CHAPIN FS III, JOHNSON DA, MC KENDRICK JD (1980) Seasonal movement of nutrients in plants of differing growth form in an Alaskan tundra ecosystem: implications for herbivory. Journal of Ecology 68: 189-210

GOSZ JR, LIKENS GE, BORMANN FH (1973) Nutrient release from decomposing leaf and branch litter in the Hubbard Brook Forest, New Hampshire. Ecological Monographs 43: 173-191

GRUBB PJ (1977) Control of forest growth and distribution on wet tropical mountains with special reference to mineral nutrition. Annual Review of Ecology Systematics 8: 83-107

GULMON SL, CHU CC (1981) The effects of light and nitrogen on photosynthesis, leaf characteristics, and dry matter allocation in the chaparral shrub, Diplacus aurantiacus. Oecologia 49: 207-212

KUMMEROW J, KRAUSE D, JOW W (1977) Root systems of chaparral shrubs. Oecologia 29: 163-177

KVET J, ONDOK JP, NECAS J, JARVIS PG (1971) Methods of growth analysis. In: SESTAK Z, CATSKY J, JARVIS PG (eds) Plant photosynthetic production - a manual of methods. Junk, The Hague, pp. 343-391

LARCHER W (1975) Physiological plant ecology. Springer-Verlag, Berlin. 252pp

LINCOLN DE (1980) Leaf resin flavonoids of Diplacus aurantiacus. Biochemical Systematics and Ecology 8: 397-400

LINCOLN DE, NEWTON TS, EHRLICH PR, WILLIAMS KS (1982) Coevolution of the checkerspot butterfly Euphydryas chalcedona and its larval food plant Diplacus aurantiacus: larval response to protein and leaf resin. Oecologia. In press

LOUSIER JD, PARKINSON D (1978) Chemical element dynamics in decomposing leaf litter. Canadian Journal of Botany 56: 2795-2812

MACLEAN DA, WEIN RW (1978) Weight loss and nutrient changes in decomposing litter and forest floor material in New Brunswick forest stands. Canadian Journal of Botany 56: 2730-2749

MAGGS J, PEARSON CJ (1977) Litter fall and litter layer decay in coastal scrub at Syndney, Australia. Oecologia 31: 239-250

MILLER PC, STONER WA (1979) Canopy structure and environmental interactions. In: SOLBRIG OT, JAIN S, JOHNSON GB, RAVEN PH (eds) Topics in plant population biology. Columbia University Press, New York, pp 428-460

MOONEY HA, FERRAR PJ, SLATYER RO (1978) Photosynthetic capacity and carbon allocation patterns in diverse growth forms of Eucalyptus. Oecologia 36: 103-111

MOONEY HA, HARRISON AT, MORROW PA (1975) Environmental limitations of photosynthesis on a California evergreen shrub. Oecologia 19: 293-301

MOONEY HA, EHRLICH PR, LINCOLN D, WILLIAMS K (1980) Environmental controls on the seasonality of a drought-deciduous shrub, Diplacus aurantiacus and its predator, the checkerspot butterfly Euphydryas chalcedona. Oecologia 45: 143-146

MOORE P (1980) The advantages of being evergreen. Nature 285: 535

NATR L (1972) Influence of mineral nutrients on photosynthesis of higher plants. Photosynthetica 6: 80-99

OECHEL WC, LAWRENCE W (1981) Carbon allocation and utilization. In: MILLER PC (ed) Resource use by chaparral and matorral: a comparison of vegetation function in two mediterranean-type ecosystems. Springer-Verlag, Berlin, pp 185-236

OECHEL WC, LAWRENCE W, MUSTAFA J, MARTINEZ J (1981) Energy and carbon acquisition. In: MILLER PC (ed) Resource use by chaparral and matorral: a comparison of vegetation function in two Mediterrranean type ecosystems. Springer-Verlag, Berlin, pp 151-184

RADIN JW, PARKER LL (1979a) Water relations of cotton plants under nitrogen deficiency. I. Dependence on leaf structure. Plant Physiology 64: 495-498

RADIN JW, PARKER LL (1979b) Water relations of cotton plants under nitrogen deficiency. II. Environmental interactions on stomata. Plant Physiology 64: 499-501

RAPP M (1969) Production de litière et apport au sol d'éléments minéraux dans deux écosystèmes Méditerranéens: le fôret de Quercus ilex L. et la garigue de Quercus coccifera L. Oecologia Plantarum 4: 377-410

SCHLESINGER WH, CHABOT BF (1977) The use of water and minerals by evergreen and deciduous shrubs in Okefenokee Swamp. Botanical Gazette 138: 490-497

OECHEL WC, LAWRENCE W, MUSTAFA J, MARTINEZ J (1981) Energy and carbon acquisition. In: MILLER PC (ed) Resource use by chaparral and matorral: a comparison of vegetation function in two mediterranean-type ecosystems. Springer-Verlag, Berlin, pp 151-184

SCHLESINGER WH, HASEY MM (1981) Decomposition of chaparral shrub foliage: Losses of organic and inorganic constituents from deciduous and evergreen leaves. Ecology 62: 762-774

SMALL E (1972) Photosynthetic rates in relation to nitrogen recycling as an adaptation to nutrient deficiency in peat bog plants. Canadian Journal of Botany 50: 2227-2233

TAKEDA T (1961) Studies on the photosynthesis and production of dry matter in the community of rice plants. Japanese Journal of Botany 17: 403-437

WINNER WE (1981) The effect of SO_2 on photosynthesis and stomatal behaviour of mediterranean-climate shrubs and trees. In: MARGARIS NS, MOONEY HA (eds) Components of productivity of mediterranean-climate regions. Basic and applied aspects. Junk, The Hague

WINNER WE, MOONEY HA (1980) Ecology of SO_2 resistance: I. Effects of fumigations on gas exchange of deciduous and evergreen shrubs. Oecologia 44:290-295

YOSHIDA S, CORONEL V (1976) Nitrogen nutrition, leaf resistance and leaf photosynthetic rate of the rice plant. Soil Science and Plant Nutrition 22: 207-211

Section III. Nutrient Cycling

Since there are similarities in both climate and vegetation throughout the mediterranean regions of the world, it would be extremely useful to investigate nutrient cycling in these regions on a comparative basis. Considerable problems arise in formulating integrated research programmes as it may be difficult to obtain comprehensive data on inputs, turnover and losses of nutrients. Majority of these ecosystems have a heterogeneous vegetation and the biotic interactions that exist between the soil, micro-organisms and plants may be extremely complex. The first nutrient budgets of mediterranean-type ecosystems were published on the garrigue (Lossaint and Rapp 1971; Lossaint 1973). Recently, Gray and Schlesinger (1981) compiled a comprehensive comparative review for Californian and Chilean sclerophyllous shrublands, although less is known on nutrient cycling than, for example on carbon balance studies.

In this section, GROVES has assessed the pool sizes of phosphorus and nitrogen in Australian heath and South African fynbos. Both heath and fynbos contain species which regenerate from underground organs after fire. Many of the heath plants have higher root:shoot ratios for carbon, phosphorus and nitrogen and a lower proportion of these elements in the litter. Groves has attempted to identify the limitations in our knowledge of nutrient cycling and in particular the few measures of the rates of nitrogen fixation and losses of phosphorus and nitrogen after fire.

READ and MITCHELL provide information on the quantitative and qualitative aspects of litter production, litter decomposition and nutrient mobilization in mediterranean and related types of heath. Litter production in mediterranean-type ecosystems is generally lower than that of other ecosystems e.g. savanna and tropical forest. Sclerophyllous vegetation produces a poor quality litter which consequently decomposes slowly. Read and Mitchell review the different decomposition models based upon environmental variables and litter chemical composition. These have been used to predict decomposition rates in the chaparral vegetation (Schlesinger and Hasey 1981).

There is good evidence of efficient internal cycling of phosphorus and nitrogen by the vegetation and mineralization of nitrogen in the litter will only occur when its carbon:nitrogen ratio drops below 30:1. Available phosphorus content is extremely low in soils of mediterranean-type ecosystems especially fynbos and Australian heath. Microbial release of phosphorus from organic complexes is of vital importance for the maintenance of ecosystem productivity during periods between fires. The seasonal events of the mediterranean climate are responsible for the complex patterns of microbial activity which provide flushes of nutrient release in the soil (Schaeffer 1973). Mediterranean-type plants contain

---

This introduction was prepared by D T Mitchell

special root adaptations (Lamont, this volume) and these efficiently capture available soil nutrients (Read and Mitchell, this volume).

Fire has been considered as the major perturbation in mediterranean-type ecosystems and is important in combusting the slowly decomposing litter. RUNDEL has stressed the complexity of studying the effects of fire on nutrient cycling. Fire may cause large losses of nutrients from the ecosystem by volatilization and erosion but is important in making soil nutrients more available for uptake by plants. Rundel also discusses the luxury uptake of nutrients by the vegetation during the early post-fire phases of regeneration.

There have been few rigorous experimental studies on the role of fire in mediterranean-type ecosystems. TRABAUD investigated the nutrient levels in the garrigue after a fire and provides general trends. There is no doubt that fire is an important physical mineralizing agent but decomposition as a biological mineralization process operating between fires must not be overlooked.

References

GRAY JT, SCHLESINGER WH (1981) Nutrient cycling in mediterranean-type ecosystems. In: MILLER PC (ed) Resource use by chaparral and matorral. A comparison of vegetation function in two mediterranean-type ecosystems. Springer-Verlag, Berlin, pp 259-285.

LOSSAINT P (1973) Soil-vegetation relationships in mediterranean ecosystems of southern France. In: DI CASTRI F, MOONEY HA (eds) Mediterranean-type ecosystems: origin and structure. Springer-Verlag, Berlin, pp 199-210

LOSSAINT P, RAPP M (1971) Repartition de la matière organique, productivité et cycles des elements mineraux dans des écosystemes de climat méditerranéen. In: DUVIGNEAUD P (ed) Productivity of forest ecosystems. UNESCO Paris, pp 597-617

SCHAEFFER R (1973) Microbial activity under seasonal conditions of drought in mediterranean climates. In: DI CASTRI F, MOONEY HA (eds) Mediterranean-type ecosystems: origin and structure. Springer-Verlag, Berlin, pp 191-198

SCHLESINGER WH, HASEY MM (1981) Decomposition of chaparral shrub foliage: losses of organic and inorganic constituents from deciduous and evergreen leaves. Ecology 62: 762-774

10. Nutrient Cycling in Australian Heath and South African Fynbos

R.H. Groves

10.1 Introduction

A review completed over three years ago (Groves 1981) assessed the state of knowledge on nutrient cycling in heaths in northwestern Europe and Australia. That review had two aims: firstly, to quantify the levels of major inputs and outputs for the nutrients phosphorus and calcium and their flow through the soil-plant system, and secondly, to identify some deficiencies in knowledge of nutrient cycling in these two vegetation types. It went on to state that "no results are available on nutrient cycling in South African fynbos - a heathland type closely comparable with the wetter end of the Australian heathland spectrum" (Groves 1981). This review seeks to update and refine the material on Australian heath in that earlier review, to incorporate the slightly changed position regarding the lack of fynbos data identified in 1977 and to point to apparent deficiencies in research effort and results on nutrient cycling in sclerophyllous shrublands both in southwestern South Africa and in southern and eastern Australia.

Firstly, I wish to discuss the choice of particular nutrients for study and the method of presentation of data for this review. The effects on nutrient cycles of two environmental factors, fire and soil moisture, will be evaluated. Results on nutrient cycling in heath and fynbos will be compared with those already available for other vegetation and this review concludes by identifying some priorities for further research.

10.2 Choice of nutrients

If nutrients in runoff water are of prime concern then the choice of elements for analysis may be very different from those selected to study the effects of fire on shrubland nutrient budgets. All too often it seems that the main basis for choice is the ease with which elements can be analysed by an atomic absorption spectrophotometer or similar automated equipment. In this way we read of reports of the effects on shrub communities of prescribed low intensity fires which ignore significant nutrients such as sulphur, but include analyses for seemingly less significant elements such as copper.

A more rational basis for selecting nutrients was provided by Garten (1978), who used principal component analysis to correlate the concentrations of many elements reported for 110 species occurring in North America. Garten (1978) showed that concentrations of elements such as N, P, Fe, Cu and S in plant tissues were correlated; together they formed a set of elements related to the metabolism of nucleic acids and proteins. Concentrations of other elements such as Ca, Mg and Mn formed what Garten termed a "structural-photosynthetic" set and a third group comprising Mn, K and Mg was identified as an "enzymatic" set. This finding, if generally valid, puts elements into different functional groups and may help to narrow the choice of nutrients for a particular study, provided the purpose of the study is clearly defined. Gill (1981b) has already pointed out that the results of Lee and Correll (1978) and O'Connell et al. (1978) on the nutrient composition of litter in open forests of Eucalyptus of southern Australia could be arranged into similar groups of elements.

For heath occurring in southeastern South Australia Specht (1963) reported a measurable growth response when only phosphorus was added. Additions of nitrogen, copper or zinc provoked no response. These are the elements, other than phosphorus, deficient for the growth of introduced, herbaceous plants on which a grazing/cropping programme can be based if the indigenous vegetation is cleared (Riceman 1948). The response of heath vegetation to phosphorus addition reported by Specht (1963) has since been verified for heath in other regions of Australia (Groves 1965; Jones 1968; Specht et al. 1977), and in the long term for southeastern South Australia by Heddle and Specht (1975).

On a species basis, rather than on a community basis, results for heath species show the overriding importance to growth of phosphorus addition (Specht and Groves 1966). Grundon (1972) reported a growth response to nitrogen and potassium as well as phosphorus in some heath species. If cut shoots of one fynbos species (Leucadendron xanthoconus) are subjected to short-term feeding of either nitrate- or ammonium-N, ammonium was preferred as a source of nitrogen to nitrate (Lewis and Stock 1978). Additions of nitrogen and potassium may ameliorate the adverse effects of adding high levels of phosphorus to some species. Other nutrients are probably significant in special ecological situations, for instance iron in seasonally waterlogged species (Groves 1981). Responses for heath and fynbos species may be compared with those of other sclerophyll species, e.g. Eucalyptus, which are promoted by the additions of phosphorus and nitrogen in combination (Groves and Keraitis 1976) and by phosphorus and sulphur in combination (Groves, unpublished 1979). The form of nitrogen added, whether as nitrate or ammonium, may also be significant (Moore and Keraitis 1971).

There can be no doubt from the evidence available that phosphorus is of major importance to a consideration of nutrient cycling in heathlands. Nitrogen is important because of its high level of interaction with phosphorus in a plant nutritional sense and also because the nitrogen cycle involves a "gaseous" phase as well as the "sedimentary" phase applying to phosphorus cycling. In a previous review (Groves 1981) calcium was also considered because it represented a nutrient which was relatively immobile in the vegetation and thus contrasted with cycling of the more mobile phosphorus; calcium also represents a different functional group of elements in Garten's terms. In this review, however, only phosphorus and nitrogen cycling will be discussed because of the paucity of data for calcium in Australian heath and South African fynbos. In other shrublands, such as the Mediterranean "garrigue", calcium is, however, of prime importance (Lossaint 1973).

10.3 Presentation of data

Nutrients such as phosphorus enter the heathland ecosystem from many sources - as ions in rain, from weathering of parent rock over geological time, from runon of nutrients in solution from adjoining areas, from ash deposited after fire, from the activities of herbivores, both invertebrate and vertebrate, and as alluvial products of erosion from outside the area. A nutrient such as nitrogen also enters from the atmosphere by microbial fixation.

Once in the soil-plant system nutrients then undergo complex chemical changes to enter the soil solution, the rates of which are dependent on a range of physical factors such as temperature, as well as the modifying biotic influence of many micro-organisms and specialized symbionts. Nutrient fluxes in soil can also occur vertically between horizons in the profile. Once taken up by the root system nutrients are subject to many metabolic changes involving complex interactions between "sources" and "sinks" governed by the rates of development and senescence of plant parts. Senescent plant parts are shed as litter, which are then subject to complex chemical changes and biotic interactions by micro-organisms before entering the soil nutrient pool.

Nutrients are lost to the system as soil particles eroded from the area either in air as dust or as solid particulate matter or as solutes in runoff water or deep drainage. When the vegetation is burned, nutrients may be volatilized or lost as smoke or ash. Herbivores, especially vertebrate grazers and browsers, may also be responsible for nutrient removal from an area.

The heathland soil-plant ecosystem as described by Groves (1981) should be termed more appropriately the soil-micro-organism-plant system, but the interactions between plants and micro-organisms and soil are extremely complex (Bowen 1981). The level of complexity makes it difficult to present nutrient cycling data simply and to produce straightforward answers to questions concerning the likely sources of nutrient loss in the management of heathland. In this review the pool sizes of the various components of the heath or fynbos nutrient cycle for phosphorus and nitrogen will be quantified as far as possible, and these are shown in Figs 1 and 2.

10.4 Phosphorus cycle

Levels of phosphate ions in rain at a forest site in southern Victoria were extremely low (Attiwill 1966). In areas of heath and fynbos receiving rainfalls as high as 1000 mm or more a year, the input level of phosphorus may still only be a "trace", although for heaths close to the sea values may be higher. For example, the phosphorus level in rain falling on British heathlands ranges from 0.002 to 0.009 g m^{-2} 100 mm^{-1} of rain (Allen et al. 1969). An unknown amount of phosphorus enters as the product of weathering and leaching of parent rock over geological time, but the amount added per year is probably very small. Phosphorus levels entering the system in other forms as ash, dust, in overland flow or as the waste products of herbivore activity are unknown and will vary only locally rather than regionally.

Some phosphorus may be lost in either smoke from a fire or water draining from the site, the latter applying in both the short (one fire cycle) and long (geological leaching) terms. Although there is no published evidence, these losses may be dismissed as minor. The levels of phosphorus lost because of erosion, ash deposition away from the site and herbivore movement may be assumed to equal the inputs. The magnitudes of the input and output of phosphorus therefore appear to be very small. Regular inputs from rain and weathering of parent rock may be insufficient to balance some irregular losses of phosphorus, e.g. as smoke or as particles of surface soil eroded from steep sites in those instances when heavy rain falls soon after a fire.

The phosphorus pool in parent rock will differ greatly according to region. For instance, for heath and sclerophyllous woodland around Sydney occurring on Hawkesbury Sandstone and for mountain fynbos in the Cape, parent rock phosphorus is low (Beadle (1962) quotes 0.006% in Hawkesbury Sandstone). The source of parent rock for those shrublands occurring on aeolian sands is displaced both geographically and temporally over recent geological time. Hence the phosphorus level in parent rock in Fig. 1 may be an overestimate for some shrublands on both continents.

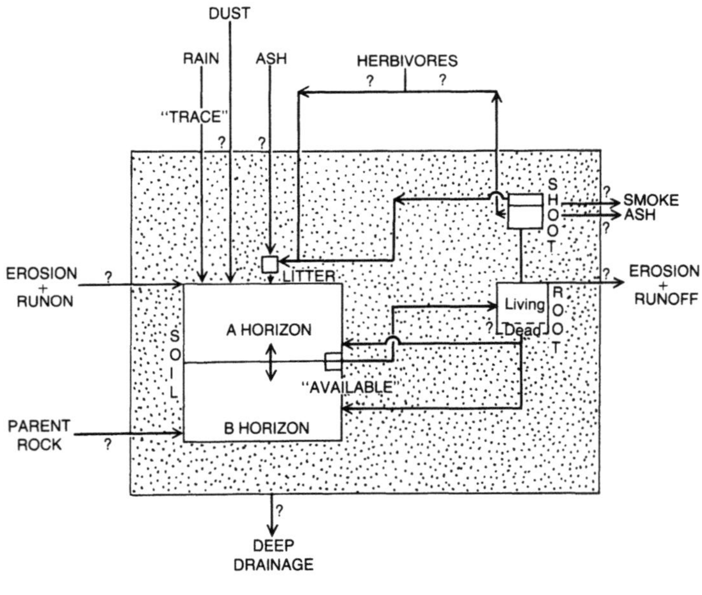

Fig. 1. Pool sizes of phosphorus (g m^{-2}) for a generalized heath/fynbos community showing some inputs and outputs to the soil-plant system (shaded). Pool sizes are based on the data for 15-year-old heath of Specht et al. (1958) and drawn to scale; they are soil 37.5, roots 0.62, shoots 0.25, and litter 0.06. ? indicates lack of data.

Fig. 1 shows total phosphorus in the two soil horizons and the level of phosphorus available to heath/fynbos plants was estimated to be about one hundredth of the total level for both horizons, although more "available" phosphorus probably occurs in the topsoil. The concept of availability of phosphorus to heath plants may not be as useful as it has been for agronomically important plants. By far the greatest proportion of phosphorus in the system occurs in the soil even though only a small amount of it is available for uptake by roots over a finite period.

The level of phosphorus in roots is at least equal to, if not higher than, that in shoots. This partly reflects the much higher proportion of biomass underground than above ground, which can be at least 3:1 (Groves 1965) and is sometimes higher. Although Fig. 1 divides the pool for roots into "living" and "dead", no information is available to show the extent of the difference between these two categories.

For shoots, phosphorus in "living" and "dead" leaves can be differentiated because phosphorus is known to be internally cycled. Using mature leaves of the Australian heath plant Banksia ornata, Specht (1981b) showed that 100 leaves on the shrub contained 3.11 mg phosphorus compared with 100 leaves recently fallen which contained 0.37 mg phosphorus, i.e. about 90% of the phosphorus had been exported from the leaf before abscission.

Although not presented in Fig. 1, ageing shoots contain an increasing but hitherto unquantified amount of phosphorus in woody tissues, both in fruiting structures and stems. Seeds of heath plants may be relatively high in phosphorus (up to ca 3.5% in B. ornata - Groves 1964). For some heath and fynbos plants (for example, species of Banksia, Hakea, Protea) in which seeds are released from woody fruits only after fire (Gill 1976; Gill and Groves 1981), the proportion of total phosphorus present as seed phosphorus in mature plants may be of considerable significance to the regeneration of those species. In a similar way, buried regenerative woody tissues (such as rootstocks, lignotubers, burls) may also contain amounts of phosphorus significant for shoot regeneration. There are preliminary data on nutrient storage in lignotubers of the so-called "mallee" form of Eucalyptus gummifera which grows intermixed with heath on soils of low nutrient status derived from Hawkesbury Sandstone (Mullette and Bamber 1978).

The mean phosphorus level in litter under heath in southeastern South Australia is 0.06 g m^{-2} for 15-year-old vegetation (Specht 1953). Phosphorus in litter remains at a fairly constant level after its initial deposition in summer for up to two years; it is then released during the following (third) autumn (Specht 1981b). This pattern for phosphorus release contrasts with that for biomass of litter which decreased by up to 50% over the first two years from litter fall.

10.5 Nitrogen cycle

The significance of phosphorus to heath plants can be greatly modified by nitrogen status. The data presented in Fig. 2 are largely those of Specht et al. (1958) and some from Hannon (1956, 1958, 1961), although the Australian community Hannon studied was not strictly a heath but a low open-forest of Eucalyptus with a heathy understorey. It does, however, occur on nutrient-deficient soil derived from Hawkesbury Sandstone.

One of the main sources of nitrogen entering the soil-plant system is rain. Although the amount entering may be up to 0.6 g N m^{-2} yr^{-1} for Calluna heaths in Britain (Robertson and Davies 1965), Hannon (1961) showed that the nitrogen in rainwater may not necessarily be completely absorbed during its passage through the soil. Of other sources of nitrogen, especially the quantitative role of free-living organisms capable of nitrogen fixation, little is known. Some data are available for symbiotic nitrogen fixation involving host genera such as Acacia (a legume),

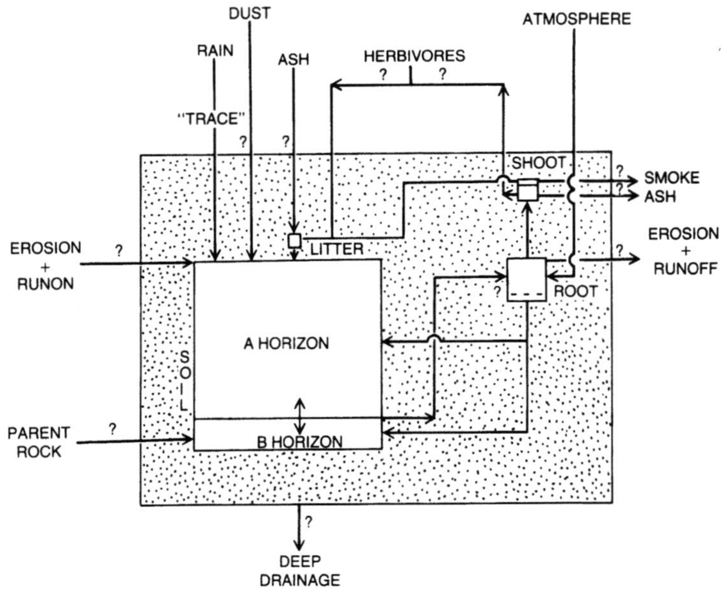

Fig. 2. Pool sizes of nitrogen (g m^{-2}) for a generalized heath/fynbos community showing some inputs and outputs to the soil plant system (shaded). Pool sizes are based on the data for 15-year-old heath of Specht et al. (1958) and drawn to scale, although a different scale from Fig. 1; they are soil 780.0, roots 29.6, shoots 9.9, and litter 3.7. ? indicates lack of data. Some data are those of Hannon (1956, 1958, 1961).

Casuarina (a non-legume) and Macrozamia (a cycad), all common components of the Australian sclerophyll flora, although Macrozamia is not common in heath. Halliday and Pate (1976) showed that symbionts associated with Macrozamia riedlei fixed about 1.9 g N m^{-2} yr^{-1}; A C Lawrie (personal communication 1980) showed a range in levels of nitrogen fixed from 0.0004-0.075 g N m^{-2} yr^{-1} for ten individual legumes growing at three sites in Victoria, one of which was a heathland. Inputs for the communities were higher (0.038-0.079 g N m^{-2} yr^{-1}). The latter study is valuable in that it shows different seasonal patterns of fixation by different species in the community.

Legumes appear to be poorly represented in the fynbos flora. Identification of the sources of input of nitrogen, especially its fixation, is one of the most obvious deficiencies in present knowledge of the fynbos ecosystem (see Chapter 13, this volume).

Nitrogen as well as phosphorus may be lost as smoke from heath or fynbos fires. From data for Calluna heaths, fire intensity and fire frequency (and thus age of heath) interact to influence the magnitude of nitrogen lost as smoke (Groves 1981). Hannon (1961) estimated that the level of nitrogen lost in a fire was low, as was the level lost by natural erosion and deep drainage. It may be that the loss from erosion of soil particles occurring soon after fire was, and presumably still is, the major source of nitrogen loss to the heath/fynbos system. Grazing of new shoots by both vertebrates and invertebrates may further deplete the system of nitrogen soon after a fire. Denitrification seems to occur only locally in swamps and may not be a significant output regionally (Hannon 1961).

Hawkesbury Sandstone contains about 180 μg N g^{-1} (Hannon 1956) and the soil derived from it contains about double that concentration at the surface. Nitrogen level in soil decreased with depth. Hannon (1956) showed, as seems to be the case for phosphorus, that "the very great mass of rock and soil...renders it the chief reserve of nitrogen (and other nutrients) in these communities". This comment was based on the relative proportion of nitrogen in rock plus soil, plants, and litter, the ratio being approximately 200:35:1.

Nitrogen in root and shoot tissues varies considerably between heath species depending on whether or not the species is associated with nitrogen-fixing organisms. Hannon (1956) showed about a twofold difference in leaf nitrogen concentration between legumes and non-legumes, a two- to threefold difference for fruit, and a fourfold difference between seeds of the two groups. For the community as a whole, living (i.e. non-woody?) root and shoot tissue comprised 5% of the total biomass but 15% of the total nitrogen level, compared with woody tissue which made up most of the biomass and a lesser proportion of the nitrogen pool.

Nitrogen levels in litter are low relative to those in other pools (see above). Internal cycling of nitrogen from senescing leaves of heath plants such as Banksia ornata is not as efficient as for phosphorus however. Whereas almost 90% of the phosphorus was recycled before leaf abscission only 30% of the nitrogen was (Specht 1981).

From the limited results presented in Figs 1 and 2 it may be concluded that the amounts of phosphorus and nitrogen cycled annually through the soil-plant system are low and that the cycling processes seem to be efficient. It is therefore unrealistic to consider phosphorus and nitrogen cycles separately, as in this review, as apparently they interact in a highly significant way. Hannon (1961) concluded from her study that in these nutrient-deficient plant communities, phosphorus levels interact with and have always controlled the economy of other nutrients, including nitogen. For Australian heath the levels of nitrogen input, especially through symbiotic fixation in the rhizosphere, interact with and may control the level of phosphorus cycled around the soil-plant system.

10.6 Nutrient cycles, fire and soil moisture

10.6.1 FIRE REGIMES

Although considered elsewhere (see Chapter 12, this volume) a review on nutrient cycling in Australian heath and South African fynbos cannot ignore fire, as it represents the major perturbation to the cycling of nutrients in the two ecosystems. Gill and Groves (1981) reviewed results to 1976 on the plant ecological effects of heathland fires in terms of fire frequency, fire intensity and seasonality of fire, what Gill (1975) had earlier termed the "fire regime". Reviews written subsequently (Bands 1977; Groves 1977; Kruger 1977; Specht 1981a) also attest to the major place of fire in the ecology of these communities.

In Britain, losses of the nutrients phosphorus and nitrogen as a result of burning of Calluna heath at 12-year intervals are usually recouped, especially from ions in rain (Groves 1981). The same may apply to heath and fynbos depending primarily on the fire frequency pattern. In all heath areas fire heats the surface soil to a depth determined by the fire's intensity and the thermal diffusivity of soil and thereby changes soil chemistry. Fires of even low intensity consume the litter. Both effects increase the level of availability of nutrients such as phosphorus and nitrogen for plant growth. Fire deposits ash from the combusted shoots onto the soil surface, the extent of which depends on fire intensity, and the ash is a richer source of nutrients for seedling establishment or new shoot growth on burnt areas than those unburnt. Intense fires may partially sterilize the soil and the micro-organisms colonizing the burnt soil and litter may be different, both taxonomically and functionally, from the population before the fire.

Fire can affect every nutrient flux in the cycle. Its potential to enhance substrate nutrient status for new growth of both seedlings and regenerating shoots, and to change the populations of micro-organisms in the soil and rhizosphere, seem to be two major and positive effects which may determine floristic composition and growth of heath or fynbos species until the next fire. On the other hand, frequent fires or intense rainfall soon after a fire, or both acting together, may deplete the nutrient status of heath or fynbos communities.

10.6.2 SOIL MOISTURE REGIMES

Nutrient cycles for heath or fynbos occurring on soils in which the surface permeable layer overlies either clay or parent rock may be influenced considerably by seasonal soil water regimes. Such texture-contrast soils are found under some heaths in Australia (the so-called "wet heaths") and under most fynbos in South Africa. Denitrification may occur from soils with a high water table. Such soils may also have a high carbon:nitrogen ratio. The availability of phosphorus and other nutrients such as iron may be influenced strongly by the seasonally fluctuating soil moisture regimes, a feature of mediterranean-type rainfall patterns, although the level of total phosphorus is usually low in all heath soils.

Locally, species may be distributed on the basis of a continuum of soil moisture regimes from permanently wet to seasonally dry. In regions where this has been examined, for example in southern England, Loach (1966) found that of three species common to the region *Molinia caerulea* was more abundant on nutrient-richer and drier sites, although it also occurred on poorer, wetter sites, whereas at the other end of the continuum *Erica tetralix* was more abundant on the site which was markedly phosphorus deficient with a permanently high water table. *Calluna vulgaris* was more abundant at the site intermediate between the two extremes of soil moisture/nutrient availability. The significance of soil moisture regimes to heath ecology may thus be more evident locally than the geographically wider effects of fire regimes. But both factors strongly influence nutrient cycling in heath and fynbos communities.

10.7 Nutrient cycles in heath and fynbos in relation to those in other vegetation types

Minimal data on the quantities of phosphorus and nitrogen are presented in Table 1 for heath and three other vegetation types occurring on nutrient-deficient soils. Although these results are limited, they are presented so that some of the claims made for the nutrition of heath (and fynbos) vegetation may be evaluated further. The absolute amounts of nutrients in the two Australian vegetation types (heath and eucalypt woodland) are low (Table 1). The data presented show that the ratio of root mass to shoot mass is higher in heath than in the other examples, even in the lignotuberous eucalypt woodland. The levels of phosphorus and nitrogen in roots of the heath are relatively high as well. Specht et al. (1958) showed that levels of nitrogen and phosphorus in roots of a 15-year-old heath are not very different from those in recently burnt vegetation (their 0-year-old heath) and root nitrogen level was lower and only increased significantly 20 years after the last fire. This time was after the species capable of fixing atmospheric nitrogen had died out after about 10 years in the pyric succession.

Table 1. Quantities (g m^{-2}) of dry matter, phosphorus and nitrogen in four vegetation types occurring on low-nutrient soils in different regions.

Location	Vegetation type	Dry matter	P	N
Southeastern South Australia	15 yr-old heath[a]			
	- tops	1853	0.25	9.9
	- litter	741	0.07	3.7
	- roots	6422	0.62	29.6
Rankin's Springs, central New South Wales, Australia	ca 55 yr-old *Eucalyptus socialis* woodland[b]			
	- tops	4017	0.50	8.8
	- litter	1137	0.31	5.4
	- roots	2854	0.67	10.4
Hubbard Brook, New Hampshire, USA	ca 55 yr-old temperate deciduous forest[c]			
	- tops	6048	3.56	37.5
	- litter	2160	6.72	110.4
	- roots	1298	5.32	18.6
Thetford Chase, England	35 yr-old *Pinus sylvestris* plantation[d]			
	- tops	26800	8.3	91.2
	- litter	33700	31.2	348.2
	- roots	-	-	-

[a] Specht et al. 1958 [b] Burrows 1976
[c] Whittaker et al. 1979 [d] Ovington 1959 (root fraction not measured)

The data of Burrows (1976) are for 50-year-old eucalypt woodland on soils of moderate nutrient status rather than heaths on nutrient-deficient soils (Table 1). Burrows (1976) also established phosphorus and nitrogen budgets for a 15-year-old stand growing at the same site. Nutrient levels were similar in 15- and 50-year-old communities, implying a rapid uptake of nutrients in the 0 to 15 year period and then reduced uptake and increased recycling as the vegetation grows and matures. The more complete set of data of Specht et al. (1958) also showed this feature. Nutrient cycling in heath and fynbos and other sclerophyll shrublands and woodlands may indeed be different from that in a temperate deciduous forest or a plantation of Pinus sylvestris on what was once a heathland soil.

Some earlier hypotheses concerning cycling of nutrients such as phosphorus for heath species may now be modified. Specht and Groves (1966) postulated an efficient mobilization of organic phosphorus from senescing leaves prior to abscission for perennial heath species. Yet Whittaker et al. (1979) showed in a very different vegetation type that nitrogen and phosphorus are both retracted from dying leaves of deciduous hardwood species at Hubbard Brook - a process they called "negative throughflow". The proportions of phosphorus and nitrogen in litter are higher relative to the total for Hubbard Brook vegetation than for the Australian heath (Table 1). Cycling of limiting nutrients such as phosphorus and nitrogen in Australian heath and eucalypt woodland, and presumably fynbos, may be efficient. This may be as much a reflection of the evergreenness and perenniality of these communities as of their occurrence on nutrient-deficient soils. An assessment of nutrient cycling in other vegetation types (chaparral, matorral, garrigue) having the same community characteristics on soils of a different nutrient status should provide a further insight on the nutrient cycling processes in mediterranean regions of the world.

10.8 Concluding discussion

A recurring theme throughout this review on nutrient cycling is the complexity of the biotic interactions existing between micro-organisms and the soil-plant system. Such complexity at least equals if not exceeds the complex chemistry of nutrient availability for ions such as phosphate. In describing nutrient cycling, the rhizosphere has been briefly mentioned. From the rhizosphere of five Victorian heath plants 115 fungal species belonging to 45 genera were isolated (Thrower 1954). The number of species confined to the rhizosphere was 27 and there were 52 species common to both the rhizosphere and the free soil. The functional significance of the microflora may be as considerable as the taxonomic diversity. The fluxes in phosphorus and nitrogen across the soil-rhizosphere-root hair zone are determined by complex interactions, both physical and microbiological, and they are modified considerably by the seasonal fluctuations in soil moisture levels. This overall aspect of ion uptake in the rhizosphere is a major deficiency in our understanding of nutrient cycling in heath and fynbos, the solution to which seems fairly intractable in the medium term, especially its quantitative aspects.

At least for Australian heaths, nitrogen fixation occurs as a result of the symbiotic association between Rhizobium spp. and leguminous hosts mostly belonging to the cross-inoculation group lumped as the "tropical legume miscellany" (Bowen 1981). Nitrogen is also fixed in the association between widespread non-legumes, such as Casuarina, and actinomycetes. Since leaves, seeds and fruits of legumes have a higher nitrogen level (Hannon 1956), presumably because of this sort of association, and because nitrogen is withdrawn from senescing leaves to a lesser extent than phosphorus (Specht 1981b), the litter from these hosts will presumably be higher in nitrogen, thereby influencing the dynamics of decomposition by micro-organisms. It may be that in communities containing host plants capable of associating with nitrogen-fixing organisms the enhanced extent of nitrogen cycling is "driving" the cycles of other nutrients such as phosphorus. More quantitative data on these aspects of nitrogen cycling in heaths are needed. In this regard, the apparent paucity of nitrogen-fixing associations in the fynbos ecosystem is indeed puzzling.

The modifications to uptake of ions such as phosphate in mycorrhizal symbioses between fungi and roots have been omitted from this review. These associations are diverse and also widespread in heathland-type communities (Bowen 1981; Chapters 12 and Chapter 14, this volume).

Another gross deficiency in our knowledge of heath/fynbos ecology is that we do not know the extent to which phosphorus and nitrogen are lost from the ecosystem after a fire or a series of fires. Some information is available for European heaths dominated by Calluna (Groves 1981a) which shows the major roles of fire in affecting the nutrient economy. To assess the nutrient losses, fire intensity must be quantified in physical terms for southern African fynbos and Australian heaths using techniques known for forests and grasslands (Gill 1981a).

Finally, this review has considered cycling of the two nutrients which seem to be most important in Australian heathlands. Most studies of biologically essential elements concern the limits they place on ecosystem productivity and continued stability. Pomeroy (1970) attributed this attitude to Liebig when he wrote (p. 71) "Investigations of essential elements are now in a transition period between Liebig-dominated studies, often of single elements, and ecosystem analysis". Although this review has taken the former approach it is hoped that other reviews will use the approach of ecosystem analysis, or at least be more advanced in the transition between the two approaches.

Acknowledgements

I thank Drs W H Burrows, A M Gill, A C Lawrie, K J Mullette and Professor R L Specht for useful comment on an earlier draft of this review and I acknowledge financial support from the CSIR, South Africa and the Australian Academy of Science.

References

ALLEN SE, EVANS CC, GRIMSHAW HM (1969) The distribution of mineral nutrients in soil after heather burning. Oikos 20: 16-25

ATTIWILL PM (1966) The chemical composition of rainwater in relation to cycling of nutrients in mature Eucalyptus forest. Plant and Soil 24: 390-406

BANDS DP (1977) Prescribed burning in Cape fynbos catchments. In: MOONEY HA, CONRAD CE (technical coordinators) Proceedings of the symposium on the environmental consequences of fire and fuel management in mediterranean ecosystems. USDA Forest Service General Technical Report WO-3, Washington, DC, pp 245-256

BEADLE NCW (1962) An alternative hypothesis to account for the generally low phosphate content of Australian soils. Australian Journal of Agricultural Research 13: 434-442

BOWEN GD (1981) Coping with low nutrients. In: PATE JS, MCCOMB AJ (eds) The biology of Australian plants. Perth, University of Western Australia Press, pp 33-64

BURROWS WH (1976) Aspects of nutrient cycling in semi arid mallee and mulga communities. PhD Thesis. Australian National University

GARTEN GT (1978) Multivariate perspectives on the ecology of plant mineral element composition. American Naturalist 112: 533-544

GILL AM (1975) Fire and the Australian flora: a review. Australian Forestry 38: 4-25

GILL AM (1976) Fire and the opening of Banksia ornata F. Muell. follicles. Australian Journal of Botany 24: 329-335
GILL AM (1981a) Fire: hazard and tool. Parks and Wildlife, Heaths in New South Wales, pp 50-56
GILL AM (1981b) Patterns and processes in open-forests of Eucalyptus in southern Australia. In: GROVES RH (ed) Australian vegetation. Cambridge University Press, pp 152-176
GILL AM, GROVES RH (1981) Fire regimes in heathlands and their plant ecological effects. In: SPECHT RL (ed) Ecosystems of the world, vol 9B. Heathlands and related shrublands. Elsevier, Amsterdam, pp 61-84
GROVES RH (1964) Experimental studies on heath vegetation. PhD Thesis. University of Melbourne
GROVES RH (1965) Growth of heath vegetation. II. The seasonal growth of a heath on ground-water podzol at Wilson's Promontory, Victoria. Australian Journal of Botany 13: 281-289
GROVES RH (1977) Fire and nutrients in the management of Australian vegetation. In: MOONEY HA, CONRAD CE (technical coordinators) Proceedings of the symposium on the environmental consequences of fire and fuel management in mediterranean ecosystems. USDA Forest Service General Technical Report WO-3, Washington, DC, pp 220-229
GROVES RH (1981) Nutrient cycling in heathlands. In: SPECHT RL (ed) Ecosystems of the world, vol 9B. Heathlands and related shrublands. Analytical studies. Elsevier, Amsterdam, pp 151-163
GROVES RH, KERAITIS K (1976) Survival and growth of seedlings of three sclerophyll species at high levels of phosphorus and nitrogen. Australian Journal of Botany 24: 681-690
GRUNDON NH (1972) Mineral nutrition of some Queensland plants. Journal of Ecology 60: 171-181
HALLIDAY J, PATE JS (1976) Symbiotic nitrogen fixation by coralloid roots of the cycad Macrozamia riedlei: physiological characteristics and ecological significance. Australian Journal of Plant Physiology 3: 349-358
HANNON NJ (1956) The status of nitrogen in the Hawkesbury Sandstone soils and their plant communities in the Sydney district. I. The significance and level of nitrogen. Proceedings of the Linnean Society of New South Wales 81: 118-143
HANNON NJ (1958) The status of nitrogen in the Hawkesbury Sandstone soils and their plant communities in the Sydney district. II. The distribution and circulation of nitrogen. Proceedings of the Linnean Society of New South Wales 83: 65-85
HANNON NJ (1961) The status of nitrogen in the Hawkesbury Sandstone soils and their plant communities in the Sydney district. III. The source of loss of nitrogen. Proceedings of the Linnean Society of New South Wales 86: 207-216
HEDDLE EM, SPECHT RL (1975) Dark Island heath (Ninety-Mile Plain, South Australia). VIII. The effects of fertilizers on composition and growth, 1950-1972. Australian Journal of Botany 23: 151-164
JONES R (1968) Productivity studies on heath vegetation in southern Australia. The use of fertilizer in studies of production processes. Folia Geobotanica et Phytotaxonomica (Praha) 3: 355-362
KRUGER FJ (1977) Ecology of Cape fynbos in relation to fire. In: MOONEY HA, CONRAD CE (technical coordinators) Proceedings of the symposium on the environmental consequences of fire and fuel management in mediterranean ecosystems. USDA Forest Service General Technical Report WO-3, Washington, DC, pp 230-244

LEE KE, CORRELL RL (1978) Litter fall and its relationship to nutrient cycling in a South Australian dry sclerophyll forest. Australian Journal of Ecology 3: 243-252

LEWIS OAM, STOCK WD (1978) A preliminary study of the nitrogen nutritional status of members of the South African Proteaceae. Journal of South African Botany 44: 143-151

LOACH K (1966) Relations between soil nutrients and vegetation in wet heaths. I. Soil nutrient content and moisture conditions. Journal of Ecology 54: 597-608

LOSSAINT P (1973) Soil-vegetation relationships in Mediterranean ecosystems in southern France. In: DI CASTRI F, MOONEY HA (eds) Mediterranean type ecosystems: origin and structure. Springer-Verlag, Berlin, pp 199-212

MOORE CWE, KERAITIS K (1971) Effect of nitrogen source on growth of eucalypts in sand culture. Australian Journal of Botany 19: 125-141

MULLETTE KJ, BAMBER RK (1978) Studies of the lignotubers of Eucalyptus gummifera (Gaertn. and Hochr.). III. Inheritance and chemical composition. Australian Journal of Botany 26: 23-28

O'CONNELL AM, GROVE TS, DIMMOCK GM (1978) Nutrients in the litter on jarrah forest soils. Australian Journal of Ecology 3: 243-252

OVINGTON JD (1959) The circulation of minerals in plantations of Pinus sylvestris L. Annals of Botany N.S. 23: 229-239

POMEROY LR (1970) The strategy of mineral cycling. Annual Review of Ecology and Systematics 1: 171-190

RICEMAN DS (1948) Mineral deficiency in plants on the soils of the Ninety-Mile Plain in South Australia. II. Effect of zinc, copper, and phosphate on subterranean clover and lucerne grown on Laffer sand, near Keith. Bulletin of the Council for Scientific and Industrial Research, Australia 234. 45 pp

ROBERTSON RA, DAVIES GE (1965) Quantities of plant nutrients in heather ecosystems. Journal of Applied Ecology 2: 211-220

SPECHT RL (1953) The ecology of the heath vegetation in the upper south-east of South Australia. PhD Thesis. University of Adelaide

SPECHT RL (1963) Dark Island heath (Ninety-Mile Plain, South Australia). VII. The effect of fertilizers on composition and growth 1950-1960. Australian Journal of Botany 11: 67-94

SPECHT RL (1981a) Responses of heathlands and related shrublands to fires. In: GILL AM, GROVES RH, NOBLE IR (eds) Fire and the Australian biota. Australian Academy of Science, Canberra, pp 395-415

SPECHT RL (1981b) Nutrient release from decomposing leaf litter of Banksia ornata, Dark Island heathland, South Australia. Australian Journal of Ecology 6: 59-63

SPECHT RL, CONNOR DJ, CLIFFORD HT (1977) The heath-savannah problem. The effect of fertilizer on sand-heath vegetation of North Stradbroke Island, Queensland. Australian Journal of Ecology 2: 179-186

SPECHT RL, GROVES RH (1966) A comparison of the phosphorus nutrition of Australian heath plants and introduced economic plants. Australian Journal of Botany 14: 201-221

SPECHT RL, RAYSON P, JACKMAN ME (1958) Dark Island heath (Ninety-Mile Plain, South Australia). VI. Pyric succession: changes in composition, coverage, dry weight, and mineral nutrient status. Australian Journal of Botany 6: 59-88

THROWER LB (1954) The rhizosphere effect shown by some Victorian heathland plants. Australian Journal of Botany 2: 246-267

WHITTAKER RH, LIKENS GE, BORMANN FH, EATON JS, SICCAMA TG (1979) The Hubbard Brook ecosystem study: forest nutrient cycling and element behaviour. Ecology 60: 203-220

11. Impact of Fire on Nutrient Cycles in Mediterranean-Type Ecosystems with Reference to Chaparral

P. W. RUNDEL

11.1 Introduction

Although there is a large literature describing the effects of fire on biogeochemical cycles of nutrients, no studies have quantified the ecosystem effects of fire on intersystem fluxes of nitrogen or other nutrients through a full successional cycle (see reviews of fire-nutrient interactions in Viro 1974; Woodmansee and Wallach 1980; Rundel 1981, 1982). The nitrogen balance of ecosystems with major portions of biomass below ground or with low above-ground coverage (grasslands, tundra, deserts) is not generally effected by fire. Low-intensity ground fires in many forest ecosystems likewise do not have a major impact on nutrient cycles. Chaparral and other mediterranean-type ecosystems, however, have characteristic high-intensity fires at relatively frequent intervals. These fires burn off much of the above-ground biomass and have very large immediate effects on output fluxes of nutrients from the ecosystem and smaller secondary effects on both outputs and inputs of nutrients through successional cycles. There are virtually no quantitative data available for the site and dynamics of ecosystem pools of nutrients during such fires.

Much of the older fire-nutrient literature is based on loosely controlled experimental investigations. Because of the wide variation in design of sampling and in the methods of chemical analysis used it is frequently impossible to compare different studies directly. Such problems have led to many apparent contradictions, and result in a data base which might be used to support virtually any possible conclusions about the effects of fire on soil nutrients. Recent investigations, using good experimental designs, however, have done much to resolve these inconsistencies. This review describes the status of our current understanding of fire-nutrient interactions in mediterranean-climate ecosystems and points out important areas for future research.

Fire has three types of effect on nutrient cycling (Woodmansee and Wallach 1980). Primary effects are due directly to the fire itself and consist of ash deposition and volatilization of nutrients (see for example Chapter 10, this volume, Figs 1 and 2). On a longer time-scale, however, secondary abiotic effects and tertiary biotic effects may be extremely significant. Abiotic factors include surface runoff and erosion, dissolution of ions, nutrient retention on soil colloids, leaching of ions out of the root zone, volatilization of nitrogen and precipitation inputs from particulate matter in smoke. Important biotic processes influenced by fire are microbial activity, plant nutrient uptake, quantity and composition of post-fire litter production, and variable nitrogen transformations (nitrification, denitrification and nitrogen fixation).

Since nitrogen and phosphorus are generally considered to be the most significant nutrients limiting productivity in these ecosystems, this review will concentrate on the impact of fire on the cycling and availability of these nutrients. The major emphasis will be on the primary effects of ash deposition and volatilization of nutrients, reflecting the availability of such data, but secondary abiotic and tertiary biotic effects will be considered as well.

Nutrient capital in above-ground canopy and litter compartments is the primary source of nutrient returned to soils as ash following fire. Although knowledge of biomass and nutrient pool sizes in mediterranean-type ecosystems is still fragmentary, representative data are available for sites within five regions (Table 1; see Chapters 10, 12, 15 and 16, this volume). For all areas where comparative data are available on nutrient pools in soil, litter and above-ground biomass, the soil contains the largest pools for both nitrogen and phosphorus. However, much of these nutrients is in forms unavailable for plant growth. Studies of nitrogen and phosphorus in soils beneath stands of *Adenostoma fasciculatum* in southern California found only 2.4% of the total soil nitrogen in available forms and only 1.5% of the total soil phosphorus was "available" as acid-soluble nutrient (Mooney and Rundel 1979).

11.2 Fire temperatures and degree of pyrolysis

Fire temperatures are very important in nutrient cycling. One aspect is the influence of temperature in determining the amount of fuel which is consumed by a fire. In light chaparral fires in California only the leaves and finest of branches within the canopy undergo pyrolysis. Even in moderately hot fires branches greater than 10 mm diameter rarely burn. A prescribed fire in spring of moderate intensity in a pure stand of *Adenostoma fasciculatum* consumed virtually 100% of the leaves and finest branches less than 5 mm diameter but only 45% of the 5-10 mm branch size and 15% of the 10-20 mm branches (Rundel et al., unpublished). No branches larger than 20 mm were consumed by the fire.

A second aspect is the effect of temperature on the degree of volatilization of nutrients, notably nitrogen and sulphur, from soil or biomass compartments. Higher temperatures increase volatilization from canopy and litter and reduce fluxes to the soil compartments (Fig. 1).

Maximum ground surface temperatures during fires in mediterranean-climate shrublands and related heathlands are high in comparison to other vegetation types. Such temperatures range from 170°C to 245°C in grasslands and from 220°C to 300°C in coniferous forests, but reach higher levels in the shrublands due to the greater mass of flammable biomass near the ground surface (Fig. 1). The highest temperatures recorded at ground surface are 700°C for wildfires in California chaparral (DeBano et al. 1977). Only slash or log pile fires reach comparable temperatures.

Table 1. Comparative aboveground biomass and nitrogen pool size in mediterranean-type ecosystems.

	Age (yrs)	Above-ground biomass (g m^{-2})	Above-ground nitrogen (g m^{-2})	Litter nitrogen (g m^{-2})	Soil nitrogen (g m^{-2})
CALIFORNIA					
Chamise chaparral (Mooney and Rundel 1979)	27	2127.0	13.5	2.5	90.0
Mixed chamise (Specht 1969)	37	4929.1	14.5	--	0.3230
Redshank chaparral (DeBano and Conrad 1978)	25	3039.4	13.4	14.7	0.1143
Ceanothus chaparral (Schlesinger and Gill 1980)	12	4158.0	26.4	23.0	--
Ceanothus chaparral (Gray and Schlesinger 1981)	20	3937.1	25.1	--	--
Ceanothus chaparral (Schlesinger and Gill 1980)	21	6337.0	41.1	--	--
Chamise chaparal (Rundel and Parsons 1980)	37	3100.0	8.3	--	0.2580
Coastal sage scrub (Gray and Schlesinger 1981)	20	1033.5	6.1	4.7	--
FRANCE					
Quercus garrigue (Specht 1969)	13	4969.3	24.0	--	--
Quercus garrigue (Lossaint and Rapp 1971)	17	2350.0	16.0	--	0.6000
CHILE					
Matorral (estimates from Mooney et al. 1977)	14	738.4	5.6	--	--
AUSTRALIA					
Heath - S. Australia (Specht 1969)	25	2717.0	11.6	4.2	
Coastal scrub - NSW (Maggs and Pearson 1977)	28	5040.0	30.5	17.8	0.4280
SOUTH AFRICA					
Fynbos (Rundel and Kruger unpublished)	22	946.6	3.4	--	--

Table 1 (Continued)

	Above-ground phosphorus ($g\ m^{-2}$)	Litter phosphorus ($g\ m^{-2}$)	Soil phosphorus ($g\ m^{-2}$)
CALIFORNIA			
Chamise chaparral (Mooney and Rundel 1979)	1.9	0.4	75.0
Mixed chamise (Specht 1969)	1.0	-	168.0
Redshank chaparral (DeBano and Conrad 1978)	1.1	2.2	53.6
Ceanothus chaparral (Schlesinger and Gill 1980)	2.0	1.2	-
Ceanothus chaparral (Gray and Schlesinger 1981)	1.8	-	-
Ceanothus chaparral (Schlesinger and Gill 1980)	2.9	-	-
Chamise chaparral (Rundel and Parsons 1980)	1.7	-	78.0
Coastal sage scrub (Gray and Schlesinger 1981)	1.1	-	-
FRANCE			
Quercus garrigue (Specht 1969)	1.8	-	-
Quercus garrigue (Lossaint and Rapp 1971)	0.9	-	55.0
CHILE			
Matorral (estimates from Mooney et al. 1977)	0.5	-	-
AUSTRALIA			
Heath - S. Australia (Specht 1969)	0.6	0.1	-
Coastal scrub - NSW (Maggs and Pearson 1977)	1.6	0.7	-
SOUTH AFRICA			
Fynbos (Rundel and Kruger unpublished)	0.14	-	-

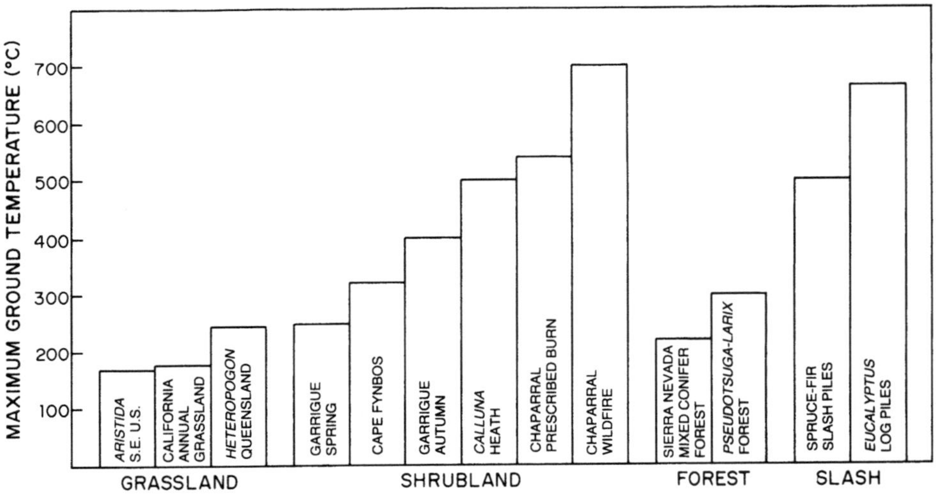

Fig. 1. Mean maximum ground surface temperatures (°C) during fires in a variety of ecosystem types.

Above-ground temperatures within the canopy during intense chaparral fires may reach 1000°C or more. Within the soil, however, maximum temperatures drop rapidly with depth (Fig. 2). Even under conditions of intense wildfire, soil temperatures reach only to 200°C at 25 mm depth and are only slightly elevated at 50 mm. Thus the effect of fire on soil nutrients is restricted to a very shallow horizon. The water content of soil is obviously a very important variable in determining soil temperatures during fire (Aston and Gill 1976; DeBano et al. 1976).

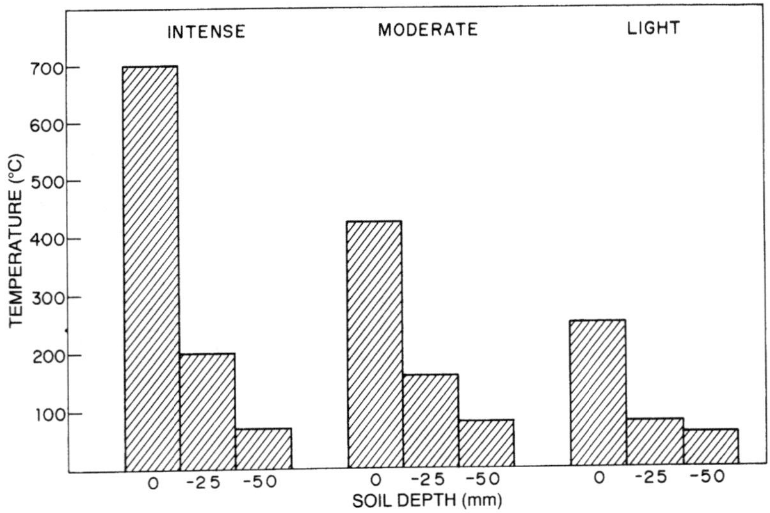

Fig. 2. Gradients of soil temperature with depth in three intensities of chaparral fires. (From DeBano et al. 1977).

11.3 Nitrogen

Laboratory studies of forest fuel types showed that 58-85% of the total nitrogen was volatilized by burning (DeBell and Ralston 1970). White et al. (1973) working with litter of *Pinus ponderosa* reported that 100% of the total nitrogen could be lost to the atmosphere at temperatures above 500°C. Temperatures of 400-500°C produced losses of 75-100%, 300-400°C produced 50-75% losses and 200-300°C caused losses up to 50%. There was no volatilization below 200°C.

DeBano et al. (1979) have produced the best data on nitrogen volatilization from soil and litter in chaparral communities from a detailed study using burns over soil slabs under laboratory conditions (Fig. 3). For soils, nitrogen losses from a 55-mm-deep layer varied directly with temperature recorded at 10 mm depth. Fifty per cent of nitrogen in the upper 20 mm of soil was volatilized at 200°C at 10 mm: this is equivalent to the effect of an intense fire. Nitrogen losses from litter were also linearly related to temperature at the surface, with losses exceeding 90% in intense fires. Overall, an intense experimental fire where litter surface temperatures reached 825°C resulted in a loss of 80% of the total nitrogen contained in the litter and upper 20 mm of the soil. Under field conditions, DeBano and Conrad (1978) estimated that 14.6 g m^{-2} of nitrogen volatilized in a prescribed burn in chamise-redshank chaparral. This is approximately 10% of the total nitrogen from plant, litter and upper soil layers. D Hollinger (unpublished) measured a loss of 24.1 g N m^{-2} during a fire in chamise chaparral in northern California, 7% of the system total. Total loss of nitrogen in a spring fire in chamise chaparral in Sequoia National Park amounted to 11.9 g m^{-2} (Rundel, unpublished). This was 7% of the total system nitrogen, but 64% of the nitrogen in litter and above-ground biomass. Although the absolute losses are quite variable in these studies as one might expect with very different preburn biomass the relative losses of nitrogen from the system are quite similar in each case.

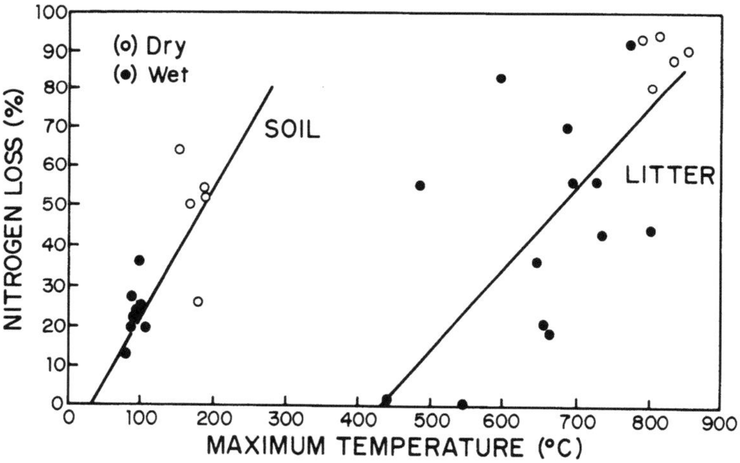

Fig. 3. Volatilization of nitrogen from soil and litter compartments of chaparral systems in relation to maximum fire temperature. (From DeBano et al. 1977).

High levels of nitrogen volatilization have also been reported for fires in Calluna heaths in Europe. Allen (1964) reported that approximately 70% of the nitrogen in the plant canopy was volatilized in an intense heath fire. Artificial fires using canopy and litter of Calluna caused a loss of 95% of the total nitrogen (Chapman 1967). The extent of nitrogen volatilization by fires in other types of vegetation have been summarized by Rundel (1982).

While the volatilization of nitrogen in shrubland fires results in a major loss to the atmosphere, nitrogen in forms available for plant uptake are commonly higher on burned than on adjacent unburned control stands (Christensen and Muller 1975; St. John and Rundel 1976; DeBano and Conrad 1978; DeBano et al. 1979). These increased concentrations probably result from a combination of increased levels of nitrate and ammonium due to direct chemical alteration of organic matter during fire and possibly to post-fire stimulation of free-living fixers (DeBano et al. 1979; Dunn et al. 1979). Unlike the situation with unburned controls, the addition of nitrogen fertilizer to recently burned chaparral soils does not increase productivity (Vlamis and Gowans 1961; DeBano and Conrad 1974).

The dynamics of soil ammonium- and nitrate-nitrogen in the first few months following a autumn prescribed burn in Adenostoma fasciculatum show characteristic patterns (Fig. 4). Ammonium-nitrogen increases 300% in the upper 100 mm of mineral soil following the fire and continues to increase over the following two months as ash residues are mineralized. Nitrate-nitrogen concentrations are unchanged by the fire but begin to increase in February as warmer temperatures promote microbial activity and thus increase rates of nitrification.

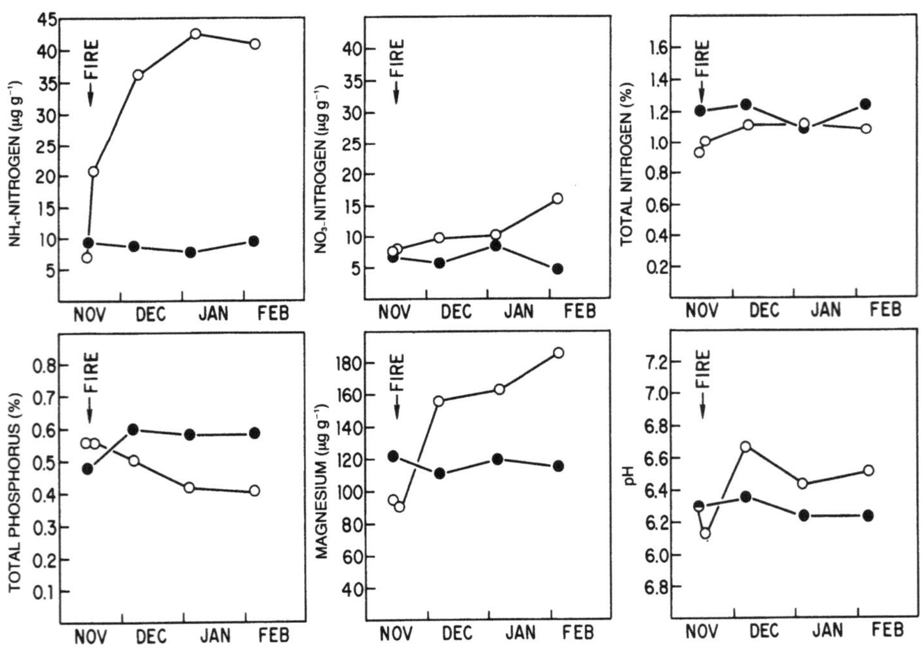

Fig. 4. Comparative soil nutrient content and chemical characteristics of control (solid circles) and prescribed burn plots (open circles) of Adenostoma fasciculatum in Sequoia National Park, California, following an autumn burn. Data for 0-50 mm depth of mineral soil (Rundel et al., unpublished).

11.3.1 NITROGEN BALANCE OVER THE COURSE OF THE FIRE CYCLES

Losses of nitrogen through dry erosion and surface runoff following fire may be considerable in mediterranean-type ecosystems. DeBano and Conrad (1976) measured losses in the first year following fire in chaparral ranging from 0.75 g N m^{-2} (20° slope) to 1.51 g N m^{-2} (50° slope), principally owing to erosion. These rates of debris and soil erosion are significant, but lower than earlier reports for steep chaparral slopes in the San Gabriel Mountains of southern California (Sinclair 1954; Krammes 1965).

The direct effects of fire on nitrogen outputs from chaparral ecosystems are very large when volatilization, erosion and runoff losses are summed. An important question then is how is ecosystem nitrogen balance restored? Inputs of nitrogen in rainfall and through dry deposition in chaparral are very small, totalling no more than 0.1-0.2 g m^{-2} yr^{-1} (Schlesinger and Hasey 1980). At this rate it would take more than 100 years to replace the losses of nitrogen due to fire, a period longer than the natural fire period. An alternative source is nitrogen fixation. Post-burn soil conditions generally favour increased soil microbial activity and favour nitrogen fixation by free-living micro-organisms (Ahlgren and Ahlgren 1965; Christensen and Muller 1975; Dunn et al. 1979). Thus, total soil nitrogen rapidly returns to prefire levels. The increased microbial activity after fire is thought to be due to both the breakdown of allelopathic compounds and the shift of pH from more acidic to neutral levels (Viro 1974; Christensen and Muller 1975; St. John and Rundel 1976). Recent studies have found that nitrifying bacteria are more sensitive to soil heating than typical heterotrophic bacteria. Since intense burning over moist soil has been found to inhibit both *Nitrosomonas* and *Nitrobacter* for periods up to 12 months, the increased levels of available forms of nitrogen following fires may be due to the action of other heterotrophic micro-organisms (Dunn et al. 1979). However, ecosystem inputs of nitrogen from free-living fixers is probably no more than about 0.1 g m^{-2} yr^{-1} in chaparral.

Shrub species with symbiotic nitrogen fixers in root nodules provide another important mechanism for restoring pre-fire levels of total stand nitrogen. In California chaparral species of *Ceanothus*, *Cercocarpus*, *Lotus*, *Lupinus*, *Pickeringia* and *Chamaebatia* are all nitrogen fixers which readily regenerate following fire. Nitrogen fixation by *Ceanothus velutinus* in coniferous forest stands in Oregon has been estimated to be as high as 10.8 g m^{-2} yr^{-1} (Youngberg and Wollum 1976) and various *Ceanothus* species in northern California are estimated to fix up to 6.0 g m^{-2} yr^{-1} (Delwiche et al. 1965). Detailed studies of fixation by other genera of nodulated chaparral shrubs have not been made. *Cercocarpus ledifolius*, related to chaparral species, is estimated to fix 0.69 g m^{-2} yr^{-1} in montane forests (Lepper and Fleschner 1977). Recent estimates of nitrogen fixation by *Ceanothus* in chaparral areas of southern California, however, are only about 0.01 g m^{-2} yr^{-1} (Kummerow et al. 1978). These low levels are due not to reduced nodule activity but more likely to low nodule density, a short period of activity due to long drought periods, and phosphorus limitations. For many chaparral stands in California, nitrogen fixation by *Lotus scoparius* in the first 5-10 years following fire is probably the predominant factor in restoring ecosystem levels of nitrogen. This short-lived semi-woody shrub may fix as much as 1.0-1.5 g N m^{-2} yr^{-1} (E Nilsen, unpublished).

In Chilean matorral Trevoa trinervis, a common shrub colonizing disturbed areas, is an active nitrogen fixer (Rundel and Neel 1978). Three other related genera in the Rhamnaceae are frequent in matorral vegetation and may well prove to be nitrogen fixers - Retanella, Colletia and Talguenea. Sophora macrocarpa, a nodulated legume, is also a widespread shrub on disturbed matorral sites.

Nitrogen fixation by nodulated shrubs in other mediterranean-climate regions has been little investigated. Coriaria myrtifolia (Coriariaceae) often forms dense stands under evergreen oaks in southern France and Spain, and may be an important source of soil nitrogen. A variety of woody and herbaceous legumes are common in the mediterranean region and Myrica species may be locally important.

A number of woody legumes, particularly Aspalathus, are present in many fynbos areas following fire. These may be important for nitrogen fixation if low soil phosphorus availability does not limit the process (see Chapters 12 and 14, this volume). Fixation in these shrubs has not been studied, however. Many fynbos stands certainly lack any known fixers. Non-leguminous fixers are restricted to species of Myrica. Blue-green algae colonies associated with the coralloid roots of cycads, Encephalartos and Stangeria are also involved in nitrogen fixation (Grobbelaar et al. 1971).

Preliminary data from a fynbos of Leucadendron laureolum in South Africa suggest fire losses of nitrogen range from 3.5-5.5 g m^{-2} (Rundel and Kruger, unpublished). This is relatively low in comparison with data from Californian chaparral because the nitrogen pool sizes are much lower. Thus, nitrogen inputs of 0.2-0.4 g m^{-2} yr^{-1} from atmospheric deposition and free-living fixation would be sufficient to restore ecosystem nitrogen pools in 15-20 years without any symbiotic fixation.

In Australian heathlands, known symbiotic nitrogen fixers are generally more common than in fynbos and related South African heathlands. Species of Acacia and other legumes and Casuarina are important heathland elements in Australia, and a number of other woody legumes are widespread. A common genus of cycads, Macrozamia, has also been shown to be an active nitrogen fixer (Halliday and Pate 1976).

11.4 Phosphorus

While phosphorus is a significant limiting nutrient in the mediterranean-climate regions of South Africa and Australia, the impact of fire on phosphorus cycling has been much less studied than in the case of nitrogen. Data from California chaparral indicates that only about 5-10% of the total ecosystem phosphorus is contained in plant and litter compartments, while in Australia and South Africa the biotic components may contain 20% or more (Table 1). DeBano and Conrad (1978) reported only a 2% loss of total phosphorus from the combined canopy, litter and upper soil pools following a chaparral fire. They suggested that volatilization of phosphorus did not take place and that virtually 100% of the phosphorus in the plants and litter was deposited as ash. Studies of phosphorus loss from plants and litter in Calluna heaths have shown high losses (Lloyd 1971), moderate losses of 26% (Chapman 1967), and no effect (Allen 1964). Recent studies in Adenostoma chaparral in California suggest that there is no change in upper soil concentrations of total phosphorus following fire, but a gradual reduction in phosphorus level occurs in subsequent months as ash is mineralized (Fig. 4).

Some studies in other ecosystems have shown considerable losses of phosphorus from plant biomass during burning, apparently by transport as fine ash particles. Studies of pine savannas in the southeastern United States have indicated a 46% loss of phosphorus during combustion of standing biomass (Christensen 1977). At the other extreme, post-fire forest soils in northern Idaho have been reported to pick up enough nutrients from ash to increase their total phosphorus concentrations by 200-500%.

The effect of fire on soluble forms of phosphorus in soils is also not clear from the literature. This can however, be attributed partly to the difficulty of comparing results in the literature in the absence of a standard method of measuring "available" phosphorus. Most studies have reported small increases of available phosphorus following fire (Viro 1974; St. John and Rundel 1976; Campbell et al. 1977). Christensen and Muller (1975) found quite large increases in available phosphorus in a burned chaparral ecosystem, but their pattern of change they reported is difficult to interpret. Other studies from coniferous forests have reported extremes from very large increases following fire (Wagle and Kitchen 1972) to decreases (Isaac and Hopkins 1934). The processes leading to increased levels of available phosphorus following fire have not been carefully studied, but the release of phosphorus from occluded sources following soil heating may be important (Humphreys and Lambert 1965).

Because of the great significance of phosphorus availability to productivity in South African and Australian heathlands, better studies of the effect of fire on phosphorus cycling are badly needed. The data in the literature appear to be somewhat contradictory, but this might be explained by differences in phosphate bonding or the fixing capacity of different soils, by variable immobilization of phosphorus by soil microbes, and by variable fire intensity.

11.5 Exchangeable cations

Since cations have not been shown to limit plant productivity, fire effects on cation cycling in mediterranean-climate shrublands appear to be less important than those on nitrogen and phosphorus. Since cations have a very low volatility during combustion, soil levels of potassium, calcium and magnesium invariably increase following fire due to ash deposition. Levels of mineral magnesium are not changed immediately after fire, but a steady increase to twice the initial concentration occurs during a three-month period as the ash is mineralized (Fig. 5).

Studies in chaparral have uniformly shown increases of 50-100% in soil potassium after fire. DeBano and Conrad (1978), however, reported that over 50% of the canopy and litter potassium was lost through volatilization in one of their experimental fires, most probably in smoke particles. Chapman (1967) found a 21% loss of potassium from *Calluna* heath through volatilization. Intermediate levels of loss have been reported for fire studies in other ecosystems (Christensen 1977). Fire temperatures above 500°C are thought to be a major factor in causing high levels of potassium loss.

The effects of fire on calcium and magnesium resemble those on potassium. DeBano and Conrad (1978) found little loss of these cations in chaparral fires, but Chapman (1967) found losses of 26% and 23% from heathlands. As with other nutrients, slow solubilization of cations from ash provides a gradual increase in burned soils over a period of 12 months or more (e.g. see Christensen 1977).

While most studies have shown increases in soil cation concentrations following fire a few have shown no change or an actual decrease following fire (Viro 1974, Christensen and Muller 1975). These results can be best explained by decreases in the cation exchange capacity of upper soil layers following fire due to a reduction in the soil organic matter content of coarse-textured soils. Excess cations solubilized from the ash under these conditions may be leached out of the rooting zone (Finn 1943; Stark 1977).

11.6 pH

The loss of cations released by the combustion of organic matter has a second effect in causing increases in soil pH. The degree to which pH is raised and the duration of time where such increases remain is a function of several variables including the amount and chemical composition of ash produced by a fire, soil organic matter content and its buffering capacity, soil pH, and regional precipitation. Such pH changes are relatively small in chaparral fires, averaging 0.2-0.3 units (Fig. 5), due to low soil organic matter contents. Hot fires in rich forest soils cause much greater changes. Shifts to a more neutral soil pH from acidic conditions commonly encourages soil microbial activity and promotes increased post-fire rates of ammonification and/or nitrification. The availability of phosphorus and some cations for plant uptake is also influenced by soil pH and thus may be affected by fire.

11.7 Biological relationships

Evidence has already been presented that fire-induced fluxes increase nitrogen and phosphorus availability in the soil. This increased availability may be highly significant for fire-adapted plants in their early stages of growth. Foliar nutrient changes in *Adenostoma fasciculatum* associated with a fire-induced age gradient suggest that nutrient limitation may be a significant factor in stand senescence (Rundel and Parsons 1980).

Although nutrient pools in ash and upper soil layers are subject to erosion, uptake in some cases has been shown to be rapid enough to prevent nutrient loss in surface runoff (McColl and Grigal 1977). Well-documented watershed studies of post-fire nutrient losses from coniferous forest ecosystems show a pattern of a small increase in nutrient loss compared to unburned forests for a period of months (Clayton 1976; Helvey et al. 1976; Schindler et al. 1978), but this increased loss is frequently within the range of normal annual variability (Wright 1980). Post-fire nutrient loss from steep chaparral slopes in Southern California may be relatively high, however (DeBano and Conrad 1976).

While resprouts and seedlings of dominant shrub species have relatively high concentrations of nutrients their total biomass is too small in the first year after fire to successfully tie up a significant percentage of the available nutrients deposited in ash. The dense post-fire herbaceous growth, however, does play a very important ecological role in immobilizing such nutrients and protecting them against loss from the ecosystem (Fig. 5). Blank et al. (1980) have discussed a similar role for herbaceous species in deciduous forest ecosystems. At the end of the spring growing season, following an autumn burn in 1979 at Sequoia National Park, the native herb biomass on burnt plots was more than twice that of *Adenostoma* resprouts. *Adenostoma* seedlings were an insignificant part of the total biomass. Most of the biomass pool of the major nutrients was also contained in the herbs, demonstrating the significance of herbaceous vegetation in sequestering nutrients immediately after fire. Herb biomass on clip plots, where introduced European annuals dominated, was more than twice as great as on burn plots and sequestered considerably more nutrients.

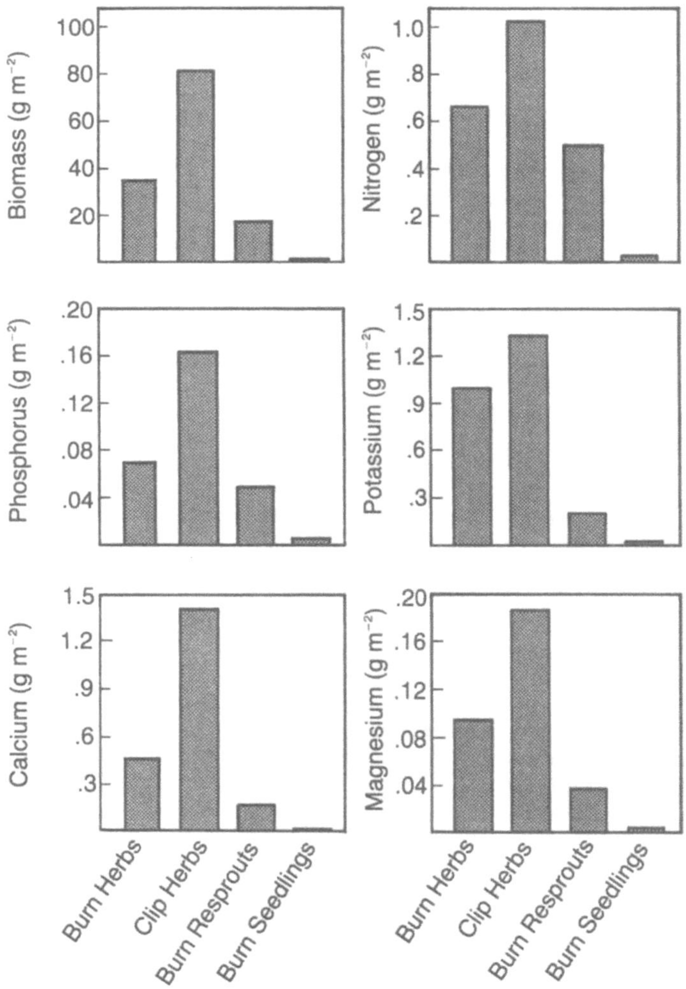

Fig. 5. Biomass and nutrient pools in above-ground tissues after one year of growth following an autumn 1979 burn in chamise chaparral at Sequoia National Park. Clip data are for an adjacent stand of chamise clipped and cleared at the same time.

Luxury uptake of nutrients (absorption in excess of those immediately required for metabolism) appears to occur in the post-fire growth of many species (Lay 1957; Steward and Ornes 1975; Rundel and Parsons 1980). Although this excess uptake and the possibility and nature of storage has not been carefully investigated, there are clear ecological advantages in low-nutrient environments to plant species which are capable of luxury uptake. Recent studies on the storage of polyphosphates in the roots of Australian heath plants (Coleman and Specht 1981) provide an example of a possible mechanism for storage of excess phosphorus.

The importance of fire-regulated nutrient cycling in mediterranean-climate ecosystems may be such that natural selection has operated to favour shrub structures and foliage secondary chemistry which provide increased potential flammability. Ecosystems with low nutrient availability may select for shrub structural forms and leaf chemical characteristics which increase flammability and thus act to recycle limiting nutrients. If fire-induced fluxes between nutrient compartments increase nitrogen and/or phosphorus availability for fire-adapted plants in their early stages of productive growth, plant fitness should be increased by greater flammability.

Preliminary studies (Rundel 1981) have indicated that Californian chaparral shrubs (with low nutrient availability in underlying soils) have structural and chemical characteristics which have higher degrees of flammability than Chilean matorral shrubs with moderate nutrient availability in the underlying soils. Mediterranean-climate heathlands in South Africa with extremely poor nutrient availability show a complex pattern of community flammability attributes. The ericaceous layer is highly flammable in both its structural and chemical characteristics and seems to depend on regular fires for recyclng nutrients from above-ground plant biomass and litter. The upper proteaceous layer, while well-adapted to fire, typically has no structural or chemical characteristics promoting flammability and apparently relies on physiological tolerance to low foliage nutrient levels to maintain growth during intervals between fires.

11.8 Conclusions

Though recent studies in many mediterranean-type ecosystems have provided new data on the immediate impact of fire on nutrient cycles our understanding of the dynamics of nutrients over the natural fire cycle in the ecosystem is still poor. Basic aspects of nutrient cycling have not been studied in mediterranean-type ecosystems to the same degree of detail as in other ecosystems. Such studies for nitrogen and phosphorus are particularly important since these nutrients are commonly limiting for net primary production. Almost no quantitative data are available for microbial nitrogen transformations, yet fire must have a major impact on these processes. These current gaps in our knowledge of fire impacts on nutrient cycling in mediterranean-type ecosystems should provide important targets for research over the next decade.

References

AHLGREN IF, AHLGREN CE (1965) Effects of prescribed burning on soil micro-organisms in a Minnesota jack pine forest. Ecology 46: 304-310

ALLEN SE (1964) Chemical aspects of heather burning. Journal of Applied Ecology 1: 347-367

ASTON AR, GILL AM (1976) Coupled soil moisture heat and water vapour transfers under simulated fire conditions. Australian Journal of Soil Research 14: 55-56

BLANK JL, OLSON RK, VITOUSEK PM (1980) Nutrient uptake by a diverse spring ephemeral community. Oecologia 47: 96-98

CAMPBELL RE, BAKER MB, FOLLIOTT PF, LARSON FR, AVERY CC (1977) Wildfire effects on a ponderosa pine ecosystem: an Arizona case study. USDA Forest Service Research Paper RM-191: 12 pp

CHAPMAN SB (1967) Nutrient budgets for a dry heath ecosystem in the south of England. Journal of Ecology 55: 677-689

CHRISTENSEN NL (1977) Fire and soil-plant nutrient relations in a pine-wiregrass savanna on the Coastal Plain of North Carolina. Oecologia 31: 27-44

CHRISTENSEN NL, MULLER CH (1975) Effects of fire on factors controlling plant growth in Adenostoma chaparral. Ecological Monographs 45: 29-55

CLAYTON JL (1976) Nutrient gains to adjacent ecosystems during a forest fire: an evaluation. Forest Science 22: 162-166

COLEMAN RG, SPECHT RL (1981) Mineral nutrition of heathlands: the possible role of polyphosphate in the phosphorus economy of heathland species. In: SPECHT RL (ed) Ecosystems of the world, vol 9B. Heathlands and related shrublands. Analytical studies. Elsevier, Amsterdam, pp 197-207

DEBANO LF, CONRAD CE (1974) Effect of a wetting agent and nitrogen fertilizer on establishment of ryegrass and mustard on a burn watershed. Journal of Range Mangement 27: 57-60

DEBANO LF, CONRAD CE (1976) Nutrients lost in debris and runoff water from a burned chaparral watershed. Proceedings of the Third Interagency Sedimentation Conference 3: 13-27

DEBANO LF, CONRAD CE (1978) The effect of fire on nutrients in a chaparral ecosystem. Ecology 59: 489-497

DEBANO LR, DUNN PH, CONRAD CE (1977) Fire's effect on physical and chemical properties of chaparral soils. In: MOONEY HA, CONRAD CE (technical coordinators) Proceedings of the symposium on the environmental consequences of fire and fuel management in mediterranean ecosystems. USDA Forest Service General Technical Report WO-3, pp 65-74

DEBANO LF, EBERLEIN GE, DUNN PH (1979) Effects of burning on chaparral soils. I. Soil nitrogen. Proceedings of the Soil Science Society of America 43: 504-509

DEBANO LF, SAVAGE SM, HAMILTON DA (1976) The transfer of heat and hydrophobic substances during burning. Soil Science Society of America Proceedings 40: 779-782

DEBELL DS, RALSTON CW (1970) Release of nitrogen by burning in light forest fuels. Proceedings of the Soil Science Society of America 34: 936-938

DELWICHE CC, ZINKE PJ, JOHNSON CM (1965) Nitrogen fixation by Ceanothus. Plant Physiology 40: 1045-1047

DUNN PH, DEBANO LF, EBERLEIN GE (1979) Effects of burning on chaparral soils. II. Soil microbes and nitrogen mineralization. Proceedings of the Soil Science Society of America 43: 509-514

FINN RF (1943) The leaching of some plant nutrients following the burning of forest litter. Black Rock Forest Paper 1: 128-134

GRAY JT, SCHLESINGER WH (1981) Nutrient cycling in mediterranean-type ecosystems. In: MILLER PC (ed) Resource use by chaparral and matorral. Springer-Verlag, Berlin. pp 259-285

GROBBELAAR N, STRAUSS JM, GROENEWALD EG (1971) Non-leguminous seed plants in southern Africa which fix nitrogen symbiotically. In: LIE TA, MULDER CC (eds) Biological nitrogen fixation in natural and agricultural habitats. Martinus Nijhoff, The Hague. pp 325-334

HALLIDAY J, PATE JS (1976) Symbiotic nitrogen fixation by coralloid roots of the cycad Macrozamia riedlei: physiological characteristics and ecological significances. Australian Journal of Plant Physiology 3: 349-358

HELVEY JD, TIEDEMANN AR, FOWLER WB (1976) Some climatic and hydrologic effects of wildfire in Washington state. Proceedings of the Tall Timbers Fire Ecology Conference 15: 201-222

HUMPHREYS FR, LAMBERT MF (1965) Soil temperature profiles under slash and log fires of various intensities. Australian Forest Research 1: 23-29

ISAAC LA, HOPKINS HG (1934) The forest soil of the Douglas-fir region and the changes wrought upon it by logging and slash burning. Ecology 18: 264-279

KRAMMES JS (1965) Seasonal debris movement from the steep mountainside slopes in southern California. Proceedings of the Third Interagency Sediments Conference. USDA Miscellaneous Publication 970, pp 85-90

KUMMEROW S, ALEXANDER JV, NEEL JW, FISHBECK K (1978) Symbiotic nitrogen fixation in Ceanothus roots. American Journal of Botany 65: 63-69

LAY DW (1957) Browse quality and the effects of prescribed burning in southern pine forests. Journal of Forestry 55: 342-349

LEPPER MG, FLESCHNER M (1977) Nitrogen fixation by Cercocarpus ledifolius (Rosaceae) in pioneer habitats. Oecologia 27: 333-338

LLOYD PS (1971) Effects of fire on the chemical status of herbaceous communities of the Derbyshire Dales. Journal of Ecology 59: 261-273

LOSSAINT P, RAPP M (1978) La forêt méditerranéenne de chênes verts (Quercus ilex L). In: LAMOTTE M, BOURLIÈRE F (eds) Problèmes d'écologie. Structure et fonctionnement des écosystèmes terrestres. Masson, Paris, pp 129-185

MAGGS J, PEARSON CJ (1977) Minerals and dry matter in coastal scrub and grasslands at Sydney, Australia. Oecologia 31: 227-237

MCCOLL JG, GRIGAL DF (1977) Nutrient changes following a forest wildfire in Minnesota: effects in watersheds with different soils. Oikos 28: 105-112

MOONEY HA, KUMMEROW J, JOHNSON AW, PARSONS DJ, KEELEY S, HOFFMANN A, HAYS RI, GILIBERTO J, CHU C (1977) The producers - their resources and adaptive responses. In: MOONEY HA (ed) Convergent evolution in Chile and California. Mediterranean-climate ecosystems. Dowden, Hutchinson and Ross, Stroudsburg, Pennsylvania, pp 1-12

MOONEY HA, RUNDEL PW (1979) Nutrient relations of the evergreen shrub, Adenostoma fasciculatum, in the California chaparral. Botanical Gazette 140: 109-113

RAPP M, LOSSAINT P (1981) Some aspects of mineral cycling in the garrigue of southern France. In: DI CASTRI F, GOODALL DW, SPECHT RL (eds) Ecosystems of the world, vol 11. Mediterranean-type shrublands. Elsevier, Amsterdam, pp 289-301

RUNDEL PW (1981) Structural and chemical components of flammability. In: MOONEY HA (ed) The role of past and present frequency and intensity of fire on ecosystem development and management. USDA Forest Service General Technical Report WO-26, pp 183-207

RUNDEL PW (1982) Fire as an ecological factor. In: LANGE OL, NOBEL PS, OSMOND CB, ZIEGLER H (eds) Plant physiological ecology I. Responses to the physical environment. Springer-Verlag, Berlin. In press

RUNDEL PW, NEEL JW (1978) Nitrogen fixation by Trevoa trinervis (Rhamnaceae) in the Chilean matorral. Flora 167: 127-132

RUNDEL PW, PARSONS DJ (1980) Nutrient changes in two chaparral shrubs along a fire-induced age-gradient. American Journal of Botany 67: 51-58

SCHINDLER DW, NEWBURY RW, BEATY KL, PROKOPOWICH J, RUSZCZYNSKI T, DALTON JA (1978) Effects of a windstorm and forest fire on chemical losses from forested watersheds and on the quality of receiving streams. Journal of the Fisheries Research Board of Canada 35.

SCHLESINGER WH, GILL DS (1980) Biomass production and changes in the availability of light, water and nutrients during the development of pure stands of the chaparral Ceanothus megacarpus after fire. Ecology 61: 781-789

SCHLESINGER WH, HASEY MM (1980) The nutrient content of precipitation dry fallout and intercepted aerosols in the chaparral of southern California. American Midland Naturalist 103: 114-122

SINCLAIR JD (1954) Erosion in the San Gabriel Mountains of California. American Geophysical Union Transactions 35: 264-268

SPECHT RL (1969) A comparison of the sclerophyllous vegetation characteristic of mediterranean-type climates in France, California, and southern Australia. II. Dry matter, energy and nutrient accumulation. Australian Journal of Botany 17: 293-308

STARK NM (1977) Fire and nutrient cycling in a Douglas-fir/larch forest. Ecology 58: 16-30

STEWARD KK, ORNES WH (1975) The autecology of sawgrass in the Florida everglades. Ecology 56: 162-171

ST JOHN TV, RUNDEL PW (1976) The role of fire as a mineralizing agent in a Sierran coniferous forest. Oecologia 25: 35-45

VIRO PJ (1974) Effects of forest fire on soil. In: KOZLOWSKI TT, AHLGREN CE (eds) Fire and ecosystems. Academic Press, New York, pp 7-45

VLAMIS J, GOWANS KO (1961) Availability of nitrogen, phosphorus and sulphur after brush burning. Journal of Range Management 14: 38-40

WAGLE RF, KITCHEN JH (1972) Influence of fire on soil nutrients in a ponderosa pine type. Ecology 53: 119-125

WHITE EM, THOMPSON WW, GARTNER FR (1973) Heat effects on nutrient release from soils under ponderosa pine. Journal of Range Management 26: 22-24

WOODMANSEE RG, WALLACH LG (1980) Effects of fire regimes on biogeochemical cyles. In: MOONEY HA, BONNICKSEN TM, CHRISTENSEN NL, LOTAN JE, REINERS WS (eds) Fire regimes and ecosystem properties. USDA Forest Service General Technical Report WO-26, pp 379-400

WRIGHT HE (1980) The role of fire in land-water interactions. In: MOONEY HA, BONNICKSEN TM, CHRISTENSEN NL, LOTAN JE, REINERS WA (eds) Fire regimes and ecosystem properties. USDA Forest Service General Technical Report WO-26, pp 421-444

YOUNGBERG CT, WOLLUM AG (1976) Nitrogen accretion in developing Ceanothus velutinus stands. Soil Science Society of America Proceedings 40: 109-112

12. Decomposition and Mineralization Processes in Mediterranean-Type Ecosystems and in Heathlands of Similar Structure

D. J. READ and D. T. MITCHELL

12.1 General background

The major resources of plant mineral nutrients in all soils exist as complex molecular aggregates from which they must be released before they become available for utilization by the vegetation. Since the rate of nutrient release often limits the primary productivity of an ecosystem, it is of fundamental importance to understand the major processes involved through decomposition and mineralization in the maintenance of nutrient cycling phenomena.

Plant nutrients are released from both mineral and organic complexes in soil as a result of either physical or biological processes. The major physical process is oxidation, either directly by fire or indirectly by heat from solar radiation. The physical process, and in particular fire, can be very important in mediterranean climates which characteristically experience seasonal drought, and is the subject of a separate paper (Chapter 11, this volume). Biological processes involve degradation by the soil macro- and microfauna and decomposition by the microflora.

While nutrient liberation both to the plant and to the soil and atmosphere occurs as a dramatic event in the course of burning, it is clear that the nutrient turnover necessary to maintain productivity of the ecosystem must be sustained for long periods between fires by other means. The fire cycle may last up to forty years in fynbos (Kruger 1979) and a similar length of time may elapse between fire in Californian chaparral (Kittredge 1955) and South Australian heath (Specht et al. 1958). During the intervening period litter fall and root breakdown will contribute the main annual increment of nutrients to the soil and for some minerals, e.g. nitrogen, these supplies of organic material may represent the only significant potential source of nutrient. Release of such a nutrient in the decomposer cycle can thus limit ecosystem development. The availability of the major elements such as phosphate from mineral sources varies, but it is evident from the work of Specht et al. (1958) and Kruger (1979) that many mediterranean-type heathlands have very low total and available phosphorus levels in the mineral horizons, probably because of the great age of the soils and the long history of leaching. Under such circumstances the major reservoir of phosphorus and nitrogen in the mediterranean ecosystem is the soil organic matter. Since the latter represents the ultimate source of some of the most important plant nutrients it is necessary to examine litter and root production and their turnover in the ecosystem. Measurements are required of litter and root production, litter quality, their decay rates, and of incorporation of organic residues into the soil. Some examples of such determinations for mediterranean-type ecosystems are discussed before we consider the process of mineralization of specific elements within the litter and soil organic matter in detail. Despite recent studies of root production in chaparral vegetation (Kummerow et al. 1978) little is yet known of the dynamics of root turnover in mediterranean-type ecosystems. Reports from other vegetation types (Santantonio et al. 1977) indicate that a considerable proportion of the total nutrient turnover takes place in the below-ground system. Research on root turnover in mediterranean-type ecosystems is urgently needed.

12.2 Litter production

Annual litter production for different sclerophyllous shrub communities and for contrasting communities are shown in Table 1. In mediterranean systems regular disturbance by fire means that the steady state of litter dynamics is difficult to identify. Different types of plant community approach the steady state at different times after a fire. Litter fall for several years after burning is inevitably small (Kittredge 1955; Specht et al. 1958; Chapman 1967). Kittredge (1955) showed that in chaparral dominated by different species, the number of years required to reach deposits equal to those before fire varied considerably, from six for *Quercus wislizenii* to seven for *Arctostaphylos glandulosa* and 14 for *Ceanothus leucodermis* and *Arcostaphylos glauca*. In stands of *Calluna* on dry heaths of England, equilibrium levels are approached only 20 years after burning (Chapman 1967).

Values in Table 1 have been selected to represent near-equilibrium conditions and hence maximum rates of deposition. They show that the highest levels of litter production in mediterranean-type ecosystems are of the order of 300-400 g m^{-2} yr^{-1}, while the lowest are close to 100 g m^{-2} yr^{-1}. The highest figures are thus comparable with those obtained in temperate deciduous woodland (Lutz and Chandler 1946), but are considerably less than those obtained in savanna (Russell 1973) or tropical forest (Nye 1961). When the whole period between fires is taken into account the mean annual figures of litter production are inevitably considerably lower. Thus litter-fall values obtained only for one period within the fire cycle must be interpreted cautiously. They also indicate that over a complete fire cycle the nutrient fund contained in the annual litter fall will be small compared to that found in ecosystems not exposed to regular burning.

While the quantities of litterfall to some extent indicate the *potential* availability of nutrients, their *actual* availability depends upon litter quality and upon environmental conditions at and below the soil surface, since both these factors have more direct effects upon decomposition processes.

12.3 Litter quality

Leaves constitute the largest proportion of the above-ground litter fall, making up 69% in scrub-heath of eastern Australia (Maggs and Pearson

Table 1. Total litter production (g m^{-2} yr^{-1}) in different mediterranean-type ecosystems and in other contrasting vegetation types.

Vegetation type	Dominant plant	Region	Litter production	Authority
Chaparral	*Adenostoma fasciculatum*	California	87	Kittredge 1955
"	*Garrya veatchii*	"	322	" "
"	*Arctostaphylos* spp.	"	384	" "
Garrigue	*Quercus coccifera*	France	230	Lossaint 1973
"	*Quercus ilex*	"	390	" "
Heath	Mixed	Victoria, Australia	154	Jones 1968
Coastal-scrub heath	*Banksia*	Sydney, Australia	490	Maggs and Pearson 1977
Fynbos	Mixed	Bains Kloof, Cape, South Africa	200	E J Moll, unpublished
Dry heath	*Calluna*	Southern England	100 (10 yr after fire)	Chapman 1967
" "	"	"	300 (20 yr after fire)	" "
Dry savanna	*Prosopis*	India	500	Rodin and Bazilevich 1967
Moist savanna		Equatorial Africa	120	" " "
Tropical forest	*Celtis-Triplochiton*	Ghana	123	Nye 1961
Temperate forest	*Fagus-Quercus*	Germany	350-650	Rodin and Bazilevich 1967

1977) and 78% in chaparral (Mooney 1977). Leaf characteristics are probably therefore the most important factor influencing litter quality. Many of the most important plant species of this type of vegetation are sclerophyllous (Beadle 1966; Kummerow 1973; Kruger 1979; Mooney and Rundel 1979). This feature may have evolved originally in a humid environment as a response to low phosphate levels (Specht 1979) but it appears to have been selected subsequently as an adaptation to periodic drought as well as nutrient stress. While the sclerophyllous habit may significantly improve the water and nutrient economy of living plants Monk (1966) has also suggested that it aids the conservation of nutrients of the ecosystem because such tissues as litter appear to decompose relatively slowly.

Nutrient conservation begins in all plants prior to leaf fall. Thus, in deciduous species, shortening day length which stimulates the production of abscisic acid and later the abscission layer also leads to reabsorption of significant quantities of soluble nutrients in the storage tissues. Recent work of Schlesinger and Hasey (1981; see Table 2) indicates that this "internal" nutrient cycling may be particularly well developed in mediterranean evergreens, significant amounts of nitrogen and phosphorus being translocated out of the leaf before drought-induced abscission. Such translocation will greatly increase the carbon:nitrogen and carbon:phosphorus ratios of the leaves and this will further increase ecosystem nutrient conservation by reducing the rate of microbial decomposition. As the tissues of the leaf die, there is, by definition, a loss of membrane integrity and a consequent breakdown of compartmentation within the cells. Even at this early stage considerable differences can occur between patterns of nutrient loss from mesophytic and sclerophyllous leaves. Nykvist (1963) has demonstrated that rainfall can leach material equivalent to 20% of the dry mass from freshly fallen leaves of mesophytic species (eg Fraxinus) whereas leaves of sclerophyllous plants (eg Pinus) lose only 1%. Most of the loss is in the form of simple organic nitrogen compounds. These high losses are very much smaller if the leaves are leached several days after the fall. Thus the pattern of nutrient loss is determined partly by leaf constitution and partly by the timing of the leaching event. In mediterranean ecosystems leaf shedding is usually associated with summer drought. Thus both sclerophylly and the absence of leaching at the time of leaf fall may combine to minimize early release of nutrients. The period from abscission to the first rainfall may nonetheless be very important in sclerophyllous vegetation because during this time transformations occur which lead to complexing of potentially leachable nutrients within the framework of the leaf and the conservation of these materials as organic complexes with secondary metabolism (Swift et al. 1979). This means that when the first rains are experienced leaching losses are likely to be minimal.

The basic structural differences between sclerophyllous and mesophytic leaves (Kummerow 1973) are paralleled by differences in their organic and inorganic constitution which may be expected to influence nutrient conservation and litter decomposition (Loveless 1961, 1962). Sclerophyllous leaves generally have lower concentrations of mineral elements such as nitrogen and phosphorus and higher concentrations of organic polymeric substances such as phenolics and and lignin (Handley 1954). Thus the level of total nitrogen in freshly fallen leaves of ericaceous shrubs is normally low. In Rhododendron ponticum, for example, it is only 0.40% compared with levels in deciduous mesophytes such as Fraxinus of 1.55% (Bocock 1964). At the same time leaves of evergreen ericaceous plants (eg Rhododendron, Erica and Calluna) have exceptionally high levels of total extractable phenolic material (Handley 1954; Ballester et al. 1975; Read and Jalal 1980). The effect of these characteristics is

Table 2. The relationship between the major features of litter quality and decomposition rate in the evergreen sclerophylls *Ceanothus megacarpus* and *Calluna vulgaris*, the drought-deciduous shrub, *Salvia mellifera* and some mesophytic species. Data selected from Heal et al. (1978), and Schlesinger and Hasey (1981).

Species		%C	%N	%P	% soluble carbohydrates	% tannin	% cellulose	% lignin	% mass loss 1st year	% mass loss 2nd year
Calluna vulgaris	roots	53	0.69	0.034	2.1	3.9	51	49	5	11.4
" "	stems	51	0.56	0.638	1.3	2.6	68	31	7.5	13.2
" "	shoots	55	1.38	0.069	4.2	3.0	42	39	19.7	28.2
Eriophorum vaginatum	leaves	49	0.97	0.041	3.5	4.3	61	22	22.0	29.0
Eriophorum angustifolium	"	50	0.94	0.050	5.7	-	56	23	17.6	32.5
Nardus stricta	"	46	0.53	0.031	2.2	2.1	72	12	22.8	38.9
Rubus chamaemorus	"	50	1.31	0.066	8.2	27.0	34	6	36.1	50.1
Ceanothus megacarpus	before abscission	-	1.60	0.111	10.4	7.3	27	24	15-19	-
	after abscission	-	0.63	0.028	9.7	9.9	21	31		
Salvia mellifera	before abscission	-	0.83	0.089	7.5	6.8	29	14	20-24	-
	after abscission	-	0.65	0.133	6.2	8.0	31	16		

to produce high carbon:nitrogen ratios in sclerophyllous tissue, and these in turn have significant negative correlations with rates of decomposition (Van Cleve 1974). Carbon:nitrogen ratios, however, provide only a general, indirect measure of the relationship between constitution of the leaf and decomposition. Recent studies have shown that a more complete understanding of the interaction between litter quality and decomposition can only be obtained by means of systematic simultaneous analysis of a wide range of organic and inorganic leaf components. For example, Heal et al. (1978) compared analyses of the major leaf constituents of species of wet heath at the time of litter fall to determine which features of leaf quality were most closely associated with rates of decomposition. The sclerophyllous Calluna vulgaris had considerably higher lignin content than most of its more mesophytic associates, but no major differences were detected in the levels of other organic or inorganic constitutents (Table 2). In view of the high levels of tannin reported by other workers (see above) the low "soluble" tannin levels reported by Heal et al. (1978) are surprising. They serve to indicate that great care must be taken in interpretation of figures for "soluble" and "insoluble" levels or organic materials, in particular "tannins" (this subject is discussed in more detail in the concluding section of this review). Schlesinger and Hasey (1981) showed in their study (Table 2) that in addition to differences between the sclerophyllous Ceanothus megacarpus and the drought-deciduous Salvia mellifera in terms of the nitrogen and phosphorus contents of their litter, there are differences in lignin content. In Ceanothus lignin content is almost double that of Salvia at the time of litter fall. Decomposition studies have shown that such differences between sclerophyllous and more mesophytic leaves can greatly influence their rates of breakdown.

12.4 Litter decomposition

The method of employing litter bags has been the most widely adopted experimental approach to the analysis of decomposition rates and hence provides the best opportunity for comparative studies within and between ecosystems. Introduced by Bocock et al. (1960), it has since been used for analysis of decomposition in many ecosystems including tundra (Heal and French 1974), temperate deciduous forest (Anderson 1973; Gosz et al. 1973) and coniferous forest (Fogel and Cromack 1977) and has now been employed in mediterranean ecosystems (Yeilding 1977; Schlesinger and Hasey 1981). Known quantities of litter are confined in nylon mesh bags, and loss of total mass or of mass of litter constituents separately is determined periodically in the natural environment. The question of mesh size has been the subject of some debate (Singh and Gupta 1977) since one must strike a balance between excessively small mesh which could inhibit gaseous diffusion (Witkamp and Olson 1963) and movement of detritivores (Singh and Gupta 1977), and a large mesh which would permit loss of fragments of litter and entry of contaminating material (Gosz et al. 1973). Most workers have accepted a 1 mm mesh size as an adequate compromise.

In her preliminary study Yeilding (1977) found that evergreen chaparral foliage decomposed at the rate of 11-24% per year. The more detailed study of Schlesinger and Hasey (1981) revealed that loss rates over a one-year period in chaparral were 15 to 19% for the evergreen Ceanothus and 20 to 24% for Salvia. The lowest figure was obtained at a high altitude site (910 m), and the highest figure at a low altitude site (350 m).

Analysis of specific leaf constituents revealed that levels of soluble carbohydrates, "total" phenolics, and potassium were reduced by more than

70% in one year whereas losses of calcium, magnesium and cellulose were smaller. No release of nitrogen or lignin was apparent during this time. The high initial lignin content of Ceanothus combined with the absence of lignin degradation in both species suggests that lignin may be the major factor controlling the rate of decomposition. The results of Heal et al. (1978) further emphasize this possibility. In the range of species examined by these workers the only organic compound to show a significant negative linear relationship with mass loss was lignin. Since mediterranean-type ecosystems are dominated by sclerophylls with high lignin contents it may be expected that decomposition would be slow. Analysis of turnover of organic matter in such systems, however, indicates that rates are now as low as might be predicted from consideration of lignin alone.

12.5 Litter turnover rates

When mass loss of decomposing litter is plotted against time a negative exponential curve is obtained which suggests that there is a constant fractional loss rate (k) with time (Jenny et al. 1949; Olson 1963), leading to a decay model which will estimate the rate of annual weight loss:

$$X/X_o = e^{-kt} \qquad (1)$$

X is mass of litter or litter constituent remaining at time t, X_o is the original mass, e is the base of natural logarithms, and k is the decomposition constant or fraction of mass lost. This model has been successfully applied in a number of ecosystems to obtain values of k, and from k to determine the mean resident or turnover time (1/k) for organic matter (Jenny et al. 1949; Olson 1963). Values of k obtained for various types of ecosystems on the assumption that they approximate a steady state are shown in Fig. 1. They are closely related to the latitude of the ecosystem, which suggests a strong climatic influence on decomposition, varying from k = 1 or greater in the moist tropics to k = 0.01 in tundra and k = 0 in waterlogged anaerobic systems such as bogs.

Values obtained for chaparral by Schlesinger and Hasey (1981) are in the range of 0.20 to 0.23. The mean turnover time for litter of the chaparral shrubs was 2.5 years for Salvia and 4.6 years for Ceanothus. Values in the same range were obtained from mass loss data of Kittredge (1955) for chaparral, Lossaint (1973) for garrigue, and Maggs and Pearson (1977) for Banksia scrub in Australia.

When such calculations are applied to the turnover of individual elements, a mobility series for the major inorganic and organic constituents of the litter can be obtained. The mobility series described by Schlesinger and Hasey (1981) was potassium > magnesium > calcium > nitrogen for inorganic nutrients and phenolic > soluble carbohydrate > cellulose > lignin for organic constituents; there was no consistent pattern of mobility of the phosphate ion. These results, including the mobility series, are similar to those reported in a recent review by Lousier and Parkinson (1978). The mobility series for inorganic elements recorded by Lossaint (1973) in garrigue was potassium > phosphorus > magnesium > nitrogen > calcium and by Maggs and Pearson (1977) for Australian heath potassium > phosphorus > calcium > magnesium > nitrogen. With the exception of the phosphate ion these figures are what might be predicted given our knowledge of the relative affinities of the elements for organic complexes. The susceptibility of potassium and soluble carbohydrates to very rapid leaching by rainfall means that the constant

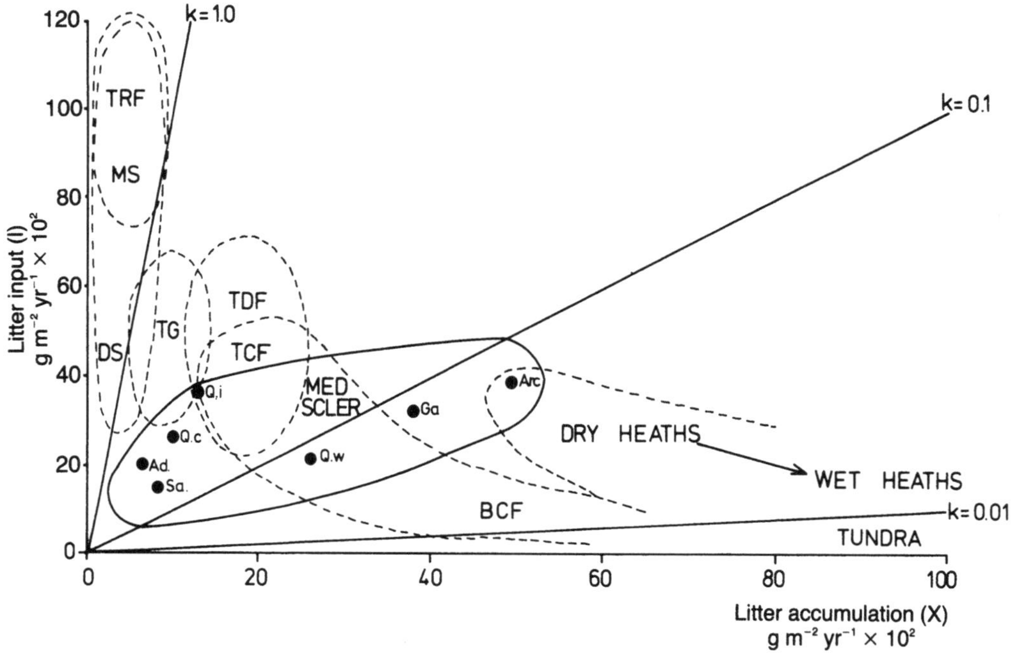

Fig. 1. The relationship between litter input (I), litter accumulation (X) and rate of decomposition (k) in a range of different ecosystems. The plot is designed to indicate the position of mediterranean-type sclerophyllous vegetation (MED SCLER) which is delimited by a continuous line, relative to other types which are delimited by a dashed line. The points within the mediterranean-type vegetation are derived from measurements of Kittredge (1955) and of Lossaint (1973). Data for other vegetation types were derived from Jenny et al. (1949), Nye (1961), Olson (1963), Chapman (1967), Rodin and Bazilevich (1967) and Heal et al. (1978). Abbreviations are TRF: tropical rainforest, MS: moist savanna, DS: dry savanna, TG: temperate grassland, TDF: temperate deciduous forest, TCF: temperate coniferous forest, BCF: boreal coniferous forest, Q.i. *Quercus ilex*, Q.c. *Quercus coccifera*, Q.w. *Quercus wislizenii*, A.f. *Adenostoma fasciculatum*, S.m. *Salvia mellifera*, G.v. *Garrya veatchii*, A.g. *Arctostaphylos glauca*.

fractional loss rate k may not appropriately describe their loss (Minderman 1968), and an instantaneous fractional loss rate is more applicable in these cases. Bunnell and Tait (1974) have employed a two-phase exponential expression to describe the pattern of loss more accurately:

$$W(t) = W_1 e^{-k_1 t} + W_2 e^{-k_2 t} \quad (2)$$

where $W(t)$ is mass at time t, and W_1 and W_2 are the proportion of materials lost rapidly and slowly, respectively. Schlesinger and Hasey (1981) have suggested that this equation may be used to describe the loss of the most soluble litter components in the chaparral vegetation.

12.6 Model systems for predicting decomposition

It is evident from the preceding discussion that certain features of litter quality, for example lignin content, and of climate are likely to be more important than others in influencing litter decomposition rates. Recently, Meentemeyer (1978) has used multiple linear regression analysis to link lignin content of litter and soil moisture status (expressed in terms of the surrogate, actual evapotranspiration AET) in a formula which can be used to predict decomposition rates in different parts of the world. The generalized model (Fig. 2) has been tested by Schlesinger and Hasey (1981) using their own data for Ceanothus and Salvia. A close relationship was observed thereby indicating that the model has some potential for practical use in mediterranean-type ecosystems. Thus with AET of 300 mm yr^{-1} Meentemeyer's formula predicts a 16.5% loss of ash-free dry mass for Ceanothus litter compared to the actual loss of 15% at an altitude of 910 m and 19% at 350 m.

In fynbos vegetation AET is likely to be higher than in chaparral and may exceed 500 mm yr^{-1} (Wicht 1971). On the assumption that plant lignin contents are similar to those of Ceanothus and that AET is 500 mm yr^{-1} rates of decomposition in fynbos would be considerably higher than in chaparral, reaching figures in excess of 30% per year. However, communities dominated by ericaceous and proteaceous plants may have higher

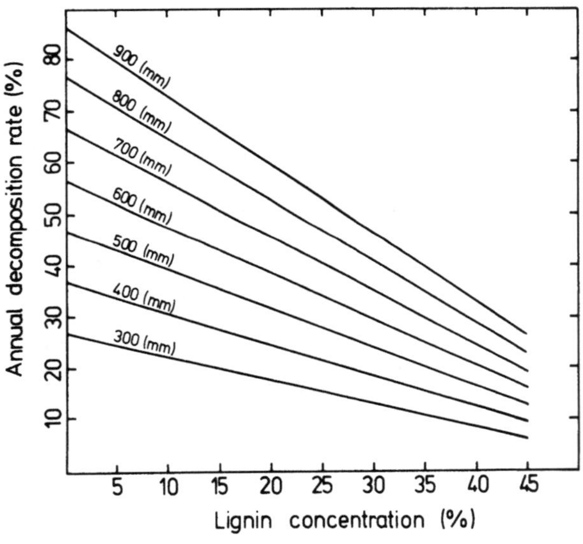

Fig. 2. Model of the relationship between annual decomposition rate (%) and lignin concentration (%) at different values of annual actual evapotranspiration (AET). The slopes were obtained using the following linear equation:

$$Y = (-3.446 + 0.1\ X_1) - (0.013 + 0.002\ X_1)X_2$$

where Y = annual decomposition rate (%), X_1 = annual AET (mm), and X_2 = lignin concentration (%)

(Figure reproduced by permission from Meentemeyer (1978). Copyright 1978, Duke University Press.)

levels of lignin (see for example Calluna levels in Table 2), and if so, this would be expected to reduce the rate to levels comparable with those of chaparral. Information on the organic constituents of members of the Proteaceae would be of great interest in this connection.

The other plant quality factor important in controlling decomposition rates is leaf nitrogen content, although not considered by Meentemeyer (1978). A model of decomposition based on nitrogen and lignin content of litter has been derived by Aber and Melillo (1980) using data from deciduous tree litter. For a number of tree litters a good inverse linear regression was obtained when percentage of original litter biomass remaining was plotted as a function of nitrogen content of the residual material. A similar relationship was found by Schlesinger and Hasey (1981) for Salvia litter, though that of Ceanothus did not respond in the predicted manner. In subsequent work Melillo et al. (1982) related the parameter k (constant fractional loss rate) to the ratio of initial lignin content to initial nitrogen content as an index of litter quality. They recorded high rates of decay in litters with low lignin:nitrogen ratios. Schlesinger and Hasey (1981) indicate that, while such a relationship may apply to mesophytic leaves in temperate climates, other factors may be involved in sclerophyllous vegetation in mediterranean regions since despite the nearly identical initial nitrogen content of the abscised leaves of their two test species, differences in decomposition rates between them were marked and were probably due to differences in lignin content.

All estimates of mobility show that nitrogen is the least readily mobilized of the nutrient elements. Thus like lignin, leaf nitrogen content has the greatest capacity to influence the rate of decomposition of litter. We shall now consider nitrogen mineralization in more detail.

12.7 Transformation of nitrogen compounds in mediterranean soils

The slow mobilization of nitrogen in litter was recognized in some of the earliest decomposition studies. Bocock (1963, 1964) found that litter nitrogen levels actually increased in the first year after leaf fall and ascribed the increases to absorption of nitrogen from rainfall and the incorporation of faecal remains of soil animals. Since no mineralization occurs in the first year the inputs produce a net gain in nitrogen content. A similar situation was found in the chaparral shrubs analysed by Schlesinger and Hasey (1981). The nitrogen concentration in the Salvia litter bags almost doubled during the year of observation, while that in Ceanothus litter remained constant. The absence of mineralization of nitrogen during the first year is presumably due to the unfavourable carbon:nitrogen ratios for the growth of micro-organisms capable of mineralization. Carbon:nitrogen ratios are at their highest when litter is fresh and it has been reported that nitrogen mineralization occurs only at carbon:nitrogen ratios of 30:1 or less (Lutz and Chandler 1946; Dutt et al. 1972). Carbon:nitrogen ratios of litter were shown to be about 80:1 in Ceanothus and 77:1 in Salvia at the time of abscission but to decline steadily through the year, reaching the lowest level at 44:1 in Salvia after one year (Schlesinger and Hasey 1981).

With the exception of communities dominated by leguminous species, microbial decomposition of organic matter is the most important process by which nitrogen is made available for plant growth (Rosswell 1976; Lee and

Stewart 1978). The quantity of organic matter thus provides an indirect measure of potentially mineralizable nitrogen (Table 3). Measurements of soil nitrogen content in a range of soils from mediterranean-type ecosystems indicate that nitrogen reserves of these soils are generally low (Table 3). Australian heath and South African fynbos soils have particularly low levels as would be expected from their low organic matter contents. Mineralization commences with the well-established chain of events involving proteolysis, then ammonification. Ammonium is the major nitrogen source for plants of acidic soils but in soils with pH above 4.5 and in those of semi-arid zones nitrification becomes an increasingly important process and the nitrate ion may be the major nitrogen source (Ellenberg 1977). The products of proteolysis are mainly amino acids. These have been detected in soils (Dadd et al. 1953) but, because they are not in equilibrium with a large source of supply, they have been regarded as an insigificant nitrogen source for plants (Kowlenko 1978). Recent research indicates that under some circumstances they may be of importance (see below).

Measurements of total or available nitrogen levels provide little information on the dynamics of release processes (Harmsen and Van Schreven 1955; Ellenberg 1977). A more useful method of evaluating nitrogen mobilization is to measure release of NH_4^+ and NO_3^- over a period during which soil cores are incubated under controlled conditions. The rates of release under these circumstances have been shown to be closely correlated with plant nitrogen uptake (Stanford et al. 1973; Ellenberg 1977). The results of such incubations can be used to produce simulation models which have provided reliable quantitative estimates of nitrogen mineralization in agricultural situations (Stanford et al. 1973; Smith et al. 1977; Watts and Hanks 1978). Nitrogen availability and the possible application of simulation models have recently been examined in chaparral ecosystems by Marion et al. (1982). These workers found that when the values for accumulated nitrogen release (N_t) were fitted to the three-parameter equation, i.e.:

$$\log(N_o - N_t) = \log N_o - k_t^b \qquad (3)$$

where N_o is potentially mineralizable nitrogen, k is the rate constant, t is time and b is the time exponent, a model was obtained which accurately predicted the rate of nitrogen release. Also, there was a significant linear relationship between potential mineralizable nitrogen and per cent soil nitrogen. The potentially mineralizable nitrogen for chaparral soils was shown to be very low compared to agricultural soils (a feature which they assumed to be due to poor litter quality). Thus at 0.15% nitrogen content the chaparral soil would release 43 $\mu g\ g^{-1}$ of nitrogen, whereas at the same content an agricultural soil would release 223 $\mu g\ g^{-1}$ (Marion et al. 1982). Only 0.1 to 4.4% of the total soil nitrogen is mineralizable in chaparral against 4 to 41% in agricultural soils. Marion et al. (1982) showed that the nitrate ion was the most important nitrogen source in the chaparral sites, thereby indicating that nitrification is an important process in this ecosystem. A similar situation prevails in garrigue for which Lossaint (1973) reported that 90% of the mineralized nitrogen is in the nitrate form.

Major peaks of mineralizing activity are found in autumn and spring. Two peaks of activity were also reported by Marion et al. (1982). In fynbos soils of the Clovelly type (yellow-brown, well-drained sandy soil of aeolian origin), Stock and Lewis (unpublished) have shown that NH_4^+ levels are higher than those of NO_3^-, ranging from 2 $\mu g\ N\ g^{-1}$ at the surface to 1 $\mu g\ N\ g^{-1}$ at 1.0 m depth. Mineralization and periodicity phenomena in fynbos soils remain to be investigated.

Table 3. Comparison of organic matter, nitrogen and phosphorus contents of soils of mediterranean-type ecosystems and of related vegetation types. All figures are expressed as % of dry mass.

Vegetation type	Dominant plant	Area	Organic matter	Total N	Total P
Chaparral	*Adenostoma fasciculatum*	California	3.9[a]	0.04[a]	
	Ceanothus greggii	"	4.0[a]	0.08[a]	
	Salvia spp.	"	9.0[a]	0.18[a]	
	Eriogonum fasciculatum		5.9[a]	0.10[a]	
Chaparral					
1-yr old	*Adenostoma fasciculatum*	California		0.17[b]	0.12[b]
9-yr old	" "	"		0.18[b]	0.04[b]
37-yr old	" "	"		0.33[b]	0.07[b]
Heath A_1 horizon	*Banksia-Xanthorrhoea-Casuarina*	Keith, South Australia	0.2-1.6[c]	0.02[b]	0.001[b]
Mallee-broombrush A_1 horizon	*Eucalyptus incrassata-Melaleuca uncinata*	Keith, South Australia		0.04[b]	0.006[b]
Garrigue					
1-yr old	*Quercus coccifera*	Montpellier, France		0.40[b]	0.031[b]
11-yr old	" "	Montpellier, France		0.48[b]	0.038[b]
Coastal fynbos	*Leucospermum parile*	Pella, Cape	1.0-3.0[e]	0.02-0.03[d]	0.002-0.003[e]
Upland heath	*Calluna*	Yorkshire, England	40-90[g]	0.98[f]	0.10[g]

References: [a] Marion et al. (1982). [b] Specht (1969). [c] Specht et al. (1958). Figure obtained by multiplication of values for carbon. [d] Stock and Lewis (unpublished). [e] Brown, Jongens and Mitchell unpublished). [f] Stribley and Read (1974). [g] Read and Jalal (unpublished).

12.8 Transformation of phosphorus compounds by soil micro-organisms

Although there is a vast literature on transformations of phosphorus compounds by soil micro-organisms, very little has been reported from mediterranean regions of the world. Data of total pool sizes in leaf litter and soil are available (Table 3) but there is no information on the types of phosphate. The phosphorus content of annual litter fall is small for *Adenostoma fasciculatum*, a Californian chaparral dominant (Mooney and Rundel 1979), and for *Quercus coccifera*, a garrigue dominant (Lossaint 1973), compared with that of plants from Californian annual grasslands (Jones and Woodmansee 1979) and heather ecosystems (Robertson and Davies 1965), but in view of the effective internal cycling of phosphorus demonstrated in *Ceanothus* by Schlesinger and Hasey (1981) this feature may not necessarily indicate a low level of phosphorus in the living plant.

The pool of available phosphorus (Bray no. 1 extract) in the surface soil of a chaparral ecosystem (Rundel and Parsons 1980) is considerably lower than that of an annual grassland (Jones and Woodmansee 1979). The soil pool beneath a canopy of *A. fasciculatum* to a depth of 100 mm is 75.0 g m^{-2} (or 95.6% of the total), whereas available phosphorus may be as little as 1.5% of the total (Mooney and Rundel 1979). Available phosphorus in Californian chaparral and Chilean mattoral soils at the surface ranges from 1.8 to 51 μg P g^{-1} dry mass (Bray no. 1 extract: Thrower and Bradbury 1977), and in the Clovelly soil form of South African coastal fynbos is 2.5-4.5 μg P g^{-1} dry mass (Bray no. 2 extract: Brown, Jongens-Roberts and Mitchell, unpublished). Either mineralization of organic phosphates by soil micro-organisms in mediterranean ecosystems is a slow process or any accumulation of available phosphorus in the soil is immediately absorbed by the root systems or alternatively precipitates as insoluble inorganic minerals.

As in the case of nitrogen, the rate of mineralization of phosphorus will depend upon the nature of the litter. It is probably inhibited by high carbon:phosphorus ratios as well as by the presence of secondary metabolites in sclerophyllous litter.

Soil organic phosphates consist of a combination of inositol phosphates, phospholipids, nucleic acids and sugar phosphates (Cosgrove 1967). Phospholipids, nucleic acids and sugar phosphates are rapidly decomposed whereas inositol phosphates, as insoluble salts of calcium, iron, magnesium and aluminium, are more resistant to breakdown. Organic phosphates can be hydrolyzed by bacteria (Greaves et al. 1967), free-living soil fungi (Dox and Golden 1911), mycorrhizas (Bartlett and Lewis 1973; Ho and Zak 1979) and nonmycorrhizal roots (Woolhouse 1969).

The enzymes which are involved in the catalytic release of orthophosphate from organic phosphates are generally known as phosphatases. These tend to be cell-surface-bound enzymes which are usually active under acid conditions but are inhibited by the presence of free orthophosphate. The chemical environment of the soil may be very important in determining the activity of total soil phosphatase. Free orthophosphate levels in mediterranean soils are very low and are not likely to inhibit phosphatase activity. The sandy soils of coastal fynbos of South Africa have a pH of 3.5-4.8 (Brown, Jongens-Roberts and Mitchell, unpublished) which should favour the fungal and mycorrhizal phosphatase enzymes. Plant phosphatases have an optimum pH at levels in the range pH 5.0-6.0 (Cosgrove 1967).

The presence of insoluble inorganic forms of phosphate in the soil must not be overlooked. Insoluble inorganic phosphates are mainly present as salts of iron and aluminium in acid soils, whereas calcium is more frequently bound to inorganic phosphates in neutral to alkaline soils (Hayman 1975). Insoluble inorganic phosphates of the Clovelly soil form from coastal fynbos are largely bound to iron (Brown, Jongens-Roberts and Mitchell, unpublished). Inorganic phosphate may also be weathered from rock phosphate sources such as apatite which are very insoluble under alkaline conditions. Many of the soils of southern France and Spain are neutral to alkaline (Lossaint 1973; Barea et al. 1975) and under these conditions the release of phosphorus from rock phosphate is thought to occur only as a result of bacterial action (Barea et al. 1975). Two bacterial genera, *Pseudomonas* and *Agrobacterium*, have been found to solubilize rock phosphate as well as secrete plant growth-regulating substances (Azcon et al. 1976). Lavender plants (*Lavandula spica* var. *vera*) inoculated with vesicular-arbuscular (VA) mycorrhizal fungi and phosphate-solubilizing bacteria were grown in Mediterranean alkaline soils amended with rock phosphate (Azcon et al. 1976). Plants with mycorrhizas and bacteria grew better and took up more phosphorus than plants not inoculated with either the bacteria or the VA mycorrhizas (Azcon et al. 1976). The bacteria were able to solubilize rock phosphate and although some of the dissociated phosphate was taken up by the bacteria, a sufficient amount was made available to stimulate plant growth. As mycorrhizal plants have been shown to take up more phosphate than nonmycorrhizal ones the solubilized phosphate released by the bacteria was probably transferred to the root systems via the VA mycorrhizas (Azcon et al. 1976).

12.9 Biological activity and litter decomposition

12.9.1 SOIL FAUNA

Soil animals are thought to play an important though largely indirect role in organic matter decomposition. They have been shown to influence the decomposition process mainly by increasing fragmentation of the litter (Edwards and Heath 1963). This has a beneficial physical effect in providing increased surface areas for attacks by the microflora. Careful studies of chemical breakdown of litter within the gut of soil arthropods such as millipedes, however, have revealed that carbon:nitrogen ratios can be considerably reduced after digestion. Thus feeding has some capacity to increase humification, and will certainly reduce the size of ingested polymers, thereby making them more susceptible to attack by the microflora (Marcuzzi 1970). While earthworm activity is generally much reduced in drought-susceptible habitats (Lofty 1974), arthropods, in particular the strongly chitinized species, are well represented (Di Castri 1973). Di Castri (1973) has stressed the need for more detailed investigation of the role of soil animals in decomposition processes of mediterranean areas.

12.9.2 SOIL MICROFLORA

Studies of biological activity, mass loss and CO_2 evolution in soil organic matter have made it clear that most of the dead plant tissues are decomposed by the microflora (Witkamp 1966; Witkamp and Crossley 1966).

Most of the major genera of both fungi and bacteria involved in litter decomposition are of cosmopolitan distribution, including the genera Penicillium and Aspergillus which have been reported to be particularly prevalent in the Mediterranean Basin (Schaeffer 1973). Coley and Mitchell (1980) found that in fynbos soils representatives of the genera Absidia, Mortierella, Penicillium and Trichoderma were common, a list of soil fungal saprophytes which are not very different from those reported from European heathlands (Sewell 1959; Burgeff 1961; Latter et al. 1967).

The bacterial flora may have an unusually large number of sporulating types and Bacillus subtilis, B. licheniformis and B. megatherium have been isolated (Schaeffer 1973). In the relatively alkaline soils of the Mediterranean Basin Azotobacter and other free-living N_2-fixing bacteria have been recorded (Sasson and Daste 1963). No studies have been made on bacteria involved in litter decomposition in fynbos soils, but in his search for free-living N_2-fixing organisms Becking (1959) could find neither Azotobacter nor the related N_2-fixing bacterium Beijerinckia which normally occurs in more acid soils, though they were present in forest soils of the area.

12.10 The effect of climatic factors on microbial activity

12.10.1 TEMPERATURE

Most of the fungi and bacteria involved in normal terrestrial decomposition processes are mesophiles (Pugh 1974). They have temperature optima for activity between 25 and 35°C and a minimum temperature for growth between 5 and 20°C. This is true even for decomposer organisms of the tundra (Flanagan and Veum 1974). This explains the fact that, providing other environmental factors are not limiting, the rate of decomposition in climatic zones with longer warm seasons is normally higher than that in cooler climates. Differences between mediterranean-type areas can thus be predicted on the basis of temperature. The fynbos areas of the western Cape with mean annual temperatures of 12°C to 18°C and a winter temperature mean of 10°C to 12°C (Kruger 1979) may expect little inhibition of microbial activity because of low temperature. In the south of France, however, frosts are not uncommon in winter and the mean minimum January temperature of 1°C (Lossaint 1973) suggests that decomposition may be significantly inhibited.

12.10.2 WATER AVAILABILITY

Marked seasonality of rainfall in mediterranean climates means that water availability is normally the major climatic factor which can limit microbial activity. Lossaint (1973) has shown that in mediterranean regions of France maximum levels of microbial activity, as indicated by CO_2 evolution, are found in the warm summer months, with depressed activity in July and August when extreme drought conditions prevail. Thus there are normally two peaks of microbial activity in this type of climate. The first occurs in spring and may be primarily a response to increasing soil temperatures; the second is in autumn when the first rains break the summer drought. The flush of activity is associated with the logarithmic phase of growth of the microbial population. These growth flushes, of great significance for the processes of nutrient release, are largely responsible for the seasonal changes in pattern of nitrogen mineralization described earlier. Areas with markedly seasonal climates, especially those characterized by sudden episodes of rainfall, are prone to suffer severe nutrient losses through leaching and in such environments mycorrhizas may become particularly prevalent. These structures reduce the risk of nutrient loss by providing direct communication between the root and microsites at which mineralization is occurring.

12.11 Mycorrhizal infection and nutrient absorption

Two major types of mycorrhizal association are recognized, namely sheathing or ectomycorrhizas in which the fungus develops outside the root, and endomycorrhizas in which the fungus develops among the root cells. Sheathing mycorrhizas, usually associated with wide areas of seasonal climate, are found on trees of boreal, temperate and seasonal subtropical zones, for example, *Pinus*, *Quercus* and *Eucalyptus*. It has been suggested that the prevalence of this type of mycorrhiza in such climatic zones is associated with the capacity of the fungal sheath, which constitutes up to 40% of the dry mass of the root (Harley 1969), to accumulate and store nutrients released during the short periods when soil conditions permit microbial mineralization. It has been shown in the case of *Eucalyptus* and *Fagus* mycorrhizas, for example, that mineral phosphate absorbed from solution can be stored in excess as polyphosphate and slowly released to the root during subsequent periods of low phosphorus supply (Ashford et al. 1975; Chilvers and Harley 1980).

Endomycorrhizas are characteristic of areas with similar climate but lower nutrient status. Thus when forest vegetation gives way to scrub or heath as a result of nutrient stress the dominant species in the more stressed environment normally have endo- rather than ectomycorrhizal associations. A good example of this transition is seen at the tree line in alpine and arctic regions where low temperatures directly inhibit mineralization (Haselwandter and Read 1980). A similar situation may be found in passing from mallee-broombush to heath in Australia, or in passing from forest to fynbos in South Africa. In all these cases the heath species which become dominant beyond the forest boundary lack the ectomycorrhizal association, and a proportion at least of the dominants, for example members of the Ericaceae or Epacridaceae, are characterized by endo-infection (Read, unpublished).

In chaparral an intermediate situation is found in which the frequently dominant genus *Arctostaphylos* is infected by the same fungi as those which form ectomycorrhizas with forest trees, but which in this case penetrate the cortical cells to form a so-called ect-endomycorrhizal infection (Zak 1976).

As yet the reason for the transition of mycorrhizal type is a matter for speculation. It is evident that the relatively massive external proliferation of fungal material in ectomycorrhizal systems is a considerable drain on the carbon resources of the host plant. As the association ages much of this carbon may be lost to the soil microflora. Endo-infection provides greater conservation of nutrients because most of the fungal biomass is within the cortical cells where it eventually breaks down as the useful physiological life of the association is completed. The nutrients contained in the fungal biomass can then be recycled within the host.

It has been suggested that the major nutritional function of ericaceous mycorrhizas in heathland is to enhance nitrogen nutrition (Read and Stribley 1973; Read 1978). Higher nitrogen levels in mycorrhizal plants result from enhanced uptake of mineral nitrogen sources when these are

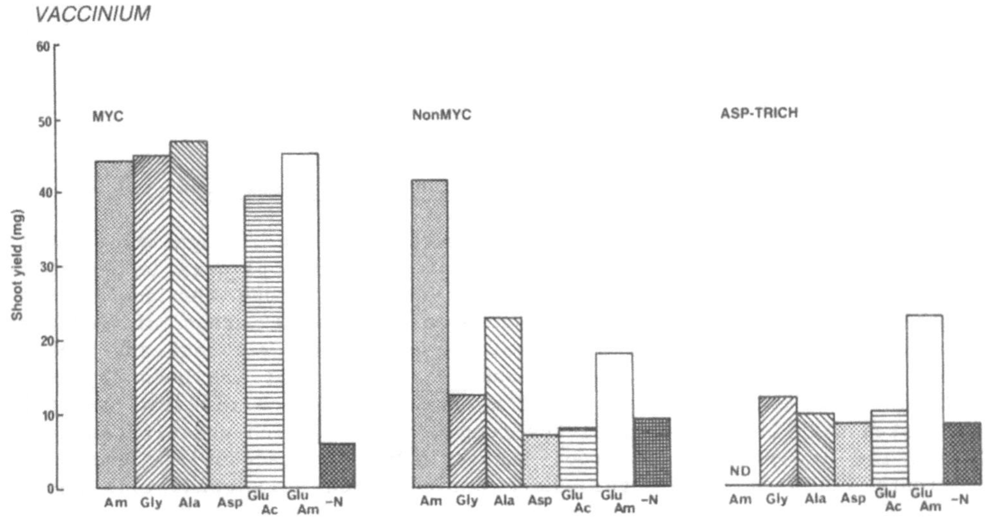

Fig. 3. Yield of the ericaceous plant Vaccinium macrocarpon when grown on mineral or simple organic nitrogen sources in the mycorrhizal (MYC), nonmycorrhizal (NonMYC) or saprophyte-inoculated (ASP-TRICH) condition. Abbreviations are Am: Ammonium, Gly: Glycine, Ala: Alanine, Asp: Asparagine, Glu-Ac: Glutamic acid, Glu-Am: Glutamine, -N: Without nitrogen. Yields of all MYC plants in the organic N treatments were significantly greater than the equivalent NonMYC or ASP-TRICH plants at $P < 0.001$ (Stribley and Read 1980).

present in otherwise growth-limiting concentrations (Stribley and Read 1976). The capacity to utilize simple organic nitrogen compounds is dependent upon mycorrhizal infection (Fig. 3; Stribley and Read 1980). Phosphate levels of shoots are also enhanced by ericoid infection and the mycorrhizal fungus can utilize complex organic and inorganic phosphate sources (Mitchell and Read 1981) as well as store phosphorus in the form of polyphosphate (Straker and Mitchell, unpublished). It therefore appears that mycorrhizas in the heathland ecosystems may function in a way similar to those of forest systems, and that the difference of type is associated with improved nutrient conservation through tighter cycling processes. Nutrients absorbed by mycorrhizal hyphae and translocated to the host cortical cells during flushes of soil microbial activity are stored in the intracellular hyphal complexes and released slowly to sustain shoot nutrient levels during the subsequent summer drought (Read 1978). This feature may help to explain the capacity of some fynbos Erica species to flower and set seed during the driest time of the year.

12.12 Other strategies for enhanced nutrient capture

While mycorrhizal plants have adopted a strategy whereby carbon is invested to maintain a fungal associate which enhances nutrient capture from soil, others have made a similar investment which makes them independent of soil nitrogen. Nodulated members of the genus Ceanothus are a very important component of chaparral vegetation whose presence would be expected to affect beneficially the nitrogen economy of communities which they dominate. The success of such species in chaparral may indicate that nitrogen is the major limiting nutrient for plant growth. By the same argument one may suggest that the relative absence of nitrogen-fixing plants in the mature vegetation of some Australian heaths and fynbos may indicate that nitrogen is not the limiting nutrient in these communities. Reduction of the N_2 molecule is an energy- and nutrient-demanding process and the major nutrient requirement is for phosphate in the form of ATP which acts as both reductant and energy supply. This phosphate, of which 12-15 molecules are required for every molecule of $N \equiv N$ reduced to NH_3, must be obtained from soil or diverted from other assimilative processes in the plant. Thus, environments extremely deficient in phosphate can support only low levels of N_2-fixing activity, and the absence of plants with nitrogen-fixing symbionts can be as much a measure of phosphorus deficiency as of a sufficiency of nitrogen. Nitrogen-fixing plants such as Phyllota and Casuarina in Australian heaths (Specht et al. 1958) and Aspalathus in fynbos (Kruger 1979) are often dominants in the earlier postfire phases of succession when soil nutrients such as phosphorus are more plentiful and available for uptake. The genera such as Podalyria, Psoralea and Aspalathus, which are probably independent of soil nitrogen supply, then fail to dominate in fynbos soils during the later stages of succession after fire, which supports the view that supplies of other nutrients are limiting their development. Similarly, the disappearance of Casuarina and Phyllota in Australian heath may be attributable to the reduction of phosphorus reserves released by fire to a level at which nitrogen fixation can no longer be sustained.

Harley (1971) has indicated that the strategy likely to be most successful for plants colonizing nutritionally marginal soils is to have both nodulation and mycorrhizal infection. The status of symbiotic relationships in heathland soils should be examined in more detail. Very little is known about the status of the VA endomycorrhizal association in these communities despite the fact that it has been shown that they can considerably enhance phosphate nutrition of infected plants (Hattingh et al. 1973; Mosse 1973). At the same time one must note that the roots of most members of the Proteaceae, the family dominating the poorer heaths, have neither nodules nor mycorrhizas. The success of these species is of great interest, as is the question of the initiation and function of their proteoid roots (see Chapter 14, this volume).

12.13 Concluding observations

Since the quantities and qualities of litter differ even between comparable ecosystems as a result of different species composition, careful measurements of both features are required as an initial step in understanding the nutrient cycling process. Internal as well as external cycling of nutrients clearly occurs, making measurements of leaf characteristics before, during and after abscission necessary to complete a balance sheet. As full a range of individual inorganic elements as possible should be examined, though nitrogen and phosphorus should be emphasized as the most likely elements to control rates of growth in the major mediterranean ecosystems. Adequate and easily applied methods exist for determining amounts of these elements (Allen et al. 1974).

Analysis of organic constituents of leaves is also necessary, since some of these materials interact so strongly with inorganic elements and directly or indirectly influence their release. Standard analytical procedures recommended by King and Heath (1967) and Allen et al. (1974) can be used since this makes comparison between studies possible. The interpretation of the results is not, however, straightforward. A particular problem concerns the quantitative and qualitative aspects of the fractions grouped under the headings "soluble" or "total" phenolics and "lignin". Phenolic materials in living tissue are often present in large quantities but mostly as molecules of relatively low molecular weight, with low reactivity in healthy tissue. As tissues senesce these form complexes with protein (Feeny and Bostock 1968; Feeny 1970) and become more condensed (Ribereau-Gayon 1972). In addition to rendering the substrate more toxic to bacterial (Byrde et al. 1960; Benoit and Starkey 1968) and fungal decomposers (Harrison 1971) the processes of complexing and condensation also decrease the solubility of the phenolic material. For this reason extractions with organic solvents like methanol give apparently decreasing levels of "soluble" phenolic acid when they are carried out over a long time period.

Decreases in phenolic content of litter extracts may therefore not represent loss, but precipitation (Suberkropp et al. 1976). This is supported by the observation that free phenolic acids are in extremely low levels in soils. Most of the reported occurrences of such substances refer to extraction following alkaline hydrolysis. Schlesinger and Hasey (1981) suggest that the apparent decline in phenolic content of Salvia litter after rainfall may be attributable to condensation rather than leaching. If, as Ribereau-Gayon (1972) suggests, such condensation is a common phenomenon it may also help to explain the parallel increase of nitrogen and lignin content in litter bags observed by Suberkropp et al. (1976) since some of the precipitated protein-polyphenolic complexes would be resistant to acid hydrolysis and may even become further condensed to form insoluble phlobophenes in the course of acid treatment. They would then be included in the fraction identified as lignin. Such addition would account for the apparent increase of lignin in Salvia litter reported by Schlesinger and Hasey (1981). This type of explanation is more probable than the alternative, which is that the complicated biochemical pathways involved in lignin synthesis are operating in decaying litter. More work is clearly necessary to identify the chemical nature of the various polymers produced in decaying litter, which are especially important since they are probably responsible for the retention of nitrogen in complexed form.

Access to these complexes and hence organic matter decomposition is essentially achieved by heterotrophs upon which the primary producers are dependent. The most effective heterotrophic system from the point of view of primary production is likely to be of a symbiotic nature since such associations provide the plant with direct access to nutrient sources. They thus "short-circuit" microbial transfer chains and reduce the chances of nutrient loss through leaching processes. More emphasis in future research should be placed on studies of those specialized plant-microbial associations which have evolved in mediterranean ecosystems in response to specific local circumstances of nutrient economy.

Acknowledgements

Thanks are due to Drs O W Heal, P C Miller and W Schlesinger, all of whom kindly provided access to data in manuscripts as yet unpublished. We also acknowledge the assistance of Mr F J Kruger, Prof E J Moll and Mrs M L Jarman who gave valuable background information. Finally thanks are extended to Miss Jane Bird for patiently typing the manuscript and to Mrs Dianne Freeman for assistance in preparing the figures.

References

ABER JD, MELILLO J M (1980) Litter decomposition: measuring relative contributions of organic matter and nitrogen to forest soils. Canadian Journal of Botany 58: 416-421

ALLEN SE, GRIMSHAW M, PARKINSON J, QUARMBY C (1974) Chemical analysis of ecological materials. Wiley, New York, 565 pp

ANDERSON JM (1973) The breakdown and decomposition of sweet chestnut (Castanea sativa Mill) and beech (Fagus sylvatica L.) leaf litter in two deciduous woodland soils. I. Breakdown leaching and decomposition. Oecologia 12: 251-274

ASHFORD AE, LING LEE M, CHILVERS GA (1975) Polyphosphate in eucalypt mycorrhizas: a cytochemical demonstration. New Phytologist 74: 447-453

AZCON R, BAREA JM, HAYMAN DS (1976) Utilization of rock phosphate in alkaline soils by plants inoculated with mycorrhizal fungi and phosphate-solubilizing bacteria. Soil Biology and Biochemistry 8: 135-138

BALLESTER A, VIEITEZ E, MANTILLA JLG (1975) Sustansias quimicas inhibidoras del crecimiento y la germinacion presentes en Ericaceas. Anales del Instituto Botanico AJ Cavinilles 32: 235-243

BAREA JM, AZCON R, HAYMAN DS (1975) Possible synergistic interactions between Endogone and phosphate-solubilizing bacteria in low-phosphate soils. In: SANDERS FE, MOSSE B, TINKER PB (eds) Endomycorrhizas. University of Leeds Symposium, July 1974. Academic Press, New York, pp 409-417

BARTLETT EM, LEWIS DH (1973) Surface phosphatase activity of mycorrhizal roots of beech. Soil Biology and Biochemistry 5: 249-257

BEADLE NCW (1966) Soil phosphate and its role in molding segments of the Australian flora and vegetation with special reference to xeromorphy and sclerophylly. Ecology 47: 992-1007

BECKING JH (1959) Nitrogen-fixing bacteria of the genus Beijerinckia in South African soils. Plant and Soil 11: 193-206

BENOIT RE, STARKEY RL (1968) Inhibition of decomposition of cellulose and some other carbohydrates by tannin. Soil Science 105: 291-296

BOCOCK KL (1963) Changes in the amount of nitrogen in decomposing leaf litter of sessile oak (Quercus petraea). Journal of Ecology 51: 555-566

BOCOCK KL (1964) Changes in the amounts of dry matter, nitrogen, carbon and energy in decomposing woodland leaf litter in relation to the activities of the soil fauna. Journal of Ecology 52: 273-284

BOCOCK KL, GILBERT O, CAPSTICK C, TWINN D, WARD JS, WOODMAN MJ (1960) Changes in leaf litter when placed on the surface of soils with contrasting humus types. I. Losses in dry weight of oak and ash leaf litter. Journal of Soil Science 11: 1-9

BUNNELL FL, TAIT DEN (1974) Mathematical simulation models of decomposition processes. In: HEAL OW, MACLEAN SF, FLANAGAN PW (eds) Soil organisms and decomposition in tundra. Tundra Biome Steering Committee, Stockholm, pp 207-226

BURGEFF H (1961) Mikrobiologie des Hochmoors. Gustav Fischer, Jena, 197 pp

BYRDE RJW, FIELDING AH, WILLIAMS AH (1960) The role of oxidized polyphenols in the varietal resistance of apples to brown rot. In: PRIDHAM JB (ed) Polyphenolics in plants in health and disease. Pergamon, Oxford, pp 95-99

CHAPMAN SB (1967) Nutrient budgets for a dry heath ecosystem in the South of England. Journal of Ecology 55: 677-689

CHILVERS GA, HARLEY JL (1980) Visualization of phosphate accumulation in beech mycorrhiza. New Phytologist 84: 319-326

COLEY PGF, MITCHELL DT (1980) The distribution of soil fungi in Cape Erica heathland community. South African Journal of Science 76: 185

COSGROVE DJ (1967) Metabolism of organic phosphates in soil. In: MCLAREN A, PETERSON GH (eds) Soil biochemistry. Dekker, New York, pp 216-228

DADD CC, FOWDEN L, PEARSALL WH (1953) An investigation of the free amino-acids in organic soil types using paper partition chromatography. Journal of Soil Science 4: 67-71

DI CASTRI F (1973) Soil animals in latitudinal and topographical gradients of Mediterranean ecosystems. In: DI CASTRI F, MOONEY HA (eds) Mediterranean-type ecosystems: origin and structure. Springer-Verlag, Berlin, pp 171-190

DOX AW, GOLDEN R (1911) Phytase in lower fungi. Journal of Biological Chemistry 10: 183-186

DUTT GR, SHAFFER MJ, MOORE WJ (1972) Computer simulation model of dynamic bio-physicochemical processes in soil. Technical Bulletin 196. University of Arizona, Tucson

EDWARDS CA, HEATH GW (1963) The role of soil animals in breakdown of leaf material. In: DOEKSEN J, VAN DER DRIFT J (eds) Soil organisms. North Holland, Amsterdam, pp 76-83

ELLENBERG VH (1977) Stickstoff als Standortsfaktor in für mitteleuropäische Pflanzengesellschaften. Oecologia Plantarum 12: 1-22

FEENY PP (1970) Seasonal changes in oak leaf tannins and nutrients as a cause of spring feeding by winter moth caterpillars. Ecology 51: 565-581

FEENY PP, BOSTOCK H (1978) Seasonal changes in the tannin content of leaves. Phytochemistry 7: 871-880

FLANAGAN PW, VEUM AK (1974) Relationships between respiration, weight loss, temperature and moisture in organic residues in tundra. In: HOLDING AJ, HEAL O, MACLEAN SF, FLANAGAN PW (eds) Soil organisms and decomposition in tundra. IBP Tundra Biome, Stockholm, pp 249-278

FOGEL R, CROMACK K (1977) Effect of habitat and substrate quality on Douglas fir litter decomposition in western Oregon. Canadian Journal of Botany 55: 1632-1640

GOSZ JR, LIKENS GE, BORMANN FH (1973) Nutrient release from decomposing leaf and branch litter in the Hubbard Brook forest, New Hampshire. Ecological Monographs 43: 173-191

GREAVES MP, ANDERSON G, WEBLEY DM (1967) The hydrolysis of inositol phosphates by Aerobacter aerogenes. Biochemica Biophysica Acta 132: 412-418

HANDLEY WRC (1954) Mull and moor formation in relation to forest soils. Forestry Commission Bulletin No. 23. HMSO, London, pp 1-115

HARLEY JL (1969) The biology of mycorrhiza. Leonard Hill, London, 334 pp

HARLEY JL (1971) Fungi in ecosystems. Journal of Ecology 59: 653-668

HARMSEN CW, VAN SCHREVEN DA (1955) Mineralization of organic nitrogen in soil. Advances in Agronomy 7: 299-398

HARRISON AF (1971) The inhibitory effect of oak leaf litter tannins on the growth of fungi in relation to litter decomposition. Soil Biology and Biochemistry 3: 167-172

HASELWANDTER K, READ DJ (1980) Fungal associations of roots of dominant and sub-dominant plants in high alpine vegetation systems with special reference to mycorrhiza. Oecologia 45: 57-62

HATTINGH MJ, GRAY LE, GERDEMAN JW (1973) Uptake and translocation of ^{32}P-labelled phosphate to onion roots by endomycorrhizal fungi. Soil Science 116: 383-390

HAYMAN DS (1975) Phosphorus cycling by soil micro-organisms and plant roots. In: WALKER N (ed) Soil microbiology. Butterworth, London, pp 67-91

HEAL OW, FRENCH DD (1974) Decomposition of organic matter in tundra. In: HOLDING AJ (ed) Soil organisms and decomposition in tundra. Tundra Biome Steering Committee, Stockholm, pp 279-310

HEAL OW, LATTER PM, HOWSON G (1978) A study of the rates of decomposition of organic matter. In: HEAL OW, PERKINS DF (eds) Production ecology of British moors and montane grasslands. Springer-Verlag, Berlin, pp 136-159

HO I, ZAK B (1979) Acid phosphatase activity of six ectomycorrhizal fungi. Canadian Journal of Botany 57: 1203-1205

JENNY H, GESSEL SP, BINGHAM FT (1949) Comparative study of decomposition rates of organic matter in temperate and tropical regions. Soil Science 68: 419-432

JONES MB, WOODMANSEE RG (1979) Biogeochemical cycling in annual grassland ecosystems. Botanical Review 45: 111-144

JONES R (1968) Estimating productivity and apparent photosynthesis from differences in consecutive measurements of total living plant parts of an Australian heathland. Australian Journal of Botany 16: 589-602

KING HGC, HEATH GW (1967) The chemical analysis of small samples of leaf material and the relationship between the disappearance and composition of leaves. Pedobiologia 7: 192-197

KITTREDGE J (1955) Litter and forest floor of the chaparral in parts of the San Dimas Experimental Forest, California. Hilgardia 23: 563-596

KOWLENKO CG (1978) Organic nitrogen, phosphorus and sulphur in soils. In: SCHNITZER M, KHAN SU (eds) Soil organic matter. Developments in Soil Science 8: Elsevier, Amsterdam, pp 95-136

KRUGER FJ (1979) South African heathlands. In: SPECHT RL (ed) Ecosystems of the world, vol 9A. Heathlands and related shrublands. Descriptive studies. Elsevier, Amsterdam, pp 19-80

KUMMEROW J (1973) Comparative anatomy of sclerophylls of mediterranean climatic areas. In: DI CASTRI F, MOONEY HA (eds) Mediterranean type ecosystems: origin and structure. Springer-Verlag, Berlin, pp 157-167

KUMMEROW J, KRAUSE D, JOW J (1978) Seasonal changes of fine root density in the southern Californian chaparral. Oecologia 37: 201-212

LATTER P, CRAGG JB, HEAL OW (1967) Comparative studies on the microbiology of moorland soils in the northern Pennines. Journal of Ecology 55: 445-464

LEE JA, STEWART G (1978) Ecological aspects of nitrogen assimilation. Advances in Botanical Research 6: 1-43

LOFTY JR (1974) Oligochaetes. In: DICKINSON CH, PUGH GFJ (eds) Biology of plant litter decomposition, vol 2. Academic Press, London, pp 467-488

LOSSAINT P (1973) Soil-vegetation relationships in mediterranean ecosystems of southern France. In: DI CASTRI F, MOONEY HA (eds) Mediterranean type ecosystems: origin and structure. Springer-Verlag, Berlin, pp 199-210

LOUSIER JD, PARKINSON D (1978) Chemical element dynamics in decomposing leaf litter. Canadian Journal of Botany 56: 2795-2812

LOVELESS AR (1961) A nutritional interpretation of sclerophylly based upon differences in the chemical composition of sclerophyllous and mesophytic leaves. Annals of Botany (London) New Series 25: 168-184

LOVELESS AR (1962) Further evidence to support a nutritional interpretation of sclerophylly. Annals of Botany (London) New Series 26: 551-561

LUTZ HJ, CHANDLER RF (1946) Forest soils. Wiley, New York, pp 150

MAGGS J, PEARSON CJ (1977) Litter fall and litter layer decay in coastal scrub at Sydney, Australia. Oecologia 31: 239-250

MARCUZZI G (1970) Experimental observations on the role of Glomeris spp. (Myriopoda:Diplopoda) in the process of humification of litter. Pedobiologia 10: 401-406

MARION GM, KUMMEROW J, MILLER PC (1982) Predicting nitrogen mineralization in chaparral soils. Ecology. In press

MEENTEMEYER V (1978) Macroclimate and lignin control of litter decomposition rates. Ecology 59: 465-472

MELILLO JM, ABER JD, MURATORE JF (1982) Nitrogen and lignin control of hardwood leaf litter decomposition dynamics. Ecology. In press

MINDERMAN G (1968) Addition, decomposition and accumulation of organic matter in forests. Journal of Ecology 56: 355-362

MITCHELL DT, READ DJ (1981) Effect of inorganic and organic phosphates on the growth of the mycorrhizal endophytes of Vaccinium macrocarpon and Rhododendron ponticum. Transactions of the British Mycological Society 76: 255-260

MONK CD (1966) An ecological significance of evergreenness. Ecology 47: 504-505

MOONEY HA (1977) Southern coastal scrub. In: BARBOUR MG, MAJOR J (eds) Terrestrial vegetation of California. Wiley, New York, pp 471-489

MOONEY HA, RUNDEL PW (1979) Nutrient relations of the evergreen shrub Adenostoma fasciculatum, in the Californian chaparral. Botanical Gazette 140: 109-113

MOSSE B (1973) Advances in the study of vesicular-arbuscular mycorrhizas. Annual Review of Phytopathology 11: 171-189

NYE PH (1961) Organic matter and nutrient cycles under moist tropical forest. Plant and Soil 13: 333-346

NYKVIST N (1963) Leaching and decomposition of water soluble organic substances from different types of leaf and needle litter. Studia Forestalia Suecica 3: 3-29

OLSON JS (1963) Energy storage and the balance of producers and decomposers in ecological systems. Ecology 44: 322-331

PUGH GJF (1974) Terrestrial fungi. In: DICKINSON CH, PUGH GJF (eds) Biology of plant litter decomposition. Academic Press, London, pp 303-336

READ DJ (1978) The biology of mycorrhiza in heathland ecosystems with special reference to the nitrogen nutrition of the Ericaceae. In: LOUTIT MW, MILES JAR (eds) Microbial ecology. Springer-Verlag, Berlin, pp 24-328

READ DJ, JALAL MAF (1980) The physiological basis of interaction between Calluna vulgaris, forest trees and other plant species. In: Proceedings of the conference on weed control in forestry. Association of Applied Biologists, London, pp 21-33

READ DJ, STRIBLEY DP (1973) Effect of mycorrhizal infection on nitrogen and phosphorus nutrition of ericaceous plants. Nature 244: 81

RIBEREAU-GAYON P (1972) Plant phenolics. Oliver and Boyd, Edinburgh, pp 254

ROBERTSON RA, DAVIES GE (1965) Quantities of plant nutrients in heather ecosystems. Journal of Applied Ecology 2: 211-219

RODIN LE, BAZILEVICH NI (1967) Production and mineral cycling in terrestrial vegetation. Oliver and Boyd, Edinburgh, pp 288

ROSSWELL T (1976) The internal nitrogen cycle between micro-organismic vegetation and soil. Ecological Bulletin (Stockholm) 22: 157-167

RUNDEL PW, PARSONS DJ (1980) Nutrient changes in two chaparral shrubs along a fire induced age gradient. American Journal of Botany 67: 51-58

RUSSELL EW (1973) Soil conditions and plant growth. 10th edn. Longmans, London. 849 pp

SANTANTONIO D, HERMAN RK, OVERTON WS (1977) Root biomass studies in forest ecosystems. Pedobiologia 17: 1-31

SASSON A, DASTE P (1963) Observations nouvelles concernant l'ecologie de l'Azotobacter dans certains sols arides du Maroc. Comptes Rendus de l'Academie de Science 257: 3516

SCHAEFFER R (1973) Microbial activity under seasonal conditions of drought in mediterranean climates. In: DI CASTRI F, MOONEY HA (eds) Mediterranean type ecosystems: origin and structure. Springer-Verlag, Berlin, pp 191-198

SCHLESINGER WH, HASEY MM (1981) Decomposition of chaparral shrub foliage: losses of organic and inorganic constituents from deciduous and evergreen leaves. Ecology 62: 762-774

SEWELL GWF (1959) Studies on fungi in a Calluna heathland soil. I. Vertical distribution in soil and on root surfaces. Transactions of the British Mycological Society 42: 343-350

SINGH JS, GUPTA SR (1977) Plant decomposition and soil respiration in terrestrial ecosystems. Botanical Review 43: 449-528

SMITH SJ, YOUNG LB, MILLER GE (1977) Evaluation of soil nitrogen mineralization potentials under modified field conditions. Journal of the Soil Science Society of America 41: 74-76

SPECHT RL (1969) A comparison of the sclerophyllous vegetation characteristic of mediterranean type climates in France, California and southern Australia. II. Dry matter, energy and nutrient accumulation. Australian Journal of Botany 17: 293-308

SPECHT RL (1979) Heathlands and related shrublands of the world. In: SPECHT RL (ed) Ecosystems of the world, vol 9A. Heathlands and related shrublands. Descriptive studies. Elsevier, Amsterdam, pp 1-18

SPECHT RL, RAYSON P, JACKMAN ME (1958) Dark Island heath (Ninety-Mile Plain, South Australia). VI. Pyric succession: Changes in composition, coverage, dry weight and mineral nutrient status. Australian Journal of Botany 6: 59-88

STANFORD G, LEGG JO, SMITH SJ (1973) Soil nitrogen availability evaluations based on nitrogen mineralization potentials of soils and uptake of labelled and unlabelled nitrogen by plants. Plant and Soil 39: 113-124

STRIBLEY DP, READ DJ (1974) The biology of mycorrhiza in the Ericaceae. IV. The effect of mycorrhizal infection on uptake of ^{15}N from labelled soil by Vaccinium macrocarpon Ait. New Phytologist 73: 1149-1155

STRIBLEY DP, READ DJ (1976) The biology of mycorrhiza in the Ericaceae. VI. The effects of mycorrhizal infection and concentration of ammonium nitrogen on growth of cranberry in sand culture. New Phytologist 77: 63-72

STRIBLEY DP, READ DJ (1980) The biology of mycorrhiza in the Ericaceae. VII. The relationship between mycorrhizal infection and the capacity to utilize simple and complex organic nitrogen sources. New Phytologist 86: 365-371

SUBERKROPP K, GODSHACK GL, KLUG MJ (1976) Changes in the chemical composition of leaves during processing in a woodland stream. Ecology 57: 720-727

SWIFT MJ, HEAL OW, ANDERSON JM (1979) Decomposition in terrestrial ecosystems. Studies in Ecology, vol 5. Blackwell, Oxford, 372 pp

THROWER NJW, BRADBURY DE (eds) (1977) Chile-Californian mediterranean scrub atlas. Dowden, Hutchinson and Ross, Stroudsburg, Pennsylvania, 237 pp

VAN CLEVE K (1974) Organic matter quality in relation to decomposition. In: HOLDING AJ, HEAL OW, MACLEAN SF and FLANAGAN P (eds) Soil organisms and decomposition in tundra. Tundra Biome Steering Committee, Stockholm, pp 311-324

WATTS DG, HANKS RJ (1978) A soil-water nitrogen model for irrigated corn on sandy soils. Journal of the Soil Science Society of America 42: 492-499

WICHT CL (1971) The influence of vegetation in the South African mountain catchments on water supplies. South African Journal of Science 67: 201-209

WITKAMP M (1966) Decomposition of leaf litter in relation to environment, microflora and microbial respiration. Ecology 47: 194-201

WITKAMP M, CROSSLEY DA (1966) The role of arthropods and microflora in the breakdown of white oak litter. Pedobiologia 6: 293-303

WITKAMP M, OLSON JS (1963) Breakdown of confined and non-confined oak litter. Oikos 14: 138-147

WOOLHOUSE HW (1969) Differences in the properties of the acid phosphatases of plant roots and their significance in the evolution of edaphic ecotypes. In: RORISON IH (ed) Ecological aspects of the mineral nutrition of plants. Blackwell, Oxford, pp 357-380

YEILDING L (1977) Decomposition in chaparral. In: MOONEY HA, CONRAD CE (eds) Symposium on the environmental consequences of fire and fuel management in mediterranean ecosystems. USDA Forest Service, Washington DC, pp 419-425

ZAK B (1976) Pure culture synthesis of bearberry mycorrhizae. Canadian Journal of Botany 54: 1297-1305

13. The Effects of Different Fire Regimes on Soil Nutrient Levels in *Quercus coccifera* Garrigue

L. TRABAUD

13.1 Introduction

Fire is known to modify the physical and chemical properties of soils, resulting in loss of organic matter and changes in availability of nutrients to plants. The major changes take place in the upper soil layer (i.e. 0-50 mm of mineral soil: DeBano and Conrad 1978; Ueckert et al. 1978).

The effect of fire on soil constituents of the vegetation of the French mediterranean region has not been studied. Investigations have been undertaken in the Californian chaparral (Christensen and Muller 1975; DeBano et al. 1977; Dunn and DeBano 1977; Zinke 1977; DeBano and Conrad 1978), and these studies as well as those of other climatic regions of the world (e.g. Ahlgren and Ahlgren 1960; Daubenmire 1968; Wells 1971; Veilleux 1972; Viro 1974; Wells et al. 1979) show that fire modifies the levels of soil nutrients. Sometimes observations differ and this may be due to differences between areas in terms of soil type and phytocoenoses.

The purpose of this study was to observe the effects of experimental fires of different frequencies and seasons on the principal nutrients of the upper soil layers of a vegetation type very frequently encountered in the French mediterranean region. Fire has passed regularly through this type of vegetation for centuries (Braun-Blanquet 1936; Kuhnholtz-Lordat 1938).

13.2 Study area and methods

13.2.1 STUDY AREA

A complex experiment was set up in a Quercus coccifera garrigue located on a hill called Puech-du-Mas-du-Juge about 10 km north of Montpellier. This area has been used in previous studies (Long et al. 1961, 1967; Trabaud 1962, 1974, 1977; Poissonet 1966). The initial vegetation was a garrigue of pyrophytic origin (the dates of the last natural fires were September 1943 and August 1951). The four main species, by decreasing frequency, are *Q. coccifera*, *Brachypodium ramosum*, *Dorycnium suffruticosum* and *Rubia peregina*. The shrub layer, 1.5 m high, is formed almost exclusively of *Q. coccifera* and the grass layer is dominated by *B. ramosum*.

The experimental plots were located on the southwestern side of the hill, the general slope of the ground being 10-15% and elevation 120 to 160 m above sea level. According to Emberger's classification (1971) the climate is a mediterranean humid type, with cool winters. The mean temperature is about 15°C, the mean of the coldest month (January) equals 2.4°, the mean of the warmest month (July) is 27.5 °C; the mean annual rainfall for the past ten years (1969-1978) is 1102 mm. The season of the first pluviometric maximum is autumn, and the second maximum occurs in spring with the lowest minimum during summer.

The soils are classified as calcimorph soils (brown, calcareous, brown soils, and rendzinas) formed on compact Eocene limestone. They generally vary in depth but are always shallow (100-150 mm). They are pebbly, with a medium-fine texture, a neutral pH (6.5 to 7.5) and a calcic mull humus layer.

The site had been cultivated a long time ago but was abandoned to pastures and has now evolved into a garrigue due to repeated fires started, deliberately or not, by shepherds to regenerate the pastures and to open the vegetation to allow the passage of flocks.

13.2.2 EXPERIMENTAL DESIGN

The study area was divided into plots of 50 m^2 and fires were applied in different combinations of two seasons and three frequencies, respectively, as follows:

T - no burning, control plots
P - plots burnt at the end of spring
A - plots burnt at the beginning of autumn
6 - plots burnt every six years
3 - plots burnt every three years
2 - plots burnt every two years

Each treatment was replicated six times. Unburnt plots were used as controls.

Two soil samples were collected from each plot before and after the fires at different times during the years of experimentation (1969-1974). The samples were taken from the first 50 mm of the mineral soil layer. Before the burns the pedogenetic characteristics of the superficial soil were similar to those of another stand of garrigue and a *Quercus ilex* coppice (Table 1).

13.2.3 CHEMICAL ANALYSIS AND DETERMINATIONS

For each treatment twelve samples were analysed yearly for organic matter, carbon and nitrogen content, and pH. Only four samples per treatment were analysed to determine total cation exchange capacity, exchangeable bases (calcium, magnesium, potassium and sodium) phosphorus (available and total P_2O_5) and potassium (K_2O). In all cases classical techniques of chemical analysis were used (Anonymous 1967; Milhomme 1979).

Upper soil organic matter content was determined by mass loss on ignition at 700°C. The wet combustion method was used for total organic

Table 1. Comparison between the pedogenetic characteristics of three related phytocoenoses.

	Depth mm	Organic matter %C x 1.72	%C	%N	C:N	Total CEC (meq 100 g^{-1})	Exchangeable cations (meq 100 g^{-1}) Ca	Mg	K	Na
Quercus ilex coppice (Rapp 1971)	10-70	8.92	5.19	0.33	15.5	21.4	40.95	1.59	1.29	0.06
Quercus coccifera garrigue (Rapp 1971)	10-100	5.66	3.29	0.32	10.3	32.4	57.57	1.41	0.68	0.06
Quercus coccifera garrigue (present research)	0-50	9.34	5.43	0.37	14.7	31.2	47.43	2.06	1.33	0.10

carbon determination. Total nitrogen was determined by the Kjeldahl technique modified to include nitrates, then samples were analysed colorimetrically using the indole-phenol blue method. Values of pH were determined using a glass electrode in distilled water suspension.

Total cation exchange capacity and soluble cations of calcium, magnesium, sodium and potassium were replaced with ammonium acetate and the concentration of each ion in the leachate measured by atomic absorption. Potassium concentrations were analysed colorimetrically by the vanadate-molybdate-yellow method. The amounts of K_2O were measured by flame photometry after precipitation of the other cations.

13.3 Results

Data are presented as means of either 12 or four observations. Significant differences between treatments were determined by variance analysis and studentized range test (Snedecor and Cochran 1971).

13.3.1 ORGANIC MATTER AND ORGANIC CARBON

Results for these components were similar and only organic matter is discussed in detail. After the first burn (1969) and immediately after every burn, organic matter content increased in the upper 50 mm of the soil (Fig. 1). One year after each burn the organic matter content reached the same level as before the burn or that measured in the control plots. In 1974, at the end of the first cycle of the experiment, the organic matter level was similar to that at the beginning except that the upper layer of the soil on plots burned every two years in autumn (2A) possessed a higher organic matter content. No matter which treatment was used, the organic matter content was close to the original five years after the start of the study.

The parallel increase of carbon during the months following each burn may have been due to deposits of ash and unburnt charcoal incorporated just below the soil surface. This phenomenon may be explained by the increase of the thermal balance and the greater heat of the blackened soil surface which stimulate decomposer populations (Ahlgren and Ahlgren 1960; Ueckert et al. 1978).

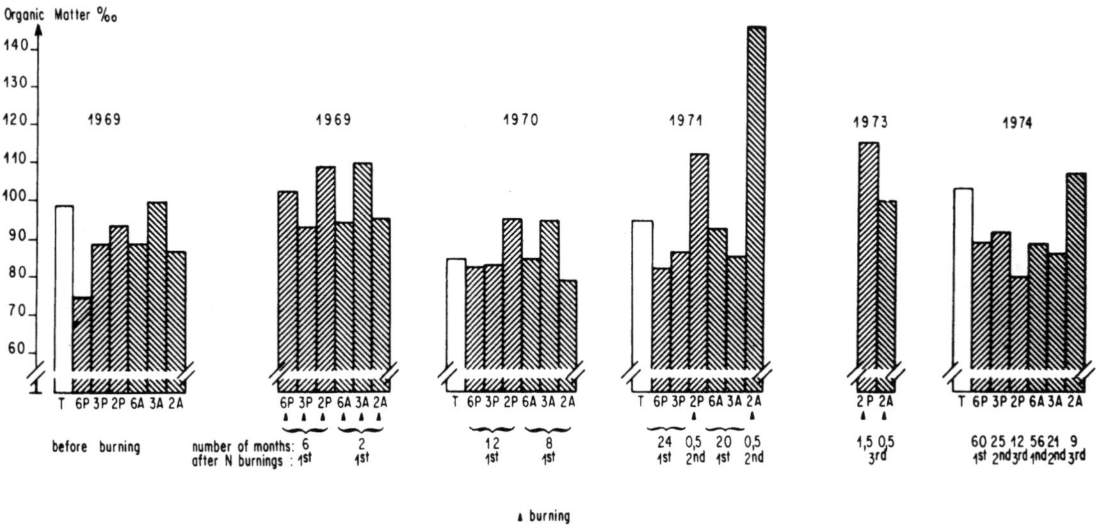

Fig. 1. Changes over time of the organic matter levels of the upper soil layer according to the frequency and the season of burning.

13.3.2 NITROGEN

Fifteen days to six months after every burn there was an increase in nitrogen in the upper layer of soil (Fig. 2A). Thereafter, the levels were similar to those of the control plots (T). However total nitrogen declined at the end of 1974, but this decrease was less in the control plots and in the soil burnt every two years during autumn (2A).

Living vegetation and litter can be completely consumed by fire. Most mineral nutrients in plants and litter remain on site in the form of ashes, whereas nitrogen is mainly volatilized. The more intense a fire, the greater is the nitrogen loss (Wells 1971). Fire temperatures in the present study exceeded 800°C and large quantities of nitrogen are known to volatilize at temperatures over 200°C (Allen 1964; Dunn and DeBano 1977; DeBano and Conrad 1978).

13.3.3 CARBON:NITROGEN RATIO

The carbon:nitrogen ratios fluctuated during the years of study (Fig. 2B), but in 1974, as the nitrogen levels were extremely low, the ratios were highest for all the treatments (including the unburnt soil, T). Lowest carbon:nitrogen ratios were found in the soils burnt in autumn (significantly different at $P < 0.05$ from the spring-burnt soils).

13.3.4 pH AND CATION EXCHANGE CAPACITY

The pH of burnt soils was higher than that of control plots and, except in 1974, the pH values of the spring-burnt soils were not significantly different from those of the autumn-burnt soils ($P > 0.05$). These differences were marginal on the whole, not exceeding about 0.3 pH units.

Fire releases basic ions from plant tissues and destroys organic matter, thus reducing the formation of organic acids in soils (Wells

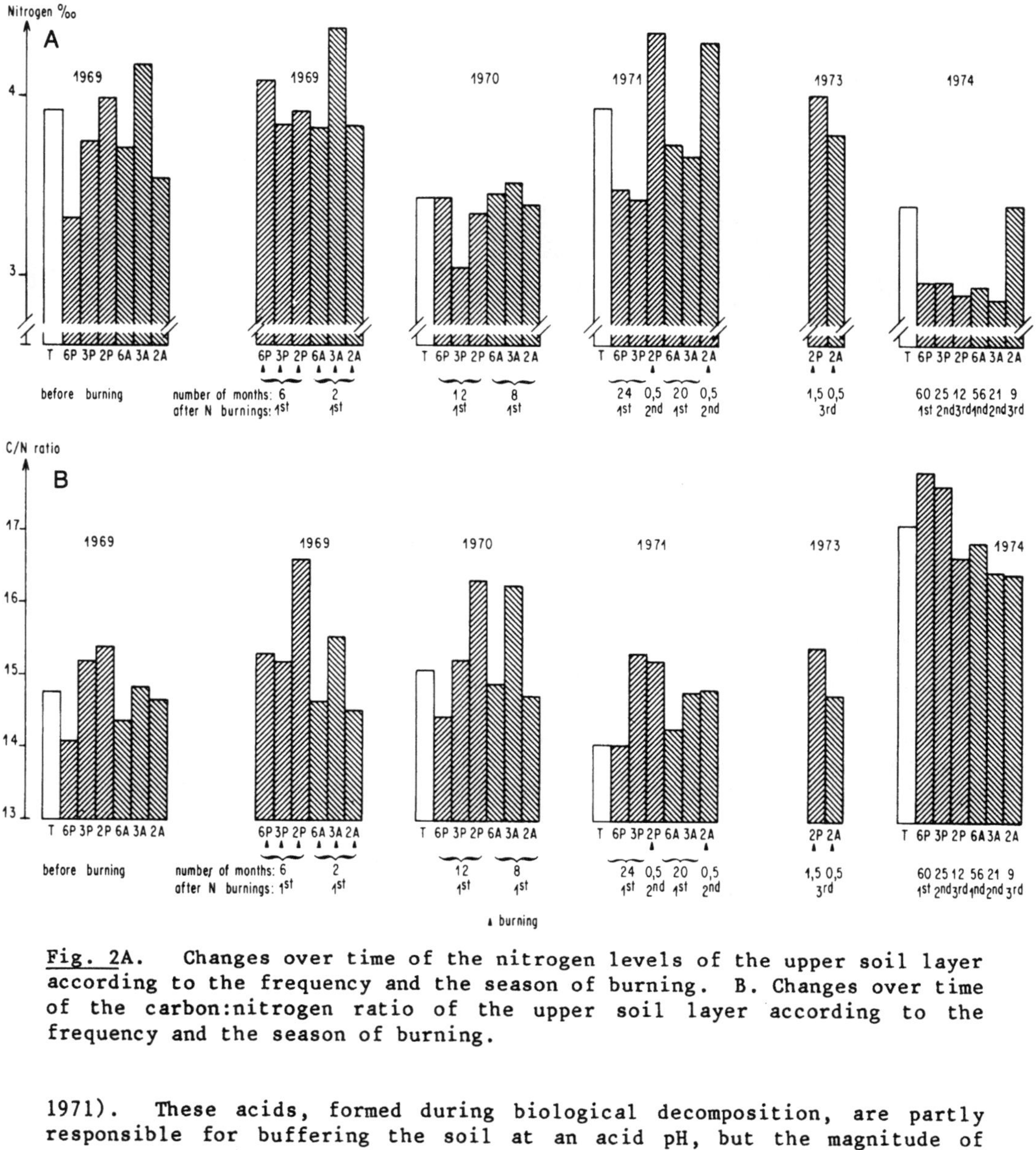

Fig. 2A. Changes over time of the nitrogen levels of the upper soil layer according to the frequency and the season of burning. B. Changes over time of the carbon:nitrogen ratio of the upper soil layer according to the frequency and the season of burning.

1971). These acids, formed during biological decomposition, are partly responsible for buffering the soil at an acid pH, but the magnitude of change of pH will depend on the amount of ash, the quantity of calcium, magnesium and potassium in the ash, and the texture and organic content of the soil (Wells 1971).

Although cation exchange capacity values were extremely variable they were generally higher after a fire than before (Fig. 3).

13.3.5 EXCHANGEABLE CATIONS

Soils in this study were saturated with calcium and exchangeable calcium levels were thus determined from total cation exchange capacity values minus the values for the other cations. There was an increase in calcium after fire, but values seem to reach their maximum only one year after a burn.

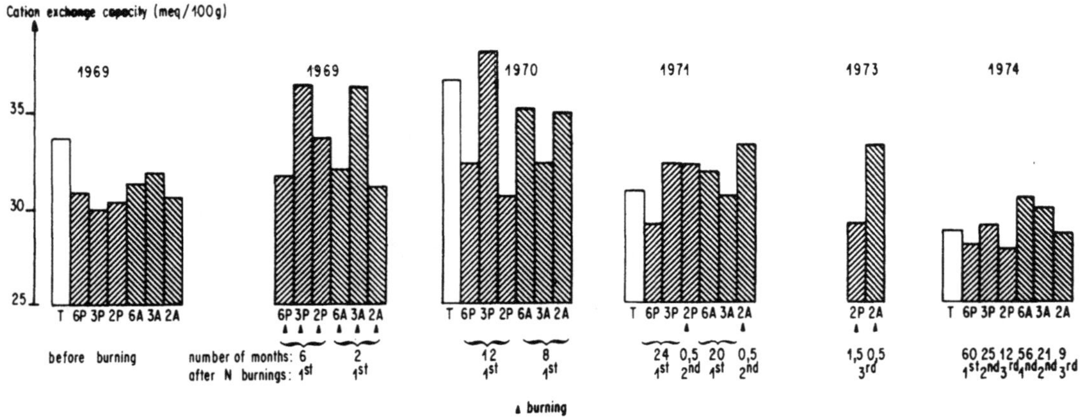

Fig. 3. Changes over time of the cation exchange capacity levels of the upper soil layer according to the frequency and the season of burning.

Immediately after the first burn potassium levels increased (Fig. 4) but after about 10 months returned to those before the burn. At the end of 1974, potassium levels of the unburnt soil had increased, but the highest values were observed on the most frequently burnt plots (2P and 2A), though they did not differ significantly from the control. The same trend was observed for K_2O: the highest K_2O content was consistently measured in plots on treatment 2A. K_2O also increased under biennial spring burns, but decreased under all other treatments.

Magnesium values increased after each burn but then declined (Fig. 5A). At the end of the study period, Magnesium values were lower in soils submitted to repeated fires than in the unburnt soils and those burnt once.

Fire decreased the sodium content of the soil (Fig. 5B). Sodium released by combustion of plants is very soluble and is leached down the profile or carried away by surface runoff.

13.3.6 PHOSPHORUS

During the period of study (1969-1974) fire increased available phosphorus in the upper soil layer (Fig. 6). Values are higher ($P < 0.01$) in those soils burnt during autumn. There appeared to be no loss of available phosphorus by volatilization. Fire produced a slight effect on total P_2O_5 levels. However, total phosphorus increased with repeated burns.

13.4 Discussion and conclusion

The results presented in the study do not provide a clear picture of the effects of fire on soil elements. This may have been due to the edaphic heterogeneity of the experimental site. There appear to be seasonal changes, upon which rainfall probably has a far greater influence than fire. Climatic influences on the levels of different soil elements have been recorded by Billes et al. (1971, 1975) and Ueckert et al. (1978). It is however possible to give general tendencies and present a provisional phytotrophic synthesis for the garrigue subjected to repeated fires.

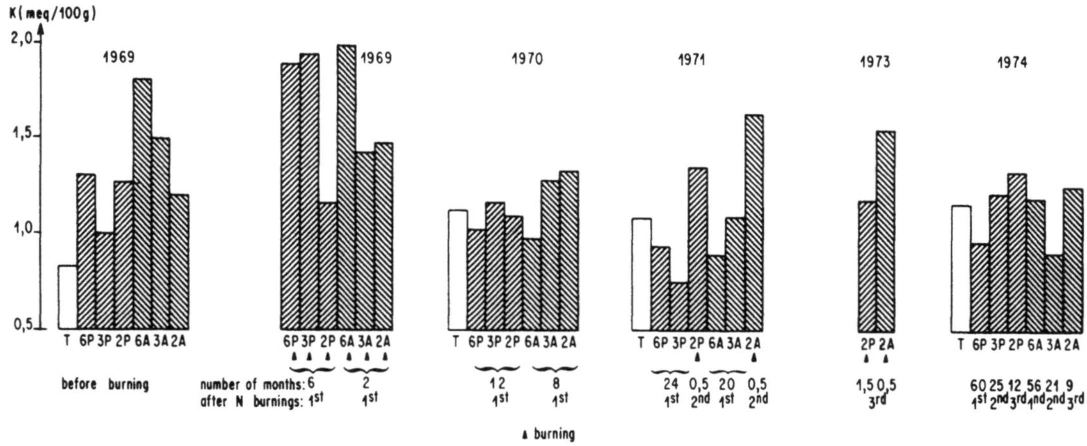

Fig. 4. Changes over time of the potassium levels of the upper soil layer according to the frequency and the season of burning.

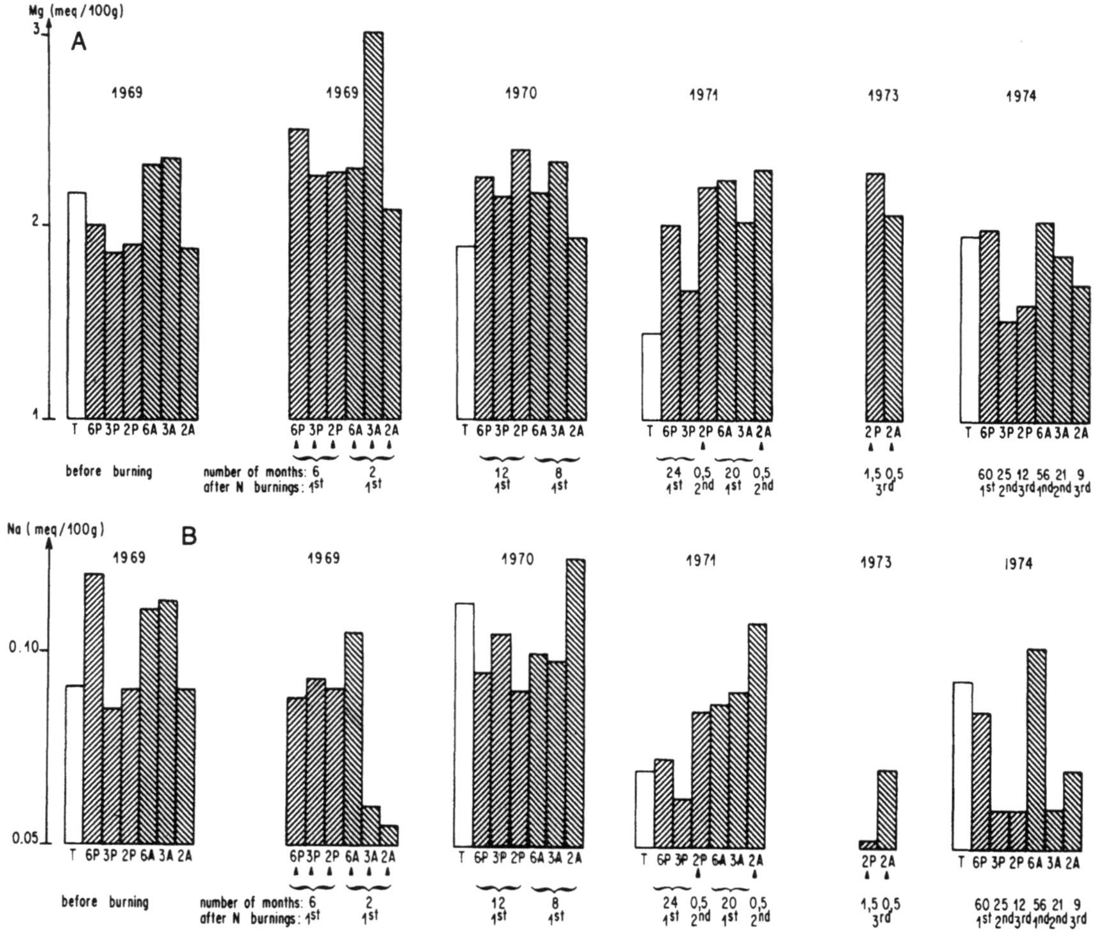

Fig. 5A. Changes over time of the magnesium levels of the upper soil layer according to the frequency and the season of burning. B. Changes over time of the sodium levels of the upper soil layer according to the frequency and the season of burning.

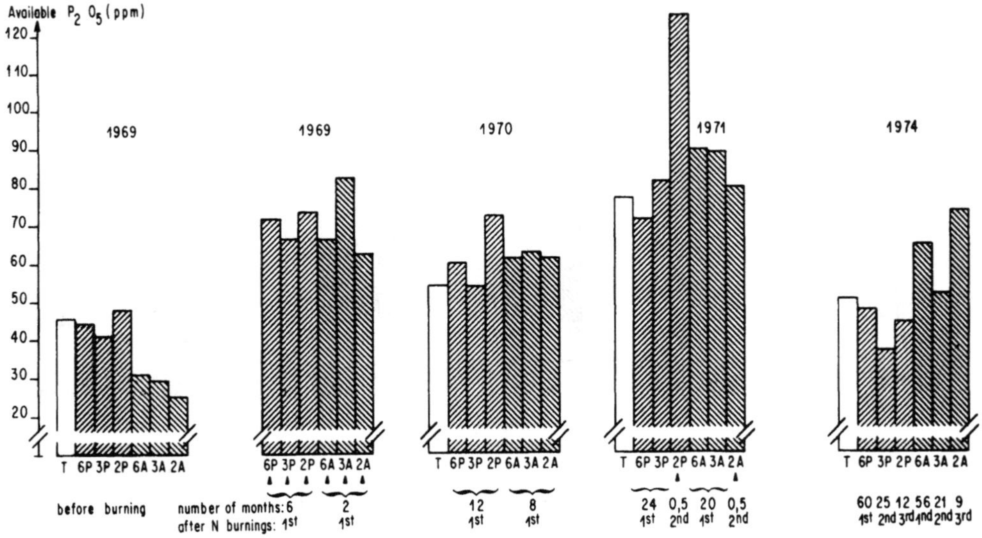

Fig. 6. Changes over time of the available phosphorus (P_2O_5) levels of the upper soil layer according to the frequency and the season of burning.

Immediately after each burn in 1969 there were increases of all the elements studied (except sodium). Subsequently these increases were very slight or non-existent. The reason lies in the high frequency of burning. Vegetation has no time to develop and litter cannot accumulate on the ground, whereas before the first burns in 1969, vegetation and litter had accumulated undisturbed in the 18 years since the last fire.

The burning season (either spring or autumn) seems to have a great effect upon the nutrient levels in the upper soil layers, since in 1974, five years after the beginning of the experiments, amounts in the plots burnt in autumn were generally higher than in others.

Whatever the season and frequency of burn, organic matter content remained approximately the same throughout the period of study. This result is partly due to charcoal deposits and organic material coming from the decay of subterranean organs located near the soil surface and killed by fire.

The decrease of nitrogen levels is important. Nitrogen is mainly volatilized by high temperatures in the form of NH_3 or the oxides NO, NO_2 and N_2O (Bush et al. 1969; Crutzen et al. 1979) or reaches the soil as ammonium which can be rapidly absorbed by plants or lost by leaching (St John and Rundel 1976; DeBano et al. 1977, 1979; Dunn et al. 1979). Though fire generally increases soluble nitrogen amounts, total nitrogen is decreased. In this sense the present study is in agreement with the statements of the authors quoted above. Because nitrogen levels were greatly decreased by burning, carbon:nitrogen ratios increased.

There was a general progressive decrease in pH over the period of study for all the treatments and particularly for the unburnt soil; but repeated burning (every two years) reduced this decrease. There was also a decrease of the cation exchange capacity for all treatments including the unburnt soil. Though sodium, magnesium, and potassium decreased overall during the period of study, treatment did have an effect on trends in potassium in that burns every two years tended to increase relative levels of potassium. Though Christensen and Muller (1957) reported no change in potassium concentration after burning, the results of the present study are in accordance with those of St John and Rundel (1976) who found that exchangeable potassium increased with burning. The amount of available P_2O_5 increased but mainly in the soils burnt in autumn. This increase can solely be due to phosphorus brought by mineral elements from ash, because the plots were not fertilized in any way. Similar results were recorded by Christensen and Muller (1975) and St John and Rundel (1967).

In conclusion, fire is an important mineralizing agent but only general trends appeared during this study, and further years of research are needed to understand the impact of repeated fires on the nutrient cycle in Quercus coccifera garrigue soil.

Acknowledgements

I thank Dr G Long for critically reviewing the text and Mr Milhomme for helpful discussion. The chemical analyses were made by the "Section des Analyses de Séries" of the Centre d'Etudes Phytosociologiques et Ecologiques L. Emberger.

I am very grateful to the technicians for their help during the burns.

References

ANONYMOUS (1967) Méthodes et techniques des analyses de terre. Centre National de la Recherche Scientifique, Centre d'Etudes Phytosociologiques et Ecologiques, Document no 34, 153 pp

AHLGREN IF (1974) The effect of fire on soil organisms. In: KOZLOWSKI TT, AHLGREN CE (eds) Fire and ecosystems. Academic Press, New York, pp 47-72

AHLGREN IF, AHLGREN CE (1960) Ecological effects of forest fires. Botanical Review 26: 483-533

AHLGREN IF, AHLGREN CE (1965) Effects of prescribed burning on soil micro-organisms in a Minnesota jack pine forest. Ecology 46: 304-310

ALLEN SE (1964) Chemical aspects of heather burning. Journal of Applied Ecology 1: 347-367

BILLES G, CORTEZ J, LOSSAINT P (1975) Etude comparative de l'activité biologique des sols sous peuplements arbusifs et herbacés de la garrigue méditerranéenne. II. Minéralisation de carbone et de l'azote. Revue d'Ecologie et de Biologie du sol 12: 115-139

BILLES G, LOSSAINT P, CORTEZ J (1971) L'activité biologique des sols dans les écosystèmes méditerranéens. II. Minéralisation de l'azote. Revue d'Ecologie et de Biologie du sol 8: 533-552

BRAUN-BLANQUET J (1936) La forêt d'yeuse languedocienne (Quercion ilicis). Monographie phytosociologique. Mémoires de la Société d'Etudes et Sciences naturelles de Nimes 5: 1-147

BUSH AF, LEONARD JJ, YUNDT WH (1969). Gas analyses in large fire experiments. In: Project Flambeau; an investigation of mass fire (1964-1969). Final Report, vol 3: appendixes. USDA Forest Service, Pacific Southwest Forest and Range Experimental Station, pp 33-90

CHRISTENSEN NL, MULLER CH (1975) Effect of fire on factors controlling plant grown in Adenostoma chaparral. Ecological Monographs 45: 29-55

CRUTZEN PJ, HEIDT LE, KRASNEC JP, POLLOCK WH, SEILER W (1979) Biomass burning as a source of atmospheric gases CO, H_2, N_2O, NO, CH_3Cl and COS. Nature 282: 253-256

DAUBENMIRE R (1968) Ecology of fire in grasslands. In: CRAGG JB (ed) Advances in Ecological Research 5. Academic Press, London, pp 209-266

DEBANO LF, CONRAD CE (1978) The effect of fire on nutrients in a chaparral ecosystem. Ecology 59: 489-497.
DEBANO LF, DUNN PH, CONRAD CE (1977) Fire's effect on physical and chemical properties of chaparral soils. In: MOONEY HA, CONRAD CE (eds) Proceedings of the symposium on the environmental consequences of fire and fuel management in mediterranean ecosystems. USDA Forest Service General Technical Report WO-3, pp 65-74
DEBANO LF, EBERLEIN GE, DUNN PH (1979) Effects of burning on chaparral soils. I. Soil nitrogen. Soil Science Society of America Journal 43: 504-509
DUNN PH, DEBANO LF (1977) Fire's effect on biological and chemical properties of chaparral soils. In: MOONEY HA, CONRAD CE (eds) Proceedings of the symposium on the environmental consequences of fire and fuel management in mediterranean ecosystems. USDA Forest Service General Technical Report WO-3, pp 75-84
DUNN PH, DEBANO LF, EBERLEIN GE (1979) Effects of burning on chaparral soils. II. Soils microbes and nitrogen mineralization. Soil Science Society of America Journal 43: 509-514
EMBERGER L (1971) Considérations complémentaires au sujet des recherches bioclimatologiques et phytogéographiques-écologiques. In: Travaux de Botanique et d'Ecologie. Masson, Paris, pp 291-301
KUHNHOLTZ-LORDAT G (1938) La terre incendiée. Essai d'agronomie comparée. La Maison Carrée, Nimes, 361 pp
LONG G, VISONA L, RAMI J (1961) La végétation du domaine de Coulondres (Herault). Relation avec les problèmes de mise en valeur. Bolletino del'Instituto Botanico del Universita di Catania 3: 5-52
LONG G, FAY F, THIAULT M, TRABAUD L (1967) Essais de détermination expérimentale de la productivité d'une garrigue de Quercus coccifera. Centre d'Etudes Phytosociologiques et Ecologiques, Centre National de la Recherche Scientifique, Montpellier. Document 39, 28 pp
MILHOMME H (1979) Méthodes et techniques d'analyses courantes par colorimétrie automatisée. Centre National de la Recherche Scientifique, Centre d'Etudes Phytosociologiques et Ecologiques. Document 88, 81 pp
MOORE JM, WEIN RW (1977) Viable seed populations by soil depth and potential site recolonization after disturbance. Canadian Journal of Botany 55: 2408-2412
POISSONET P (1966) Etude méthodologique en écologie végétale a partir des photographies aériennes. Third cycle thesis (ecology), Faculty of Sciences, University of Montpellier, 107 pp
RAPP M (1971) Cycle de la matière organique et des éléments minéraux dans quelques ecosystemes méditerranéens. Centre National de la Recherche Scientifique, Paris, 184 pp
SNEDECOR G, COCHRANE W (1971) Méthodes statistiques. ACTA, Paris, 650 pp
ST JOHN TV, RUNDEL PW (1976) The role of fire as a mineralizing agent in a Sierran conifer forest. Oecologia 25: 35-45
TRABAUD L (1962) Monographie phytosociologique et écologique de la region de Grabels-St Gély du Fesc. Third cycle thesis (ecology), Faculty of Sciences, University of Montpellier, 131 pp
TRABAUD L (1974) Experimental study of the effects of prescribed burning on a Quercus coccifera L. garrigue. Proceedings of the Annual Tall Timbers Fire Ecology Conference 13: 97-129
TRABAUD L (1977) Comparison between the effects of prescribed fires and wildfires on the global quantitative development of the kermes scrub oak (Quercus coccifera L.) garrigues. MOONEY HA, CONRAD CE (eds) Proceedings of the symposium on the environmental consequences of fire and fuel management in mediterranean ecosystems. USDA Forest Service General Technical Report WO-3, pp 271-282

UECKERT DN, WHIGHAM TL, SPEARS BM (1978) Effects of burning on infiltration, sediments and other soil properties in a mesquite-tobosa grass community. Journal of Range Management 31: 420-425

VEILLEUX JM (1972) Effet d'un brûlage contrôlé sur les propriétés physico-chimiques de l'humus. Ministère des Terres et Forêts du Québec, Mémoire 9: 1-31

VIRO PJ (1974) Effects of forest fire on soil. In: KOZLOWSKI TT, AHLGREN CE (eds) Fire and ecosystems. Academic Press, New York, pp 7-45

WELLS CG (1971) Effects of prescribed burning on soil chemical properties and nutrient availability. Proceedings of the symposium on prescribed burning. USDA Forest Service, Southeastern Forest Experimental Station, pp 86-99

WELLS CG, CAMPBELL RE, DEBANO LF, LEWIS CE, FREDRICKSEN RL, FRANKLIN EC, FROELICH RC, DUNN PH (1979) Effects of fire on soil. USDA Forest Service General Technical Report WO-7, pp 1-34

ZINKE PJ (1977) Mineral cycling in fire type ecosystems. In: MOONEY HA, CONRAD CE (eds) Proceedings of the symposium of environmental consequences of fire and fuel management on mediterranean ecosystems. USDA Forest Service General Technical Report WO-3, pp 385-394

Section IV. Plant Nutrition

Even though the mediterranean-type vegetation of California and Chile may have evolved by convergence on the basis of similar environmental conditions (Di Castri and Mooney 1973; Cody and Mooney 1978), the seasonal cycling of nutrients and mobile carbon pools of plant species from the same mediterranean regions showed little convergence (Shaver 1981). In this section, SHAVER provides further data of mineral, nonstructural carbohydrates and fat content, and weighted concentrations and standing stocks were calculated and presented in the same way as that of Whittaker et al. (1979). In California, the photosynthetic carbon fixation of plants exceeds carbon use during winter and summer whereas during spring carbon use is greater than fixation. The nutrient concentrations of Californian plant species were similar to those of other mediterranean-type ecosystems whereas concentrations in the matorral vegetation were higher. Shaver also showed a more marked seasonal pattern of nitrogen and phosphorus uptake by the Californian plant species than the matorral species. Nutrients in the Californian soils may be more readily available for plant uptake during the wet winter months than at other times of the year.

The different adaptations that most plants from mediterranean South Africa and Western Australia possess for maximizing nutrient uptake, are described by LAMONT. These include mycorrhizas, haustoria of parasitic angiosperms, trapping mechanisms of carnivorous plants and various other root morphologies such as proteoid, capillaroid and dauciform roots. Most of the specialized root systems are located in the uppermost 150 mm of the soil and are prevalent during the moister periods of the year. Lamont provides examples of nodulated nitrogen-fixing legumes which are mycorrhizal and indicates the importance of legumes as the initial colonizers of burnt areas (also considered by Groves, Rundel and Read and Mitchell, this volume).

OECHEL and HASTINGS investigated the effects of burning on the resprouting of <u>Adenostoma fasciculatum</u>, <u>Arctostaphylos glandulosa</u> and <u>Quercus dumosa</u>. Photosynthetic rates of post-fire resprouted plants were higher than the controls and hand-cleared plants. Since there was no change in photosynthetic rates of the resprouts of hand-cleared plants and there were marked increases in soil ammonia-nitrogen in the burnt plots, they concluded that nutrient enhancement in the soil was the major cause of increased photosynthetic rates in the post-fire resprouts. More studies on plant physiological responses to changes in the availability of scarce water and nutrients following fire are necessary for a proper understanding of the ecological effects of fire in mediterranean-type ecosystems.

This introduction was prepared by D T Mitchell

References

CODY ML, MOONEY HA (1978) Convergence versus nonconvergence in mediterranean-type ecosystems. Annual Review of Ecology and Systematics 9: 265-321

DI CASTRI F, MOONEY HA (eds) (1973) Mediterranean-type ecosystems: origin and structure. Springer-Verlag, Berlin, 405 pp

SHAVER GR (1981) Mineral nutrient and nonstructural carbon utilization. In: MILLER PC (ed) Resource use by chaparral and matorral. A comparison of vegetation function in two mediterranean-type ecosystems. Springer-Verlag, Berlin, pp 237-257

WHITTAKER RH, LIKENS GE, BORMANN FH, EATON JS, SICCAMA TG (1979) The Hubbard Brook ecosystem study: forest nutrient cycling and element behaviour. Ecology 60: 203-220

14. Strategies for Maximizing Nutrient Uptake in Two Mediterranean Ecosystems of Low Nutrient Status

BYRON B. LAMONT

14.1 Introduction

The problems of nutrient uptake by plants in mediterranean ecosystems are centred largely on restriction of effective rainfall to only part of the year, usually under six months, with that part also being the coolest. These limitations are intensified in those ecosystems with strongly leached soils derived from inherently infertile parent material (see Chapter 7, this volume). As prime examples, the mediterranean regions of South Africa and Western Australia are recognized as the most nutrient-impoverished of the six mediterranean ecosystems (Beadle 1966; Specht 1979; Cowling and Campbell 1980). Plant losses through drought (Marloth 1915; Hnatiuk and Hopkins 1980) and nutrient-deficiency symptoms (Schütte 1960; chlorosis on some soils) are indicators of the constraints of the edaphic environment in South African fynbos and Australian heath.

When nutrient supply is the major determinant of plant growth and survival there may either be an increase in efficiency of uptake, or tapping of novel sources of nutrients. Both strategies are examined in this review. The actual benefits of each strategy are emphasized rather than its structure or biochemistry. There are no claims that the phenomena described are peculiar to these, or other, mediterranean ecosystems, only that they are well-developed in the Cape and southwestern Australia. It is true however that many taxa possessing these strategies have their world centres of distribution in these two regions.

14.2 Root distribution and seasonality

From extensive studies of sclerophyll vegetation in mediterranean Australia, and to a lesser extent South Africa (Specht and Rayson 1957; Lamont 1976b, 1981c; Low 1979), and of their component species (Specht and Rayson 1957; Purnell 1960; Lamont 1973, 1974; Jongens-Roberts et al. 1980; Dodd and Heddle 1981) clear patterns in root distribution emerge. Most dicotyledons develop a tap root system, sometimes supplemented by sinker roots from extensive laterals when the tap root is weak or atrophied (e.g. *Casuarina pusilla*, *Eucalyptus marginata*). Monocotyledons typically have a fibrous root system associated with rhizomes. Some dicotyledons also reproduce vegetatively from roots (e.g. Jacksonia floribunda, Banksia elegans) or rhizomes (e.g. Protea acaulos, Nuytsia floribunda). Such vegetative reproduction can be viewed as devices for extending the root system through the soil. On the other hand, the less well-documented occurrence of autografts of roots (e.g. Eucalyptus wandoo) and rhizomes (e.g. N. floribunda) not only gives stability to the shoot system but probably improves the overall distribution of water and mineral nutrients through the plant (Lamont 1981e).

The extent to which the root system extends beyond the canopy and down towards the water table depends on size of the plant. The root systems of plants under 0.25 m high are usually confined to the uppermost 0.15-0.25 m of humus-bearing soil. The root systems of larger plants are still concentrated in the uppermost 0.15 m and decrease exponentially with depth. Where litter has accumulated, the greatest concentration per unit depth may actually occur in the decomposing litter between the raw litter and humus horizons (Lamont 1973). Grasstrees (Xanthorrhoeaceae) and many eucalypts are exceptional in that there may be few roots in the uppermost 0.20 m, but prolific development in the less "fertile" subsoil between 0.20 m and 0.70 m (Specht and Rayson 1957; Muir 1977; Lamont 1981b,f), a device perhaps for avoiding summer desiccation of roots. Where tree roots approach the water table the concentration of fine roots may build up again (Kimber 1974; Lamont and Lange 1976). A special feature of the grasstree Kingia australis, prominent in heaths and forests of southwestern Australia, is its mantle of aerial roots concealed between the stem and leaf bases. ^{32}P injected among the leaf bases in summer is readily absorbed by the abundant rootlets located here where water and nutrient availability may exceed that in the soil (Lamont 1981d).

The pattern of root distribution by shrubs through the soil profile corresponds closely to the pattern of nutrient availability, especially nitrogen (Lamont 1973; Jongens-Roberts et al. 1980; Stock and Lewis 1980; Specht 1981) and a causative relationship has been demonstrated between these for two Hakea species in Western Australia (Lamont 1973). Nutrients are released from surface soil and litter particles at the start of the wet season. This is accompanied by new root growth, usually of adventitious origin - which ensures access to nutrients from the plant's own litter (Lamont 1972a, 1973, 1976a). Root growth continues in moist patches, depending on temperature and rainfall pattern, until start of the summer drought.

Phenology of the subsoil roots has not been studied. It is possible that growth near the water table may continue throughout the year, and could explain the almost unabated transpiration of some trees in southwestern Australia (Grieve and Hellmuth 1970). Lamont (1976a) showed that watering patches around individual shrubs of Hakea prostrata in the middle of summer could induce prolific local root growth in an otherwise dormant root system. Other experiments, with ryegrass in mediterranean South Australia, have demonstrated absorption of micronutrients from dry soil, provided the subsoil remained moist (Nambiar 1976, 1977). Here, water is diverted into the surface roots and secreted into the rhizosphere along with mucigel to maintain root growth and nutrient uptake. This system overcomes the usual unavailability of surface nutrients during the mediterranean summer. "Anomalous" summer root growth with abundant water in the rhizosphere has recently been observed in the field, especially Leucospermum hypophyllocarpodendron in coastal fynbos of the western Cape (Lamont 1980a). But transpiration approaches zero in many shrubs during summer and this suggests that water and thus nutrient uptake from anywhere in the root system is usually negligible at this time (Grieve and Hellmuth 1970, Colquhoun 1979).

14.3 Role of root hairs

Mathematical models have been devised for predicting nutrient (Nye 1979) and water uptake (So 1979; Weatherley 1979). A major determinant is absorptive surface area, which appears in these formulae as root length and "effective" root radius. "Effective" allows for the inclusion of any root hair zone, which has been shown by Drew and Nye (1969) and Bhat and Nye (1973) to have a major part to play in absorption, especially when diffusion controls uptake. Root hairs are often very long and abundant on plants of mediterranean South Africa and Western Australia (Lamont 1972b, 1974, 1980a, 1981b; Lamont and Lange 1976; Barrow 1977). For example, the root hairs of *Leucadendron laureolum*, common in South African fynbos, increase the epidermal surface of the ultimate rootlets by over 5.8 times. Under high nutrient conditions this increase was reduced to less than twice, indicating suppression of root hairs. The abundance of root hairs is associated with the presence of very fine ultimate parent roots (<0.2 mm wide) and contrasts with the coarse (> 0.5 mm wide) and hairless "magnolioid" roots of Baylis (1975). There are exceptions with no root hairs (*Podocarpus*, *Acacia*, Cycadaceae, Orchidaceae, Ericales, Santalales) but these have alternative mechanisms for enhancing uptake which will be discussed later.

Root hairs simultaneously increase the area for absorption and shorten the diffusion path for nutrients of low mobility (e.g. phosphorus) or in short supply. This should partly offset the retarding effect on uptake of low winter temperatures here (Walters and Jooste 1980) and the hydraulic resistance of dry soil, especially at the end of the wet season. Root hair secretions are also a major source of enzymes, metabolites and mucigel for rhizosphere-enhanced nutrient release (Head 1964; Kana and Tjepkema 1978; Boerd and Thien 1979) and plugging of gaps that develop as the soil dries (Nambiar 1976). During temporary dry periods, roots and soil particles shrink. Root hairs serve to maintain contact with water films at this time and facilitate rapid recovery following rewetting (Greenland 1979). Their production and elongation is actually enhanced by sub-optimal water availability. Reid and Bowen (1979) found *Medicago truncula*ta growing in South Australian loam had no root hairs in waterlogged soils but a maximum root hair number and length near permanent wilting point. Their appearance on new roots precedes that by the hyphae of endomycorrhizas (Baylis 1972) without retarding their growth rate, unlike sheathing mycorrhizas (Chilvers and Pryor 1965).

Clearly, root hairs have a selective advantage in seasonal climates, and their special role in prolonging the period of nutrient uptake in mediterranean regions should not be underestimated. The proliferation of species with hairy root clusters in the Cape and southwestern Australia, discussed below, is no doubt a reflection of the inherent advantages of root hairs.

14.4 Specialized roots lacking symbionts

Two root modifications fit into this category, the haustoria of parasistic plants and root clusters. The latter are dense bunches of hairy rootlets which are only produced at relatively low soil nutrient levels. They are considered to enhance nutrient uptake, but, unlike mycorrhizas and root nodules (see later), this is without the aid of microsymbionts. Current research indicates that root clusters may be a special feature of mediterranean South Africa and Western Australia, reaching their peak occurrence in the most impoverished sands, laterites and sandstones.

14.4.1 PROTEOID ROOTS

The most studied clusters are proteoid roots with the densely packed rootlets in longitudinal rows (Purnell 1960; Lamont 1972a). They are characteristic of the family Proteaceae (Table 1) including 13 endemic genera, with all but one genus (*Persoonia* sensu lato) bearing them. Rootlets in this primitive genus, however, usually bear a dense cover of root hairs up to 6 mm long (Pathmaranee 1974; Lamont, unpublished). Proteoid roots are concentrated in the top 100 mm of soil, especially in contact with actively decomposing litter (Lamont 1973; Jongens-Roberts et al. 1980). The proteoid roots of *Banksia* and *Dryandra* are the most complex of all, and form a dense, continuous mat at the surface when they are the dominant species (Jeffrey 1967; Lamont, unpublished). The location of proteoid roots at the sites of greatest nutrient and water input to the soil suggests a special role in absorption.

Table 1. Plants with clusters of hairy roots in mediterranean South Africa and Western Australia. Number of species with positive records to date are given in parentheses after each genus.

Type	Family	Genus
PROTEOID	Proteaceae	*Adenanthos* (6), *Aulax* (2), *Banksia* (16), *Brabejum* (1), *Conospermum* (9), *Diastella* (3), *Dryandra* (6), *Franklandia* (1), *Faurea*, (2), *Grevillea* (40), *Hakea* (81), *Isopogon* (9), *Lambertia* (8), *Leucadendron* (10), *Leucospermum* (6), *Mimetes* (2), *Orothamnus* (1), *Paranomus* (2), *Petrophile* (3), *Protea* (14), *Serruria* (8), *Sorocephalus* (2), *Spatalla* (3), *Stirlingia* (1), *Strangea* (2), *Synaphea* (1), *Xylomelum* (3)
	Fabaceae	*Lupinus* (1), *Viminaria* (1)
	Mimosaceae	*Acacia* (1)
DAUCIFORM	Cyperaceae	*Carex* (11), *Cladium* (1), *Cyathochaete* (1), *Gahnia* (2), *Kobresia* (1), *Lepidosperma* (5), *Schoenus* (5)
CAPILLAROID	Restionaceae	*Elegia* (1), *Hypolaena* (1), *Restio* (1), *Staberoha* (1), *Willdenowia* (2)
STALAGMIFORM	Myrtaceae	*Eucalyptus* (1)

Jeffrey (1967) found excised proteoid roots of *Banksia ornata* absorbed 2.4 times as much ^{32}P from water culture as non-proteoid roots on a dry-mass basis, while those of *Leucadendron salignum* absorbed twice as much sodium and rubidium (Walters and Jooste 1980). Relative uptake of ^{32}P varied from 0.9 for intact root clusters of *Banksia ericifolia* to 2.3 for *B. aspleniifolia* (Siddiqi and Carolin 1976). *Banksia marginata* seedlings absorbed 12.7 times more ^{32}P via their proteoid roots than their non-proteoid roots on a root-mass basis (Green 1976), with 16% of this due to enhanced extracellular (capillary-held) uptake. Malajczuk and Bowen (1974) obtained a 3.8 times increase by proteoid roots of the largest *Banksia* in southwestern Australia, *B. grandis*, on an epidermal surface basis. Though phosphorus uptake is active, the enhanced absorption noted here cannot be attributed to greater metabolic activity per unit total (epidermal plus root hair) surface. For example, Gullan (1975) calculated root hairs increased the total surface of proteoid roots of *B. marginata* twelvefold. As a result, absorption per unit total surface area is less than for non-proteoid roots (Jeffrey 1967). Enhanced uptake is thus largely due to the much greater surface area per unit root weight, e.g. 4.5 times for *B. ornata* (Jeffrey 1967) and five times for *Leucadendron laureolum* (Lamont 1980a).

Because of their enhanced uptake of ^{32}P, apparent formation of insoluble polyphosphates and negligible immediate transfer of ^{32}P to the shoots, it has been postulated that proteoid roots store phosphorus to be mobilized later during shoot growth (Jeffrey 1967; Green 1976; Siddiqui and Carolin 1976). This would fit in with the well-known phenomenon of summer growth in many proteaceous plants. However, proteoid roots are short-lived, losing their cortex in 6-10 weeks, and lack storage tissue (Lamont 1972a). In addition, using more specific tests, Anne Ashford (personal communication 1980) was unable to detect polyphosphate granules in mature proteoid or non-proteoid roots of two *Hakea* spp. grown in nutrient solution.

Proteoid roots are initiated soon after the onset of the mediterranean winter and their formation continues until late spring (Lamont 1976a; Green 1976). Experiments with *Hakea francisiana* showed proteoid, relative to non-proteoid, root production peaked at 1-2 times field capacity, diminishing, but not ceasing, with waterlogging and at the standard permanent wilting point (Lamont 1976a). The proteoid roots of *Viminaria juncea*, a legume, may even develop in flooded soil, along with pneumatophores and root nodules (Lamont 1972a). In the presence of sufficient soil water they will also develop during summer in the field (Purnell 1960; Lamont 1976a). In addition, Lamont (1980a) recently observed young proteoid roots on a *Leucospermum* sp. in dry sand during summer. In contrast to adjacent soil, the rhizosphere soil water, apparently secreted by the roots themselves (Nambiar 1976; Lamont 1981b), was above field capacity. This can be considered as a mechanism for prolonging nutrient release and absorption, in an otherwise highly seasonal climate and deserves closer study.

Although proteoid root formation is favoured by nutrient-rich (surface) portions of otherwise infertile soil, they are suppressed in fertile soils (Lamont 1981a). This apparent paradox is explained by differences in nutrient status of the plant. Plants with a low nitrogen concentration will form proteoid roots in abundance, but preferentially in humus-rich pockets. Plants with relatively high nitrogen may not form any proteoid roots (though abundant "normal" roots) even when repotted in leached sand (Lamont 1973). While the contribution by proteoid roots varies markedly between species, it appears to increase with plant age. For example, 80% of the dry mass of new season's surface roots of mature *Leucadendron argenteum* may be proteoid. Green (1976) did however note failure of the development of new proteoid roots by *Banksia ornata* following an intense fire, but this also appeared to involve suppression of parent roots. Apart from nutrient levels which are too low even for significant proteoid root production, there is clearly a close relationship between the extent of their contribution to the root system and plant demand for nutrients.

14.4.2 OTHER ROOT CLUSTERS

Dauciform roots are carrot-shaped rootlets, densely covered in long root hairs when mature (Lamont 1974). They occur in clusters but are not as discrete nor as dense as proteoid roots. These roots are characteristic of the Cyperaceae, with seven of nine genera in mediterranean South Africa and Western Australia examined so far possessing them (Table 1). The only genus that has been studied extensively, Carex, is unlike the others in that only half its species appear to have dauciform roots. The profile distribution, physiology and function of these roots appear very similar to those of proteoid roots (Lamont 1981a). There is as yet no direct evidence of how their rate of nutrient uptake compares with non-dauciform roots, though they are also eliminated at moderately high levels of nitrogen (Lamont 1974).

Clusters of rootlets densely covered with long root hairs have also been noted on some members of the Restionaceae (Campbell 1964, 1981; Gullan 1975; Low 1980; Table 1). Like the Proteaceae, the centres of distribution of the restios is mediterranean South Africa and Western Australia. My preliminary study of these root clusters (Lamont 1981e) shows them to be abundant in South African fynbos - both in bogs and on deep sand. Two types of lateral may arise from the rhizome - a thick, sparsely branched one which penetrates deeply into the soil, and a thin, highly branched one which is best developed near the surface. Gullan (1975) described the latter parent roots in Hypolaena fastigiata as 2 mm wide, bearing five, 6.5 mm long, laterals per 10 mm length of parent root. The root hairs were on average 2 mm long and they increased the surface area of the laterals by 26 times. Thus, though these rootlets contributed only 4% of the volume of the root system of this species, they accounted for almost 90% of its surface area.

This dimorphic root system appears to be best developed on rejuvenating rhizomes, after disturbance or fire. Under waterlogged conditions the lateral branches divide many times and twine among the organic particles to form a spongy mat at the soil surface. According to Campbell (1981) this mat may hold 15 times its dry mass in water. It is proposed that this final type of root cluster be called "capillaroid" to emphasize the unique water-holding capacity of root clusters in general, considered here as an integral part of their nutrient uptake strategy.

There is less well-defined root clustering in other major groups here, especially eucalypts in southwestern Australia (Shea and Dell 1981). This is usually in association with mycorrhizas (see later). The stalagmite-like aggregations of rootlets of Eucalyptus gomphocephala beneath drip-points in limestone caves however are sparsely mycorrhizal (Lamont and Lange 1976). These "stalagmiform" roots seem to play a small part in prolonging nutrient uptake into summer, as the surface soil above, where most rootlets are located, dries out first.

14.4.3 HAUSTORIA

Haustoria function like roots in obtaining water and nutrients from the substrate, except they bypass the soil by tapping host plants. Most of the world's parasitic angiosperms (11 out of 14 families) are represented in mediterranean South Africa and Western Australia (Table 2). There are more species of root parasite here than elsewhere in the two sub-continents. Most root-attaching haustoria occur at depths of 100-200 mm on long lateral branches which usually lack root hairs (Herbert

1925). The laterals of *Nuytsia floribunda*, a species widespread on the sandplains of southwestern Australia, in fact are rhizomes up to 110 m long (Herbert 1919). *Nuytsia* affixes onto any root in its path, including proteoid roots (Lamont 1977), but especially those of sedges and Restionaceae. Such insignificant hosts are surprising, since *Nuytsia* is probably the world's largest parasite, reaching a height of 12 m, with trunks arising from a large lignotuber (Gobel 1975) which is usually linked by rhizomes to adjacent clumps.

Table 2. Parasitic angiosperms in mediterranean South Africa and Western Australia. Numbers refer to all species recorded, and in some cases their parasitic habit has been assumed.

Type	Family	Genus
ROOT HEMIPARASITE	Loranthaceae	*Nuytsia* (1)
	Olacaceae	*Olax* (2)
	Santalaceae	*Choretrum* (4), *Colpoon* (1), *Exocarpus* (3), *Leptomeria* (8), *Santalum* (3), *Thesidium* (8), *Thesium* (86)
STEM HEMIPARASITE	Lauraceae	*Cassytha* (7)
	Loranthaceae	*Amyema* (8), *Lysiana* (1), *Moquinella* (1), *Septulina* (1), *Tapinanthus* (1)
	Viscaceae	*Viscum* (4)
ROOT HOLOPARASITE	Balanophoraceae	*Mystropetalon*[a] (2), *Sarcophyte*[a] (1)
	Hydnoraceae	*Hydnora* (1)
	Orobanchaceae	*Orobanche* (3)
	Rafflesiaceae	*Cytinus*[a] (2)
	Scrophulariaceae	*Harveya* (9), *Hyobanche* (2)
STEM HOLOPARASITE	Convolvulaceae	*Cuscuta* (2)
	Rafflesiaceae	*Pilostyles*[a] (1)

[a] Endoparasite.

Hemiparasites can be defined as those substantially independent of hosts for carbohydrate requirements. Most root hemiparasites, such as *Nuytsia*, have a wide range of hosts, and self-parasitism is common. Those which appear host-specific, such as *Thesidium fragile* on *Metalasia muricata* (South Africa) and *Leptomeria spinosa* on *Eremaea* spp. (Western Australia), also avoid self-parasitism. There are a few examples of epiparasitism, such as *Colpoon compressum* on *Thesidium fragile* and *Santalum acuminatum* on *Exocarpus sparteus* (but not the reverse). Most root hemiparasites can be grown, at least as seedlings, in the absence of a host. However, in southwestern Australia, Herbert (1925) recorded death within 12 months for *Santalum spicatus* lacking hosts, compared with vigorous growth of those attached to hosts. Autrophic plants of Thesidium fragile are more chlorotic and stunted than heterotrophic plants (L Ayris, unpublished).

Little work on nutrition of root parasites has been done in these two regions. In Rhinanthus serotinus (Scrophulariaceae, with 11 parasitic species in mediterranean South Africa), top growth increased six times on attachment compared with the autotrophic plants, with no change in root growth (Klaren and Janssen 1976). This difference was eliminated simply by adding inorganic nutrients to the unattached plants.

Stem parasites are not as conspicuous as root parasites in these two mediterranean regions (Table 2). However many major species are parasitized by mistletoes, especially in Western Australia, which have become more abundant following clearing for agriculture (Lamont and Southall 1982a). In contrast to root hemiparasites, mistletoes (Loranthaceae, Viscaceae) here possess one (primary) haustorium which, of necessity, is attached to a single host. There is considerable host specificity, usually operating at genus level. For example, Amyema lynophyllum parasitizes Casuarina spp. preferentially, Viscum continuum only Acacia spp. Amyema miraculosum is an epiparasite on its usual host Santalum acuminatum (Barlow and Wiens 1977).

Not only do mistletoes transpire at 2-4 times the rate of their hosts (Hartel 1937; Hellmuth 1971; Lamont and Southall 1982a) they also possess a higher concentration of most nutrients (Lamont 1982a; Lamont and Southall 1982b). This is consistent with the reduced sclerophylly of all parasites compared with their hosts. In particular, the ratio of potassium to calcium is much higher, e.g. five times for Amyema preissii on Acacia acuminata in southwestern Australia (Lamont and Southall 1982b). This selective uptake rules out greater transpiration as the sole cause of higher concentrations in the parasite. In view of the contrasting remobilization properties of these two nutrients, non-xylem feeding is implicated as well. Ansiaux (1958) however considered such findings to be a feature only of holoparasites, which have phloem links in their haustoria, in contrast to hemiparasites which at best have xylem links. Clearly more attention needs to be given to non-xylem uptake by the haustoria of stem hemiparasites (Lamont 1982a).

Preferential transport of mineral nutrients and carbohydrates across the haustoria of the holoparasites Orobanche and Cuscuta is well established (Whitney 1972; Tsivion 1978). This is effected via contact cells with a highly convoluted plasmalemma giving maximum absorptive area (Dörr 1975). Though rare, the remarkable endo-holoparasites are better represented in mediterranean South Africa and Western Australia than in the remainder of their continents. The vegetative body of species in the Balanophoraceae and Rafflesiaceae consists of septate filaments closely associated with the host's cambium (Kuijt 1969). No doubt because of their high nutrient content, pea legumes are the preferred hosts (obligately so for Pilostyles), though Marloth (1913) recorded other hosts for the South African holoparasites.

14.5 Specialized roots bearing symbionts

Nutrient-gathering devices of the symbiotic type include nitrogen-fixing nodules and the fungal associations with rootlets termed mycorrhizas. Some of these systems are special features of either or both these nutrient-impoverished regions, e.g. ericoid mycorrhizas, while others, though widespread here, may be better represented elsewhere, e.g. legume nodules in South Africa.

14.5.1 NITROGEN-FIXING SYSTEMS

About 1500 legumes, including 11 small endemic genera, occur in mediterranean South Africa and Western Australia. Nearly all the species in genera represented here possess root nodules (Table 3). The only legume family poorly represented in these two regions (though abundant in the

Table 3. Nitrogen-fixing symbiotic systems in mediterranean South Africa and Western Australia. Figures in brackets refer to number of species (including some non-mediterranean) at present known to possess nitrogen-fixing structures.

		Macrosymbiont	
Structure	Microsymbiont	Family	Genus
ROOT NODULE	Bacteria (*Rhizobium*)	Mimosaceae	*Acacia* (131), *Albizia* (1)
		Fabaceae	*Amphithalea* (1), *Argyrolobium* (20), *Aotus* (5), *Aspalathus* (59), *Bossiaea* (11), *Brachysema* (4), *Burtonia* (1), *Chorizema* (5), *Cyclopia* (6), *Daviesia* (10), *Dillwynia* (5), *Dolichos* (6), *Eutaxia* (5), *Gastrolobium* (8), *Gompholobium* (6), *Hardenbergia* (1), *Hovea* (3), *Indigofera* (121), *Jackonsia* (7), *Kennedia* (6), *Latrobea* (1), *Lebeckia* (7), *Lotononis* (36), *Melolobium* (7), *Medicago* (16), *Mirbelia* (5), *Oxylobium* (5), *Podalyria* (8), *Priestleya* (6), *Psoralea* (19), *Pultenaea* (5), *Rafnia* (9), *Rhynchosia* (32), *Sphaerolobium* (3), *Templetonia* (2), *Viminaria* (1), *Virgilia* (1), *Wiborgia* (3)
ROOT NODULE	Actinomycete (*Frankia*)	Casuarinaceae	*Casuarina* (24)
		Myricaceae	*Myrica* (9)
CORALLOID ROOT	Blue-green algae (*Nostoc*)	Zamiaceae	*Encephalartos* (7), *Macrozamia* (2)
THALLUS	Lichen (e.g. *Leptogium*)	-	Woody stems (substrate)
INFLORESCENCE (NECTAR)	Blue-green algae	Proteaceae	*Banksia* (3)

summer-rainfall parts of the two continents) is Caesalpiniaceae, whose species usually lack nodules. The common legumes accept a range of nodule-inducing rhizobial bacteria, especially those with wide host and low pH tolerances (Bowen 1981).

Four alternative nitrogen-fixing symbiotic systems are also found (Table 3). The massive coralloid roots of the two cycads are inhabited by blue-green algae (usually *Nostoc*, Bowyer and Skerman 1968) while the well-represented Casuarina and Myrica usually have clustered root nodules, especially in moist soils, containing an actinomycete. Two "indirect" systems appear worthy of more research, in view of the scarcity of nitrogen in the soils of southwestern South Africa and Western Australia: blue-green algae are the symbionts in some epiphytic lichens (Marloth 1913) and in nectar of several Banksia spp. (Lamont 1980b), and could augment soil nitrogen via throughfall and stemflow following rain.

Verification of nitrogen fixation in these two mediterranean ecosystems, using the acetylene reduction technique, has been obtained for Albizia lophantha (Tacey 1977), Podalyria calyptrata, Acacia cyclops and A. saligna (Haxen 1978), Acacia pulchella, Mirbelia dilatata and Bossiaea aquifolium (Shea and Kitt 1976), Viminaria juncea, Oxylobium lanceolatum and Jacksonia furcellata (Finlayson and McComb 1978), Kennedia prostrata (Lawrie 1981) and K. coccinea (Hingston et al. 1982). Though the conversion factor into actual units of N_2 fixed varies greatly between species and experimental conditions, the annual legume Trifolium pratense, from the Mediterranean Basin, appears to have much higher rates than those above (Hardy et al. 1973). Among non-legume N_2-fixers, verification of fixation has been obtained for Encephalartos altensteinii and E. ferox (Grobbelaar et al. 1971), Macrozamia riedlei (Halliday and Pate 1976; Grove et al. 1980), Casuarina cunninghamiana (indigenous to eastern Australia, Tyson and Silver 1979) and the lichen Leptogium (Bond and Scott 1955). Within the limits of the techniques, legumes appear to have higher rates of acetylene reduction than do non-legumes. This may be related to annual renewal of nodules in legumes, whereas those of non-legumes are perennial structures containing much moribund tissue from previous seasons.

In southwestern Australia Lange (1959) observed that 94% of legume species had active (healthy) nodules in winter, but only 3% had them in summer. Associated with this senescence is a marked drop in fixation rates during the summer period (Halliday and Pate 1976; Lawrie 1981; Hingston et al. 1982). Since most nodules are located in the upper 0.15 m of soil (Engin and Sprent 1973; Tyson and Silver 1979) such depressive effects on nodule longevity and fixation by drought and high temperatures are to be expected in these mediterranean ecosystems. Even so, survival of both perennial and seasonal nodules is dependent on uptake of internal water from deeper roots rather than through direct absorption (Sprent 1972). Environmental factors besides drought which probably limit N_2 fixation in these two regions are seasonal waterlogging, phenolic compounds from litter, low winter temperatures, salinity, defoliation (fire, drought, herbivory), sometimes shading, low pH and low nutrient availability. However, insufficient studies have been done to identify the relative importance of each factor.

Nodule formation, including proliferation of rhizobia, requires the presence of potassium, calcium, zinc, copper and boron (Lie and Mulder 1971; Nutman 1976). The N_2-fixation process itself involves iron, molybdenum, cobalt, zinc, phosphorus and to a lesser extent copper, sulphur and calcium (Becking 1970; Bergerson 1974; Hewitt and Smith 1975; Snowball et al. 1980). Some combined nitrogen is necessary for maximum nodule production (Stewart 1963, Rodriguez-Barrueco et al. 1970) and sometimes their optimal efficiency as well (Gibson 1976). Clearly therefore N_2-fixation imposes additional nutrient demands on the plant. This is highlighted by the fact that only N_2-fixers require molybdenum and cobalt for growth. Not only do legumes have a high nutrient demand in general, the nutrients associated with nodule activity, such as phosphorus, iron, calcium, molybdenum, copper, cobalt and zinc, are poorly mobile, unavailable and/or leached out of most soils in mediterranean South Africa and Western Australia. This could explain why very few N_2-fixers are trees here, and their scarcity in climax vegetation on highly impoverished substrates, especially fynbos (personal observations).

However, the rapid establishment of hard-seeded legumes following fire is well-known in these two regions (Levyns 1935; Shea and Kitt 1976; Baird 1977). In jarrah forest, stands of shrub legumes two years after a "hot" fire were estimated to contribute 0.2-1.5 g N m^{-2} yr^{-1} (Hingston 1977; Malajczuk and Grove 1977). Six years after a fire, the leaves of two legumes contained 2.5 times the nitrogen concentration of the eucalypt overstorey (Malajczuk and Grove 1977). Grove et al. (1980) observed six times increase in acetylene reduction by patches of Macrozamia riedlei in the same forest following a fire. Fertilizer trials suggest that part of the fire effect is enhancement of nodulation and fixation efficiency through nutrient accession, especially phosphorus (Shea and Kitt 1976; Hingston et al. 1982). In the absence of fire, nutrient uptake could be enhanced instead by other specialized roots. Mosse et al. (1976) found that three legumes failed to nodulate in phosphorus-impoverished soil if they were not also bearing mycorrhizas. It is of particular interest therefore that all species of 11 N_2-fixing genera (except Viminaria, which has proteoid roots, Lamont 1972c) examined so far in mediterranean South Africa and Western Australia (Tables 3, 4; Lamont 1982b) have mycorrhizas as well as nodules.

14.5.2 GENERAL MYCORRHIZAS

The term "general" is used here to refer to the so-called vesicular-arbuscular (VA) mycorrhizas, to distinguish them from the three "host-specific" groups: ericoid, orchid and sheathing. Over 90 genera of 50 families in these two regions have species with VA mycorrhizas (Table 4). VA mycorrhizas have been noted in various habitats of southwestern South Africa (Low 1980). Of 19 of the most common species in a patch of eucalypt woodland on laterite in southwestern Australia, Titze et al. (1980) found that 14 consistently carried these general mycorrhizas. Infection was highest in herbs and small shrubs (Boronia, Logania, Opercularia, Senecio, Stylidium and Hibbertia). Collation of Sward's (1978) records for two infertile sites in Victoria (with almost all their genera, if not species, in common with mediterranean Western Australia) shows 44 species were VA mycorrhizal, 43 possessed other specialized modes of nutrition while 20 appeared to lack any specialized structures. These figures are consistent with evidence from other ecosystems where greatest incidence of VA mycorrhizas occurs at sites where the soils are shallowest, least fertile or with most interspecific competition (Read et al. 1976).

Plants of the following genera present in these two mediterranean ecosystems (but not necessarily the same species) grew much larger in infertile soils when bearing VA mycorrhizas than did the controls which lacked them: Podocarpus, Pittosporum (Baylis 1967), Atriplex (Williams et al. 1974), Leptospermum and Metrosideros (Hall 1975), Viola (Mosse 1975), Pteridium and Solanum (Cooper 1975) and Poa (Powell 1975). This was accompanied by a higher total content of inorganic nutrients, and usually a higher concentration of phosphorus (e.g. Hall 1975). Differences in concentration between infected and control plants can be attributed to selective properties of the microsymbiont (Cooper and Tinker 1978), and the special advantages of mycorrhizas may be the correction of nutrient imbalances rather than just increased uptake of ions in general. Growth responses to VA mycorrhizas increase up to light saturation (Johnson 1973;

Table 4. Selected genera with mycorrhizas in mediterranean South Africa and Western Australia. Some records in parentheses include non-mediterranean species.

Type	Family	Genus
VESICULAR-ARBUSCULAR	Aizoaceae	Galenia (1)
	Asteraceae	Arctotheca (1), Eriocephalus (1), Gnaphalium (2), Helichrysum (6), Senecio (4), Stoebe (1)
	Boraginaceae	Myosotis (1)
	Bruniaceae	Staavia (1)
	Casuarinaceae	Casuarina (3)
	Celastraceae	Cassine (1)
	Cunoniaceae	Cunonia (1)
	Cupressaceae	Widdringtonia (1)
	Cyperaceae	Carex (4), Schoenus (3)
	Dennstaedtiaceae	Pteridium (1)
	Dilleniaceae	Hibbertia (11)
	Ebenaceae	Diospyros (1)
	Euphorbiaceae	Amperea (2), Phyllanthus (1)
	Fabaceae	Aotus (2), Bossiaea (1), Daviesia (1), Dillwynia (2), Dolichos (1), Indigofera (3), Kennedia (1), Oxylobium (1), Pultenaea (1), Trifolium (7)
	Goodeniaceae	Dampiera (2), Goodenia (2)
	Hydrocotylaceae	Platysace (1), Trachymene (1), Xanthosia (1)
	Lamiaceae	Prostanthera (1), Teucrium (1)
	Liliaceae	Asparagus (1)
	Linaceae	Linum (1)
	Loganiaceae	Logania (1)
	Mimosaceae	Acacia (9)
	Myrtaceae	Eucalyptus (5), Leptospermum (7), Melaleuca (1)
	Plantaginaceae	Plantago (1)
	Poaceae	Agrostis (3), Poa (4), Danthonia (1)
	Podocarpaceae	Podocarpus (13)
	Polygalaceae	Comesperma (1)
	Polygonaceae	Polygonum (2)
	Ranunculaceae	Ranunculus (1)
	Rhamnaceae	Phylica (1)
	Rosaceae	Prunus (1)
	Rubiaceae	Anthospermum (1), Galium (3), Opercularia (1)
	Rutaceae	Boronia (5), Eriostemon (3)
	Solanaceae	Solanum (3)
	Thymelaeaceae	Pimelea (4)
ERICOID	Ericaceae	Erica (19), Grisebachia (1), Simocheilus (1)
	Epacridaceae	Astroloma (2), Leucopogon (7), Monotoca (1)
ORCHID	Orchidaceae	Acianthus (3), Caladenia (20), Diuris (8), Elythranthera (2), Eriochilus (1), Gastrodia (1), Lyperanthus (1), Microtis (2), Prasophyllum (2), Pterostylis (5), Thelymitra (6)
SHEATHING	Casuarinaceae	Casuarina (2)
	Fabaceae	Dillwynia (2)
	Lobeliaceae	Lobelia (3)
	Myrtaceae	Eucalyptus (157), Leptospermum (4), Melaleuca (3)

Hayman 1974). Since light is rarely limiting in these two regions (closed forests are extremely rare) a high response to infection on these grounds alone is therefore to be expected, but has yet to be confirmed.

Especially relevant is the limited evidence that VA mycorrhizas enhance recovery from wilting (Safir et al. 1972), as do root hairs. Indeed, because they are usually narrower, longer and more versatile in their direction of growth, hyphae should be more efficient than root hairs in absorption of water and nutrients. But in contrast to root hairs, rate of infection falls markedly once soil moisture is below field capacity (Reid and Bowen 1979). This preference for moist soils suggests severe limits to the value of VA fungi in the dry sands and rocks so common in mediterranean South Africa and Western Australia. Since many plants at these sites appear to be VA mycorrhizal, the possibility of special drought-adapted endophytes may warrant investigation. Many soils in these two regions may have inadequate levels of phosphorus (in particular), nitrogen, calcium and/or potassium for optimal infection rates and growth responses (Abbott and Robson 1979; Daft 1979; Malajczuk and Lamont 1981). Excessive levels of nutrients for VA formation (e.g. Chambers et al. 1980) on the other hand are most unlikely.

14.5.3 HOST-SPECIFIC MYCORRHIZAS

Though not as widespread taxonomically as VA mycorrhizas, this group contains some of the most characteristic components of mediterranean South Africa and Western Australia (Table 4). These regions share almost 1000 species in the Ericales (580 in the South African genus *Erica*) with 26 endemic genera. Though few of these species have been examined for ericoid mycorrhizas, rootlets of all those studied so far have clusters of hyphae in their hypodermal cells (Malajczuk and Lamont 1981). Terrestrial, rather than epiphytic, orchids are a feature of these two regions, with over 340 species. Only Western Australian species have been examined and cortices in all bear coils of hyphae characteristic of orchid mycorrhizas (Malajczuk and Lamont 1981).

Most woodlands and forests in southwestern Australia are dominated by *Eucalyptus* species, which possess sheathing (ecto-) mycorrhizas. Unlike other mycorrhizal fungi, these are intercellular and form a dense sheath up to 40 μm wide around the rootlet. While *Eucalyptus* has speciated strongly in southwestern Australia (with 107 species), another 400 are spread over the Australian continent (Chippendale 1981) and probably these are all also ectomycorrhizal (Chilvers and Pryor 1965; Lamont 1979). Sheathing mycorrhizas appear optional on other prominent myrtaceous genera (*Melaleuca* and *Leptospermum*) and *Casuarina* in mediterranean Western Australia, alternating or co-occurring with VA mycorrhizas, as in some Eucalypts (Lamont and Lange 1976). The apparent absence of ectomycorrhizal species in South Africa is notable (see also Chapter 12, this volume).

Host-specific mycorrhizas are concentrated in the uppermost 100 mm of soil and the decomposing litter layer above it (P G F Coley, personal communication 1980; Malajczuk and Hingston 1981). Hot fires suppress sheathing mycorrhizas at least, possibly through both loss of inoculum and litter (Renbuss et al. 1972; Malajczuk and Hingston 1981). There is a wide range of ectomycorrhizal fungi which are specific to eucalypts. Malloch et al. (1980) argue that such host-specificity gives a competitive advantage to plants in species-poor vegetation strata (as with eucalypt trees in southwestern Australia). However, any potential rain forest (VA mycorrhizal) competitors have long since been eliminated in southwestern

Australia. There may also be a nutritional benefit in the presence of a mosaic of different symbionts on the one root system, as found in eucalypts, but this has yet to be established experimentally. In contrast are ericoid mycorrhizas where a single, well-adapted fungus with wide host tolerance is found in species-rich heathlands of South Africa (Robinson 1973). This has the advantage of facilitating infection from adjacent root systems when they are more than likely to belong to different species (Read et al. 1976).

There are no reports of the benefits of ericoid and orchid mycorrhizas in these two regions, but their nutritional advantages are well established in related species and vegetation elsewhere (Stribley and Read 1976; Smith 1966). Malajczuk et al. (1975) obtained a 30% increase in dry mass and phosphorus concentration when seedlings of Eucalyptus calophylla (a major tree in southwestern Australia) were allowed to develop sheathing mycorrhizas. The low phosphorus status of the laterite soil, from the species' habitat, was confirmed by the continuation of mycorrhizal enhancement even when fertilizer application was increased to an equivalent of 1.8 g P m^{-2}. At this level 73% of the rootlet apices were mycorrhizal, whereas previously (at 0.3 g P m^{-2}) only 47% of the apices were mycorrhizal. Consistent with its early release from organic matter, ammonium is taken up preferentially to nitrate by host-specific mycorrhizas (Carrodus 1967; Harley 1969; Stribley and Read 1976). However, some eucalypts, grown experimentally without mycorrhizas, may absorb nitrate preferentially to ammonium (Moore and Keraitis 1971).

Apart from making use of soluble phosphorus, there is growing evidence of the capacity of host-specific (and to some extent VA) mycorrhizas to mineralize normally unavailable sources of phosphorus. This is effected through phosphatases and related enzymes on the hyphal surfaces (Bartlett and Lewis 1973; Williamson 1973; Harley 1975; Mitchell and Read 1980). Since phosphatase activity is inversely proportional to phosphorus availability (Beresford 1979) its rate should be at a maximum in these two ecosystems. Several factors work against this however: levels of other nutrients (e.g. magnesium, potassium) may not be adequate for maximum phosphatase synthesis, the basal metabolic rates of most sclerophylls are low (in turn related to nutrient availability), and a major component of fossil soils in southwestern Australia is aluminium, which inhibits phosphatase activity. However, 18% of families in this region, including Myrtaceae, Proteaceae and Rutaceae, accumulate aluminium without harmful effects (Webb 1954; Lamont, unpublished). There is some evidence that roots and mycorrhizas may secrete organic acids which chelate and detoxify aluminium and iron released during solubilization of laterite particles (Mullette et al. 1974).

Despite little direct evidence, sheathing mycorrhizas in particular have several features especially adaptive in a mediterranean environment. The first relates to their storage capacity and slow rate at which nutrients are released from the hyphae to the root (Harley 1975) and the second to their longevity, which may span two growing seasons. Together they suggest mechanisms for sustaining growth over summer in some eucalypts, and rapid root growth and mycorrhizal re-establishment at the start of winter (B Rochel, personal communication 1980). Thirdly, they tend to bunch together in the fashion of root clusters (Ashton 1976; Shea and Dell 1981, Fig. 1) suggesting a similar role in water retention and nutrient release. The sheath probably protects the root against desiccation and maintains close root-soil contact. To what extent this is responsible for increased heat tolerance of ectomycorrhizal pines (Marx and Bryan 1971) remains unexplored in the mediterranean context.

While sheathing mycorrhizas are a considerable carbohydrate drain on their hosts (Harley 1975), orchid and ericoid mycorrhizas maintain a degree of saprophytism, including absorption of various organic nitrogen products of litter decomposition (Harley 1969; Stribley and Read 1975). Indeed, in mediterranean Western Australia, several orchids (Gastrodia, Rhizanthella) must be holosaprophytic as they lack chlorophyll. They usually occur under sclerophyllous shrubs (e.g. R. gardneri with Melaleuca uncinata, George 1980) which are no doubt mycorrhizal (Table 4). As shown by Campbell (1963), such orchids may thus be partly epiparasitic on these shrubs, via a common symbiont.

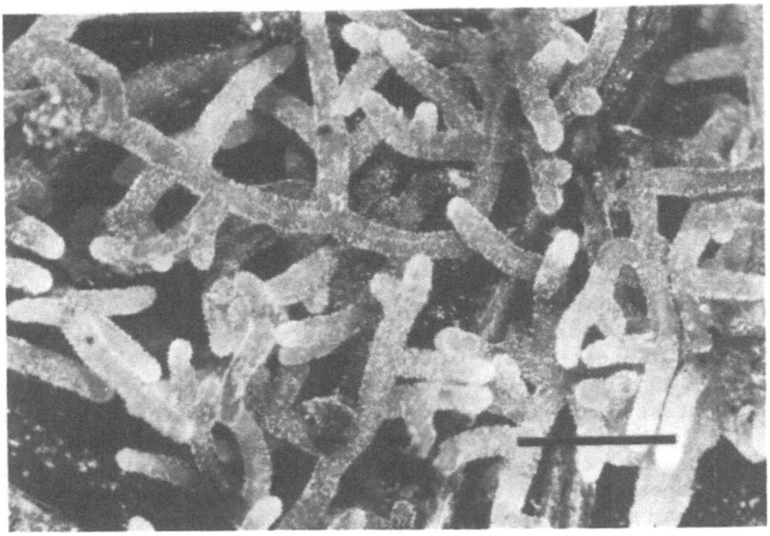

Fig. 1. Rootlets of the major tree in southwestern Australia, Eucalyptus marginata (jarrah), bearing sheathing (ecto-) mycorrhizas. Note the extensive clumping of the rootlets reminiscent of non-symbiotic root clusters. Scale = 1 mm. Courtesy Dr N Malajczuk, CSIRO, Perth.

14.6 Leaves specialized for carnivory

Another feature of mediterranean South Africa and Western Australia is the abundance of carnivorous plants (Carlquist 1974; Table 5). Three types of "trap" occur here: glandular hairs, pitchers and bladders. Southwestern Australia is the world centre of distribution for Drosera, which has stalked glands that capture small arthropods. In sand culture, increases of up to 50% in dry mass and higher concentrations of nitrogen, phosphorus, potassium, sulphur and zinc have been demonstrated for plants fed vinegar-flies compared with the controls (Chandler and Anderson 1976a; Pate and Dixon 1978). Dixon et al. (1980) showed arthropods contributed up to 17% of net uptake of nitrogen by D. erythrorhiza, widespread in coastal sandplains of southwestern Australia. Uptake by the glands of labelled ammonium, phosphorus, sulphur, calcium, various amino acids and glucose have been observed directly (Chandler and Anderson 1976b; Juniper et al. 1977; Dixon et al. 1980). That the prey serves essentially as a source of mineral nutrients was demonstrated by Chandler and Anderson (1976c) on the grounds that (a) in the presence of a complete nutrient solution droseras grew equally whether or not supplied with insects, and (b) insect-fed plants grew poorly in the dark compared with those in the light. In fact, the tentacles are probably a net carbohydrate drain on the plant, as starch was immediately mobilized from the leaf of D. capensis when fed insects or mineral nutrients (Giessler 1928).

Table 5. Carnivorous plants in mediterranean South Africa and Western Australia. Records refer to the number of species in these two regions.

Type	Family	Genus
Glandular hairs	Byblidaceae	Byblis (1)
	Droseraceae	Drosera (54)
	Proteaceae	Grevillea (1)[a]
	Roridulaceae	Roridula (2)[a]
Pitchers	Cephalotaceae	Cephalotus (1)
Bladders	Lentibulariaceae	Utricularia (8, excluding aquatic spp.), Polypompholyx (3)

[a] Prey not digested

Unlike Drosera, Byblis has separate glands for capture and digestion of the prey. Plants grown for three months in their own sand in southwestern Australia and fed insects had 2.5 times the wet mass and 2.3 times the nitrogen content of those kept free of insects (K Loveridge and B Lamont, unpublished). The differences were negligible when both sets of plants were watered with 1.5 mM NH_4NO_3 rather than deionized water. Cephalotus follicularis is the only member of its family and is restricted to peat swamps in the wettest parts of southwestern Australia. Though the nutritional role of the pitcher has received little attention (Dakin 1919), its structure and functioning appear remarkably similar to those in unrelated pitcher plants (Lüttge 1971; Adams and Smith 1977). Movement of fluid to (with agents of digestion) and from (with products of digestion) the pitcher appears to be effected through sunken glands and permanently open stomates in certain parts of the pitcher.

Unlike mediterranean Western Australia, terrestrial bladderworts (Lentibulariaceae) are better represented outside the mediterranean portion of South Africa. Utricularias whose bladders received protists grew larger than those not receiving prey (Sorenson and Jackson 1968), while flowering only occurred in sterile plants which received peptone and meat extract (Pringsheim and Pringsheim 1962). Unlike the other carnivorous systems already described, digestive glands are absent, though internal stellate hairs appear to pump out fluid from the bladder chamber. In common with the other carnivorous systems are (a) an apparent involvement of micro-organisms in the digestive process, and (b) a preference for nitrogen in the form of ammonium rather than nitrate.

Deserving of more attention are Roridula, endemic to the Cape and some members of Grevillea in southwestern Australia. Resinous glands, which catch but do not digest arthropods, are widespread in these species

(Marloth 1925; Lamont 1982c). Despite decomposition and eventual return of nutrients to soil beneath the plant, the animals appear to make an insignificant contribution to the nitrogen content of *Grevillea leucopteris*, a major colonizer in some sandplain scrub of mediterranean Western Australia. Their major role may lie in discouraging herbivory (relevant to nutrient conservation rather than uptake).

14.7 Discussion

A wide array of nutrient-uptake mechanisms has been demostrated in the floras of mediterranean Australia and South Africa. Not only are all known specialized modes of nutrition among higher terrestrial plants represented, but they are invariably more abundant (on a species per unit area basis) in these regions than in the surrounding (more fertile) semi-arid to subtropical areas and apparently in the other (more fertile) mediterranean regions (Table 6). This applies in particular to root clusters (proteoid, dauciform and capillaroid roots), root-parasitic haustoria, non-legume nitrogen-fixing systems, host-specific mycorrhizas (ericoid, orchid and sheathing) and carnivorous leaves. The only exceptions are stem-parasitic haustoria, legume nodules and general (vesicular-arbuscular) mycorrhizas which, while abundant in the Cape, may be better represented in the subtropical forests to the north-east. These exceptions can probably be attributed to greater moisture and/or nutrient requirements of the parent plants (e.g. stem parasites, legumes) or of the uptake strategies (e.g. preference of VA mycorrhizas for moist soils, extra nutrient requirements for nodule formation and function).

Table 6. Incidence of special nutrient-uptake strategies in mediterranean South Africa and Western Australia and their abundance, on a species per unit area basis, relative to elsewhere in the respective subcontinents and the remaining mediterranean ecosystems.

Nutrient-uptake strategy	Relative number of species with each strategy		
	Med. vs non-med. Africa	Med. vs non-med. Australia	Med. S. Afr. and W. Aust vs other med. ecosystems[a]
ROOT CLUSTERS	>	>	>
HAUSTORIA			
Root	>	>	>?
Stem	<	>	>?
ROOT NODULES			
Legume	<	>	<?
Non-legume	>	>	>
MYCORRHIZAS			
General	?	>	<?
Host-specific	>	>	>
CARNIVORY	>	>	>

[a] Statements very tentative and require detailed investigation.

plants (e.g. stem parasites, legumes) or of the uptake strategies (e.g. preference of VA mycorrhizas for moist soils, extra nutrient requirements for nodule formation and function).

Almost all taxonomic groups that dominate floristically or physiognomically in these two regions appear to have supplementary mechanisms for acquiring nutrients. Plants in the tree layer have mycorrhizas (VA in South Africa, sheathing in Western Australia) or proteoid roots; plants in the shrub layer have mycorrhizas (VA or ericoid), proteoid roots or N_2-fixing nodules; while plants in the herb layer have mycorrhizas (VA) or non-proteoid root clusters. An exception appears to be the grasstree *Xanthorrhoea* in Western Australia, which has scores of thick, wiry roots that are renewed at intervals: perhaps the slow growth rate reduces demand to a level where enhanced uptake is unnecessary (Lamont and Downes 1979). While a few *Casuarina* spp. (nodules) and *Nuytsia floribunda* (root haustoria) may be codominant in some tree strata in Western Australia, other hemiparasites (shrubs), holoparasites (herbs) and carnivorous plants (herbs) are never floristically or physiognomically dominant here.

That the various types of mycorrhizas and root clusters are essentially strategies for enhanced uptake of water and limiting nutrients, especially phosphorus and micronutrients, is supported by the fact that they rarely co-occur on the same root system. The exceptions are *Leucopogon virgatus* with ericoid and VA mycorrhizas (Gullan 1975), *Acacia mucronata* with proteoid roots and VA mycorrhizas (Sward 1978), and Eucalypts with sheathing and VA mycorrhizas (Lamont and Lange 1976; Titze et al. 1980). In Eucalypts, the possibility that VA mycorrhizas are associated with short-term, and sheathing with longer-term, uptake is consistent with their functions when kept independent (see text) and is worth further study. On the other hand, the invariable co-occurrence of N_2-fixing nodules with VA mycorrhizas (Tables 3 and 4) confirms that they have complementary functions, although much of the early mycorrhizal uptake feeds the N_2-fixing process itself (see text). Referring to Fig. 2, it is as if mycorrhizas increased nutrient availability along the x axis, with nodule production (= specialized roots) rising exponentially. A similar function for proteoid roots has been implicated for *Viminaria juncea*, one of the few legumes with root clusters (Pathmaranee 1974; Lamont 1981a). That root clusters are usually important in nitrogen uptake as well is suggested by their general absence from legumes and suppression at moderately high levels of added nitrogen (Lamont 1972b, 1974), but direct evidence is needed. The usual absence of other uptake strategies from plants with root haustoria and carnivorous leaves attests to their ability to contribute to a balanced nutrient uptake. The special feature of haustoria is their preferential uptake of potassium and sodium which appear more related to the maintenance of a positive water balance than to the metabolic needs of the parasite.

Nutrient-uptake strategies operate essentially through two mechanisms. The absorptive area per nutrient cost is increased enormously, either directly through increased fineness of the root system and proliferation of long root hairs (root clusters), or indirectly through the absorptive capacity of innumerable hyphal extensions from rootlets into the soil (mycorrhizas). Alternatively, they tap novel sources of nutrients, such as atmospheric N_2 (nodules), insect prey (carnivorous leaves) or intact leaf bases (aerial root of *Kingia australis*). Haustoria do both, by tapping nutrients already acquired by other plants through their own root system. That root clusters, mycorrhizas, nodules and root haustoria (and perhaps root hairs, Lamont 1980a) are strategies of particular relevance to infertile soils is indicated by their peak production at soil nutrient levels well below that required for optimal plant growth in the absence of other constraints such as drought (Fig. 2). They may be eliminated altogether at nutrient levels required for optimal growth of "normal" roots. These patterns should also be viewed in perspective of the much lower nutrient requirements of the sclerophylls bearing these structures than of mesophytes (Beadle 1966; Specht and Groves 1966).

The nutrient-depleted parent material of much of mediterranean South Africa and Western Australia results in the decomposing litter layer beneath the plant's canopy and ash after fire providing the major source of nutrients. As described in the text, both new season's "normal" and specialized roots are concentrated in or just below this nutrient-enriched layer (Lamont 1973; Specht 1981). Nutrient capture is enhanced by the tendency for the rootlets to cluster, increasing their capacity to act as a water and nutrient trap by capillary action, and prolonging the period of uptake. This close association between sites of nutrient release and absorption favours rapid uptake (nitrogen as ammonium rather than nitrate) and minimizes nutrient escape. There is evidence from tropical ecosystems that mycorrhizas obtain phosphorus directly from litter, bypassing the soil (Herrera et al. 1978). A parallel situation may exist in Australia and

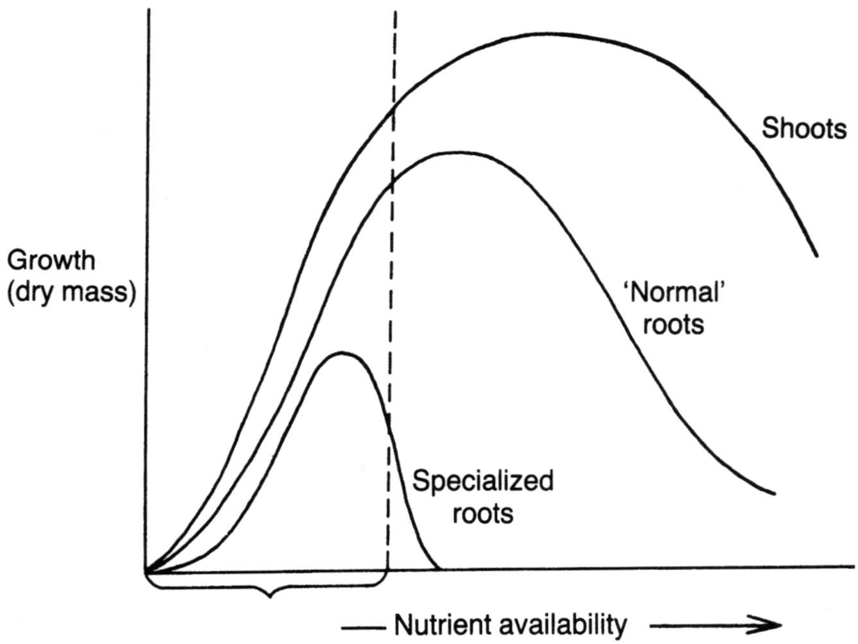

Fig. 2. Generalized response of specialized and "normal" root and shoot growth to increasing soil nutrient availability. Based on Stewart (1963), Rodriguez-Barrueco et al. (1970), Lamont (1972b, 1974, unpublished), Abbott and Robson (1979), Daft (1979), Malajczuk and Lamont (1981). The broken line tentatively separates soils of mediterranean South Africa and Western Australia on the left from more fertile soils on the right, as part explanation for the abundance of specialized modes of nutrition in these two regions.

South Africa for root clusters which cling to litter and soil particles (Specht 1981).

It was stated in the introduction that the major constraints to nutrient uptake in mediterranean ecosystems in general, and the Cape and southwestern Australia in particular, are (a) low nutrient availability (b) low winter temperatures and (c) summer drought. In general the strategies which overcome (a) also overcome (b) and they also tend to prolong the period of uptake, partly overcoming (c). For small plants that cannot reach sub-soil water throughout the year, summer drought continues to be the major constraint on growth and nutrient uptake. Of those that can, the specialized modes of nutrition of some species continue to function. There are even records of proteoid root growth in dry surface soil (Lamont 1980a) which deserve further study.

From this review it is evident that much work has been done or is underway on the mineral nutrition of plants from the infertile soils of mediterranean Australia and South Africa. However the remaining gaps have had to be filled from literature outside these regions or simply left as hypotheses. There is a need for more comparative research in the various mediterranean regions on the occurrence and mode of operation of all uptake strategies discussed here (Table 6). There is as yet no satisfactory answer to why, if most of these can be described as "low demand" plants (Chapin 1980), they possess such efficient means of nutrient acquisition.

References

ABBOTT LK, ROBSON AD (1979) A quantitative study of the spores and anatomy of mycorrhizas formed by a species of Glomus, with reference to its taxonomy. Australian Journal of Botany 27: 363-375

ADAMS RM, SMITH GW (1977) An SEM survey of the five carnivorous pitcher plant genera. American Journal of Botany 64: 265-272

ANSIAUX JR (1958) Sur l'alimentation minérale des phanérogemes parasites. Bulletin de l'Académie royal de Belgique, Classe des Sciences 5 séries 44: 787-793

ASHTON DH (1976) Studies on the mycorrhizae of Eucalyptus regnans F. Muell. Australian Journal of Botany 24: 723-774

BAIRD AM (1977) Regeneration after fire in King's Park, Western Australia. Journal of the Royal Society of Western Australia 60: 1-22

BARLOW BA, WIENS D (1977) Host-parasite resemblance in Australian mistletoes: the case for cryptic mimicry. Evolution 31: 69-84

BARROW NJ (1977) Phosphorus uptake and utilization by tree seedlings. Australian Journal of Botany 25: 571-584

BARTLETT EM, LEWIS DH (1973) Surface phosphatase activity of mycorrhizal roots of beech. Soil Biology and Biochemistry 5: 249-257

BAYLIS GT (1967) Experiments on the ecological significance of phycomycetous mycorrhizas. New Phytologist 66: 231-243

BAYLIS GT (1975) The magnolioid root and mycotrophy in root systems derived from it. In: SANDERS FE, MOSSE B, TINKER PB (eds) Endomycorrhizas. Academic Press, London, pp 373-389

BEADLE NC (1966) Soil phosphate and its role in molding segments of the Australian flora and vegetation, with special reference to xeromorphy and sclerophylly. Ecology 47: 992-1007

BECKING JH (1970) Plant-endophyte symbiosis in non-leguminous plants. Plant and Soil 32: 611-654

BERESFORD RT (1979) Nutrient imbalances in tomato plants and acid phosphatase activity in the leaves. Journal of Science, Food and Agriculture 30: 275-280

BERGERSON FJ (1974) Formation and function of bacteroids. In: QUISPEL A (ed) The biology of nitrogen fixation. North-Holland Publishing Company, Amsterdam, pp 473-498

BHAT KK, NYE PH (1973) Diffusion of phosphate to plant roots in soil: 1. Quantitative autoradiography of the depletion zone. Plant and Soil 38: 161-175

BOERD G, THIEN S (1979) Phosphatase activity and phosphorus availability in the rhizosphere of corn roots. In: HARLEY JL, RUSSELL RS (eds) The soil-root interface. Academic Press, London, pp 231-242

BOND G, SCOTT GD (1955) An examination of some symbiotic systems for fixation of nitrogen. Annals of Botany 19: 67-77

BOWEN GD (1981) Coping with low nutrients. In: PATE JS, MCCOMB AJ (eds) The biology of Australian plants. University of Western Australia Press, Perth, pp 33-64

BOWYER JW, SKERMAN VB (1968) Production of axenic cultures of soil-borne and endophytic blue-green algae. Journal of General Microbiology 54: 299-306

CAMPBELL EO (1963) Gastrodia minor Petrie, an epiparasite of Manuka. Transactions of the Royal Society of New Zealand 2: 73-81

CAMPBELL EO (1964) The restiad peat bogs at Motumaoho and Moanatuatua. Transactions of the Royal Society of New Zealand 2: 219-227

CAMPBELL EO (1981) The water relations of heathlands: morphological adaptation to waterlogging. In: SPECHT RL (ed) Ecosystems of the world, vol 9B. Heathlands and related shrublands. Analytical studies. Elsevier, Amsterdam. pp 107-109

CARLQUIST S (1974) Island biology. Columbia University Press, New York, pp 253-330

CARRODUS BB (1967) Absorption of nitrogen by mycorrhizal roots of beech. II. Ammonia and nitrate as sources of nitrogen. New Phytologist 66: 1-4

CHAMBERS CA, SMITH SE, SMITH FA (1980) Effects of ammonium and nitrate ions on mycorrhizal infection, nodulation and growth of Trifolium subterraneum. New Phytologist 85: 47-62

CHANDLER GE, ANDERSON JW (1976a) Studies on the nutrition and growth of Drosera species with reference to the carnivorous habit. New Phytologist 76: 129-141

CHANDLER GE, ANDERSON JW (1976b) Uptake and metabolism of insect metabolites by leaves on tentacles of Drosera species. New Phytologist 77: 625-634

CHANDLER GE, ANDERSON JW (1976c) Studies on the origin of some hydrolytic enzymes associated with the leaves and tentacles of Drosera species and their role in heterotrophic nutriton. New Phytologist 77: 51-62

CHAPIN FS (1980) The mineral nutrition of wild plants. Annual Review of Ecology and Systematics 11: 233-260

CHILVERS GA, PRYOR LD (1965) The structure of eucalypt mycorrhizas. Australian Journal of Botany 13: 245-259

CHIPPENDALE GM (1981) Distribution density of Eucalyptus species in Australia. Search 12: 131-133

COLQUHOUN IJ (1979) The water relations of two plant species on the Swan Coastal Plain (abstract). Symposium on the biology of native Australian plants, Aug 11-13, University of Western Australia, Perth, p 47

COOPER KM (1975) Growth responses to the formation of endotrophic mycorrhizas in Solanum, Leptospermum, and New Zealand ferns. In: SANDERS FE, MOSSE B, TINKER PB (eds) Endomycorrhizas. Academic Press, London, pp 391-408

COOPER KM, TINKER PB (1978) Translocation and transfer of nutrients in vesicular-arbuscular mycorrhizas. New Phytologist 81: 43-52

COWLING RM, CAMPBELL BM (1980) Convergence in vegetation structure in the mediterranean communities of California, Chile and South Africa. Vegetatio 43: 191-198

DAFT MJ (1979) Effects of calcium, phosphorus and potassium on mycorrhizal plants. In: HARLEY JL, RUSSELL RS (eds) The soil-root interface. Academic Press, London, pp 420-421

DAKIN WJ (1919) The West Australian pitcher plant (Cephalotus follicularis) and its physiology. Journal and Proceedings of the Royal Society of Western Australia 4: 37-53

DIXON KW, PATE JS, BAILEY WJ (1980) Nitrogen nutrition of the tuberous sundew Drosera erythrorhiza Lindl. with special reference to catch of arthropod fauna by its glandular leaves. Australian Journal of Botany 28: 283-297

DODD J, HEDDLE EM (1981) Root systems of some swamp and banksia woodland plants of the Swan Coastal Plain, Western Australia (abstract). Bulletin of the Ecological Society of Australia 11: 4

DöRR I (1975) Development of transfer cells in higher parasitic plants. In: ARONOFF S, DAINTY J, GORHAM PR, SRIVASTAVA LM, SWANSON CA (eds) Phloem transport. Plenum Press, New York, pp 177-186

DREW MC, NYE PH (1969) The supply of nutrient ions by diffusion to plant roots in soil. II. The effects of root hairs on the uptake of potassium by roots of rye grass (Lolium multiflorum). Plant and Soil 31: 407-424

ENGIN M, SPRENT JI (1973) Effects of water stress on growth and nitrogen-fixing activity of Trifolium repens. New Phytologist 72: 117-126

FINLAYSON M, MCCOMB AJ (1978) Nitrogen fixation in wetlands of southwestern Australia. Search 9: 98-99

GEORGE AS (1980) Rhizanthella gardneri R.S. Rogers - the underground orchid of western Australia. American Orchid Society Bulletin 49: 631-646

GIBSON AH (1976) Recovery and compensation by nodulated legumes to environmental stress. In: NUTMAN PS (ed) Symbiotic nitrogen fixation in plants. Cambridge University Press, Cambridge, pp 405-420

GIESSLER A (1928) Einfluss von Salzlösungen auf die Stärheverarbeitung bei Drosera. Flora 23: 133-190

GOBEL T (1975) Some field observations on Nuytsia floribunda (Labill.) R. Br. West Australian Naturalist 29: 50-60

GREEN PS (1976) Ecological and nutritional aspects of proteoid roots. Hons Thesis. University of Adelaide, South Australia

GREENLAND DJ (1979) The physics and chemistry of the soil-root interface: some comments. In: HARLEY JL, RUSSELL RS (eds) The soil-root interface. Academic Press, London, pp 83-98

GRIEVE BJ, HELLMUTH EO (1970) Eco-physiology of Western Australian plants. Oecologia Plantarum 5: 33-68

GROBBELAAR N, STRAUSS JM, GROENEWALD EG (1971) Non-leguminous seed plants in southern Africa which fix nitrogen symbiotically. Plant and Soil. Special vol, pp 325-334

GROVE TS, O'CONNELL AM, MALAJCZUK N (1980) Effects of fire on the growth, nutrient content and rate of nitrogen fixation of the cycad Macrozamia riedlei. Australian Journal of Botany 28: 271-281

GULLAN PK (1975) Vegetation at Cranbourne, vol II. PhD Thesis. Monash University, Victoria

HALL IR (1975) Endomycorrhizas of Metrosideros umbelleta and Weinmannia racemosa. New Zealand Journal of Botany 13: 463-472

HALLIDAY J, PATE JS (1976) Symbiotic nitrogen fixation by coralloid roots of the cycad Macrozamia riedlei: physiological characteristics and ecological significance. Australian Journal of Plant Physiology 3: 349-358

HARDY RW, BURNS RC, HOLSTEN RD (1973) Applications of the acetylene-ethylene assay for measurement of nitrogen fixation. Soil Biology and Biochemistry 5: 47-81

HARLEY JL (1969) The biology of mycorrhiza. Leonard Hill, London, 334 pp

HARLEY JL (1975) Problems in mycotrophy. In: SANDERS FE, MOSSE B, TINKER PB (eds) Endomycorrhizas. Academic Press, London, pp 1-24

HARTEL O (1937) Über den Wasserhaushalt von Viscum album L. Bericht der Deutschen botanischen Gesellschaft 59: 136-148

HAXEN P (1978) Aspects of nodule physiology of some southwestern Cape leguminous species. Hons Thesis. University of Cape Town

HAYMAN DS (1974) Plant growth responses to vesicular-arbuscular mycorrhiza. New Phytologist 73: 71-80

HEAD GC (1964) A study of "exudation" from the root hairs of apple roots by time-lapse cine-photomicrography. Annals of Botany 28: 495-498

HELLMUTH EO (1971) Eco-physiological studies in plants in arid and semi-arid regions in Western Australia. IV. Comparison of the field physiology of the host, Acacia grasbyi and its hemiparasite, Amyema nestor under optimal and stress conditions. Journal of Ecology 59: 351-363

HERBERT DA (1919) Nuytsia floribunda (the Christmas tree) - its structure and parasitism. Journal of Proceedings of the Royal Society of Western Australia 5: 72-88

HERBERT DA (1925) The root parasitism of Western Australian Santalaceae. Journal of the Royal Society of Western Australia 11: 127-149

HERRERA R, MERIDA T, STARK N (1978) Direct phosphorus transfer from leaf litter to roots. Naturwissenschaften 65: 208-209

HEWITT EJ, SMITH TA (1975) Plant mineral nutrition. English University Press, London, 298 pp

HINGSTON FJ (1977) Sources of, and sinks for, nutrients in forest ecosystems. In: Nutrient cycling in indigenous forest ecosystems. CSIRO, Division of Land Resources Management, Perth. pp 41-53

HINGSTON FJ, MALAJCZUK N, GROVE TS (1982) N_2-fixation (ascetylene reduction) by legume seedlings following an intense fire in jarrah forest. Journal of Applied Ecology (in press)

HNATIUK RJ, HOPKINS AJ (1980) Western Australian species-rich kwongan (sclerophyllous shrubland) affected by drought. Australian Journal of Botany 28: 573-586

JEFFREY DW (1967) Phosphate nutrition of Australian heath plants. I. The importance of proteoid roots in Banksia (Proteaceae). Australian Journal of Botany 15: 403-411

JOHNSON PN (1973) Mycorrhizae of coniferous-broadleaved forest. PhD Thesis. Otago University, New Zealand, 301 pp

JONGENS-ROBERTS SM, BROWN GJ, MITCHELL DT (1980) Studies on phosphorus cycling processes in the fynbos biome. CSIR, Fynbos Biome Annual Report No 2

JUNIPER BE, GILCHRIST AJ, ROBINS AJ (1977) Some features of secretory systems in plants. Histochemical Journal 9: 659-680

KANA TM, TJEPKEMA JD (1978) Nitrogen fixation associated with Scirpus atrovirens and other non-nodulated plants in Massachusetts. Canadian Journal of Botany 56: 2636-2640

KIMBER PC (1974) The root system of jarrah (Eucalyptus marginata) Research Paper No 14. Forests Department, Perth, 5 pp

KLAREN CH, JANSSEN B (1976) Physiological changes in the hemiparasite Rhinanthus serotinus before and after attachment. Physiologia Plantarum 42: 151-155

KUIJT K (1969) The biology of flowering parasitic plants. University of California Press, Berkeley, 246 pp

LAMONT B (1972a) The morphology and anatomy of proteoid roots in the genus Hakea. Australian Journal of Botany 20: 155-174

LAMONT B (1972b) The effect of soil nutrients on the production of proteoid roots by Hakea species. Australian Journal of Botany 20: 27-40

LAMONT B (1972c) 'Proteoid' roots in the legume Viminaria juncea. Search 3: 90-91

LAMONT B (1973) Factors affecting the distribution of proteoid roots within the root systems of two Hakea species. Australian Journal of Botany 21: 165-187

LAMONT B (1974) The biology of dauciform roots in the sedge Cyathochaete avenacea. New Phytologist 73: 985-996

LAMONT B (1976a) The effects of seasonality and waterlogging on the root systems of a number of Hakea species. Australian Journal of Botany 24: 691-702

LAMONT B (1976b) A biological survey and recommendations for rehabilitating a portion of Reserve 31030 to be mined for heavy minerals during 1975-81. WAIT-AID Ltd., Perth, 61 pp

LAMONT B (1977) Root parasitism of Hakea sulcata by Nuytsia floribunda. West Australian Naturalist 13: 201-202

LAMONT B (1979) Root systems of the Myrtaceae. Australian Plants 10: 74-78

LAMONT B (1980a) Proteoid roots in the South African Proteaceae. CSIR, Fynbos Biome Annual Report No 2, 31 pp

LAMONT B (1980b) Blue-green algae in nectar of Banksia aff. sphaerocarpa. West Australian Naturalist 14: 193-194

LAMONT B (1981a) Specialized roots of non-symbiotic origin in heathlands. In: SPECHT RL (ed) Ecosystems of the world, vol 9B. Heathlands and related shrublands. Analytical studies. Elsevier, Amsterdam, pp 183-195

LAMONT B (1981b) Morphometrics of the aerial roots of Kingia australis (Liliales). Australian Journal of Botany 29: 81-96

LAMONT B (1981c) Autografting of roots and stems in Eucalyptus and of rhizomes in Nuytsia floribunda. West Australian Naturalist 15: 26-28

LAMONT B (1981d) Availability of water and inorganic nutrients in the persistent leaf bases of the grasstree Kingia australis and the uptake and translocation of labelled phosphate by the embedded aerial roots. Physiologia Plantarum 52: 181-186

LAMONT B (1981e) A botanist in South Africa, 1980: study leave report. School of Biology, Western Australian Institute of Technology, 22 pp

LAMONT B (1981f) Understorey suppression by Eucalyptus wandoo (abstract). Conference on the Biology of Eucalyptus. King's Park and Botanic Garden, Perth. p 4

LAMONT B (1982a) Mineral nitrition of mistletoes. In: CALDER DM, BERNDHART P (eds) The biology of mistletoes. Academic Press, London. In press

LAMONT B (1982b) Specialised roots in the genus Acacia: a review. Mulga Research Centre Annual Report 5: 9-11

LAMONT B (1982c) The reproductive biology of Grevillea leucopteris (Proteaceae), including reference to its glandular hairs and colonizing potential. Flora 172: 1-20

LAMONT B, LANGE BJ (1976) "Stalagmiform" roots in limestone caves. New Phytologist 76: 353-360

LAMONT B, DOWNES S (1979) The longevity, flowering and fire history of the grasstrees Xanthorrhoea preissii and Kingia australis. Journal of Applied Ecology 16: 893-899

LAMONT B, SOUTHALL KJ (1982a) Biology of the mistletoe Amyema preissi on road verges and undisturbed vegetation. Search 13: 87-88

LAMONT B, SOUTHALL KJ (1982b) Distribution of mineral nutrients between the mistletoe Amyema preissi and its host Acacia acuminata. Annals of Botany 49: 721-725

LANGE RT (1959) Additions to the known nodulating species of Leguminosae. Antoni van Leeuenhoek Journal 25: 272-276

LAWRIE AC (1981) Nitrogen fixation by native Australian legumes. Australian Journal of Botany 29: 143-157

LEVYNS M (1935) Veld burning experiments at Oakdale, Riversdale. Transactions of the Royal Society of South Africa 23: 231-243

LIE TA, MULDER EG (eds) (1971) Biological nitrogen fixation in natural and agricultural habitats. Plant and Soil. Special vol, 580 pp

LOW AB (1979) Phytomass and litter studies on the Cape Flats. CSIR, Fynbos Biome Annual Report

LOW AB (1980) Preliminary observations on specialized root morphologies in plants of the western Cape Province. South African Journal of Science 76: 513-516

LÜTTGE U (1971) Structure and function of plant glands. Annual Review of Plant Physiology 22: 23-44

MALAJCZUK N, BOWEN GD (1974) Proteoid roots are microbially induced. Nature 251: 316-317

MALAJCZUK N, MCCOMB AJ, LONERAGAN JF (1975) Phosphorus uptake and growth of mycorrhizal and uninfected seedlings of Eucalyptus calophylla R. Br. Australian Journal of Botany 23: 231-238

MALAJCZUK N, GROVE T (1977) Legume understorey biomass, nutrient content and nitrogen fixation in eucalypt forests of southwestern Australia. In: Nutrient cycling in indigenous forest ecosystems. Division of Land Resources Management, CSIRO, Perth, pp 36-39

MALAJCZUK N, HINGSTON F (1981) Ectomycorrhizae associated with jarrah. Australian Journal of Botany 29: 453-463

MALAJCZUK N, LAMONT BB (1981) Specialized roots of symbiotic origin in heathlands. In: SPECHT RL (ed) Ecosystems of the world, vol 9B. Heathlands and related shrublands. Analytical studies. Elsevier, Amsterdam, pp 165-182

MALLOCH DW, PIROZYNSKI KA, RAVEN PH (1980) Ecological and evolutionary significance of mycorrhizal symbiosis in vascular plants (A review). Proceedings of the National Academy of Sciences of the USA 77: 2113-2118

MARLOTH R (1913) The flora of South Africa, vol 1. Darter Brothers and Company, Cape Town, 264 pp

MARLOTH R (1915) The effects of droughts and of some other causes on the distribution of plants in the Cape Region. South African Journal of Science 12: 383-390

MARLOTH R (1925) The flora of South Africa, vol 2 (pt 1). Darter Brothers and Company, Cape Town, 124 pp

MARX DH, BRYAN WC (1971) Influence of ectomycorrhizae on survival and growth of aseptic seedlings of loblolly pine at high temperature. Forest Science 17: 37-41

MITCHELL DT, READ DJ (1981) Utilization of inorganic and organic phosphates by the mycorrhizal endophytes of Vaccinium macrocarpon and Rhododendron ponticum. Transactions of the British Mycological Society 76: 255-260

MOORE CW, KERAITIS K (1971) Effect of nitrogen source on growth of eucalypts in sand culture. Australian Journal of Botany 19: 125-141

MOSSE B (1975) Specificity in VA mycorrhizas. In: SANDERS FE, MOSSE B, TINKER PB (eds) Endomycorrhizas. Academic Press, London, pp 469-484

MOSSE B, POWELL CL, HAYMAN DS (1976) Plant growth responses to vesicular-arbuscular mycorrhiza. IX. Interactions between VA mycorrhiza, rock phosphate and symbiotic nitrogen fixation. New Phytologist 76: 331-342

MUIR BG (1977) Biological survey of the Western Australian Wheatbelt. 2. Vegetation and habitat of Bendering Reserve. Records of the Western Australian Museum. Supplement No 3

MULLETTE JJ, HANNON NJ, ELLIOT AG (1974) Insoluble phosphorus usage by Eucalyptus. Plant and Soil 41: 199-205

NAMBIAR EK (1976) Uptake of Zn^{65} from dry soil by plants. Plant and Soil 44: 267-271

NAMBIAR EK (1977) The effects of drying of the topsoil and of micronutrients in the subsoil on micronutrient uptake by an intermittently defoliated ryegrass. Plant and Soil 46: 185-193

NUTMAN PS (ed) (1976) Symbiotic nitrogen fixation in plants. Cambridge University Press, Cambridge, 584 pp

NYE PH (1979) Soil properties controlling the supply of nutrients to the root surface. In: HARLEY JL, RUSSELL RS (eds) The soil-root interface. Academic Press, London, pp 39-49

PATE JS, DIXON KW (1978) Mineral nutrition of Drosera erythrorhiza Lindl. with special reference to its tuberous habit. Australian Journal of Botany 26: 455-464

PATHMARANEE N (1974) Observations of proteoid roots. MSc Thesis. University of Sydney, New South Wales, 123 pp

POWELL CL (1975) Rushes and sedges are non-mycotrophic. Plant and Soil 42: 481-484

PRINGSHEIM EG, PRINGSHEIM O (1962) Axenic culture of Utricularia. American Journal of Botany 49: 898-901

PURNELL HM (1960) Studies of the family Proteaceae. I. Anatomy and morphology of the roots of some Victorian species. Australian Journal of Botany 8: 38-50

READ DJ, KOUCHEKI HK, HODGSON J (1976) Vesicular-arbuscular mycorrhiza in natural vegetation systems. I. The occurrence of infection. New Phytologist 77: 641-653

REID CP, BOWEN GD (1979) Effects of soil moisture on VA mycorrhiza formation and root development in Medicago. In: HARLEY JL, RUSSELL RS (eds) The soil-root interface. Academic Press, London. pp 211-219

RENBUSS MA, CHILVERS GA, PRYOR LD (1972) Microbiology of an ashbed. Proceedings of the Linnean Society of New South Wales 97: 302-316 and plate

ROBINSON RJ (1973) Mycorrhiza in certain Ericaceae native to Southern Africa. Journal of South African Botany 39: 123-129

RODRIGUEZ-BARRUECO C, MACKINTOSH AH, BOND G (1970) Some effects of combined nitrogen on the nodule symbioses of Casuarina and Ceanothus. Plant and Soil 33: 129-139

SAFIR GR, BOYER JS, GERDEMANN JW (1972) Nutrient status and mycorrhizal enhancement of water transport in soybean. Plant Physiology 49: 700-703

SCHÜTTE KH (1960) Trace element deficiencies in Cape vegetation. Journal of South African Botany 26: 45-49

SHEA SR, KITT RJ (1976) The capacity of jarrah forest native legumes to fix nitrogen. Forests Department of Western Australian Research Paper 21

SHEA SR, DELL B (1981) Structure of the surface root system of Eucalyptus marginata Sm. and its infection by Phytophthora cinnamomi Rands. Australian Journal of Botany 29: 49-58

SIDDIQUI MY, CAROLIN RC (1976) Studies on the ecology of coastal heath in New South Wales. II. The effects of water supply and phosphorus uptake on the growth of Banksia serratifolia, B. aspleniifolia and B. ericifolia. Proceedings of the Linnean Society of New South Wales 101: 38-52

SMITH SE (1966) Physiology and ecology of Orchis mycorrhizal fungi with reference to seedling nutrition. New Phytologist 65: 488-499

SNOWBALL K, ROBSON AD, LONERAGAN JF (1980) The effect of copper on nitrogen fixation in subterranean clover (Trifolium subterranean). New Phytologist 85: 63-72

SO HB (1979) Water potential gradients and resistances of a soil-root system measured with the root and soil psychrometer. In: HARLEY JL, RUSSELL RS (eds) The soil-root interface. Academic Press, London, pp 99-113

SORENSEN D, JACKSON WT (1968) Utilization of paramecium by Utricularia gibba. Planta 83: 166-170

SPECHT RL (1979) Heathlands and related shrublands of the world. In: SPECHT RL (ed) Ecosystems of the world, vol 9A. Heathlands and related shrublands. Descriptive studies. Elsevier, Amsterdam, pp 1-19

SPECHT RL (1981) Nutrient release from decomposing leaf litter of Banksia ornata. Dark Island heathland, South Australia. Australian Journal of Ecology 6: 59-64

SPECHT RL, GROVES RH (1966) Comparison of the phosphate nutrition of Australian heath plants and introduced economic plants. Australian Journal of Botany 14: 201-221

SPECHT RL, RAYSON P (1957) Dark Island heath (Ninety-Mile Plain, South Australia). III. The root systems. Australian Journal of Botany 5: 103-114

SPRENT JI (1972) The effects of water stress on nitrogen-fixing root nodules. IV. Effects on whole plants of Vicia faba and Glycine max. New Phytologist 71: 603-611

STEWART WD (1963) The effect of combined nitrogen on growth and nodule development of Myrica and Casuarina. Zeitschrift für allgemeine Mikrobiologie 3: 152-156

STOCK W, LEWIS OAM (1980) An investigation on cycling and processing of nitrogen in the fynbos biome. CSIR, Fynbos Biome Annual Report No 2

STRIBLEY DP, READ DJ (1975) Some nutritional aspects of the biology of ericaceous mycorrhizas. In: SANDERS FE, MOSSE B, TINKER PB (eds) Endomycorrhizas. Academic Press, London, pp 195-208

STRIBLEY DP, READ DJ (1976) The biology of mycorrhiza in the Ericaceae. VI. The effects of mycorrhizal infection and concentration of ammonium nitrogen on growth of cranberry (Vaccinium macrocarpon Ait.) in sand culture. New Phytologist 77: 63-72

SWARD RJ (1978) Studies on vesicular-arbuscular mycorrhizas of some Australian heathland plants. PhD Thesis, Monash University, Victoria

TACEY WH (1977) Nitrogen fixation rate of Albizia lophantha. In: Nutrient cycling in indigenous forest ecosystems. CSIRO, Division of Land Resources Management, Perth, pp 59-63

TITZE JF, CRAIG G, LAMONT BB (1980) Vesicular-arbuscular mycorrhizae in jarrah forest - a preliminary note. Mulga Research Centre Annual Report 3: 29-33

TSIVION Y (1978) Loading of assimilates and some sugars into the translocation system of Cuscuta. Australian Journal of Plant Physiology 5: 851-857

TYSON JH, SILVER WS (1979) Relationship of ultrastructure of acetylene reduction (N_2 fixation) in root nodules of Casuarina. Botanical Gazette 140 (suppl.): 44-48

WALTERS CM, JOOSTE JH (1980) Aspekte van die minerale voeding van lede van die Proteaceae as verteenwoordigers van die fynbosgemeenskap. CSIR, Fynbos Biome Annual Report No 2
WEATHERLEY PE (1979) The hydraulic resistance of the soil-root interface - a cause of water stress in plants. In: HARLEY JL, RUSSELL RS (eds) The soil-root interface. Academic Press, London, pp 275-286
WEBB LJ (1954) Aluminium accumulation in the Australian-New Guinea flora. Australian Journal of Botany 2: 176-196
WHITNEY PJ (1972) The carbohydrate and water balance of beans (Vicia faba) attacked by broomrape (Orobanche crenata). Annals of Applied Biology 70: 59-66
WILLIAMS SE, WOLLUM AG, ALDON EF (1974) Growth of Atriplex canescens (Pursh) Nutt. improved by formation of vesicular-arbuscular mycorrhizae. Proceedings of the Soil Science Society of America 38: 962-965
WILLIAMSON B (1973) Acid phosphatase and esterase activity in orchid mycorrhiza. Planta 112: 149-158

15. The Effects of Fire on Photosynthesis in Chaparral Resprouts

W. C. OECHEL and S. J. HASTINGS

15.1 Introduction

The success and rapid growth of vegetation regenerating after fire is probably due to a complex set of factors. In sprouting shrubs it may be due to the utilization of carbohydrate reserves in the roots and lignotubers. It may also be due in part to increased photosynthetic rates, as Radosevich et al. (1977) have reported for resprouting Adenostoma fasciculatum. Christensen and Muller (1975) point out the enhancing effects of improved nutrient status of the chaparral on growth following fire. Photosynthesis and the rate of regrowth could also be increased due to improved water relations of resprouts. A reduction in the leaf area index after top removal would decrease the total transpirational surfaces and presumably decrease water loss. Greater tissue water potentials following fire may result in lowered stomatal resistance to water vapour and carbon dioxide flux. Christensen and Muller (1975) reported increased soil nutrients in chaparral areas after a fire. Higher nutrient levels may result in greater enzyme concentrations, higher respiration rates and greater photosynthetic rates. Numerous studies have shown the positive correlation between nitrogen availability and photosynthetic rate (Longstreth and Nobel 1980) and between leaf nitrogen and photosynthetic performance (Natr 1970). Most essential nutrients, including nitrogen, phosphorus, potassium, magnesium, sulphur, calcium, iron, manganese, copper, boron, zinc and molybdenum, in deficient amounts may limit photosynthesis (Bottrill et al. 1970; Longstreth and Nobel 1980; Spiller and Terry 1980; Terry 1980).

This study was undertaken to determine if sprouts showed enhanced photosynthesis following fire. Photosynthesis, leaf conductance and water potential were measured over time in the field on mature plants and on sprouts after fire and hand-clearing of Arctostaphylos glandulosa, Quercus dumosa, and Adenostoma fasciculatum over time. Total nonstructural carbohydrate (TNC) was determined in leaves and stems of mature and sprouting plants. The hypothesis was that sprouting plants would exhibit increased photosynthetic rates due to improved water status.

15.2 Methods

15.2.1 SITE DESCRIPTION

The research location is within the Laguna-Morena Demonstration Area of the Cleveland National Forest in San Diego County (southern California) and is located on the eastern slopes of the Laguna Mountains at about 1400 m. The experiment sites are situated on an east-facing slope along Kitchen Creek, south of Kitchen Valley, about 76 km east of the coast at San Diego.

The experimental area was approximately 250 m wide, 670 m long, and encompassed 16.5 ha. It consisted of a uniform slope with an average incline of 45% supporting a mixed chaparral community composed of *Adenostoma fasciculatum*, *Ceanothus greggii*, *Ceanothus leucodermis*, *Arctostaphylos* spp., *Garrya* spp., *Quercus dumosa*, *Q. wislizenii*, *Rhus ovata*, *Rhamnus crocea*, *Prunus ilicifolia*, and *Eriogonum fasciculatum*. Nomenclature follows Munz (1974). The community has remained unburned for 34 years. The experimental area was located on the lower 60% of the slope.

15.2.2 EXPERIMENTAL METHODS

Burning of chaparral is reported to increase available soil nutrients. These nutrients may originate from the ash and char of the above-ground plant material and litter and from decomposing root systems. An increase in nutrients could elevate photosynthetic rates of sprouts after fire above those of mature vegetation. Due to a decrease in leaf area of the sprouts, improved water status could also result in higher photosynthetic rates. To distinguish these effects, two field manipulations were carried out. The first was hand-clearing where shoots were cut and removed from the plots. Soil nutrients in the hand-cleared areas should be similar to the control plots, but presumably water status of the plants would be improved. The second manipulation involved controlled burning of mature vegetation, allowing nutrient inputs following fire. Thus, the burned plots would have both higher levels of nutrients and improved water status. Seven strips were cleared between 30 May and 13 June 1979. Each strip, which ran up the slope, was 250 m long. The northern and southern boundaries of the experimental area were cleared to a width of about 30 m; smaller cleared areas, which were 2 m wide, separated the plots to be burned. Three controlled burns were conducted in November 1979 (weather conditions delayed burning). Each burned plot was approximately 40-50 m wide and 250 m long.

Maximum photosynthetic rates were determined on attached, fully expanded leaves of resprouts or on one-yr-old leaves on mature plants. Carbon-14 dioxide techniques were used as previously described (Tieszen et al. 1974; Mustafa 1978; Oechel and Mustafa 1979) except that a methanol killing-fixing procedure was employed (Lawrence and Oechel, unpublished). Dawn (04h00-06h00) and midday (11h00-12h00) measurements of water potential and leaf conductance were made using a Scholander pressure chamber and a null balance porometer (Bingham and Coyne 1977). Sprouts and mature plants were measured near the peak period of photosynthesis (10h00-12h00) every four to six weeks on *Q. dumosa*, *A. glandulosa* and *A. fasciculatum*. In addition, full diurnal measurements were made on the control vegetation. Twelve replicate measurements were made on each treatment and species.

To measure tissue carbohydrate content of control vegetation and sprouts from burned and hand-cleared areas, stems and leaves of *A. glandulosa*, *Q. dumosa* and *A. fasciculatum* were sampled on the same day in June 1980 within 1.5 hours of noon and immediately frozen on dry ice. On the same day, samples were directly transferred from dry ice to a drying oven and dried for one week at 80°C, ground in a Thomas-Wiley intermediate mill (40 mesh) and stored in sealed plastic bags until analysed. Total nonstructural carbohydrates (in the form of polysaccharides, disaccharides and monosaccharides) were determined at the US Forest Service Fire Laboratory in Riverside, California using a modification of the automated technique of Hudson et al. (1976). The colour reagent was p-hydroxybenzoic acid hydrazide which is specific for monosaccharides. Disaccharides were determined from the difference in the monosaccharide levels before and after sample treatment with invertase. TNC was determined using an amyloglucosidase-invertase enzyme mix to convert the nonstructured carbohydrates to monosaccharides. Polysaccharides were calculated as TNC minus mono- and disaccharides.

All measures of concentration and rate have been expressed in terms of units of plant dry mass.

15.3 Results

15.3.1 CARBOHYDRATES

In general, mono- and disaccharides were considerably more abundant in leaf and stem tissue than were polysaccharides (Table 1). These data indicate a much higher TNC in the leaves and young stems in *A. glandulosa* than in *A. fasciculatum* or *Q. dumosa* but little difference in the leaves of *A. fasciculatum* and *A. glandulosa* and the stems of *A. fasciculatum* between control vegetation and sprouts from burned and hand-cleared plots. However, the proportions of polysaccharides to mono- and disaccharides varied in some cases. Where differences in TNC levels did occur as a result of treatment, they were a decrease in the TNC levels in the postfire sprouts (*Q. dumosa* leaves and stems and *A. glandulosa* stems) or an increase in leaf TNC levels in the hand-cleared plots (*Q. dumosa*) relative to the other treatments.

15.3.2 PHOTOSYNTHESIS

Mature *A. fasciculatum* from the control plots showed photosynthetic rates of about 0.03 μmol CO_2 g^{-1} leaf s^{-1} in late April (Fig. 1). The photosynthetic rate gradually increased to about 0.05 μmol CO_2 g^{-1} leaf s^{-1} in late June. These values are similar to maximum photosynthetic rates reported at 1000 m for Echo Valley, California, where 0.019-0.022 μmol CO_2 g^{-1}leaf s^{-1} was found using carbon-14 dioxide techiques and about 0.036 μmol CO_2 g^{-1} leaf s^{-1} was determined by infrared gas analysis (Oechel et al. 1981). Radosevich et al. (1977), however, reported photosynthetic rates of 0.0057 and 0.011 for *A. fasciculatum* in March and June 1977, respectively, from the Hopland Field Station, Mendocino County.

The photosynthetic rate of postfire sprouts of *A. fasciculatum* increased markedly compared to mature control vegetation. In late April they exhibited a photosynthetic rate of 0.14 ± 0.026 μmol CO_2 g^{-1} leaf s^{-1}, which is over four times the rate of the control vegetation. Photosynthetic rates of postfire sprouts increased to 0.21 in May. However, these rates fell to 0.080 ± 0.0009 in June, about 1.7 times the June control values. The photosynthetic rates for sprouts from the hand-cleared plots were virtually identical to the rates for the control

Table 1. Mean carbohydrate contents and standard errors of leaves and stems (mg g^{-1} dry mass) in control plants and sprouts after fire and hand-clearing. N = 3 unless given. ND = No data.

Carbohydrate	*Adenostoma fasciculatum*			*Arctostaphylos glandulosa*			*Quercus dumosa*		
	control	fire	hand-cleared	control	fire	hand-cleared	control	fire	hand-cleared
LEAVES									
monosaccharides	77.5 ± 8.4	160.9 ± 38.1	144.5 ± 3.6	256.8 ± 2.4	226 ± 13.5 N=2	246.33 ± 33.0	102.2 ± 6.1	70.7 ± 9.5	83.9 ± 14.7
disaccharides	2.7 ± 1.5	2.9 N=1	ND	23.5 N=1	ND	12.9 N=1	13.1 ± 0.6	5.8 ± 1.6	0.6 N=1
nonstructural polysaccharides	54.8 ± 4.4	16 N=1	38.2 ± 7.7	18.7 ± 3.8 N=2	61.6 ± 18.8 N=2	47.8 ± 3.8	34.7 ± 7.9 N=2	22.8 ± 1.7 N=2	9.5 ± 1.5
total nonstructural carbohydrates	134.9 ± 6.4	169.9 ± 45.0 N=2	182.6 ± 6.2	286.3 = 16.3 N=2	287.6 ± 32.3 N=2	282.6 ± 22.4	152.7 ± 2.7 N=2	90.1 ± 11.2	93.6 ± 13.6
STEMS									
monosaccharides	46.0 ± 3.0	52.1 ± 6.1	38.8 ± 1.8	79.5 ± 4.0	113.6 N=1	121.7 N=1	37.5 ± 1.6 N=2	45.8 ± 2.4	47.4 ± 3.3
disaccharides	7.1 ± 0.8	6.8 ± 0.1 N=2	7.8 ± 1.2	18.7 ± 2.6	8.5 N=1	9.4 N=1	7.6 ± 0.6 N=2	7.2 ± 0.4 N=2	5.0 ± 0.8
nonstructural polysaccharides	26.9 ± 3.8	16.4 ± 1.2 N=2	14.2 ± 4.0	80.3 ± 24.2	ND	7.3 N=1	38.2 ± 13.6 N=2	9.6 N=1	55.6 ± 7.4
total nonstructural carbohydrates	80.0 ± 3.9	75.3 ± 7.4 N=2	60.8 ± 3.7	178.6 ± 28.4	122.1 N=1	171.2 N=1	83.3 ± 15.8 N=2	54.2 ± 6.4	108.0 ± 7.4

vegetation. The rates for the hand-cleared plots for March, May and June were 0.025 ± 0.004, 0.040 ± 0.001 and 0.042 ± 0.004 μmol CO_2 g^{-1} leaf s^{-1}, respectively.

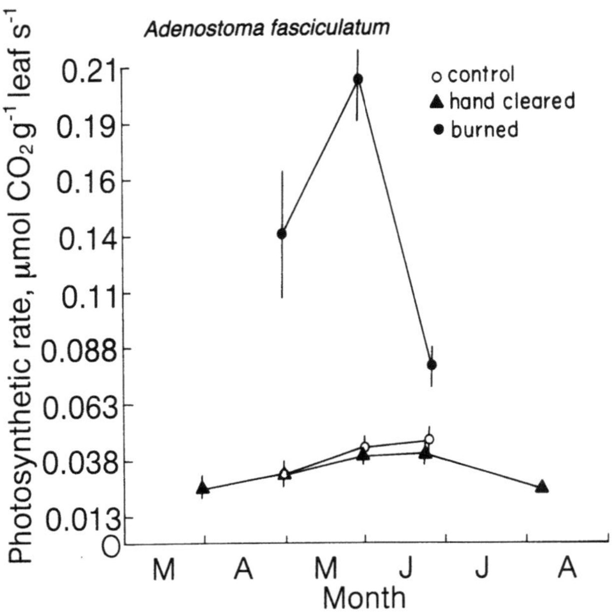

Fig.1. Photosynthetic rate (μmol CO_2 g^{-1} leaf s^{-1}) of Adenostoma fasciculatum versus time in mature plants (control vegetation) and in sprouts following fire and hand clearing. Vertical lines show standard error.

Photosynthesis was similarly enhanced in sprouts of Q. dumosa after fire in April and May (Fig. 2) and in A. glandulosa burn sprouts (Fig. 3). In Q. dumosa, the control vegetation showed photosynthetic rates of about 0.038 mol CO_2 g^{-1} leaf s^{-1} in April and June, whereas postfire sprouts had photosynthetic rates of about 0.063 to 0.069 μ mol CO_2 g^{-1} leaf s^{-1} for April and May. Photosynthesis in postfire sprouts of Q. dumosa fell to about half that of control values in June and August. This appears to be due to the effects of a fungal disease caused by Sphaerotheca lanestris, (Harkin), family Erysiphaceae, which is common on Q. dumosa sprouts following fire, and resulted in considerable dieback. Photosynthetic rates in the Q. dumosa sprouts from the hand-cleared plots were similar to those from the control areas. In late June, plants from the hand-cleared plots had photosynthetic rates of 0.042 ± 0.0044 μmol CO_2 g^{-1} leaf s^{-1}, but the fungal infection in these sprouts was not as severe as in the postfire sprouts.

In early and late June, photosynthetic rates in undisturbed A. glandulosa plants were 0.033 ± 0.0032 and 0.021 ± 0.0019 μmol CO_2 g^{-1} leaf s^{-1} respectively. The postfire sprouts measured in early June had photosynthetic rates of 0.076 ± 0.0063 μmol CO_2 g^{-1} leaf s^{-1}, i.e. 2.3 times the control values. A. glandulosa was the only species in which photosynthesis in sprouts from the hand-cleared plots was markedly enhanced. In early June, sprouts from the hand-cleared plots had photosynthetic rates of 0.067 ± 0.0044 μmol CO_2 g^{-1} leaf s^{-1}, a value about twice that of the control vegetation and almost as high as the sprouts from the burned plots. However, by late June the sprouts from the hand-cleared plots had fallen to 0.028 ± 0.0013, a value slightly larger (30%) than that of the control.

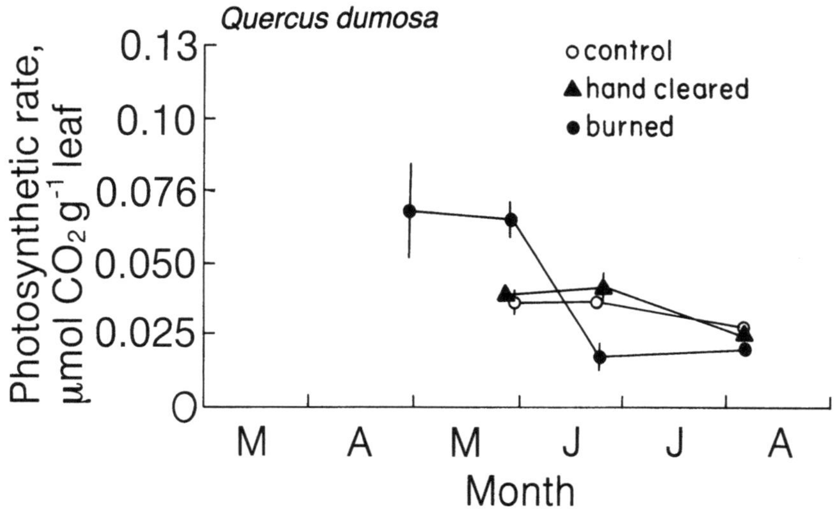

Fig. 2. Photosynthetic rate (μmol CO_2 g^{-1} leaf s^{-1}) of *Quercus dumosa* versus time in mature plants (control vegetation) and in sprouts following fire and hand clearing. Vertical lines show standard error.

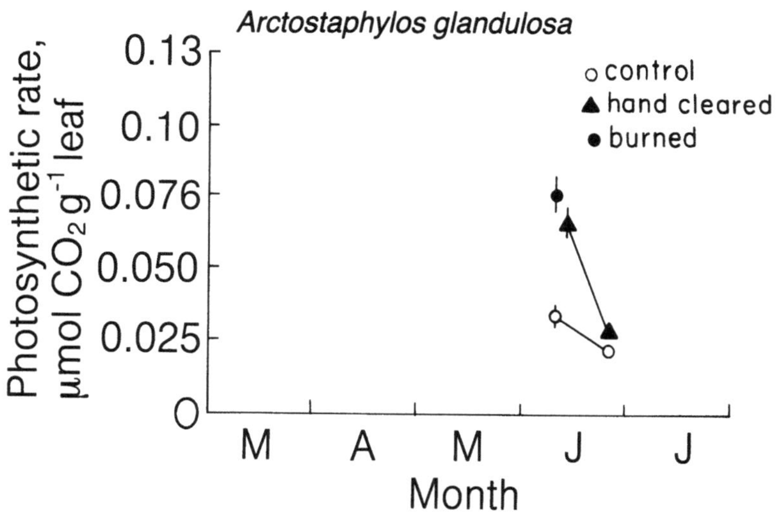

Fig. 3. Photosynthetic rate (μmol CO_2 g^{-1} leaf s^{-1}) of *Arctostaphylos glandulosa* versus time in mature plants (control vegetation) and in sprouts following fire and hand clearing. Vertical lines show standard error.

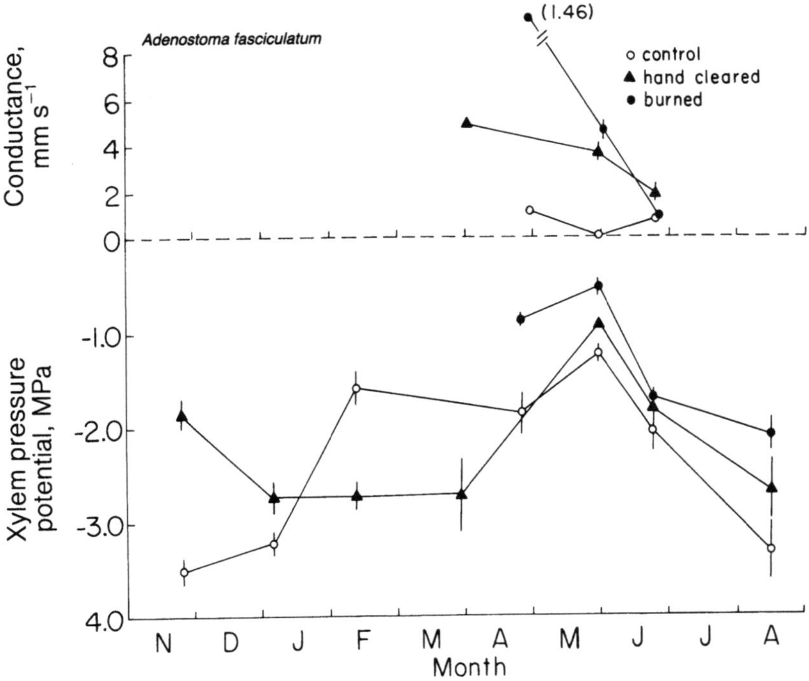

Fig.4. Minimum (midday) conductance (mm s^{-1}) (TOP) and minimum (midday) xylem pressure potentials (MPa) (BOTTOM) for Adenostoma fasciculatum versus time in mature plants (control vegetation) and in sprouts following fire and hand clearing. Vertical lines indicate standard errors.

15.3.3 WATER RELATIONS

In A. fasciculatum minimum (midday) water potentials of the control vegetation were -1.8, -1.3 and -2.6 MPa for April, May and June respectively (Fig. 4). Late spring rains raised the tissue water potentials during the late May sampling and temporarily reversed the normal seasonal drying pattern (Poole and Miller 1975; Oechel and Mustafa 1979). The tissue water potentials were highest in the sprouts on the burned plots and were -0.8, -0.5 and -1.8 MPa in April, May and June respectively. In May 1980, water potentials of sprouts in the hand-cleared plots were intermediate between those of plants in the control and burned plots. Minimum water potentials of sprouts due to fire and hand clearing were similar in June.

Tissue water potentials of mature plants and postfire sprouts of Q. dumosa were generally higher than those of A. fasciculatum (Fig. 5). The control vegetation showed midday water potentials of -1.0 MPa in late May and -1.4 MPa in June. Again the postfire sprouts showed substantially higher water potentials than did the mature vegetation. The midday water potentials of sprouts in late May averaged -0.6 MPa.

Arctostaphylos glandulosa control and sprouting plants showed less variation in midday water potential than did A. fasciculatum. Control vegetation showed a water potential of -1.6 and -2.1 MPa in early and late June, and postfire sprouts varied between -1.4 and -1.8 MPa over the same period. In late June, water potential values of sprouts from the hand-cleared plots fell between the control and postfire sprouts. However, in early June, the cut sprouts actually had a lower water potential than the control plants.

Minimum (midday) leaf conductances in control A. fasciculatum varied from a high value of 4.8 mm s^{-1} in late March to 0.1 mm s^{-1} in late May (Fig. 4). The conductances of sprouts on hand-cleared and burned plots were generally higher than or equal to the control vegetation. In late May, the conductances of the control vegetation and sprouts from hand-cleared and burned plots were 1.0, 3.6 and 4.6 mm s^{-1}, respectively. The highest conductance was recorded for postfire sprouts in late April with a value of 14.6 mm s^{-1}.

Conductances for Q. dumosa control vegetation were 1.0 and 0.9 mm s^{-1} for late May and June. Conductances for sprouts of burned vegetation were higher, i.e. 6.3, 2.3 and 1.3 mm s^{-1} for April, May and June, respectively. Q. dumosa sprouts from the hand-cleared plots had higher conductances than the postfire resprouts or the control vegetation, 3.0 mm s^{-1} (Fig. 5).

In late June, A. glandulosa showed the typical pattern with the lowest conductances in the control plants, intermediate conductances in the sprouts from the hand-cleared plots, and the highest conductances in the postfire sprouts (Fig. 6). In early June, the pattern was somewhat altered as conductances in sprouts on hand-cleared and burned plots were lower than in controls.

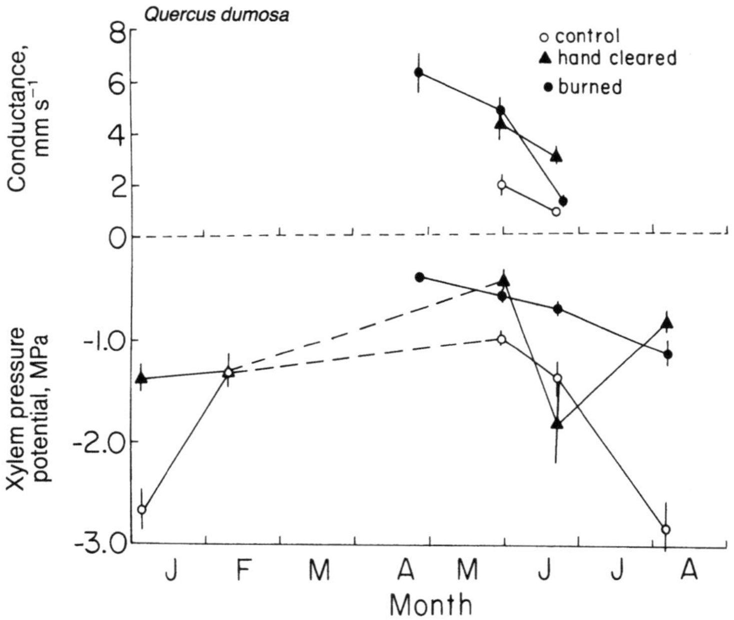

Fig. 5. Minimum (midday) conductance (mm s^{-1}) (TOP) and minimum (midday) xylem pressure potentials (MPa) (BOTTOM) for Quercus dumosa versus time in mature plants (control vegetation) and in sprouts following fire and hand clearing. Vertical lines indicate standard errors.

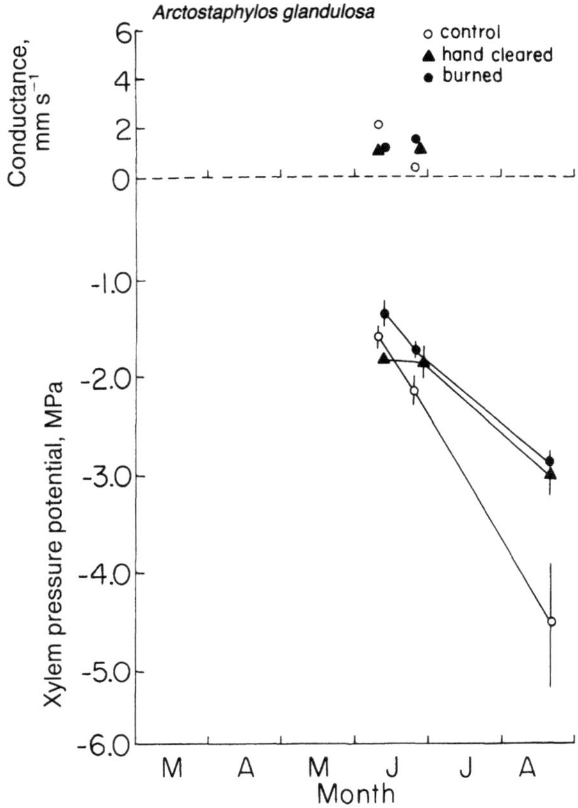

Fig. 6. Minimum (midday) conductance (mm s^{-1}) (TOP) and minimum (midday) xylem pressure potentials (MPa) (BOTTOM) for Arctostaphylos glandulosa versus time in mature plants (control vegetation) and in sprouts following fire and hand clearing. Vertical lines indicate standard errors.

15.4 Discussion

The lower TNC levels in postfire sprouts when compared to control vegetation may reflect the combined effects of the loss of carbohydrates and photosynthetic potential through shoot removal and the effect of enhanced nutrient levels and growth rates on reserve carbohydrate pools. The sprouts from hand-cleared plots did not show the same depression of TNC levels, possibly because of the greater period of time since top removal (about one year) and the lack of nutrient enhancement of the site which could cause depletion of TNC due to increased growth rates (Shaver 1981; Chapter 16, this volume). The high TNC levels in the leaves of Q. dumosa sprouts from hand-cleared plots may result from movement of carbohydrate from the roots and lignotubers to the shoots.

The weak suppression of TNC levels in the other sprouts studied is surprising. Following shoot removal, the plants have a large, respiring below-ground biomass to support and little, if any, photosynthetic tissue. High growth rates following shoot removal should accentuate the depression of available TNC. That such depressions do not occur or are not greater may be due to mobilization of organic nutrients from the root system and lignotubers. These structures need to be analysed to complete the picture of carbon balance in sprouting plants.

In late June, leaf conductances of A. fasciculatum sprouts after either a burn or hand clearing were similar. Water potential values of the sprouts after hand clearing were intermediate between the control and the postfire sprouts. However, the sprouts after hand clearing showed no enhancement of photosynthesis relative to the controls. Yet the postfire plants showed a marked enhancement of photosynthesis, 170% of the control value.

Similar results were found for Q. dumosa where photosynthesis in postfire plants was markedly enhanced in April and May, while photosynthesis in the cut sprouts was similar to the control despite higher leaf conductances. The cut sprouts of A. glandulosa showed a tendency to increased photosynthesis, indicating the possibility of interspecific differences. The results of Radosevich et al. (1977) showed similar trends. In their study, water potentials did not differ significantly between the unburned and burned chamise in February despite differences in the photosynthetic rates. The evidence therefore suggests that directly enhanced photosynthesis in sprouts is not due to either increased stomatal conductance or to indirect effects arising from altered internal mesophyll area to leaf area ratios resulting from improved moisture status.

Increased photosynthesis may result from increased availability of macro- or micronutrients following fire. This may lead to altered membrane permeability, increased RUBP carboxylase activity, or increased activity in other enzymes which results in increased photosynthetic performance.

The fire resulted in substantial increases in nitrogen as ammonium. On 18 May 1979 soil ammonium-N and nitrate- plus ammonium-N levels were 1.2 $\mu g\ g^{-1}$ and 5.8 $\mu g\ g^{-1}$, respectively. The day after the burn (15 November 1979) ammonium-N increased to 11.5 $\mu g\ g^{-1}$, while nitrate- plus ammonium-N was 15 $\mu g\ g^{-1}$. These data indicate a sizeable increase in nitrogen availability following fire which correlates with the enhanced photosynthesis.

15.5 Conclusions

These results provide some information on the causes of increased photosynthetic rates in resprouting vegetation. Increased photosynthesis in postfire sprouts may result either from nutrient enhancement following fire (Vlamis and Gowans 1962; DeBano and Conrad 1974; DeBano et al. 1979) or from improved moisture relations.

Photosynthesis in postfire sprouts of A. fasciculatum was markedly higher than in the controls. Sprouts following hand clearing did not show the same rise in photosynthesis rates despite higher water potential values and higher leaf conductances than the control vegetation. In this study, the increases in photosynthetic rates in the postfire sprouts were due to increased nutrient supply resulting from fire and not due to the altered water relations following top removal. Future research must determine the whole plant TNC and nutrient balance after a fire. The specific causes of the altered photosynthetic rates following fire have to be evaluated properly and will include altered component resistances to carbon dioxide flux in sprouts from burned and hand-cleared plots.

Acknowledgements

The authors thank Phil Riggan (USFS) for the design and establishment of the field plots, Judd Weaver (USFS) for a portion of the water potential data and Phil Riggan, Ernie Lopez and Judd Weaver (USFS) for soil nitrogen values. We also thank William Collins (USFS Riverside, California) for the carbohydrate analysis, Patsy Miller for drafting the figures and Marti Poole, Beth Sigren and Rhonda Watson for assistance in preparation of the manuscript. Support was provided by grants from the US Forest Service and the National Science Foundation.

References

BINGHAM GE, COYNE PI (1977) A portable, temperature-controlled, steadystate porometer for field measurements of transpiration and photosynthesis. Photosynthetica 11: 148-160

BOTTRILL DE, POSSINGHAM JV, KREIDEMANN PE (1970) The effect of nutrient deficiencies on photosynthesis and respiration in spinach. Plant Soil 32: 424-438

CHRISTENSEN NL, MULLER CH (1975) Effects of fire on factors controlling plant growth in Adenostoma chaparral. Ecological Monographs 45: 29-55

DEBANO LF, CONRAD CE (1974) Effect of a wetting agent and nitrogen fertilizer on establishment of ryegrass and mustard on a burned watershed. Journal of Range Management 27: 57-60

DEBANO LF, RICE RM, CONRAD CE (1979) Soil heating in chaparral fires: Effects on soil properties, plant nutrients, erosion, and runoff. US Forest Service Research Paper PSW-145. 21 pp

HUDSON GJ, JOHN PMV, BAILEY BS, SOUTHGATE DAT (1976) The automated determination of carbohydrate. Development of a method for available carbohydrates and its application to foodstuffs. Journal of the Science of Food and Agriculture 27: 681-687

LONGSTRETH DJ, NOBEL PS (1980) Nutrient influences on leaf photosynthesis. Plant Physiology 65: 541-543

MUNZ PA (1974) A flora of southern California. Berkeley, University of California Press. 1086 pp

MUSTAFA J (1978) The effect of growth and species specific variability on photosynthesis along an elevational gradient in the chaparral. MSc Thesis. McGill University.

NATR L (1970) Gas exchange of barley leaves as influenced by mineral deficiency. Scientia Agriculturae Bohemoslavaca 2: 211-218

OECHEL WC, MUSTAFA J (1979) Energy utilization and carbon metabolism in mediterranean scrub vegetation of Chile and California. II. The relationship between photosynthesis and cover in chaparral evergreen shrubs. Oecologia 41: 305-315

OECHEL WC, LAWRENCE W, MUSTAFA J, MARTINEZ J (1981) Energy and carbon acquisition. In: Miller PC (ed) Resource use by chaparral and matorral. Springer-Verlag, Berlin, pp 151-184

POOLE DK, MILLER PC (1975) Water relations of selected species of chaparral and coastal sage communities. Ecology 56: 1118-1128

RADOSEVICH SR, CONRAD SG, ADAMS DR (1977) Regrowth responses of chamise following fire. In: Mooney HA, Conrad CE (eds) Proceedings of the symposium on the environmental consequences of fire and fuel management in mediterranean ecosystems. US Department of Agriculture, Washington, DC, pp 378-382

SHAVER GR (1981) Mineral nutrient and nonstructural carbon utilization. In: Miller PC (ed) Resource use by chaparral and matorral. Springer-Verlag, Berlin, pp 237-258

SPILLER S, TERRY N (1980) Limiting factors in photosynthesis. II. Iron stress diminishes photochemical capacity by reducing the number of photosynthetic units. Plant Physiology 65: 121-125

TERRY N (1980) Limiting factors in photosynthesis. I. Use of iron stress to control photochemical capacity in vivo. Plant Physiology 65: 114-120

TIESZEN LL, JOHNSON DA, CALDWELL MM (1974) A portable system for the measurement of photosynthesis using 14-carbon dioxide. Photosynthetica 3: 151-160

VLAMIS J, GOWANS KD (1961) Availability of nitrogen, phosphorus and sulfur after brush burning. Journal of Range Management 14: 38-40

16. Mineral Nutrient and Nonstructural Carbon Pools in Shrubs from Mediterranean-Type Ecosystems of California and Chile

G. R. SHAVER

16.1 Introduction

Mediterranean-type ecosystems are frequently used as examples of convergent evolution, based on similarities in climate, geography, physiognomy and community structure (e.g. Cody and Mooney 1978). In Californian and Chilean examples, most interpretations of controls on primary production and plant growth form have emphasized the interaction between temperature, water availability and photosynthesis through the year (Specht 1969a,b; Mooney et al. 1970; Mooney and Dunn 1970; Miller and Mooney 1974; Parsons 1976). In South Africa and Australia, however, low soil nutrient availability is of considerable importance in determining the structure and distribution of mediterranean-type vegetation (Specht 1963; Cody and Mooney 1978). Relatively little is known of the mechanisms whereby plants of the Californian chaparral and Chilean matorral have adapted to nutritional factors in their environment, or of the interactions between carbon and nutrient use in any mediterranean-type vegetation. Similarity in the patterns of nutrient and mobile carbon accumulation would constitute additional evidence of convergence in vegetation and ecosystem function.

The first aim of this paper is to compare the overall concentrations and standing stocks of mineral nutrients, nonstructural carbohydrates and crude fat in two mediterranean-type vegetations of California and Chile. Similar information is available for nutrients at several sites in California, Australia and the Mediterranean region (Gray and Schlesinger 1981), but total nutrient stocks in the Chilean vegetation have not been determined. Total non-structural carbohydrate (TNC) and fat pools have not been determined for any mediterranean-type ecosystem, though Oechel and Hastings (Chapter 15, this volume) report TNC concentrations for some chaparral plants.

The second aim is to evaluate the hypothesis that nutrient uptake during the winter, prior to any above-ground growth, is a major source of nutrients for the whole vegetation in Chile and California. The chaparral species *Adenostoma fasciculatum* (Mooney and Rundel 1979) and south Australian vegetation (Specht 1973) seem to follow this pattern. On the basis of individual tissue concentrations, wintertime nutrient uptake in

Chile appears to be of little importance (Shaver 1981).

Thirdly, annual cycles and standing stocks of mineral nutrients are compared with cycles and stocks of mobile, nonstructural carbon pools.

16.2 Research sites

The research sites were situated at Echo Valley, near San Diego, California, and at Fundo Santa Laura, 35 km north of Santiago, Chile. Community structure, soils, climate, physiography and geology of the sites have been described extensively (Mooney 1977; Thrower and Bradbury 1977; Miller 1981). Each site, dominated by evergreen sclerophyllous shrubs, is representative of inland mediterranean-type vegetation. Annual rainfall at Fundo Santa Laura is higher than at Echo Valley (550 versus 475 mm yr^{-1}), but a lower proportion of the rain falls in late spring or summer in Chile (Miller et al. 1977). The soils are more fertile at Fundo Santa Laura and have higher levels of available phosphorus and exchangeable cations (Miller et al. 1977). On the other hand, nitrate-N levels at Echo Valley are about twice those at Fundo Santa Laura. The soil is derived from andesites at the Chilean site, compared to a quartz diorite or a gabbro in California. The species and growth form composition of the Chilean matorral are more diverse than in the chaparral, but shrub biomass and primary production are twice as high in California (Parsons 1976; Mooney et al. 1977; Kummerow 1981; Steward and Webber 1981). The low shrub productivity in Chile is complemented by a diverse and productive annual herb community (Keeley and Johnson 1976) although total primary production is still lower than in California.

Shrub growth phenology in California is much more uniform among species than in Chile. Whereas production of new leaf and stem tissue is primarily restricted to spring months in California, in Chile some shrub growth occurs throughout the year. The major period of plant growth in Chile extends from mid-winter to late summer, although no single species grows throughout that period (Hoffmann et al. 1977; Mooney et al. 1977; Mooney and Kummerow 1977). The more uniform phenology of growth in California might cause a more intense demand for nutrients over a shorter period of time than in Chile.

16.3 Presentation of data

Weighted-mean concentrations and standing stocks of nitrogen, phosphorus, potassium, calcium, magnesium, TNC and "crude fat" were calculated for the whole vegetation and for individual species at Echo Valley and Fundo Santa Laura. The concentration of each component in leaves, stems or roots was multiplied by the proportion of biomass represented by that tissue and species. The resulting values were then summed for a mean concentration, weighted according to the relative biomass of tissues and species (Whittaker et al. 1979).

Above-ground biomass data used in the calculations were taken from Mooney et al. (1977) and Kummerow (1981), and include leaf and stem masses by species for each site (Table 1). Root biomass was estimated using the root:shoot ratios of Kummerow et al. (1977), Miller and Ng (1977) and Hoffmann and Kummerow (1978). Total biomass, by species, was the sum of the measured above-ground biomass and the estimated below-ground biomass.

Where there were no root:shoot biomass ratios known, a value of 0.7 was assumed. This value was the mean for all species from both sites for which root:shoot ratios were available. The assumption was necessary to estimate

Table 1. Cover and biomass of the dominant shrub species at Echo Valley and Fundo Santa Laura, from Miller and Ng (1977); Mooney et al. (1977) and Hoffmann and Kummerow (1978). Values in parenthesis denote estimates including undetermined root:shoot biomass ratios, assumed to be 0.7.

	Per cent cover	Above-ground biomass g m^{-2}	Stem biomass g m^{-2}	Leaf biomass g m^{-2}	Root: shoot ratio	Root biomass g m^{-2}	Total biomass g m^{-2}
ECHO VALLEY (CALIFORNIA)							
Adenostoma fasciculatum	52.2	901.1	749.7	151.4	0.6	540.7	1441.8
Arctostaphylos glauca	5.0	168.0	121.3	46.7	0.9	151.2	319.2
Ceanothus greggii	25.4	859.2	708.0	151.2	0.3	251.8	1117.0
Rhus ovata	1.1	23.3	18.2	5.0	(0.7)	(16.3)	(39.6)
Sum (4 species)	83.7	1951.6	1597.2	354.3		(966.0)	(2917.6)
Total community (shrubs only)	100.0	2307.9	1892.5	415.4		(1142.4)	(3450.3)
FUNDO SANTA LAURA (CHILE)							
Colliguaya odorifera	4.2	58.2	47.7	10.5	0.6	34.9	93.1
Cryptocarya alba	16.8	284.8	156.4	128.4	1.4	398.7	683.5
Lithraea caustica	6.6	57.1	37.6	19.5	2.9	165.6	222.7
Quillaja saponaria	0.3	3.8	2.7	1.1	(0.7)	(2.7)	(6.5)
Satureja gilliesii	7.9	60.7	51.7	9.0	0.7	42.5	103.2
Trevoa trinervis	5.0	41.3	33.5	7.9	(0.7)	(28.9)	(70.2)
Kageneckia oblonga	2.2	29.5	18.6	10.9	(0.7)	(20.7)	(50.2)
Sum (7 species)	43.0	535.4	348.2	187.3		(694.0)	(1229.4)
Total community (shrubs only)	59.3	738.4	531.7	206.8		(957.1)	(1695.5)

standing stocks of nutrients in the whole vegetation. However, the species for which root:shoot ratios were unknown accounted for only 1-2% of shrub biomass in California and 10% in Chile (Table 1).

Concentrations of mineral nutrients, TNC and "crude fat" were taken from Shaver (1981) and included a breakdown of roots, leaves and stems into several age-classes of tissue. Because the biomass data did not distinguish between tissue age-classes, a simple mean concentration for all age-classes and sample dates was used in calculating standing stocks and overall concentrations of nutrients and mobile carbon pools (Table 2). There were no significant differences between age-classes in concentration of the relevant components except in new tissues (Shaver 1981).

Levels of nitrogen, phosphorus, potassium, calcium and magnesium were total concentrations in the tissue (acid digest followed by colorimetric determination with a Technicon autoanalyser for nitrogen and phosphorus, atomic absorption spectroscopy for potassium, calcium and magnesium; Chapin et al. 1980; Shaver and Chapin 1980). TNC included monosaccharides, oligosaccharides and starch (acid hydrolysis and alcohol extraction, colorimetric determination on the autoanalyser; Smith 1969; Technicon method #280-73A). "Crude fat" included all the lipids extracted by chloroform-methanol extract (Randall 1974).

Data were not available from all shrub species at each site and only the dominant species were sampled (Table 1). In California, the four species listed in the table had a combined cover of 84% and comprised 85% of above-ground shrub biomass. The seven species listed for Chile had a cover of only 43% but comprised 72% of above-ground shrub biomass. In extrapolating nutrient standing stocks to the whole community, weighted-mean concentrations for the species sampled were multiplied by whole shrub community biomass rather than biomass of only the species sampled.

16.4 Results

16.4.1 STANDING STOCKS OF MINERAL NUTRIENTS, NONSTRUCTURAL CARBOHYDRATE AND CRUDE FAT

Mean concentrations of nutrients and nonstructural carbon pools (Table 2) showed a pattern similar to the specific comparisons of individual age classes and sample dates (Shaver 1981). Unweighted community mean nutrient concentrations were higher in Chile. TNC and fat concentrations were higher in leaves and stems in California. TNC concentrations in roots were higher in Chile, and mean fat concentrations were about equal.

Relative overall nutritional status of individual species was compared using "nutrient profiles" (Woodwell et al. 1975; Whittaker et al. 1979). For each species, mean whole-plant concentration was plotted as a per cent deviation from the mean for all 11 species (Fig. 1). In each case the mean whole-plant concentrations were weighted according to relative tissue masses within a species but not according to the relative abundance of the species. In California, the results were similar to the comparison of simple means, with three exceptions: (a) Adenostoma fasciculatum was low in overall TNC concentration relative to the other ten species; (b) Ceanothus greggii, which is an N-fixer, was high in nitrogen and (c) Rhus ovata was extremely high in potassium, particularly in stems (Table 2). The Chilean species were much more variable in overall nutrient and nonstructural carbon concentration (Fig. 1).

Table 2. Mean concentrations (per cent of dry mass) of mineral nutrients, TNC and crude fat in leaves, stems and roots of the dominant shrub species at the Californian sites and Chilean sites, by species, and simple un-

	N			P			K		
	Leaves	Stems	Roots	Leaves	Stems	Roots	Leaves	Stems	Roots
ECHO VALLEY (CALIFORNIA)									
Adenostoma fasciculatum	0.91	0.65	0.46	0.13	0.11	0.12	0.52	0.37	0.16
Arctostaphylos glauca	0.81	0.46	0.27	0.12	0.10	0.10	0.44	0.54	0.22
Ceanothus greggii	1.32	0.87	0.59	0.11	0.10	0.10	0.48	0.50	0.17
Rhus ovata	0.89	0.48	0.38	0.14	0.12	0.15	0.75	1.87	0.40
Mean	0.98	0.62	0.43	0.13	0.11	0.12	0.55	0.82	0.24
FUNDO SANTA LAURA (CHILE)									
Colliguaya odorifera	1.22	0.99	0.77	0.18	0.15	0.14	0.73	1.18	0.59
Cryptocarya alba	0.95	0.61	0.56	0.17	0.15	0.13	0.85	0.63	0.47
Lithraea caustica	0.84	0.49	0.48	0.14	0.13	0.11	0.60	0.96	0.59
Quillaja saponaria	1.00	0.83	0.74	0.15	0.15	0.13	0.71	0.53	0.61
Satureja gilliesii	1.72	0.81	0.95	0.26	0.21	0.18	1.72	0.88	0.78
Trevoa trinervis	2.73	1.38	1.00	0.20	0.14	0.15	0.65	0.46	0.36
Kageneckia oblonga	1.76	1.04	0.77	0.26	0.19	0.14	0.98	0.59	0.41
Mean	1.46	0.88	0.75	0.19	0.16	0.14	0.89	0.75	0.54

Despite the variation among species, weighted means for the vegetation at each site showed the same relationships between the two sites (Table 3) as the comparison of simple means (Table 2). Nutrient concentrations were consistently higher in Chile, and TNC and fat were higher in leaves and stems in California and in roots in Chile. Overall, the shrubs from Echo Valley and Fundo Santa Laura differed most in potassium, calcium, magnesium and fat concentration and least in nitrogen and TNC concentration.

The total quantities (g m^{-2}) of nutrients and mobile carbon present in the shrub vegetation were higher in California in every case (Table 4). Only calcium and potassium content in Chile came within 25% of the content in California. However, considering roots alone the amounts of nitrogen, potassium, calcium and TNC were similar in the two sites, and quantities of potassium, magnesium and fat in roots were higher in Chile. The reason for the larger quantities in California despite low concentrations is the much higher total and above-ground biomass in California (Table 1). Root biomass was similar in the two sites because root:shoot ratios of the two most abundant species, Cryptocarya alba and Lithraea caustica, were exceptionally high in Chile. The high root:shoot ratios in those two species probably resulted from periodic cutting of the Chilean shrubs for firewood, and regrowth from the stumps (Hoffmann and Kummerow 1978).

16.4.2 SEASONAL VARIATION IN CONCENTRATIONS

In California, seasonal changes in community leaf nitrogen concentration followed a pattern very similar to that in Adenostoma fasciculatum (Mooney and Rundel 1979), with an increase during the winter weighted means for each site. Data from Shaver (in press), calculated as means for all age classes and sample dates over 15 months in 1975 and 1976.

Ca			Mg			TNC			Fat		
Leaves	Stems	Roots	Leaves	Stems	Roots	Leaves	Stems	Roots	Leaves	Stems	Roots
0.73	0.32	1.31	0.16	0.09	0.07	17.38	5.72	6.82	9.19	5.20	0.49
0.84	0.44	0.32	0.21	0.10	0.06	17.40	11.14	8.79	7.97	1.85	0.33
1.13	0.68	1.24	0.23	0.10	0.07	15.06	8.13	5.81	5.44	2.70	0.93
1.06	0.81	1.05	0.20	0.13	0.11	9.93	9.52	11.69	3.43	3.45	1.94
0.94	0.56	0.98	0.20	0.11	0.08	14.94	8.63	8.28	6.51	3.30	0.92
1.06	0.87	0.67	0.30	0.18	0.13	4.68	4.09	14.49	1.74	2.32	1.15
2.25	1.72	1.38	0.28	0.07	0.11	7.93	6.40	8.23	4.76	0.85	1.56
1.37	1.53	1.48	0.34	0.17	0.18	3.32	5.89	9.18	1.80	1.49	1.18
1.70	1.14	1.33	0.64	0.45	0.25	7.82	4.61	13.55	1.00	0.77	0.76
1.45	0.30	0.66	0.34	0.14	0.23	8.90	4.05	3.28	2.02	0.60	0.67
3.65	1.30	2.11	0.26	0.09	0.15	9.60	6.00	10.65	0.22	0.34	0.58
1.97	1.92	1.48	0.34	0.22	0.27	6.20	4.60	7.00	3.27	1.72	0.58
1.92	1.25	1.30	0.36	0.19	0.19	6.92	5.09	9.48	2.12	1.16	0.93

Table 3. Weighted mean nutrient and nonstructural carbon concentrations (per cent dry mass) for each site. Means were weighted at each site according to the relative biomass of each tissue for each species (Whittaker et al. 1979).

	N	P	K	Ca	Mg	TNC	Fat
ECHO VALLEY							
Leaves	1.07	0.12	0.50	0.92	0.20	16.29	7.35
Stems	0.73	0.11	0.46	0.49	0.10	7.24	3.81
Roots	0.46	0.11	0.18	1.13	0.07	6.94	0.61
Total plant	0.69	0.11	0.37	0.75	0.10	8.23	3.18
FUNDO SANTA LAURA							
Leaves	1.11	0.18	0.87	2.00	0.30	7.28	3.85
Stems	0.78	0.16	0.76	1.56	0.12	5.53	1.08
Roots	0.60	0.13	0.52	1.33	0.14	8.55	1.32
Total plant	0.73	0.15	0.64	1.50	0.16	7.50	1.64

followed by a drop during rapid growth in spring (Fig. 2A). There was a small decline in stem nitrogen concentrations during winter which then increased slightly during spring. Root nitrogen concentration remained constant in winter and then more than doubled between March and September. The net result was that whole-plant nitrogen concentration did not change

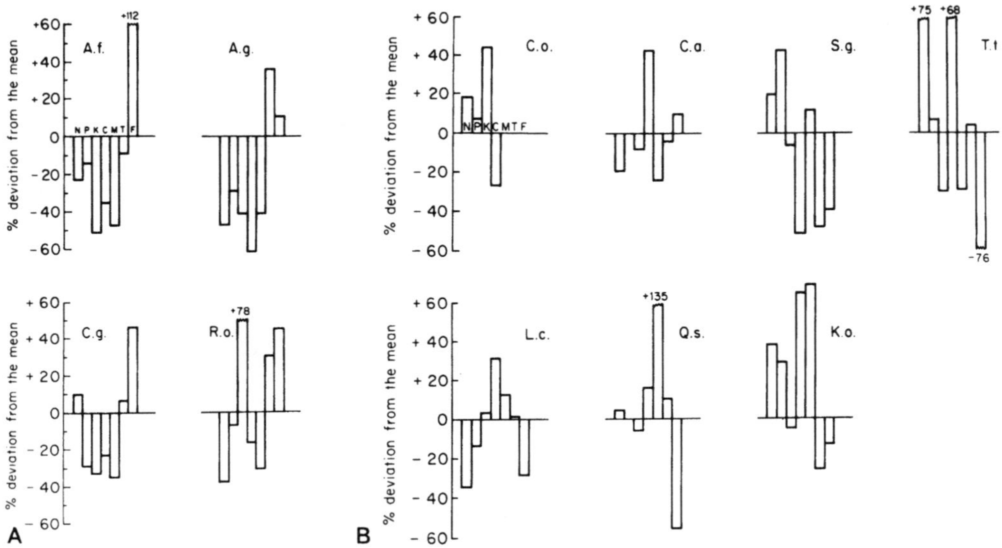

Fig. 1. Nutrient profiles for the dominant shrub species of Echo Valley and Fundo Santa Laura. Bars represent percentage deviation of component concentrations from the simple mean value (unweighted for species abundance) for all 11 species. Within each species the concentration was weighted according to the relative biomass of leaf, stem and root tissues. Species are indicated by the first letters of their generic and specific names. N, P, K, C, M, T and F stand for nitrogen, phosphorus, potassium, calcium, magnesium, TNC and crude fat, respectively. Where there is no bar visible the species value was equal to the overall mean (eg for Mg, TNC and fat in Colliguaya odorifera in Chile). A: Californian species, B: Chilean species.

over the winter in California because the amounts lost from the large stem biomass were compensated for by the amounts gained by the small leaf biomass. Whole-plant nitrogen concentration increased throughout the spring and summer, primarily due to the increase in root nitrogen.

Seasonal changes in whole-plant nitrogen concentration also varied among individual species, but only slightly (Fig. 2B).

Phosphorus concentrations in California changed much more uniformly through the year, both between tissues and between species (Fig. 2A). The concentration increased in winter and spring, then dropped steeply through the summer. Phosphorus concentration of the whole vegetation reached a maximum of 0.20%, 2.5 times the initial value of 0.08%.

TNC in the Californian plants followed different seasonal patterns in different tissues (Fig. 3A). In leaves, TNC decreased during the winter, remained low during spring growth and increased in summer. Stems, roots and the whole vegetation, however, increased in TNC concentration in winter and summer, and decreased during the spring. Whole plants of individual species in general behaved as the whole vegetation did, the only exception being C. greggii in winter (Fig. 3B).

Fat concentrations in California hardly varied through the year in leaves and roots, but in stems decreased during winter, increased in spring and decreased over the summer (Fig. 3A). Individual species all followed a pattern similar to that for the stems and for the whole vegetation (Fig. 3B).

Table 4. Nutrient and nonstructural carbon content (g m^{-2}) of the woody shrubs at Fundo Santa Laura and Echo Valley.

	N	P	K	Ca	Mg	TNC	Fat
ECHO VALLEY							
Leaves	4.44	0.50	2.08	3.82	0.83	67.67	30.53
Stems	13.82	2.08	8.71	9.27	1.89	137.02	72.10
Roots	5.26	1.26	2.06	12.91	0.80	79.28	6.97
Total plant	23.46	3.80	12.77	25.88	3.45	283.96	109.72
FUNDO SANTA LAURA							
Leaves	2.30	0.37	1.80	4.14	0.62	15.06	7.96
Stems	4.15	0.85	4.04	8.29	0.64	29.40	5.74
Roots	5.74	1.24	4.98	12.73	1.34	81.83	12.63
Total plant	12.38	2.54	10.85	25.43	2.71	127.16	27.81

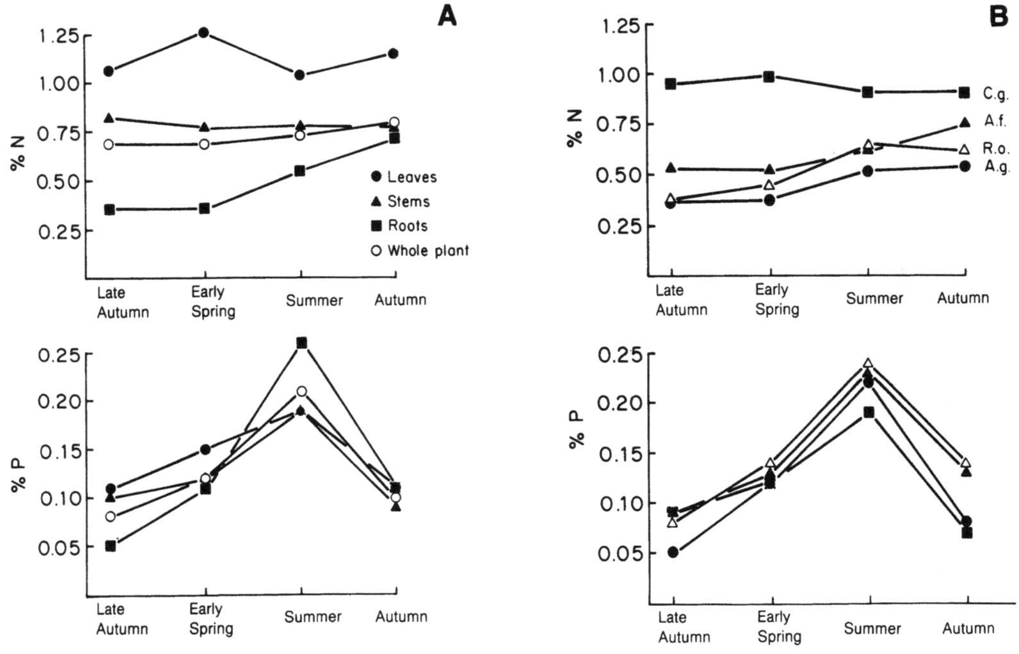

Fig. 2. Changes in leaf, stem, root and whole-plant concentrations of nitrogen and phosphorus in California. Sampling dates were: 17-20 November 1975 (late autumn), 19 March 1976 (early spring), 27 June to 20 July 1976 (summer) and 30 September to 10 October 1976 (autumn). A. Weighted means for the whole vegetation, B. Means weighted according to the relative biomass of leaf, stem and root tissues. Species identified by first letters of their generic and specific names.

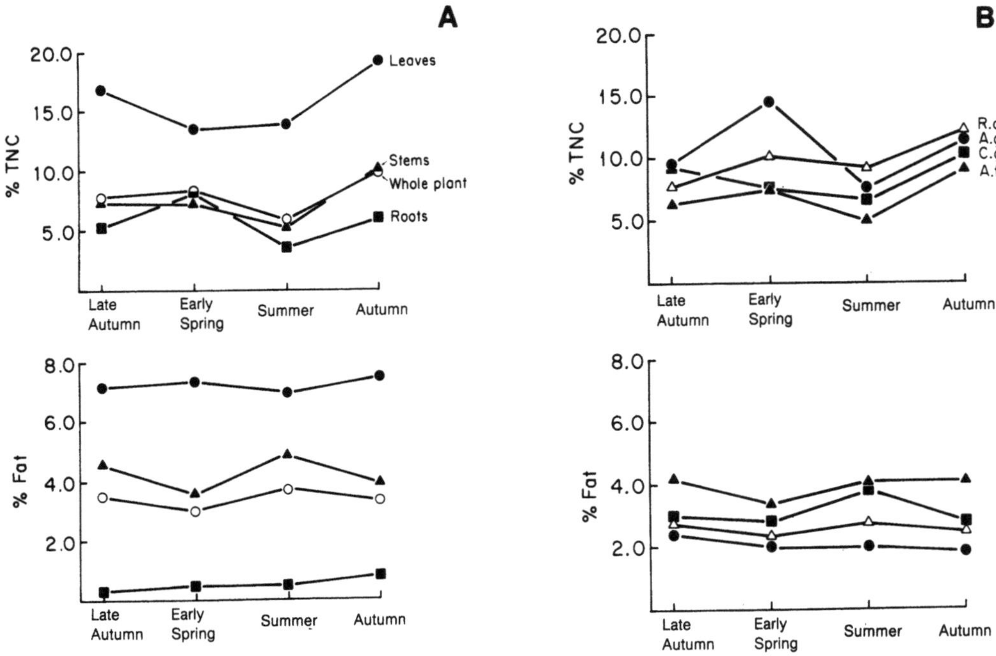

Fig. 3. Changes in leaf, stem, root and whole-plant concentration of TNC and crude fat in California. Sample dates as in Fig. 2. Species identified by first letters of their generic and specific names. A. Weighted means for the whole vegetation, B. Means weighted according to relative biomass of leaf, stem and root tissues.

Comparison of concentration changes in the Chilean shrubs was restricted to four species at two dates. Weighted means for Chile were calculated according to the relative tissue biomass of these four species (Tables 5, 6). In both California and Chile the interval between samplings was the winter.

The shrubs from California and Chile did not behave similarly over the winter (Fig. 4). For the whole shrub vegetation in California, phosphorus and TNC concentrations increased while nitrogen did not change and fat concentrations decreased. In Chile, vegetation nitrogen, phosphorus and TNC concentrations decreased while fat increased. However, except for phosphorus in California the changes over the winter were never uniform among all species at a site. The lack of uniformity among species was particularly noticeable for TNC in Chile, where there were large changes for all species in opposite directions (Fig. 4).

16.5 Discussion

Overall, weighted mean nitrogen, phosphorus and potassium concentrations in Chile were higher than those of other mediterranean-type sites reported by Gray and Schlesinger (1981). The results for Echo Valley in California were slightly above Gray and Schlesinger's mean nitrogen,

Table 5. Weighted-mean nutrient and nonstructural carbon concentrations, Fundo Santa Laura, Chile. Dates of sampling were 5-26 May 1976 (autumn) and 21 September 1976 (early spring).

	Late autumn	Early spring
NITROGEN		
Leaves	1.11	0.98
Stems	1.18	0.94
Roots	0.61	0.72
Whole plant	0.85	0.82
PHOSPHORUS		
Leaves	0.18	0.17
Stems	0.17	0.15
Roots	0.13	0.16
Whole plant	0.15	0.16
TNC		
Leaves	3.8	3.1
Stems	6.3	4.7
Roots	11.8	10.1
Whole plant	9.0	7.5
FAT		
Leaves	1.85	1.97
Stems	2.24	1.95
Roots	0.94	1.19
Whole plant	1.45	1.93

Table 6. Weighted-mean whole-plant concentrations of nutrients and nonstructural carbon, Fundo Santa Laura. Dates of sampling were 5-26 May 1976 (autumn) and 21 September 1976 (early spring).

	Late autumn	Early spring
NITROGEN		
Colliguaya odorifera	1.09	1.10
Lithraea caustica	0.54	0.62
Quillaja saponaria	0.83	0.77
Trevoa trinervis	1.51	1.08
PHOSPHORUS		
Colliguaya odorifera	0.18	0.16
Lithraea caustica	0.13	0.16
Quillaja saponaria	0.17	0.15
Trevoa trinervis	0.17	0.15
TNC		
Colliguaya odorifera	4.0	8.9
Lithraea caustica	8.3	6.5
Quillaja saponaria	10.4	14.3
Trevoa trinervis	9.5	6.2
FAT		
Colliguaya odorifera	2.63	2.41
Lithraea caustica	1.18	2.25
Quillaja saponaria	0.98	1.30
Trevoa trinervis	0.58	0.95

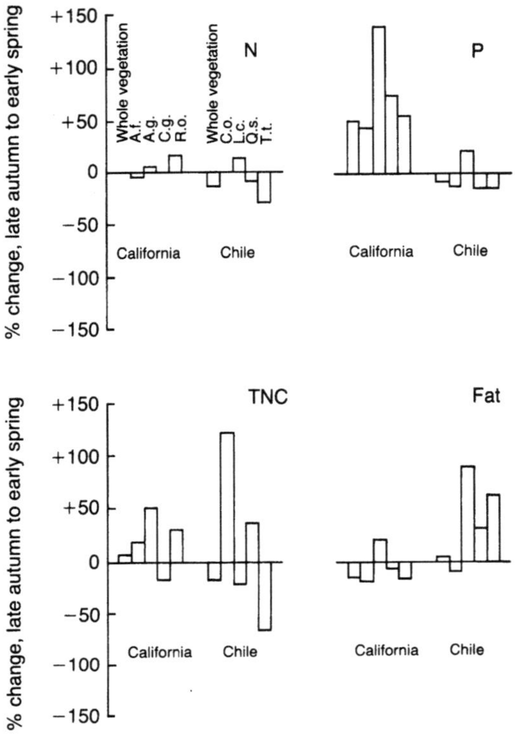

Fig. 4. Relative changes in N, P, TNC and fat concentrations from late autumn to early spring sample dates in Chile and California. Bars are percentages, calculated as: ((early spring concentration divided by late autumn concentration) - 1) x 100. "Whole plant" values taken from Figs 2A and 3A for California and Table 5 for Chile. Individual species values, denoted by genus and species first letters, taken from Figs 2B and 3B for California and Table 6 for Chile. There was no change in whole-plant N concentration in California.

phosphorus and potassium concentrations (0.54%, 0.05% and 0.36% respectively), but were well within the range of values they reported. Gray and Schlesinger (1981) did not calculate mean calcium and magnesium concentrations but their data suggest that weighted means for these nutrients also were exceptionally high in Chile and near average in California in comparison with other mediterranean-type vegetations.

The low concentrations of TNC and fat in Chile contrast strongly with the high nutrient concentrations there. The reasons for this contrast are not clear, but similar inverse relationships between nutrient and carbohydrate status of plants have been observed in both agricultural and natural ecosystems (Fleming 1973; Smith 1973; Evans 1975; Shaver and Chapin 1980). Elucidation of the mechanisms responsible will require further research.

Some of the differences in component concentrations and standing stocks may be due to specific site factors at Echo Valley and Fundo Santa Laura. For example, very different results might have been found on other soil parent materials. The ages of both stands were about 20 years but the vegetation in Chile had been cut for firewood while that in California had been burned (Aschmann and Bahre 1977). Even with these minor differences the similarity in climate, geology and topography between the sites was remarkable.

The hypothesis that nutrient uptake during the winter is a major nutrient source in mediterranean-type vegetations was not supported by this investigation. Except for phosphorus in California, nutrient concentrations in the whole plant were either constant or declined through the winter in the two sites. Furthermore, the greatest increase in phosphorus concentrations in California was in spring, while the plants were growing rapidly. Comparison of the nutrient data from 1975-76 at Echo Valley (this paper) with results from 1972-73 (Mooney et al. 1977; Mooney and Rundel 1979) indicates considerable year-to-year variation. Although leaf nitrogen concentration in California showed very similar seasonal changes in 1975-76 and 1972-73, leaf phosphorus concentrations behaved very differently in the two years. In 1972-73 leaf phosphorus concentrations followed a pattern similar to that for nitrogen. Root nitrogen concentrations in California also ended up in 1976 at over twice the initial value in 1975. Thus nutrient uptake during the winter, and storage in older tissues, may occur in some years but not necessarily every year.

The two sites also differed in the seasonal behaviour of TNC and fat. In California, seasonal changes in whole plant TNC concentration were consistent with the hypothesis that photosynthetic carbon accumulation exceeds carbon utilization in the winter and summer. In spring, when growth is rapid, carbon use exceeds fixation and TNC levels decline. A similar pattern was found in Heteromeles arbutifolia grown in pots (Mooney and Chu 1974). In contrast to the pattern of TNC concentrations in whole plants, that in leaves was very different but could not be correlated with overall growth. Fat concentrations in California also were not well correlated with growth. In Chile, the large interspecific differences in TNC changes over the winter may reflect interspecific phenological variation.

In sum, this investigation has shown large differences in plant nutrition between Echo Valley and Fundo Santa Laura. Concentrations of mineral nutrients were exceptionally high in the matorral, but in mobile carbon pools were small. There was less variation among species in California, both seasonally and in overall concentrations and standing stocks. Functional or evolutionary convergence between the two sites was not strongly apparent. However, comparisons involving mineral nutrients in particular must usually be more site-specific than geographic comparisons of vegetation structure or temperature, light and water relations.

Acknowledgements

F S Chapin III, R A Houghton, H A Mooney and W H Schlesinger kindly reviewed this paper in manuscript. The research was supported by the National Science Foundation and The Ecosystems Center.

References

ASCHMANN H, BAHRE C (1977) Man's impact on the wild landscape. In: MOONEY HA (ed) Convergent evolution in Chile and California: mediterranean climate ecosystems. US IBP Synthesis Series, vol 5. Dowden, Hutchinson and Ross, Stroudsberg, Pennsylvania, pp 73-84

CHAPIN FS, JOHNSON DA, MCKENDRICK JD (1980) Seasonal movement of nutrients in plants of differing growth form in an Alaskan tundra ecosystem: Implications for herbivory. Journal of Ecology 68: 180-209

CODY ML, MOONEY HA (1978) Convergence vs nonconvergence in mediterranean-climate ecosystems. Annual Review of Ecology and Systematics 9: 265-321

EVANS LT (1975) The physiological basis of crop yield. In: EVANS LT (ed) Crop physiology. Cambridge University Press, Cambridge, pp 327-356

FLEMING GA (1973) Mineral composition of herbage. In: BUTLER GW, BAILEY RW (eds) Chemistry and biochemistry of herbage, vol I. Academic Press, New York, pp 529-566

GRAY JT, SCHLESINGER WH (1981) Nutrient cycling in mediterranean-type ecosystems. In: MILLER PC (ed) Resource use by chaparral and matorral: a comparison of vegetation function in two mediterranean-type ecosystems. Ecological Studies Series, vol 39. Springer-Verlag, Berlin, pp 259-285

HOFFMANN A, KUMMEROW J (1980) Root studies in the Chilean matorral. Oecologia 32: 57-69

HOFFMANN A, MOONEY HA, KUMMEROW J (1977) Qualitative phenology. In: THROWER NJW, BRADBURY DE (eds) Chile-California mediterranean scrub atlas. US IBP Synthesis Series, vol 2. Dowden, Hutchinson and Ross, Stroudsberg, Pennsylvania, pp 102-120

KEELEY SC, JOHNSON AW (1976) A comparison of herb and shrub growth in comparable sites in Chile and California. American Midland Naturalist 97: 120-132

KUMMEROW J (1981) Biomass, phenology, and growth. In: MILLER PC (ed) Resource use by chaparral and matorral: a comparison of vegetation function in two mediterranean-type ecosystems. Springer-Verlag, Berlin, pp 69-96

KUMMEROW J, KRAUSE D, JOW W (1977) Root systems of chaparral shrubs. Oecologia 29: 163-177

MILLER PC (ed) (1981) Resource use by chaparral and matorral: a comparison of vegetation function in two mediterranean-type ecosystems. Springer-Verlag, Berlin, 455 pp

MILLER PC, BRADBURY DE, HAJEK E, LAMARCHE V, THROWER NJW (1977). Past and present environment. In: MOONEY HA (ed) Convergent evolution in Chile and California: Mediterranean climate ecosystems. US IBP Synthesis Series, vol 5. Dowden, Hutchinson, and Ross, Stroudsberg, Pennsylvania, pp 27-72

MILLER PC, MOONEY HA (1974) The origin and structure of American arid-zone ecosystems. The producers: Interactions between environment, form, and function. In: Proceedings of the First International Congress of Ecology, The Hague, Netherlands, pp 201-209

MILLER PC, NG E (1977) Root:shoot biomass ratios in shrubs in southern California and central Chile. Madrono 24: 215-223

MOONEY HA, DUNN EL (1970) Photosynthetic systems of mediterranean-climate shrubs and trees of California and Chile. American Naturalist 104: 447-453

MOONEY HA, DUNN EL, SHROPSHIRE F, SONG L (1970) Vegetation comparisons between the mediterranean climate areas of California and Chile. Flora 159: 480-496

MOONEY HA, CHU C (1974) Seasonal carbon allocation in Heteromeles arbutifolia, a Californian evergreen shrub. Oecologia 14: 295-306

MOONEY HA, KUMMEROW J (1977) Quantitative phenology. In: THROWER NJW, BRADBURY DE (eds) Chile-California mediterranean scrub atlas. US IBP Synthesis Series, vol 2. Dowden, Hutchinson and Ross, Stroudsberg, Pennsylvania, pp 121-125

MOONEY HA, KUMMEROW J, JOHNSON AW, PARSONS DJ, KEELEY S, HOFFMANN A, HAYS RI, GILIBERTO J, CHU C (1977) The producers - their resources and adaptive responses. In: MOONEY HA (ed) Convergent evolution in Chile and California mediterranean climate ecosystems. US IBP Synthesis Series, vol 5. Dowden, Hutchinson and Ross, Stroudsbertg, Pennsylvania, pp 85-143

MOONEY HA, RUNDEL PW (1979) Nutrient relations of the evergreen shrub, Adenostoma fasciculatum, in the California chaparral. Botanical Gazette 140: 109-113

PARSONS DJ (1976) Vegetation structure in the mediterranean scrub communities of California and Chile. Journal of Ecology 64: 435-437

RANDALL EL (1974) Improved method for fat and oil analysis by a new process of extraction. Journal of the Association of Official Analytical Chemists 57: 1165-1168

SHAVER GR (1981) Mineral nutrient and nonstructural carbon utilization. In: MILLER PC (ed) Resource use by chaparral and matorral: a comparison of vegetation function in two mediterranean-type ecosystems. Springer-Verlag, Berlin, pp 237-257

SHAVER GR, CHAPIN FS (1980) Response to fertilization by various plant growth forms in Alaskan tundra: Nutrient accumulation and growth. Ecology 61: 662-675

SMITH D (1969) Removing and analyzing total nonstructural carbohydrates from plant tissue. University of Wisconsin, College of Agriculture and Life Sciences Research Report 41. University of Wisconsin, Madison

SMITH D (1973) The nonstructural carbohydrates. In: BUTLER GW, BAILEY RW (eds) Chemistry and biochemistry of herbage, vol 1. Academic Press, New York, pp 106-156

SPECHT RL (1963) Dark Island heath (Ninety-Mile Plain, South Australia). VII. The effects of fertilizers on composition and growth, 1950-1960. Australian Journal of Botany 11: 67-94

SPECHT RL (1969a) A comparison of the sclerophyllous vegetation characteristic of mediterranean-type climates in France, California and southern Australia. I. Structure, morphology, and succession. Australian Journal of Botany 17: 277-292

SPECHT RL (1969b) A comparison of the sclerophyllous vegetation characteristic of mediterranean-type climates in France, California and southern Australia. II. Dry matter, energy, and nutrient accumulation. Australian Journal of Botany 17: 293-308

SPECHT RL (1973) Structure and functional response of ecosystems in the mediterranean climate of Australia. In: DI CASTRI F, MOONEY HA (eds) Mediterranean type ecosystems: origin and structure. Springer-Verlag, Berlin, pp 113-120

STEWARD D, WEBBER PJ (1981) The plant communities and their environments. In: MILLER PC (ed) Resource use by chaparral and matorral: a comparison of vegetation function in two mediterranean-type ecosystems. Springer-Verlag, Berlin, pp 43-68

THROWER NJW, BRADBURY DE (eds) (1977) Chile-California mediterranean scrub atlas. US IBP Synthesis Series, vol 2. Dowden, Hutchinson and Ross, Stroudsberg, Pennsylvania, 273 pp

WHITTAKER RH, LIKENS GE, BORMANN FH, EATON JS, SICCAMA TG (1979) The Hubbard Brook ecosystem study: Forest nutrient cycling and element behaviour. Ecology 60: 203-220

WOODWELL GM, WHITTAKER RH, HOUGHTON RA (1975) Nutrient concentrations in plants in the Brookhaven oak-pine forest. Ecology 56: 318-332

17. Comparative Phenology of Mediterranean-Type Plant Communities

J. KUMMEROW

17.1 Introduction

Phenological studies were defined in 1751 by Carolus Linnaeus in his Philosophia Botanica as the observation and compilation of the different phenophases of flowering plants with respect to climatological data. Thus, phenology can deal with the phenophases of individuals of a single species in a limited area, with individual plants covering a wide geographical range, or with representative members of a larger number of species in a defined ecosystem. In this way, phenology serves as a tool in autecology, for exploring environmental conditions over the geographical range of a species, and in the comparative analysis of ecosystems. Since organisms adapt to environmental parameters according to their genetic constitution, an array of genetically determined responses will characterize the temporal sequence of phenophases in an ecosystem (Anderson 1974). It is this part of phenology which will be the major focus of this study.

The physiognomic similarity of the world's mediterranean-climate ecosystems (Walter 1968; Specht 1980), i.e., the Mediterranean Basin, southern California, central Chile, southeastern Australia, and South Africa, make it probable that similar trends of phenophase sequences exist in these regions. This predicted similarity in the general phenology of the mediterranean-climate ecosystems does not necessarily imply uniform growth phenologies among the species of a given community, e.g., that all the species develop leaves or extend their shoots at the same time. One may predict that effective coexistence in a vegetation rich in species and growth-forms requires some staggering of resource use in time and space. Thus, a comparison of the phenology in these areas should show the strongest similarities in phenology to exist between comparable growth forms. For example, drought-deciduous shrubs might be expected to shed their foliage at similar times in all the mediterranean areas, triggered by the seasonal onset of soil-moisture stress. An interesting possible exception to synchronous phenology may be found in reproductive physiology (Chapter 6, this volume). Thus flowering may be less dependent on basic resources such as water and mineral nutrients. If this is indeed the case, the phylogeny of species of similar growth form and co-evolution with pollinators may be more decisive factors regulating the time of flowering rather than specific conditions of the physical environment.

It seems feasible to develop an effective model to predict the timing of phenophases of major species in the mediterranean areas. Such a development would be a key for integration of other biological processes in an ecosystem, including micro-organisms and consumers, and would be a useful aid for refining land use and management, for example, in determining appropriate times for prescribed burns. The purpose of this contribution is to compare the phenology of mediterranean-climate areas, to identify the main factors which control the rates of plant development, and to analyse the possibilities for phenophase prediction.

17.2 Phenophase development and moisture availability

Abundant evidence has been accumulated that points towards the decisive influence of rainfall patterns on the timing of the vegetative growth phases in the mediterranean vegetation. This is obvious for annual grasses and forbs and valid also for that fraction of the perennial vegetation which does not have access to ground water sources. Since precipitation is markedly seasonal, it can be expected that plant species have adapted in order to cope with this condition. However, relating the sequence of phenophases to the amounts of precipitation can be misleading. Storms with high rainfall intensity produce substantial runoff, especially on slopes, while more evenly spaced moderate rainfalls can charge the soil profile with more moisture than short and intensive downpours. It is necessary to keep in mind that soil moisture and not rain is the source of water for plant growth. Unfortunately, only in a few cases have phenological observations been related to soil moisture (Kummerow et al. 1981). Our discussion is, therefore, based on the assumption that after the initiation of winter rains, the soil profile has stored enough water to allow development of the characteristic phenophases in a given ecosystem.

In the following paragraphs evidence for moisture-controlled phenological events will be presented.

17.2.1 SEASONAL AND SHOOT BUD FORMATION

Since bud formation is dependent on leaf growth, it might be concluded that seasonal leaf development implies seasonal bud growth. This reasoning may be faulty however because at the termination of the spring growth-flush bud growth may continue. It seems relevant therefore, for phenological purposes, to observe whether fully grown buds are present or absent at the end of the shoot growth period. However, the distinction between a growing and a resting bud is not an easy task under field conditions. A second problem rests in the difficulty of differentiating between vegetative and flower buds. Only observers with sufficient field experience may be able to make the distinction.

The timing of vegetative growth was shown with a series of common eastern Mediterranean chamaephytic shrubs (*Sarcopoterium spinosum* (= *Poterium spinosum*), *Cistus salvifolius*, *Coridothymus capitatus* (= *Thymus capitatus*), *Teucrium polium*, *Artemisia monosperma*; Orshan 1964). The detailed analysis of the phenophases produced a complex picture because not only different leaf, but also long- and short-shoot generations follow each other in the course of the growing season. A fraction of the buds formed during the growing season in late winter and spring remains dormant until the following winter when rainfall triggers the new growth flush. Another group of buds, determined by their position on the stem, does not enter this period of summer inactivity but develops in only a few weeks into short shoots characterized by limited elongation growth (Orshan 1964).

A case of complex adaptation of vegetative growth to summer drought is that of Trevoa trinervis, a spiny summer-deciduous matorral shrub in central Chile (Hoffmann 1972). In midwinter (July), provided that winter rains had begun to fall as usual in May, buds burst and leaves develop. Short lateral thorn shoots grow from axillary buds at the base of stem-spines. Between the base of the thorn and its respective leaf, a group of buds develops (Fig. 1). These buds do not sprout during the summer but remain dormant until the next winter, protected by the modified foliar bases. The following year's rainy season triggers development of these buds. Thus, in T. trinervis morphological features have evolved which ensure that preformed leaves expand quickly when water becomes available. The critical role that water plays in this system becomes evident in a simple field experiment: a Trevoa shrub in the matorral was watered at regular intervals during the summer and did not drop its foliage (Hoffmann 1972).

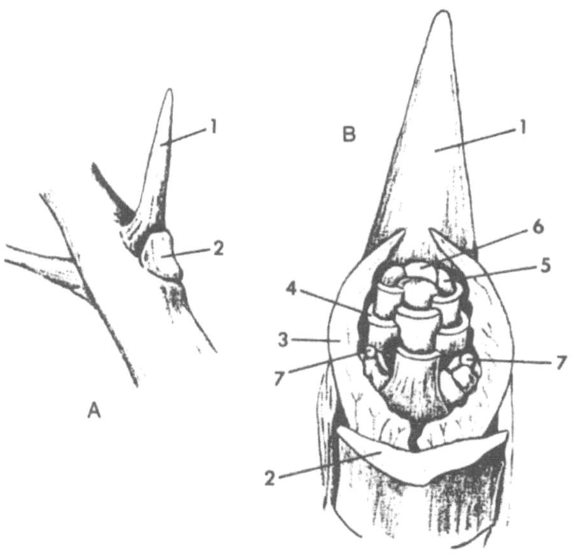

Fig. 1. Winter bud of Trevoa trinervis. A. The bud is positioned below the thorn (1) and protected by the scale leaf (2). B. Bud with the scale leaf removed. 1: thorn, 2: severed base of the scale leaf, 3: oldest pair of bud leaves, 4: basal portion of young leaves cut off in this preparation, 5: leaf primordium, 6: apical meristem, 7: lateral reserve meristem. (After Hoffmann 1972).

A somewhat less extreme case of seasonal bud behaviour was found in Colliguaya odorifera, another matorral shrub from equatorward slopes in central Chile. This shrub is a sclerophyllous evergreen with a shallow root system (Miller and Ng 1977, Hoffmann and Kummerow 1978); its growing season extends from early winter to the beginning of summer. When the dry season begins, partly predetermined vegetative buds develop. These buds remain dormant during the drought period and resume activity soon after the first winter rain (Hoffmann and Hoffmann 1976).

An example from southern California confirms bud dormancy from late spring until late winter also for the chaparral. In Ceanothus greggii var. perplexans, a sclerophyllous evergreen shrub, vegetative and flower buds are differentiated during the spring growing season and remain dormant until the following winter (Fishbeck and Kummerow 1976). These few examples show that bud formation is followed in many species by apparent inactivity for 6-9 months. This adapatation to a summer-dry climate can also be expected to exist in heath and fynbos vegetation, although documentation is absent.

Stem elongation and leaf expansion are the most obvious expressions of growth. A detailed analysis of the growth patterns of individual shoots was made with six Chilean matorral shrubs (Montenegro et al. 1979). Three species, Satureja gilliesii, Trevoa trinervis and Colliguaya odorifera initiate growth in midwinter and continue to grow into the summer until late December, thus maintaining growth activity for six months. In contrast, Quillaja saponaria, Cryptocarya alba, and Lithraea caustica begin to grow later in October and November, towards the end of spring. Satureja gilliesii, T. trinervis, and C. odorifera are shallow rooters while Q. saponaria and L. caustica have root systems of several metres deep. Cryptocarya alba is not deep-rooted but grows in mesic habitats at the bottom of ravines (Hoffmann and Kummerow 1978; Giliberto and Estay 1978). The two species S. gilliesii and T. trinervis are both summer-deciduous and C. odorifera is semi-deciduous (Hoffmann and Hoffmann 1976). Thus, patterns in the timing of growth initiation, shoot elongation, and leaf expansion are correlated with seasonal availability of soil moisture, whereas growth rates may be influenced most by temperature. This behaviour may be controlled by soil moisture and temperature but recent investigations by Margaris (1975) indicate that a photoperiodic control of the timing of phenophases cannot be excluded.

Similar information was obtained for four evergreen sclerophyllous chaparral shrubs. Ceanothus greggii and Adenostoma fasciculatum are relatively slow-growing shrubs which initiate their shoot growth in early spring. The stems of C. greggii show a rather linear length increase from April to May; the same pattern is found to a lesser degree in A. fasciculatum. This growth behaviour contrasts with that of Arctostaphylos glauca and Rhus ovata. Both shrubs initiate growth late in spring; however, stem elongation and leaf growth takes only 48 and 31 days respectively versus 68 and 69 days needed by the first-mentioned species pair (Kummerow unpublished; Fig. 2).

Shoot growth in three of four shrub species in the fynbos of the Cape Province showed the same general pattern; Protea nitida (= P. arborea), P. repens, and P. neriifolia displayed the characteristic spring growth pattern while Leucospermum tottum flushed in summer and autumn (Kruger 1981). In general it was stated for the fynbos vegetation that shrubs grow actively any time from late winter into the summer with some species extending the growth period well into the autumn (Kruger 1979; 1981). In the Dark Island heath of south Australia the main growth period of the dominant sclerophyllous shrubs Banksia ornata and B. marginata fell in midsummer from December to March (Specht and Rayson 1957a). The new shoots that were formed during the summer gradually matured during the cooler autumn and winter months.

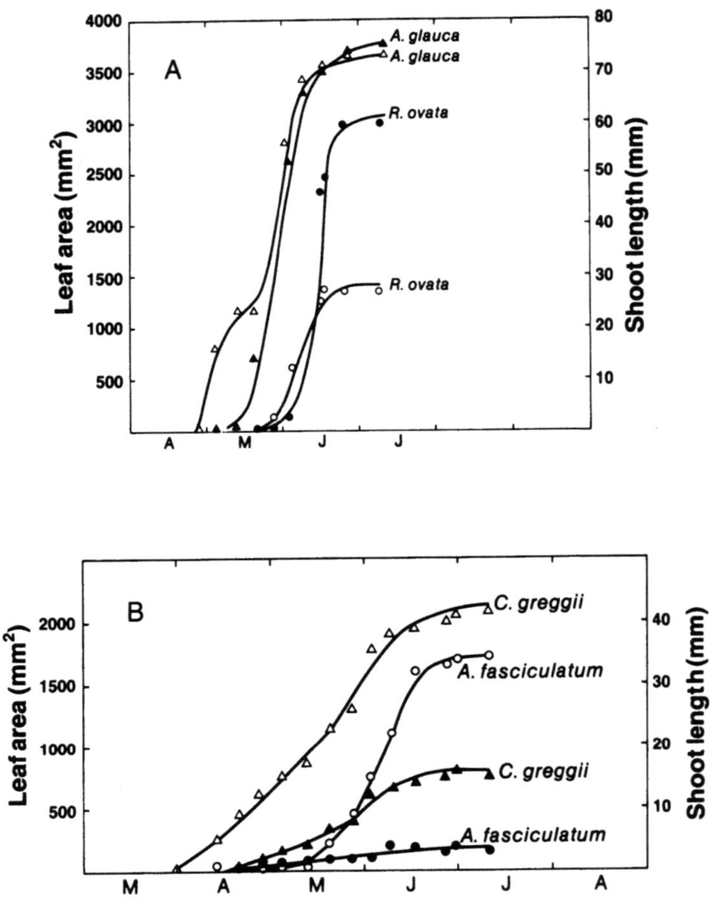

Fig.2. Leaf area expansion (closed symbols) and shoot elongation (open symbols) of a representative shoot of Arctostaphylos glauca and Rhus ovata (A), and Adenostoma fasciculatum and Ceanothus greggii (B). Each data point indicates the accumulated leaf area or shoot length of the sample shoot at the recording data. Note the difference in slope between the two species pairs. (After Kummerow, unpublished)

The examples of time-courses in shoot and leaf growth used here demonstrate that in spite of similar climatic conditions there are important differences in the shoot phenology of dominant shrub species (see also Levyns 1964; Cody and Mooney 1978). In Chile, the summer-deciduous shrubs initiated shoot growth in early winter and continued to grow until midsummer, presumably until exhausting the available soil moisture. Evergreen sclerophylls with deeper root systems flushed in spring. Lithraea caustica, the last species to flush, prolongs its growth well into the late summer. In California seasonality is more pronounced but shoot phenology is broadly similar (Mooney et al. 1977). In South Africa, although only limited information has become available, it is obvious that there are shrub species with the main growth period in summer; and data from southern Australia suggest that summer growth is a common feature (Specht and Rayson 1957a; Specht et al. 1981).

The late-growing species Lithraea caustica, Banksia ornata and B. marginata appear to have deep root systems with access to a longer-lasting water supply, thus allowing for a later growth period in comparison with the shallow rooters (Specht and Rayson 1957b; Hoffmann and Kummerow 1978). The unanswered question remains why these shrubs grow during a period of extreme water stress.

17.2.2 CANOPY GROWTH

Studies of seasonal canopy growth, where seasonal shoot and leaf production values of entire plant communities are measured, obscure the differences between individual species but have the advantage of providing more general information for comparisons between ecosystems.

Canopy growth (g m^{-2} $month^{-1}$) of eight major chaparral and seven Chilean matorral species over a period of four years demonstrated that community growth in California is less protracted than in Chile but that growth culminated in spring (Mooney and Kummerow 1977). Measurements of four plant associations including leaf and stem growth in Israel show a smooth seasonal curve with a maximum in the spring months (Orshan and Diskin 1968). Two consecutive years were observed and in spite of differences in the absolute values for both years the curves run parallel with the maximum in April-May and the minimum in September-October. Very similar results were reported for a Greek phrygana ecosystem (Margaris 1976).

17.2.3 CAMBIAL ACTIVITY

Primary growth, as shown above, has a pronounced seasonal character. However, secondary growth, i.e., the thickening growth of stems, does not follow this pattern strictly, as shown by comparative studies reported by Avila et al. (1975). The two evergreen sclerophylls Rhus ovata and Lithraea caustica from California and Chile respectively show a spring maximum in cambial activity followed by continuous but less intense activity throughout the year; this behaviour contrasts with that of Quercus dumosa (chaparral) and Quillaja saponaria (matorral), both of which have a three-month period of cambial growth, from spring to early summer (Fig. 3). During the rest of the year the vascular cambium remains dormant. This pattern is also found among four dominant shrubs of the Greek phrygana formations (Fig. 3; Margaris and Papadogianni 1978).

The four shrubs from California and Chile all have deep root systems. Xylem water potential measurements have shown that these shrubs do not pass through extremely severe drought stress (Poole and Miller 1957). Nevertheless they exhibit two different patterns of cambial activity. Therefore, it might be concluded that genetically fixed traits are, at least in part, responsible for the particular rhythms of vascular cambium growth.

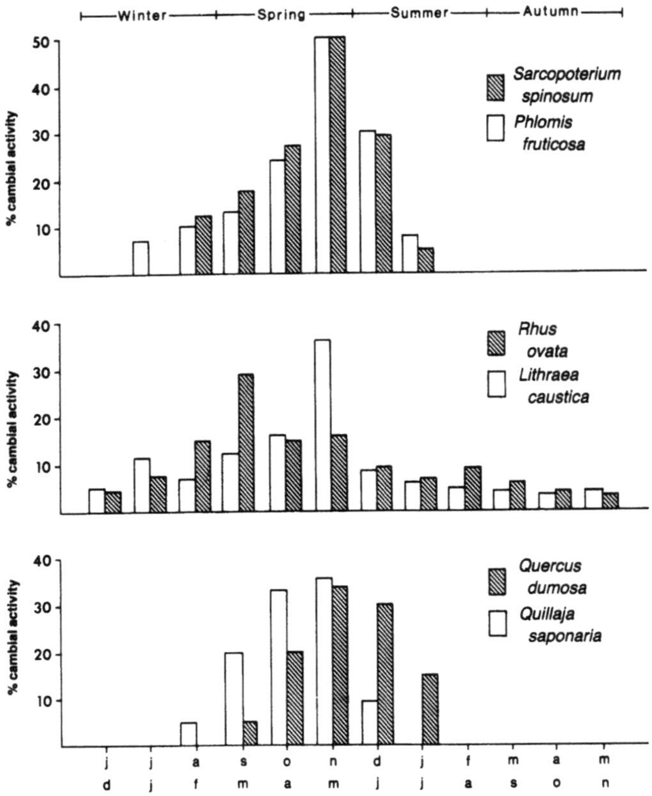

Fig.3. Seasonal fluctuations of the cambial activity of two Greek phrygana shrubs (Sarcopoterium spinosum, Phlomis fruticosa), two California chaparral shrubs (Rhus ovata, Quercus dumosa), and two Chilean matorral shrubs (Lithraea caustica, Quillaja saponaria). Cambial activity values expressed as per cent of total yearly xylem increase. (Phrygana species after Margaris and Papadogianni 1978; chaparral and matorral species after Avila et al. 1975)

17.2.4 SEASONAL FINE-ROOT GROWTH

The relative wealth of information regarding the seasonality of canopy growth contrasts with the scarcity of data on root growth. But available evidence indicates that the root systems of shrubs in summer-dry areas extend horizontally beyond the crown radius and vertically downward according to prevailing soil conditions (Evenari 1938; Hellmers et al. 1955; Specht and Rayson 1957b; Hanes 1965; Kummerow et al. 1977; Hoffmann and Kummerow 1978). These root systems, observed and harvested by means of mechanical or hydraulic excavations, are comparable to the framework of branches above ground and seasonal fluctuations in their cambial activity can be expected. However, the main question is whether the absorbing fine roots (rootlets < 1 mm in diameter), most of which are lost with conventional excavation techniques, show seasonal fluctuations analogous to the seasonal fluctuations of the foliage area.

Observations of seasonal changes of root biomass in the Greek phrygana revealed a biomass maximum of 2700 g m^{-2} in April-May and a minimum of 1400 g m^{-2} in September-October (Margaris 1976). These observations

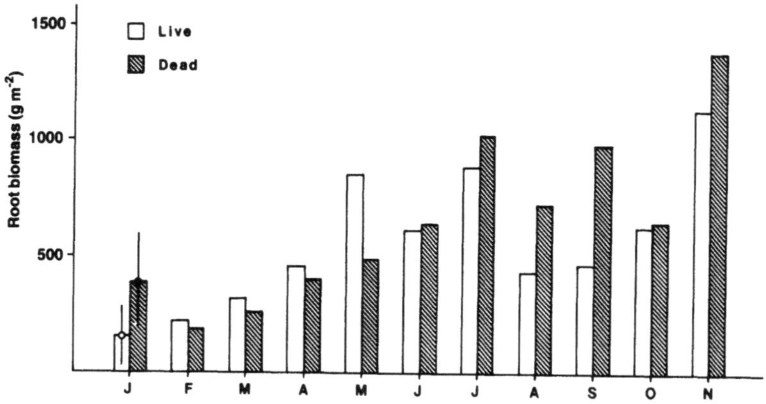

Fig.4. Seasonal course of live and dead fine-root dry masses in the California chaparral. Each column represents the mean value of 38 cores of 125 mℓ each. The cores were extracted from the 0.1-0.2 m layer, the zone of the highest fine root densities. Total monthly biomass was estimated from samples of cores 0.5 m deep. Samples collected from a dense, 23-year-old chaparral stand with Adenostoma fasciculatum, Arctostaphylos glauca, Ceanothus greggii, and Rhus ovata. Vertical bars indicate 95% confidence intervals. (After Kummerow et al. 1978)

were made on an extremely shallow soil of only 0.3 m depth. The data do not discriminate between roots of different size classes but it seems likely that the difference between minimum and maximum root biomass is due mainly to seasonal changes in the ephemeral rootlet population.

Studies in south-eastern Australia (Victoria) have shown that root elongation in Leptospermum heathland was initiated in spring (September) and continued until the end of May, although growth rates declined as soil moisture decreased (Haines 1967). In the southern California chaparral the total fine-root biomass of Adenostoma fasciculatum increased from 260 g m^{-2} in early spring to 1580 g m^{-2} in late summer (Fig. 4). Three other shrub species from the same area, Rhus ovata, Arctostaphylos glauca, and Ceanothus greggii, showed similar changes. Ratios of late summer rootlet biomass to late winter rootlet biomass ranged from 2.4 to 6.0 (Kummerow et al. 1978). Present evidence indicates that marked seasonal fluctuations in the fine-root biomass occurs in mediterranean regions of the world but more information is required to substantiate these findings.

17.2.5 SEASONAL FLOWERING

In general, flower development does not require the energy and moisture needed for the vegetative growth processes. Flower production in eight chaparral shrubs and seven matorral shrubs amounted to an average of 5.3% and 2.3% respectively of the total annual terminal shoot production (Mooney et al. 1977). Consequently, a higher degree of independence from moisture conditions can be expected. However, flowering, more than any other growth process, seems to be controlled by temperature.

Flowering and fruiting of the main shrub species of chaparral and matorral have been recorded from 1971-1974 (Hoffmann et al. 1977). In the chaparral flowering reached the peak between May and July and during autumn and winter little flowering was observed. The flowering phenophase was followed by fruiting with the expected time lag. In Chile, a different situation was observed. The flowering period extended over the entire year with only a slightly discernible maximum in November-December. However, the corresponding curve for fruiting showed a protracted maximum in late summer and autumn in each year of observation (Fig. 5).

The monthly distribution of flowering of the Palestinian flora shows a peak in April but 3-5% of the vegetation flowers as late as November and December (Fig. 6; Zohary 1962). These data were obtained from the observations on about 1600 species, including more than 1000 annuals. However, most trees and shrubs flowered in March and April and matched observations for shrubs in other areas of Palestine (Zohary 1962).

A similar periodicity of flowering was found in the fynbos of the Cape Province (Rycroft 1953 and van der Merwe 1966 in Kruger 1981). Records of nearly 750 species in the mountain fynbos of the Jonkershoek valley showed maximum flowering in early spring and minimum in autumn. However, some flowering could be observed at all times (Fig. 6). Likewise, a spring flower peak was observed for Australian heath (Specht and Rayson 1957a). This peak, based on 103 vascular plants, was well pronounced and, as in the other areas, some flowering was observed during the entire year.

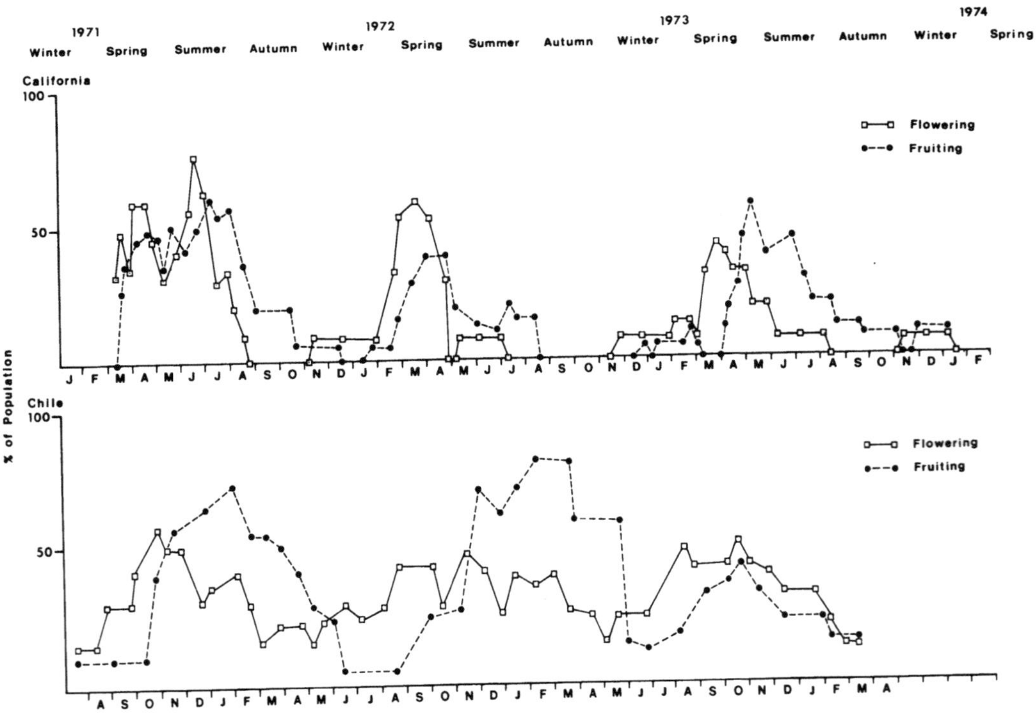

Fig.5. Seasonality and year-to-year variation in flowering and fruiting in representative chaparral and matorral stands. Values are numbers of active individuals as per cent of the numbers of shrubs under observation in vegetation of the two research sites, Echo Valley in California and Fundo Santa Laura in Chile, in the respective phenophase. (After Hoffmann et al. 1977)

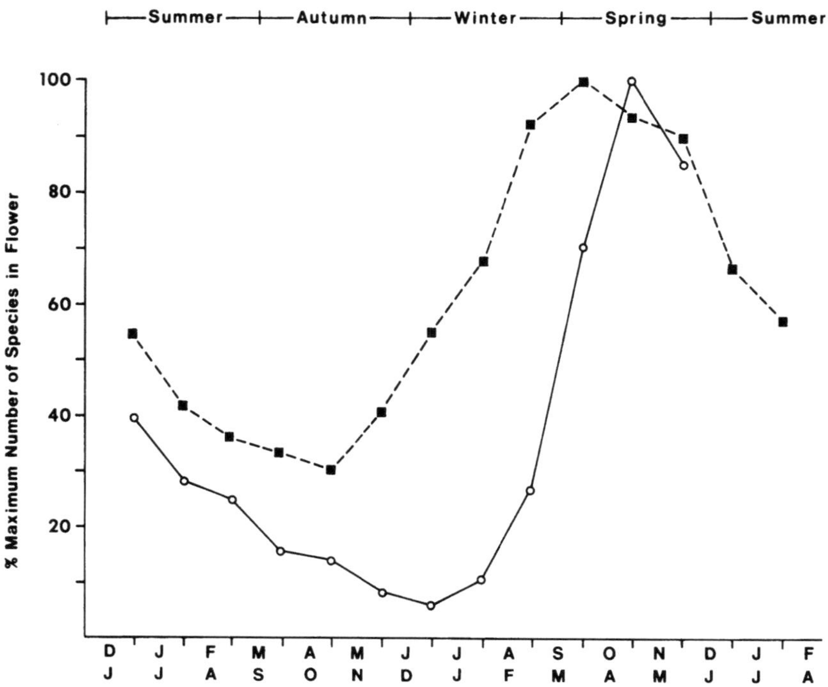

Fig.6. Seasonal course of flowering in the fynbos flora of Swartboschkloof (black squares) in the Jonkershoek Valley, Cape Province (448 species observed) and in Palestine (circles) based on 1600 observed species. (Fynbos after van der Merwe in Kruger 1981; Palestine after Zohary 1962)

Thus, considering entire ecosystems, similarity in the general timing of flowering and fruiting of mediterranean-climate ecosystems is unquestionable, especially when contrasted with a cold-temperate or tropical climate region. Nevertheless, there is diversity in flowering patterns of any specific ecosystem. Limited information in this regard is available but a few examples may illustrate the point.

The dominant shrub species of large chaparral areas is Adenostoma fasciculatum. This shrub flowers relatively late in midsummer. Since A. fasciculatum develops inflorescences from terminal buds of the current year's shoots it follows that flowering cannot take place before shoot elongation and leaf formation is completed (Jow et al. 1980). High spring temperatures and restricted soil moisture will have an accelerating effect. The genera Ceanothus and Arctostaphylos, represented by several species in the same ecosystem and growing in mixed stands with A. fasciculatum, are characterized by early blooming. Arctostaphylos glauca may open its first flower buds in January. In both genera, flower buds are differentiated as soon as the current year's leaves are formed. By May fully developed buds enter a dormancy period until the initiation of the next year's growing season (Fishbeck and Kummerow 1976). Thus, timing of flowering within the season is preconditioned by morphology and shoot growth phenology. Similar, and even more complex relations between morphology and flowering time were shown for Israeli shrub species (Orshan 1964).

That information on the flowering patterns of individual species is important can be shown by the following examples. The tall shrubs *Protea repens* and *P. neriifolia* that dominate in some fynbos communities flower "off season" in autumn and early winter, thus producing abundant flowers when most of the species-rich vegetation has already passed through the flowering phase. This has a significant impact on the fynbos consumer biology (Mostert et al. 1980), an effect not evident from simply counting numbers of species flowering at a given time since these two shrubs form an insignificant fraction of the total flora (Kruger 1981). Similar observations have been made in the southern Australian heath. Although peak flowering occurs in spring (between September and November), the more prominent shrubs like *Banksia ornata* and *B. marginata* do not flower until late summer, autumn, or even mid-winter (Specht and Rayson 1957a). This situation is far less obvious in California where the dominant *Adenostoma fasciculatum* flowers late in June, i.e., in midsummer, towards the end of the flowering season of the chaparral. In Israel, *Coridothamnus capitatus* begins to flower at the end of May and the flowering season lasts until October. Some mediterranean half-shrubs, especially Chenopodiaceae, flower in autumn under conditions of maximum water stress (G Orshan, personal communication). This phenomenon has not been reported for central Chile.

17.2.6 SEASONAL LITTER PRODUCTION

Lossaint and Rapp (1978) monitored monthly litter production in a *Quercus ilex* forest in southern France, an area with high annual precipitation (770 mm) including some summer rain, from 1965 to 1972. A significant maximum rate of fall was found in May and June. At this time about 100 g m^{-2} $month^{-1}$ of mostly old leaves dropped. A minimum of 12-18 g m^{-2} $month^{-1}$ was recorded from January to March. Comparable absolute values have been found for the strictly summer-dry chaparral and matorral (Mooney and Kummerow 1977). July and August in California and January and February in Chile are the months of highest litter fall.

The effect of soil moisture on litter fall was pointed out by Specht and Rayson (1957a) for the south Australian heathlands. Leaves remain attached to their twigs for four years in several shrub species after a sequence of wet years although they may drop after only two years under drier conditions. Regardless of the soil moisture, the bulk of the leaves fall during the summer or somewhat earlier after a spring drought period.

Although other phenophases vary to a lesser or greater extent among individual species of the fynbos, leaf shedding seems highly synchronous. Of 12 species monitored, 11 were found to drop the bulk of their leaves and twigs between December and March, the period of the lowest soil moisture potential (Kruger 1981).

These few data indicate that litter fall is probably the best synchronized phenophase in plants of all mediterranean-climate areas. Moisture dependence seems obvious but the role of photoperiodism cannot be excluded without experimental work (see Margaris 1975).

17.2.7 THE PHENOLOGY OF ANNUAL GRASSES AND FORBS

The moisture dependence of the annual vegetation is obvious. Rainfall in combination with specific temperatures triggers germination. The extension of the life cycle is determined by the length of time that sufficient soil moisture is available. In matorral and chaparral it was found that variations in annual rainfall were reflected in variations of herb biomass production (Sierra Rafols and Hays 1977). Seasonal fluctuations of nutrient availability, probably related to variations in annual rainfall, might also influence the time course of the patterns of annual plant development.

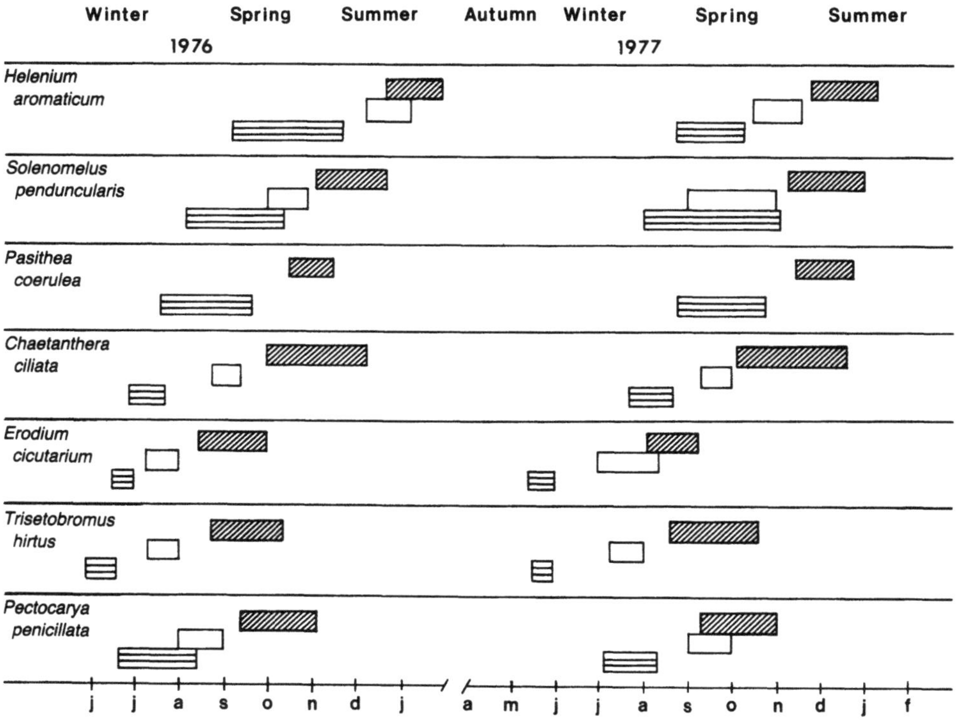

Fig.7. Phenology of seven herbaceous matorral species in 1976 and 1977. Bars with horizontal lines: leaves full grown, open bars: first leaves senescent, bars with oblique hatching: flowering. (After Montenegro et al. 1978)

Precise observations in the Chilean matorral with seven floristically important herbaceous plants may be indicative of patterns in mediterranean-climate areas (Montenegro et al. 1978). Growth began only a few weeks after the first rainfall in May and lasted about six months in 1976 and 1977, years with 374 and 676 mm rainfall, respectively. In spite of this significant difference in precipitation each of the species observed had a specific time sequence of development (Fig. 7). Thus, the Chilean data show that the annual vegetation exhibits characteristic phenophases for each different species, although the time for completion of the life cycle of any one is only six months. Since the matorral climate is more maritime than that of other mediterranean areas it can be expected that annual plant development would take even less time in the latter. More studies are needed to confirm that these patterns are general.

Table 1. Flower initiation and onset of stem elongation for a given percentage of the population as a function of warming hours (Σ hr$\geq$15°C) in four chaparral species. All regressions are significant at P = 0.005 except the flower initiation regression for *Ceanothus greggii* (0.20<P>0.10). (From Kummerow et al. 1981)

SPECIES	Minimum days to initiation	Range of initiation period (days)	Regression equations[a]	n	r^2
FLOWER INITIATION					
Adenostoma fasciculatum	193	50	$Y_1 = -144.6 + (0.13)X_1$	5	0.940
Arctostaphylos glauca	0	122	$Y_1 = -25.6 + (0.19)X_1$	6	0.896
Ceanothus greggii	77	87	$Y_1 = -66.9 + (0.20)X_1$	4	0.748
Rhus ovata	134	54	$Y_1 = -69.4 + (0.11)X_1$	6	0.985
ONSET OF STEM ELONGATION					
Adenostoma fasciculatum	127	61	$Y_1 = -111.5 + (0.20)X_1$	7	0.931
Arctostaphylos glauca	76	118	$Y_1 = -41.1 + (0.12)X_1$	8	0.935
Ceanothus greggii	96	77	$Y_1 = -50.0 + (0.17)X_1$	6	0.943
Rhus ovata	163	65	$Y_1 = -81.8 + (0.10)X_1$	6	0.961

a Regression for rate of accumulation of warming hours: $\log X_1 = 2.16923 + (0.0045535)X_2$; n = 16; $r^2 = 0.985$, where
Y_1 = per cent of population in which flower initiation or stem elongation has been initiated.
X_1 = Σ hr during which temperature $\geq$ 15°C from time soil moisture $\geq$ 0.1 g mℓ^{-1} in the preceding fall.
X_2 = days from date in preceding fall when soil moisture $\geq$ 0.1 g mℓ^{-1}.

In order to predict when a given per cent of the population has initiated flowering or stem elongation (in terms of accumulated warming hours) use the following form of the equation:
X_2 (days to flowering or elongation) = (log[(% population flowering or elongating − Y_1 − intercept)b_1] − 2.16923)/b_2, where Y_1 − intercept is the y intercept of the regression for a given species; b_1 is the slope of the regression for a given species; and b_2 is the slope of the regression for the accumulation of warming hours.

17.2.8 PREDICTION OF PHENOPHASES

Possibilities and methods of seasonality and phenology modelling have been summarized recently by Lieth (1974). Regular patterns of phenophase sequences have become apparent after several years of phenological observations in the chaparral (Kummerow et al. 1981). Therefore, trial predictions of the timing of specific phenophases for key species were made from statistical models (Kummerow et al. 1981). These trials focused on four evergreen sclerophyllous shrubs where biweekly observations were available for four to eight years, and flower initiation and the onset of stem elongation were used as response variables. Day length, the sum of precipitation and of warming hours (with two thresholds, 0°C and 15°C) for the season up to the time of the phenological event, and soil moisture content were the independent variables tested. Flowering of early spring bloomers can extend over several weeks and the earliest observed flowers may not be adequate to describe the blooming of a species. Therefore, per cent flowering and shoot elongation of a stand of the respective species were the data used in these initial models. Regression analysis showed rainfall to have no, and warming hours above 0°C to have little predictive value. However, a decisive variable was the day in the preceding autumn when soil moisture reached the threshold value of 0.1 g $m\ell^{-1}$. The sums of hours with temperatures 15°C or warmer, following this date, were used in regressions on per cent of shrub stand that had flowered, and that had initiated stem elongation and the equations so obtained were used to predict the initiation of these phenophases (Table 1). Good fits were obtained but predictions of flowering and flushing in the spring of 1980 for all four species erred considerably; for example, flowering dates were recorded three to four weeks earlier than predicted. Further study is clearly necessary. The result could be due to the unusually mild winter experienced in 1980.

17.3 Discussion

A brief survey of the mediterranean-climate areas of the world shows the overall similarity in the patterns of plant development in these regions, although a closer scrutiny demonstrates some important differences (Cody and Mooney 1978). The less pronounced seasonality of the climate in central Chile is reflected by a less obvious seasonal plant development, at least with respect to the shrubs. The summer growth of larger shrubs in southern Australia, occurring to a lesser degree also in the fynbos, has been documented only to a minor degree for Chile and the Mediterranean area. The reason for this difference is not clear. It has been suggested (Specht and Rayson 1957a) that the phenomenon could be connected with the evolutionary history of the vegetation. Heath species might have evolved during the Pliocene with a warmer and moister climate which favoured a summer growth period. Only those heath elements that developed mechanisms for water conservation to reduce the moisture loss during a summer growth period without rainfall, would have survived.

A striking difference in the physical environment is the remarkably low nutrient level of the Australian heath soils (Specht and Rayson 1957a). The impact of low nutrient levels on seasonal plant development is discussed by Specht et al. (Chapter 7, this volume).

With the exception of the summer-growing species, the patterns of shoot growth in the mediterranean-climate areas are strikingly similar. Spring growth flushes are most frequent and for each region a range of flushing periods from late winter to late spring can be established. Limited information on the timing of flower bud formation is available, but the broad range of flowering periods for the various species in each ecosystem is suggestive of the possibility that blooming is one of the more conservative phenological characteristics. Late flowering in summer as well as early blooming in spring are morphologically conditioned and one is tempted to speculate that the late-blooming species in which flower buds differentiate from the current year's meristems, are those which have evolved in a more tropical climate, e.g. Adenostoma fasciculatum in the rose family. The patterns of cambial activity may also reflect the evolutionary history of the species. The two Anacardiaceae, Rhus ovata in the chaparral and Lithraea caustica in the matorral, have continuous vascular cambial activity (Avila et al. 1975). Both the species are late blooming (Hoffmann et al. 1977) and may have evolved under moister and warmer conditions (Raven and Axelrod 1974). Acacia raddiana in the eastern mediterranean region and A. caven in central Chile show the same year-round cambial growth (Fahn et al. 1968). These Anacardiaceae and Acacia species have been found to be deep-rooted (Halevi and Orshan 1973; Avila et al. 1975). From this point of view it would be attractive to study the relation between the seasonality of growth, flowering and cambial activity for the summer-growing Proteaceae in heath and fynbos.

The seasonality of fine-root growth appears to be controlled by soil moisture. The influence of this growth process on the phenology of the ecosystems is difficult to assess because even the deep-rooting species like Quercus dumosa in the chaparral or Lithraea caustica in the matorral have a high fine-root density in the upper 300 mm of the soil and are, therefore, exposed to the extreme soil moisture fluctuation (Kummerow and Mangan 1981; G Avila, unpublished). However, it is clearly important to note that the seasonal fine-root turnover in summer-dry areas reaches values well above 1000 g m^{-2} yr^{-1}, a below-ground litter production which is more significant than the litter produced by the above-ground plant fraction (Kummerow et al. 1978; Margaris 1976). Information on nutrient cycling resulting from this process is not yet available.

All mediterranean-type vegetation seems to behave similarly with regard to leaf litter-fall, in which seasonality is clear-cut and definitely moisture-dependent. In annual grasses and forbs where life cycles develop according to the available soil moisture, specific differences are nevertheless found among species of the matorral, despite the short growth cycle (Montenegro et al. 1978). Work in Australia has shown that changes in nutrient availability may accelerate or delay the growing season of such plants (Specht 1963).

The obvious soil moisture dependence of the phenology in the mediterranean climate regions has stimulated attempts to predict the timing of specific phenophases. The results of these efforts are not yet satisfactory; however, they demonstrate clearly that once a soil moisture threshold has been reached, accumulating warming hours seem to trigger the initiation of leaf growth and stem elongation.

17.4 Conclusions

A general similarity of seasonal growth and development of the vegetation of mediterranean-climate areas has been documented. However, specific differences exist which make each of these areas unique. The main controlling factor of plant growth is the soil moisture. Above a soil moisture threshold value, temperature controls timing and rates of specific phenophases. The broad mosaic of phenological behaviour of the species-rich plant communities can be understood as an expression of a resource allocation system which must be expected in ecosystems that are characterized by a wide variety of micro-habitats. The role of seasonal fine-root growth for nutrient turnover in the ecosystem is a topic which requires more studies before quantitative assessments can be attempted. Phenophase prediction in a mediterranean-climate area has been attempted but is not yet fully satisfactory.

Acknowledgements

I thank William Jow and James Lopez for the provisions of unpublished data. Dr David Rayle helped with the manuscript preparation and Shari Sitko prepared the illustrations. The work was supported by NSF-Grant DEB 77-13944.

References

ALJARO ME, AVILA G, HOFFMANN A, KUMMEROW J (1972) The annual rhythm of cambial activity in two woody species of the Chilean "matorral." American Journal of Botany 59: 879-885

ANDERSON RE (1974) Seasonality in terrestrial primary producers. In: LIETH H (ed) Phenology and seasonality modelling. Springer-Verlag, Berlin, pp 103-111

AVILA G, ALJARO ME, ARAYA S, MONTENEGRO G, KUMMEROW J (1975) The seasonal cambium activity of Chilean and Californian shrubs. American Journal of Botany 62: 473-478

CODY ML, MOONEY HA (1978) Convergence versus nonconvergence in mediterranean-climate ecosystems. Annual Review of Ecology and Systematics 9: 265-321

DI CASTRI F, GOODALL DW, SPECHT RL (ed) (1981) Ecosystems of the world, vol 11. Mediterranean-type shrublands. Elsevier, Amsterdam

EVENARI M (1938) Root conditions of certain plants in the wilderness of Judaea. Journal of the Linnean Society (Bot.) 51: 383-388

FAHN A, WAISEL Y, BENJAMINI L (1968) Cambial activity in Acacia raddiana Savi. Annals of Botany 32: 677-686

FISHBECK K, KUMMEROW J (1976) Flower bud differentiation in Ceanothus greggii. Abstract. Bulletin of the Ecological Society of America 57: 14

GILIBERTO J, ESTAY H (1978) Seasonal water stress in some Chilean matorral shrubs. Botanical Gazette 139: 236-240

HAINES WM (1967) Experimental approaches to the study of seasonal root activity and phosphorus nutrition of Australian heath plants. MSc Thesis. University of Melbourne, Victoria

HALEVI G, ORSHAN G (1973) Ecological studies in Acacia species in the Negev and Sinai. Israel Journal of Botany 22: 120-138

HANES TL (1965) Ecological studies on two closely related chaparral shrubs in southern California. Ecological Monographs 35: 213-235

HELLMERS H, HORTON JS, JUHREN G, O'KEEFE J (1955) Root systems of some plants in southern California. Ecology 36: 667-678

HOFFMANN A (1972) Morphology and histology of Trevoa trinervis (Rhamnaceae), a drought deciduous shrub from the Chilean matorral. Flora 161: 527-538

HOFFMANN AJ, HOFFMANN AE (1976) Growth pattern and seasonal behaviour of buds of Colliguaya odorifera, a shrub from the Chilean mediterranean vegetation. Canadian Journal of Botany 54: 1767-1774

HOFFMANN A, KUMMEROW J (1978) Root studies in the Chilean matorral. Oecologia 32: 57-69

HOFFMANN A, MOONEY HA, KUMMEROW J (1977) Qualitative phenology. In: THROWER NJW, BRADBURY DE (eds) Chile-California mediterranean scrub atlas. Dowden, Hutchinson and Ross, Stroudsburg, Pennsylvania, pp 102-120

JOW WM, BULLOCK SH, KUMMEROW J (1980) Leaf turnover rates of Adenostoma fasciculatum (Rosaceae). American Journal of Botany 67: 256-261

KRUGER FJ (1979) South African heathlands. In: SPECHT RL (ed) Ecosystems of the world, vol 9A. Heathlands and related shrublands. Descriptive studies. Elsevier, Amsterdam, pp 19-80

KRUGER FJ (1981) Seasonal growth and flowering rhythms: South African heathlands. In: SPECHT RL (ed). Ecosystems of the world, vol 9B. Heathlands and related shrublands. Analytical studies. Elsevier, Amsterdam, pp 1-4

KUMMEROW J, MANGAN R (1981) Root systems in Quercus dumosa Nutt. dominated chaparral in southern California. Acta Oecologica, Oecologia Plantarum 2: 177-188

KUMMEROW J, KRAUSE D, JOW W (1977) Root systems of chaparral shrubs. Oecologia 29: 163-177

KUMMEROW J, KRAUSE D, JOW W (1978) Seasonal changes of fine root density in the southern California chaparral. Oecologia 37: 201-212

KUMMEROW J, MONTENEGRO G, KRAUSE D (1981) Biomass, phenology and growth. In: MILLER PC (ed). Resource use by chaparral and matorral. Springer-Verlag, Berlin, pp 69-96

LEVYNS (1964) Migration and origin of the Cape flora. Transactions of the Royal Society of South Africa 37: 85-107

LIETH H (ed) (1974) Phenology and seasonality modeling. Springer-Verlag, Berlin, 444 pp

LOSSAINT P, RAPP M (1978) La fôret méditerranéenne de chêne verts. In: LAMOTTE M and BOURLIERE F (eds) Problemes d'écologie. Structure et fonctionnement des écosystèmes terrestres. Masson, Paris, pp 129-185

MARGARIS NS (1975) Effect of photoperiod on seasonal dimorphism of some mediterranean plants. Berichte der Schweizerischen Botanischen Gesellschaft 85: 96-102

MARGARIS NS (1976) Structure and dynamics in a phryganic (East Mediterranean) ecosystem. Journal of Biogeography 3: 249-259 MARGARIS NS, PAPADOGIANNI P (1978) Cambial activity in some plants dominating phryganic formations in Greece. Phyton (Buenos Aires) 36: 1-5

MILLER PC, NG E (1977) Root:shoot biomass ratios in shrubs in southern California and central Chile. Madrono 24: 215-223

MONTENEGRO G, RIVERA O, BAS F (1978) Herbaceous vegetation in the Chilean matorral. Dynamics of growth and evaluation of allelopathic effects on some dominant shrubs. Oecologia 36: 237-244

MONTENEGRO G, ALJARO ME, KUMMEROW J (1979) Growth dynamics of Chilean matorral shrubs. Botanical Gazette 140: 114-119

MOONEY HA, KUMMEROW J (1977) Quantitative phenology. In: THROWER NJW, BRADBURY DE (eds) Chile-California mediterranean scrub atlas. Dowden, Hutchinson and Ross, Stroudsburg, Pennsylvania, pp 121-125

MOONEY HA, KUMMEROW J, JOHNSON AW, PARSONS DJ, KEELEY S, HOFFMANN A, HAYS RI, GILIBERTO J, CHU C (1977) The producers - their resources and and California. Dowden, Hutchinson and Ross, Stroudsburg, Pennsylvania, pp 85-143

MOSTERT DP, SIEGFRIED WR, LOUW GN (1980) Protea nectar and satellite fauna in relation to the food requirements and pollinating role of the Cape sugarbird. South African Journal of Science 76: 409-412
ORSHAN G (1964) Seasonal dimorphism of desert and Mediterranean chamaephytes and their significance as a factor in their water economy. In: RUTTER AJ, WHITEHEAD FH (eds) The water relations of plants. Blackwell, Oxford, pp 206-222
ORSHAN G, DISKIN S (1968) Seasonal changes in productivity under desert conditions. In: Proceedings of the UNESCO symposium on functioning of terrestrial ecosystems at the primary production level. Copenhagen, pp 191-201
POOLE DK, MILLER PC (1975) Water relations of selected species of chaparral and coastal sage communities. Ecology 56: 1118-1128
RAVEN PH, AXELROD DI (1974) Angiosperm biogeography and past continental movements. Annals of the Missouri Botanical Garden 61: 539-673
SIERRA RAFOLS E, HAYS RI (1977) Herb productivity. In: THROWER NJW, BRADBURY DE (eds) Chile-California mediterranean scrub atlas. Dowden, Hutchinson and Ross, Pennsylvania, p 126
SPECHT RL (1963) Dark Island heath (Ninety-Mile Plain, South Australia). VII. The effect of fertilizers on composition and growth, 1950-1960. Australian Journal of Botany 11: 67-94
SPECHT RL, RAYSON P (1957a) Dark Island heath (Ninety-Mile Plain, South Australia). I. Definition of the ecosystem. Australian Journal of Botany 5: 52-85
SPECHT RL, RAYSON P (1957b) Dark Island heath (Ninety-Mile Plain, South Australia). III. The root systems. Australian Journal of Botany 5: 103-114
SPECHT RL, ROGERS RW, HOPKINS AJM (1981) Seasonal growth and flowering rhythms: Australian heathlands. In: SPECHT RL (ed) Ecosystems of the world, vol 9B. Heathlands and related shrublands. Analytic studies. Elsevier, Amsterdam pp 5-13
WALTER H (1968) Die Vegetation der Erde, Bd. II. Die gemässigten und arktischen Zonen. Fischer Verlag, Jena, 1001 pp
ZOHARY M (1962) Plant life in Palestine. Ronald Press, New York, 262 pp

Section V. Community Patterns and Diversity

Mediterranean-type shrublands, as a rule, are richer in plant species than are communities of adjoining plant formations; certain samples indicate that some of the richest plant communities in the world are found in these zones. As for the communities, so for the regional floras. The Californian Floristic Province, the South West Botanical Province of Western Australia, and especially the Cape Floristic Kingdom have notably rich floras. Our appreciation of community dynamics and the maintenance of species diversity depends on understanding the continental patterns of diversity, their evolution, and the manner in which species are assembled and coexist in communities.

This section covers a spectrum of topics, including aspects of theoretical models to explain diversity patterns, empirical studies of diversity patterns, and reviews of our knowledge of spatial patterning in plant communities as well as the role of fire in structuring both plant and animal communities. Elsewhere in this volume, MOONEY and KUMMEROW have indicated how competition for resources among plants may result in evolution of diverse phenologies and hence temporal partitioning of resources.

Here, TILMAN applies his theory of resource competition to plant communities, indicating how diversity may be predictably related to different availabilities and patterning of limiting resources as well as to disturbance. These relationships show maximum diversity at intermediate to low levels of nutrient availability, and at intermediate but rather low frequencies of disturbance. These model patterns match those found in classical field studies, and TILMAN suggests specific experiments to test his theory. BOND reports on patterns of species diversity in plant communities of the fynbos in the constant-rainfall zone of the southern Cape Province. Although more data are required generally to allow valid comparisons, it is clear from this work that these shrublands are not uniformly richer than the richest communities elsewhere, and that the patterns of richness within regions are at least as interesting as the comparisons between regions. He presents a relationship between plant biomass and alpha diversity that is as predicted from current community diversity theory. His analyses of the different components of plant species diversity and of the way that these relate to the length and diversity of environmental gradients and to biogeographic patterns suggest appropriate explanations for observed differences in diversity between continents.

The paper by CODY on bird community diversity extends his earlier exposition of a theory for continental diversity patterns (Cody 1975). He proposes specific relationships between the different components of species diversity and their evident controlling factors such as habitat gradients

---

This introduction was prepared by F J Kruger

and the area, contiguity and patchiness of habitats, to explain inter- and intra-continental variations in components of diversity. Data from different continents are used to assess these hypotheses. However, the focus is on avifaunas of the Cape Province, where bird diversity patterns and community composition and structure vary among the geographic transects sampled, despite similar habitat gradients. In this area geography and the distribution of habitats provide ready testing grounds for continental diversity theory.

SIEGFRIED and CROWE have investigated communities in contrasting geographic situations (coastal versus montane, nutrient-rich versus nutrient-poor) for parallels and contrasts in patterns of fynbos bird and plant species diversity and composition. These detailed analyses provide results encompassed by continental theory but point to the need for further studies on succession, local habitat and productivity gradients, and seasonal migratory patterns to explain diversity patterns in finer detail.

Community theory has been developed in a large degree by animal ecologists because the resources consumed by different animal species can be readily distinguished and different patterns of consumption measured. This theory is not easily applicable to plant communities, since plant species compete for essentially the same resources of light, nutrients, and space. WESTMAN addresses the question of the measure to which spatial structuring of vegetation reflects resource partitioning and in doing so provides a comprehensive review of the literature available on spatial patterning at different scales in mediterranean-type ecosystems. KRUGER reviews the information available on the ecological effects of fire on plant communities and the possible role of resource partitioning over successional time in maintaining plant species diversity. Differences between Californian chaparral on the one hand and Australian heathland and South African fynbos on the other with regard to fire regime and pyric succession are apparent and evidently correlated in part with variations in plant community diversity and structure. However, although there is a large literature on vegetation response to fire in these regions, as well as on plant life-cycles, critical studies that precisely document diversity patterns and processes involved in pyric succession on all of the different continents are still too few to allow robust generalizations. Intercontinental differences and similarities in plant life-history strategies are apparently related to different patterns of resource availability during pyric succession in certain cases. For fynbos and Australian heathlands, shrub fitness in the low-nutrient fire environment appears to be linked with unusual features of plant carbon balance and life-history constraints that include, for example, short lifespans (see also MOONEY, this volume); demographic studies in this context could be particularly helpful in understanding the basis of vegetation dynamics.

FOX synthesizes information on the responses of small-mammal communities to fire in Australian heathland. He finds consistent patterns of succession after fire, and emphasizes the importance of fire-adapted mammals which exploit the recently burned habitat. Such succession, together with fire-determined shifts in habitat preference in some species, evidently allow richer small-mammal communities than would otherwise be the case. FOX explains patterns of diversity in small-mammal communities of the eastern Australian heathlands in these terms. NEWSOME and CATLING also focus on the effects of fire on Australian mammal communities but address the problem of explaining the variations in patterns of pyric succession reported in the literature. They use different models that simulate population dynamics and species responses to variations in resources such as food and shelter and to factors such as migration and predation to analyse and explain the various responses to fire that they have recorded among different mammals.

References

CODY ML (1975) Towards a theory of continental diversities: bird distributions over mediterranean habitat gradients. In: CODY ML, DIAMOND JM (eds) Ecology and evolution of communities. Harvard University Press, Cambridge, Massachusetts, pp 214-257

18. Some Thoughts on Resource Competition and Diversity in Plant Communities

D. TILMAN

18.1 Introduction

The most diverse terrestrial plant communities in the world, i.e. those that are richest in species, occur on some of the most nutrient-poor soils. The fynbos of the Cape Province of South Africa and the heaths of Australia are two such cases, and both occur on soils that are very poor in nutrients (Goldblatt 1978; Taylor 1978; Kruger 1979; Specht 1979; George et al. 1979; Naveh and Whittaker 1980; Chapter 19, this volume). Although possibly not as diverse as the fynbos (Kruger and Taylor 1979), tropical forests are also relatively rich in species and also occur on relatively poor soils (Holdridge et al. 1971; Huston 1979). This association of high plant diversity with nutrient-poor habitats is not limited to mediterranean and tropical climates, nor to terrestrial plant communities. In a broad variety of plant communities - in temperate and tropical regions, in grasslands, forests, lakes, rivers, estuaries, and the oceans - plant species diversity within a geographical area is locally higher in habitats relatively poor in nutrients, and lower in sites very poor, or richer in nutrients (Huston 1979). For instance, the nutrient-poor, unproductive waters of the Sargasso Sea support a very diverse phytoplankton assemblage, whereas the more productive and nutrient-rich temperate oceans support a much less diverse plant community (Fischer 1960; Guillard and Kilham 1977). Similarly, nutrient-rich marine upwelling regions tend to be much less diverse than nearby regions that do not experience nutrient upwelling (Blasco 1971; Dugdale 1972; Smayda 1975). The addition of nutrients (cultural eutrophication) to coastal marine areas, estuaries, rivers and lakes has consistently led to decreased species diversity (Patrick 1963, 1967; Williams 1964; Schleske and Stoermer 1971).

These observations fit a larger pattern of the dependence of plant species diversity on nutrient availability. Both Al-Mufti et al. (1977) and Grime (1973) report that plant communities have maximum diversity on soils of intermediate fertility but are less diverse on soils that are richer or poorer in nutrients. Reports by Beadle (1966) for Australian xeromorphic and rain forest genera, by Ashton (1977) for Malaysian rain forest, and by Holdridge et al. (1971) for Costa Rican forests also show such a relationship between species richness and local soil nutrient availability. Thus, a graph of species richness against nutrient availability often gives a humped curve.

This brief summary suggests that the availability of those resources which limit plant growth influences plant community diversity in a consistent manner. In this paper I briefly present a simple, graphical theory of plant competition for limiting resources (Tilman 1980), extend it to include multispecies competition in spatially heterogeneous environments, and then show that such a theory predicts that the most diverse plant communities should occur in habitats relatively poor in resources. This analysis is done in two parts. First plant communities in which two nutrient resources are potentially limiting are considered, then plant communities in which open sites and a nutrient are the limiting factors are discussed. For both cases the analyses show that a graph of species richness against the supply rate of a limiting resource gives a humped curve. The supply rate of mineral resource is a measure of the richness of the soil. The supply rate of open sites is related to the rate of disturbance in a community. The graphical techniques used in this paper show the similarity of the effect of any potentially limiting resource - be it a nutrient or open sites - on the structure of plant communities.

18.2 A graphical theory of resource competition

The graphical theory of resource competition used in this paper is an extension of the work of MacArthur (1972), Maguire (1973), Leon and Tumpson (1975) Petersen (1975), Taylor and Williams (1975), and Tilman (1980). Because the justification and development of this approach are fully presented in Tilman (1980) I will concentrate on extensions of the approach to the influence of resources on the structure of plant communities.

The resource requirements of a plant can be represented by a series of growth isoclines (Fig. 1A). Such isoclines show all the combinations of two limiting resources for which a plant has a particular reproductive rate. The three isoclines of Fig. 1A show the requirements of a hypothetical plant species for R_1 and R_2 at three different reproductive rates (0.1, 0.2, and 0.3 of its maximum reproductive rate, r). The rectangular intersection shown for these growth isoclines means that the two resources are non-interactive essential resources (Tilman 1980). Curved isoclines will be considered later. A species with a rectangular isocline is limited by either one or the other resource. The broken line through the corner of an isocline defines the availabilities of R_1 and R_2 for which a species is equally limited by R_1 and R_2. Below the broken line of Fig. 1A, the species is limited by R_2; changes in R_1 will have no effect on its growth rate. Above the broken line the species's growth-rate is limited by R_1.

In a particular habitat a species will experience a certain mortality rate. For that species to maintain a stable population in that habitat at equilibrium (i.e. for dN/dt = 0) it must have sufficient resources for its reproductive rate to balance its mortality rate. The growth isocline along which the reproductive rate of a species exactly balances its mortality rate represents such sufficient availabilities of the limiting resources. This is an isocline of zero net growth and is called the <u>z</u>ero <u>n</u>et <u>g</u>rowth <u>i</u>socline, or ZNGI. If the species of Fig. 1A were to experience a mortality rate of 0.2r, its ZNGI would be that shown in

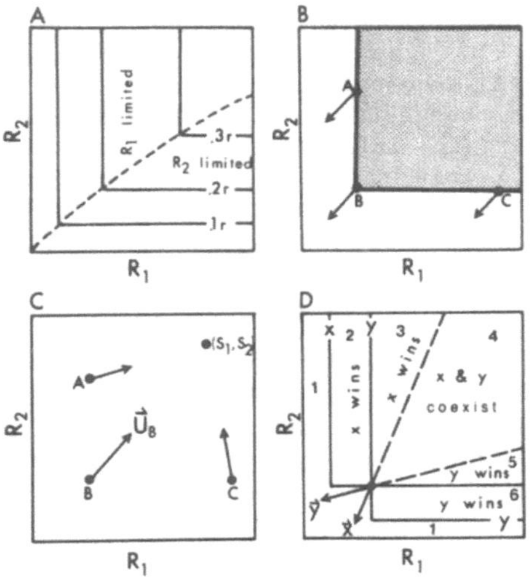

Fig. 1. Illustration of a graphical theory of resource competition. A. Growth isoclines. B. Illustration of the zero net growth isocline. C. A resource supply point and resource supply vectors. D. Two-species equilibria for different hypothetical habitats. See text for explanation.

Fig. 1B. If a habitat had resource availabilities that fell in the shaded region outside the ZNGI the population density of this species would increase, whereas the population density would decrease for habitats with resource availabilities in the unshaded region inside the growth isocline. Population density would remain unchanged (i.e. equilibrium would occur) for any habitats that had resource availabilities that fell on the growth isocline.

We need to know the rates of supply and consumption of nutrients, as well as the ZNGIs for the competing species, to determine the outcome of interspecific competition. Nutrient consumption is represented as a vector (Fig. 1B). The slope of the vector is the ratio of the consumption rate of R_2 to the consumption rate of R_1. The length of the vector is proportional to the instantaneous rate of consumption of these resources by the total population of this species. Here, I will assume that a species consumes resources in the proportion at which the resources are required for growth, i.e. in the proportion at which they equally limit the growth of a species (Droop 1974). This means that the consumption vector for a species will be parallel to the broken line separating the region in which a species is limited by R_1 or R_2.

I assume also that resources are supplied in an equable mode (Stewart and Levin 1973; Tilman 1980), which can be considered to be a linear approximation to many other processes of resource supply. For equable resources the resource supply process can be represented on graphs such as those of Fig. 1 as follows. A single point (S_1,S_2), termed the resource supply point, is that where S_1 is the maximal amount of resource 1 and S_2 the maximal amount of R_2 that can exist in that habitat. The rate of supply of the resources is assumed to be proportional to the deviation of a habitat away from its resource supply point. Thus, if a habitat had environmental availabilities of resources which were R_1 and R_2, and the resource supply point was at S_1 and S_2, the instantaneous rate of supply of each resource would be

$$dR_1/dt = a(S_1 - R_1) \text{ and}$$
$$dR_2/dt = a(S_2 - R_2)$$

where a is a constant giving the rate of supply for a given deviation. The instantaneous rates of supply defined by these two equations can be represented as a vector (Fig. 1C). The resource supply vector always points from the actual environmental availability of the two resources, the point (R_1,R_2), towards the supply point, (S_1,S_2), with a magnitude proportional to the distance between the points. U_B is the resource supply vector for habitat B of Fig. 1C.

This information can be used to predict the equilibrium outcome of resource competition between several species. Fig. 1D shows the ZNGI for two species. These growth isoclines cross at one point, shown with a dot. This point is a two-species equilibrium point. It shows the environmental availabilities of these resources at which these two species may coexist. Lines through this equilibrium point with the slopes of the consumption vectors of these two species and the ZNGI for these two species define six regions in the R_1-R_2 resource quarter-plane (Fig. 1D). Habitats which have resource supply points that fall in region 1 have insufficient resources for the survival of either species. Habitats which have resource supply points that fall in region 2 have sufficient resources for the survival of species X but insufficient resources for the survival of species Y. Species Y will tend towards extinction in such habitats, whereas species X will approach a stable equilibrium population density. In region 3 there are sufficient resources for the equilibrium survival of either species if growing in the absence of the other species. When both species co-occur in a habitat which has a resource supply point that falls in region 3, species X will consume resources down to its ZNGI, and there will be insufficient resources for the equilibrium survival of species Y. Thus, resource supply points in region 3 lead to dominance by species X and competitive exclusion of species Y. Species X and Y will stably coexist, at equilibrium, in habitats which have a resource supply point that falls in region 4. This occurs because the consumption vectors of these species combined with the resource supply process will focus resource availabilities into the two-species equilibrium point. The two-species equilibrium point shown in Fig. 1D is locally stable because each species consumes proportionately more of the resource which limits its growth at equilibrium (Leon and Tumpson 1975). For all cases considered here the assumption that species consume resources in the proportion required for balanced growth means that all two-species equilibrium points are locally stable (Tilman 1980). Habitats which have resource supply points in region 5 will be dominated by species Y, with species X approaching extinction. In region 6 there are sufficient resources for the survival of species Y but insufficient resources for the survival of species X; species Y will dominate and X go extinct. Thus, the ZNGI and consumption vectors of two species define the types of habitats in which neither species can survive, in which one will be competitively dominant, and in which both will coexist at equilibrium. This is illustrated by experimental results. Fig. 2 illustrates a case of competition between two plants for limiting soil nitrogen and phosphorus, as derived from the results of the long-term fertilization experiments performed at Rothamsted, England (Brenchley and Warington 1958; Thurston 1969; Lawes and Gilbert 1980; Tilman, in preparation). Of the 20 Rothamsted plots, the three fertilized with all nutrients except nitrogen were dominated by _Lathyrus pratensis_, a legume. Plots receiving all nutrients except phosphorus were dominated by _Agrostis tenuis_, a grass. This suggests that these two species have ZNGIs as illustrated in Fig. 2.

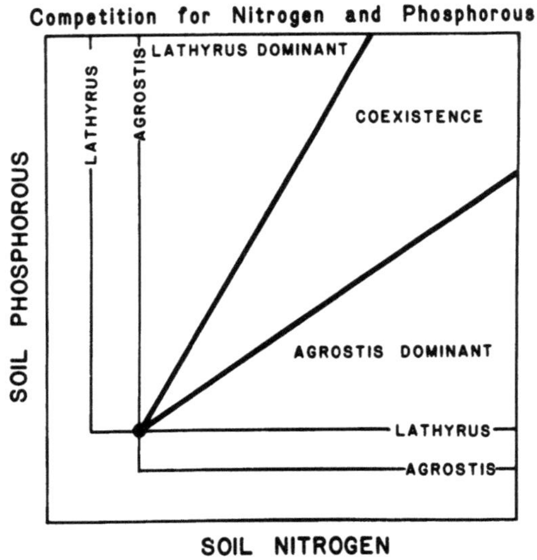

Fig. 2. Competition between Agrostis tenuis and Lathyrus pratensis for nitrogen and phosphorus. Lathyrus, a legume, has a lower nitrogen requirement but a higher phosphorus requirement than Agrostis, a grass. This leads to ZNGI with the relative positions shown. See text for explanation.

18.2.1 MULTISPECIES COMMUNITIES

This graphical approach to the theory of resource competition is easily extended to numerous species competing for two limiting resources. Let us first consider situations in which several plant species compete for two non-interactive, essential resources, such as the mineral nutrients required by plants (Droop 1974; Rhee 1978).

Fig. 3 shows a case of six species competing for two essential resources. The growth isoclines for these six species were chosen on the assumption that a species which is a slightly superior competitor for one of the resources would be a slightly inferior competitor for the other resource. The consumption vectors shown assume that each species consumes the two resources in the proportion required for balanced growth. These ZNGIs and consumption vectors define numerous regions in which either one species will be dominant or in which various pairs of species will coexist. If a habitat had a resource supply point where the number 3 is located in Fig. 3, species C and D would eventually coexist in that habitat, and the other species would be competitively displaced. A habitat with a resource supply point where the number 4 is located would be dominated by species E and F and a habitat with a resource supply point where the number 5 is located would be dominated by species A and B, with the other species becoming extinct. Thus, for spatially homogeneous habitats, no more species can coexist than there are limiting resources. Although a particular habitat will have an average rate of resource supply, it seems unlikely that all the individual plants living in a habitat will experience the same rates of resource supply; rather, some microsite-to-microsite variation in the resource supply processes seems likely. Such variation could allow the stable coexistence of more species than the use of the average resource supply point would predict. For instance, let the circle around the number 1 represent the range of microsite-to-microsite variation in the resource supply point for habitat 1. The circle could be considered to be the 0.99 probability contour for a probability distribution of resource supply points in habitat 1. As drawn in Fig. 3, this spatial heterogeneity in resource supply includes microsites in which all six species can potentially coexist. Because the probability distribution of resource supply points for these microsites is weighted around the mean, species C and D will be the most abundant species in habitat 1. However, the spatial variation in resource supply allows species A, B, D, and F to stably coexist with species C and D. A diagram such as this is easily used to predict which species should dominate in any given habitat, and which other species should be able to coexist with them.

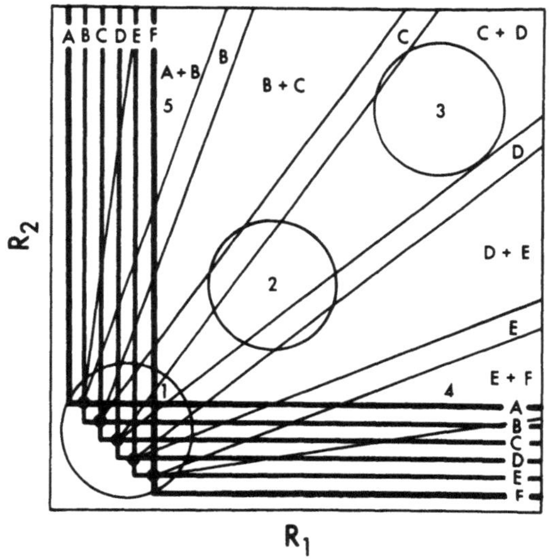

Fig. 3. Competition among six species for two essential resources is shown, with the ZNGI for the species labelled A to F. Regions in which various pairs of species coexist are labelled, with A+B meaning that only species A and B can coexist in habitats which have resource supply points in this region, etc. The circles represent spatial variation in resource supply (see text).

This diagram also illustrates how resource richness may influence the species diversity of a habitat. The circle around the number 2 of Fig. 3 shows a habitat with the same amount of spatial heterogeneity as habitat 1, but which is richer in resources. Inspection shows that only four species are able to coexist in such a habitat. Similarly, in the equally heterogeneous but even richer habitat 3 only two species are able to coexist. This implies that more species are able to coexist in resource-poor habitats than in resource-rich habitats, at a given level of heterogeneity. Obviously, no species can exist in habitats very poor in resources (i.e., like those of region 1 of Fig. 1D). Thus, a graph of species richness against resource richness should give a humped curve with maximal species richness in habitats which have just enough resources to support stable populations in the absence of competition. For such habitats the average resource supply points would be near the corners of the ZNGIs of the species.

If the probability distribution about the average resource supply point is explicitly defined we may determine analytically the expected density of each of the species for a particular degree of spatial heterogeneity and a given average resource supply point and thus explore more thoroughly the response of community diversity to various levels of resource richness. This has been done using numerical simulation techniques. For these simulations, a community of 200 competing plant species was used. These were assumed to have ZNGIs and consumption vectors as illustrated in Fig. 4. The microsite-to-microsite variations about the resource supply points were assumed to follow bivariate normal distributions with equal variances and no correlation.

In order to solve for the expected community structure for any given spatially heterogeneous habitat, 4000 microsites were chosen from the bivariate normal probability distribution. For each of these microsites, the computer programme would compute the expected equilibrium density of the one or two species that could coexist in that microsite. The densities of the species were summed over all the microsites to give the total expected equilibrium density of all the species in the spatially heterogeneous habitat. Such calculations were made for a whole range of habitats, each differing in their average resource supply point, to determine how resource richness would affect species richness. To summarize the results of these calculations graphically, I define resource richness to be $(S_1 + S_2)/2$.

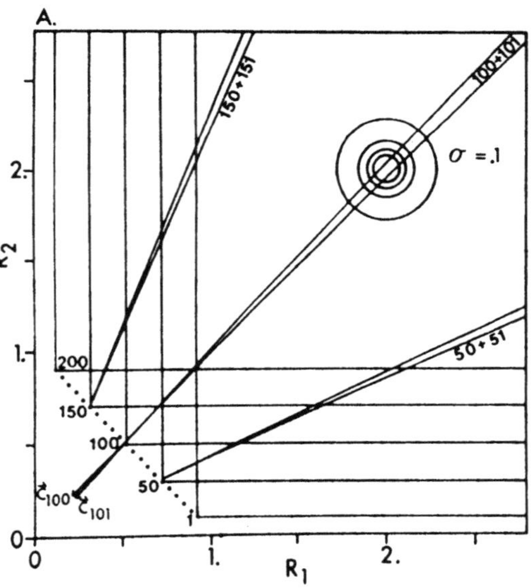

Fig. 4. This figure illustrates the constants used in the computer simulation of 200 species competing for two limiting resources. The ZNGIs of species 1, 50, 100 and 200 are shown, as are the regions in which species 50 and 51, species 100 and 101, and species 150 and 151 can coexist. The set of circles schematically represents the bivariate probability cloud used in the simulations.

Fig. 5 shows the simulated dependence of species richness and species diversity on resource richness for six different levels of spatial heterogeneity. Here spatial heterogeneity is defined as the standard deviation of the resource supply point probability distribution cloud, and is represented by σ. The graph of species richness against resource richness is a humped curve for each level of spatial heterogeneity. Maximum species richness occurs in relatively nutrient-poor habitats. These same data were used to calculate H', the Shannon-Wiener index of species diversity. This index also shows a humped relationship between species diversity and resource richness. Using either measure of diversity the ascending portion of the humped curve is much steeper than the descending portion. Thus, an explicit numerical solution of the equations defining the equilibrium density of species competing for two resources in a spatially heterogeneous environment predicts that maximal species richness should occur in nutrient-poor habitats, just as was suggested by the graphical model of Fig. 3.

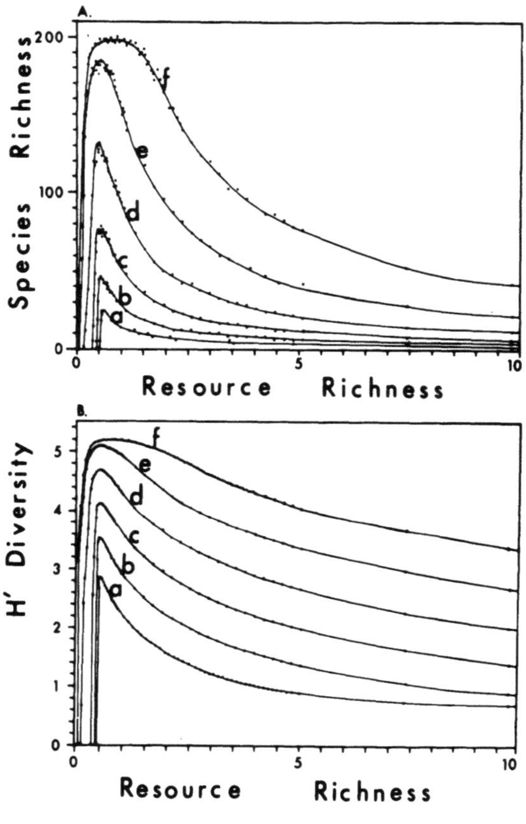

Fig. 5. Results of the simulations of 200 species competing for two resources in a spatially heterogeneous environment. Curves a to f show results for six different levels of spatial heterogeneity (σ of 0.025, 0.05, 0.1, 0.2, 0.4 and 0.8, respectively). The upper figure (A) shows the predicted dependence of equilibrium species richness on resource richness. The lower graph shows the predicted dependence of the Shannon-Wiener species diversity index, H', on resource richness.

18.2.2 SPACE AS A RESOURCE

Thus far I have been discussing nutrient resources required by plants. Numerous experimental and correlational studies in a variety of plant communities indicate that open sites are an essential resource for plants. Most plants require some sort of opening in order to become established. Open site limitation most likely reflects light limitation of the seedling stage of a plant, not limitation by "space" itself. The population growth rate of a particular species of space-limited plant will increase as the amount of open space is increased, at least until inhibitory effects come into play. As the population grows it reduces the amount of open space available for potential colonization by other species. Thus, competition for open sites is qualitatively similar to competition for other nutrient resources. In both cases increased resource availability leads to increased growth rate and the resource is consumed so that it is no longer available to competitors. Space may be a more complex resource than mineral elements. The size of an open site and the time of the season that a site becomes open may influence which species can successfully colonize it. Moreover, the very processes that create an open site must do so by killing the plant species that had occupied the site. Thus, unlike nutrient resources, the rate of supply of space depends directly on mortality rate. This complicates an analysis of the effects of various rates of disturbance on the structure of plant communities.

Let us consider plants living in a community with two potentially limiting resources - open space and a nutrient - and explore qualitatively how the rate of disturbance may affect the structure of plant communities. These two resources are essential with respect to each other because a plant must have both the nutrient and open space in order to survive. This can be illustrated by growth isoclines (Fig. 6A). The isocline is shown with a curve instead of a right angle because some interaction between nutrients and space in determining growth rate seems likely. For instance, a plant population might be able to attain a certain growth rate in a habitat with less available open space if the open sites were richer in nutrients. If that habitat happened to have nutrient and open space availabilities that fell in the shaded region of Fig. 6A, the population density of this plant would increase, etc., as for Fig. 1B. The outcome of multispecies competition in a particular habitat can be predicted as for nutrient resources, as shown in Fig. 6B. Again, there are regions in which various pairs of species may coexist when competing for these two resources.

Because mortality rate changes as the rate of supply of new open sites changes, it is not possible to superimpose probability clouds of resource supply points on this diagram as was done in Fig. 3 for two nutrient resources. Rather, each resource supply point has associated with it a particular set of growth isoclines. The higher the rate of supply of open space - i.e., the higher the disturbance rate - the further from the origin the ZNGIs will be. This is because higher disturbance rates lead to higher mortality rates. Species require more resources to maintain stable populations when they experience higher mortality rates. Thus, it is necessary to use a different set of ZNGIs for each microsite to calculate how the species richness of nutrient- and space-limited plant communities may be influenced by changes in disturbance rate.

This raises a second problem. Do all species experience the same mortality rates, as disturbance rate is increased, or are some species more susceptible to disturbance-caused mortality than others? Does disturbance that most affects the species which is the best competitor for open space lead to a different sort of community than disturbance that most affects the species which is the worst competitor for open space? To explore these questions I modelled several cases of competition for limiting space and nutrient resources among 20 species. For all cases, the same parameters were used to describe the requirements of the 20 species for these resources. These parameters are not illustrated in a figure but are qualitatively similar to those shown in Fig. 6B for four species. Species 1 was the superior competitor for limiting open space and the inferior competitor for limiting nutrient similar to species A of Fig. 6B. Species 20 was the superior competitor for limiting nutrient, similar to species D of Fig. 6B. The other 18 species formed a series intermediate between species 1 and 20 in their requirements for space and nutrient, much as do species B and C of Fig. 6B. As before, a spatially heterogeneous habitat was used, with different rates of supply of the nutrient and different disturbance rates for the microsites in the habitat.

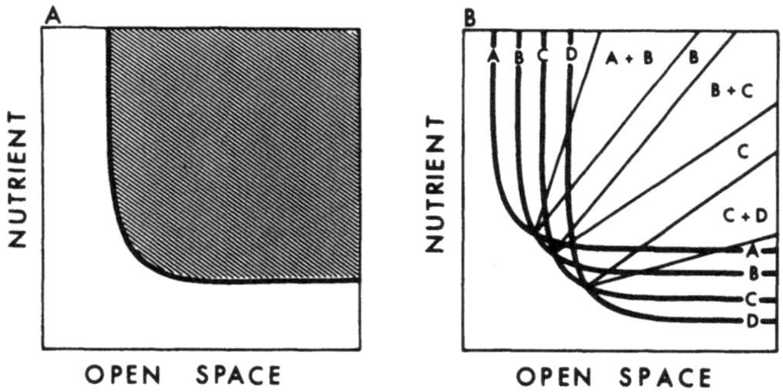

Fig. 6. The graph theory where space is a limiting resource. A. Illustration of the principle of zero net growth isocline. B. Illustration of multispecies interactions. See text for explanation.

Three different cases were explored. In case 1 all species were equally affected by the disturbance rate. Thus, an increase in disturbance rate for case 1 would shift the ZNGIs of all species equally away from the origin. In case 2 the species which was the best competitor for open space (species 1) was assumed to suffer the least disturbance-caused mortality, with the species that was the best competitor for the limiting nutrient (species 20) suffering most of the mortality caused by increases in the disturbance rate. The other species suffered intermediate increases in mortality rate with increases in disturbance rate. In case 3 the best competitor for open space (species 1) suffered most of the mortality from disturbance, whereas the species which was the worst competitor for open space (species 20) suffered the least mortality from disturbance. In all cases a background mortality rate of 0.33 of the maximum reproductive rate was assumed and all species were assumed to have the same maximum reproductive rate. Half of the increase in mortality caused by increased disturbance was assumed to substitute for the background mortality rate. Fig. 7 shows how changes in the disturbance rate for cases 1, 2 and 3 influence the equilibrium species richness of these communities. The disturbance rate shown in

Fig. 7 is the average of disturbance rates experienced by all the species for a given spatially heterogeneous habitat. For all three cases the species richness-disturbance rate curve is humped, with maximum diversity in habitats with moderately low disturbance rates. Species diversity tends to fall more slowly with disturbance rate when the species with the highest requirement for open space suffers the most mortality (case 2), and tends to fall more rapidly when the species with the highest requirement for open space suffers the least mortality (case 3), than where all species experience the same mortality rate (case 1). These deviations seem intuitively to make good sense. In all three cases, changes in the supply rate of the resource space affect species richness just as do changes in the supply rate of nutrient resources (compare Figs 5 and 7).

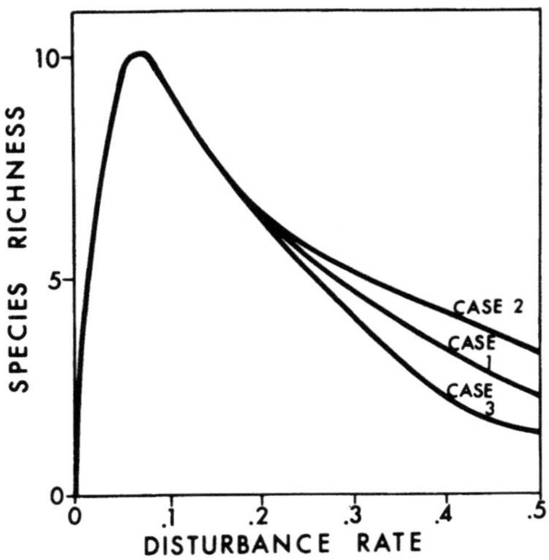

Fig. 7. Illustration of responses in community diversity to disturbance rate, given different assumptions about species's competitive abilities. See text for explanation.

Many others have suggested the periodic disturbance, including predation, may be important in maintaining the species diversity of prairies (Platt and Weis 1977), tropical forests and coral reefs (Connell 1978; Huston 1979), intertidal sessile animal and plant communities (Paine 1966; Lubchenco 1978), and plant communities in general (Grubb 1977). The usual explanation for the effect of disturbance on species richness is that disturbance prevents competitive displacement by opening up areas which are colonized by inferior competitors capable of reproduction before displacement can occur. In this view, disturbance ameliorates competition and allows a mosaic pattern of hopscotch coexistence. Connell (1978) and Lubchenco (1978) have suggested that disturbance, acting in this fashion, should lead to communities with a humped diversity curve when species richness is plotted against disturbance rate. The results briefly presented in this paper also suggest that a graph of species richness against disturbance rate should give a humped curve, but for different reasons than suggested by Connell, Lubchenco, and others.

According to the model presented here only the very best competitors for open space can survive in habitats with very low disturbance rates. These are the species which can establish in almost closed communities. Increases in disturbance rate allow invasion by other species which are superior competitors for nutrients but inferior competitors for open space (i.e., species which require more open space to become established, such as species C and D of Fig. 6B). Thus, communities with moderately low rates of disturbance can have numerous species coexisting. However, further increases in the disturbance rate preclude those species which are the best competitors for open space because the species which are the best competitors for nutrients reduce the level of nutrients below the point at which the superior competitors for open space can survive. This view of competition for open space and nutrients thus differs from that of Connell (1978) and Lubchenco (1978) in that disturbance is not considered to be a process that periodically interrupts competition; rather, it considers open space to be one of several resources for which competition occurs.

According to this model changes in disturbance rates affect community structure because they change the availability of one essential resource relative to another. Thus disturbance is seen as a process that supplies the essential resource, space. Disturbance affects plant community structure just as the supply rate of other essential resources affects it. Although space is a complex resource, this view suggests that the major qualitative effect of multispecies competition for open space is identical to the major qualitative effect of competition for any other essential resource: a graph of species richness against the supply rate of the resource should give a humped curve.

18.3 Overview

This outline represents the initial steps in developing a theory of multispecies competition for limiting resources in spatially heterogeneous environments. A much more thorough exploration of the effects of resource richness and disturbance rate on components of plant community structure, including species-area curves, diversity-resource richness curves, and dominance-diversity curves, is in preparation. However, a general and robust conclusion of all the analyses done to date is that the species diversity of such communities should depend on the rates of supply of the limiting resources, with maximum diversity in moderately resource-poor habitats and diversity decreasing in either more resource-rich or more resource-poor habitats. As the analyses indicate, such humped diversity curves are expected both for mineral resources and for space as a limiting resource in plant communities.

Correlational and experimental data for a variety of plant communities seems to be consistent with these theoretical predictions. As mentioned in the introduction, the most diverse plant communities in the world occur in the more nutrient-poor localities. Within a geographic area, the plant communities with locally higher diversity occur on the poorer soils. Addition of nutrients to plant communities often leads to decreased species diversity. For instance, the Park Grass Plots at Rothamsted, England, have been subjected to various levels and types of fertilization since 1856. The numbers of species in plots receiving complete fertilization have decreased from an initial level of about 35 species to two to five species today, whereas the unfertilized plots have maintained their original levels of diversity (Brenchley and Warington 1958; Thurston 1967; Lawes and Gilbert 1980). Similar decreases in species diversity following fertilization have been reported for other terrestrial plant communities (Milton 1947; Willis 1963), and for freshwater aquatic communities (Patrick 1963; Williams 1964). The work of Platt and Weiss (1977), Connell (1978), Lubchenco (1978) and others has strongly suggested that disturbance rate also greatly affects plant community structure. This suggests that studies of plant community structure should look at the combined effects of competition for one or two limiting nutrient resources and competition for open sites. The theory suggests that such studies may allow us to determine how it is that very highly diverse communities are maintained, and why it is that such communities tend to occur in habitats poor in resources.

To test these ideas thoroughly, and to be able to distinguish the hypothesis presented here from others that have been developed or will be developed, it will be necessary to re-analyse the data base that has already been collected and to greatly extend it. Intensive studies are needed of species-rich plant communities such as the fynbos to document thoroughly the relationship between small-scale differences in species richness and differences in resource supply rates, including the rate of disturbance. For these same localities, long-term experiments are needed to test the effects of fertilization and disturbance, as well as glasshouse studies of the resource-dependence in growth and reproduction of the major species. Such thorough experiments are necessary to test the theory outlined in this paper.

Acknowledgements

This research was supported by National Science Foundation grant DEB-79-04250 and by funds from the Graduate School and the Computer Center of the University of Minnesota. I thank the College of Biological Sciences, the Graduate School and the Office of International Programs of the University of Minnesota for funds which allowed me to attend this conference. I thank P Morrow for comments on this manuscript and H Mooney for encouraging me to attend this conference. I thank Ricka Robb for assistance in preparation of this manuscript.

References

AL-MUFTI MM, SYDES C, FURNESS SB, GRIME JD, BAND SR (1977) A quantitative analysis of shoot phenology and dominance in herbaceous vegetation. Journal of Ecology 65: 759-792

ASHTON PS (1977) A contribution of rain forest research to evolutionary theory. Annals of the Missouri Botanical Gardens 64: 694-705

BEADLE NCW (1966) Soil phosphate and its role in molding segments of the Australian flora and vegetation, with special reference to xeromorphy and sclerophylly. Ecology 47: 992-1007

BLASCO D (1971) Composicion y distribucion del fitoplancton en la region del afloramiento de las costas peruanas. Investigaciones Pesques 35: 61-112

BRENCHLEY W, WARINGTON K (1958) The Park Grass Plots at Rothamsted. Rothamsted Experimental Station, Harpenden, United Kingdom

CONNELL JH (1978) Diversity in tropical rain forests and coral reefs. Science 199: 1302-1310

DROOP MR (1974) The nutrient status of algal cells in continuous culture. Journal of the Marine Biological Association, United Kingdom 54: 25-855
DUGDALE RC (1972) Chemical oceanography and primary productivity in upwelling regions. Geoforum 11: 47-61
FISCHER A (1960) Latitudinal variation in organic diversity. Evolution 14: 54-81
GEORGE AS, HOPKINS AJM, MARCHANT NG (1979) The heathlands of Western Australia. In: SPECHT RL (ed) Ecosystems of the world, vol 9A. Heathlands and related shrublands. Descriptive studies. Elsevier, Amsterdam, pp 211-230
GOLDBLATT P (1978) An analysis of the flora of southern Africa: its characteristics, relationships and origins. Annals of the Missouri Botanical Gardens 65: 369-436
GRIME J (1973) Control of species diversity in herbaceous vegetation. Journal of Environmental Management 1: 151-167
GRUBB P (1977) The maintenance of species richness in plant communities: the importance of the regeneration niche. Biological Review 52: 107-145
GUILLARD RRL, KILHAM P (1977) The ecology of marine planktonic diatoms. In: WERNER D (ed) The biology of diatoms. Blackwell, Oxford, pp 327-469
HOLDRIDGE LR, GRENKE WC, HATHEWAY WH, LIANG T, TOSI JA (1971) Forest environments in tropical life zones: a pilot study. Pergamon Press, Oxford
HUSTON M (1979) A general hypothesis of species diversity. American Naturalist 113: 81-101
KRUGER FJ (1979) South African heathlands. In: SPECHT RL (ed) Ecosystems of the world, vol 9A. Heathlands and related shrublands. Descriptive studies. Elsevier, Amsterdam, pp 19-80
KRUGER FJ, TAYLOR HC (1979) Plant species diversity in Cape fynbos: gamma and delta diversity. Vegetatio 41: 85-93
LAWES J, GILBERT J (1980) Agricultural, botanical and chemical results of experiments on the mixed herbage of permanent grassland, conducted for many years in succession on the same land. Philosophical Transactions of the Royal Society 171: 189-416
LEON J, TUMPSON D (1975) Competition between two species for two complementary or substitutable resources. Journal of Theoretical Biology 50: 185-201
LUBCHENCO J (1978) Plant species diversity in a marine intertidal community: importance of food preference and algal competitive ability. American Naturalist 112: 23-29
MACARTHUR RH (1972) Geographical ecology. Harper and Row, New York, 269 pp
MAGUIRE B (1973) Niche response structure and the analytical potentials of its relationship to the habitat. American Naturalist 107: 213-246
MILTON W (1947) The yield, botanical and chemical composition of natural hill herbage under manuring, controlled grazing and hay conditions. I. Yield and botanical. Journal of Ecology 35: 65-89
NAVEH Z, WHITTAKER, RH (1980) Structural and floristic diversity of shrublands and woodlands in northern Israel and other mediterranean areas. Vegetatio 41: 171-190
PAINE RT (1966) Food web complexity and species diversity. American Naturalist 100: 65-75
PATRICK R (1963) The structure of diatom communities under varying ecological conditions. Annals of the New York Academy of Science 108: 359-365

PATRICK R (1967) The effect of varying amounts and ratios of nitrogen and phosphate on algae blooms. Proceedings of the 21st annual waste water conference. Purdue University, Lafayette, Indiana, pp 41-51

PETERSEN R (1975) The paradox of the plankton: an equilibrium hypothesis. American Naturalist 109: 35-49

PLATT W, WEIS IM (1977) Resource partitioning and competition within a guild of fugitive prairie plants. American Naturalist 111: 479-513

RHEE G-YULL (1978) Effects of N:P atomic rations and nitrate limitation on algal growth, cell composition, and nitrate uptake. Limnology and Oceanography 23: 10-25

SCHELSKE CL, STOERMER E (1971) Eutrophication, silica depletion, and predicted changes in algal quality in Lake Michigan. Science 173: 423-424

SMAYDA T (1975) Net phytoplankton and the greater than 20 micron phytoplankton size fraction in upwelling waters off Baja California. Fisheries Bulletin 73: 38-50

SPECHT RL (1979) The sclerophyllous (heath) vegetation of Australia: the eastern and central states. In: SPECHT RL (ed) Ecosystems of the world, vol 9A. Heathlands and related shrublands. Descriptive studies. Elsevier, Amsterdam, pp 125-210

STEWART FM, LEVIN BR (1973) Partitioning of resources and the outcome of interspecific competition: a model and some general considerations. American Naturalist 107: 171-198

TAYLOR P, WILLIAMS P (1975) Theoretical studies on the coexistence of competing species under continuous flow conditions. Canadian Journal of Microbiology 21: 90-98

TAYLOR HC (1978) Phytogeography and ecology of Capensis. In: WERGER MJA (ed) The biogeography and ecology of South Africa. Junk, The Hague, pp 171-229

THURSTON J (1969) The effect of liming and fertilizers on the botanical composition of permanent grasslands, and on the yield of hay. In: RORISON I (ed) Ecological aspects of the mineral nutrition of plants. Blackwell, Oxford

TILMAN D (1980) Resources: a graphical-mechanistic approach to competition and predation. American Naturalist 166: 362-393

WILLIAMS L (1964) Possible relations between plankton-diatom species numbers and water quality estimates. Ecology 45: 809-823

WILLIS A (1963) Braunton Burrows: the effects on the vegetation of the addition of mineral nutrients to the dune soils. Journal of Ecology 51:

19. On Alpha Diversity and the Richness of the Cape Flora: A Study in Southern Cape Fynbos

W. Bond

19.1 Introduction

The Cape floristic region is very rich in species with a high proportion of endemics (Goldblatt 1978; Taylor 1978). Most speculation on the origin of fynbos (and Australian heathland) diversity has had a biogeographic perspective dealing with total floras and the effects on speciation of geomorphic rejuvenation, climatic change, topographic diversity, age of taxa and other broad-scale influences (Levyns 1964; Raven 1973; Axelrod and Raven 1978; Goldblatt 1978; Taylor 1978, 1980; Hopper 1979). Few studies have dealt with lower levels of diversity with the recent and notable exception of Kruger and Taylor's (1979) comparative study on gamma and delta diversity of fynbos[a]. Alpha and/or beta diversity patterns have been studied by Werger (1972), Campbell and van der Meulen (1980) and Naveh and Whittaker (1980).

Rich floras do not necessarily imply rich communities. High species turnover rates along geographic or environmental gradients will result in greater overall diversity than very rich communities in monotonous landscapes. Indeed most variation in the richness of floras can probably be attributed to geographic and historical factors and their effects on species replacement rates along gradients. If ecological conditions favour high α diversity, however, more species can be accommodated within these gradients resulting in appreciably higher overall diversity.

Adamson (1927) characterized fynbos communities as very rich in species, lacking dominance, but with a striking conformity of growth forms. Taylor (1972) recorded 121 species of flowering plants in a 100 m^2 quadrat of Mountain Fynbos; this figure has been much quoted as substantiation for the richness of fynbos communities (Kruger 1977; Taylor 1978). High α diversity has also been recorded in Australian heathlands (Parsons and Cameron 1973; George et al. 1979). Lamont et al. (1977) provided diversity statistics for Western Australian heath which exceeded nearly all values for terrestrial vegetation reported in the literature.

---

[a] Diversity terminology follows Whittaker (1972, 1977) - alpha: within habitat, beta: between habitats, gamma: landscape, i.e. the sum of species in several communities, delta: geographic differentiation, species turnover between geographic regions.

In this study I report α diversity figures for fynbos in the non-seasonal rainfall areas of the southern Cape. I compare these with adjacent non-fynbos vegetation types and with other heathlands and shrublands of the world to assess whether they are particularly rich, whether special explanations should be sought for this richness and whether high α diversity may contribute significantly to the richness of Capensis. I have also investigated possible controls on diversity in fynbos by correlative methods.

19.2 Study area

Samples were collected from mountain fynbos (Acocks's (1953) Veld Type 70) of the coastal and inland mountain ranges of the southern Cape, and from adjacent shrublands and forests. Mean annual rainfall for the data set varies from 300 to 1000 mm and is distributed fairly evenly throughout the year. Seasonal differences in evapotranspiration, however, result in maximum soil moisture in winter months (Bond 1980).

Soils of the fynbos samples were all derived from quartzites and sandstones of the Table Mountain Group and were generally coarse-textured, acid, and with a low base status. Karoid shrublands, renosterveld (Acocks Veld Type 43) and valley bushveld (Acocks Veld Type 23) occurred on more fertile, finer-textured soils derived from shales and conglomerates.

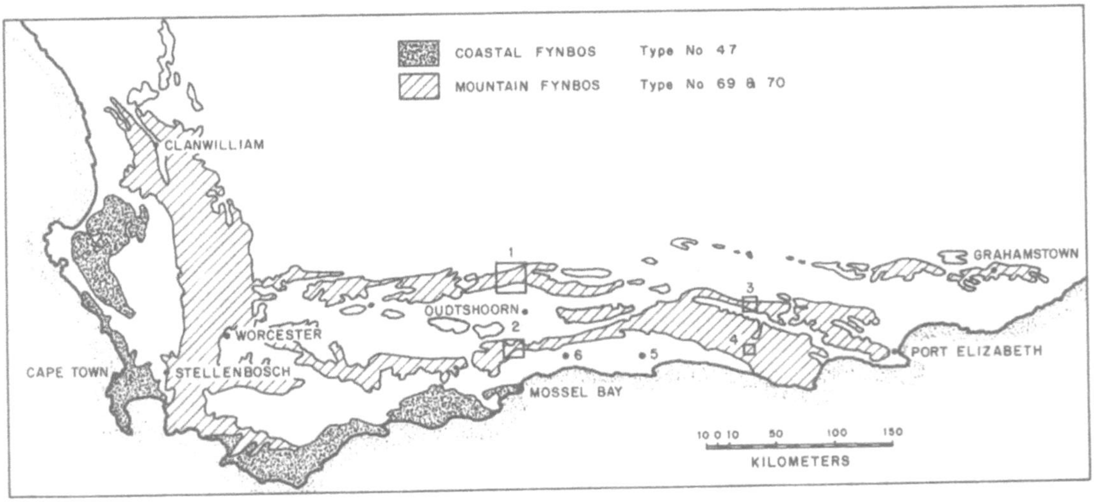

Fig. 1. Map of the fynbos biome (from Acocks 1953) showing study areas. 1: Swartberg, 2: Outeniquas, 3: Baviaanskloof, 4: Tsitsikamma, 5: Fynbos "island" (Dirk se Eiland), 6: Moist forest.

19.3 Methods

Fynbos and other woody communities were sampled from the Outeniqua, Tsitsikamma, Baviaanskloof and Swartberg mountain ranges (Fig. 1). A fynbos "island" in the indigenous forests was sampled on the Knysna coastal plateau as were the forests themselves. Sample plots were selected (originally for rodent studies - Bond et al. 1980) to cover a range of altitude and rainfall. All plots were sampled in reproductively mature vegetation showing no signs of senescence and negligible human disturbance.

Plot layout was similar to George et al. (1979), Whittaker et al. (1979), and Naveh and Whittaker (1980). Plots of 20 x 50 m were placed in visibly homogeneous vegetation with the long axis perpendicular to the contour. Nested subsamples of 5 x 1 m^2, 10 m^2, 50 m^2, 100 m^2, 500 m^2 and 1000 m^2 were assessed. All vascular plant species were recorded and their height and cover estimated. Species overlapping, but not rooted in, sample plots were counted (cf. Naveh and Whittaker 1980). Deciduous geophytes and annual species are probably underrepresented since plots could not be revisited.

Aspect, slope and altitude were recorded and rainfall was estimated from isohyetal maps. Soil pits were excavated in a number of plots and samples from the various horizons analysed for chemical and physical properties.

Surface rock cover in four diameter classes (0-0.2 m, 0.21-0.75 m, 0.76-2.50 m and 2.50 m+) and percentage bare soil were estimated by point quadrats and foliage profiles were sampled as described in Bond et al. (1980). Two surrogate measures of biomass were calculated - the area under the foliage profile and total "phytovolume", the sum of the product of cover and height for each species in a sample.

Rainfall is a poor estimator of moisture availability because of the complicating effect of topography on evapotranspiration. A moisture index was calculated to combine effects of rainfall and potential evapotranspiration (PET). Penman PET estimates were made for agricultural research station records at George, Oudtshoorn and the Langkloof (representative of the north slopes of the Outeniquas and Tsitsikammas). Aspect effects on different slopes were modelled by entering adjusted radiation (Swift 1976) into the Penman equation and calculating the ratio of PET on slopes to PET on a horizontal surface. Thornthwaite moisture balance (Israelson and Hansen 1965) was calculated for each site from the nearest representative weather station records and adjusted for slope and aspect by this ratio. Rainfall data from isohyetal maps (assuming uniform monthly distribution) were used to calculate monthly moisture surplus or deficit from the adjusted Thornthwaite estimates of PET from which the moisture index (Schulze and McGee 1978) was derived.

19.4 Diversity indices

Peet (1974) and Whittaker (1972, 1977) have reviewed diversity measures. The indices used in this study are listed in Table 1.

Species-area curves were fitted to the numbers of species in progressively larger plot sizes by least squares regression using $S = b + d\log_{10}A$ and $\log S = b + d\log_{10}A$ where S is total number of species, A is area and b and d are regression constants.

19.5 Results

19.5.1 SPECIES RICHNESS

Diversity measures are listed in Tables 2 and 3. Correlation analysis showed progressively weaker relationships between untransformed species number in 1 m^2 plots and larger plot sizes ($r = 0.92$ for 5 m^2, $r = 0.78$ for 100 m^2, $r = 0.65$ for 1000 m^2) suggesting different controls on diversity at large and small scales.

Table 1. Diversity indices used in this study.

Symbol	Index	Function	Highest value in southern Cape fynbos
S	Richness	No. of species in plot of given size	
H'	Shannon-Weiner	$-\sum_{i=1}^{S} p_i \log_2 p_i$	4.93
C	Simpson's	$\sum_{i=1}^{S} p_i^2$	0.29 (lowest 0.05)
Ec	Logvariance	$4[\sum_{i=1}^{S} (\log_{10}p_i - \log_{10}\bar{P})^2/S]^{\frac{1}{2}}$	3.69
	Whittaker's	S/Ec	29.6
b	Intercept	of species-log-area curve $S = b + d\log_{10}A$	31.2
d	Slope	" " "	26.3

S: total species, A: area in m^2, p_i: relative importance value which is percentage cover for H' and C and the product of cover and height for Ec and Whittaker's measure, $\bar{P}$: geometric mean.

The largest mean number of species in 1 m^2 plots was 27.6 in a mesic, south-facing site in the Baviaanskloof Mountains (B4) and the smallest only 5 in a fynbos island in the Knysna forests supporting very tall, dense fynbos on forest fringes. The mean for all fynbos plots was 14.9 (n = 21, s = 4.64).

A maximum of 104 species for a 0.1 ha plot was recorded in waboomveld, a vegetation category characterized by Protea nitida (Taylor 1978), followed by a closed ericoid heath community with 90 species at high elevations in the Swartberg. The lowest number of species was 21, again from the fynbos island community, with a next lowest of 41 from a proteoid-restioid community on the north slopes of the Swartberg. The mean for all fynbos plots was 63.1 (n = 21, s = 17.58).

Species number in 1 m^2 plots was appreciably lower in renosterveld (6.2), succulent karoo (4.2) and karoo (8.2) samples than in most fynbos plots, but comparatively high in a rich sample of valley bushveld (13.6). A similar trend was found at the 0.1-ha scale with renosterveld the poorest (28) followed by succulent karoo (36), karoid shrublands (49), but a rich valley bushveld (79), richer than all but two of the fynbos samples.

19.5.2 DOMINANCE

Dominance was measured by Simpson's Index (C), based on cover values, the proportional contribution of an importance value (cover x height) for the three most abundant species and Whittaker's logvariance (Whittaker 1972, 1977). Values for Simpson's Index and Whittaker's logvariance are

Table 2. Plant species diversity measures in fynbos of the southern Cape Mountains. Cover was used as importance value for calculations. The product of cover and height was used in calculation of logvariance.

Site No	Locality	Structural formation	Total Species				Species-log-area parameters		H'	C	Ec
			1 m²	100 m²	500 m²	1000 m²	b	d			
S1	Swartberg	Tall open proteoid shrubland - heath understorey	13.2	43	58	65	14.1	16.2	3.14	0.19	3.45
S2	Swartberg	" "	18.6	58	68	76	23.2	17.5	3.91	0.13	3.29
S3	Swartberg	" "	16.8	37	46	54	19.7	10.4	2.93	0.26	3.61
S4	Swartberg	Low closed *Erica* spp. heath	16.8	60	76	90	12.6	23.6	4.32	0.10	3.04
S5	Swartberg	Tall open proteoid shrubland, heath understorey	15.8	54	65	74	19.6	17.8	3.64	0.16	2.93
S6	Swartberg	Tall open proteoid shrubland, restioid heath	10.2	26	34	41	10.3	9.1	3.19	0.18	3.27
S7	Swartberg	Tall open proteoid shrubland, restioid heath	12.4	37	45	51	13.3	12.2	3.95	0.12	2.78
S9	Swartberg	Low open heathland. Arid fynbos	10.6	42	56	59	11.1	15.8	4.76	0.05	2.59
S11	Swartberg	Arid fynbos-renosterveld transition	9.8	35	43	49	7.7	13.2	3.61	0.13	3.07
B2	Baviaanskloof	Tall open proteoid shrubland-heath understorey	13.6	40	55	64	10.4	16.4	4.02	0.11	3.17
B4	Baviaanskloof	Tall closed proteoid shrubland-heath "	26.6	63	74	77	31.2	15.9	3.73	0.14	3.28
B5	Baviaanskloof	Tall proteoid shrubland-grassy heath "	13.2	51	66	75	11.75	20.4	4.18	0.10	2.88
B6	Baviaanskloof	Low open grassy heathland	16.0	47	55	65	16.6	15.2	3.26	0.23	3.15
O3	Outeniquas	Closed proteoid scrub-heath understorey	13.0	43	52	55	10.8	15.0	3.76	0.13	3.49
O4	Outeniquas	Tall open proteoid shrubland-heath understorey	16.8	55	61	65	19.0	15.9	2.92	0.29	3.69
O6	Outeniquas	Tall open proteoid shrubland-heath understorey	20.2	50	59	66	22.2	14.3	3.87	0.15	3.04
O8	Outeniquas	Low open heathland-arid fynbos	10.4	43	54	60	9.9	16.4	4.61	0.07	2.11
O9	Outeniquas	Tall open shrubland-heath understorey (waboomveld)	18.0	68	91	104	21.6	26.3	4.93	0.05	2.70
T1	Tsitsikama	Tall open shrubland-heath understorey	15.6	35	40	45	17.2	9.0	3.43	0.23	3.15
T2	Tsitsikama	Tall open shrubland-grassy heath understorey	20.0	54	61	69	24.7	14.6	4.65	0.06	2.64
E	Knysna "island"	Open scrub	5.0	15	19	21	4.2	5.4	2.43	0.28	3.83
Mean[a]			15.4	47.1	57.95	65.2	16.4	15.8	3.83	0.14	3.07
s[a]			4.15	10.82	13.4	15.1	6.2	4.3	0.61	0.07	0.38

[a] Excluding E, Knysna "island" plot.

Table 3. Plant species diversity measures in non-fynbos of southern Cape Mountains. Importance values as for Table 2. Structural formation is given with Acocks (1953) Veld Type in parentheses.

Site No	Locality	Structural formation	S: 1 m^2	S: 100 m^2	S: 500 m^2	S: 1000 m^2	Species-log-area parameters: b	Species-log-area parameters: d	H'	C	Ec
S8	Swartberg	Succulent karoo (25)	4.2	24	30	36	0.95	11.04	3.54	0.13	2.83
B7	Baviaanskloof	Karoid shrubland (26(b))	8.2	33	39	49	7.0	12.9	3.62	0.16	3.01
S10	Swartberg	Renosterveld (43)	6.2	21	25	28	5.85	7.3	1.55	0.62	3.13
B1	Baviaanskloof	Valley bushveld. Short, spinescent thicket (23(b))	13.6	55	71	79	14.8	20.9	4.26	0.12	2.78
F1	George	Moist forest (4)	5.8	38	49	52	2.0	16.8	3.47	0.11	4.86

given in Tables 2 and 3. Mean C for fynbos was 0.14 with lowest dominance (0.05) in arid fynbos (Swartberg) and waboomveld (Outeniquas) and highest in the fynbos island community (0.28) and a proteoid community in the Outeniquas (0.29).

Naveh and Whittaker (1980) give comparative values of C for a number of mediterranean-type communities. Mean C of fynbos is lower than nearly all others except for grazed Israeli woodlands and coastal Chilean matorral.

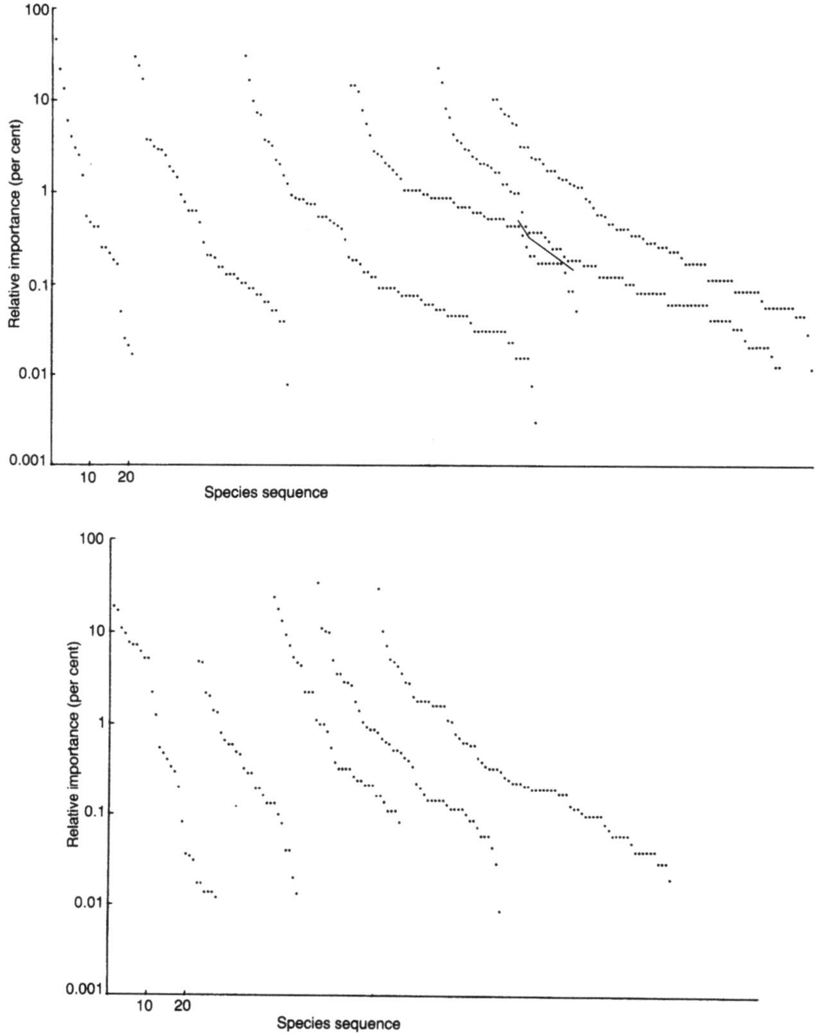

Fig. 2. Dominance-diversity curves for southern Cape woody communities. The y axis is the logarithm of cover x height values for each species as a proportion of total stand value. TOP. Fynbos; from left to right: 1: Fynbos island, 2: Proteoid shrubland with restioid understorey - S6, 3: Proteoid shrubland with heath understorey - B4, 4: Arid fynbos - 08, 5: High elevation ericoid heath - S4, 6: Waboomveld - 09. BOTTOM. Non-fynbos woody communities; from left to right: 1: Moist forest, 2: Renosterveld - S10, 3: Succulent karoo - S8, 4: Karoid shrubland - B7, 5: Succulent and spinescent thicket (= valley bushveld) - B1.

The renosterveld community is exceptional in showing strong single-species dominance by Elytropappus rhinocerotis.

Dominance-diversity curves (Whittaker 1972, 1975) for a range of communities are shown in Fig. 2. The product of cover and height for each species has been used as an importance value. The curve for tall fynbos in islands adjoining indigenous forests follows a geometric series, in accordance with the niche pre-emption hypothesis, with a few species occupying most of the niche space and less successful species occupying the remainder (Whittaker 1972). Renosterveld and a proteoid/restioid community of the Swartberg also share this "feudal" system with competition along relatively few resource gradients (May 1976). Most fynbos (and non-fynbos) communities show a lognormal curve or, in May's (1976) graphic terms, "a preponderance of middle class species" where a multitude of ecological dimensions become relevant to community composition. The waboomveld approaches MacArthur's "broken stick" model, although the slope is steeper than predicted. Very rich communities in Western Australian heathlands also approximate a MacArthur distribution (Lamont et al. 1977) which sometimes fits "small samples of taxonomically related (animals) from narrowly defined, homogeneous communities...primarily for higher animals with contest competition, stable populations and relatively long lives" (Whittaker 1975).

Whilst Adamson's (1927) observation of lack of dominance in fynbos tends to be supported by Naveh and Whittaker's (1980) comparative data, it is neither universally true nor a phenomenon restricted to fynbos in Cape vegetation. Indeed only the renosterveld community showed strong single-species dominance. The considerable variation in dominance-diversity curves indicates a range of dominance patterns in fynbos and the causes of this variation are of greater interest than the weak generalization of multiple species dominance.

19.5.3 SPECIES-AREA CURVES

The model $S = b + d\log_{10}A$ consistently gave the best fit. Both models overestimated the number of species at 1 m^2 but otherwise gave reasonably good fits (Fig. 3). Only the untransformed species number against log-area curve is discussed hereafter.

The intercept is highly correlated with species numbers in 1 m^2 plots in fynbos samples ($r = 0.94$, $n = 20$). Slope and intercept values are given in Tables 2 and 3. Mean fynbos intercept is 16.4, three times higher than Australian mallee, suggesting dense species packing at microscales. The intercept is consistently low in xeric communities including arid fynbos, karoid shrublands and renosterveld. Highest values were found in mesic proteoid fynbos and waboomveld.

The slope of the species-area curve is independent of the intercept in fynbos samples ($r = 0.104$, NS). High species richness at 0.1 ha may thus be derived from a high intercept and shallow slope (e.g. plot 4, Baviaanskloof) or low intercept and steep slope (plot 4, Swartberg). The slope of the species-area curve is a measure of the rate of species addition with expanding area. It is thus a measure of patchiness or scale of pattern induced either by patchy microsites or as a consequence of species' interactions. It can be used as a measure of Whittaker's (1977) pattern diversity synonymous with "internal β diversity" or "the change between parts of intracommunity pattern".

Mean slope for fynbos plots was 15.8 (Table 2), comparable to mallee (15.28: Whittaker et al. 1979).

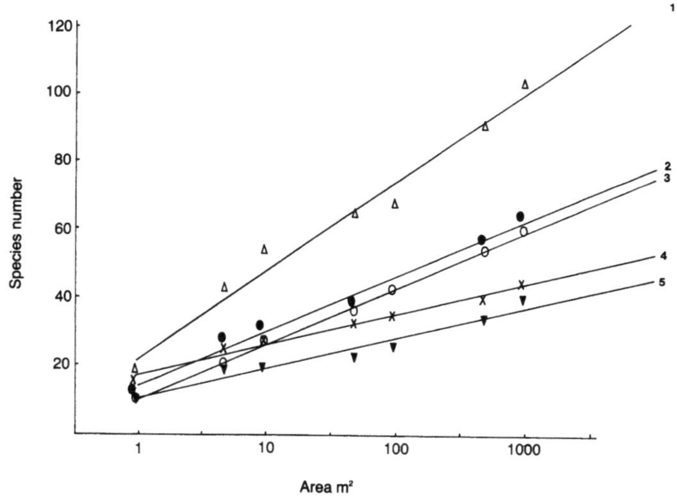

Fig. 3. Species-area curves for southern Cape fynbos. Lines were fitted by least squares regression to the model $S = b + d\log_{10}A$ where S is number of species and A is area in m^2. The curves demonstrate the variable contribution of slope and intercept to diversity in 0.1 ha plots. 1. Waboomveld, 09; 2. Proteoid shrubland with heath understoreys, S1; 3. Arid fynbos, 08; 4. Proteoid shrubland with heath understorey, T1; 5. Proteoid shrubland with restioid understorey, S6.

19.5.4 DIVERSITY PATTERNS

Diversity patterns were sought by correlation analysis of species diversity, site and surrogate biomass variables for the fynbos sample set only. Foliage profile and rock cover values were not available for plot numbers S9, S10, S11, 09 and the fynbos island site. Foliage profile values were not available for 04 and T1.

Species richness. There was no consistent relationship between species number on 0.1 ha plots and environmental variables. Species richness is often high in low, windswept, ericoid heath at high elevations (S4, S5, B4), but also in *Protea nitida* communities (09 and unpublished 50 m^2 relevé data). The latter usually occupy relatively dry talus slopes, often mixed with mountain renosterveld species; their richness may thus be partially due to ecotonal species mixing.

A tall, dense, proteoid shrubland (03 in Table 2) was excluded as an outlier in calculations. This plot had unusually low species densities in small plots, possibly because of species reduction by competition for light. Species richness in 1 m^2 plots of the remaining sites was correlated with foliage density below 0.6 m ($r = 0.74$, $n = 14$, $P < 0.01$) and total area under the curve of foliage profiles ($r = 0.78$, $n = 14$, $P < 0.01$). Both canopy parameters are correlated with the moisture index ($r = 0.72$, $P < 0.01$ for 0-0.6 m; $r = 0.75$, $P < 0.01$ for 0-2 m), but species richness is only weakly correlated with this variable ($r = 0.59$, $P < 0.05$).

Species richness of 1 m^2 plots in fynbos was inversely related to soil fertility measured as the S value in milliequivalents 100 g^{-1} soil (S value is the sum of exchangeable Ca, Mg, Na and K: $r = -0.56$, $n = 17$, $P < 0.05$). Potassium and calcium were the two most important cations correlated with species richness, accounting for 42.2% and 42.0% of the variance respectively ($n = 15$ with O3 and the fynbos island excluded, $P < 0.01$; $r^2 = 0.307$ and 0.362, $P < 0.05$, respectively with only the island excluded). Neither S value nor potassium and calcium are independent of the moisture regime and all are significantly correlated with the moisture index so that the roles of moisture and nutrients in controlling diversity is difficult to separate in this study.

In general, small areas are rich in species where the site and especially its available moisture can support dense foliage at low heights or where soil fertility, measured by potassium or calcium, is low. In arid areas with sparse foliage, richness is usually low. Note, however, that only perennially visible species were counted - diversity in annuals or deciduous geophytes is probably subject to quite different constraints.

Equitability and dominance. Dominance, measured by the proportional contribution of the three species with highest importance values and by Simpson's Index, was correlated with phytovolume ($r = 0.70$, $P < 0.01$ and $r = 0.65$, $P < 0.01$, $n = 20$ respectively). Phytovolume is weakly correlated with the moisture index ($r = 0.46$, $P < 0.05$). This suggests that the major limitation on standing crop of mature fynbos is moisture and that sites with large biomass show resource concentration in a few dominant species.

Whittaker's (1972) logvariance index, based on the slope of the dominance-diversity curve, offers an alternative measure of the division of community resources (Whittaker 1972; Peet 1974). The index, which assumes a lognormal distribution, is an estimate of the logvariance of importance values and should, theoretically, be less sensitive to extreme values of dominants and rare species (Peet 1974). Logvariance was strongly correlated with phytovolume ($r = 0.77$, $n = 20$, $P < 0.01$), area under the foliage profile ($r = 0.78$, $n = 14$, $P < 0.01$) and with the moisture index ($r = 0.66$, $P < 0.01$). The range in importance values for a stand is thus a function of biomass and its determinants, especially available moisture, stands of high biomass having a greater variance of importance values and strong dominance concentration.

Species-area curves. If factors affecting the slope and intercept of species-area curves were known, diversity could be predicted for a range of plot sizes. Since species richness in 1 m^2 plots is correlated with predicted intercept values ($r = 0.94$, $n = 20$, $P < 0.01$) intercept will also be related to biomass at low heights and factors affecting it. The slope of the curve, as a measure of pattern diversity, was less tractable to correlation analysis. Pattern diversity may be related to the number of microsites in a heterogeneous edaphic environment. This was tested by correlation analysis with frequencies of different rock diameter classes ane diversity measures of rock size. No significant correlations were found with slope (d) values, but intercepts and dominance/equitability measures were negatively correlated with small stones of diameter less than or equal to 0.13 m. This probably does not reflect a causal relationship since small stones increase in frequency in arid areas and thus parallel the moisture gradient.

No other measured variables were correlated with slope values.

19.6 Discussion

19.6.1 ARE FYNBOS COMMUNITIES RICH IN SPECIES?

Sufficient published α diversity data, particularly that of Whittaker and his colleagues, have accumulated to allow at least superficial comparisons between a variety of temperate vegetation types. Data on variations in diversity within types are scarce, inevitably leading to comparisons of unrepresentative or extreme values. Sampling methods vary and may also distort comparisons.

Species richness summed from a number of separated plots, for example, will usually be higher than contiguous plots because of heterogeneity of large areas. Diversity indices incorporating species proportions are very sensitive to different importance values so that species richness is the least ambiguous measure. Richness is, however, area-dependent and standardization on a single plot size, whilst assisting comparative studies, will obscure diversity patterns. A further difficulty lies in the selection of plant taxa. Most European studies include bryophytes and lichens, while these have been ignored by other workers. Studies also vary in the inclusion or exclusion of annuals and geophytes. Given these limitations, comparative studies should be treated cautiously, except where the data have been accumulated by a single investigator.

I have compared species richness in southern Cape fynbos communities with mediterranean-type shrublands, heathlands, South African arid shrublands, and various temperate and subtropical grasslands, woodlands and forests (Table 4). The South African data are mostly drawn from Werger (1972) and the Mediterranean and heathland data from Naveh and Whittaker (1980), Whittaker et al. (1979) and from personal observations in the French maquis, Californian chaparral and Mojave Desert shrublands (Table 5). Most of the comparisons are based on 1 m^2 or 0.1 ha plot sizes although Werger's (1972) data had to be extrapolated from semilogarithmic species-area curves for the latter, despite their rather poor fit.

Species richness in 1 m^2 southern Cape fynbos plots (mean = 14.9, maximum 27.6) is greater than most southern African arid shrublands, chaparral, mallee and maquis. It is comparable to grasslands, but greatly exceeded by some herbaceous European communities. Van der Maarel (1971) reported densities of more than 36 species per 1 m^2 in species-rich dune vegetation in Holland and Grime (1979) has recorded up to 40 species per 1 m^2 in grazed Derbyshire grasslands. The inclusion of bryophytes and lichens probably accounts for some of this richness. British heathlands, for example, support a mean of 17 species per m^2 of which only about four species are vascular plants (Gimingham et al. 1979).

Species richness on 0.1 ha plots (mean = 63.1, maximum 104) in this study is apparently slightly lower on average than western Cape fynbos, comparable to Australian heathlands, and higher than chaparral, mallee and karoid shrublands. It is considerably exceeded by some Israeli shrublands and woodlands sampled by Naveh and Whittaker (1980). Several temperate woodlands (Peet 1978) and European grasslands (van der Maarel 1971; Dony 1977) are comparable to or richer than many fynbos communities.

Naveh and Whittaker (1980) claimed that the Gondwana shrublands with some Israeli communities are the most species-rich temperate vegetation in the world. Although a detailed study of α diversity in the floristically rich southwestern Cape Province has yet to be made, I conclude that there is little evidence for this generalization.

Nevertheless fynbos communities rank as rich in the world's vegetation. Perennial species are five to ten times richer in small plots

Table 4. Comparisons of diversity measures. Values exceeding southern Cape fynbos are underlined.

	n	S, 0.1 ha^{-1} Mean	Min	Max	S per metre-square
GONDWANA HEATHLANDS					
S. Cape fynbos	20	65.2(58)	41	104	15,4
S. Cape fynbos - extreme values					26.6
W. Cape fynbos	10	75	52	128	16.7
Heath, SW Australia	7	65	46	82	13.3
Heath, SW Australia	25	59.5	37	92	
N. HEMISPHERE HEATHLAND					
Tennessee heathland	2	21.5	20	23	
Heath forest, Tennessee	2	6	5	7	
British heathland (including bryophytes and lichens)	5	39.6(9.2)	30(7)	49(11)	17.0(4.3)
MEDITERRANEAN SHRUBLANDS					
Chaparral, *Adenostoma*	5	24	16	34	7.3
Chaparral, *Ceanothus*	5	32	27	34	3.9
Maquis, Israel	1	65			14
Mallee, Australia	6	49	39	68	6.1
Mallee, Australia	10	52.7	49	62	8.0
Batha, Mt Carmel	1	65			14
Open shrubland, Israel	3	138.7	118	179	
Renosterveld, S. Cape	1	28			6.2
ARID SHRUBLANDS					
Karoid open scrub	9	42.9(43.1)	23	62	6.2
Arizona spinose, suffrutescent scrub	1	41			
Arizona desert scrub	1	33			
Succulent Karoo	2	42.5	36	49	6.2
Sonoran desert scrub	?	62.6			
MEDITERRANEAN AND S. CAPE WOODLANDS					
Knysna Forest	1	52			5.8
Valley Bushveld	1	79			13.6
Grazed woodland, Israel	3	136.7	135	138	
Woodland, Santa Barbara	6	47	35	53	10.7
Woodland, California, protected	6	58	40	68	13.8
Woodland, Carmel, grazed	5	64	50	89	15.6
Dune scrub, Holland	?	90			
TEMPERATE WOODLANDS, FORESTS					
Mesic forest, N. Carolina		85-95			
Pinus taeda forest, N. Carolina		95-115			
Cove forest, Great Smoky Mountains		61-68			
GRASSLANDS					
New South Wales	1	55			
Alpine Themeda-Festuca, S. Africa	1	37.8			12(0.8 m^2)
Dry grassveld, S. Africa	1	50.6(53)			14(0.8 m^2)
British grassland	1	88.9			
Dutch dune grassland	1	130(0.2 ha)			36+ (in some cases)

Sources - 1: This study, 2: Naveh and Whittaker 1980, 3: George et al. 1979, 4: Whittaker 1979, 5: Gimingham et al. 1979, 6: van der Maarel 1971, 7: Werger 1972, 8: Whittaker and Niering 1975, 9: Peet 1978, 10: Whittaker et al. 1979,

C	Species-log-area parameters b	d	Source
0.08	16.4	15.8	1 (0.05 ha values in brackets)
0.03	31.2	26.3	1
0.22	17.3	19.0	2
0.21	14.1	17.1	2
			3 (Value for 0.05 ha)
			4
			4
	19.2(5.1)	6.8(1.3)	5 (Vascular plants only in brackets)
0.58	8.0	5.6	2
0.56	3.1	9.2	2
0.46	15.9	16.9	2
0.30	3.8	14.5	2
	5.3	15.3	10
0.46	15.9	16.9	2
			2
0.62	5.9	7.3	1
	3.8	13.1	7
			8
			8
0.14	4.0	12.0	1
			9
0.11	2.0	16.8	1
0.12	14.8	20.9	1
0.07			2
0.23	9.1	12.1	2
0.25	12.0	13.9	2
0.16	12.2	15.7	2
			6 (0.02 ha only)
			9
			9
			9
	18.5	12.8	10
	11.4	8.8	7
	14.1	12.2	7
			11
			6

11: Dony 1977. Values for Werger's South African data have been extrapolated from species-log area curves. Value for 0.08 ha are given in parentheses.

Table 5. Plant species diversity measures in French Mediterranean and Californian shrublands[a]. Cover was used as importance value for calculations. The product of cover and height was used in calculation of logvariance (Ec).

Site No	Locality	Structural formation	Total S				Species-log-area parameters		H'	C	Ec
			1 m^2	100 m^2	500 m^2	1000 m^2	b	d			
M1	France	*Erica arborea* maquis	12.2	33	40.8	43.9[a]	12.5	10.5	2.8	0.28	4.77
M2	France	*Quercus suber*, *Cistus salvaefolia* maquis	5.6	21	25.1[a]	27.3[a]	4.8	7.5	2.31	0.30	5.37
C1	California, Sky Oaks	Chamise (*Adenostoma fasciculatum*) chaparral	1.4	7	10	10.4	0.98	3.1	0.17	0.97	6.35
C2	California, Sky Oaks	Redshank (*Adenostoma sparsifolium*) chaparral	1.4	5	7	7.1	0.66	2.16	1.19	0.56	7.4
D1	Mojave Desert Granite Mountains	*Yucca schidigera*, *Opuntia* spp., *Eriogonum fasciculatum* shrubland - 1220 m	2.0	20	24	26	0.55	8.7	3.6	0.11	4.17
D2	Mojave Desert	*Larrea tridentata*, *Franseria dumosa* shrubland 777 m	0.4	5	8	12	-1.4	3.7	1.97	0.33	4.70
D3	Mojave Desert	*Larrea tridentata*, *Franseria dumosa* shrubland 375 m	0.6	2	2	3	0.51	0.72	1.01	0.51	4.0
D4	Mojave Desert Joshua Tree N.P.	*Yucca* spp., *Coleogyne ramosissima*, *Opuntia* spp. shrubland - 1200 m	1.4	12	17	20	-0.26	6.4	2.37	0.28	4.28

[a] The same methods were used as in the southern Cape set. The French samples (up to 10 x 10 m only) were taken in maquis on siliceous substrates in Les Moures l'Esteral. The chaparral samples (500 m^2 only) were taken near Sky Oaks, southern California in relatively arid chaparral. The Mojave Desert samples (D1-D3) were taken on a gentle altitudinal gradient on coarse-textured soils on plains south of the Granite Mountains. D4 was taken on similar substrates in Joshua Tree National Park.

in fynbos than in chaparral and pattern diversity, measured by the slope of the species-area curve, is two to five times greater. Given environmental gradients (such as moisture) of equivalent length with equivalent rates of species turnover, α diversity alone can account for the much higher richness of the Capensis flora relative to the Californian Floristic Province (or at least its shrubland component).

Australian heaths have similar life-form distribution and diversity statistics to fynbos (George et al. 1979; Naveh and Whittaker 1980). The greater richness of the Cape flora relative to south-west Australia must thus be sought in the length and diversity of environmental gradients and species turnover rates along them (β diversity) or in differences in species turnover between geographic areas with similar environments (delta diversity).

19.6.2 DIVERSITY PATTERNS IN FYNBOS

Species-area curves, despite their limitations in fitting field data (Hopkins 1955; Dony 1977), are a useful approach to explaining species richness. The intercepts, or richness in 1 m^2 plots, describe "point diversity" (Whittaker 1977), where "determinants of density are those which allow plants of different biology to establish and survive in close proximity" (Grime 1979). Point diversity is possibly the least ambiguous measure of species richness. The slope of the species-area curve is a measure of heterogeneity or "pattern diversity" (Whittaker 1977). (Though Gould (1979) has criticized use of the intercept as a measure of richness where it is a distant extrapolation from actual data and use of slope and intercept as separate diversity measures where they may be statistically interdependent, neither criticism is valid in this data set.) Even the most uniform terrain may contain a complex of microhabitats imposed by the physical environment, faunal activities or the plants themselves and their spatial arrangement. The task of explaining this patchiness is largely independent of the problem of coexistence in very small plots. Finally, both measures of richness will respond to the size of the regional species reservoir and the time elapsed for species diffusion.

Several authors (Harper 1969; Connell 1978; Grime 1979; Huston 1979; Pickett 1980) have suggested that coexistence of plants is non-equilibrial and is dependent on factors reducing dominance or preventing the attainment of competitive equilibrium (climax). Competitive exclusion may be limited in stressed environments (such as nutrient-poor soils) by reduced growth rates and in all environments by the frequency of disturbance (Grime 1979; Huston 1979). Al-Mufti et al. (1977) and Grime (1979) have proposed an explicit "humpbacked" model for the control of species richness in herbaceous vegetation of small areas. Highest species richness occurs at moderate intensities of stress or disturbance with reduction in numbers of species where vigorous dominants are favoured or, at the other end of the scale, in stressed or very frequently disturbed environments. Rephrased, where rates of competitive exclusion are low and disturbance moderately frequent, more species can coexist. Al-Mufti et al. (1977) successfully tested the model on British herbaceous vegetation using a biomass gradient as an estimator of the relative stress or dominance, or rate of competitive displacement in Huston's (1979) terms.

I have shown that point diversity is correlated with biomass which is largely controlled by available moisture. Few species occur in stressed, arid environments of low biomass, but many species coexist in relatively moist environments with high biomass. Few plots were sampled in very high biomass stands. Unfortunately foliage profiles were not recorded for the fynbos "island" plot, a tall dense scrub of high biomass. In Fig. 4 I have inserted a foliage profile value for a similarly structured community on the foothills of the Outeniquas. The available data appear to support the humpbacked model of diversity with few species in stressed, arid habitats or high biomass sites with strongly expressed dominance (Fig. 2) and many species in intermediate conditions.

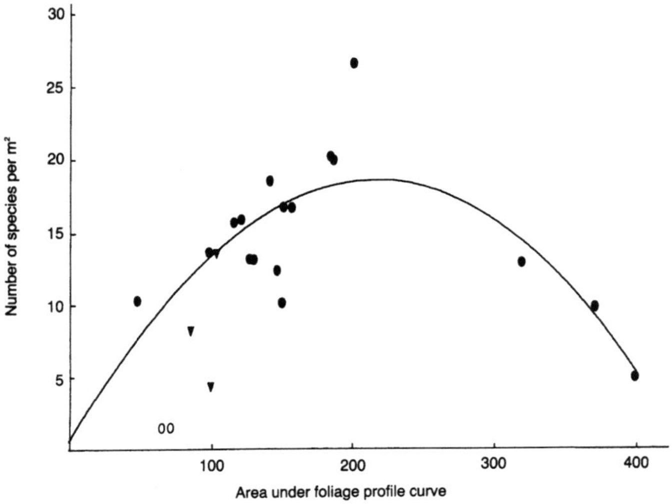

Fig. 4. The relationship between species density in 1 m^2 plots and area under the foliage profile (a measure of biomass). The curve was fitted to the fynbos data set only (closed circles) by a quadratic equation, $y = a + bx + cx^2$ where y is species number, x is area under the foliage profile and a, b and c are regression constants. Point diversity (9.9 spp m^{-2}) of a dense Protea aurea fynbos from Ruitersbos-Outeniquas, not listed in the tables, was used as an additional datum for large biomass samples. Chaparral (C1, C2 - open circles), and arid Cape shrublands data (B1, B7, S8 - triangles) are included for comparison. Area under the foliage profile curve measured in arbitrary units.

Fire acts as a very effective generalized non-selective defoliator of communities, thus reducing dominance. The "humpbacked" hypothesis would predict low point diversity in frequently fired communities and communities burned at long intervals, with greatest diversity at intermediate fire frequencies (Huston 1979). Casual observations of regularly burnt firebreaks partially support the hypothesis, but it has yet to be rigorously tested.

Tilman (1980 and Chapter 18, this volume) has derived a model of species coexistence with similar predictions, i.e. a humpback form over a disturbance or nutrient gradient, but based on a deterministic, mechanistic and essentially equilibrial view of species interactions. Although based on very different premises, his model cannot be explicitly tested by the correlative studies reported here and the role of competition and disturbance in structuring fynbos communities remains an open question.

Use of the humpback model is limited by the difficulty of assessing the scale at which it operates. If slopes of semilogarithmic species-area curves are similar (parallel) in a given vegetation type, the effective area will be large. Where slopes differ, however, pattern effects will predominate in communities with steep slopes and the model will only apply to small areas. I have examined the relationship between the area under foliage profiles (as a biomass estimator) and species density for progressively larger plot sizes to determine the scale of stress-dominance effects. Using only the low to medium biomass sections of the hump, the area appears to be about 5 m^2 (correlation of species density to area under foliage profiles, 5 m^2: $r = 0.66$, $P < 0.01$, $n = 14$; 10 m^2: $r = 0.65$, $P < 0.05$; 50 m^2: $r = 0.46$, NS; 100 m^2: $r = 0.38$, NS). This scale will vary for different growth forms and would have to be reassessed in different vegetation types.

Species richness on 0.1 ha plots is controlled in complex ways by a combination of point diversity and heterogeneity within the stand (internal beta diversity - Whittaker 1977), where the latter is reflected in the slope of the species-area curve. Samples with steep slopes included waboomveld (O9), high montane ericoid heath (S4), several proteoid shrublands (B5, S2 and 5) and arid fynbos (O8). Particularly low values occurred in proteoid-restioid fynbos on north aspects (S6 and 7, T1), tall proteoid heath on mesic south slopes on deep soils (S3) and open scrub in fynbos islands in indigenous forest. No common factor appears to link either the low or high group of plots besides the expected high dominance concentration in plots with flat slopes.

Community patchiness may be controlled by a multitude of factors of varying importance to different synusiae. Microsite variation is probably important in montane environments with variable soil depth and fertility (Harner and Harper 1976). Grace (1977) has documented the local spatial variability of wind and its significance to plant growth. Local variation in insolation, precipitation, interception and infiltration (Specht 1957) are also significant in creating microsites. Differences in regenerative strategies may affect pattern (Grime 1979). Fynbos communities are usually even-aged with populations failing to regenerate between fires. Cyclic fluctuations and gap replacement typical of European heathlands (Watt 1947; Grubb 1977; Gimingham et al. 1979) will therefore contribute little to spatial heterogeneity.

Janzen (1970) has suggested a mechanism for faunal controls on pattern by selective seed predation. But target-specific seed dispersal by animals is probably more significant in fynbos, for example through the effects of ants on microdistribution of elaiosome-bearing seeds (Berg 1975).

Patchiness is also dependent on plant size and a variety of mechanisms controlling competitive displacement. One example of the latter is allelopathy which appears to reduce the pool of potentially coexisting species in chaparral (Christensen and Muller 1975) and may be significant in renosterveld in its purer forms which has species-areas curves with anomalously low slopes.

Both patchiness and point diversity are affected by the size of the species reservoir (Connell 1978; Grime 1979). Thus fynbos islands in indigenous forests may be species-poor because of their relative isolation (thus weakening the test of the humpback model which used data from this plot!). Conversely the richness of waboomveld may be due to the availability of both fynbos and renosterveld species pools. British heathlands, although physiognomically similar to fynbos and also occurring on nutrient-poor soils, are extremely poor in vascular plant species. In contrast to fynbos, communities are more diverse on base-rich soils through the addition of woodland species. British heathlands have a short history (less than 2000 years) and were derived from a depauperate glaciated flora (Gimingham et al. 1979). They provide a striking example of the influence of the size of the regional species pool on α diversity.

19.7 Conclusions

Recent hypotheses on the control of species diversity by stress, competition and frequency of disturbance (Connell 1978; Grime 1979; Huston 1979) appear to be of theoretical and practical value in predicting and maintaining point diversity. The same cannot be said for factors controlling the patchiness component of α diversity (Whittaker and Levin 1977) for which no comparable generalized model has yet been developed.

Species richness in 0.1 ha plots, commonly used in plant diversity studies, is significantly affected by relative patchiness which may have a multitude of causes. It seems that, because of this, few useful general principles will emerge from large plot studies. Point diversity studies, however, are likely to yield useful information on mechanisms controlling coexistence of multiple plant growth forms.

I have not discussed the relative contribution of different growth forms to α diversity. However, it seems some of the problems of patchiness will be best resolved by imitating the zoologists. Their success in the study of guilds and their ability to predict species diversity patterns (MacArthur 1964; Cody 1970; Pianka 1973) is to be envied by their botanical colleagues (Peet 1978). Defining guilds which utilize similar resource axes is of critical importance. Selection of taxocenes, e.g. Proteaceae in fynbos, seems a useful starting point (Cody 1978; Yeaton and Cody 1979).

Acknowledgements

I thank my colleagues at Saasveld, especially C J Geldenhuys, J C van Daalen, R America and J Breytenbach for assistance in the field and critical comment and F J Kruger and B M Campbell for stimulus and discussion. M L Cody gave generous hospitality and field assistance in California. L Trabaud, R Hobbs and A Malek are thanked for stimulating discussions on Mediterranean shrublands and British heathlands respectively. This work forms part of the conservation research programme of the Department of Environment Affairs and Fisheries.

References

ACOCKS JPH (1953) Veld types of South Africa. Memoirs of the Botanical Survey of South Africa 28: 1-128

ADAMSON RS (1927) The plant communities of Table Mountain. I. Preliminary account. Journal of Ecology 15: 278-309

AL-MUFTI MM, SYDES CL, FURNESS SB, GRIME JP, BAND SR (1977) A quantitative analysis of shoot phenology and dominance in herbaceous vegetation. Journal of Ecology 65: 759-792

AXELROD DI, RAVEN PH (1978) Late Cretaceous and Tertiary vegetation history of Africa. In: WERGER MJA (ed) Biogeography and ecology of southern Africa. Junk, The Hague, pp 77-130

BERG RY (1975) Myrmecochorous plants in Australia and their dispersal by ants. Australian Journal of Botany 23: 475-508

BOND WJ (1980) Periodicity in fynbos of the non-seasonal rainfall belt. Journal of South African Botany 46: 343-354

BOND W, FERGUSON M, FORSYTH G (1980) Small mammals and habitat structure along altitudinal gradients in the southern Cape Mountains. South African Journal of Zoology 15: 34-43

CAMPBELL BM, VAN DER MEULEN F (1980) Patterns of plant species diversity in fynbos vegetation, South Africa. Vegetatio 43: 43-47

CHRISTENSEN NL, MULLER CH (1975) Effects of fire on factors controlling plant growth in Adenostoma chaparral. Ecological Monographs 45: 29-55

CODY ML (1978) Distribution ecology of Haplopappus and Chrysothamnus in the Mojave Desert. I. Niche position and niche shifts on north-facing granitic slopes. American Journal of Botany 65: 1107-1116

CODY ML (1970) Chilean bird distribution. Ecology 51: 455-463

CONNELL JH (1978) Diversity in tropical rain forests and coral reefs. Science 199: 1302-1309

DONY JG (1977) Species-area relationships in an area of intermediate size. Journal of Ecology 65: 475-484

GEORGE AS, HOPKINS AJM, MARCHANT NG (1979) The heathlands of Western Australia. In: SPECHT RL (ed) Ecosystems of the world, vol 9A. Heathlands and related shrublands. Descriptive studies. Elsevier, Amsterdam, pp 211-230

GIMINGHAM CH, CHAPMAN SB, WEBB NR (1979) European heathlands. In: SPECHT RL (ed) Ecosystems of the world, vol 9A. Heathlands and related shrublands. Descriptive studies. Elsevier, Amsterdam, pp 365-413

GOLDBLATT P (1978) An analysis of the flora of South Africa. Its characteristics, relationships and origins. Annals of the Missouri Botanic Gardens 65(2): 369-436

GOULD SJ (1979) An allometric interpretation of species-area curves: the meaning of the coefficient. American Naturalist 114: 335-343

GRACE J (1977) Plant response to wind. Academic Press, London

GRIME JP (1979) Plant strategies and vegetation processes. Wiley, Chichester

GRUBB PJ (1977) The maintenance of species richness in plant communities: the importance of the regeneration niche. Biological Review 52: 107-145

HARNER RF, HARPER KT (1976) The role of area, heterogeneity, and favourability in plant species diversity of pinyon-juniper ecosystems. Ecology 57: 1254-1263

HARPER JL (1969) The role of predation in vegetational diversity. In: Diversity and stability in ecological systems. Brookhaven Symposium Biology 22, pp 48-62

HOPKINS B (1955) The species-area relations of plant communities. Journal of Ecology 43: 409-426

HOPPER SD (1979) Biogeographical aspects of speciation in the southwest Australian flora. Annual Review of Ecology and Systematics 10: 399-422

HUSTON M (1979) A general hypothesis of species diversity. American Naturalist 113: 81-101

ISRAELSON OW, HANSEN VF (1965) Irrigation principles and practices, 3rd edn. Wiley, New York

JANZEN DH (1970) Herbivores and the number of tree species in tropical forests. American Naturalist 104: 501-528

KRUGER FJ (1977) Ecology of Cape fynbos in relation to fire. In: MOONEY HA, CONRAD CE (eds) Symposium on environmental consequences of fire and fuel management in mediterranean ecosystems. USDA Forest Service General Technical Report WO-3, pp 230-244

KRUGER FJ, TAYLOR HC (1979) Plant species diversity in Cape fynbos: gamma and delta diversity. Vegetatio 41: 85-93

LAMONT BB, DOWNES S, FOX JED (1977) Importance value curves and diversity indices applied to a species-rich heathland in Western Australia. Nature 265: 438-441
LEVYNS MR (1964) Migrations and origin of the Cape flora. Transactions of the Royal Society of South Africa 37: 85-107
MACARTHUR RH (1964) Environmental factors affecting bird species diversity. American Naturalist 98: 387-397
MAY RM (1976) Patterns in multi-species communities. In: MAY RM (ed) Theoretical ecology: principles and applications. Blackwell, Oxford, pp 147-162
NAVEH Z, WHITTAKER RH (1980) Structural and floristic diversity of shrublands and woodlands in northern Israel and other Mediterranean areas. Vegetatio 41: 171-190
PARSONS RF, CAMERON DG (1973) Maximum plant species diversity in terrestrial communities. Biotropica 6: 202-203
PEET RK (1974) The measurement of species diversity. Annual Review of Ecology and Systematics: 285-307
PEET RK (1978) Forest vegetation of the Colorado front range: patterns of species diversity. Vegetatio 37: 65-78
PIANKA ER (1973) The structure of lizard communities. Annual Review of Ecology and Systematics 4: 53-74
PICKETT STA (1980) Non-equilibrium coexistence of plants. Bulletin of the Torrey Botanical Club 107: 238-248
RAVEN PH (1973) The evolution of Mediterranean floras. In: DI CASTRI F, MOONEY HA (eds) Mediterranean-type ecosystems: origin and structure. Springer-Verlag, Berlin, pp 213-224
SCHULZE RE, MCGEE OS (1978) Climatic indices and classifications in relation to the biogeography of southern Africa. In: WERGER MJ (ed) Biogeography and ecology of southern Africa. Junk, The Hague, pp 19-52
SPECHT RL (1957) Dark Island heath (Ninety-Mile Plain, South Australia). IV. Soil moisture patterns produced by rainfall interception and stem-flow. Australian Journal of Botany 5: 137-150
SWIFT LW (1976) Algorithm for solar radiation on mountain slopes. Water Resources Research 12: 108-112
TAYLOR HC (1972) Fynbos. Veld and Flora 2: 68-75
TAYLOR HC (1978) Capensis. In: WERGER MJA (ed) Biogeography and ecology of southern Africa. Junk, The Hague, pp 171-229
TAYLOR HC (1980) Phytogeography of fynbos. Bothalia 13: 231-235
TILMAN D (1980) Resources: a graphical-mechanistic approach to competition and predation. American Naturalist 116: 362-393
VAN DER MAAREL E (1971) Plant species diversity in relation to management. In: DUFFEY E, WATT AS (eds) The scientific management of animal and plant communities for conservation. Blackwell, Oxford pp 45-64
WATT AS (1947) Patterns and process in the plant community. Journal of Ecology 35: 1-22
WERGER MJA (1972) Species-area relationship and plot size: with some examples from South African vegetation. Bothalia 10: 583-594
WHITTAKER RH (1972) Evolution and measurement of species diversity. Taxon 21: 213-251
WHITTAKER RH (1977) Evolution of species diversity in land communities. Evolutionary Biology 10: 1-67
WHITTAKER RH, LEVIN SA (1977) The role of mosaic phenomena in natural communities. Theoretical Population Biology 12: 117-139
WHITTAKER RH, NIERING WA, CRISP MD (1979) Structure, pattern and diversity of a mallee community in New South Wales. Vegetatio 39: 65-76
YEATON RI, CODY ML (1979) The distribution of cacti along environmental gradients in the Sonoran and Mohave Deserts. Journal of Ecology 67: 529-541

20. Continental Diversity Patterns and Convergent Evolution in Bird Communities

M. L. Cody

20.1 Introduction

Convergent evolution in mediterranean-climate ecosystems has received intensive study over the last decade. At this stage, with considerable data on bird community structure available from all five continents on which mediterranean-type habitats exist, there are both similarities and dissimilarities to be interpreted and resolved. In this paper I compare patterns of alpha and beta diversity in bird communities of California, Chile, Sardinia and South Africa and point out what questions remain unanswered, how they can be refined, and what sorts of data might help to answer them. I then discuss data from South Africa which help illuminate some of the factors that control species diversity patterns within continents, especially at the levels of β and γ diversity.

The chapter uses the review of convergent evolution and α diversity by Cody and Mooney (1978) as a point of departure, then compares β diversity over matched habitat gradients, emphasizing broad differences between continents and among transects within a single region of South Africa. Birds on similar habitat gradients in three different areas of Cape Province show reasonable similarities in α diversity, but strikingly different β diversity patterns. I attribute the similarities to the matched vegetation in the three gradients, and the dissimilarities to differences both in total habitat area available to the birds and in the distribution of this habitat area over the gradient from low, open vegetation to taller and denser fynbos and woodland.

Our understanding of continental diversity patterns is still rather rudimentary. It will develop, I believe, by use of concise and quantitative definitions of different components of this diversity, and a realization that the controlling factors for each component may be quite independent. I discuss these conceptual questions, point out what I see as the major unknowns for each component of this diversity, and sketch the ways in which these components interact. In the final section of the paper I get back to the real world of ecological biogeography with presentation and discussion of data recently collected on South Africa. I ask whether the clear-cut examples of habitat patchiness within a continental setting, such as the string of patches of Afromontane woodland that stretch westward from Knysna, can be extrapolated to a more subtle effect on diversity within complex mosaics of variable habitat. A detailed analysis of bird distributions over habitats in the three areas of Cape Province mentioned above illustrates this theme. The geographic position of each site censused, relative to the distributions of its habitat type over the map - karoo, fynbos, forest etc. - affects the α diversity of birds within the habitat type, bird species turnover (β diversity) between habitat types, and turnover within habitat types between each of the three geographic areas (γ diversity).

20.2 Convergent evolution at the community level

20.2.1 MACCHIA SITES OF FOUR CONTINENTS

A detailed analysis of convergent and nonconvergent evolution in birds and plants of matched sites on four continents has been published earlier (Cody and Mooney 1978). The sites compared are from the chaparral of southern California, the matorral of central Chile, macchia in southern Sardinia and fynbos in South Africa. All sites enjoy the characteristic mediterranean temperature and precipitation regime, but there are climatic differences among sites that appear to be reflected in both the plant and bird ecologies. In addition, two control sites were used, one in Yorkshire with similar vegetation structure (successional thorn scrub) but a different climate, a second in California, evergreen oak woodland with a different vegetation structure but similar (mediterranean) climate. For details of site locations, vegetation and bird census methods, see Cody and Mooney (1978) and references therein.

The four "experimental" sites all support a thick and fairly continuous scrub cover 2-3 m high, with both broad- and narrow-sclerophyllous plants, and all four appear very similar. Both plants and birds differ taxonomically among sites; the sites share on the average 48% of their bird families, 10% of their genera and only 2% of their bird species. On 2.5-4-ha sites with homogeneous vegetation, total bird species seen in the breeding season were 24 in Sardinia, 28 in South Africa, 30 in California and 31 in Chile. The eight most abundant species at each site comprised 74% of the total bird density at Sardinia, 70% in South Africa, 79% in California and 67% in Chile, while total bird densities were assessed at 13.3, 9.3, 14.4 and 16.6 pairs ha^{-1} respectively.

At each site the birds were assembled into rough guilds according to foraging ecology. Each site had at least four species of foliage insectivores, at least six species of ground-foraging omnivores, two to four species of seed and fruit eaters, one aerial feeder, one crepuscular insectivore ("nightjar-type") and two to four of raptors and scavengers. Each site except Sardinia (with none) had one to two woodpeckers, each site except Sardinia (with a single medium-sized species) had a small and a larger flycatching species, and each site except Sardinia (with none), two to three nectarivorous species. The insular position of Sardinia may account for the absence of the first-mentioned rather more specialized species, but not for the lack of nectarivores.

Despite some overall similarity in guild structure among the four sites, differences do appear. For example, Chile has only four species (5.8 pairs ha^{-1}) of foliage insectivores whereas Sardinia has nine species (7.7 pairs ha^{-1}) compared to the average 6.2 species (5.6 pairs ha^{-1}). But Chile has 12 species of ground omnivores totalling 7.1 pairs ha^{-1} versus the average 8.3 species with 4.8 pairs ha^{-1}. And whereas Chile and California have two species each with 1.1 and 1.2 pairs ha^{-1} respectively of nectarivores, the three South African species total 2.7 pairs ha^{-1}.

We used the two control sites to calibrate these purportive similarities and differences in species numbers and densities. Both species numbers (36) and total bird densities (20.7 pairs ha^{-1}) were higher in California oak woodland. In particular, the species numbers and densities in the categories of foliage insectivores, sallying flycatchers and seed and fruit eaters are two to three times that of the average macchia site. In the Yorkshire successional scrub species numbers (31) and densities (18.6 pairs ha^{-1}) resemble those of macchia, but the distribution of species numbers and densities among guilds is different. The ground-foraging omnivores comprise the largest category with 12 species and 7.9 pairs ha^{-1}; in addition foliage insectivores are more numerous (10 versus average 6.2 species), flycatchers are rarer (0.1 versus average 0.7 pairs ha^{-1}) and seed and fruit eaters much commoner than in macchia (5 species, 4.1 pairs ha^{-1} versus an average 3 species, 0.9 pairs ha^{-1}).

20.2.2 DENSITY COMPENSATION

Community organization may be more similar than the differences in species numbers among the different macchia communities indicate. Through density compensation, fewer species can make up, at least in part, for the absence of ecologically similar competitors, and by foraging over a broader height interval or in a wider variety of ways support more pairs ha^{-1} in species-poor sites. Thus the fewer Chilean foliage insectivores are denser (1.44 pairs ha^{-1} $species^{-1}$) than the overall average for the four sites (0.90 pairs $ha^{-1}/species^{-1}$), and the larger number of ground foragers in Chile are each less dense (0.59 pairs ha^{-1} $species^{-1}$) than the average in Sardinia and California (0.73 and 0.76 pairs ha^{-1} $species^{-1}$ respectively). Yet there are instances where no density compensation is apparent: the three species of nectarivores in South African fynbos are each more common than the two nectarivores in California and Chile (0.89 versus 0.57 pairs ha^{-1} $species^{-1}$). And finally, there are major discrepancies in the total bird densities among sites, greatest between the South African and Chilean sites (80%), and in general much greater than the differences between species totals (up to 30%).

20.2.3 DIFFERENCES IN FOOD RESOURCES

Similarities in community organization (guild structure, species numbers and relative abundances) may follow from broad similarities in climate and vegetation structure among the different sites, whereas differences in bird densities may reflect differences in food resources. The only resource that I have directly measured is that which bears directly on flycatcher and foliage insectivore densities, the insect catches on Tanglefoot plaques. These catches are reasonabe estimates of the food available to insectivorous birds, since in Arizona pine-oak woodland (Cody 1981) and in Maroccan macchia (Cody 1980b) 70% and 85% of the variation in bird densities is accounted for by variation in Tanglefoot catches respectively. In the macchia sites, total bird density D is directly related to the area A under the graph of insects caught versus height above ground: $D = 8 + A/25$; South African insect catches were three to four times lower than those in California and Chile, where foliage insectivores and flycatchers are two and a half times as common. In other aspects also the resources are not well matched among sites. There are no nectar-producing plants north of the Mediterranean that are morphologically adapted to use by birds, and hence no nectarivorous birds in Sardinia. In South Africa, many nectar-producing plants in the families Proteaceae and Ericaceae are coevolutionarily associated with bird consumers. No plants comparable to the Proteaceae exist in Chile or California; their evolution appears tied to nitrogen- and phosphorus-poor soils, and their flowering and leafing strategies result in more nectar and fewer insects (on the relatively well-defended leaves).

The other major anomaly of high diversity and abundance of ground foragers in Chile is attributable to the matorral having a more open canopy and more summer rainfall. Therefore spring and summer annual plants are important components of that vegetation, and this presumably results in more and various bird food near and on the ground.

20.2.4 COMMUNITY COMPARISONS IN PHENOTYPES AND FORAGING ECOLOGY

Factor analysis of bird phenotype was used to explore questions about the range of morphotypes and of feeding ecologies, and the interrelations among the two sets of observations by cross-continent comparisons. The following conclusions were reached.

(a) The same range of phenotypes is present in macchia on each continent. The principal components of the seven morphological characters used are similarly constituted on each continent, and species morphotypes from each continent are distributed with statistically indistinguishable means and variances on the morphological plane of the first two principal components.

(b) Foraging behaviours are similar among the four communities. The first two principal components of foraging behaviour are similarly constituted on each continent, account for similar amounts of behavioural variation (89 ± 4%), and there are no significant differences among continents in the means and variances of species in this behavioural plane.

(c) The components of foraging ecology covary similarly on the four continents. The principal components of foraging behaviour and foraging height show that (i) faster species forage higher and (ii) species with long pauses in foraging sequence are more variable in foraging height. Species from each continent show statistically similar ranges and means on these two variables, but the South African species exploit a narrower range in relation (ii) above. Canonical correlation between morphology and feeding height first segregates long-winged and short-tailed birds with high foraging ranges from larger, longer-legged and -tailed species that forage on the ground. This relation is not significantly different among continents, despite the fact that in the South African community the composition of the canonical variables is somewhat different. A similar analysis of morphology versus foraging behaviour shows that there are no differences among continents in the way these variables are related. In general, a given foraging behaviour is conducted at a given height in a certain vegetation density similarly on all continents, and phenotypes vary with foraging behaviour in the same way on all continents.

20.3 Bird turnover rates along habitat gradients

20.3.1 ALPHA AND BETA DIVERSITY ON MEDITERRANEAN HABITAT GRADIENTS

Whereas α diversity measures the number of species that pack into and coexist in a given habitat type, a second and potentially independent diversity component is β diversity, the rate at which these species are replaced by others as habitat structure varies. Most habitat variability follows a gradient based on the structure of the vegetation, and thus habitat can be considered a continuous variable H. As H is sampled from low to high values bird species are added to a cumulative list whilst others are lost. The functions g(H) and l(H) describe the way in which species are gained and lost over habitats H respectively. I have defined α diversity as the difference between the species gain and loss functions, g(H) - l(H), itself a function of H, and β diversity as the average rate at which species are gained and lost at a given H: $d/dH(\frac{1}{2}[g(H) + l(H)])$; H is calculated as the logarithm of the principal component of vegetation height and vegetation half-height (Cody 1975).

Bird diversity along similar habitat gradients in the mediterranean-climate zones of California, Chile and South Africa was the subject of an earlier paper (Cody 1975). I have added bird censuses from Sardinia (Appendix A; discussed in Cody 1980a) and two additional transects from South Africa, one in the Outeniqua and Swartberg Mountains in the southern Cape Province and the other from the Cedarberg in the northwestern Cape Province. The three South African transects are the subject of detailed analysis in following sections; here I simply use their α and β diversity curves in a comparative fashion, and give further details of the sites later.

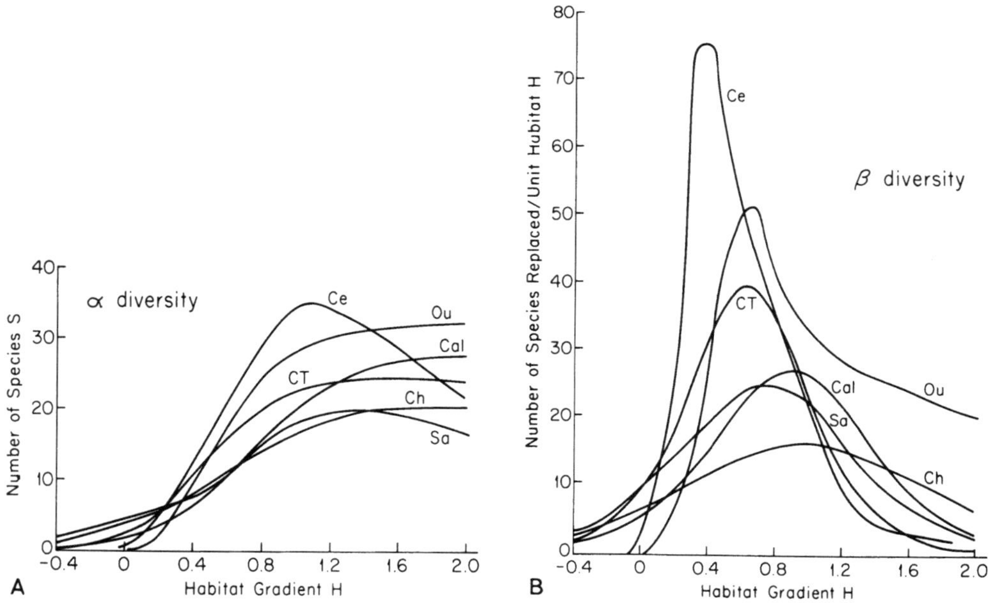

Fig. 1. Plots of α and β diversity over a gradient of mediterranean-climate habitats in six different areas. Four continents are represented. Ch: Chile, Sa: Sardinia, Cal: California, Ce: Cedarberg, Ou: Outeniqua Mountains, CT: Cape Town area, the last three in South Africa. A. α diversity. B. β diversity.

Fig. 1A shows bird α diversity plotted against habitat; fynbos habitats are located at around $0.8<H<1.2$, with shorter, more open and drier habitats to the left and taller, mesic woodlands to the right. The number of bird species increases with increasing H in an approximately sigmoid fashion, but in habitats taller than H = 1 species numbers increase at a decreasing rate, and may decrease in woodlands. Fig. 1B gives comparable data on β diversity. Unlike curves of α diversity, the species turnover rates on the gradient are quite different between transects. β diversities are highest overall in South Africa, where they peak at lower H values. California and Sardinia are similar with maximum β diversities of 25 species per unit H around H = 1.0. Chile has the lowest β diversity, and its maximum of 16 species per unit H is nearly five times lower than figures from the Cedarberg, and three times lower than those in the Outeniqua-Swartberg Mountains. These differences contrast to differences in α diversity among transects that are at most a factor of two and usually much less.

20.3.2 DIVERSITY AND HABITAT AREA

There are two reasonable and empirically acceptable sorts of relationships between α diversity and habitat area. The first is illustrated by the peninsula phenomenon: species diversity in general decreases from the proximal to the distal end of peninsulas. Less obviously, α diversity declines in more isolated or peripheral locations, and in areas isolated by topographic or vegetational discontinuities. The second sort of relationship takes account of the abundance in a region of a particular habitat. Common habitats are in general richer in species (other things being equal) than are rare habitats. Thus it is the rain forests of South America that are rich in bird species, while its savannas are relatively impoverished, but in Africa the reverse is true. These two types of relationship refer to the shape and area of habitat respectively.

The relationships between β diversity and habitat area are more obscure, but some might be extrapolated from the foregoing. For example, if two habitat types adjacent on H are both common, then species are more likely uniquely adapted to each and consequently a high species turnover will occur between the two habitats, even though they are adjacent on the map. If a habitat type occurs as isolated patches, a consequent reduction in α diversity might permit species in habitats adjacent both on H and on the map to move into the patches with a resulting reduction in β diversity. Relations like these are largely speculative, and must be explored.

Table 1 shows for each of six transects, the total "source area" of habitat within 1600 km of the sample sites and its distribution over H. Habitat "shadowed" from the transect locality by mountains in excess of 1500 m is not included as source area. This method is explained in Cody (1975); Acock's (1975) map of veld types is used for South Africa, and a vegetation map by Sahab (1966) gives the information for the mediterranean region. The figures of habitat availability are related to α and β diversity in Fig. 2. In Fig. 2A the diversity data define curves for each transect in the α - β plane, with the H values corresponding to the given α and β diversity values shown for each curve. On Fig. 2B habitat isoclines join the same habitat points on different transects, and the figures on these isoclines show the cumulative habitat area reached by a given H value on the gradient. Thus from Table 1, by H = 0.6 on the Californian transect $.39 \times 24.5 = 9.6$ units (10^5 km^2) of habitat area are accumulated, so this value appears on Fig. 2B, second from the left on the third curve up, where

α = 11 and β = 24. Notice that increased habitat availability in general enhances diversity; at low H more habitat influences chiefly β diversity, while more area at higher H promotes greater α diversity.

Table 1. Habitat characteristics around bird census transects in California, South Africa, Chile and Sardinia-Corsica. Total habitat is given in both km^2 (first column) and mi^2 (second column). The area under the $\alpha - \beta$ curve is calculated from Fig. 2A and is in arbitrary units.

Transect	Total Habitat Area (10^5)		Area under $\alpha - \beta$ curve	Cumulative proportion of habitat up to H = -0.2	0.2	0.6	1.0	1.4	1.8	2.2
California	24.5	9.5	160	.03	.21	.39	.58	.66	.81	1.0
Cape Town	22.5	8.7	191	.05	.22	.51	.79	.95	.98	1.0
Cedarberg	25.0	9.7	419	.10	.30	.55	.81	.96	.99	1.0
Outeniqua	24.1	9.3	350	.05	.20	.49	.78	.95	.99	1.0
Chile	3.4	1.3	69	.15	.29	.50	.71	.88	.96	1.0
Sardinia	40.2	15.5	94	.07	.39	.68	.77	.81	.87	1.0

Because the lines of Fig. 2B reflect internally consistent effects, rough algorithms can be generated as in Figs 2C and 2D. In Fig. 2C assume that the distribution of habitat over H is uniform, but that larger habitat areas are available, from 5 x 10^5 km^2 in distribution (i) to 30 x 10^5 km^2 in distribution (vi). Increasing habitat area produces the set of $\alpha - \beta$ curves shown in the figure. Fig. 2D shows the diversity curves produced by habitat distributions that vary from a preponderance of low habitat (vi) to mostly tall habitat (i), leaving total habitat area constant at 20 x 10^5 km^2. In each circumstance a different $\alpha - \beta$ curve will result, conforming to the known relationships as they are currently understood.

Figures for habitat availability on the Sardinia-Corsica transect are included on Fig. 2B inside dashed circles. While habitat availability in this region is high, the transects were made on islands and thus a major proportion of this available habitat is isolated from the transect by the Mediterranean Sea. The Sardinia-Corsica data allows us to infer, from their positions relative to continental sites, that these islands have an effective area of available habitat around one-third that measured in my conventional way. Thus we can estimate the efficacy of sea barriers compared to overland barriers.

To return to the data shown in Fig. 2A, my a posteriori explanation for a low α and β diversity over the Chilean transect is that the total habitat area there is small; the Andean cordillera reduces available source habitats to the strip between the mountains and the coast. This restricts immigrant species, reduces α diversity and thus favours habitat generalists and low β diversity. The California and Cape Town transects have a similar area of source habitat, but more of this habitat is woodland and forest in California, where consequently α diversity is higher at higher H. Around Cape Town there is more habitat around $0.6<H<1.0$, and this is where Cape

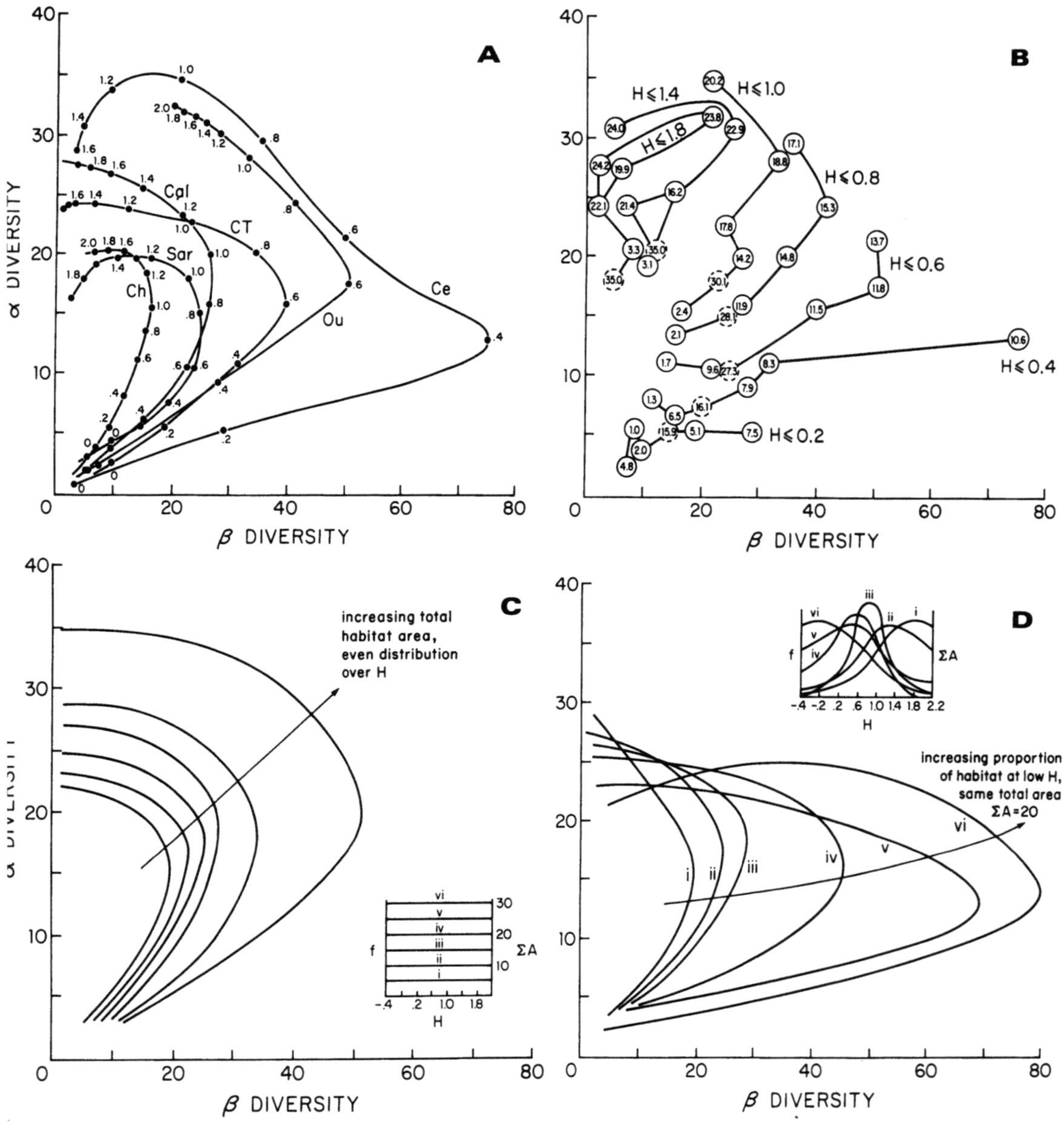

Fig. 2. The α and β diversity data of Fig. 1 are replotted in the diversity plane α - β. The figures on each curve represent the values of habitat H on the gradient by which the given α and β diversity values are reached (A). In B lines are drawn which connect points at the same H-values in different regions. The encircled figures on these lines are cumulative habitat area (km^2) reached in a given region under curves that describe how much habitat area of each type is available there (see text). If habitat availability is described by a uniform distribution over H, as in the inset graph of C, larger total areas of habitat (i) through (vi) will produce α - β curves as shown in C from innermost for smallest total habitat area to outermost for largest total habitat area. In D total habitat area is held constant, while its distribution over habitat type H varies. This produces the α - β curves shown. See text for further discussion.

Town β diversity is highest. Comparing the remaining two South African transects to that near Cape Town, both the Outeniqua and the Cedarberg transects have larger source areas (since they are not centred at the tip of the continent) and higher total bird diversities (area under the α - β curve of Fig. 2A). The Cedarberg has a variety of habitat with low H values - strandveld, renosterveld and karoid bushveld - each available in the source area over large and topographically simple areas. This gives a maximum β diversity value at H = 0.4, and high α diversity values up to H = 1.0, but thereafter the lack of habitat with H>1.4 causes α diversity to plunge. In the Outeniqua Mountains transect there is none of the low-H strandveld habitat, only renosterveld and the taller valley bushveld, and thus β diversity peaks at higher H. But this transect is located close to quite large forest areas to the east and north-east, and this influence is seen in a continually rising α diversity and a β diversity that remains high at the largest H values.

20.3.3 CONVERGENT EVOLUTION AND DIVERSITY ON HABITAT GRADIENTS

Both theory and common sense lead us to expect that competition among consumer species for limited resources will place an upper limit on α diversity and will shape the ecologies and morphologies of these consumers by jointly minimizing interspecific overlap and maximizing resource utilization. To the extent that habitats present to consumers similar densities and distributions of resources, we can expect convergent evolution to produce parallel α diversities, consumer densities, and community organization. But to the extent that these habitats differ among continents in their total areas, in the continuity of these areas, and in the accessibility of the habitat to potential colonists from similar habitat nearby, then α diversity might be expected to vary. In particular, in discontinuous or scarce habitats α diversity might equilibrate at levels below those permitted by competition among the members of a large and accessible species pool.

Since α diversity and β diversity are never entirely independent of each other, the factors that produce convergent evolution at the community level are likely to have some influence on β diversities. But this influence appears slight; β diversity is controlled more by topography, habitat continuity and contiguity than by such predictable biological factors as competition and adaptation to habitat-specific challenges of food supply and predators. Thus there is very little reason a priori to expect convergent evolutions in β diversities, and only in such circumstances of parallel distributions of habitat area and habitat availability is such convergence likely to be seen.

20.4 Effects of habitat contiguity

20.4.1 HABITAT PATCHES AS ISLANDS

Real islands are discrete parcels of land surrounded by water, but to express the difficulty of reaching the island as simply proportional to the width of the water barrier is a simplification that ignores the fact that a certain stretch of water might be far more of a barrier to one species than to another (Case and Cody 1982). These problems are magnified where isolation is by different habitats rather than by water gaps. Rosenzweig (1975) talks about "habitat taboos" in his discussion of continental analogies to island biogeography theory, and Cody (1975) related this notion to β diversity; it is epitomized by the striking fact that exotic habitats go virtually unoccupied by birds in regions where the β diversity in native vegetation is high, but are well populated in regions of low β diversity. Thus in the mediterranean zones of Chile and South Africa, pine and eucalyptus plantations support diverse and dense bird communities in the former, but are virtual deserts in the latter country.

In continental landscapes habitats occur naturally as mosaics, since vegetation structure as well as plant species vary with varying soil type, slope, aspect, fire and management history as well as climate. Thus patches of like habitat are often separated by intervening habitat of a different type that constitutes both a physical barrier, in terms of a stretch of alien terrain, and a biological barrier, since it is likely to contain competitor species that render the habitat additionally inhospitable. Here the black and white of land versus water becomes an intermediate grey, and we can rely on only empirical facts to estimate the importance of habitat barriers to the distribution of birds within continents.

These two aspects of continental distributions, the natural occurrence of a certain habitat type in a patchwork, and the varying degrees to which intervening habitats constitute a barrier to would-be colonists, have received very little attention from biogeographers and ecologists, yet it seems that their understanding is fundamental to both theoretical and analytical studies of distribution. The outcome of adaptation and competition among species that inhabit a patchwork environment is basically different from that in broad continuous habitat (Levins and Culver 1971; Horn and MacArthur 1972), and the same is true for habitats that are patchy in time (Levins 1968). Thus where habitats are patchy and/or rare there may be an increasingly important role for species with good invasion but poor competitive capacities which prevent the evolution of equilibrial communities. Such species would increase α diversity and decrease β diversity in patchy, marginal or rare habitat, and might through temporary overexploitation of resources be found in unexpectedly high densities (Diamond 1974; Case et al. 1979). This and other real-island phenomena are likely to show up in modified form on continents with habitat mosaics, but both the concepts and the data base are poorly developed. We know that North American mammals of pine and fir woodlands have not crossed desert barriers of modest width in thousands of years (Brown 1971), that many forest birds on the New Guinea coast will not cross water gaps of a few metres whereas others routinely fly thousands of miles to reach new islands (Diamond 1975), that isolated granitic buttes serve as habitat islands to saxicolous plants (Cody 1978) and so do thistleheads for phytophagous insects (Brown and Kodric-Brown 1977), but facts such as these have yet to be organized into a theory of continental diversity to parallel that developed for real islands.

20.4.2 THE AFROMONTANE WOODLAND PATCHES IN CAPE PROVINCE

When habitats of a distinct type occur in very small and very isolated patches, then the effects of these island-like characteristics on the resultant communities are readily observed; the Afromontane forest of the southwestern part of Cape Province is a good example. This distinctive 20-30 m evergreen vegetation is fairly continuous east of Knysna (White 1978), but between George and Cape Town exists in patches that become smaller and more isolated from the main forest block as one moves west. A detailed analysis of the plant communities in these forest patches is given by MacKenzie (1978), and of the bird communities by Cody (1982); here I will refer only to some of the most obvious results of habitat patchiness.

The distribution of the largest patches of Afromontane forest between George and Cape Town is shown in Fig. 3. Censuses of the birds were made at five sites: 2.8 ha at Kirstenbosch on Table Mountain, 4.2 ha in the Langeberg near Riviersonderend, 5.2 ha in Grootvadersbos near Heidelberg, 10.0 ha near Diepwalle, Knysna, and lastly 7.8 ha in the Alexandria forest east of Port Elizabeth. Both the vegetation structure and the density and distribution of insects vary somewhat among the five sites: mean vegetation height from a low of 22 m in Alexandria to a high of 28 m in Knysna, area under the foliage profile from a low of 40.1 units in Alexandria to a high of 82.6 at Kirstenbosch, and mean insect densities caught on Tanglefoot boards from a low of 7.4 insects per plaque per day at Grootvadersbos and Knysna to a high of 13.5 at Riviersonderend. Yet these forests are generally very similar in structure (see also MacKenzie 1978) and presumably also in the resources they provide for insectivorous birds. The Riviersonderend and Grootvadersbos areas were censused by Martin et al. in 1979; the correspondence in results is high, with 20 out of 22 and 23 out of 28 of my species totals respectively in their lists, and two species at each site recorded by them but not by me. Bird species totals decrease from Alexandria to Cape Town, from 42 to 15. Yet despite this nearly threefold difference in α diversity, the total bird densities are extremely similar (average difference 5%, largest 11%). These figures indicate

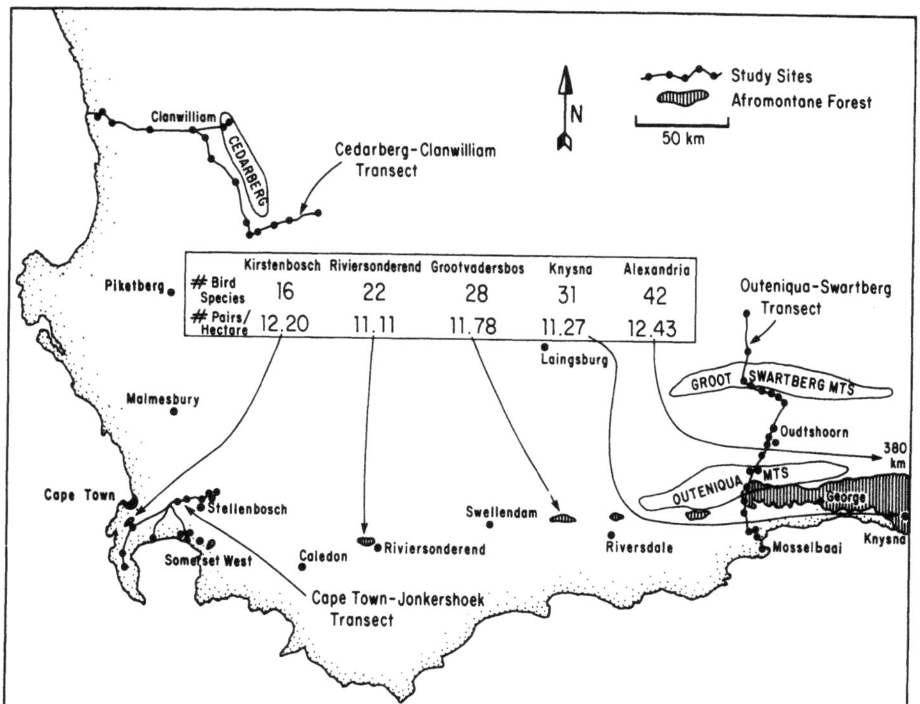

Fig. 3. Map of part of Cape Province, South Africa, showing the locations of five sites at which birds were censused in Afromontane woodland, and locations of three transects along which vegetation structure and bird diversity and density were measured.

considerable density compensation in the bird communities; the absence of bird species from the smaller and more isolated forest patches to the west, up to 65% of the Alexandria community by Kirstenbosch, is compensated by expanded foraging niches and increased densities of the remaining species.

The decrease in species numbers from Alexandria to Cape Town is very largely a filtering out of more and more species with increasing isolation, and not a process of species turnover. A few bird species with wider habitat tolerance make use of the forest habitat only towards the western end where α diversity is low. These are *Serinus canicollis* and *Streptopelia capicola* which occur in forest only from Grootvadersbos west, and *Onychognathus morio* and *Cinnyris chalybeus*, which occur in forest only from Knysna west. These species appear to be replaced by other canaries, doves and sunbirds to the east.

Density compensation is most readily observed in those species which are dominant insectivores and occur from Cape Town across to Alexandria. *Zosterops pallidus* decreases in density from west to east across the five sites: 2.5, 2.5, 1.5, 1.4 and 0.8 pairs ha^{-1} respectively; *Muscicapa adusta* with similarly decreasing densities: 1.1, 0.7, 0.4, 0.3 and 0.3 pairs ha^{-1} respectively; and *Cossypha caffra*, which barely makes it in forest to Alexandria: 1.0, 0.5, 0.4, 0, and 0.03 respectively. There is just one species, the insectivore *Apalis thoracica*, which is reasonably common in all forests east of Kirstenbosch, quite common in shorter habitats such as kloof woodland and tall fynbos near Kirstenbosch, but virtually absent from the Kirstenbosch forest. It appears that the density compensation has all gone to *Zosterops*, the only foliage insectivore at Kirstenbosch, to the complete exclusion of *Apalis*. I observed putative equilibrium densities of the four foliage insectivores at the Knysna forest of *Zosterops pallidus* (Z): 1.36 pairs ha^{-1}; *Apalis thoracica* (A): 0.44 pairs ha^{-1}; *Seicerus ruficapilla* (S): 0.94 pairs ha^{-1}; *Camaroptera brachyura* (C): 1.46 pairs ha^{-1}. Competitive interactions among these four species are estimated from the overlaps between their normalized foraging height distributions using

	Z	A	S	C
Z	1.00	0.31	0.70	0.22
A	0.31	1.00	0.10	0.50
S	0.70	0.10	1.00	0.00
C	0.22	0.50	0.000	1.00

Carrying capacities at Knysna can be derived from the following equilibrium condition:

$$\begin{bmatrix} K_Z \\ K_A \\ K_S \\ K_C \end{bmatrix} = \begin{bmatrix} 1.00 & 0.31 & 0.70 & 0.22 \\ 0.31 & 1.00 & 0.10 & 0.50 \\ 0.70 & 0.10 & 1.00 & 0.00 \\ 0.22 & 0.50 & 0.00 & 1.00 \end{bmatrix} \begin{bmatrix} 1.36 \\ 0.44 \\ 0.94 \\ 1.46 \end{bmatrix}$$

whence $K_Z = 2.48$, $K_A = 1.69$, $K_S = 1.94$, and $K_C = 1.98$. We notice immediately that the carrying capacity of *Zosterops* at Knysna is just the equilibrium density of the species where it occurs alone at Kirstenbosch, strong evidence that we have identified the relevant competitors and that the resources are basically similar in those two forests. Yet if we remove *Seicercus* and *Camaroptera* from the four-species assembly, we predict that

Zosterops and *Apalis* can coexist at elevated densities via density compensation for the missing two species:

$$\begin{bmatrix} 2.48 \\ 1.69 \end{bmatrix} = \begin{bmatrix} 1.00 & 0.31 \\ 0.31 & 1.00 \end{bmatrix} \begin{bmatrix} X_Z \\ X_A \end{bmatrix}$$

whence the new equilibrium densities are $X_Z = 2.15$ and $X_A = 1.06$. Yet X_A is observed to be zero. If overall resource levels are elevated at Kirstenbosch by a factor P, then the observed equilibrial densities should be equally enhanced by the factor P. I know that the insect densities are very similar, but carrying capacity for *Zosterops* might be increased at Kirstenbosch by more available fruit, and by *Zosterops*' shifting its foraging niche from that at Knysna. If the carrying capacity of *Zosterops* at Kirstenbosch is in fact changed by a factor k, then

$$\begin{bmatrix} 2.48k \\ 1.69 \end{bmatrix} = \begin{bmatrix} 1.00 & 0.31 \\ 0.31 & 1.00 \end{bmatrix} \begin{bmatrix} X_Z \\ X_A \end{bmatrix}$$

Solving for equilibrium density of *Apalis* by Cramer's Rule,

$$X_A = \frac{\begin{vmatrix} 1.00 & 2.48k \\ 0.31 & 1.69 \end{vmatrix}}{\begin{vmatrix} 1.00 & 0.31 \\ 0.31 & 1.00 \end{vmatrix}} = 1.87 - 0.85k$$

Thus *Zosterops* can exclude *Apalis* from the Kirstenbosch forest if it can exploit new resources or shift its foraging niche and thus its carrying capacity there by a factor (1.87/0.85) of just over two.

Alternatively, *Zosterops* might exclude *Apalis* from the Kirstenbosch forest by niche shifts that increase the competition coefficient α_{AZ} above 0.31. In that case

$$\begin{bmatrix} 2.48 \\ 1.69 \end{bmatrix} = \begin{bmatrix} 1.00 & 0.31 \\ \alpha_{AZ} & 1.00 \end{bmatrix} \begin{bmatrix} X_Z \\ X_A \end{bmatrix}$$

at equilibrium, whence

$$X_A = \frac{\begin{vmatrix} 1.00 & 2.48 \\ \alpha_{AZ} & 1.69 \end{vmatrix}}{\begin{vmatrix} 1 & 0.31 \\ \alpha_{AZ} & 1.00 \end{vmatrix}} = \frac{1.69 - 2.48\,\alpha_{AZ}}{1 - 0.31\,\alpha_{AZ}}$$

If the competitive effect of *Zosterops* on *Apalis* were increased to $\alpha_{AZ} = 1.69/2.48 = 0.68$, slightly more than double 0.31, then at equilibrium X_A would be zero. These questions are discussed in more detail elsewhere (Cody 1982) and have received some theoretical attention. The likelihood and stability of alternative equilibria in multispecies competition communities are known from simulation (Case and Casten 1979). Its importance to us here is that communities that are lower in α diversity because of habitat patchiness or isolation may be subjected to additional (internal) factors that will further reduce α diversity.

20.4.3 SPECIES TURNOVER WITHIN A HABITAT BETWEEN DIFFERENT REGIONS

When a given habitat is fragmented into patches or sections that are separated by either physical barriers (e.g. cordilleras or water bodies) or on a more local level by intervening habitat of a different type, then there arises the possibility that different species might fill similar ecological roles in different parts of the range of the habitat. These ecological vicars or ecological analogues constitute a third diversity component that I call γ diversity (γ for "geography"). Here the species turnover is independent of habitat differences and is measured over geographic area within a given habitat type, H.

In the Afromontane woodland censuses discussed above there are few replacements of ecological counterparts between sites. One reason might be that as recently as 10,000 years ago many of these forest patches were contiguous or nearly so such that there would have been more interchange of birds between them. Although now there is circumstantial evidence of genetic isolation in the bulbul Andropadus importunus populations, whose songs vary little within but considerably between the western forest patches, there are no indigenous species or races there.

If ecological counterparts each occupy habitats that differ markedly in area, then the species in the more restricted area is likely to be replaced, over time, by that of the larger habitat as long as the geographic isolation of the two areas does not preclude some threshold level of immigration. Thus the point at which ecological analogues replace each other on the map should be affected by the relative areas of habitats on H (see Trombulak and Cody 1980).

Examples of γ diversity in fynbos and related vegetation are discussed below; others are to be found among flycatchers (Batis), sugarbirds (Promerops), seedeaters (Serinus), robins (Erythropygia) and rock-thrushes (Monticola).

20.5 Ecological biogeography and convergent evolution

20.5.1 EVOLUTION OF GAMMA DIVERSITY

In this section of the paper I deal with concepts necessary for understanding firstly, continental diversity patterns and secondly, convergent and non-convergent evolution. These subjects are closely related, since the major unknowns in continental diversity patterns are those factors that control γ diversity, and it is between γ diversity counterparts that we expect to see the products of convergent evolution.

A basic fact of ecological biogeography is a statement such as: Species X_1 occurs in region R_1 where it occupies habitats $f_1(H)$ in which it pursues a certain ecological role. Juxtaposed with this fact might be a second statement: X_1 fails to occur in region R_2. As ecological biogeographers, we want to know why.

Perhaps the two statements can be reconciled because there is no suitable habitat in R_2. Unless some part of $f_1(H)$ is represented in the second region, then we do not expect X_1 to occur there in the first place. But if $f_1(H)$ includes habitat present in R_2, then perhaps the absence of X_1 is explained in terms of access. X_1 may be excluded by major physical barriers such as oceans or mountain chains or by biotic barriers such as very different habitat; these biotic barriers might be inhospitable per se such as a desert for woodland birds or inhospitable because of competitor species that live in the habitat barrier. If there is no obvious barrier then we must look to the habitats $f_1(H)$ in R_2 for the explanation. The answer may be simple competition in the form of a single species X_2, the ecological counterpart of X_1. X_2 might be quite unrelated to X_1 in which case its role as ecological counterpart or γ diversity counterpart is a product of convergent evolution. Alternatively regions R_1 and R_2 might have been divided by a physical or biotic barrier in the past, and the two species have a recent common ancestor. Sets of γ diversity counterparts in this category were mentioned above.

If no single ecological counterpart prevents access to R_2 by X_1, perhaps diffuse competition is the answer. Communities structured by competition may have alternative stable equilibria, some of which might include X_1 and some not. In this case the presence of several competitor species jointly serve to exclude X_1. Again, this competition has several aspects, for close competitors might be species either with intolerably large ecological overlap with X_1 or might be species that have a larger carrying capacity in R_2 than in R_1 where they coexist with X_1. The case of Zosterops and Apalis in Kirstenbosch could have this explanation.

Another possibility is that despite the duplication of f_1 habitats in R_2 and the absence of strong competitors, resources in these habitats might differ. Resources and consumers can coevolve tight and specific interrelations that subordinate the effects of habitat structure. If X_1 feeds only from flowers with 25-28 mm corollas or eats only hard seeds of diameters 7.5-8.6 mm, then we must be sure that such resources are present in R_2 before we wonder at X_1's absence.

And finally the constraints of resources, access and competitors might all be satisfied, yet the areal extent of f_1 habitats in R_2 might be very small. X_1 is unlikely to persist in such small patches, and even though it might occasionally reach and use these patches we might not find it there at a given time. There are other but less tangible or less interesting possibilities why X_1 is restricted to R_1. One might be the evolution of a sedentary tendency in X_1 such that it does not even try to reach R_2, or strange new predators eat it when it arrives, or perhaps X_1 needs the social facilitation of its home population; for small, mobile birds oriented towards resource competition, these possibilities seem distant.

20.5.2 FIGURE 4: EFFECTS OF HABITAT PATCHINESS

Figure 4 is an attempt to clarify my perspective of some of the factors that influence continental species diversity and involves many of the foregoing concepts. I imagine that subject species X occupies a geographic range G, but within this range is able to occupy only those habitats where $H>0.5$ and does rather poorly in habitats $H>1.5$. Within G I have drawn contours of H showing where there is occupiable habitat. The largest habitat block is at (a) where X concentrates its f(H) around H = 1. Other species restrict X to this range of habitats from below and above on H. Here X's food niche F can be translated into a density, distributed over H as indicated by the hatched lines. X has access to habitats in block (b) but because habitats here are commoner in the range $H>1.5$, X is subjected to stronger competition from the species of taller vegetation and it shifts its f(H) lower on H. This in turn affects the availability of foods to which its morphology and behaviour are best suited, and thus its density is less in lower H.

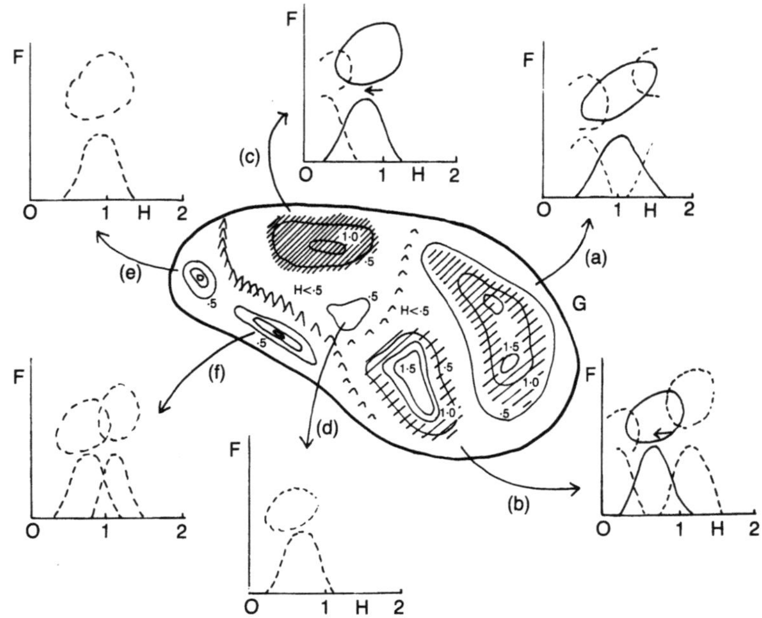

Fig. 4. Schematic representation of the availability of habitat type H, patchily distributed over a geographic region G. Contours represent changing habitat H; hatched lines and frequency distributions on the inset graphs show habitat occupancy of a given species, which changes as does habitat availability. A second niche axis F represents foraging ecology, relative to competitor species (dashed lines).

A third but smaller habitat block (c) is also colonized by X, and here the high-H competitor is absent. Thus X can both occupy all of H>0.5 and make up for the missing competitor over median H values. Its density in (c) is thus high; this in turn might enable X to push more strongly into lower habitats, to which it might become somewhat better adapted by virtue of the small area and relative isolation of (c) from (a) and (b). There is also accessible habitat in patch (d) where H>0.5, but the patch is too small for X to inhabit regularly, and species from the surrounding lower-H habitat annex the patch and its resources. Two additional patches of suitable habitat occur on the edge of X's range, but these are rendered less accessible by the presence of a barrier. Either patch might be invasible for X, depending on the extent to which the species that use X's resources and habitats in its absence evolve to become close ecological counterparts. With reduced accessibility of these patches to X, the time necessary for this convergent evolution is better assured.

20.5.3 EVOLUTION OF NON-CONVERGENCE

At this point I wish to restate some of the major problems in studies of convergent evolution. The important message from physiological plant ecology for bird ecologists is that soils differ in the sorts and amounts of nutrients they provide for plant growth, and these differences have been part of the evolutionary scene for millions of years. Thus plants differ in their life histories, in the rates at which they fix carbon, replace leaves, in their defence budgets and in their modes of reproduction. In many respects the typical fynbos Protea and the typical chaparral Ceanothus are at opposite poles, with the former using a few, thick, unpalatable and long-lived leaves, producing copious nectar to attract large vertebrate pollination vectors and setting a few very large seeds. The latter has many smaller, edible and often augmented leaves, is bee-pollinated and produces many smaller seeds that are of much less significance in its population cycle. Thus despite the similar climate of fynbos and chaparral, insects and insectivorous birds are scarcer in fynbos and nectarivorous birds are scarcer in chaparral.

The importance of coevolution between resources and consumers has not been accorded the attention it deserves, and recent (unpublished) papers by A V Milewski point this out. In particular the resources that plant parts constitute for seed-, fruit-, leaf- and nectar-eating birds, and the consumers of these resources, interact to provide reciprocal feedback that affects the course of each other's evolution. Especially where these interactions are mutual, as in pollination and seed dispersal, and where it behooves both the plant and its helpful consumer to preclude the involvement of other consumers, then the two halves of the system may together trace a divergent and distinct evolutionary pathway unlikely to have a counterpart elsewhere. To some extent we can control for this effect in different communities by more accurate measurement of kinds of resources and how they differ between continents. In this way our expectations of convergence will be more modest and realistic.

The other variable I have stressed is the ecological and biogeographic consequences of habitat areas. Habitat area and the extent to which it is patchy or continuous strongly affects α and β diversity. We cannot expect to find close ecological counterparts in communities that differ in α diversity by factors of two or more as in the woodlands discussed earlier, and clearly there will be little one-to-one correspondence between the species in such communities.

We cannot control for differences in habitat area very easily, but we can at least conduct studies that will elucidate the role of habitat area and contiguity in diversity. Moreover there is a large and varied category of events and circumstances generally referred to as "historical factors" of which our understanding is even poorer. These may be short-term, and include such varied effects on diversity as the date of the last fire, the last time pipe-wood was removed from the macchia, or the last time natural browsers were present. They may also be the sorts of factors that affect speciation, the long-term contiguity and subdivision of habitats, the shifts over the map of a particular habitat type with changing climate to alter its relative accessibility from a source region, and even changes in the maps themselves. Clearly we are hardly able to ask the appropriate sorts of questions of these factors, and even further from the answers that bear on convergent evolution.

20.6 Bird diversity in three areas of Cape Province

20.6.1 SITES, DATA COLLECTION, HABITAT ORDINATION

For further insights into the interrelations among α, β and γ diversity in South African birds I studied three areas of the Cape Province in which a similar sequence of habitats was assessed for vegetation structure and over which bird diversity and density were measured. These three areas are in the Cape Town-Jonkershoek region, the Clanwilliam-Cedarberg area to the north-west, and the Outeniqua-Swartberg area in the southeastern part of the province (Fig. 3). These study sites and their characteristics are listed in Appendix B. For the Cape Town area I use the 13 sites reported in Cody (1975), located between the Cape Peninsula and the Jonkershoek valley. These sites cover a sequence of habitats between the Cape Flats in coastal strandveld to renosterveld, fynbos, kloof woodland and finally the patch of Afromontane woodland on the side of Table Mountain at Kirstenbosch. In the Cedarberg-Clanwilliam area I collected data from 2-22 October 1979 in strandveld vegetation, in renosterveld and fynbos in the Cedarberg, and in sites in karoo, karoid broken veld and succulent karoo (14 sites). I selected 17 sites in the Outeniqua-Swartberg area and worked on them from 3 November to 12 December 1979. These are located between the coast near Mosselbaai north across the Outeniquas at Robinson Pass, near Oudtshoorn and across Swartberg Pass to Prince Albert in the Karoo. The sites include coastal renosterveld and valley bushveld near the coast, mountain renosterveld and fynbos in both mountain ranges, various types of karoid vegetation in the Little Karoo and north of Swartberg, kloof woodland near Robinson's Pass, and Knysna forest at Diepwalle, north of Knysna. In general climatic variation is as great within as between transects. Rainfall on the Outeniqua-Swartberg transect shows at least a four-fold variation (George: 851 mm yr^{-1}, Oudtshoorn: 237 mm yr^{-1}); Clanwilliam is similar to Oudtshoorn with 203 mm yr^{-1}, Stellenbosch with 704 mm yr^{-1} approaches the George total. Nevertheless, wetter sites increase from the northwestern to southeastern Cape Province, and the incidence of summer rainfall follows the same trend.

Vegetation measurements. Each site consisted of relatively homogeneous and undisturbed vegetation; although site areas are variable, from a low of 0.74 ha to a high of 30.1 ha, over one half of the sites are within the range 2-5 ha (see Appendix B). Site area was determined by the extent and continuity of the vegetation, and by bird density and distribution: taller, denser habitats with high bird density, especially those such as kloof woodland that occur naturally as patches, were sampled over small areas, while low, open and continuous vegetation with low bird density, such as the karoo sites, were sampled over much larger areas.

Within each delimited site 25 points were chosen haphazardly, and from each point orthogonal axes were established oriented according to the direction of the sun from the observer. Vegetation densities were then measured, using the inverse of the distance from the observer to points of 50% vegetation cover, along each of the four axes at a series of height intervals: 0.15 m, 0.30 m, 0.60 m, 1.2 m, 2 m, 3 m, 4 m, 6 m and thence at 2-m intervals until the vegetation was exceeded. From these 100 vegetation density measures a mean vegetation profile was constructed, and from this site-specific profile the following six variables are derived: (a) $\log(ht_{.01})$, the logarithm of the height above the ground at which vegetation density falls below .01 (reciprocal of 100 feet, = 30 m); (b) $\log(1 + \frac{1}{2}ht)$, the logarithm of 10x the height at which half the vegetation density of the profile is below and half is above (the half-height of Cody, 1975); (c) $\log(ht_{.1})$, the logarithm of the height above ground where vegetation density diminishes to 0.1 (reciprocal of 10 feet, = 3m); (d) logA, the logarithm of the area under the foliage profile, in units of an interval of 1 foot (= 0.3 m) on the ordinate and an interval of 0.1 (reciprocal of 10 feet, = 3.0 m) on the abscissa; (e) the ratio $\log(ht_{.01})/\log A$, a measure of the openness of the vegetation; and finally (f) $\log(2 + \frac{1}{2}d)$, the logarithm of 100x the vegetation density at one-half the vegetation height ($ht_{.01}$). The values of these six variables for each of the 44 sites on three transects are given in Appendix B. The data on vegetation structure are used to define a habitat plane. Since my goal is to compare bird distributions and densities between the three transects, I ordinate the habitats using vegetation data alone, with no input from bird data or distributions. I use factor analysis on the vegetation data to collapse the six variables into their first two principal components. These two variables, PCI and PCII, are statistically independent, are very similarly constructed from original measurements in each of the three data sets independently, and between them account for some 80% of the overall variation in vegetation structure. The variables are derived via the following steps: (a) construct a 44 x 6 data matrix, using all 44 sites and all six vegetation variables described above; (b) normalize the columns to zero means and unit variance; (c) derive the 6 x 6 correlation matrix; and (d) calculate the eigenvectors of this matrix. The first two eigenvectors are given in Table 2, and when the data matrix is postmultiplied by these vectors, the first two principal components are produced (last two columns of Appendix B).

Table 2. First two eigenvectors used to derive the principal components of vegetation structure of sites in three areas of Cape Province.

Q1	Q2
0.387	-0.505
0.382	-0.232
0.543	-0.080
0.518	0.035
-0.265	-0.598
0.265	0.571

Bird censuses. I conducted bird censuses that were designed to answer the question: Given that a site of these vegetational characteristics in this region is surveyed at this time of year, which bird species are encountered at the site and with what frequency? The censuses do not establish which parts of a site a species uses, what its population structure is, what territorial or other social system operates in each species, what the demographies of the various populations are, or what foraging ecology a species employs. Bird census results for sites in the three areas are given in Appendices C, D and E. The species include those that foraged within the sites during my census visits, and exclude all raptors, noctural species, brood parasites (cuckoos and honeyguides) and aerial foragers (swallows and swifts). The great majority of these species foraged regularly within the sites and were breeding at the time of the censuses. Many species are monogamous, defended small territories vigorously, and sang consistently (e.g. Cisticola spp., Emberiza capensis, Bradypterus victorinii, Seicercus ruficapilla, Sphenoeacus afer), and from this activity their presence and density were accurately determined. A minority of species foraged in small, narrow-ranging groups (Phoeniculus, Zosterops), others are gregarious and bred outside the study sites (Spreo, Onychognathus, many doves); Euplectes is polygynous (2-3 females per male at my sites), and a variety of species are less flamboyant in their territorial activities (Stenostira, Malurus, Telophorus, Francolinus, many sunbirds). Species in these last categories are less readily counted. At just one site (no. 17, Outeniqua-Swartberg) a north-temperate migrant

Hippolais icterina, was included in the census as a regular site user.

Most sites were visited 3-6 times, and each visit lasted 2-6 hours; during a single day I censused 3-5 sites on a transect, visiting each site at different times on different days. I moved slowly through the study area, recording on a map the location and movements of each bird encountered (by sight or sound), aided by flagging tape on the vegetation at regular intervals. At most sites successive visits produced the same list of bird species, but at others some species were observed just once or twice. Some birds had wide foraging ranges of which the study site was but a small part, and an effort was made to determine what part from observations beyond the site boundaries. Around 15 hours were spent at an average site, but considerably more detailed observations were made in the fynbos (site 11) at Jonkershoek (see Cody and Mooney 1978) and in the woodland sites at Kirstenbosch (no. 13) and Knysna (no. 7, see Cody 1982).

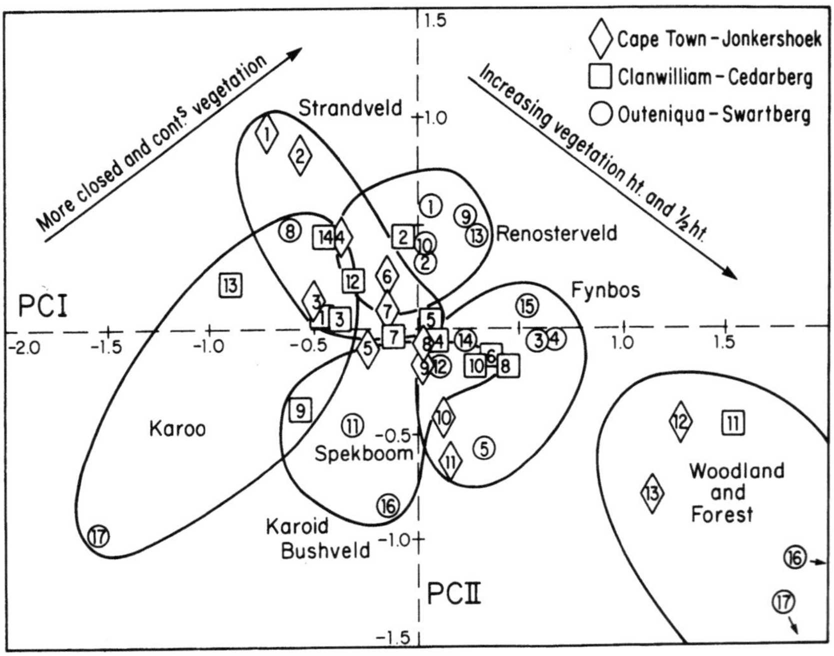

Fig. 5. Distribution of 44 sites on three transects of similar habitat, ordinated by the first two principal components of vegetation structure, PCI and PCII, which allows the sites to be represented in the habitat plane of this figure. The symbols refer to transects, and the numbers enclosed refer to the site numbers in Appendix B; the appendix lists the original habitat variables at each site, and the PCI-II figures computed for the site. Sites group according to veld type, and the axes are interpreted biologically.

At each site the census data were converted to bird density estimates in pairs ha^{-1}. These figures are given in the appendices (C-E), and while most reflect the number of breeding pairs per ha, others reflect comparable utilization of the sites by birds breeding elsewhere or at some other time. Low density figures are obtained for those species observed only intermittently, and/or have wide foraging ranges within or beyond the study site. Although some estimates are crude, the bird density figures are used here only to compare between sites and among transects. The census data permit such conclusions as: Emberiza capensis, observed at all sites in the Clanwilliam-Cedarberg area, has a much wider habitat range there than in the other areas; Apalis thoracica is widely distributed over habitat in the Outeniqua Mountains, more restricted in the Cape Town-Jonkershoek area and, despite a range that reaches the Cedarberg, must be extremely restricted there as it was not observed at any Cedarberg site.

The sites in the Cape Town-Jonkershoek area are those of Cody (1975), but several (nos. 8, 11, 12, 13) were recensused in the last week of October 1979. The additional surveys account for some discrepancies between Appendix D of Cody (1975) and Appendix C here: add Serinus albogularis to site 8, Tersiphone viridis to sites 12 and 13, Serinus canicollis and Onychognathus morio to site 13. Anthus novaeseelandiae was omitted from site 3 and Corvus omitted for four sites in the 1975 list by error and policy respectively.

The distribution of the census sites in the plane of PCI and PCII is shown in Fig. 5. PCI correlates positively with the vegetation height and total vegetation density, whereas PCII correlates positively with high vegetation density at one-half height, and negatively with tall vegetation that is of overall low density. It is easier to interpret these axes in reference to vegetation types as named by Acocks (1975). Fig. 5 shows that the Cape Flats sites are upper left on the habitat plane, and the woodland and forest sites are to the bottom right. Between these two poles are, sequentially from the upper left, strandveld, renosterveld and fynbos. The low, sparse and open karoo sites are to the left on PCI, and the taller bushveld and karoid broken veld sites are low on PCII around the center of PCI. Each transect contributes comparable sites to most sections of the habitat plane, with the exception that no karoo sites are available on the Cape Town-Jonkershoek transect.

20.6.2 COMPARISONS OF THE BIRD FAUNAS

A total of 63 species was recorded on the Cedarberg transect, 66 species in the Outeniqua-Swartberg area exclusive of a further 17 species found in the Knysna forest, and 53 species for the Cape Town-Jonkershoek transect. Altogether 109 species were found in the combined censuses.

Some 30 species are found on all three transects, and the differences among the three lists are due chiefly to (a) a pool of eastern and/or forest species present in the Outeniqua-Swartberg area, few of which reach the Cape area and even fewer of which extend further north to the Cedarberg, and (b) a pool of northwestern and/or karoo species that fail to reach the southern and eastern habitats. Among the first category are the 17 species (see Appendix B, site 17) of the Knysna forest that show up nowhere else, and species such as Dicrurus adsimilis and Serinus scotops that extend beyond the forest but only in the Outeniquas. Other forest-associated birds that reach the Cape but are not recorded in the Cedarberg censuses are Tersiphone viridis, Andropadus importunus, Muscicapa adusta, Batis capensis, Apalis thoracica and Cisticola fulvicapilla. Twenty-six species are restricted to the Outeniqua-Swartberg transect, a further nine are shared with the Cape only, and a further 16, including many karoo species censused in the Swartberg area, are shared with the Cedarberg region only.

A small pool of eight species is found mainly on the Cape Flats and was not censused elsewhere; amongst these are Macronyx capensis, Calandrella cinerea, Cisticola textrix, Anthus novaeseelandiae and A. leucophrys. The 13 species found only in the Cedarberg area includes such karoo specialities as Oenanthe monticola and Serinus alario, and species common in strandveld and succulent karoo such as Parus afer, Batis pririt, Eremomela icteropygialis, Myrmecocichla formicivora and Melaenornis infuscatus. Clearly the location of each region has greatly influenced the species recorded in the transects despite the similarity of the habitats involved. There is an infusion of karoo and coastal strandveld species into the northwestern sites, and infusion of forest species into the eastern sites, and few of either in the Cape area where the karoo is further inland, the strandveld has almost petered out down the west coast, and the forest connection is tenuous.

20.6.3 COMPARISONS OF BIRD DIVERSITIES

Some mention has already been made of α and β diversities on the three Cape Province transects (see above, Figs 1A and 1B). I have looked at α diversity in a somewhat different light by plotting cumulative species numbers against cumulative area on the habitat plane PCI-PCII. The result confirms what was apparent earlier, that the avifauna of the Cape Town-Jonkershoek transect is somewhat depauperate in comparison to the other two. Taking nested samples from the origin of the habitat plane, an area of 1.0 on the plane produces 51 and 52 species in the Cedarberg and Outeniqua-Swartberg respectively, but only 43 species in the Cape Town area. The species-area curve for the habitat plane is worth plotting only if there is an equal density of samples over the plane from each region, and so I defer further comment until I compare specific habitat types from the three regions below.

I next analyse bird species turnover between different sites within each region. As usual I measure species turnover as the average of the number of species gained and the number of species lost from one census to another. Thus turnover from a census of five species to a census of seven species with three in common is $\frac{1}{2}$(no. spp gained + no. spp. lost) = $\frac{1}{2}(4 + 2) = 3$. In this way each pair of censuses on the habitat plane is characterized by a turnover value, and census sites can be grouped according to these values. In Figs 6A-C I show the contours of faunal turnover on the habitat plane for each region. These figures show clearly the extent to which there is species turnover between different habitats. In particular the contours delimit habitat sets with relatively constant bird composition (low β diversity) and those where there is high turnover (high β diversity). In the Outeniqua-Swartberg area (Fig. 6C) there are four main sets of bird species: one each in the karoo, the renosterveld and karoid broken veld, the centrally positioned valley bushveld, and the fynbos and woodland to the centre and lower right. By comparison, in the Cedarberg area (Fig. 6B) the woodland species set extends not nearly as far into the fynbos habitats in the centre of the habitat plane, there is no valley bushveld, and there is a strong segregation between the birds of the strandveld plus renosterveld and those of the karoid bushveld. Here again the karoo sites are distinct. In the Cape Town area (Fig. 6A) the fynbos and woodland sites again cluster together at low turnover levels, while the second major bird grouping is based in low strandveld with the renosterveld group annexed. These diagrams illustrate at a glance which habitats are occupied by certain cohesive faunal groups, and how these groupings shift over the habitat plane between different regions.

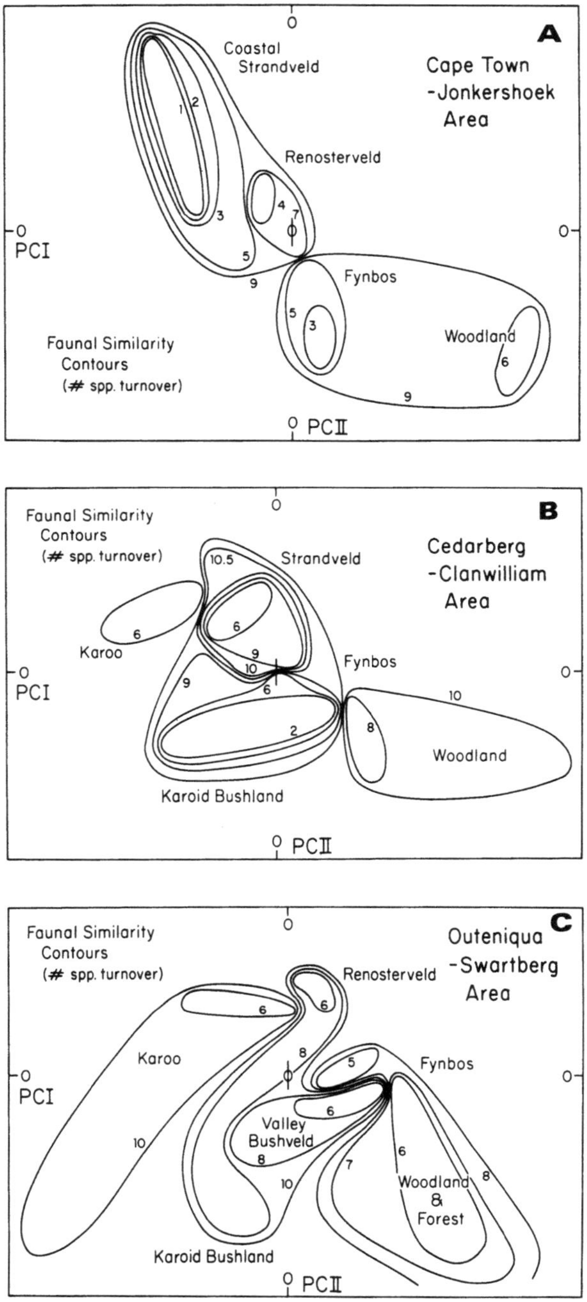

Fig. 6. Faunal similarity contours for the habitat transects. The contours include sites between which β diversity is at the magnitude indicated along the contour. Sites most similar in species lists (low β diversity) are grouped first, and the contours expanded to include additional sites that support increasingly different bird species. Thus censuses are similar within contours, but are increasingly different across contours.

20.6.4 COMPARISON OF INDIVIDUAL SPECIES DISTRIBUTIONS AMONG TRANSECTS

Next we ask to what extent do species that occur in all three regions censused occupy the same range of habitats. Taking first the ubiquitous Zosterops pallidus, its distribution on the habitat plane is shown in Fig. 7A. Not only does this species occupy a broad range of habitats, but the pattern indicates that it supports higher densities in certain habitat types from which in all directions on the habitat plane its density declines. Furthermore, there is no evidence that either the range of habitats utilized or the density in any particular habitat varies among transects. These data indicate that what is represented on the habitat plane is certainly something relevant to Zosterops. Further evidence that the habitats are arranged by combinations of variables that relate to bird habitat selection is shown in Fig. 7B where three warbler species with restricted habitat ranges are pictured. These three species are each represented several times, on two or more transects, and each occupies a discrete sector of the habitat plane.

Another widely-ranging and often abundant species is Prinia maculosa, and its distribution is shown in Fig. 7C. Apart from the one outlying point that shows the species in high density in Cedarberg kloof woodland (see below) there is little to indicate that the species occupies different habitat ranges on the three transects. However, there are differences among the densities it reaches in different regions, and in the Cape censuses Prinia numbers are double those counted elsewhere. Cisticola subruficapilla is another habitat generalist and although one cannot suggest from the data that its densities vary among transects this species occupies a more restricted range of habitats in the Cape area than it does in either of the other transects (Fig. 7D). And the last broadly distributed species considered is the bunting Emberiza capensis (Fig. 7E), recorded in all transects. This species is both more restricted in habitat range and present in lower densities in the Cape area, and reaches the highest densities where it occupies the most habitat types, in the Cedarberg.

In Fig. 7F I show the habitat distribution of Apalis thoracica, a species that favours taller and denser vegetation. It is in the eastern transect, towards the region where woodland and forest become comon habitats, that this species reaches its maximum distribution over habitat types. Here it occurs in quite low habitat in the centre of the habitat plane, habitat that is avoided in the Cape area where the species accepts nothing shorter or more open than tall fynbos. It is absent from the Cedarberg transect entirely. In Fig. 7G the distributions of two species of bulbuls are shown. That which favours forest and taller scrub habitats, Andropadus importunus, fails to reach the Cedarberg, but Pyconotus capensis occurs on all three transects. In the Outeniqua-Swartberg area (specifically in the Outeniqua censuses only) Andropadus extends its habitat range into fynbos and even renosterveld near the coast, as well as using the taller kloof woodland and forest. Pycnonotus on this transect is most abundant in the open bushveld and is absent from fynbos. In the Cape Andropadus shows up only in the Kirstenbosch forest habitat, and Pycnonotus is present throughout the fynbos. In the Cedarberg there are no Andropadus, and here Pycnonotus extends through tall fynbos and into the kloof woodland.

Finally I show the distribution of two canary species in Fig. 8 to illustrate once more that species distribution over habitat can change markedly between different regions. In the Cedarberg the turnover point between the two species distributions occurs around PCI = -0.25, whereas in

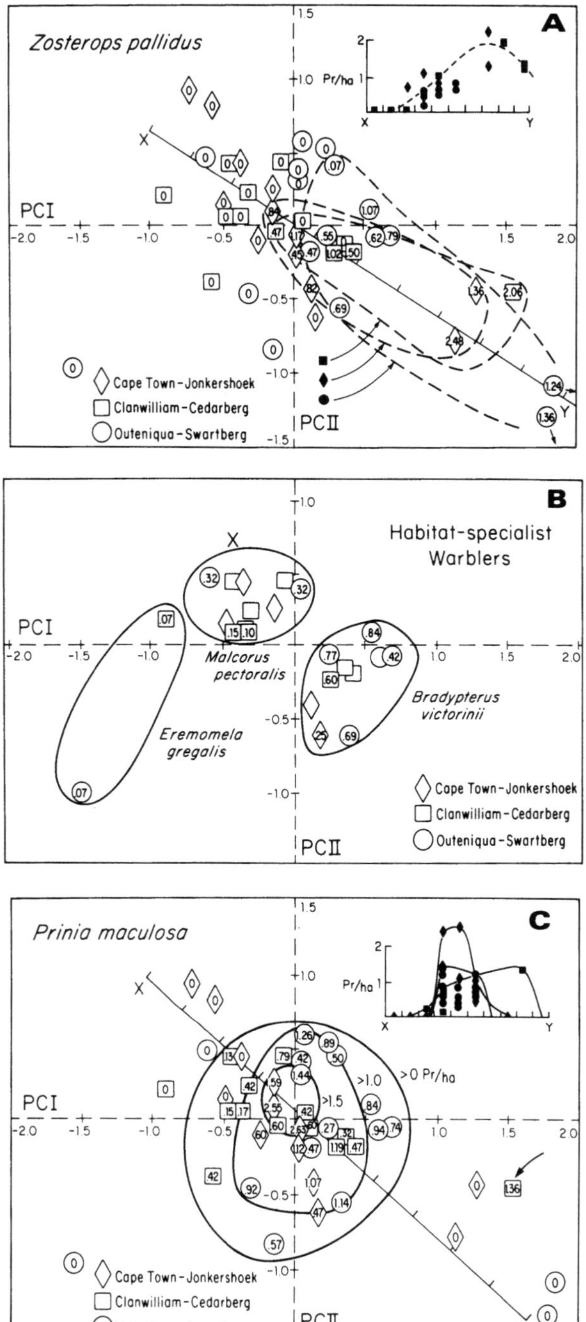

Fig. 7 (legend see p. 383)

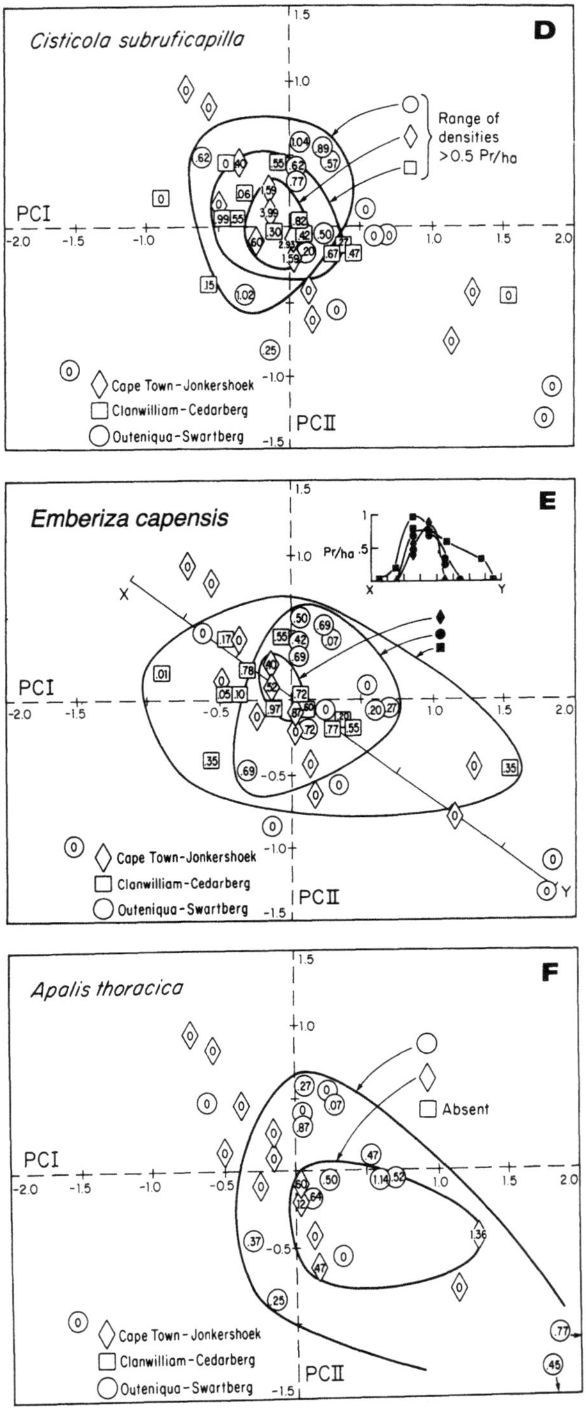

Fig. 7 (legend see p. 383)

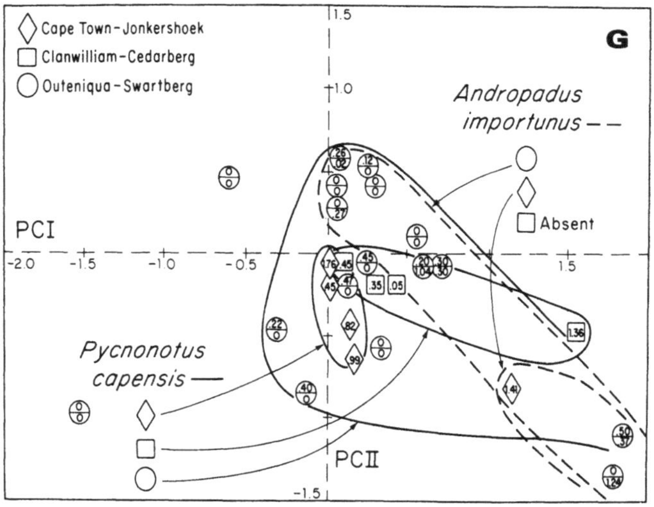

Fig. 7. Distribution of representative species in the habitat plane. The figures enclosed by the symbols are observed densities (pairs ha^{-1}). The white-eye (Zosterops pallidus) (A) is a species of tall fynbos and woodland, and shows no appreciable differences in either the range of habitats occupied or the densities it reaches in comparable habitats among the three transects. The three warbler species shown (B) are restricted in the habitats they occupy, with the karoo green warbler at left, the rufous-eared warbler above, and Victorin's scrub warbler to the right. Each is patchily distributed within a narrow range of habitats; empty symbols indicate absence from censused habitat. The karoo prinia (Prinia maculosa) (C) is a broadly distributed species that occupies all sites within its habitat range. Its range differs little among transects; densities of prinias are highest in the centre of its habitat range, but are also higher in the Cape Town area, and in taller habitat in the Cedarberg. The grey-backed cisticola (D) is a warbler that occupies a narrower habitat range in the Cape Town-Jonkershoek region and a wider range in the Clanwilliam-Cedarberg area compared to that of the third and intermediate region. Cape buntings (Emberiza capensis) (E) have both a narrower habitat range and lower densities in the Cape Town area, and occupy a broader habitat range at higher densities in the Cedarberg. The warbler (Apalis thoracica) (F), occupies a large habitat range in the Outeniqua-Swartberg area where its preferred habitats, tall scrub and woodland, are common, a reduced range in the Cape Town area, and was not found at sites in the Cedarberg area where its preferred habitat is rare. Cape bulbuls (Pycnonotus capensis) (G) were censused on all three transects, but are variable in their habitat ranges among transects. Sombre bulbuls (Andropadus importunus) are absent from the Cedarberg, occur only in tall woodlands in the Cape Town area, but occupy a wide range of habitat in the Outeniqua-Swartberg region.

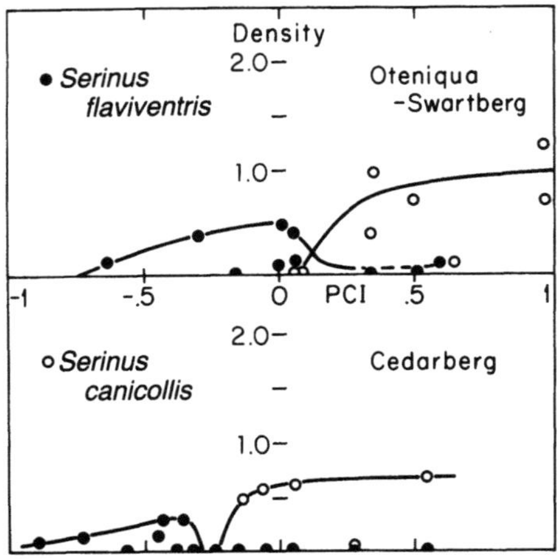

Fig. 8. The yellow canary (Serinus flaviventris) is found in shorter and more open habitats than the Cape canary (Serinus canicollis) but the turnover point between the two differs between the Cedarberg and the Outeniqua-Swartberg.

the Outeniquas and Swartberg it is around PCI = 0.15. This case is particularly interesting, for since S. canicollis is the species of taller and denser vegetation, one might predict that it would push the second species, C. flaviventris, into shorter vegetation where it is at a habitat-area advantage in the east, and the reverse would happen in the Cedarberg where shorter vegetation predominates. In fact the reverse happens; perhaps the presence of S. scotops in even taller (eastern) habitat might be involved, but the example does show that we understand poorly the factors that control species turnover on habitat gradients.

20.6.5 COMPARISON OF THREE SPECIFIC HABITATS BETWEEN TRANSECTS

Renosterveld. Unlike other habitats renosterveld is both widely distributed over the Cape Province and is dominated by a single low shrub, Elytropappus rhinocerotis. The habitat provides a relatively constant vegetation structure, and because it is so widespread hundreds of bird species must have access to it. It is instructive therefore to compare the bird communities of renosterveld in four sites: site 6 in Devon Valley of the Cape transect, site 1 near Hartenbos in the Outeniquas, site 13 in the Swartberg and site 2 at Kromrivier in the Cedarberg. Note that these sites are adjacent on the habitat plane (Fig. 5).

All four sites are similar in α diversity, with 14, 13, 15 and 14 species. Yet a total of 27 bird species is found on the four sites, and consequently there is a considerable species turnover between sites. This is γ diversity, measured again as the average number of species gained and lost between censuses, and values are obtained as shown in Table 3. Here two values for γ diversity are given, the first taking account simply of species names, and the second of the densities of these species. Computation of this second figure is straightforward: if census 1 has a total of 8 pairs ha^{-1}, and census 2 a total of 9 pairs ha^{-1}, and the species in common have a summed density of 5 pairs ha^{-1} (using the lower density figure of the two values from the two sites) then the turnover in bird density is ½(4 + 3) = 3.5 pairs ha^{-1}. Table 3 shows that species turnover is 5-8 species between different renosterveld censuses, and density differences are greater between the Cape and the other sites.

Table 3. Gamma diversities of species numbers and species densities (pairs ha^{-1}) in three habitat types among four areas of Cape Province. ◆= Cape Town-Jonkershoek, ○ = Outeniqua Mountains, ● = Swartberg and ■ = Cedarberg

Habitat	Transects compared					
	◆-○	◆-●	◆-■	○-●	○-■	●-■
RENOSTERVELD						
Species:	6.5	5.5	5.0	8.0	7.5	6.5
Density:	3.5	4.2	3.7	2.5	2.5	2.2
FYNBOS						
Species:	7.0	7.0	8.0	9.0	16.0	14.0
Density:	4.5	4.6	6.1	3.0	4.0	4.5
KLOOF WOODLAND						
Species:	4.5	---	6.0	---	7.5	---
Density:	3.8	---	7.9	---	7.7	---

Gamma diversities of 5-8 spp. between sites are misleading in that they do not account for the relative abundances of the species in question. In fact the renosterveld bird communities are similarly composed and organized in that they share core or dominant species with similar relative abundances (Table 4). There are five spp. present at every site, and these shared species account for between 55 and 70% of the total bird density. In Fig. 9A I have plotted cumulative relative abundance of the "core" species against their cumulative rank order overall, adding to the commonest species (Prinia maculosa) the second commonest (Cisticola subruficapilla) and so on, and I find that a set of eight commonest species, species that occur in at least three of the four sites, account for between 65% and 72% of the total bird densities. Thus I conclude that these communities share the same basic organization.

There remain two ways in which the communities differ. Firstly, there are 5-7 species that differ between censuses; but these are peripheral species using a minor proportion of the birds' resources. Secondly, there are major differences in total bird densities among sites (factor of almost four). This illustrates again that matching vegetation structure assures a reasonable match in α diversity, but does not control for bird density. It remains untested that food resources differ in abundance between sites, perhaps in response to rainfall differences, and that in consequence bird densities differ accordingly.

Fynbos. In each of four ranges, Cedarberg, Jonkershoek, Outeniquas and Swartberg, I select two fynbos sites, one that is more open and dominated by Protea nitida, and the other a more closed and dense vegetation dominated by other Protea spp. These sites are nos. 10 and 11 in Jonkershoek, 4 and 5 in the Outeniquas, 7 and 10 in the Cedarberg and 14 and 15 in the Swartberg. These censuses comprise the basis for a comparison of fynbos bird communities in the four different regions (see Table 5). Looking first at γ diversity, the values for species turnover

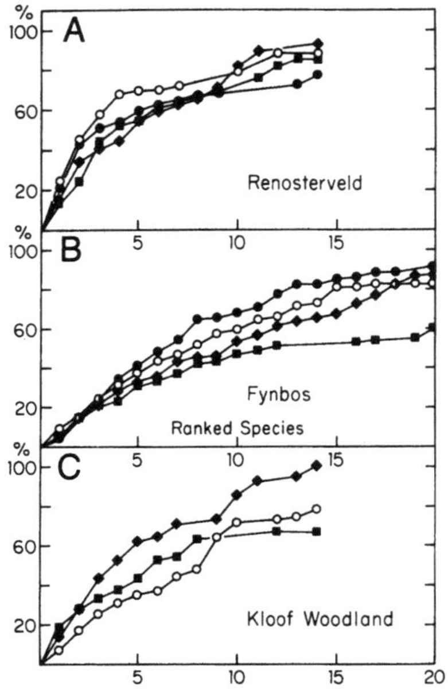

Fig. 9. For three habitat types, (A) renosterveld, (B) fynbos and (C) kloof woodland, species are ranked in order of decreasing density (abscissa). The curves show that the cumulative total bird density of these ranked species is similar among transects for the first two habitats, especially in renosterveld. Symbols are as before, with Outeniqua open and Swartberg closed circles.

Table 4. Renosterveld censuses at four sites in Cape Province.

Ranked bird species	Densities (pairs ha^{-1}) and cumulative percentages			
	Cape Town-Jonkershoek	Outeniqua	Swartberg	Clanwilliam-Cedarberg
Prinia maculosa	1.59(15%)	1.26(23%)	0.50(20%)	0.79(15%)
Cisticola subruficapilla	1.59(34%)	1.04(42%)	0.57(45%)	0.55(25%)
Erythropygia coryphaeus	0.69(36%)	0.62(54%)	0.15(49%)	0.94(43%)
Emberiza capensis	0.40(46%)	0.50(63%)	0.07(52%)	0.55(53%)
Streptopelia capicola	0.99(57%)	0.07(64%)	0.17(59%)	0.12(55%)
Streptopelia senegalensis	0.40(61%)	----(64%)	0.05(61%)	0.40(63%)
Nectarinia chalybea	0.40(66%)	0.07(66%)	0.07(64%)	----(63%)
Telophorus zeylonus	0.30(69%)	----(66%)	0.07(67%)	0.12(65%)
Francolinus spp.	0.40(73%)	----(66%)	----(67%)	0.07(67%)
Serinus flaviventris	0.99(84%)	0.42(73%)	----(67%)	----(67%)
Euplectes capensis	0.79(92%)	----(73%)	----(67%)	0.55(77%)
Nectarinia famosa	----(92%)	0.42(81%)	----(67%)	0.27(82%)
Mirafra apiata	----(92%)	----(81%)	0.07(70%)	0.27(87%)
Numida meleagris	0.20(96%)	----(81%)	0.12(75%)	----(87%)
Additional species:	2	5	5	3
Total bird species:	14	13	15	14
Total bird density:	9.24	5.42	2.47	5.32

and density turnover are given in Table 3. The Cape sites, with their position intermediate between the northern Cedarberg and the eastern Outeniquas and Swartberg, are about equally dissimilar (or similar) to both, a 7-8 spp. difference. A similar difference distinguishes the Outeniquas from the Swartberg, but the Cedarberg and the Outeniqua-Swartberg have twice this γ diversity. In terms of density turnover, all six comparisons are roughly similar.

The γ diversity figures again reflect the fact that many species can occur in fynbos in very low density, and generally these rarer species are different from one region to another. Yet the important core species are those that occur both in high density and in all four areas. This is shown in Table 5 and Fig. 9B. Although a total of 55 species are listed at the eight sites under consideration (two sites in each area), a small proportion of these constitute a major proportion of the total fynbos bird density. Species are ranked according to overall abundance in Table 5. Some thirteen species occur at all four localities, including four foliage insectivores (nos. 1,3 6 and 9 in the list) and two nectarivores (nos. 2 and 4 in the list). A robin, finch, dove, shrike, bulbul, widow-bird and francolin complete the list of ubiquitous species (listed nos 5, 7, 8, 11,

Table 5. Fynbos censuses at four sites in Cape Province.

Ranked bird species	Densities (pairs ha^{-1}) and cumulative percentages			
	Cape Town-Jonkershoek	Outeniqua	Swartberg	Clanwilliam-Cedarberg
Prinia maculosa	0.77(06%)	0.94(10%)	0.56(06%)	0.90(09%)
Promerops cafer	0.63(11%)	0.54(16%)	0.95(15%)	0.51(14%)
Zosterops pallidus	0.83(17%)	0.74(24%)	0.81(23%)	0.75(21%)
Nectarinia violacea	0.96(24%)	0.65(31%)	1.02(33%)	0.42(25%)
Cossypha caffra	0.58(29%)	0.65(38%)	0.67(40%)	0.75(33%)
Sphenoeacus afer	0.67(34%)	0.57(44%)	0.79(47%)	0.34(36%)
Serinus canicollis	0.72(39%)	0.26(47%)	0.60(53%)	0.34(40%)
Streptopelia capicola	0.47(43%)	0.39(52%)	0.39(57%)	0.38(43%)
Bradypterus victorinii	0.13(44%)	0.56(58%)	0.81(65%)	0.30(46%)
Nectarinia famosa	0.88(51%)	0.23(60%)	----(65%)	0.34(50%)
Laniarius ferruginea	0.46(54%)	0.39(64%)	0.14(67%)	0.13(51%)
Pycnonotus capensis	0.91(61%)	0.15(66%)	0.23(69%)	0.18(53%)
Cisticola subruficapilla	0.21(63%)	0.51(72%)	0.71(76%)	----(53%)
Onychognathus morio	0.66(68%)	----(72%)	0.16(77%)	0.26(55%)
Euplectes capensis	0.20(69%)	0.26(74%)	0.36(81%)	0.18(57%)
Apalis thoracica	0.24(71%)	0.26(77%)	0.49(86%)	----(57%)
Serinus leucopterus	0.60(76%)	----(77%)	0.37(89%)	----(57%)
Francolinus spp.	0.60(80%)	0.10(78%)	0.10(90%)	0.14(60%)
Batis capensis	0.34(83%)	0.58(85%)	----(90%)	----(60%)
Serinus sulphuratus	0.64(88%)	0.08(85%)	----(90%)	0.09(61%)
Nectarinia chalybea	----(88%)	0.16(87%)	0.14(91%)	0.42(65%)
Additional species:	5	8	7	22
Total bird species:	25	27	25	39
Total bird density:	13.11	9.20	10.13	10.44

12, 15 and 18 respectively). Additional species of foliage insectivores (nos 13, 16), sunbirds (nos 10, 21), a starling (no. 14) and a finch (no. 20) occur at three localities. Cumulative percentages of total species density, of species in the rank order given, remain within 15% of each other throughout the cumulative ranking, and end within 3% of each other in the first three comparisons. The Cedarberg sites are somewhat different (39 versus 25-27 species), and have 22 other species that comprise a combined one-third of the total bird density.

The outcome, as seen in Fig. 9B, is that fynbos communities in different regions have the same basic constitution, the same species with the same relative abundances, over at least the first 50-60% of their bird numbers and densities. The 21 spp. ranked in Fig. 9B total between 65% and 91% of the total bird density. The other 34 species are best termed peripheral species, and their occurrence in fynbos seems an artifact of the character of the surrounding habitat and its extent. For example, the foliage insectivores Cisticola fulvicapilla and Apalis thoracica were not found in fynbos in the Cedarberg area; both species become much rarer in lower habitat of the sort that surrounds the Cedarberg. These species are fifth and sixth in abundance among the foliage insectivores, and there are insectivores in the Cedarberg fynbos, such as Sylvietta rufescens, Parisoma subcaeruleum and P. layardi, characteristic of shorter habitat that fill a comparable role as auxiliary foliage gleaners. In a similar way, the core dove of the fynbos is Streptopelia capicola, and whereas in the Cape and Outeniqua areas it is backed up by Streptopelia semitorquata typical of taller and denser habitat, in the Cedarberg the backup role is played by Oena capensis and Streptopelia senegalensis, both typical of lower and more open habitat.

I have recorded 10 species of seed-eating birds in fynbos, but only two of these are listed in the core species list, Euplectes capensis and Serinus canicollis. Of the remainder, many species appear to be opportunistic in their use of a particular patch of fynbos. I recorded Serinus sulphuratus and Serinus leucopterus in Jonkershoek, the former plus Serinus albogularis in the Cedarberg, the former again plus Serinus gularis in the Outeniquas, and just Serinus leucopterus in the Swartberg. The largest category of peripheral species is that of the open-country ground-foraging omnivores that enters fynbos in the Cedarberg. Besides the two core species, Cape robin and boubou shrike, there are starlings, chats, shrikes, larks and robins, all typical of lower and drier habitat, and occurring unpredictably in low densities in fynbos. Despite their contribution to α and γ diversity, such species are of only minor importance in the fynbos community, which in its essentials remains unchanged over many hundreds of miles of territory.

Kloof woodland Renosterveld is of very wide distribution and fynbos is of more restricted occurrence in the southern Cape Province. The habitat I consider next is of even more restricted distribution, being confined to valley bottoms where there is a permanent watercourse. Kloof woodlands were censused in the Outeniquas, in the Jonkershoek valley and in the Cedarberg; these are sites 6, 12 and 11 respectively in Appendix B. In this habitat there is more potential for dramatic differences between the bird communities, for this is the habitat that becomes abundant in eastern Cape Province and is almost nonexistent in the Cedarberg.

The species censused in the kloof woodland are given in Table 6; the totals vary from 18 in the Outeniquas to 13 in both the Cape and the Cedarberg. There is a close correspondence between Cape and Outeniquas with only two species present in the former and absent from the latter. But the Outeniqua census contains six species not present at the Cape, four

Table 6. Kloof woodland censuses at three sites in Cape Province.

Ranked bird species	Densities (pairs ha^{-1}) and cumulative percentages		
	Cape Town-Jonkershoek	Outeniqua	Clanwilliam-Cedarberg
Cossypha caffra	1.56(12%)	0.99(08%)	2.75(20%)
Zosterops pallidus	1.36(23%)	1.24(19%)	2.06(34%)
Laniarius ferruginea	1.56(35%)	0.87(26%)	0.69(39%)
Serinus canicollis	0.97(42%)	0.62(31%)	0.69(44%)
Turdus olivaceus	0.97(50%)	0.62(36%)	0.69(49%)
Streptopelia capicola	0.25(52%)	0.25(38%)	1.36(58%)
Nectarinia famosa	0.67(57%)	0.74(44%)	0.17(60%)
Batis capensis	2.70(78%)	1.98(61%)	----(60%)
Apalis thoracica	1.36(88%)	0.77(67%)	----(60%)
Pycnonotus capensis	----(88%)	0.50(71%)	1.36(69%)
Sigelus silens	0.67(93%)	----(71%)	0.22(71%)
Onychognathus morio	----(93%)	0.25(74%)	0.69(76%)
Tersiphone viridis	0.45(97%)	0.45(77%)	----(76%)
Streptopelia semitorquata	0.27(99%)	0.25(79%)	----(76%)
Additional species:	1	5	3
Total bird species:	13	18	13
Total bird density:	12.91	12.01	14.10

of these being typical forest or woodland birds (Andropadus importunus, Serinus scotops, Dicrurus adsimilis and a replacement flycatcher for Sigelus, Muscicapa adusta). The Cedarberg census shares nine species with the Outeniquas, but includes in addition three species not at all typical of this woodland habitat: Euplectes capensis, Prinia maculosa, Emberiza capensis (and once again Sigelus silens). It appears that in the absence of birds more appropriate to this habitat in the Cedarberg outpost, the only patch of woodland I could find in the area, non-woodland species are able to utilize it.

The γ diversity figures between paired sites are given in Table 3. These figures are notable only to the extent that species densities vary considerably among sites. Rather surprising is the fact that the highest total bird density is observed in the Cedarberg. It appears that opportunistic species in the Cedarberg kloof are able to function very well there, a situation reminiscent of that mentioned above with island supertramps.

From Fig. 9C it is seen that, accumulating species in order of their mean abundances over the three sites, there is little correspondence in the community structure in different kloofs beyond the first four or five species and little evidence of convergent evolution in the patches. The high densities of ecologically inappropriate species in the Cedarberg kloof emphasizes the classical ecological question of the extent to which bird densities are regulated by food resources directly rather than indirectly via behavioural mechanisms.

Acknowledgements

The recent field work in South Africa reported in the paper was conducted with financial assistance from the Guggenheim Foundation, while the author was a Fellow, and from the Percy Fitzpatrick Institute of African Ornithology through the kind consideration of its director W R Siegfried. The Office of Academic Computing at the University of California, Los Angeles provided computing facilities. My wife Daryl Ann Cody helped with the field work. Peter Frost laboured to improve the manuscript, but is not sure his efforts were well spent. My colleague Hartmut Walter assisted in the collection of Sardinian bird data.

References

ACOCKS JPH (1975) Veld types of South Africa. Memoirs of the Botanical Survey of South Africa 40

BROWN JH (1971) Mammals on mountaintops: non-equilibrium insular biogeography. American Naturalist 105: 467-478

BROWN JH, KODRIC-BROWN A (1977) Turnover rates in insular biogeography: effect of immigration on extinction. Ecology 58: 445-449

CASE TJ, CASTEN RG (1979) Global stability and multiple domains of attraction in ecological systems. American Naturalist 113: 705-714

CASE TJ, GILPIN ME, DIAMOND JM (1979) Overexploitation, interference competition, and excess density compensation in insular faunas. American Naturalist 113: 843-854

CASE TJ, CODY ML (1982) The theory of island biogeography. Chapter 12. In: CASE TJ, CODY ML (eds) Biogeography of islands in the sea of Cortez. University of California Press, Berkeley. In press

CODY ML (1975) Towards a theory of continental diversities: bird distributions over mediterranean habitat gradients. In: CODY ML, DIAMOND JM (eds) Ecology and evolution of communities. Harvard University Press, Cambridge, pp 214-257

CODY ML (1978) Distribution ecology of Haplopappus and Chrysothamnus in the Mohave Desert. 1. Niche position and niche shifts on north-facing granitic slopes. American Journal of Botany 65: 1107-1116

CODY ML (1980a) Evolution of habitat use: geographic perspectives. Proceedings of the Seventeenth International Ornithological Congress. Berlin. Verlag der Deutschen Ornithologen-Gesellschaft. V.II: 1013-1018.

CODY ML (1980b) Species-packing in insectivorous bird communities: density, diversity, and productivity. Proceedings of the Seventeenth International Ornithological Congress. Berlin. Verlag der Deutschen Ornithologen-Gesellschaft. V.II: 1071-1077

CODY ML (1981) Habitat selection in birds: the roles of vegetation structure, competitors and productivity. Bioscience 31: 107-113

CODY ML (1982) Bird diversity and density in the afromontane woodlands of Cape Province. MS in review

CODY ML, MOONEY HA (1978) Convergence versus nonconvergence in mediterranean-climate ecosystems. Annual Review of Ecology and Systematics 9: 265-321

DIAMOND JM (1974) Colonization of exploded volcanic islands by birds: the supertramp strategy. Science 184: 803-806

DIAMOND JM (1975) Assembly of species communities. In: CODY ML, DIAMOND JM (eds) Ecology and evolution of communities. Harvard University Press, Cambridge. pp 342-444

HORN HS, MACARTHUR RH (1972) Competition among fugitive species in a harlequin environment. Ecology 53: 749-752

LEVINS R (1968) Evolution in changing environments. Princeton Monographs in Population Biology, vol 2. Princeton, New Jersey. 120 pp

LEVINS R, CULVER D (1971) Regional coexistence of species, and competition between rare species. Proceedings of the National Academy of Science of the United States of America 68: 1246-1248

MACKENZIE B (1978) A quantitative and qualitative study of the indigenous forests of the southwestern Cape. MSc Thesis. Unversity of Cape Town

MARTIN (1979)

MILEWSKI AV (1982) A theory of coadaptation with frugivores and nectarivores to explain differences in the vegetation of mediterranean Australia and South Africa. Unpublished MS

ROSENZWEIG MR (1975) On continental steady states of species diversity. In: CODY ML, DIAMOND JM (eds) Ecology and evolution of communities. Harvard University Press, Cambridge, Massachussetts, pp 121-140

SAHAB (1966)

TROMBULAK S, CODY ML (1980) Distributions of *Pinus edulis* and *P. monophylla* in the New York Mountains, eastern Mojave Desert. Madrono. In press

WHITE F (1978) The afromontane region. In: WERGER MJA (ed) Biogeography and ecology of southern Africa, vol 1. Junk, The Hague, pp 463-513

Appendix A. Avifauna census data for Sardinia and Corsica. "+" indicates species' presence, asterisks its absence.

Study site:	San Sal-vatore	Sant' Antioco	Pabillo-nis	Campa-nasissa	Narcao	Capo Car-bonara	Bau Pressiu	Ter-rubia	Terra-seo	Porto Pino	San Bene-detto	Zonza
AREA(ha):	2.22	2.04	1.85	3.08	2.64	2.31	1.94	3.96	2.58	4.72	1.90	2.94
VEG. HT. (m):	0.42	0.60	0.80	0.96	1.44	2.41	2.38	3.79	7.20	9.06	35.45	-?
VEG.½HT(m):	0.07	0.17	0.11	0.17	0.39	0.42	0.91	1.21	2.97	1.70	4.55	10.91
H:	0.054	0.251	0.327	0.420	0.630	0.820	0.885	1.067	1.375	1.399	1.501	2.034
No. SPP. GAINED:	6	9	11	14	20	26	30	38	38	38	40	46
No. SPP. LOST:	1.24	2.64	2.19	4.74	6.48	7.61	12.29	20.01	20.55	20.55	21.37	30.64
Tot. SPP:	6	5	6	10	12	15	14	20	19	14	19	19
SPECIES: Tot. DENSITY (pairs ha^{-1}):	2.47	2.65	5.41	9.36	7.07	8.08	8.46	13.40	15.28	5.29	18.88	13.22
1. Cisticola juncidis	0.45											
2. Alauda arvensis	0.67	**	0.74									
3. Anthus campestris	0.45	**	**	0.25								
4. Calandrella cinerea	0.45	0.15	4.22	2.68								
5. Coturnix coturnix	+	**	**	0.97								
6. Saxicola torquata	0.45	1.46	0.10	0.97	0.50	0.12						
7. Emberiza calandra		0.67	+	2.28	1.34	**	0.20					
8. Sylvia sarda		0.20	**	**	0.77	**	0.42					
9. Sylvia melanocephala		0.17	**	0.25	1.59	2.60	0.77	0.52	0.12	0.79	0.27	
10. Melanocorypha calandra			0.10									
11. Sylvia conspicillata			0.25	0.50								
12. Alectoris barbara				0.32	**	**	+	+				
13. Acanthis cannabina				0.50	0.15	0.87	**	**	**	0.50		
14. Carduelis carduelis				0.64	0.37	1.02	1.56	0.60	1.17	0.64		
15. Emberiza cirlus					0.57	0.87						
16. Lanius senator					0.37	0.87						
17. Lullula arborea					0.37	**	**	**	**	0.22		
18. Sylvia undata					0.77	0.64	2.33	0.64	+			
19. Streptopelia turtur					+	+	**	**	**	+		
20. Parus major					0.27	0.25	0.52	0.60	0.77	0.32	1.41	0.37
21. Monticola solitarius						0.42						
22. Merops apiaster						+	+	+				
23. Upupa epops						+	**	**	+	**	+	
24. Sylvia cantillans						0.42	0.52	1.46	1.74	**	1.84	
25. Serinus serinus						0.42	0.25	0.30	0.47	0.84	1.07	
26. Luscinia megarhynchos						+	**	**	0.05	**	0.27	
27. Lanius collurio							0.52	0.05	**	**	0.67	
28. Chloris chloris							0.25	0.06	0.50	0.42	**	0.40
29. Turdus merula							0.72	0.45	0.20	0.35	1.31	0.35
30. Troglodytes troglodytes							0.40	0.72	1.07	0.22	0.79	0.74

31. Sylvia atricapilla	1.17	2.23	+	1.31	0.82
32. Erithacus rubecula	2.93	1.36	**	1.14	2.38
33. Parus caeruleus	0.35	0.30	**	0.79	0.45
34. Muscicapa striata	0.30	1.04	0.22	0.79	0.17
35. Regulus ignicapillus	0.87	2.13	0.42	1.59	1.36
36. Parus ater	1.46	0.57	**	0.52	1.64
37. Fringilla coelebs	0.72	1.36	0.35	3.77	1.31
38. Garrulus glandarius	0.20	0.20	**	0.27	0.25
39. Passer hispaniolensis				+	
40. Coccothraustes coccothraustes				1.07	0.37
41. Parus aegithaloides					0.74
42. Endrocopous major					0.62
43. Regulus regulus					0.55
44. Certhia familiaris					0.35
45. Sitta whiteheadi					0.35
46. Columba palumbus					+

Appendix B. Site data for transects in the Cape Province. "+" indicates

Location of site	Habitat	Dominant plants
CAPE TOWN-JONKERSHOEK		
1. Gordon's Bay	Lower Cape Flats	Restiads, Asteraceae
2. Gordon's Bay	Low Cape Flats	Asteraceae, Iridaceae
3. Cape Peninsula	Coastal fynbos	Restiads, Erica
4. Gordon's Bay	Low renosterveld	Elytropappus
5. Cape Peninsula	Coastal fynbos	Thamnochortus
6. Devon Valley	Renosterveld	Elytropappus
7. Devon Valley	Tall renosterveld	Elytropappus
8. Swartklip	Coastal fynbos	Metalasia, Euclea
9. Jonkershoek	Low fynbos	Protea repens
10. Jonkershoek	Open fynbos	Protea nitida
11. Jonkershoek	Tall dense fynbos	P. repens, P. neriifolia
12. Jonkershoek	Kloof woodland	Brabejum, Cunonia
13. Kirstenbosch	Afromontane forest	Virgilia, Cassine
CEDARBERG-CLANWILLIAM		
1. Lambert's Bay	Low strandveld	Ruschia, Helichrysum
2. Kromrivier	Renosterveld	Elytropappus, Berzelia
3. Lambert's Bay	Succulent strandveld	Ruschia, Cotyledon
4. Kromrivier	Low fynbos	Leucadendron
5. Lambert's Bay	Tall strandveld	Zygophyllum, Lycium
6. Clanwilliam	Karoid broken veld	Carissa
7. Pakhuispas	Low fynbos	Protea glabra
8. Uitkomsberge	Fynbos	Protea, Euclea
9. Keurbos, Clanwilliam	Karoid broken veld	Rhus incisa
10. Pakhuispas	Tall fynbos	Protea nitida
11. Algeria	Kloof woodland	Brabejum
12. Grootrivierpas	Succulent karoo	Cotyledon, Ruschia
13. Suurfontein	Succulent karoo	Euphorbia, Hoodia
14. Brakfonteinspruit	Low karroo	Eriocephalus
OUTENIQUA-SWARTBERG AREA		
1. Hartenbos	Coastal renosterveld	Elytropappus
2. Hartenbos	Tall renosterveld	Elytropappus
3. Hartenbos	Valley bushveld	Aloe ferox
4. Bonnievale	Fynbos	Protea repens
5. Ruitersbos	Tall fynbos	Protea coronata
6. Robinson Pass	Kloof woodland	Virgilia, Alsophila
7. Knysna	Knysna forest	Podocarpus, Olea
8. Kliprivier	Low mt. renosterveld	Elytropappus
9. Kliprivier	Mt. renosterveld	Elytropappus
10. Kliprivier	Spekboom-renosterveld	Elytropappus
11. Oudtshoorn	Spekboomveld	Portulacaria
12. Alwynkoppe	Mt. thorn scrub	Cussonia, Maytenus, Aloe
13. Swartbergpas	Mt. renosterveld	Elytropappus
14. Swartbergpas	Open fynbos	Protea nitida
15. Swartbergpas	Dense fynbos	Protea eximia, P.punctata
16. Scholtzkloof	Bushveld	Acacia karroo
17. Prince Albert	Karoid broken veld	

[a] Habitat variables 1-6 are described in the text. [b] PCI, PCII refer to derived from all 44 sites listed.

species' presence, asterisks its absence.

Site area (ha)	Habitat variables[a] 1	2	3	4	5	6	PCI[b]	PCII
2.41	-0.013	0.072	-0.097	0.903	-0.014	1.500	-0.73	0.93
1.20	0.530	0.076	-0.071	1.146	0.046	1.400	-0.57	0.80
3.33	0.267	0.140	0.176	1.380	0.194	0.700	-0.49	0.11
2.58	0.322	0.146	0.176	1.223	0.263	1.300	-0.42	0.46
1.67	0.532	0.152	0.477	1.519	0.350	0.700	-0.24	-0.10
2.52	0.596	0.260	0.415	1.322	0.451	1.400	-0.17	0.27
1.76	0.740	0.462	0.544	1.146	0.646	1.500	-0.16	0.07
1.71	0.769	0.461	0.602	1.491	0.516	1.130	0.05	-0.07
2.22	0.892	0.491	0.634	1.447	0.616	1.161	0.04	-0.18
1.86	1.037	0.322	0.653	1.724	0.602	0.778	0.11	-0.42
4.59	1.131	0.649	0.681	1.649	0.686	0.699	0.16	-0.65
0.74	1.556	1.212	1.380	1.954	0.796	1. 16	1.27	-0.44
2.81	1.888	1.305	1.602	1.681	1.123	1.600	1.12	-0.78
5.55	0.462	0.699	0.243	0.991	0.466	1.255	-0.49	0.11
3.76	0.580	1.000	0.431	1.004	0.578	1.362	-0.06	0.46
11.15	0.653	0.845	0.358	0.869	0.751	0.954	-0.38	0.07
3.33	0.785	1.041	0.439	1.265	0.621	0.892	0.06	-0.02
5.57	0.799	1.041	0.550	1.030	0.776	1.130	0.06	0.01
7.50	0.839	1.415	0.078	0.978	0.858	1.146	0.29	-0.17
4.18	0.949	0.204	0.628	1.193	0.795	0.954	-0.12	-0.03
5.02	0.962	1.301	0.708	1.196	0.806	1.161	0.43	-0.17
5.83	0.964	0.415	0.342	0.820	1.176	0.845	-0.58	-0.39
2.97	1.143	1.255	0.748	1.438	0.795	0.778	0.55	-0.53
1.46	1.362	1.996	1.227	1.575	0.865	1.516	1.56	-0.44
8.33	0.580	0.955	0.414	0.806	0.720	1.146	-0.31	0.23
30.10	0.415	0.845	0.146	0.462	0.898	1.049	-0.90	0.21
15.06	0.477	0.955	0.267	0.724	0.659	1.301	-0.44	0.46
3.54	0.580	0.903	0.420	1.199	0.484	1.415	0.05	0.59
3.63	0.690	0.954	0.435	1.140	0.605	1.255	0.02	0.31
4.77	0.940	1.362	0.782	1.305	0.720	1.250	0.61	-0.06
4.77	0.964	1.322	0.832	1.300	0.742	1.301	0.64	-0.05
3.47	1.152	1.-14	0.591	1.373	0.839	0.699	0.32	-0.57
1.20	1.757	2.643	0.086	0.429	0.395	-0.317	11.77	-1.91
20.00	2.455	3.625	0.108	0.958	0.352	-0.352	2.70	-2.47
3.23	0.431	0.699	0.230	0.724	0.595	1.130	-0.63	0.48
2.22	0.633	1.000	0.519	1.233	0.513	1.491	0.22	0.55
4.81	0.622	1.000	0.435	1.117	0.557	1.274	0.01	0.39
3.70	0.832	1.114	0.431	0.839	0.992	0.699	-0.30	-0.45
4.17	0.875	1.146	0.628	0.959	0.912	1.097	0.08	-0.18
6.02	0.699	1.114	0.571	1.173	0.596	1.498	0.27	0.43
3.63	0.875	1.000	0.556	1.283	0.682	1.000	0.20	-0.05
2.13	0.881	1.146	0.720	1.356	0.650	1.295	0.51	0.12
6.25	1.068	1.255	0.415	0.924	1.156	0.556	-0.18	-0.84
29.50	0.580	0.955	-0.208	0.279	2.079	0.556	-1.59	-0.98

the first two principal components of the habitat variables 1-6,

Appendix C. Bird census results from sites in the Cape Town-Jonkershoek Area. Data are densities as pairs ha^{-1} or equivalent for gregarious species (see text). Asterisks indicate species' absence.

SITE NUMBER:	1	2	3	4	5	6	7	8	9	10	11	12	13
H:	-0.09	-0.03	0.20	0.25	0.43	0.53	0.72	0.74	0.85	0.93	0.97	1.56	1.83
BIRD SPECIES													
1. *Calandrella cinerea*	0.30	**	0.30										
2. *Galerida magnirostris*	0.35	0.35	0.30	0.40									
3. *Anthus novaeseelandiae*	0.62	**	0.12										
4. *Cisticola textrix*	1.49	1.49	0.89	0.40	0.25								
5. *Macronyx capensis*	0.15	0.15	**	0.15	**	0.30							
6. *Anthus leucophrys*		0.84	0.25										
7. *Mirafra apiata*		0.32	**	0.77									
8. *Coturnix coturnix*				0.20									
9. *Telophorus zeylonus*				0.20	**	0.30	**	0.30					
10. *Cisticola subruficapilla*				0.40	0.60	1.59	3.99	2.93	1.59				
11. *Francolinus capensis*					0.30	0.40	0.27	0.60	0.45				
12. *Sphenoeacus afer*					0.60	**	0.27	1.17	1.81	1.34			
13. *Prinia maculosa*					0.60	1.59	2.55	2.63	1.12	1.07	0.47		
14. *Euplectes capensis*					0.12	0.79	0.57	*	0.22	0.27	0.12		
15. *Serinus flaviventris*						0.99	2.28	1.76					
16. *Erythropygia coryphaeus*						0.69	1.41	1.16					
17. *Numida meleagris*						0.20	0.57						
18. *Streptopelia senegalensis*						0.40	1.14						
19. *Emberiza capensis*						0.40	0.57	0.87					
20. *Cinnyris chalybeus*						0.40	0.57	1.17	**	**	**	**	0.10
21. *Streptopelia capicola*						0.99	1.14	0.30	0.89	0.82	0.12	0.25	0.25
22. *Sigelus silens*						0.20	**	**	0.45	0.40	0.12	0.67	
23. *Oena capensis*							0.27						
24. *Serinus canicollis*							0.27	**	0.89	1.07	0.37	0.97	0.89
25. *Cossypha caffra*							0.27	1.17	1.12	0.40	0.77	1.56	1.00
26. *Zosterops pallidus*							0.84	1.17	0.45	0.82	0.84	1.36	2.48
27. *Saxicola torquata*								0.15					
28. *Parisoma subcaeruleum*								0.60					
29. *Serinus albogularis*								0.37					
30. *Pycnonotus capensis*								1.76	0.45	0.82	0.99		

31. *Francolinus africanus*								0.60	**	0.82	0.37	0.12	
32. *Onychognathus morio*								0.60	0.25	1.07	0.25	**	0.12
33. *Apalis thoracica*								0.60	0.12	**	0.47	1.36	
34. *Corvus albicollis*								1.12	0.10	0.12	0.07		
35. *Promerops cafer*									1.59	0.82	1.44		
36. *Nectarinia violacea*									0.89	0.82	1.09		
37. *Serinus leucopterus*									0.45	0.82	0.37		
38. *Laniarius ferrugineus*									0.22	0.25	0.67	1.56	0.35
39. *Batis capensis*									0.22	**	0.67	2.70	1.41
40. *Monticola rupestris*										0.82	0.20		
41. *Nectarinia famosa*										1.61	0.15	0.67	
42. *Cisticola fulvicapilla*										0.27	0.15		
43. *Serinus sulphuratus*										1.07	0.20		
44. *Mesopicos griseocephalus*										0.12	0.20	**	0.12
45. *Streptopelia semitorquata*										0.12	0.12	0.27	0.17
46. *Bradypterus victorinii*											0.25		
47. *Turdus olivaceus*												0.97	0.94
48. *Tersiphone viridis*												0.45	0.89
49. *Andropadus importunus*													1.41
50. *Muscicapa adusta*													1.07
51. *Columba arquatrix*													0.35
52. *Bradypterus sylvaticus*													0.50
53. *Aplopelia larvata*													0.25
TOTAL BIRD SPECIES:	5	5	5	7	6	14	16	20	19	22	24	13	17
TOTAL BIRD DENSITY (pairs ha^{-1}):	2.91	3.15	1.86	2.52	2.47	9.24	16.98	21.20	13.28	15.74	10.47	12.91	12.30

Appendix D. Bird census results from sites in the Clanwilliam-Cedarberg area. Asterisks indicate species' absence.

SITE NUMBER:	1	2	3	4	5	6	7	8	9	10	11	12	13	14
H:	0.38	0.53	0.57	0.71	0.72	0.83	0.87	0.90	0.92	1.05	1.37	0.52	0.25	0.42
BIRD SPECIES														
1. *Saxicola torquata*	0.10													
2. *Malcorus pectoralis*	0.15	**	0.10											
3. *Corvus albicollis*	0.02	**	0.02											
4. *Eupodotis afra*	0.03	**	0.05	**	0.02									
5. *Mirafra apiata*	0.62	0.27	0.17	0.30	**	0.12	0.12							0.07
6. *Cercomela familiaris*	0.10	**	0.05	**	**	0.40	**	**	0.32					
7. *Passer melanurus*	0.15	**	**	**	**	**	0.25							
8. *Spreo bicolor*	0.15	**	0.20	**	0.20	**	0.05	**	0.12					
9. *Certhilauda albescens*	0.37	**	0.22	**	0.17	**	**	**	0.17			0.12	0.07	0.10
10. *Stenostira scita*	0.07	**	**	**	0.17	**	0.25					0.12		
11. *Parisoma layardi*	0.27	**	**	**	0.27	**	0.37	0.25				0.12		
12. *Serinus flaviventris*	0.27	**	**	**	0.10	**	**	0.27					0.03	0.13
13. *Cisticola subruficapilla*	0.99	0.55	0.55	0.42	0.82	0.27	0.30	0.47	0.15	0.67		0.06		
14. *Erythropygia coryphaeus*	0.07	0.94	0.27	0.25	0.30	0.12	0.60	0.25	0.25			0.06		
15. *Serinus albogularis*	0.15	**	0.22	0.15	**	0.12	0.30	0.22	0.35	0.35		0.30		
16. *Telophorus zeylonus*	0.05	0.12	**	0.40	0.32	0.20	0.55	0.20	0.17	0.35		0.12		
17. *Cinnyris chalybeus*	0.15	**	0.25	**	0.99	**	0.72	0.27	0.15	0.12				0.07
18. *Emberiza capensis*	0.05	0.55	0.10	0.60	0.72	0.20	0.97	0.55	0.35	0.77	0.35	0.78	0.01	0.17
19. *Prinia maculosa*	0.15	0.79	0.17	0.60	0.42	0.32	0.60	0.47	0.42	1.19	1.36	0.42		0.13
20. *Lanius collaris*		0.12	0.10	**	0.17	0.12	**	**	0.17			0.12	0.02	
21. *Streptopelia senegalensis*		0.40	**	**	0.12	**	**	0.07	**	0.12		0.24	0.02	
22. *Geocolaptes olivaceus*		0.02	**	**	**	**	0.05	0.02	**	0.07				
23. *Francolinus capensis*		0.07	**	0.10	**	**	0.12	0.02	**	0.15				
24. *Euplectes capensis*		0.55	**	**	**	**	**	0.07	**	0.35	1.71			
25. *Streptopelia capicola*		0.12	0.25	**	0.42	0.20	0.40	0.17	0.17	0.35	1.36	0.12		
26. *Nectarinia famosa*		0.27	**	0.15	**	**	0.25	0.12	**	0.42	0.17	0.24		
27. *Serinus canicollis*		0.55	**	0.60	**	**	**	**	**	0.67	0.69			
28. *Oena capensis*			0.07	0.30	0.10	**	0.12	0.07						
29. *Emberiza impetuani*			0.05	**	0.12								0.05	0.07
30. *Certhilauda curvirostris*			0.17											

31. Melaenornis infuscatus			0.05											
32. Creatophora cinerea			0.12											
33. Myrmecocichla formicivora			0.07											
34. Eremomela icteropygialis			0.10	0.15										
35. Corvus albus			0.02	**	0.02	0. 02	0.02	0.02	0.02	0.02				
36. Sylvietta rufescens			0.05	**	0.37	**	0.25	0.25	**	0.35				
37. Cercomela schlegelii				0.07										0.06
38. Pycnonotus capensis				0.15	**	**	**	0.05	**	0.35	1.36			
39. Parus afer					0.05							0.12		
40. Colius colius					0.10	**	0.12							
41. Parisoma subcaeruleum					0.37	**	**	**	**	0.12				
42. Ploceus capensis					0.15	**	**	**	**	0.25				
43. Sigelus silens					0.17	**	0.25	**	**	**	0.22			
44. Serinus sulphuratus							0.17							
45. Columba guinea							0.30	0.10				0.12	0.02	
46. Onychognathus morio							0.17	0.10	0.12	0.35	0.69			
47. Cossypha cafra							0.72	0.12	**	0.77	2.75			
48. Zosterops pallidus							0.47	0.50	**	1.02	2.06			
49. Bradypterus victorinii										0.60				
50. Nectarinia violacea										0.84				
51. Promerops cafer										1.02				
52. Sphenoeacus afer										0.67				
53. Numida meleagris										0.07				
54. Batis pririt										0.12				
55. Laniarius ferrugineus										0.25	0.69			
56. Turdus olivaceus											0.69			
57. Oenanthe monticola												0.12	0.02	
58. Eremomela gregalis													0.07	
59. Unidentified lark													0.12	0.17
60. Estrilda astrild													0.03	
61. Pterocles namaqua													0.02	0.05
62. Galerida magnirostris													0.02	
63. Serinus alario														0.07
TOTAL BIRD SPECIES:	19	14	24	14	24	11	26	23	14	28	13	16	13	11
TOTAL BIRD DENSITY (pairs ha^{-1}):	3.91	5.32	3.42	4.24	6.66	2.09	8.49	4.63	2.93	12.38	14.10	3.18	0.50	1.09

Appendix E. Bird census results from sites in the Outeniqua-Swartberg area. Asterisks indicate species' absence.

BIRD SPECIES	Site number: Habitat H:	1 0.51	2 0.61	3 0.89	4 0.90	5 1.19	6 1.65	7 1.97	8 0.15	9 0.62	10 0.66	11 0.95	12 1.03	13 0.79	14 0.89	15 1.04	16 1.31	17 0.52
1.	Spreo bicolor	0.12	0.12	**	**	**	**	**	0.15	**	0.12	0.12						
2.	Cisticola subruficapilla	1.04	0.77	**	**	**	**	**	0.62	0.89	0.62	1.02	0.20	0.57	0.50	**	0.25	
3.	Serinus flaviventris	0.42	0.47	0.20	**	**	**	**	0.15	**	0.12	0.40	0.17					
4.	Erythropygia coryphaeus	0.62	0.94	0.37	**	**	**	**	**	0.89	0.52	0.69	0.42	0.15	**	**	0.47	
5.	Emberiza capensis	0.50	0.69	0.20	0.27	**	**	**	**	0.69	0.42	0.69	0.72	0.07				
6.	Sphenoeacus afer	0.35	0.69	**	0.50	0.64	**	**	**	**	**	**	**	0.32	0.74	0.84		
7.	Prinia maculosa	1.26	1.44	0.94	0.74	1.14	**	**	**	0.89	0.42	0.92	0.47	0.50	0.27	0.84	0.57	
8.	Streptopelia capicola	0.07	0.42	0.32	0.30	0.47	0.25	**	**	0.22	0.20	0.40	0.37	0.17	0.42	0.35	0.15	
9.	Nectarinia famosa	0.42	0.22	0.32	0.15	0.30	0.74	**	0.32	**	0.32	0.10	0.25					
10.	Pycnonotus capensis	0.26	**	0.20	0.30	**	0.50	**	**	0.12	**	0.22	0.47	**	0.45	**	0.40	
11.	Nectarinia chalybea	0.07	0.22	0.42	**	0.35	0.62	0.22	**	**	0.37	0.55	0.47	0.07	0.27			
12.	Apalis thoracica	0.27	0.87	1.14	0.52	**	0.77	0.45	**	**	**	0.37	0.64	0.07	0.50	0.47	0.25	
13.	Andropadus importunus	0.02	0.27	1.04	0.30	**	0.37	1.24										
14.	Oena capensis		0.22	**	**	**	**	**	0.32	0.45	**	**	**	**	**	**	0.15	
15.	Emberiza impetuani	0.69																
16.	Mirafra apiata		0.12	**	**	**	**	**	**	**	0.30	**	**	0.07				
17.	Lanius collaris		0.42	**	**	**	**	**	**	**	**	**	**	**	**	**	0.15	0.02
18.	Telophorus zeylonus		0.27	0.17	**	**	**	**	0.22	0.45	0.32	0.55	0.25	0.07	**	**	0.32	
19.	Streptopelia senegalensis		0.42	0.32	**	**	**	**	**	**	**	0.60	0.42	0.05	**	**	0.32	0.02
20.	Colius striatus		0.35	0.62	**	**	**	**	**	**	0.20	0.35	**	**	0.27	**		
21.	Tchagra tchagra		0.12	0.20	**	**	**	**	**	**	0.25							
22.	Cisticola fulvicapilla		0.27	0.62	0.72	0.30	**	**	**	**	**	0.12	0.40	0.05	1.04	0.42		
23.	Euplectes capensis		0.12	**	**	0.52	**	**	**	**	**	**	**	**	0.45	0.27		
24.	Serinus sulphuratus		0.22	**	**	0.15												
25.	Cossypha caffra		0.27	0.74	0.60	0.69	0.99	0.22	**	**	**	**	0.15	0.12	0.55	0.79	0.15	
26.	Passer melanurus			0.37	**	**	**	**	**	**	**	0.10	0.10	**	**	**	0.15	0.05
27.	Sigelus silens			0.37	**	**	**	**	**	**	**	0.22	0.25					
28.	Numida meleagris			0.12	**	**	**	**	**	**	**	**	0.25	0.12	0.15			
29.	Parisoma subcaeruleum			0.42	**	**	**	**	**	0.12	0.72	0.50	0.20	**	**	**	0.57	
30.	Streptopelia semitorquata			0.12	0.20	0.15	0.25											
31.	Laniarius ferrugineus			0.42	0.42	0.35	0.87	0.27	**	**	**	**	**	**	0.27			
32.	Zosterops pallidus			0.62	0.79	0.69	1.24	1.36	**	**	**	**	0.47	0.07	0.55	1.07		
33.	Francolinus capensis				0.20	**	**	**	**	**	**	**	**	**	**	0.20		
34.	Promerops cafer				0.20	0.87	**	**	**	**	**	**	**	**	0.82	1.07		
35.	Nectarinia violacea				0.52	0.77	**	**	**	**	**	**	**	**	0.77	1.26		

36. Bradypterus victorinii	0.42	0.69	**	**	**	**	**	**	**	**	0.77	0.84		
37. Serinus canicollis	0.10	0.42	0.62	**	**	**	**	**	**	**	0.45	0.74		
38. Dicrurus adsimilis	0.20	0.07	0.25	0.07	**	**	**	0.10						
39. Batis capensis	0.60	0.55	1.98	0.50	**	**	**	**	0.12	**	**	**	0.12	
40. Serinus totta		0.30					**							
41. Serinus gularis		0.30												
42. Serinus scotops		0.47	0.74	0.07										
43. Muscicapa adusta		0.15	0.50	0.27										
44. Onychognathus morio			0.25	0.02	**	**	**	**	**	**	0.32			
45. Turdus olivaceus			0.62	0.03										
46. Terpsiphone viridis			0.45	0.42										
47. Tauraco corythaix				0.20										
48. Camaroptera brachyura				1.46										
49. Seicercus ruficapilla				0.94										
50. Nectarinia afra				0.57										
51. Cossypha dichroa				0.45										
52. Oriolus larvatus				0.40										
53. Dryoscopus cubla				0.30										
54. Coracina caesia				0.30										
55. Phyllastrephus terrestris				0.27										
56. Apaloderma narina				0.27										
57. Mesopicos griseocephalus				0.22										
58. Columba arquatrix				0.20										
59. Aplopelia larvata				0.17										
60. Sarothrura elegans				0.11										
61. Telephorus olivaceus				0.05										
62. Dendropicos fuscescens				0.05										
63. Phoeniculus purpureus				0.12										
64. Campephaga flava				0.05										
65. Serinus albogularis						0.89	**	0.27	0.25	**	0.15			
66. Stenostira scita						0.45	**	0.47	0.15	**	0.32			
67. Certhilauda albescens					0.94	**	0.52							
68. Eupodotis afra					0.07	**	0.17	**	0.05					
69. Sylvietta rufescens							0.20	0.12						
70. Pterocles namaqua					0.32	**	0.10	**	**	**	**	**	**	0.02

Appendix E (continued)

BIRD SPECIES	Site number:	1	2	3	4	5	6	7	8	9	10	11	12	13	14	15	16	17
	Habitat H:	0.51	0.61	0.89	0.90	1.19	1.65	1.97	0.15	0.62	0.66	0.95	1.03	0.79	0.89	1.04	1.31	0.52
71. *Corvus capensis*											0.10	**	0.10					
72. *Malcorus pectoralis*									0.32	**	0.32							
73. *Cercomela familiaris*											0.20	0.27	0.17	**	0.25			
74. *Cercomela schlegelii*									0.77									
75. *Estrilda astrild*												0.27						
76. *Anthus similis*												0.50						
77. *Monticola rupestris*												0.20						
78. *Saxicola torquata*															0.07			
79. *Serinus leucopterus*																0.74		
80. *Colius colius*																	0.15	
81. *Eremomela gregalis*																		0.07
82. *Certhilauda curvirostris*																		0.20
83. *Hippolais icterina*																		0.07
TOTAL BIRD SPECIES:		13	24	23	20	22	18	31	11	11	21	26	25	15	23	14	15	7
TOTAL BIRD DENSITY (pairs ha^{-1})		5.42	10.61	10.26	8.05	10.34	12.01	11.27	4.20	6.06	6.51	10.12	7.51	2.47	10.35	9.90	4.17	0.45

21. Distribution and Species Diversity of Birds and Plants in Fynbos Vegetation of Mediterranean-Climate Zone, South Africa

W. R. SIEGFRIED and T. M. CROWE

21.1 Introduction

The Fynbos Biome corresponds geographically with the Capensis Region delimited by Werger (1978) as one of the plant biogeographical regions of southern Africa. The biome is characterized by four indigenous vegetation types: strandveld, coastal renosterveld, coastal fynbos and mountain fynbos. The main physiognomic features of the vegetation are the prevalent sclerophyllous shrub form, the scarcity of trees, and the relatively minor importance of grasses and evergreen succulent shrubs (Kruger 1979). Important biogeographical features of the flora are the great concentration of species, the high degree of endemism, the characteristic distribution patterns of typical elements, despite a lack of species dominance, and the preponderance of certain families and genera (Taylor 1978).

Although birds are by far the best known group of animals in the fynbos (Bigalke 1979), work to evaluate the structure and functioning of selected avian communities in the biome against a background of the paradigms of modern ecology has only recently begun (Siegfried 1979). Here our purpose is to compare broadly the species composition and diversity of avian communities in the main vegetation types of the fynbos biome and to investigate general relationships, in accord with theoretical expectations, between bird communities and the structural and floristic complexities of their habitats. More particularly, we attempt to determine whether structure or floristic composition alone or in combination is a consistent and significant predictor of the species composition and richness of the avifauna.

21.2 Methods

21.2.1 STUDY SITES

We selected study sites representative of the four major vegetation types of the fynbos biome in the southwestern Cape (Table 1; Fig. 1). Each site (except Lebanon, see Table 1) embraced at least 50 ha of reasonably

Table 1. Characteristics of plant and bird community study sites in the southwestern Cape fynbos biome.

Site	Coordinates	Altitude (m a.s.l.)	Aspect	Slope	Annual[a] rain-fall (mm)	Geology	Area (ha)	Vegetation	
								Age (yr)	Type
Melkbosch	33°45'S, 18°27'E	0-15	W and E	0-8	570	White-grey sandy soil	50	15	strandveld
Geelbek	33°12'S, 18°08'E	0-15	W	0-8	320	Langebaan limestone	50	15	strandveld
De Grendel	33°51'S, 18°35'E	300	WSW	9-26	450	Malmesbury slates, Tygerberg formation	50	15	coastal renosterveld
Barhuis	33°14'S, 18°15'E	100	WNW	0-2	470	White to light red sand	50	23	coastal fynbos
Swartbosch-kloof	34°00'S, 18°57'E	300	SSE	9-45	1470	Cape granite, coarse grained	50	20	mountain fynbos
Sir Lowry's Pass	34°09'S, 18°50'E	600	NE	0-26	No data	TMG, Peninsula Formation, mainly quartzites	50	5	mountain fynbos
Silvermine	34°05'S, 18°25'E	1000	NW	0-10	1300	TMG, Peninsula Formation, mainly quartzites	50	21	mountain fynbos
Lebanon 1	34°12'S, 19°08'E	700	SW	0-18	950	TMG, Peninsula Formation, mainly quartzites	50	21	mountain fynbos
Lebanon 2		900	SW	15-25			5	5	
Lebanon 3		950	SW	30-35			40	10	
Lebanon 4		1000	SW	0-35			5	10	

[a] All are 25-year means except Lebanon (4 yr) and Silvermine (3 yr).

homogeneous, undisturbed vegetation of the designated type. The Silvermine site was cleared and subjected to a controlled burn in June 1978. Therefore, we have grouped our bird and plant surveys into pre- and post-fire categories. Complete lists of the plant and bird species recorded in these vegetation types, and their respective mean cover and abundance values, can be obtained from the senior author.

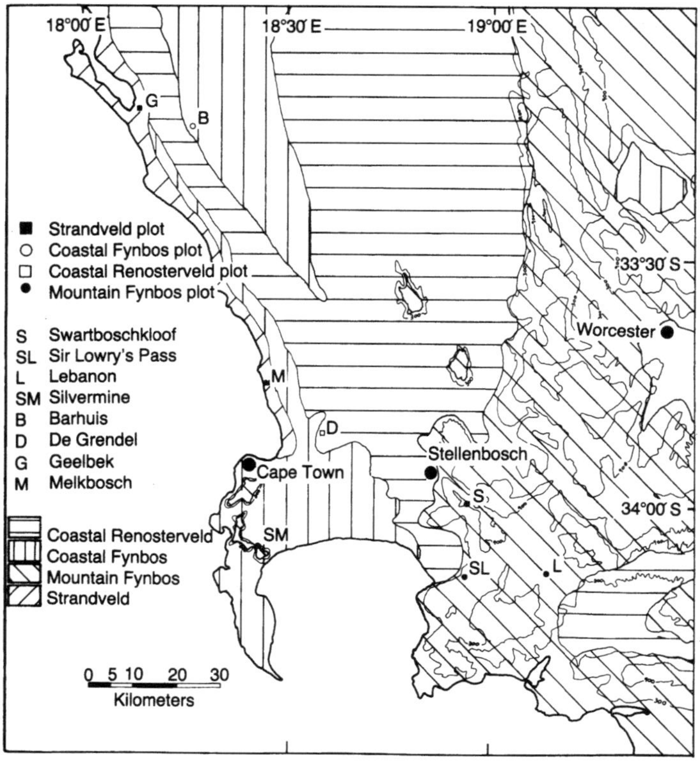

Fig. 1. Fynbos vegetation types in the southwestern Cape and sites at which bird and plant data were collected (after Acocks 1953).

21.2.2 SAMPLING PLANT COMMUNITIES

Floristic and structural characteristics of the vegetation at each study site were sampled during August and November 1979 (but see below) in at least eight 10 x 10 m plots spaced along transect segments, so as to encompass any altitudinal gradients in the vegetation. Floristic and structural composition data were collected in the form of Braun-Blanquet (B-B) estimates of the cover-abundance of all species present in the plot and included in 12 growth forms: ferns (non-vascular plants > 50 mm in height, to exclude mosses); restioids (aphyllous stem monocots); graminoids (monocots with leaf diameters < 20 mm); broad-leaf monocots (monocots with leaf diameters > 20 mm); dicot herbs (non-woody plants with reticulate venation); woody plants (divided into picophyll, leptophyll, nanophyll, microphyll and mesophyll), and succulents (divided into leaf and stem succulents). These are the most common growth forms in the study sites. Vertical stratification was represented by B-B cover-abundance estimates of vegetation in the following strata: 0-0.25 m, 0.25-0.5 m, 0.5-1 m, 1-1.5 m, 1.5-2 m and > 2 m. For each study site (except Lebanon and pre- and post-fire Silvermine, for which plots were analysed separately) mean B-B values over all plots for each site were calculated for each species, growth form and stratum, and were used in subsequent statistical analyses. Thus, units of vegetation analysis are termed "mean" vegetation plots. Overall vegetation structural diversity for each of these "mean" plots was calculated using the foliage diversity formula, i.e.

$$-\sum_{i=1}^{6} C_i \log C_i$$

(MacArthur and MacArthur 1961), in which C is the mean B-B cover value for each of the six vegetation strata.

Modifications to these procedures were employed at Geelbek and Lebanon. The vegetation at Geelbek is a mosaic of two communities each of which was surveyed separately. Plot size and shape were dependent on the bush clumps surveyed. At Lebanon, vegetation composition data were extracted from records obtained through line-point surveys made by personnel of the Directorate of Forestry. Basal and canopy strikes for each species were expressed as a percentage of 1000 points (100 points, one pace apart, on each of 10 lines 7.6 m apart) and converted to the Braun-Blanquet scale. Each species was classified by height and growth form according to the system used for the other study sites. Totals for each height stratum and growth form were calculated. Since the species' heights are mean values, the actual vegetation is probably taller than indicated. The Lebanon site was sampled during February and March 1979.

21.2.3 SAMPLING BIRD COMMUNITIES

A standard 21-ha plot in the form of a rectangle 700 x 300 m was demarcated in each study site except Lebanon (see below). The size and shape of the plot were designed to accommodate at least partial contiguous territories of virtually all the bird species representative of a given locality. Within each plot, two parallel transect lines were established, running perpendicularly to the long axis of the plot and at approximately one-third intervals along it. The transects were marked off in 20-m segments. At Lebanon a 200-m transect was laid out in each of the four plots.

Initial trials indicated that three censuses in the first five hours after sunrise gave the optimum return for effort, and that no significant benefits accrued from censusing on more than one day in a given period, except when weather extremes were encountered. Accordingly, a standardized transect-line census procedure was developed covering 3 ha per plot. Censuses, each of three 35-min "runs", were made on two days of reasonably uniform weather in every month of the year during 1978 and 1979. All birds (except aerial foragers) seen or heard within 25 m of the transect line were counted.

21.2.4 NUMERICAL METHODS

Patterns of similarity attending floristic and structural features of vegetation and bird species composition were identified by means of cluster analysis (Field and McFarlane 1968; Anderberg 1973) and multidimensional scaling (Kruskal 1964). These two methods are complementary approaches to multivariate data analysis. Cluster analysis (CA) is designed to identify patterns of hierarchical similarity (i.e. groups and subgroups) among samples, and the results of a CA are depicted in a dendrogram. Multidimensional scaling (MDS) reveals patterns of sample similarity independent of group affiliation, and the results of an MDS analysis are depicted in an ordination diagram. Thus, MDS is better than CA at detecting similarity trends among samples. Information statistic tests (Field 1969) between plots or groups of plots identified bird species which characterized ($P < 0.001$) our study sites. Characteristic bird species were ranked according to their $2\Delta I$ value in information statistic tests, and, for the sake of brevity, no more than 10 characteristic species are discussed for each plot or plot grouping. The presence of each bird species in each plot was weighted according to its mean abundance (mean number of birds per census). The Bray and Curtis (1957) measure of similarity and a group-average sorting method (Lance and Williams 1967) were used in cluster analyses.

21.3 Results and discussion

21.3.1 SPECIES RICHNESS

In marked contrast to the fynbos flora, one of the richest in the world (Taylor 1978), the avifauna of the biome is relatively poor (Table 2)

Table 2. Species richness for birds and plants and vegetation structural diversity in fynbos vegetation types in the southwestern Cape. There is a significant ($P < 0.001$; t test) difference in bird species richness between lowland (strandveld, coastal renosterveld and coastal fynbos and montane (mountain fynbos) formations, but not in plant species richness or vegetation structural diversity.

Vegetation type and study site	No. of bird species	No. of plant species	Vegetation structural diversity
STRANDVELD			
Melkbosch	39	61	1.44
Geelbek	35	85	1.58
COASTAL RENOSTERVELD			
De Grendel	34	79	1.37
COASTAL FYNBOS			
Barhuis	35	21	1.21
MOUNTAIN FYNBOS			
Swartboschkloof	15	72	1.77
Sir Lowry's Pass	23	113	1.05
Silvermine pre-fire	9	28	1.36
Silvermine post-fire	10	53	1.01
Lebanon 1	9	43	1.31
Lebanon 2	11	70	1.24
Lebanon 3	8	67	1.08
Lebanon 4	9	53	1.08

though similar in richness to those in other comparable mediterranean-type ecosystems. This is largely a function of similar vegetation structure (Cody 1974, 1975; Cody and Mooney 1978). More particularly, the scarcity of trees in the fynbos (Moll et al. 1980) is probably an important determinant of the low bird species richness, not "the fact that the flora is poor in seeds suitable for and available to birds" (Winterbottom 1968, quoting M K Rowan). This "fact" has never been demonstrated.

Our results show that the fynbos avifauna consists of two suites of species each tending to occur in habitats formed by lowland and montane vegetation formations, respectively (Fig. 2). A basic difference in a combination of structure and floristic composition of the vegetation between the lowland and montane formations parallels this dichotomy in avifauna (Fig. 3). At a less gross level, each of the four vegetation types (strandveld, coastal renosterveld, coastal fynbos and mountain fynbos), which hitherto have been recognized mainly on a qualitative basis (Kruger 1979), are distinguished quantitatively by their own associations of bird species and combinations of flora and physiognomic features (Figs 2 and 3). However, in marked contrast to MacArthur's (1965) predictions, neither vegetation structure, as determined by us, nor floristic composition alone is a consistent and significant predictor of the species richness of the bird communities of each vegetation type (Figs 4 and 5).

The lowland sites contain approximately three times the number of avian species as the montane sites (at a 3-ha level of comparison), in contradistinction to the species richness of the vegetation (Table 2; Fig. 5). Nor is the relatively enhanced species richness of the lowland avifauna associated with a markedly more complex vegetation structure (Table 2; Fig. 4). However, the lowland avifauna occurs in a greater variety of vegetation types than the montane suite which is virtually confined to mountain fynbos.

The avifaunas of strandveld and coastal renosterveld have more in common with each other than either one has with that of coastal fynbos (Table 3; Fig. 2). Only three species which characterize coastal fynbos also are indicators of the other two lowland vegetation types (Table 3). Both strandveld and coastal renosterveld typically form dense shrublands, whereas our coastal fynbos communities tend to be more open. Avian species richness is almost the same in all three lowland vegetation types which, however, differ markedly in plant species richness between coastal fynbos on the one hand and strandveld and coastal renosterveld communities on the other (Table 2). Also to be noted is that at the community level, plant species richness in mountain fynbos is not significantly higher than in strandveld and coastal renosterveld.

21.3.2 CHARACTERISTIC SPECIES

Among the ten main species which characterize the lowland avifauna, mousebirds, the Cape bulbul, Karoo scrub robin, Karoo prinia and lesser double-collared sunbird are common in strandveld communities (Table 3). The Karoo prinia and lesser double-collared sunbird are common in coastal renosterveld as well, but they do not feature among the main species characterizing coastal fynbos. The prominence of larks, the rufous-eared warbler and chat flycatcher in the coastal fynbos community points to a strong link with the avifauna of the Karoo region. Indeed, the avifauna characterizing the lowland formation as a whole is most strongly allied with communities in Karoo vegetation types, and Winterbottom (1960, 1968) has suggested that the fynbos avifauna is predominantly derived from that

Table 3. Characteristic bird species in lowland vegetation study sites in the southwestern Cape fynbos biome types. Given data are ranks of species by information statistic tests.

	Strandveld		Renosterveld	Coastal Fynbos
	Geelbek	Melkbosch	De Grendel	Barhuis
Namaqua dove *Oena capensis*				9
Speckled mousebird *Colius striatus*			10	
White-backed mousebird *Colius colius*	4	7		
Red-faced mousebird *Colius indicus*		3		
Karoo lark *Certhilauda albescens*				1
Clapper lark *Mirafra apiata*				3
Lark sp.				7
Cape bulbul *Pycnonotus capensis*	1	1		
Cape robin *Cossypha caffra*		10	2	
Karoo scrub robin *Erythropygia coryphaeus*	7	2		
Bar-throated apalis *Apalis thoracica*	6			
Grey-backed cisticola *Cisticola subruficapilla*			1	8
Rufous-eared warbler *Malacorus pectoralis*				6
Karoo prinia *Prinia maculosa*	5	5	4	
Tit babbler *Parisoma* sp.	2			
Chat flycatcher *Melaenornis infuscatus*				10
Pied starling *Spreo bicolor*	10			
L. Double-collared sunbird *Nectarinia chalybea*	3	4	5	
Cape white-eye *Zosterops pallidus*			9	
Cape sparrow *Passer melanurus*			8	
Yellow-rumped bishop *Euplectes capensis*			6	2
Cape canary *Serinus canicollis*			7	
Black-headed canary *Serinus alario*			3	
White-throated seed-eater *Serinus albogularis*		8		
Yellow canary *Serinus flaviventris*	8	6		5
Bully canary *Serinus sulphuratus*				4
Cape bunting *Emberiza capensis*	9	9		

Table 4. Characteristic bird species of montane vegetation type study sites in the southwestern Cape fynbos biome. Given data are ranks of species by information statistic tests.

	Mountain fynbos							
	Sir Lowry's Pass	Swartbosch kloof	Silvermine		Lebanon			
			Pre-fire	Post-fire	1	2	3	4
Black harrier *Circus maurus*								3
Grey-wing francolin *Francolinus africanus*				2				
Ground woodpecker *Geocolaptes olivaceus*	9							
Grey-backed finchlark *Eremopterix verticalis*	10							
Cape rockjumper *Chaetops frenatus*	1							
Cape rock thrush *Monticola rupestris*	5	8						
Sentinel rock thrush *Monticola explorator*	6							
Victorin's scrub warbler *Bradypterus victorini*					1	3	3	1
Grassbird *Sphenoeacus afer*				5	2			
Neddicky *Cisticola fulvicapilla*	3							
Fiscal flycatcher *Sigelus silens*			1					
Cape batis *Batis capensis*		3						
Fiscal shrike *Lanius collaris*				4				
Southern boubou *Laniarius ferrugineus*		5	3					
Bokmakierie *Telophorus zeylonus*				6				
Cape sugarbird *Promerops cafer*		4				1	2	4
Malachite sunbird *Nectarinia famosa*		2						
Orange-breasted sunbird *Nectarinia violacea*	2	7				2	1	2
Red bishop *Euplectes orix*			4					
Yellow-rumped bishop *Euplectes capensis*			5					
Cape siskin *Serinus tottus*	4			3				
Yellow canary *Serinus flaviventris*	8	6	2					
Protea sead-eater *Serinus leucopterus*		1						
Pigeon sp.	7							
Canary sp.				1				

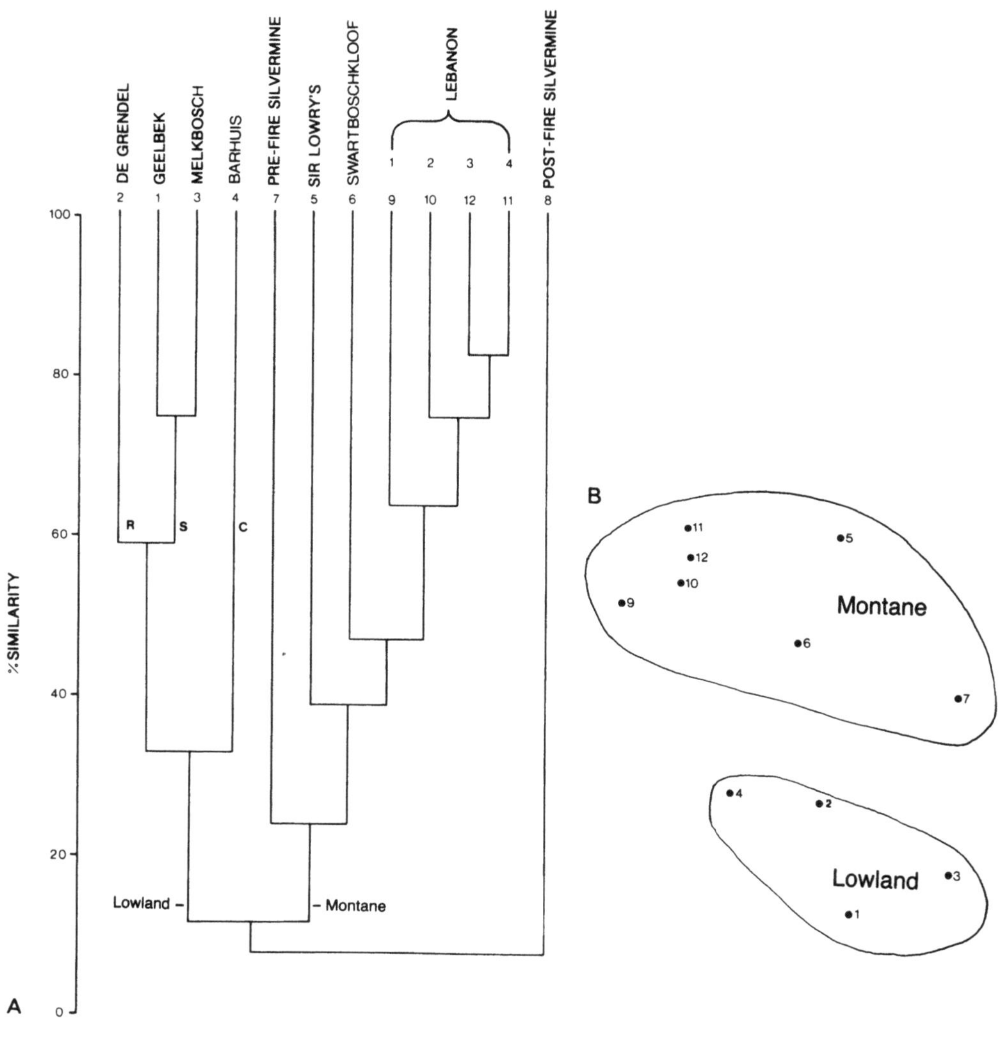

Fig. 2. The results of a cluster analysis (A) and a multidimensional scaling (B) of 12 sites according to their avifaunas in four vegetation types of the fynbos biome. See Fig. 2A for key to names of sites.

of the Karoo. The transitions of the avian elements, as well as certain plant species, tend to be gradual between the lowland fynbos and the Karoo formations (Siegfried, unpublished), unlike the more abrupt division between the lowland and montane formations in the fynbos biome. The restioid, proteoid and ericoid elements of coastal and mountain fynbos provide a link between the lowland and montane vegetation formations which is not paralleled by avian elements (Tables 3 and 4). In other words, coastal and mountain fynbos have stronger affinities through the growth forms of common plant groups (Taylor 1978) than through their birds. However, this affinity is apparently a weak one, and it is possible that another botanical link exists between coastal fynbos and Karoo vegetation types, at least in areas north of Cape Town.

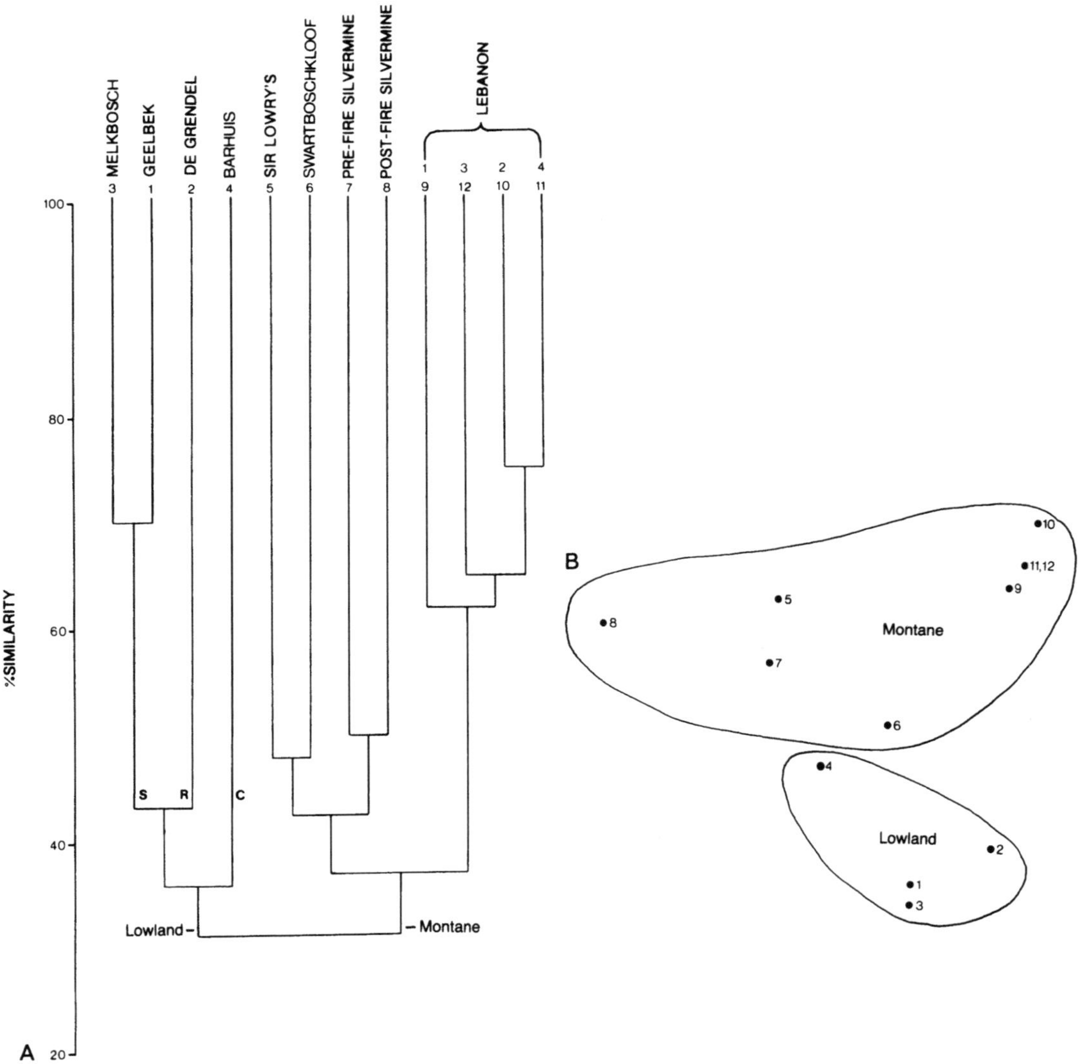

Fig. 3. The results of cluster analysis (A) and a multidimensional scaling (B) of 12 sites according to their floras and the structure of their vegetation in the fynbos biome. See Fig. 2A for key to names of sites.

The avifauna characteristic of the montane vegetation (Table 4) is more distinct than that of the lowland formation, in that it contains five species (protea seed-eater, Cape siskin, orange-breasted sunbird, Cape sugarbird and Victorin's scrub warbler) which are practically endemic to mountain fynbos. (There are no avian species endemic to the lowland vegetation formation.) The balance of the avifauna has links with communities in a very wide range of major vegetation formations.

21.3.3 DIFFERENCES BETWEEN COMMUNITIES

Bird assemblages of the various plant communities of mountain fynbos clearly differ, independently of age of the plant community. This is exemplified by the avifaunas of the approximately equal-age communities of Swartboschkloof, Silvermine (pre-fire) and Lebanon 1 (Table 2; Fig. 2). However, these differences are weaker than those in either structure or, to a lesser extent, in species composition of the plant communities concerned (Figs 4 and 5). Normally one species is very common in each community: the Cape sugarbird dominates in proteoid and the orange-breasted sunbird in ericoid communities generally. Old mountain fynbos communities apparently support a lower avian species richness than younger ones (cf. Sir Lowry's Pass and Swartboschkloof), and very young communities (e.g., post-fire Silvermine) can embrace a distinctly non-fynbos avifauna (Table 4; Fig. 2). It is also likely that bird and plant species richness decrease while structural diversity and age of vegetation increase. All these intercommunity comparisons are, however, preliminary and provisional, because seasonal changes in flowering and vegetative growth can have marked effects on the composition of the bird communities (Siegfried, unpublished). Moreover, description and categorization of the plant communities of the heterogeneous mountain fynbos vegetation are still inadequate (Kruger 1979).

In the fynbos biome, the physical environment described by climate, topography, soil type and fire regime increases in heterogeneity as one moves inland from the marine littoral. Consequently, the relatively uniform areas covered by either strandveld or coastal renosterveld embrace fewer different plant communities, with respect to both floristic and structural diversity either in seral or climax stages, than mountain fynbos. Areas of coastal fynbos are apparently intermediate in heterogeneity, but tend towards the lowland pattern of relative uniformity. In effect, in the lowland a few different plant communities

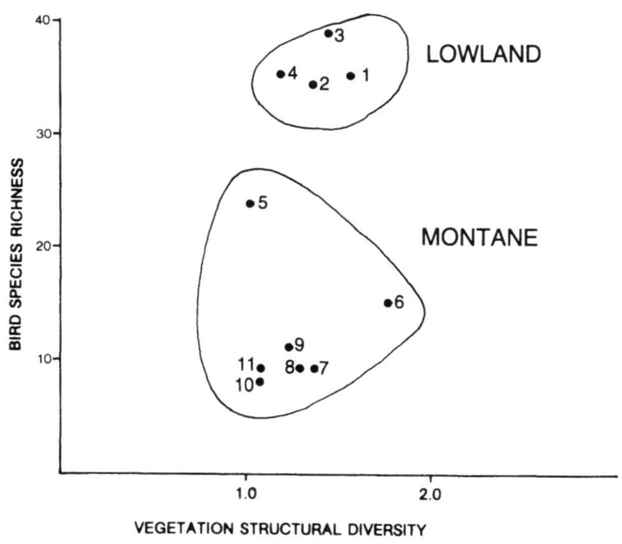

Fig. 4. Bird species richness plotted against the structural diversity of the vegetation at 11 sites (post-fire Silvermine exempted) in the fynbos biome. See Fig. 2A for key to names of sites.

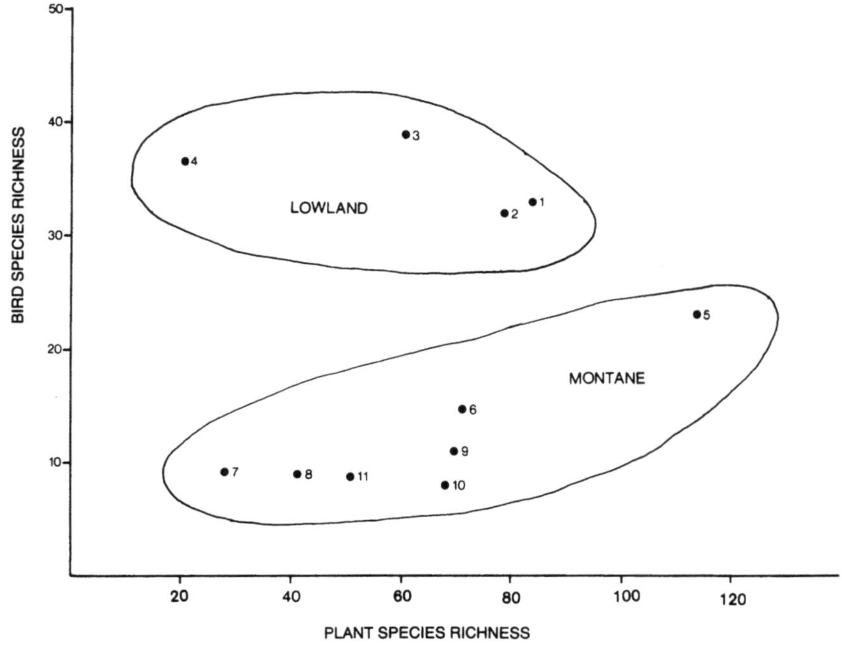

Fig. 5. Bird species richness plotted against plant species richness at 11 sites (post-fire Silvermine exempted) in the fynbos biome. See Fig. 2A for key to names of sites.

tend to cover uniformly large areas, whereas in the montane environment a much greater variety of communities is dispersed widely as relatively small entities over the landscape.

In spite of the heterogeneity of mountain fynbos, the avifauna is poor with the same small core of characteristic species occurring generally in the vegetation. This applies particularly to the orange-breasted sunbird and the Cape sugarbird which maintain their dominance in mountain fynbos by moving regularly or nomadically, over relatively long distances if necessary, between different plant communities. The Cape sugarbird migrates regularly into coastal fynbos, but the orange-breasted sunbird seldom leaves mountain fynbos (Siegfried, unpublished).

Cody (1975) found that bird species replace one another relatively rapidly as one moves from one scrub community to another in the fynbos. This implies that fynbos birds occupy narrow niches and are sterotyped to a narrow habitat range. Our findings support Cody's conclusion in so far as there are marked changes in avian species composition between major vegetation types of the fynbos. The still scanty information on differences between bird communities within a major vegetation type, however, points to a relaxation of niche breadth and increased mobility of species especially in mountain and possibly also coastal fynbos, compared to strandveld and coastal renosterveld. In other words, the relatively spatially uniform, temporally constant, and productive strandveld and coastal renosterveld allow finer partitioning of avian resources, resulting in the coexistence of more species than in coastal and mountain fynbos.

Finally, it should be noted that this study is limited geographically to the southwestern Cape and that the results are not necessarily applicable to fynbos in the eastern part of the biome and, more particularly, the southern Cape including coastal fynbos east of 19° longitude.

Acknowledgements

We thank R K Brooke, R A C Jensen, R and J Martin, P and M Netherway, J C Sinclair and G D Underhill for assisting with the collecting of bird data. Botanical data were collected by J Sommerville and members of the Department of Environment Affairs. C Boucher and F J Kruger helped to select our sampling sites. We are grateful to the Department of Environment Affairs and the Divisional Council of the Cape Province for access to study sites. Financial support was provided by the South African Council for Scientific and Industrial Research, through its Fynbos Biome Project, and the University of Cape Town.

References

ACOCKS JPH (1953) Veld types of South Africa. Memoirs of the Botanical Survey of South Africa 28: 1-192

ANDERBERG MR (1973) Cluster analysis for applications. Academic Press, London. 359 pp

BIGALKE RC (1979) Aspects of vertebrate life in fynbos, South Africa. In: SPECHT RL (ed) Ecosystems of the world, vol 9A. Heathlands and related shrublands. Descriptive studies. Elsevier, Amsterdam, pp 81-95

BRAY JR, CURTIS JJ (1957) An ordination of the upland forest communities of southern Wisconsin. Ecological Monographs 27: 325-349

CODY ML (1974) Competition and the structure of bird communities. Princeton University Press, New Jersey. 318 pp

CODY ML (1975) Towards a theory of continental species diversity: bird distribution over mediterranean habitat gradients. In: CODY ML, DIAMOND JM (eds) Ecology and evolution of communities. Bellknap Press, Cambridge. 545 pp

CODY ML, MOONEY HA (1978) Convergences and dissimilarities of mediterranean-climate ecosystems. Annual Review of Ecology and Systematics 9: 265-321

FIELD JG (1969) The use of the information statistic in the numerical classification of heterogenous systems. Journal of Ecology 57: 565-569

FIELD JG, MCFARLANE G (1968) Numerical methods in marine ecology 1. A quantitative 'similarity' analysis of rocky shore samples in False Bay, South Africa. Zoologica africana 3: 119-137

KRUGER FJ (1979) Plant ecology. In: DAY J, SIEGFRIED WR, LOUW GN, JARMAN ML (eds) Fynbos ecology: a preliminary synthesis. South African National Scientific Programmes Report No. 40. CSIR, Pretoria

KRUSKAL JB (1964) Multidimensional scaling by optimizing goodness of fit to a non-metric hypothesis. Psychometrika 29: 1-27

LANCE GN, WILLIAMS WT (1967) A general theory of classificatory programs 1. Hierarchical systems. Computer Journal 9: 373-380

MACARTHUR R (1965) Patterns of species diversity. Biological Reviews 40: 510-533

MACARTHUR R, MACARTHUR J (1961) On bird species diversity. Ecology 42: 594-598

MOLL EJ, MCKENZIE B, MCLACHLAN D (1980) A possible explanation for the lack of trees in the fynbos, Cape Province, South Africa. Biological Conservation 17: 221-228

SIEGFRIED WR (1979) Animal ecology. In: DAY J, SIEGFRIED WR, LOUW GN, JARMAN ML (eds) Fynbos ecology: a preliminary synthesis. South African National Scientific Programmes Report No. 40. CSIR, Pretoria
TAYLOR HC (1978) Capensis. In: WERGER MJA (ed) Biogeography and ecology of southern Africa. Junk, The Hague
WERGER MJA (1978) Biogeographical division of southern Africa. In: WERGER MJA (ed) Biogeography and ecology of southern Africa. Junk, The Hague
WINTERBOTTOM JM (1960) The zoogeographical affinities of the avifauna of the western Cape Province. Ibis 102: 383-393
WINTERBOTTOM JM (1968) Remarks on the possible origin of the avifauna of the South West Cape. Ibis 110: 91-93

22. Plant Community Structure - Spatial Partitioning of Resources

W. E. WESTMAN

Throw away the lights, the definitions,
And say of what you see in the dark
That it is this or that it is that,
But do not use the rotted names.
How should you walk in that space and know
Nothing of the madness of space,
Nothing of its jocular procreations?
Stevens (1937)

22.1 Introduction

The spatial partitioning of phytomass is considered in this chapter at three levels of community organization: the supra-organismal (patterning among suites of species, associations and "microcommunities"), the organismal (spatial patterning of a species colony or population) and the suborganismal (vertical and horizontal arrangement of leaves, stems and roots). The evergreen and facultatively deciduous shrublands of the five mediterranean-climate regions are considered, recognizing that the distinctive soil fertility levels and longer evolutionary histories of the Australian and South African heathlands reduce their comparability with other mediterranean-climate shrublands (Specht 1979a; Naveh and Whittaker 1980).

22.2 Spatial partitioning at the supra-organismal level

22.2.1 HERB STRATUM PATTERNING IN RELATION TO SHRUBS

The shrub and herb strata of mediterranean-climate shrubland communities appear to act, in general, as distinct synusia, establishing spatial patterns in relation to each other as units. Exceptions occur where nutrients or allelopathic substances released by particular shrub species favour or discourage the growth of particular species of herb.

Specht and Morgan (1981) demonstrated the inverse relationship between overstorey and understorey foliage projective cover in a range of community types, including Australian dry-heath and mallee-heath. They considered overstorey foliage cover to be a function of available water, but did not venture a hypothesis to explain understorey dependence on overstorey cover. Kirkpatrick and Hutchinson (1977) and Westman (1981a) found an inverse relationship between shrub and herb cover in Californian coastal sage scrub, the latter finding a correlation of $r = -0.86$ ($P < 0.01$) in 67 sites. Stratal interdependence in both chaparral and sage scrub is demonstrated by the marked decline in herb cover with successional development of the shrub stratum (Hanes 1971; Westman 1981a). In the following subsections, six factors influencing herb stratum patterns in relation to shrubs are reviewed: available moisture, light, grazing, allelopathy, substrate and soil nutrients, and air pollution.

Available moisture. There is some evidence that herb cover in Californian shrublands is moisture-sensitive. Keeley (1977) reported marked increases in spring herb cover with above-average winter precipitation levels, during the first five years following fire. Halligan (1973) found a positive correlation ($P < 0.01$) between lushness of herb growth and soil moisture in coastal sage scrub stands. In *Adenostoma* (chamise) chaparral at the I.B.P. site at Echo Valley, California, Keeley and Johnson (1977) found 99% of the herb cover under shrubs during normal to dry rainfall years. Following a winter (1977-78) of double the normal rainfall, Shmida and Whittaker (1981) found the greatest density of herbs in open areas between shrubs, in the same chaparral type in the San Jacinto Mountains about 100 km to the north. They found the cover of annuals to be 7-19% in openings, compared to less than 1% found by Keeley and Johnson (1977) in lower rainfall years. Presumably the rapid increase in density of these herb opportunists in chaparral openings in wet years is made possible by germination of seeds dispersing from adjacent areas of recently burned chaparral, and from mature coastal sage scrub sites, as well as from seed stores. Westman (1979a) found that 58% of the herb species recorded on 1-4 year-old chaparral sites could be found in the understorey of coastal sage scrub stands that had not burned in seven years or more.

Jaksic and Montenegro (1979) also reported an increase in herb biomass in matorral following a wet year, but the growth response was exhibited only by the annuals (two Chilean native grasses, *Chaetanthera ciliata*, *Trisetobromus hirtus* and a Mediterranean forb, *Erodium cicutarium*, and not by the native perennial iris studied (*Solenomelus peduncularis*)).

Light. Light is also a factor which appears to stimulate herb growth in California shrublands. Despite their occupation of higher rainfall sites than coastal sage scrub, mature chaparral stands have lower herb cover. The density of overstorey foliage in chaparral is twice that of sage scrub (leaf area index 2.61 for chaparral versus 1.31 for coastal sage scrub: Mooney et al. 1977b), and produces substantial shading. Once moisture is no longer the limiting factor, herb growth appears to be limited by light. Shmida and Whittaker (1981) sampled in an unusually wet year and found the biomass per plant to be markedly less for herbs under chaparral shrub canopies than for those in the open. *Ceanothus cuneatus*, a chaparral species, has been shown capable of suppressing growth of seedlings of *Salvia leucophylla*, a sage-scrub species, due to shading (McPherson and Muller 1967). Addition of nutrients to herbs shaded by chaparral shrubs only induces etiolation (Christensen and Muller 1975).

The tendency for native herbs to be found primarily under the canopy of shrubs in years of average or low precipitation is even more marked in matorral than in chaparral. Although herb densities are higher in Chile than in California (Keeley and Johnson 1977; Montenegro et al. 1978), native herbs are sharply restricted to sites underneath the canopies of matorral shrubs during years of normal precipitation (Carter 1973; Keeley and Johnson 1977). However, Mediterranean herbs occur below gaps in the canopy. Shmida and Whittaker (1981) also report this phenomenon in chaparral: six of the 11 herb species restricted to openings in their study were Mediterranean annuals.

Presumably the more extensive area of openings in shrublands of the Mediterranean region, induced by prolonged grazing, has selected for heliophilic annuals able in competition to displace most herbs from other regions which have retained more generalized abilities to grow in a variety of light environments. In support of this hypothesis is the study of Israeli maquis by Naveh and Whittaker (1980), in which herb richness increased fourfold from closed maquis to open and disturbed shrublands. By contrast, Jaksic and Montenegro (1979) found that native Chilean grass did not grow differently in the shade of matorral shrubs than in openings, and the native perennial iris produced more biomass (though fewer fruits) in the shade.

Grazing. The role of native rodents in grazing forbs and grasses in coastal sage scrub and chaparral (Christensen and Muller 1975; Meserve 1976), and on the border between coastal sage scrub and grassland in California has been reasonably well documented (Bartholomew 1970, 1971; Halligan 1973, 1974). Where shrubs form a continuous canopy, rodents use these for shelter and venture into adjacent grassland to graze, thus contributing to a border zone of markedly low phytomass. This "bare" zone is actually occupied by diminutive herbs much smaller than those in the adjacent grassland (Halligan 1973). The density of herbs is, in fact, lowest within the shrublands themselves (Halligan 1973; Christensen and Muller 1975). In the case of chamise chaparral, Bradford (1976) has shown that rodents are abundant within as well as at the borders of the shrub stand, with some tendency for rodent densities to increase near shrubland margins.

Domestic stock will markedly decrease the abundance of herbs in openings accessible to them. The steeper gradient of increase in herb cover from openings to shrub understorey in Chile than California is attributed to the heavier grazing pressure of the Chilean sites (Montenegro et al. 1978). Naveh and Whittaker (1980) note an intermediate grazing level at which herb species richness in Israeli open oak woodland reaches its maximum. Grazing has been heavier over a longer period in the Mediterranean than in Chile, and more so in Chile than in California (Aschmann and Bahre 1977). One might therefore expect herbs from the Mediterranean to invade both Chilean and Californian shrubland sites, the extent of invasion increasing with grazing pressure. The more marked concentration of introduced annuals in openings in Chile than in California (Carter 1973) is consistent with these expectations.

The invasion of Mediterranean annuals into Australia (Specht 1969a) and South Africa (coastal renosterveld; Taylor 1978) has also been noted. Raven (1973) and others have discussed exchanges between these regions in broad biogeographic terms, and Gulmon (1977) has compared extent of invasion of Mediterranean exotics into grassland sites in California and Chile. A quantitative study of species richness and abundances of exotic annuals in each mediterranean-climate region is needed to determine whether, in fact, the Mediterranean portions of Europe provide the most vigorous competitors among mediterranean-climate annuals, and to what extent the regions differ in their resistance to this invasion.

Allelopathy. Allelopathic substances in shrubs or small trees, potentially capable of inhibition of selected herb species, have been found among species of chaparral (Adenostoma fasciculatum; Arctostaphylos glandulosa var. zacaensis; A. glauca; Lepechinia calycina: Muller et al. 1968; McPherson and Muller 1969; Chou and Muller 1972), coastal sage scrub (Artemisia californica, Salvia apiana, S. leucophylla, S. mellifera: Muller and Muller 1964; Muller et al. 1964; C H Muller 1965; W H Muller 1965; Muller 1966; Muller and del Moral 1966; Halligan 1973, 1974, 1976), French garrigue (with Rosmarinus and Erica arborea: Deleuil 1951; Guyot 1951), Spanish bush heath or maquis (Erica arborea, E. australis: Ballester et al. 1979), Israeli deciduous oak woodland (Quercus ithaburensis: Dinoor 1962) and maquis (Yardeni and Evenari 1952), and Australian coastal dry-heath (Eucalyptus baxteri: Del Moral et al. 1978). No water-soluble allelopathic substances were isolated from leaves of five Chilean matorral shrubs tested (Montenegro et al. 1978). In a number of the cases cited, tests for allelopathic effects were performed on species not found in, or native to, the area, so that demonstration of potential allelopathic effects on native herb species was not always accomplished. Beyond this, the rapidity of degradation of these substances by micro-organisms, their proneness to leaching by rainfall, and their volatilization by high temperatures and fire remain to be quantified for most field sites. To the extent that allelopathic substances do reach sensitive herb species, their effect will be masked to a degree by moisture, light, nutrient and herbivory effects on the same herb species (e.g. Halligan 1973; Christensen and Muller 1975). Thus allelopathic controls on herb growth in the field are not definitely proven, and in most cases the effect remains only a possibility. Halligan (1973) has shown that allelopathic substances released by Artemisia californica, a coastal sage scrub species, differ in their effects on different herb species. Hypochoeris glabra and Madia sativa showed sensitivity, while Festuca megalura and Bromus rigidus did not. The implications of this for spatial patterning are significant. Were allelopathic effects of shrubs on selected herbs strong, one would expect their expression in strong negative associations between species pairs within the shrubland, that could be revealed in appropriate field data. Studies that combine such association analyses with allelopathic testing are needed.

Substrate and soil nutrients. Herbs can potentially exhibit a spatial pattern corresponding to microsite heterogeneities in soil nutrient levels. Beyond the funnelling of nutrients to particular microsites by the capture of rain as stemflow, plants may enrich microsites by nutrient accumulation or fixation and subsequent release as litter fall, insect frass, or as ash following fire. Before local enrichment could be considered potentially significant in patterning of herbs, however, it is necessary to ask whether growth of shrubland or heathland species is nutrient-limited, and if so, by which nutrients under what conditions. The question of nutrient limitation is discussed by other authors in this volume, and the evidence will be summarized only briefly here.

Specht (1979b, 1980) has reviewed soil nutrient studies in mediterranean-type shrublands and heathlands and concludes that heathlands are invariably found on less fertile soils than broad-leaved sclerophyllous shrublands, although regional variations exist. Thus the heathland soils of southwestern Australia appear less fertile than those of southern Australia or South Africa (Kruger 1979; Specht 1979b, 1980). Among shrublands, the Australian mallee and South African renosterveld occur on soils lower in both nitrogen and phosphorus (less than 0.024% total phosphorus, less than 0.13% total N) than shrublands of the other three continents. Of the latter group, French maquis has a higher nitrogen:phosphorus ratio (13:1) (0.4 to 0.5% N; 0.03 to 0.04% P) than chaparral or matorral (1.7:1). The lower phosphorus content of European shrublands is reflected in comparisons of vegetative stocks of chaparral and garrigue (Mooney and Rundel 1979).

A number of ecologists have noted the shorter life span and generally higher nutrient concentrations of mesophytic, drought-deciduous leaves relative to sclerophyllous, evergreen ones (e.g. Monk 1966; Janzen 1974; Schlesinger and Chabot 1977; Westman 1975a, 1978a; Mooney and Rundel 1979). The higher nutrient requirements of the mesophytic, drought-deciduous shrublands imply that these communities would be found on more fertile soil than their evergreen counterparts, other factors being equal. The evidence for this hypothesis at present is limited and equivocal. Westman (1981b) found surface soils of drought-deciduous coastal sage scrub (mean age since fire = 17 years) to have 3-4 times higher levels of total nitrogen, phosphorus and magnesium than evergreen chaparral of ages 9-37 years (Specht 1969b; DeBano and Conrad 1978; Mooney and Rundel 1979). On the other hand, levels of available nitrogen as ammonium and nitrate were 10% less than that for 26-year-old chaparral (Mooney and Rundel 1979), and 60% less in the case of calcium. Similarly, the surface soils of coastal matorral in the Papudo-Zapallar region of Chile have equivalent levels of nitrate-nitrogen to the evergreen matorral of Fundo Santa Laura, and higher levels of phosphorus and sodium, but levels of calcium, magnesium and available phosphorus are 20-60% lower at the coast (data of D E Bradbury in Miller et al. 1977).

The absence of a drought-deciduous counterpart to evergreen shrublands in Australia is noteworthy. It is of interest to ask whether this is attributable to the extreme infertility of the soils in the mediterranean-climate portions of that continent, and the consequent selective disadvantage of deciduousness. The Atriplex-Maireana saltbush-bluebush association in xerothermomediterranean regions of Australia is facultatively drought-deciduous, but is of dubious analogy, since it occurs in desert-margin habitats substantially drier than those of drought-deciduous shrublands in mediterranean-climate portions of other continents.

Fertilizer experiments provide some basis for improving understanding of nutrient relations. In Australia, Specht (1963) and Heddle and Specht (1975) have documented the growth stimulation to heath provided by phosphorus, with or without nitrogen addition. Specht et al. (1977) review experimental studies indicating that phosphorus fertilizer applications to Australian wet- and dry-heaths can inhibit reproduction of heath species and favour invasion of heath communities by mesophytic annual and perennial grasses and shrubs. Chamise chaparral showed growth response to nitrogen (with or without phosphorus) in field trials (Hellmers et al. 1955a). Christensen (1973) postulated that herb growth following fire in chaparral awaits activation of nitrifying bacteria by moisture, implying a nitrate limitation to herb growth in this community. Christensen and Muller (1975) found selective stimulation of herb biomass by applied ash or fertilizer (N-P-K) in chamise chaparral, once seeds had been heat-treated to break dormancy. In coastal sage scrub, Salvia mellifera showed shoot-growth responses to nitrogen (± phosphorus) which increased with increasing stand age up to about 25 years, then declined somewhat (Davis 1980). This behaviour is consistent with the hypothesis that nutrients become limiting to growth as they are stored increasingly in standing vegetation with stand age (cf. Hanes (1971) for a similar hypothesis with respect to chaparral).

In South African Cape dune scrub, the legume Psoralea fruticans is the only species which can withstand severe competition from the Australian legume, Acacia cyclops (Taylor 1972), suggesting that nitrogen stresses limit growth in the other scrub species. Lewis and Stock (1978) report that, under laboratory conditions, selected species of South African Proteaceae assimilate and utilize nitrogen only slowly. Groves and Keraitis (1976) found similar results for the protead Banksia serrata from Australian heath.

In summary, one would generally expect microsite differences in nitrogen levels to be particularly significant in North American, Chilean and South African shrublands, those for phosphorus to be particularly significant in Australian and French shrublands. The effects would be more significant in the heathlands and evergreen shrublands than in the broad-leaved, facultatively deciduous or seasonally dimorphic shrublands on more fertile soils.

Effects would be further buffered to the extent that mechanisms in the plants had evolved to enhance assimilation, retention and internal recirculation of nutrients. Thus to some extent root crowns and lignotubers may serve for storage of nutrients as well as starch storage (Specht 1980). Mullette and Bamber (1978) indicate significant nutrient contents in lignotubers of Eucalyptus gummifera, and Westman and Rogers (1977) document the role of root crowns in nutrient storage at the community level in the heath understorey of a eucalypt forest. Proteoid roots (Lamont 1980), mycorrhizae (Malajczuk and Lamont 1980) and symbiotic N_2-fixing micro-organisms additionally serve to increase assimilation and retention of nitrogen, phosphorus or both. Mooney and Rundel (1979) note the adaptive feature of evergreen leaves in providing storage for nutrients during periods of no growth. Storage of phosphorus as polyphosphates in roots during the rainy season in heath species (Jeffrey 1968) is another mechanism which enhances retention. Preferential uptake of nitrogen in the ammonium form has been shown for the fynbos species, Leucadendron xanthoconus (Lewis and Stock 1978) and indicated for Salvia mellifera in coastal sage scrub (Z W Davis and W E Westman, unpublished). This feature favours retention of nitrogen by the vegetation after fire, by enabling assimilation of nitrogen to occur before the rains stimulate microbial conversion of nitrogen to the more soluble nitrate form.

To the extent that selected species act to store and retain nutrients they provide concentrated sources of these elements upon release by leaching, litter fall, insect herbivory and fire. With respect to leaching, Specht (1957) observed differences in soil moisture over short distances in South Australian heath, due to patterns of interception and stemflow. Westman (1978a) showed as much as threefold differences in nutrient quantities arriving at the base of different eucalypt species in a eucalypt forest-heath assemblage in Queensland. Del Moral et al. (1978) found higher nutrient levels in soils under Eucalyptus baxteri canopies than in adjacent dry-heath in Victoria. Halligan (1973) found half as much throughfall under the canopy of Artemisia californica shrubs in areas with sparse than those with dense herb growth. In the latter two cases, the overstorey species also released allelopathic substances. Hamilton and Rowe (1949) observed differences in patterns of rainfall interception on three chaparral sites.

As to insect herbivory and fire, Springett (1978) has speculated that the role of insect frass in nutrient enrichment in Australian eucalypt forests can be significant. Release of nutrients by fire differentially concentrates phosphorus and micronutrients in the residue relative to nitrogen and potassium (Dunn and DeBano 1977; DeBano et al. 1977, 1979), but does provide a stimulatory ash bed. To the extent that the ash accumulates in patches by burned stumps and in site gullies, it provides a heterogeneity to which post-fire herbs and shrub seedlings might respond in spatial pattern of growth. Encelia californica seedlings have been observed to cluster just downslope of the resprouted parent in the second year after fire in sage scrub (Westman, personal observation). It is not clear whether this phenomenon is purely a result of seed dispersal from fruits of sprouts from the previous season, or whether soil surface nutrient heterogeneities also play a role. On steep slopes, repeated fires could act to redistribute substantial quantities of available nutrients downslope, inducing a gradient in biomass and composition between more and less nutrient-demanding species. At least some species appear able to respond to such fire-induced trends. Delvosalle (1954) reports that repeated burning of maquis in Corsica results in increased abundance of the herb Asphodelus microcarpus due to enrichment of the soil, especially with nitrate-nitrogen. More careful autecological studies of post-fire herbs and shrub seedlings to distinguish nutrient from light and other post-fire habitat requirements would help to test this hypothesis.

Whether any of these postulated effects register in the field will depend largely upon the extent to which more severely growth-limiting factors act in opposition to the trends induced by nutrient availability. In the few studies which consider a number of growth-limiting factors jointly, nutrient factors do not appear predominant. Herb growth under chaparral was found to be influenced more by heat-treatment of seeds, animal grazing, light levels and allelopathy than by nutrients (Christensen and Muller 1975). Allelopathy was found to be more significant than light levels, soil moisture or nutrients in suppression of heath under Eucalyptus baxteri (del Moral et al. 1978). Levels of evapotranspirative stress, and in some cases, oxidants (Westman 1979b), were found to be better predictors of species distributions in coastal sage scrub than nutrient levels (Westman 1981b). Poissonet et al. (1978) found a medium level of N-P-K fertilizer application to encourage herb richness in a Quercus coccifera garrigue after fire more than levels of application double this amount.

Air pollution. Most work on air pollution effects in mediterranean climates has concerned effects on forests rather than on shrublands (see, for example, Miller 1980). Nevertheless, there is incipient evidence that a variety of air pollutants are affecting the structure of mediterranean-climate shrublands. Winner and Mooney (1980a,b) found a drought-deciduous shrub, Diplacus aurantiacus, to be more sensitive to sulphur dioxide damage than a chaparral species (Heteromeles arbutifolia). Winner (1981) reports even greater sensitivity in the sage scrub species Salvia mellifera. Preston (1980) studied the structure of a sage scrub stand downwind of an oil refinery emitting primarily sulphur dioxide. He found injury to Salvia mellifera including significant reductions in photosynthetically active tissue and in leaf size. Stomatal resistance of leaves was decreased, with a concomitant 35% increase in transpiration rates. Flowering capacity, as measured by the number of flower whorls per flower spike, was significantly reduced. Significant decreases in the height:width ratio of Salvia shrubs was postulated to be due to inhibition of apical dominance, an SO_2-induced effect observed on other species. The thinning of the Salvia population on the site resulted in an increase in the abundance of annuals. Of the invading annual cover, 59% was comprised of introduced species (K Preston, personal communication 1980) compared to the control site. As a result of the influx of annuals, there was a net increase in species richness. Westman (1979b) found a somewhat similar effect on coastal sage scrub, apparently induced by oxidants. He noted a reduction in foliar cover of native species on 11 sites of highest oxidant concentration among 67 examined, and a trend for partial replacement of native shrubs by introduced herbs or grasses.

It would appear, in sum, that moisture availability, light and native and domesticated grazing are most frequently the factors controlling microsite herb patterns within the shrublands, with air pollution, allelopathy and substrate effects playing controlling roles in occasional instances. The true extent of such instances needs further documentation. Substrate plays a greater role in mesoscale patterning, where whole suites of dominant species may be included or excluded due to ecotope differences.

22.2.2 MESOSCALE PATTERNING

Species colonies. A mesoscale phenomenon observable in both chaparral and coastal sage scrub is the occasional occurrence of entire stands (of the order of 0.1-5 ha) in which only one or two shrub species dominate (*Ceanothus* spp., *Adenostoma fasciculatum* in the case of chaparral; *Salvia mellifera*, *S. leucophylla*, *Encelia californica* in the case of coastal sage scrub). Kirkpatrick and Hutchinson (1980) and Westman (1981b) have commented on this phenomenon in coastal sage scrub and Mooney et al. (1977b) have done so for chaparral. The preclusion of other shrub species which are sympatric in range is initially puzzling. Possible explanations include allelopathic suppression, selective effects of fire and/or grazing history, habitat changes due to substrate or topography and unknown mechanisms of competitive exclusion following chance establishment.

Allelopathic suppression remains an undocumented hypothesis since no tests to date have been conducted on potential toxic interactions between shrubs. All tests have looked at effects of shrub releases on herbs only.

The fire history hypothesis suggests that because of a particular fire intensity or periodicity or both, only a small number of species are favoured for rapid post-fire regrowth, and these then successfully preclude the establishment of additional shrub species. A combination of fire with selective grazing could act similarly. While a number of authors have shown that different fire intensities and periodicities differentially affect post-fire reproduction in chaparral and sage scrub species (Keeley 1977; Westman 1981a; Westman et al. 1981), the only documentation of a fire-related effect on mesoscale shrub dominance has been that of Bradbury (1978), who showed the persistence of a sharp boundary between a chaparral stand dominated by a single shrub species (*Adenostoma fasciculatum*, 79% cover) and a coastal sage stand dominated by two shrub species (*Salvia apiana*, 32%; *Eriogonum fasciculatum*, 27%) for at least 100 years following fire. No edaphic differences were evident across the ecotone. Bradbury (1978) found much evidence for browsing by packrat (*Neotoma* spp.) and other rodents in both the chaparral and sage stands, and suggested that these animals were playing a role in perpetuating the boundary by destroying shrub seedlings and feeding on seeds (see also Westman 1982).

A shift in species composition with change in aspect or other topographic feature is common in southern Californian shrublands. These

Fig. 1. View of associations of coastal sage scrub dominated by Salvia mellifera (right and left) and S. leucophylla (centre) meeting sharply at boundaries of uncertain origins, Leo Carillo State Park, Los Angeles County, California. Large dark spots in both associations are clumps of Rhus laurina.

shifts tend to be visually dramatic because of the high concentration of dominance in a small number of shrub species (e.g. Fig. 1).

Cole (1980) has described the occurrence of sharp boundaries between coastal sage scrub and chaparral on the south-central coast of California that appear ascribable to substrate differences rather than fire boundaries. Westman (1981b) has suggested that in the case of Salvia mellifera and S. leucophylla, sharp changes in abundance of the two species with aspect are due to their distinctive moisture preferences, S. leucophylla being the more xeric species. He noted that potentially hybridizable congeners of Salvia are effectively isolated by the mesoscale patterning phenomenon. Thus whatever the mechanism of segregation, one selective advantage is the preservation of genetic identities among closely related species.

In Israel, there is also some tendency for mesoscale (0.1 ha) shrub richness in batha to be relatively low, the concentration of dominance (Simpson's (1949) index) for batha being 0.46 compared to 0.34 for Israeli maquis, 0.15 for open shrubland, and 0.07 for grazed woodlands (Naveh and Whittaker 1980). Californian and Israeli samples have high concentrations of dominance relative to eastern Australian and South African samples (Table 1). South-west Australian heath also has low values (0.21, Naveh and Whittaker (1980); 0.03, Lamont et al. 1977). Californian coastal sage scrub values are somewhat lower than chaparral (range 0.09-0.85; a mean for 67 sites of coastal sage scrub was 0.30; Westman 1981a), while Chilean interior and coastal matorral (0.15, 0.11; Naveh and Whittaker 1980) resemble vegetation of other Southern Hemisphere regions. Presumably the higher equitability of the Southern Hemisphere shrublands is a reflection of the longer evolutionary time for speciation of shrubs in these regions, the enhanced shrub richness leading to niche specialization and reduced dominance.

Microcommunities and associations. We have noted above the mesoscale aggregation of one or two shrub dominants. A second micro- or mesoscale pattern of interest is that created by associated suites of species, or "microcommunities." The term "microcommunity" is used here to refer to aggregations of co-occurring subsets of species within a sample stand. The term differs from the "association" in the European sense, in that while the subsets are detectable statistically within a stand, the composition of subsets may differ between stands. Studies of microcommunity patterning in Australia and California (Table 1) indicate two common scales of microcommunity pattern, at 5-8 m and 16-20 m respectively. The larger scale appeared only in communities with a low, open overstorey of *Eucalyptus*. The limited sample of shrublands indicates a common scale of aggregation equivalent to 1-8 shrub diameters, in the absence of an overstorey pattern of shade and gaps. This generalization needs further testing in a range of mediterranean shrubland types.

Some European ecologists regard substrate as the most important environmental factor affecting the distribution of entire shrubland plant associations or community types in the mediterranean-climate region of France. Thus Duvigneaud (1954) notes that garrigue occurs on calcareous soils, maquis and lande on siliceous ones. He further distinguishes between garrigue types on compacted soils and those of friable calcareous soils, and considers grazing intensity a secondary factor to distinguish garrigue or maquis types in various stages of degradation on each substrate. Among submontane garrigues, he distinguishes between types occurring on marl and those on limestone. Duvigneaud notes an example in which the vegetation dominants, by acidifying the soil through their litter and leachate, can shift associated species composition. He cites the case of garrigue in eastern Provence, in which *Cistus albidus*, by acidifying the soil, will encourage the growth of *Pinus pinaster* over the calcicole *P. halepensis*, and may even encourage some plants that favour siliceous substrates (*Cistus salviifolius*, *C. monspeliensis* and *Lavandula stoechas*) to invade.

In synecological studies of Californian coastal sage scrub in relation to environmental factors, Axelrod (1978), Kirkpatrick and Hutchinson (1980) and Westman (1981b) all concluded that the major axis of floristic variation reflected temperature effects, expressed either as annual temperature range or maximum summer temperature. Kirkpatrick and Hutchinson (1980) and Westman (1981b) both concluded, however, that there was a high degree of substrate specificity among individual sage scrub species. Westman (1981b) found that 18 of the 21 dominant species in the flora showed significant preference for one among seven substrate classes. Kirkpatrick and Hutchinson (1980) found significant positive or negative correlations of sage floristic associations with one of three substrate classes for four of the 11 associations tested. The occurrence of highly patchy substrates in the Coastal Ranges of California may contribute significantly to the marked changes in floristic composition that can be observed at an intermediate scale in this region (e.g. Cole 1980). The phenomenon of microcommunity patterning is intimately related to issues of scale of patterning of component species.

22.3 Spatial patterning at the organismal level

As a generalization, the South African and Australian heathlands exhibit a greater degree of species intermingling than do the mediterranean-climate shrublands on more fertile soils. The scale of patterning in the heathlands seems more often to correspond to the relatively small size of the individuals. In the shrublands, aggregation within species is more common, and this is compounded by the fact that a single shrub clump, including resprouts, can become quite large with age.

Two kinds of measurements are available to document these

generalizations. The first are measures of pattern diversity; the second, studies of scales of patterning of species at a site.

22.3.1 PATTERN DIVERSITY

Using the pattern diversity index of Pielou (1966), Westman (1975b) found values in a closed wet-heath near Brisbane, Queensland, to depart only slightly towards aggregation from the value (1.0) for random intermingling (Table 1). Another measure of pattern diversity is the rate of turnover in species composition along a gradient - either a complex gradient measured directly in the field, or a gradient constructed by rearrangement of field samples through ordination. Whittaker (1960) first introduced this measurement as beta diversity along a coenocline, and later applied it to changes between quadrats along a belt transect ("internal beta diversity"). Results shown in Table 1 initially suggest that the heathland samples are less diverse in species and patterning. However, as Shmida and Whittaker (1981) point out, the calculation of beta diversity involves describing broad changes in slope, determined by changes in abundance of dominant species, and along which occurrences of rare species are considered "noise." Thus, in rich communities with low concentration of dominance, such as the heathlands, this approach to pattern diversity measurement fails to reflect the true richness, hence pattern diversity of the communities.

The mean number of species per small quadrat (1 m^2) is a more revealing measure, in such instances, of the degree of species intermingling. Thus in Adenostoma chaparral, the mean number of species m^{-2} was eight (Shmida and Whittaker 1981); in mallee, eight (Whittaker et al. 1979b); in Sympezia fynbos, 24 (Shmida and Whittaker 1981); in Western Australian dry-heath on laterite, 50 km south west of Eneabba, 36 (A J M Hopkins, personal communication 1979). These figures reflect the higher richness at a smaller scale in heathlands.

22.3.2 SCALE OF PATTERN OF SPECIES

A second approach to pattern analysis is to examine the mean scale of aggregation of individual species. Greig-Smith's (1964) block-size analysis technique has been applied to a wide variety of communities but not often in mediterranean-type ecosystems (Table 1). Here, major scales of pattern range from 3 m to 20 m and greater. The large size and range of scales of aggregation suggest that site heterogeneities play at least as important a role in determining pattern as any biological factors, and that further studies in these shrublands will confirm the wide range in scales reported to date.

Aggregation at scales generally larger than the size of individuals further emphasizes the role of environment in pattern determination, although the smaller scales of aggregation (3-4 m) are within the range of size of some individuals, including clump sprouts (e.g. Rhus laurina in Californian chaparral and sage scrub; Banksia collina in Australian wet-heath). Field observations in coastal sage scrub suggest that low intensity fires increase the sizes of individual clumps by favouring crown sprouting, and intense fires have the reverse effect by requiring reproduction from seed (Westman, unpublished). Presumably light grazing would expand the size of clumps of unpalatable species, while heavy grazing could increase pattern diversity by increasing the number of

Table 1. Scales of microcommunity pattern, species diversity and species pattern diversity in some mediterranean-climate or analogous shrublands. See text for values of Simpson's index in other regions.

Vegetation type	Mean scale of micro-community pattern, m	Pattern analysis technique	Species diversity: Concentration of dominance	Species diversity: Richness (no. of spp.)	Species diversity: Area m^2	Scales of patterning of species, m (number of species at this scale in parentheses)	Species pattern diversity	Pattern diversity technique
AUSTRALIA								
S.E. Queensland								
Closed wet-heath	5,8[a]	phase changes in belt transect[a]	0.065-0.085[a]	51[a]	300		0.800 ± 0.037; 0.916 ± 0.032[a]	Pielou's index[j]
Dry-heath under-storey to low, closed eucalypt forest	16[a]	block size analysis[d]	0.211[a]	12[a]	1024	16(9)	0.6[a]	half-changes along elevational transect[k]
Central New South Wales								
Mallee	5-12,20[b]	block size analysis[e,f]	0.30[h]	53[b]	1000	5-12(7);20+(3)	2.0[b]	half-changes along ordination axis(100 m)[k]
CALIFORNIA								
Adenostoma chaparral	5[c]	reciprocal averaging trace[g]	0.58[h]	30[c]	1000	3-4(14); 10-13 (8);18-20(2)	1.8[c]	half-changes along ordination axis(100 m)[k]
SOUTH AFRICA								
Sympieza fynbos			0.22[i]	90[c]	1000		0.37[c]	half-changes along ordination axis(100 m)[k]

[a] Westman 1975b. [b] Whittaker et al. 1979b. [c] Shmida and Whittaker 1981. [d] Kershaw 1957. [e] Greig-Smith 1964. [f] Hill 1973. [g] Whittaker et al. 1979a. [h] Naveh and Whittaker 1980. [i] Mean of nine samples from coastal and mountain fynbos, data of H C Taylor, F J Kruger and R H Whittaker in Naveh and Whittaker 1980. [j] Pielou 1966. [k] Whittaker 1960.

gaps in the canopy available for seedling colonization. However, no experimental studies of the effects of fire and grazing on scales of pattern in mediterranean-type vegetation have yet been reported.

Finally, it should be noted that the smaller scale of pattern in heathlands is a phenomenon due to the vegetation type, and not to the geographic region in which it occurs. Thus in Australia, the mallee woodland shows a much larger scale of pattern than the wet or dry heathlands. In South Africa Taylor (1972) reports pure stands of Rhus mucronata 20-84 m^2 in extent, 0.6-1.0 m high, possibly derived from a single coppicing individual, in inland dune fynbos. This pattern is in marked contrast to the highly intermingled species pattern reported from other types of fynbos.

Explanations for the frequently observed reduction in the scale of pattern in heathlands, as opposed to shrublands, may be sought among several somewhat interrelated factors. (a) The lengthier evolutionary history of the heathlands has led to a flora of greater species richness, composed of more highly specialized, K-adapted species of keen competitive abilities; these in turn have decreased the tendency toward dominance on an area by one or a small number of species. (b) The evergreen, sclerophyllous, perennial habit of the heathland species has resulted in slower growth rates, and decreased ability of a single population to overtake a bare area after disturbance. (c) The low fertility of heathland soils further discourages rapid growth, again inhibiting resource pre-emption by a single species over a continuous area following disturbance. (d) The smaller size of mature individuals in heathlands leads to a smaller scale of potential intermingling, and is partially a result of their slow growth rates. These factors have been discussed to some extent by Lamont et al. (1977) and Westman (1975b). Consequences of this pattern include a lower concentration of dominance and less canopy closure, leading to a higher occurrence of herbaceous species and higher alpha diversity. The high degree of intermingling of species within heath or fynbos may also result in lower levels of herbivory by host-specific animals, reducing the selective pressure for developing other defences against herbivores (spines, defensive chemicals). While the selective advantage of pattern diversity over other means of defence would thus lie in the savings in energy otherwise expended on specialized defences against herbivores, it is clear that selection for such a pattern must act at the level of the individual. Thus plants with mechanisms of seed dispersal or vegetative reproduction leading to effective dispersal, discouraging clumping, would be selected for.

To the extent that these factors operate, it is possible to derive corollary predictions. Thus the following factors should lead to larger scales of pattern diversity: (a) greater representation of fast-growing, mesophytic, drought-deciduous species in the shrub stratum; (b) a herb stratum with lower representation of annual species; (c) the application of fertilizer; (d) a decrease in the volatile oil content of plant parts, leading to a decrease in fire intensity and increased frequency of coppicing following fire; (e) grazing and predation by small rather than large herbivores, leading to increased herbivory on seedlings and seeds rather than on shrub sprouts; (f) increased effectiveness of allelopathy. While differences in the levels of these factors in mediterranean-climate shrublands and heathlands have been documented, studies of differences in scales of pattern diversity are as yet too few to test to these hypotheses statistically.

22.4 Spatial partitioning of resources at the suborganismal level

22.4.1 ABOVE-GROUND

The above-ground physiognomy of mediterranean-climate vegetation has been relatively well described qualitatively. Tables presenting the proportions of species in various growth- or life-form categories have been compiled for examples from all regions. For Australia, Western Australian heathlands are described by George et al. (1979) and Lamont et al. (1977); eastern and south-central heathlands by Specht (1979b); mallee by Whittaker et al. (1979b). For North America, chamise chaparral is characterized by Whittaker et al. (1979a) and Mooney et al. (1977b), coastal sage scrub by Westman (1981b). For South Africa, Kruger (1979) presents data for various fynbos types. For Chile, data are presented in Parsons and Moldenke (1975), Mooney et al. (1977b) and Rundel (1980). For Europe, Naveh and Whittaker (1980) present data from Israel. Growth-form representations are compared between North America, Europe and Australia by Specht (1969a) and Naveh and Whittaker (1980).

Beyond this, several authors have attempted to indicate, qualitatively or quantitatively, the environmental factors influencing growth-form representations in mediterranean-climate shrublands and heathlands. Thus Kruger (1979) provides a qualitative analysis for fynbos; Naveh and Whittaker (1980) provide some information in relation to grazing stress for Israeli samples; Westman (1981a) provides regressions of growth-form and life-cycle variables on environmental factors for coastal sage scrub; Miller (1979) models shifts in growth form with elevation and evapotranspirative stress for chaparral; and Specht (1969a) compares factors responsible for structure in France, southern California and Australia. In addition, South African ecologists have commented on the shifts in representation by major plant families with environmental factors, since these correspond to major physiognomic shifts in the Cape region: Milewski (1978a, b) and Milewski and Esterhuysen (1977) for Cape coastal flats; Werger et al. (1972) for fynbos at Jonkershoek; Taylor (1979) for mountain fynbos at Rooiberg; Rutherford (1978) for western Cape mountain fynbos; Glyphis et al. (1978) and Laidler et al. (1978) for fynbos on Table Mountain. Bridgewater (1978) and Bridgewater and Zammit (1979) have analysed southern and southwestern Australian heaths using a similar approach. One of the more significant generalizations to emerge from these studies is that the Australian and South African heathlands differ from other mediterranean-climate shrublands in having a persistent component of hemicryptophytes many decades after fire, this trait being best developed in fynbos communities (Kruger 1977).

The factors identified as influencing shifts in physiognomy depend on the scale of analysis. At a regional scale, evapotranspirative stress is generally identified as a crucial determinant of the foliar cover and proportions of evergreen to deciduous species (Flahault 1937; Specht 1969a; Kruger 1979; Taylor 1979; Miller 1979; Westman 1981b). Some generalized trends along a gradient of decreasing available moisture and soil macronutrients are summarized in Fig. 2.

Fig. 2 is derived from data from various studies. Parameters of leaf and stem morphology in chaparral and matorral observed by Mooney et al. (1977a) were correlated with leaf macronutrient data (nitrogen and phosphorus) obtained by Fishbeck and Kummerow (1977). Resulting correlations are reported in Table 2. The trend for increased branching with decreasing moisture availability was documented for *Eucalyptus* woodlands by Mooney et al. (1978).

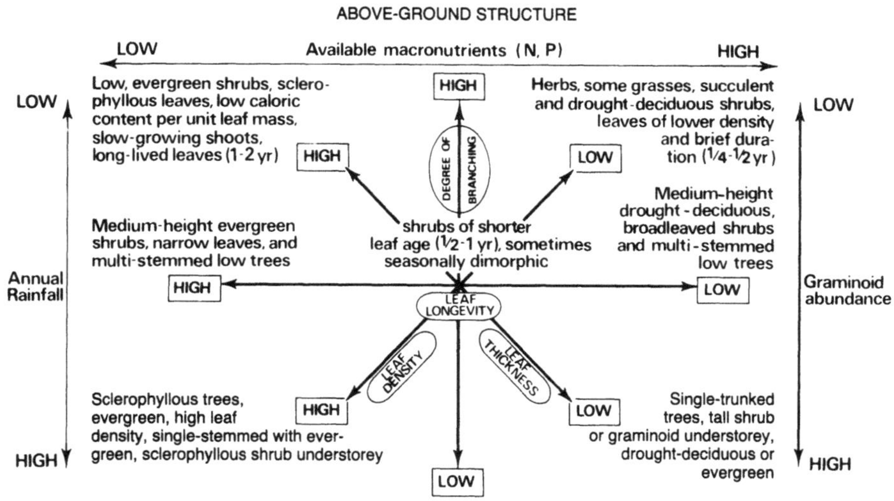

Fig. 2. Tentative, generalized trends of change in above-ground plant community structure in mediterranean-climate shrublands along gradients of moisture and macronutrient availability. Studies upon which figure is based are cited in text.

Mooney et al. (1978) found leaf specific mass in eucalypts to increase with decreasing moisture availability; at the same time, leaf nitrogen contents increased. In the chaparral and matorral, by contrast, there is a slight tendency for leaf specific mass to decrease with increasing leaf nitrogen content. In neither case is there evidence that the relationship between leaf nitrogen content and leaf specific mass is causal.

In general, it appears that available moisture is the primary determinant of physiognomy, with nutrients playing a secondary but interrelated role. Thus the effect of both moisture and nutrient deficits in encouraging sclerophylly has been noted by numerous authors (e.g. Beadle 1966; Seddon 1974; Westman 1978a,b; Specht 1979b). The role of soil depth, drainage, texture and nutrient content in determining physiognomy in Cape vegetation has been explored by Werger et al. (1972), Glyphis et al. (1978), Laidler et al. (1978), Milewski (1978a,b) and Kruger (1979).

Since the nutrient and moisture factors are both important and interactive in the field, it has been difficult to separate their influences by synecological studies. Studies such as those of Westman (1981a,b) on coastal sage scrub, which used multiple regression and multivariate approaches to test the relative influence of moisture, temperature, nutrients and disturbance factors in influencing growth form composition, confirm the conclusion that evapotranspirative stresses are more important than soil nutrient differences in affecting growth-form spectra.

Fig. 2 contains oversimplifications. Moisture and nutrient axes are shown without scales, trends are pictured as linear rather than curvilinear, shrublands and heathlands are not differentiated, and disturbance factors are not included. The greatest value of such diagrams may be in provoking researchers to collect new data to challenge its assertions. In the proceedings of the workshop following the present symposium, these relationships have been refined (Westman and Cowling 1981).

Table 2. Correlations between leaf energy and leaf nutrient contents, leaf structure and shoot growth in seven Chilean and eight Californian evergreen shrub species of the I.B.P. primary matorral and chaparral sites (Fundo Santa Laura, Chile; Echo Valley, California). Raw data from Fishbeck and Kummerow (1977) and Mooney et al. (1977a). Data for leaf nutrient contents, size and density were log-transformed to normalize the data. *Indicates $P<5\%$, **indicates $P<1\%$.

	Log % N in leaf	Log % P in leaf	Leaf energy cal g^{-1}	Shrub height	Shrub diameter	Leaf cuticle thick-ness	Leaf area index per shrub	Log leaf size	Annual terminal shoot	Log leaf density
Log % N in leaf	1.000									
Log % P in leaf	0.306	1.000								
Leaf energy	-0.147	-0.642**	1.000							
Shrub height	0.008	-0.519*	0.271	1.000						
Shrub diameter	-0.143	-0.102	-0.367	-0.100	1.000					
Leaf cuticle thickness	-0.431	-0.146	0.262	-0.292	0.134	1.000				
L.A.I. per shrub	-0.030	-0.372	0.575*	0.539*	-0.250	-0.120	1.000			
Log leaf size	-0.558*	-0.146	-0.031	0.314	0.560*	0.225	0.131	1.000		
Annual terminal shoot growth	-0.452	-0.552*	0.367	0.254	0.025	0.471	0.320	0.353	1.000	
Log leaf density	-0.182	-0.306	0.449	-0.385	0.192	0.725**	-0.065	-0.063	0.236	1.000

22.4.2 BELOW-GROUND

Factors affecting the partitioning of biomass below ground and the allocation of resources between shoots and roots, are broadly generalized in Fig. 3. Most noteworthy is that while root masses of heathlands (in Australia, at least) increase with decreasing soil fertility, the reverse is the case for North American and Chilean shrublands. Presumably the long periods of evolution on infertile soils in Australia selected for heath species capable of storing nutrients underground in order to conserve nutrients leached during rains and to recover rapidly following fire or other defoliation. On the younger, more fertile soils of North America and Chile, nutrient storage below ground has been less essential, and relatively more photosynthate is allocated to above-ground parts to increase competitive domination of light and space to enhance reproductive output. Thus, an increase in root:shoot ratio with decreasing nutrient availability has been documented in the laboratory for certain Australian heath species (Banksia ornata, Casuarina pusilla) by Specht and Groves (1966). In field studies, Westman and Rogers (1977) noted the very high below-ground mass of root crowns in eucalypt forest and heath understorey on highly infertile, sandy soils compared to more fertile ones.

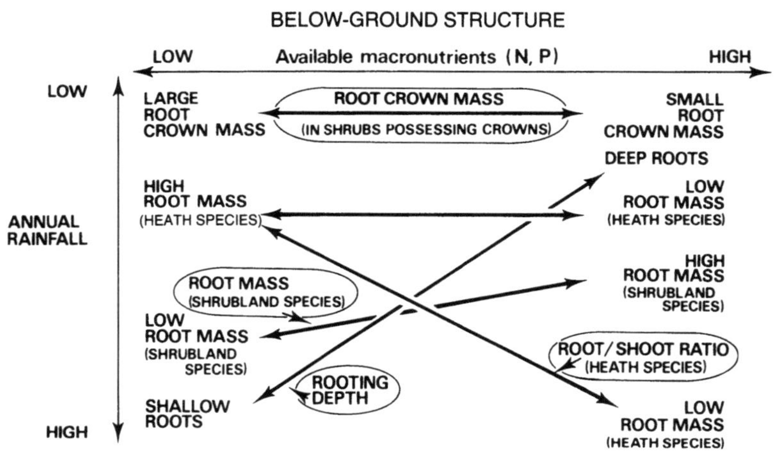

Fig. 3. Tentative, generalized trends of change in below-ground plant community structure in mediterranean-climate shrublands and heathlands along gradients of moisture and macronutrient availability. Studies upon which figure is based are cited in text.

By contrast, root:shoot biomass ratios are higher in matorral on the more fertile soils of Fundo Santa Laura, Chile, than in chaparral on a matched site in Echo Valley, southern California (Hoffmann and Kummerow 1978).

There is a good deal of variability in root:shoot ratios between species and among sites. Thus, on a smaller subset of shrubs from the same sites as the previous study, Miller and Ng (1977) did not find significant differences in root:shoot ratios between Californian and Chilean species. It is perhaps most meaningful, therefore, to examine root:shoot ratios on a community basis, where samples sizes are larger and broad trends can be seen more clearly. Available figures have been

Table 3. Community-level root:shoot ratios and soil nitrogen and phosphorus levels from four mediterranean-climate shrubland and heathland communities.

Location	Vegetation	Root:shoot ratio	Stand age, yr	%N	Total P,%	Available P, μg^{-1}	Extractant used in available P determination
AUSTRALIA							
Dark Island, S. Australia[a]	heath	6.3	25	0.018	--	7.5	butyl alcohol
CALIFORNIA							
Echo Valley	chaparral	0.5[b,c]	25[f]	0.065[d]	0.039[d]	7.1[d]	0.03N NH_4F in 0.025N HCl
CHILE							
Fundo Santa Laura	matorral	1.5[b,c,d]	16[f]	0.233[d]	0.092[d]	32.5[d]	0.03N NH_4F in 0.025N HCl
GREECE							
Mt Hymettus Athens[e]	phrygana	1.5-1.8	53+	0.143	--	--	--
SOUTH AFRICA							
Cape region	mountain fynbos			0.10-0.21	--	1-29	Bray No. 2 (strong acid)
	coastal fynbos			0.26	--	44	

[a] Specht et al. 1958. [b] Mooney et al. 1977b. [c] Miller and Ng 1977. [d] Hoffmann and Kummerow 1978. [e] Margaris 1976. [f] Aschmann and Bahre 1977. [g] Kruger 1979.

compiled or calculated for four of the five mediterranean-climate regions, and are presented in Table 3. It can be seen that while root:shoot ratios on the infertile Australian heathland soils are large, root:shoot ratios decrease with decreasing fertility on the three shrubland types on more fertile soils (California, Chile, Greece). Root:shoot ratios may also increase with increasing stand age and sequestration of nutrients in the phytomass. South African fynbos, being a heathland type on somewhat more fertile soils than the Australian, could be predicted to have community root:shoot ratios of intermediate range (1-4).

The extent of exploration of the soil by roots of mediterranean-climate heathlands and shrublands is high. Specht et al. (1958) found the density of shallow roots (to 0.3 m depth) to be the same in both vegetated and "bare" portions of a South Australian heathland. Kummerow et al. (1977) found the same in chaparral. Miller and Ng (1977) report that roots extended 1.4-7.2 times the width of the canopy of individual shrubs in five of the six chaparral and matorral species examined. Kummerow et al (1977) found root:canopy projection ratios as high as 40 for Haplopappus pinifolius, a chaparral species.

The increasing soil fertility leads to a higher root density among Chilean than Californian evergreen shrublands (Hoffmann and Kummerow 1978). Miller (1979) calculated, from nutrient diffusion rates, that absorbing roots would need to be 4 mm apart to exploit all available soil phosphate, 40 mm for nitrate, and 80 mm for water, and noted that absorbing root densities in chaparral were adequate to exploit fully the available soil water and nitrate, but not phosphate. In the case of phosphorus, higher initial levels of available soil phosphorus in Chilean soil compared to Californian at the I.B.P. sites (4.6:1; Hoffmann and Kummerow 1978), has helped lead to a much higher (16 times) Chilean root biomass which, by stimulating higher root densities, in turn has enabled more complete exploitation of available soil phosphate reserves.

Thus in the case of the shrublands evolving on more fertile soils, the ability to increase root growth with increasing soil fertility is advantageous. In the case of heathlands evolved on nutrient-poor soils, by contrast, there has apparently been a selective advantage for the opposite trend. Heddle and Specht (1975), for example, report phosphorus toxicity in seedlings of heath species on soils fertilized to normal crop levels. Possibly the advantage of this inhibition lies in preventing overstimulation of root and subsequent shoot growth following fire on otherwise infertile soils, thus preventing nutrient deficiencies from developing in the phytomass after the initial flush of post-fire nutrients has been leached or assimilated.

Species-specific differences in root structure have been documented for Australian heath by Specht and Rayson (1957); for chaparral by Hellmers et al. (1955b), Hanes (1965), Kummerow et al. (1977) and Miller and Ng (1977); and for matorral by Miller and Ng (1977) and Hoffmann and Kummerow (1978). The combined effect on below-ground community structure is similar in each case, however: a predominance of root biomass within 0.5 m of the surface, with occasional sinker roots to depths of 3 m and more. Specht and Rayson (1957) found that a quarter of the heath species possessed deep tap roots, one-half shallow tap roots and the remaining quarter, fibrous roots. They found four times more root biomass in the top 0.3 m than in the next 1.4 m of depth. Kummerow et al. (1977) and Hoffmann and Kummerow (1978) excavated chaparral and matorral sites, respectively, to 0.6 m depth. They found the bulk of large roots within this zone, with greater zonation within this depth exhibited by the chaparral, which was growing on a shallower soil.

Specialized root structures of Australian heath species have somewhat different responses to nutrient levels than the remainder of the root system. The lignotubers of Eucalyptus species, which were earlier noted to act as starch- and nutrient-storage organs (Mullette and Bamber 1978), increase in size with increasing levels of applied phosphorus (Kerr 1925; Weir 1969; Mullette and Bamber 1978), although at high levels of phosphorus, growth was sometimes inhibited (Beadle 1968). Proteoid roots in Hakea and Banksia (see Chapter 14, this volume) are stimulated to grow by increasing soil moisture up to twice field capacity (Lamont 1976) and by low levels of phosphorus and nitrogen (Lamont 1972; Siddiqi et al. 1976). Waterlogging will stimulate proteoid root initiation but suppress total proteoid root mass (Lamont 1976). There are species-specific differences in responses of these proteads to moisture and aeration (Lamont 1976; Siddiqi and Carolin 1976).

Shrublands in all five regions contain species that sprout after fire. The proportions of species that sprout to those that do not will differ from site to site, due in part to the selective influence of fire frequency and intensity (Keeley and Zedler 1978; Westman 1981a). Specht and Rayson (1957) report that 70% of the species in a South Australian heath sprout. Hanes (1971) noted that of 59 dominants characteristic of coastal chaparral, about half are sprouters. The proportion of sprouting dominants in coastal sage scrub is somewhat higher. Of 43 dominant species for which data are available (Table 3 in Westman 1981b), 25 species of shrub and low tree can sprout from root crowns or stem bases following fire. Of the 18 herb species (including eight grasses), five have been observed to sprout from underground organs (Westman, unpublished). Of this sample, then, 70% of the species are capable of sprouting, of which 83% are shrubs. In mountain fynbos, van der Merwe (1956) and Kruger (1979) report that 60-90% of the plants in different communities sprout. Of this number, however, a greater proportion appear to be geophytes, hemicryptophytes and chamaephytes than in California or Australia, while most species of dominant fynbos shrubs do not sprout (Kruger 1979). In Chile, Mooney et al. (1970) estimate that 50% of matorral shrub species are capable of sprouting. Thus at least half of the species possess sprouting ability in the four regions cited. In regions where mesophyllous and sclerophyllous shrublands co-occur, the former appear to exhibit a slightly higher sprouting ability, based on the limited data so far available, but more data are needed for confident comparisons.

While root/micro-organism relationships (mycorrhizae, proteoid roots, symbiotic and asymbiotic N_2-fixing organisms) have been studied in some mediterranean-climate shrublands and heathlands, much remains to be done to measure their role in nutrient budgets in each community type. There is reason to believe that important differences exist. For example, in California, mycorrhizae are reported from all chaparral shrubs (Kummerow et al. 1978), symbiotic N_2-fixation is important in communities of 10 to 40+ years (Hanes 1971) and asymbiotic N_2-fixation is insignificant (Dunn and Poth 1979). By contrast, in the coastal sage scrub, symbiotic N-fixation is known to become negligible 20 years after fire (Westman 1981a), and mycorrhizae and asymbiotic N_2-fixing organisms have not been studied. No studies of proteoid roots in South African fynbos have yet been published, and the net contribution of these root masses to Australian heath phosphorus budgets has not been quantified.

22.5 Concluding remarks

A review of the patterns of spatial partitioning in mediterranean-climate regions reinforces the value of distinguishing between the ancient South African and Australian heathlands on the one hand, and the more recent shrublands evolved on more fertile soils in the Mediterranean and New World on the other. At the supra-organismal level, the more fertile shrublands show distinct patterning of the shrub and herb strata in relation to each other, induced variously by moisture stress, light availability, rodent activity, allelopathy, grazing and air pollution. Substrate and nutrient effects are more evident in selecting for or against whole suites of species at a mesoscale of pattern, than in influencing herb patterning at a microscale. Mesoscale patterning by colonies of one or two shrub species is repeatedly observed in Californian shrublands. Openings between shrubs are typically colonized by annuals, especially from the Mediterranean. The canopies of heathlands by comparison are generally lower, shrub clumps are smaller and stratal differentiation in patterning is less marked. In South Africa however, the herb layer is much richer in hemicryptophytes than in Australia, especially in the early post-fire years.

At the organismal level, the greater intermingling of species in the heathlands results in higher α diversity and equitability values, and can be attributed to the high levels of competition and niche specialization existing among these slow-growing species. The shrublands, largely composed of generalists recently derived from subtropical elements on fertile soils, are more subject to pre-emption of resources by a few dominants, resulting in an enlargement of the scale of patterning and species intermingling.

At the suborganismal level the heathlands have evolved mechanisms of below-ground nutrient storage, and biomass appears to be partitioned increasingly to below-ground parts on soils of decreasing fertility. In the shrublands, by contrast, increasing soil fertility levels stimulate root growth, and root:shoot biomass ratios increase somewhat, while remaining substantially lower than those of heathlands.

Distribution and longevity of above-ground parts respond primarily to evapotranspirative stress, and secondarily to nutrient levels. Net primary production decreases and canopy branching increases with aridity, resulting in canopy gaps which promote understorey growth, hence stratal differentiation. Selective disturbance factors (e.g. grazing) will produce a similar effect. On more fertile soils, shorter-lived species (annuals, perennial herbs) are favoured, as opposed to longer-lived shrubs. Similarly, leaves are more likely to be mesophytic and short-lived (malacophyllous, drought-deciduous or seasonally dimorphic) as opposed to evergreen and sclerophyllous. The higher fertility is accompanied by a shift in emphasis from mineral nutrient retention to carbon assimilation, with an accompanying increase in growth and biomass turnover rates. The effects of moisture and nutrients on structure are only partially covariant, sometimes producing similar plant adaptations, at other times acting distinctively as selective forces. Attempts to summarize the influence of the two factors (Figs 2 and 3) inevitably lead to oversimplification.

Gaps in research in each of these areas can be identified. Studies of spatial patterning at the supra- and organismal levels have been performed primarily in California and Australia. Studies of above-ground partitioning are well advanced at a qualitative level (growth forms, life forms), but not well advanced quantitatively, especially in the

Mediterranean and South Africa. Studies of root structure are needed especially for South African shrublands and heathlands. The Fynbos Biome Project currently under way (Day et al. 1979) should provide much of this information. There has been a tendency to study evergreen more than drought-deciduous shrublands in all of the mediterranean-climate regions (except Australia, where the drought-deciduous analogue does not exist). Unfortunately the latter also tend to be smaller in extent and under more severe development pressure. Indeed, disturbance clearly plays an important part in shaping community structure, particularly in the Mediterranean and in Chile. While fire effects have been well studied (Mooney and Conrad 1977), the impact of grazing on community structure has been studied largely in the Mediterranean, and of air pollution largely in California. Much progress has been made in characterizing plant community structure in mediterranean-climate regions in the past decade. There remain ample challenges, however, in fully elucidating the "jocular procreations" of mediterranean-climate vegetation.

Acknowledgements

Previously unpublished data on coastal sage scrub of mine and of my students, Z Davis and K Preston, were obtained with the support of the Division of Environmental Biology, National Science Foundation under Grant DEB 76-81712. I also thank A J M Hopkins for supplying unpublished data. A Troeger performed calculations for Table 2.

References

ASCHMANN H, BAHRE C (1977) Man's impact on the wild landscape. In: MOONEY H A (ed) Convergent evolution in Chile and California mediterranean climate ecosystems. Dowden, Hutchinson and Ross, Stroudsberg, Pennsylvania, pp 73-84

AXELROD D (1978) The origin of coastal sage vegetation, Alta and Baja California. American Journal of Botany 65: 1117-1131

BALLESTER A, VIEITEZ AM, VIEITEZ E (1979) The allelopathic potential of Erica australis L. and E. arborea L. Botanical Gazette 140: 433-436

BARTHOLOMEW B (1970) Bare zone between California shrub and grassland communities: the role of animals. Science 170: 1210-1212

BARTHOLOMEW B (1971) Role of animals in suppression of herbs by shrubs. Science 171: 463

BEADLE NCW (1966) Soil phosphate and its role in molding segments of the Australian flora and vegetation, with special reference to xeromorphy and sclerophylly. Ecology 47: 992-1007

BEADLE NCW (1968) Some aspects of the ecology and physiology of Australian xeromorphic plants. Australian Journal of Science 30: 348-355

BRADBURY DE (1978) The evolution and persistence of a local sage/chamise community pattern in southern California. Yearbook of the Association of Pacific Coast Geographers 40: 39-56

BRADFORD DF (1976) Space utilization by rodents in Adenostoma chaparral. Journal of Mammalogy 57: 576-579

BRIDGEWATER PB (1978) Coastal vegetation of East Gippsland, Victoria: a comparison of physiognomic and floristic classifications. Phytocoenologia 4: 471-490

BRIDGEWATER PB, ZAMMIT CA (1979) Phytosociology of S. W. Australian limestone heaths. Phytocoenologia 6: 327-343

CARTER SI (1973) A comparison of the pattern of herb and shrub growth in comparable sites in Chile and California. MA Thesis. California State University, San Diego

CHOU C-H, MULLER CH (1972) Allelopathic mechanisms of Arctostaphylos glandulosa var. zacaensis. American Midland Naturalist 88: 324-347

CHRISTENSEN NL (1973) Fire and the nitrogen cycle in California chaparral. Science 181: 66-67

CHRISTENSEN NL, MULLER CH (1975) Relative importance of factors controlling germination and seedling survival in Adenostoma chaparral. American Midland Naturalist 93: 71-78

COLE K (1980) Geologic control of vegetation in the Purisima Hills, California. Madrono 27: 79-89

DAVIS ZW (1980) The effects of applied nutrients on Salvia mellifera (Black sage) stands of different ages. MA Thesis. University of California, Los Angeles

DAY J, SIEGFRIED WR, LOUW GN, JARMAN ML (eds) (1979) Fynbos ecology: a preliminary synthesis. South African National Scientific Programmes Report No 40, CSIR, Pretoria, 166 pp

DEBANO LF, CONRAD CE (1978) The effect of fire on nutrients in a chaparral ecosystem. Ecology 59: 489-497

DEBANO LF, DUNN PH, CONRAD CE (1977) Fire's effect on physical and chemical properties of chaparral soils. In: MOONEY HA, CONRAD CE (eds) Proceedings of the symposium on the environmental consequences of fire and fuel management in mediterranean ecosystems. USDA Forest Service General Technical Report WO-3, pp 65-74

DEBANO LF, EBERLEIN GE, DUNN PH (1979) Effects of burning on chaparral soils: I. Soil nitrogen. Soil Science Society of America Journal 43: 504-509

DELEUIL G (1951) Origine des substances toxiques du sol des associations sans therophytes du Rosmarino-Ericion. Compte-Rendu de l'Académie des Sciences, Paris 232: 2038-2039

DEL MORAL R, WILLIS RJ, ASHTON DH (1978) Suppression of coastal heath vegetation by Eucalyptus baxteri. Australian Journal of Botany 26: 203-219

DELVOSALLE L (1954) Aspects végétaux de la Corse. In: DE SMET S, DUVIGNEAUD P, DELVOSALLE L, LAWALREE A, SEMAL-VAN GAUSEN P, VANDEN EECKHOUDT J-P, RASMONT R (eds) Végétation et faune de la région méditerranéenne francaise. Les Naturalistes Belges, Bruxelles, pp 44-58

DINOOR A (1962) The effect of Quercus ithaburensis on its herbaceous associates. MSc Thesis. Hebrew University, Jerusalem

DUNN EL, SHROPSHIRE FM, SONG LC, MOONEY HA (1976) The water factor and convergent evolution in mediterranean-type vegetation. In: LANGE OL, KAPPEN L, SCHULZE E-D (eds) Water and plant life. Ecological Studies 19: 492-505

DUNN PH, DEBANO LF (1977) Fire's effect on biological and chemical properties of chaparral soils. In: MOONEY HA, CONRAD CE (eds) Proceedings of the symposium on the environmental consequences of fire and fuel management in mediterranean ecosystems. USDA Forest Service General Technical Report WO-3, pp 75-84

DUNN PH, POTH M (1979) Nitrogen replacement after fire in chaparral. In: GORDON JC, WHEELER CT, PERRY DA (eds) Symbiotic nitrogen fixation in the management of temperate forests. Forest Research Laboratory, Oregon State University, Corvallis, Oregon, pp 287-293

DUVIGNEAUD P (1954) Les groupements végétaux de la France m diterranéene. In: DE SMET S, DUVIGNEAUD P, DELVOSALLE L, LAWALREE A, SEMAL-VAN GAUSEN P, VANDEN EECKHOUDT J-P, RASMONT R (eds) Végétation et faune de la région méditerranéenne francaise. Les Naturalistes Belges, Bruxelles, pp 15-43

FISHBECK K, KUMMEROW J (1977) Comparative wood and leaf anatomy. In: THROWER NJW, BRADBURY DE (eds) Chile-California mediterranean scrub atlas. A comparative analysis. Dowden, Hutchinson and Ross, Stroudsberg, Pennsylvania, pp 148-161

FLAHAULT C (1937) La distribution géographique des végétaux dans la région méditerranéene francaise. Encyclopédie Biologique 18: 1-178. Lechevalier, Paris

GEORGE AS, HOPKINS AJM, MARCHANT NG (1979) The heathlands of Western Australia. In: SPECHT RL (ed) Ecosystems of the world, vol 9A. Heathlands and related shrublands. Descriptive studies. Elsevier, Amsterdam, pp 211-230

GLYPHIS J, MOLL EJ, CAMPBELL BM (1978) Phytosociological studies on Table Mountain, South Africa: 1. The Back Table. Journal of South African Botany 44: 281-289

GREIG-SMITH P (1964) Quantitative plant ecology, 2nd edn. Butterworths, London, 256 pp

GROVES RH, KERAITIS K (1976) Survival and growth of seedlings of three sclerophyll species at high levels of phosphorus and nitrogen. Australian Journal of Botany 24: 681-690

GULMON SL (1977) A comparative study of the grassland of California and Chile. Flora 166: 261-278

GUYOT L (1951) Sur un aspect du déterminisme biologique de l'évolution floristique de quelques groupements végétaux. Compte-Rendu Sommaire des Séances de la Société de Biogeographie 293: 3-14

HALLIGAN JP (1973) Bare areas associated with shrub stands in grassland: the case of Artemisia californica. BioScience 23: 429-432

HALLIGAN JP (1974) Relationship between animal activity and bare areas associated with California sagebrush in annual grassland. Journal of Range Management 27: 358-362

HALLIGAN JP (1976) Toxicity of Artemisia californica to four associated herb species. American Midland Naturalist 95: 406-421

HAMILTON EL, ROWE PB (1949) Rainfall interception by chaparral in California. California Department of Natural Resources, Division of Forestry

HANES TL (1965) Ecological studies on two closely related chaparral shrubs in southern California. Ecological Monographs 35: 27-52

HANES TL (1971) Succession after fire in the chaparral of southern California. Ecological Monographs 41: 27-52

HEDDLE EM, SPECHT RL (1975) Dark Island heath (Ninety-Mile Plain, South Australia). VIII. The effect of fertilizers on composition and growth, 1950-72. Australian Journal of Botany 23: 151-164

HELLMERS H, BONNER J, KELLEHER JM (1955a) Soil fertility: a watershed management problem in the San Gabriel mountains of southern California. Soil Science 80: 189-197

HELLMERS H, HORTON JS, JUHREN G, O'KEEFE J (1955b) Root systems of some chaparral plants in southern California. Ecology 36: 667-678

HILL MO (1973) The intensity of spatial pattern in plant communities. Journal of Ecology 61: 225-235

HOFFMANN A, KUMMEROW J (1978) Root studies in the Chilean matorral. Oecologia 32: 57-69
JAKSIC FM, MONTENEGRO G (1979) Resource allocation of Chilean herbs in response to climatic and microclimatic factors. Oecologia 40: 81-89
JANZEN DH (1974) Tropical blackwater rivers, animals and mast fruiting by the Dipterocarpaceae. Biotropica 6: 69-103
JEFFREY DW (1968) Phosphate nutrition of Australian heath plants. II. The formation of polyphosphate by five heath species. Australian Journal of Botany 16: 603-613
KEELEY JE, ZEDLER PH (1978) Reproduction of chaparral shrubs after fire: a comparison of sprouting and seeding strategies. American Midland Naturalist 99: 142-161
KEELEY SC (1977) The relationship of precipitation to post-fire succession in southern California chaparral. In: MOONEY HA, CONRAD CE (eds) Proceedings of the symposium on the environmental consequences of fire and fuel management in mediterranean ecosystems. USDA Forest Service General Technical Report WO-3, pp 387-390
KEELEY SC, JOHNSON AW (1977) A comparison of the pattern of herb and shrub growth in comparable sites in Chile and California. American Midland Naturalist 97: 120-132
KERR LR (1925) The lignotubers of eucalypt seedings. Proceedings of the Royal Society of Victoria 37: 79-97
KERSHAW KA (1957) The use of cover and frequency in the detection of pattern in plant communites. Ecology 38: 291-299
KIRKPATRICK JB, HUTCHINSON CF (1977) The community composition of Californian coastal sage scrub. Vegetatio 35: 21-33
KIRKPATRICK JB, HUTCHINSON CF (1980) The environmental relationships of Californian coastal sage scrub and some of its component communities and species. Journal of Biogeography 7: 23-28
KRUGER FJ (1977) A preliminary account of aerial plant biomass in fynbos communities of the mediterranean-type climatic zone of the Cape Province. Bothalia 12: 301-307
KRUGER FJ (1979) South African heathlands. In: SPECHT RL (ed) Ecosystems of the world, vol 9A. Heathlands and related shrublands. Descriptive studies. Elsevier, Amsterdam, pp 19-80
KUMMEROW J, KRAUSE D, JOW W (1977) Root systems of chaparral shrubs. Oecologia 29: 163-177
KUMMEROW J, KRAUSE D, JOW W (1978) Seasonal changes of fine root density in the southern California chaparral. Oecologia 37: 201-212
LAIDLER D, MOLL EJ, CAMPBELL BM, GLYPHIS J (1978) Phytosociological studies in Table Mountain, South Africa. 2. The Front Table. Journal of South African Botany 44: 291-295
LAMONT B (1972) The effect of soil nutrients on the production of proteoid roots by Hakea species. Australian Journal of Botany 20: 27-40
LAMONT B (1976) The effects of seasonality and waterlogging on the root systems of a number of Hakea species. Australian Journal of Botany 24:691-702
LAMONT B (1981) Specialized roots of non-symbiotic origin in heathlands. In: SPECHT RL (ed) Ecosystems of the world, vol 9B. Heathlands and related shrublands. Analytical studies. Elsevier, Amsterdam, pp 183-195
LAMONT B, DOWNES S, FOX JED (1977) Importance-value curves and diversity indices applied to a species-rich heathland in Western Australia. Nature 265: 438-441
LEWIS OAM, STOCK WD (1978) A preliminary study of the nitrogen nutritional status of members of the South African Proteaceae. Journal of South African Botany 44: 143-151

MALAJCZUK N, LAMONT B (1980) Specialized roots of symbiotic origin in heathlands. In: SPECHT RL (ed) Ecosystems of the world, vol 9B. Heathlands and related shrublands. Analytical studies. Elsevier, Amsterdam, pp 165-182

MARGARIS NW (1976) Structure and dynamics in a phryganic (East Mediterranean) ecosystem. Journal of Biogeography 3: 249-259

MCPHERSON JK, MULLER CH (1969) Allelopathic effect of Adenostoma fasaciculatum, "chamise," in the California chaparral. Ecological Monographs 39: 177-198

MESERVE PL (1976) Food relationships of a rodent fauna in a California coastal sage scrub community. Journal of Mammalogy 57: 300-319

MILEWSKI AV (1977) Habitat of Restionaceae endemic to western Cape coastal flats. Journal of South African Botany 43: 233-241

MILEWSKI AV (1978a) Habitat of threatened Proteaceae endemic to western Cape coastal flats. Journal of South African Botany 43: 233-241

MILEWSKI AV (1978b) Habitat of threatened species of Serruria and Protea endemic to western Cape coastal flats. Journal of South African Botany 44: 363-371

MILLER PC (1979) Quantitative plant ecology. In: HORN DJ, STAIRS GR, MITCHELL RD (eds) Analysis of ecological systems. Ohio State University Press, Columbus, pp 179-231

MILLER PC, BRADBURY DE, HAJEK E, LA MARCHE V, THROWER NJW (1977) Past and present environment. In: MOONEY HA (ed) Convergent evolution in Chile and California mediterranean climate ecosystems. Dowden, Hutchinson and Ross, Stroudsberg, Pennsylvania, pp 27-72

MILLER PC, NG E, (1977) Root:shoot biomass ratios in shrubs in southern California and Central Chile. Madrono 24: 215-223

MILLER PR (1980) (ed) Proceedings of symposium on effects of air pollutants on mediterranean and temperate forest ecosystems. USDA Forest Service General Technical Report PSW-43

MONK CD (1966) An ecological significance of evergreenness. Ecology 47: 504-505

MONTENEGRO G, RIVERA O, BAS F (1978) Herbaceous vegetation in the Chilean matorral: dynamics of growth and evaluation of allelopathic effects of some dominant shrubs. Oecologia 36: 237-244

MOONEY HA, CONRAD CE (eds) (1977) Proceedings of the symposium on the environmental consequences of fire and fuel management in mediterranean ecosystems. USDA Forest Service General Technical Report WO-3, 498 pp

MOONEY HA, DUNN EL, SHROPSHIRE F, SONG L (1970) Vegetation comparisons between the mediterranean climatic areas of California and Chile. Flora 159: 480-496

MOONEY HA, FERRAR PJ, SLATYER RO (1978) Photosynthetic capacity and carbon allocation patterns in diverse growth forms of Eucalyptus. Oecologia 36: 103-111

MOONEY HA, KUMMEROW J, CHU C, HAYS RI (1977a) Biomass, energy and mineral characteristics. In: THROWER NJW, BRADBURY DE (eds) Chile-California mediterranean scrub atlas. A comparative analysis. Dowden, Hutchinson and Ross, Stroudsberg, Pennsylvania, pp 162-164

MOONEY HA, KUMMEROW J, JOHNSON AW, PARSONS DJ, KEELEY S, HOFFMANN A, HAYS RI, GILIBERTO J, CHU C (1977b) The producers - their resources and adaptive responses. In: MOONEY HA (ed) Convergent evolution in Chile and California mediterranean climate ecosystems. Dowden, Hutchinson and Ross, Stroudsberg, Pennsylvania, pp 85-143

MOONEY HA, RUNDEL PW (1979) Nutrient relations of the evergreen shrub, Adenostoma fasciculatum, in the California chaparral. Botanical Gazette 140: 109-113

MULLER CH (1965) Inhibitory terpenes volatilized from Salvia shrubs. Bulletin of the Torrey Botanical Club 92: 38-45
MULLER CH (1966) The role of chemical inhibition (allelopathy) in negetational composition. Bulletin of the Torrey Botanical Club 93: 332-351
MULLER CH, DEL MORAL R (1966) Soil toxicity induced by terpenes from Salvia leucophylla. Bulletin of the Torrey Botanical Club 93: 130-137
MULLER CH, HANAWALT RB, MCPHERSON JK (1968) Allelopathic control of herb growth in the fire cycle of California chaparral. Bulletin of the Torrey Botanical Club 95: 225-231
MULLER CH, MULLER WH, HAINES BL (1964) Volatile growth inhibitors produced by aromatic shrubs. Science 143: 471-473
MULLER WH (1965) Volatile materials produced by Salvia leucophylla: effects on seedling growth and soil bacteria. Botanical Gazette 126: 195-200
MULLER WH, MULLER CH (1964) Volatile growth inhibitors produced by Salvia species. Bulletin of the Torrey Botanical Club 91: 327-330
MULLETTE KJ, BAMBER RK (1978) Studies of the lignotubers of Eucalyptus gummifera (Gaertn. and Hochr.). III. Inheritance and chemical composition. Australian Journal of Botany 26: 23-28
NAVEH Z, WHITTAKER RH (1980) Structural and floristic diversity of shrublands and woodlands in northern Israel and other mediterranean areas. Vegetatio 41: 171-190
PARSONS DJ, MOLDENKE AR (1975) Convergence in vegetation structure along analogous climatic gradients in California and Chile. Ecology 56: 950-957
PIELOU EC (1966) Species-diversity and pattern-diversity in the study of ecological succession. Journal of Theoretical Biology 10: 370-383
POISSONET P, ROMANE F, THIAULT M, TRABAUD L (1978) Evolution d'une garrigue de Quercus coccifera L. soumise à divers traitements: quelques résultats des cinq premières années. Vegetatio 38: 135-142
PRESTON K (1980) Effects of sulfur dioxide pollution on coastal sage scrub. MA Thesis. University of California, Los Angeles
RAVEN PH (1973) The evolution of Mediterranean floras. In: DI CASTRI F, MOONEY HA (eds) Mediterranean-type ecosystems: origin and structure. Springer-Verlag, Berlin, pp 213-224
RUNDEL PW (1981) The matorral zone of central Chile. IN: DI CASTRI F, GOODALL DW, SPECHT RL (eds) Ecosystems of the world, vol 11. Mediterranean-type shrublands. Elsevier, Amsterdam. In press
RUTHERFORD MC f1978) Karoo-fynbos biomass along an elevational gradient in the western Cape. Bothalia 12: 555-560
SCHLESINGER WH, CHABOT BF (1977) The use of water and minerals by evergreen and deciduous shrubs in Okefenokee Swamp. Botanical Gazette 38: 490-497
SEDDON G (1974) Xerophytes, xeromorphs and sclerophylls: the history of some concepts in ecology. Biological Journal of the Linnaean Society 6: 65-87
SHMIDA A, WHITTAKER RH (1981) Pattern and biological microsite effects in two shrub communities, southern California. Ecology 62: 234-251
SIDDIQI MY, CAROLIN RC (1976) Studies in the ecology of coastal heath in New South Wales. II. The effects of water supply and phosphorus uptake on the growth of Banksia serratifolia, B. aspleniifolia and B. ericifolia. Proceedings of the Linnaean Society of New South Wales 101: 38-52

SIDDIQI MY, MYERSCOUGH PJ, CAROLIN RC (1976) Studies in the ecology of coastal heath in New South Wales. IV. Seed survival, germination, seedling establishment and early growth in Banksia serratifolia Salisb., B. aspleniifolia Salisb. and B. ericifolia L.F. in relation to fire: temperature and nutritional effects. Australian Journal of Ecology 1: 175-183

SIMPSON EH (1949) Measurement of diversity. Nature 163: 688

SPECHT RL (1957) Dark Island heath (Ninety-Mile Plain, South Australia). IV. Soil moisture patterns produced by rainfall interception and stem-flow. Australian Journal of Botany 5: 137-150

SPECHT RL (1963) Dark Island heath (Ninety-Mile Plain, South Australia). VII. The effect of fertilizers on composition and growth, 1950-60. Australian Journal of Botany 11: 67-94

SPECHT RL (1969a) A comparison of the sclerophyllous vegetation characteristic of mediterranean type climates in France, California and southern Australia. I. Structure, morphology, and succession. Australian Journal of Botany 17: 277-292

SPECHT RL (1969b) A comparison of the sclerophyllous vegetation characteristic of mediterranean type climates in France, California, and southern Australia. II. Dry matter, energy, and nutrient accumulation. Australian Journal of Botany 17: 293-308

SPECHT RL (1979a) Heathlands and related shrublands of the world. In: SPECHT RL (ed) Ecosystems of the world, vol 9A. Heathlands and related shrublands. Descriptive studies. Elsevier, Amsterdam, pp 1-18

SPECHT RL (1979b) The sclerophyllous (heath) vegetation of Australia: the eastern and central states. In: SPECHT RL (ed) Ecosystems of the world, vol. 9A. Heathlands and related shrublands. Descriptive studies. Elsevier, Amsterdam, pp 125-210

SPECHT RL (1981) Primary production in mediterranean-climate ecosystems regenerating after fire. In: DI CASTRI F, GOODALL DW, SPECHT RL (ed) Ecosystems of the world, vol 11. Mediterranean-type shrublands. Elsevier, Amsterdam

SPECHT RL, CONNOR DJ, CLIFFORD HT (1977) The heath-savannah problem: The effect of fertilizer on sand-heath vegetation of North Stradbroke Island, Queensland. Australian Journal of Ecology 2: 179-186

SPECHT RL, GROVES RH (1966) A comparison of the phosphorus nutrition of Australian heath plants and introduced economic plants. Australian Journal of Botany 14: 201-221

SPECHT RL, MORGAN DG (1981) The balance between the foliage projective covers of overstorey and understorey strata in Australian vegetation. Australian Journal of Ecology 6: 193-202

SPECHT RL, RAYSON P (1957) Dark Island heath (Ninety-Mile Plain, South Australia). III. The root systems. Australian Journal of Botany 5: 103-114

SPECHT RL, RAYSON P, JACKMAN ME (1958) Dark Island heath (Ninety-Mile Plain, South Australia). VI. Pyric succession: changes in composition, coverage, dry weight, and mineral nutrient status. Australian Journal of Botany 6: 59-88

SPRINGETT BP (1978) On the ecological role of insects in Australian eucalypt forests. Australian Journal of Ecology 3: 129-139

STEVENS W (1937) The man with the blue guitar. In: MORSE SF (ed) (1959) Poems. Wallace Stevens. Vintage Books, New York, pp 73-90

TAYLOR HC (1972) Notes on the vegetation of the Cape Flats. Bothalia 10: 637-646

TAYLOR HC (1978) Capensis. In: WERGER MJA (ed) Biogeography and ecology of Southern Africa. Junk, The Hague, pp 171-230

VAN DER MERWE P (1966) Die flora van Swartboschkloof, Stellenbosch en die herstel van die soorte na 'n brand. Annals of the University of Stellenbosch 41, Series A, Number 14: 691-736

WEIR ICA (1969) Regrowth studies in uneven-aged Eucalyptus obliqua stands. MSc (Forestry) Thesis, University of Melbourne

WERGER MJA, KRUGER FJ, TAYLOR HC (1972) A phytosociological study of the Cape fynbos and other vegetation at Jonkershoek, Stellenbosch. Bothalia 10: 599-614

WESTMAN WE (1975a) Edaphic climax pattern of the pygmy forest region of California. Ecological Monographs 45: 109-135

WESTMAN WE (1975b) Pattern and diversity in swamp and dune vegetation, North Stradbroke Island. Australian Journal of Botany 23: 339-354

WESTMAN WE (1978a) Inputs and cycling of mineral nutrients in a coastal subtropical eucalypt forest. Journal of Ecology 66: 513-531

WESTMAN WE (1978b) Evidence for the distinct evolutionary histories of canopy and understorey in the Eucalyptus forest-heath alliance of Australia. Journal of Biogeography 5: 365-376

WESTMAN WE (1979a) A potential role of coastal sage scrub understories in the recovery of chaparral after fire. Madrono 26: 64-68

WESTMAN WE (1979b) Oxidant effects on Californian coastal sage scrub. Science 205: 1001-1003

WESTMAN WE (1981a) Diversity relations and succession in Californian coastal sage scrub. Ecology 62: 170-184

WESTMAN WE (1981b) Factors influencing the distribution of species of Californian coastal sage scrub. Ecology 62: 439-455

WESTMAN WE (1982) Coastal sage scrub succession. In: Proceedings of the International symposium on dynamics and management of mediterranean-climate ecosystems. Pacific Southwest Forest and Range Experiment Station, Berkeley, California. In press

WESTMAN WE, COWLING RM (1981) Plant community structure and dynamics. In: DAY J (ed) Proceedings of the workshop on nutrients as determinants of structure in mediterranean-type ecosystems. In press

WESTMAN WE, O'LEARY JF, MALANSON GP (1981) The effect of fire intensity, aspect and substrate on post-fire growth of Californian coastal sage scrub. In: MARGARIS NS, MOONEY HA (eds) Components of productivity of mediterranean regions - basic and applied aspects. Junk, The Hague. In press

WESTMAN WE, ROGERS RW (1977) Nutrient stocks in a subtropical eucalypt forest, North Stradbroke Island. Australian Journal of Ecology 2: 447-460

WHITTAKER RH (1960) Vegetation of the Siskiyou Mountains, Oregon and California. Ecological Monographs 30: 279-338

WHITTAKER RH, GILBERT LE, CONNELL JH (1979a) Analysis of two-phase pattern in a mesquite grassland, Texas. Journal of Ecology 67: 935-952

WHITTAKER RH, NIERING WA, CRISP MD (1979b) Structure, pattern and diversity of a mallee community in New South Wales. Vegetatio 39: 65-76

WINNER WE (1981) The effect of SO?2? on photosynthesis and stomatal behaviour of mediterranean climate shrubs. In: MARGARIS NS, MOONEY HA (eds) Components of productivity of mediterranean regions - basic and applied aspects. Junk, The Hague. In press

WINNER WE, MOONEY HA (1980a) Ecology of SO?2? resistance: I. Effects of fumigations on gas exchange of deciduous and evergreen shrubs. Oecologia 44: 290-295

WINNER WE, MOONEY HA (1980b) Ecology of SO?2? resistance: II. Photosynthetic changes of shrubs in relation to SO?2? absorption and stomatal behavior. Oecologia 44: 296-302

YARDENI D, EVENARI M (1952) The germination inhibiting and phytotoxic effects of certain leaves and leaf extracts. Phyton 2: 11-17

23. Plant Community Diversity and Dynamics in Relation to Fire

F. J. KRUGER

23.1 Introduction

Recurrent fires in vegetation are a general feature of mediterranean-type ecosystems and the profound effect of fire regimes on these ecosystems is generally acknowledged (Mooney and Conrad 1977).

Management of natural and semi-natural landscapes in these zones inevitably implies control of fire. This needs an understanding of vegetation dynamics both to enable fire control and to allow predictions of the ecological effects of fire control. The apparent interactions between fire regime and community (alpha) diversity raises important questions regarding both community ecology and practices for ecosystem conservation.

Here I have attempted to review and synthesize the information on the dynamics of mediterranean-climate vegetation in relation to fire regimes. Coexistence of plant species in the context of vegetation dynamics is considered with a view to our understanding of the effects of changes in regime. This is done principally for the mediterranean-climate zones of California, Australia, and South Africa where effects of fire regime are confounded less by human activities than in the Mediterranean Basin and Chile.

Cody and Mooney (1978) and Gill and Groves (1981) have recently reviewed information on the ecological effects of fire in these ecosystems. Some aspects of their findings are summarized in Table 1. Gill and Groves (1981) concluded that the pattern of change in vegetation after fire in heathlands and mediterranean-type shrublands in general "...is more a gradual elimination of individuals present from the outset than a replacement of initial species by new species..." and the classical succession model of "facilitation" (Connell and Slatyer 1977) does not apply. Another characteristic they noted is that "Pyric succession in heathlands comprises regeneration both from seed and from rootstocks." They maintained that "... the responses of heathlands to fires is probably one of the major features common to these different shrublands." Both reviews provided instances of the pronounced effects of different fire regimes on vegetation composition. But in spite of the overall similarities in vegetation response to fire there are important differences

Table 1. Summary characteristics of pyric succession in mediterranean-type ecosystems as described by Cody and Mooney (1978) and Gill and Groves (1981).

	Chaparral	Matorral	Maquis and garrigue	Australian heath	Fynbos
Annuals and short-lived fire ephemerals	Dominant post-fire	None	None?	Rare	Rare
Immigrant post-fire annuals	None?	None?	Often abundant	None?	None?
Geophytes	Some	?	Many	Prominent	Prominent
Persistent herbaceous perennials	Very few	Abundant (decrease post-fire)	Grasses stimulated post-fire. Rare in old stands.	Sedges, restios dominant post-fire and remain prominent	Grasses, sedges, restios dominant post-fire and remain prominent
Regeneration mode of shrub species	About 50% sprout	Most sprout, few by seed only	Most sprout, few by seed only	Both modes prominent	Both modes prominent
Timing of plant recruitment	Immediately post-fire	?	Immediately post-fire	Immediately post-fire	Immediately post-fire
Fire-stimulated germination in shrubs?	Yes	No	No	Yes	Yes
Redevelopment of canopy cover	100% in 10-20 yr	?	Closed canopy in 5 yr	100% in 10 yr	70-90% in 2 yr
Diversity trends	Species richness (S) decreases with age	?	S decreases with age	S decreases with age	S decreases with age (sometimes an initial increase)
Effect of long absence of fire	Stagnation	?	?	Stagnation; in some cases, trend toward *Banksia* or *Eucalyptus* woodland, or *Leptospermum* thicket	Stagnation; in some cases immigration of evergreen forest species

among mediterranean-type ecosystems that correspond with differences in nutrient status (Table 1).

23.2 Fire regimes

The degree to which the floras and vegetations of the different mediterranean regions converge with respect to their responses to fire must depend in part on the measure of correspondence between the respective fire regimes. Fire regimes include the following: fire type, intensity, seasonality, and frequency (Gill 1975; Gill and Groves 1981) and comparative data are not available on all components.

23.2.1 FIRE FREQUENCY

There are few good data on fire frequency as characterized by the mean period between fires and variance about the mean. Indirect evidence suggests that Australian heaths experience a relatively high frequency of fires with low variance. Coaldrake (1951) describes a regime of frequent (less than 10 year recurrence interval?) fires in the extensive sand-heaths of South Australia; Specht et al. (1958) could find no stand older than 25 years on their study site in the Dark Island heath and had to travel 16 km to find a 50-year-old stand. Siddiqi et al. (1976) worked in vegetation where fire had recurred fortuitously five years and 12-15 years after the previous burns. The oldest stand of heath that Russell and Parsons (1978) could find in Wilson's Promontory National Park, an area where fire suppression is practised, was 20 years old and burnt shortly after their study was completed. Fires had recurred about once in 6.3 years in one of the stands they analysed, and these were mainly wildfires. Coaldrake (1961) reported a prevalent frequency of about once in five years for the heathlands of the southern Queensland coastal lowlands, for the period after European settlement. Prolonged absence of fire (i.e. 32 and 80+ years) has been reported only occasionally (Hazard and Parsons 1977).

Some indication of low frequencies in South African mountain fynbos is available from State forests where suppression is practised (Kruger 1979a). Here, most fynbos may burn either every 20 years or less, and very rarely once in 40 years. Horne (1981) analysed fire records for the Groot-Swartberg, a southern mountain range where fire suppression has been practised, and found an average frequency for all fires of once in 18 years, and for fires of natural origin - if man-caused fires are ignored - once in about 40 years. Fire frequencies are greater than these in most fynbos areas under present conditions.

The picture for chaparral indicates a fairly long mean recurrence interval, with relatively large variance. Schlesinger and Gill (1978), reported that "...stands older than 35 to 40 years are extremely rare..." in the Santa Ynez Mountains. However, several of Hanes's (1971) study sites were in vegetation more than 80 years old, and Keeley (1977b) sampled a stand more than 90 years old - these ages have not been reported for fynbos, for example. At San Dimas Experimental Forest, eight fires occurred between 1895 and 1960 on the area of about 7000 ha but only the fires of 1920 and 1960 were large; thus most of the vegetation had reached an age of 40 years and a number of stands were 65 years old in 1960 (Mooney and Parsons 1973).

The incidence of fire in National Forests of the San Bernardino range, southern California, and in fynbos of the Cedarberg State Forest is shown in Fig. 1. In both these areas fires have been suppressed by fire-break systems and fire-fighting measures and therefore allow a fairly exact comparison between chaparral and fynbos ecosystems. The data clearly show that fynbos burns more frequently than chaparral.

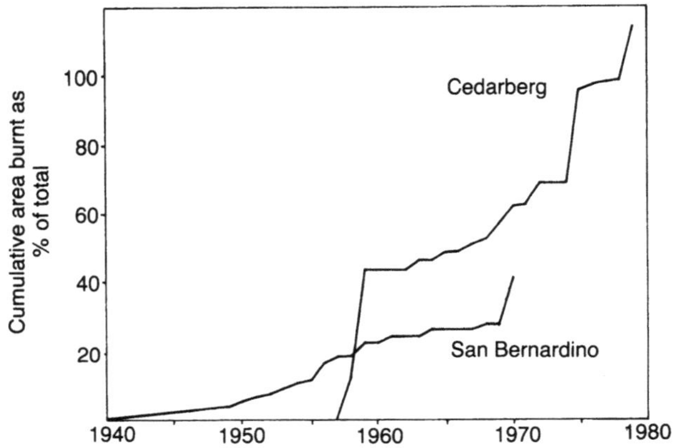

Fig. 1. Cumulative sum of area burnt from arbitrary initial dates in portion of the San Bernardino National Forest, southern California, and Cedarberg State Forest, South Africa. Derived from data in Philpot (1974) and the records of the South African Directorate of Forestry.

Fire frequencies are apparently highly variable in all of the mediterranean-type ecosystems. The upper limits of frequency are determined by vegetation structure and productivity. Thus Coaldrake (1951), Specht et al. (1958) and Kruger (1977a) have pointed out that South African fynbos and Australian heathland normally require 4-5 years before sufficient standing vegetation and litter have accumulated to carry a fire, though fires may sometimes spread into younger stands if conditions are extreme. In the chaparral, Hanes (1971) reported fire periods of 2-10 years at its lower elevational limits in the zone of transition with coastal sage, but Philpot (1977) found from an analysis of fire atlases for National Forests in the San Bernardino range of southern California that almost no large fires occurred in vegetation less than 15-17 years old. Early recurrence of fire in chaparral may depend on the nature of the herb regrowth and especially on the degree of invasion by exotic grasses. Fire appears never to be very frequent in the average natural chaparral stand. By contrast, sprouting hemicryptophytes in fynbos and Australian heath quickly re-establish a nearly continuous fuel bed that permits the frequencies of up to once in four years reported there.

The lower limits of frequency are determined by probability of ignition. Few or no cases have been reported where prolonged deferral of fire leads to changes in vegetation composition that reduce the flammability of stands, though inferred in certain instances in Australia (Jackson 1968).

23.2.2 OTHER ELEMENTS OF FIRE REGIME

Fires naturally tend to occur in the summer half-year of the mediterranean-type climate but fire seasons are in fact relatively variable. Thus, in chaparral, Horton and Kraebel (1955) reported burns that had occurred in January, March, July, August, September and November. The fire atlas for the San Bernardino National Forest shows burns from February to December, but with less than 1% of the total area burning in spring, about 30% in summer, and over 60% in autumn (Philpot 1974).

Wild-fires have been reported during any season in South African fynbos, although there appears to be a summer peak (Kruger 1979a). In the Cedarberg, 14% of the area burnt was by spring wild-fires, 52% in summer, and 32% in autumn (Andrag in Kruger 1979b). Horne (1981) found for the Groot Swartberg that 24% of wild-fires were reported in spring and 51% in summer (21% and 60% respectively for lightning fires). There is some indication of a greater incidence of large winter fires in the southern coastal mountains than elsewhere, and therefore of regional variation in the seasonality of fires in fynbos (Kruger 1979a).

Thus, in both chaparral and fynbos the seasonal timing of fire can be unpredictable from year to year with considerable variance within about a six-month range. The same should apply for Australia and A M Gill (personal communication 1980) reports maximum occurrence from spring to midsummer for heathlands in eastern Australia. This significant variability in fire season is to be expected. With regard to South Australian heath, Specht et al. (1958) have noted that "Times of high fire risk do not necessarily correspond with times of climatic stress on the vegetation. Several hot, windy days are sufficient to dry the surface and to make the fire risk high despite high soil moisture reserves." The fact that conditions for spreading fires are strongly controlled by current weather, and only partly correlated with season, has also been emphasized by McCutcheon (1977) and this observation would be valid for all mediterranean-type ecosystems.

The scant data available show that chaparral fires are distinctly more intense in terms of energy release rates (see Rothermel and Philpot 1973 cf. Bands 1977; Gill and Groves 1981) and hotter in terms of flame temperatures in the canopy and soil surface and shallow sub-surface temperatures (Rundel 1982) than fynbos or heathland fires. The deep ash-beds found after incineration of biomass in chaparral fires (e.g. Vogl and Schorr 1972) are not found after average fynbos and heathland fires (Kruger, unpublished).

Variation in the frequency of fire on any given site allows variation in the amount of fuel that accumulates and is available for each fire. This with variability in season and weather would cause considerable variation in fire intensity and in temperatures at the fire front. This has not been properly investigated, though Gill and Groves (1981) indicate the likely magnitude of variation. Gill (1981a) has also shown how intensities vary by up to a factor of five within a given fire, simply because of different rates of spread around the perimeter.

23.3 Pyric succession

23.3.1 SUCCESSION AS A SELF-REPLICATION OF STANDS: AUTOSUCCESSION

Plant succession after fire in mediterranean-type shrublands consists largely of the regeneration and development through their respective life-cycles of the species present before the fire (Specht et al. 1958; Hanes 1971), and studies have shown that the burnt stand retains its floristic identity, i.e. is more similar to the pre-fire stand than to others in the area (Russell and Parsons 1978; van Wilgen and Kruger 1981). Hanes (1971) called this "autosuccession". However, most succession studies have relied on comparisons among stands apparently alike aside from fire history, rather than on successive observation of permanent sample sites. Studies of the latter kind usually show that recovery in composition is not complete. In fynbos, Adamson (1935) found that 16 of an original 92 species had not reappeared within 6.5 years after fire. Van Wilgen and Kruger (1981) found after a repeat survey of 33 permanent plots in fynbos burnt six years previously that the floristic composition of each plot had apparently changed by an average of about 50%, even though the species richness per plot and the total flora encountered in the sample had changed little. Despite observation error in both cases, changes were real and significant. Horton and Kraebel (1955) showed floristic changes both in dominants and in overall composition of the annuals and herbaceous perennials after successive fires in chaparral of the San Bernardino mountains (some change being due to immigration of an exotic species); their records also indicated that shrubs such as Ceanothus crassifolius could respond differently to successive fires on the same site; seedlings were abundant after the first (summer) fire, absent after the second (in spring).

The relative abundance of species may also change after different fires because of differences in one or more components of fire regime. Specht et al. (1958) recorded major differences in the relative abundances after burning of seedlings of early-maturing and late-maturing species in adjacent heath stands, one six years old when burnt, and the other 10 years old; the ratio of late-maturing to early-maturing species was 3:1 in the former case, 30:1 in the latter. Bond (1980) found significant and major (up to 12-fold) reductions in Proteaceae and dicotyledon seedling densities after fire in senescent fynbos stands, compared with mature stands. Relative abundances of seeding species (those regenerating solely from seed) in fynbos are markedly affected by season of burn with the pattern often being one of successful regeneration after summer or autumn fires, but low success, or loss of populations, after winter or spring fires (Jordaan 1965, 1981; Bond 1980), a pattern apparently also found in some chaparral species (Horton and Kraebel 1955). These reports are from comparative data (same fire burning adjacent stands of different ages, for example) but show how fire-to-fire variation in period, season and other regime components on the same site can affect initial stand composition.

Thus, though sclerophyllous shrublands tend to retain their identity despite burning there is at least some progressive or fluctuational change in species composition after successive fires. Contributing to this change are (a) the effects of season of the burn, (b) interval between fire (see below), (c) the interaction between timing of the fire and subsequent weather (Horton and Kraebel 1955; Specht et al. 1958), and (d) by analogy with Australian eucalyptus forest understoreys (Floyd 1976; Gill 1981a), intensity of fire.

23.3.2 TRENDS IN COMMUNITY DIVERSITY

Most studies show that old stands contain less species than they do in the immediate post-fire phase but the reported decline in diversity is not necessarily a uniform progressive change from a maximum in the first year

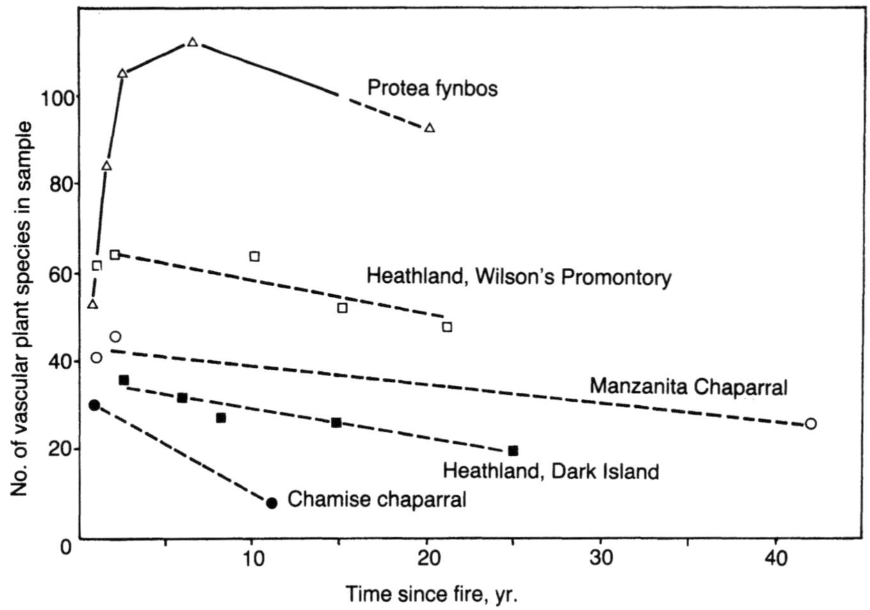

Fig. 2. Trends in community diversity with pyric succession. Data from Adamson (1935), Specht et al. (1958), McPherson and Muller (1969), Vogl and Schorr (1972), and Russell and Parsons (1978).

after fire (Fig. 2). The trend of increasing richness in the first few years to levels beyond pre-fire richness reported by Adamson (1935) has also been found in Australian heathland or heath understoreys (Bell and Koch 1980; Posamentier et al. 1981). In these cases, some or all of the increasing trend is spurious, owing to difficulties of detecting or identifying plants in the early stages of regeneration. Results are often difficult to interpret because of sampling techniques. For example, Posamentier et al. (1981) sampled plots of 0.08 ha by means of 16-20 0.2 x 0.5 m quadrats. With such small quadrats, some increase in numbers of species recorded must be owing to changes in frequency, rather than new appearance of species in the plot. In general, the progressive loss of species from the stand involves both the rapid disappearance of relatively short-lived plants, usually "fire ephemerals" (short-lived fugitive species that appear soon after fire and depend on the fire for regeneration - Gill 1981a) and gradual suppression of understorey species by overstorey species (e.g. Specht et al. 1958; Hanes 1977). The wet-heaths of Australia, richer in species than other heaths and dominated by sprouters, lose less species during succession (e.g. Russell and Parsons 1978). In Australian heathland and South African fynbos species are lost gradually during succession, but in chaparral most species that do so (about 40% of the flora) disappear relatively rapidly, in the first five years (see review in Hanes 1977; Keeley et al. 1981). This latter pattern applies to coastal sage scrub as well (Westman 1981). In fynbos, α diversity can increase later in succession (more than 30 years after fire), apparently owing to regeneration of understorey species as the canopy opens with senescence in the overstorey (van Wilgen 1981). An increase has also been reported for Californian coastal sage scrub at about 20 years after fire, but is not associated with an opening of the overstorey, rather the immigration of tolerant herbs (Westman 1981). At 40 years, α diversity is reduced once more, despite a sparser overstorey.

23.3.3 DYNAMIC RELATIONS BETWEEN UNDERSTOREY AND OVERSTOREY STRATA

Specht and coworkers (Specht and Morgan 1981; Chapter 7, this volume) have derived a useful construct to describe the successional relations between the overstorey and the understorey strata in mediterranean communities. Briefly, sclerophyllous vegetation recovers rapidly after fire until a climatically determined maximum foliage projective cover is reached, normally within two to five years. This rapid recovery is owing to the regrowth of the understorey; once the maximum foliage projective cover is reached, the slower overstorey species begin to dominate and the cover of the understorey declines sharply. An inverse linear relationship between overstorey cover and understorey cover then obtains, while the equilibrial foliage projective cover is maintained. The number of active plant species on the site is also in an inverse linear relation to the overstorey foliage projective cover.

Several points are of interest in this regular relationship. First, the overstorey apparently begins to influence the understorey already when sparse, i.e. with a foliage projective cover of 30-40%, normally much less. This may reflect unusual light sensitivity among understorey species, or the early onset of competition below ground. Second, the overstoreys in different shrublands vary considerably in their rates of recovery. In the Dark Island heath, the overstorey reached about 30% foliage projective cover in about three years and was still increasing at 20 years. In the fynbos, overstorey cover was less than 10% at three years, and increased very slowly afterward. By contrast, chaparral overstorey shrubs recover their pre-fire canopy cover within five to 7-9 years (Hanes 1977; Keeley 1977c). Though Specht and others (Specht et al. 1958; Specht and Morgan 1981; Chapter 7, this volume) argue that the relation between understorey and overstorey is competitive, the curves for fynbos show a decline in understorey cover well before the overstorey has reached the level of 30% or so that apparently caused a decline in the understorey of Australian heath, for example. This implies effects due to competition for different resources (water and/or nutrients, instead of light), or different life-history strategies (earlier senescence, in fynbos understoreys). Also, the pattern of senescence in the overstorey accompanied by recovery in the understorey reported by van Wilgen (1981, 1982) shows that the inversely linear progression may not continue indefinitely. In any case, this overstorey-understorey relationship promises a useful basis for comparison of successional patterns and processes among the different mediterranean-type ecosystems, and a basis for extrapolating from local experiments.

23.3.4 CONSEQUENCES OF PROLONGED ABSENCE OF FIRE

The potential for replacement of mediterranean-type shrublands by alternative vegetation types if fire is excluded is a controversial matter and empirical studies are few. First, prolonged absence of fire may lead to cyclic succession where shrubland dominants alternate or replace themselves as gaps occur through population senescence or death of large individuals, as in Australian heath or in fynbos (Specht et al. 1958; van Wilgen 1981, 1982). This has not been suggested for chaparral. This pattern would fit the inhibition model of succession proposed by Connell and Slatyer (1977).

Second, senescence among the dominants of the mature stand may allow less abundant species already present to propagate and gain prominence - the tolerance model (Connell and Slatyer 1977). In chaparral (of the "climax" zone) Patric and Hanes (1964) have suggested that a "scrub oak chaparral" can develop over about 50 years, through the death of certain typical chaparral species such as in the genus Ceanothus and persistence and seed-establishment by Quercus dumosa, Prunus ilicifolia, and Rhamus ilicifolia (see also Horton and Kraebel 1955). In northern California, Sampson (1944) has suggested replacement of chaparral by grass with time. In Australian heath, Hazard and Parsons (1977) deduced that in certain parts of their study area, Banksia and Leptospermum, unable to regenerate without fire, were slowly being replaced by Leucopogon (Epacridaceae), seedlings of which were able to establish in old vegetation.

In other cases, senescence of the dominants appears to allow immigration of species of an alternative cover type - perhaps the facilitation model applies (Connell and Slatyer 1977). Van Wilgen (1981) has described a measure of immigration by tree species such as Kiggelaria africana of senescent fynbos on relatively fertile granite soil at Jonkershoek after 25-35 years of protection against fire. These species that are seen as precursors of evergreen rain forests. Del Moral et al. (1978) report a slow replacement of heath by mallee eucalypts and Burrell (1969, in Hazard and Parsons 1977) and Hazard and Parsons (1977) show how typical mixed heath can sometimes be replaced by Leptospermum thicket.

In some cases, the vegetation seems simply to stagnate, where there is gradual mortality among long-lived dominants but little immigration or re-establishment. This would fit the inhibition model, and has been reported for chaparral (Horton and Kraebel 1955; Hanes 1977). Stagnation appears to occur in fynbos on quartzites. Bond (1980) could find no successful establishment in a recent sample survey in his senescent stands. A recent survey in 35-year-old vegetation in the Kogelberg showed that though 90% of the dominant Leucadendron population was dead, no seedlings or young plants were present nor were any immigrant shrubs or trees to be found (Kruger, unpublished).

Presently, no useful generalizations seem possible about the likelihood of and patterns in vegetation replacement in the absence of fire.

23.4 Secular changes in vegetation under different fire regimes

Pyric succession in the process of vegetation change between successive fires. The longer-term changes in vegetation induced by a change from one regime to another has hardly been studied, but for experimental work in garrigue by Trabaud (1973) and in fynbos transitional to grassland by Trollope and colleagues (Downing et al. 1978). Also van Wilgen (1981, 1982) reports results from comparative studies in frequently and less frequently burned vegetation.

These studies show pronounced changes in structure and diversity with different fire frequencies. Frequent fires in garrigue and transitional fynbos favour perennial grasses. Frequent fires in fynbos at Jonkershoek induce richer plant communities than otherwise, even though some shrub species are eliminated (van Wilgen 1981), and Campbell and van der Meulen (1980) have shown how where the climax is a tall-scrub, increased fire frequency increases understorey graminoid species, whereas where the climax is a graminoid fynbos, greater frequency increases both graminoid and woody species. Despite these changes, there was no great change in species composition overall.

Our empirical knowledge of the relation between fire regime and secular vegetation change is therefore very limited, though many speculative models have been inferred (e.g. Cooper 1922; Jackson 1968; Hanes 1977).

23.4.1 CHANGES IN STRUCTURE: SPROUTERS VERSUS SEEDERS

One aspect of the effects of changes in regime that has claimed attention is that of balance between sprouters and seeders. Perennial species that regenerate vegetatively when burnt (sprouters) form a relatively stable component of the stand and are relatively little affected by a single fire, but stand composition in terms of sprouters and seeders can be changed by changing fire regime. The most evident effect is that of too-frequent fire. In fynbos and Australian heath seeders tend to be eliminated because they cannot mature, and sprouters favoured because their propagation is stimulated by too-frequent fire (Specht et al. 1958; Siddiqi et al. 1976; van Wilgen 1981, 1982) and the same holds by implication for chaparral (Hanes 1971; Keeley 1977a) but considerable variation in the relative abundances of sprouting and seeding shrubs is evidently not explained by such obvious fire effects.

Table 2. Regeneration mode in some typical mediterranean-type shrub taxa (data from Wells 1968; Rourke 1972; Williams 1972).

	Number of species	
	Sprouting	Seeding
CHAPARRAL		
Arctostaphylos	16	59
Ceanothus	12	46
CAPE FYNBOS		
Bruniaceae[a]	4	8
Ericaceae[a]	4	46
Leucadendron	7	73
Leucospermum	4	36[b]

[a] Flora of Jakkalsrivier, Lebanon only: Kruger, unpublished.
[b] At least three of these species show some tendency to resprout from epicormic buds - Rourke 1972.

Among trees or shrubs about half or more of any regional flora comprises sprouting species (Keeley 1977a; Kruger 1977b; Gill and Groves 1981), though large genera characteristic of the floras tend to have more seeders (Table 2). However, mature stands vary considerably in the ratio of seeders to sprouters, both between and within regions and in terms of both floristic composition and ecologial dominance (Tables 3 and 4). In chaparral, stands may consist nearly entirely of sprouting Adenostoma fasciculatum (80% or more of cover - Hanes 1977); by contrast, Schlesinger and Gill (1978) studied several nearly pure stands of Ceanothus megacarpus, a seeder; in one 21-year-old stand C. megacarpus comprised 98% of individuals and 100% of canopy cover. In Australian heathlands, sprouters appear to predominate. Specht (1981) lists data for seven stands in which sprouters account for between 54% and nearly 100% of above-ground plant biomass, except in an atypical Leptospermum stand where sprouters were absent. Seventy-three per cent of the shrub species in vegetation surveyed by Russell and Parsons (1978) and all 26 species in the Banksia woodland analysed by Whelan and Main (1979) regenerated vegetatively. There are nevertheless numerous cases, at least in the south-east, where typical heaths are dominated by seeders such as Banksia ericifolia, Casuarina distyla, Hakea gibbosa, H. sericea and H. teretifolia (Siddiqi et al. 1976; A M Gill, personal communication 1980).

Two related ecological explanations have been proposed for this variation. Keeley (1977a) argued that the relative success of seeders depends on the availability of "post-fire openings" in the regenerating vegetation, where seedlings could establish and grow free of competition from sprouters. Thinning of a stand of sprouters is progressive in the interval between fires. Live and dead fuel accumulates, and the longer a fire is deferred, the greater its intensity and the greater the mortality in sprouters, and the more openings after fire. Chaparral seeders are relatively long-lived and maintain fecundity (see below) so that abundant seed is available for regeneration in the openings. Seed stored in the soil is well insulated against fire. Thus seeders have the advantage in the open regrowth that occurs after rare fires, as opposed to the competitive conditions in the relatively dense regrowth of sprouters found when fire recurs soon, and the ratio of seeders to sprouters therefore increases progressively with decreasing fire frequency. Coexistence of the two forms requires a variable fire frequency, including long fire-free periods.

Specht (1981) also emphasizes the competitive advantage enjoyed by sprouters during post-fire regeneration. Available moisture determines the density (foliage projective cover) of Australian heath, and therefore the more humid the environment, the less gaps for seeders, and the less their importance in the climax stand. Thus, the relative abundance of sprouters correlates positively with foliage projective cover so that seeders are virtually absent from humid sites. The relative prevalence of either mode of regeneration is a consequence of site productivity and hence availability of gaps for seeders, rather than of fire response.

Both Specht and Keeley argue for the crucial role of competition for gaps in the evolution of sprouters and seeders. In South African fynbos patterns in the relative abundances of sprouters and seeders are not readily apparent. Fynbos stands, including those on moist sites, have a number of species of both types, though often dominated by seeding shrubs that may cover half or more of the ground; sprouting shrubs are seldom if ever so dense (Kruger 1979b; Tables 3 and 4) and the pattern that Specht (1981) reports for Australian heathlands is not found. Furthermore, senescence in the dominant serotinous seeding shrubs results in reduced

Table 3. Species from representative stands of chaparral, heath and fynbos classified according to mode of post-fire regeneration.

	Chamise chaparral (McPherson and Muller 1969)	Manzanita chaparral (Vogl and Schorr 1972)	Australian heath Wilsons Promontory (Russell and Parsons 1978)[a]	*Protea* fynbos (Kruger 1979b[b])	*Sympieza* fynbos (Kruger 1979b[b])
Age, yr since fire	1,11	1 2, 42	1 - 20	17	18
Seeding trees and shrubs					
- seed storage on plant	0	0	2	1	2
- seed storage in soil	3	3	3	14	9
Sprouting trees and shrubs					
- subterranean regenerative buds only	1	8	29	16	7
- aerial regenerative buds present	0	2	1	0	0
Seeding chamaephytes or sub-shrubs	4	20	0	3	2
Sprouting chamaephytes or sub-shrubs	0	0	10	2	1
Hemicryptophytes (sprouting perennial herbs)	0	1	16	14	22
Geophytes	2	2	15	18[c]	8[c]
Annuals	19	25	0	1[c]	0[c]
Uncertain and other	0	0	8	1	1
Total	29	61	84	70	51

a Species regeneration modes determined from Specht (1972) where necessary.
b Species regeneration modes determined from Van der Merwe (1969) and Kruger, personal observation.
c More species would probably have been found by sampling soon after fire.

Table 4. Relative importance of species of different modes of regeneration, recorded for various stands of chaparral, fynbos and Australian heath, and expressed as relative contribution of each category to cover or estimated biomass: data for most abundant species, representing a sum of 90% of total cover or biomass.

	Manzanita chaparral (Vogl and Schorr 1972)	Chamise chaparral (Mooney et al. 1977)[a]	*Protea nerifolia* fynbos (Kruger 1979b)[b]	*Protea nitida* woodland (Kruger 1979b)[b]	*Erica* fynbos (Kruger 1979b)[b]	*Sympieza* fynbos (Kruger 1979b)[b]	Dark Island dry heath-land (Specht 1979)[c]	*Leptospermum* heath (Russell and Parsons 1978)
Age, yr since fire	$\geq$ 42	>20?	17	7	19	18	9	20
Seeding trees and shrubs								
- seed storage on plant	0	0	48	0	0	0	18	2
- seed storage in soil	7	33	0	11	36	54	16	0
Sprouting trees and shrubs								
- subterranean regenerative buds only	89	60	43	24	9	2	38	76
- aerial regenerative buds present	0	0	0	37	0	0	0	0
Herbaceous perennials - seeding	0	ND	0	0	0	0	0	12
Herbaceous perennials - sprouting	0	ND	0	19	47	35	11	6
Uncertain	0	0	0	0	0	0	9	5

[abc] Information on regeneration mode from Hanes (1971), Van der Merwe (1969) and Kruger, personal observation, and Specht (1972) and Specht et al. (1958), respectively.

seed banks and a decline in daughter populations after fire, with prolonged deferral of fire (Bond 1980); the balance between seeders and sprouters can therefore respond strongly to both an increase (see above) and a decrease in fire frequency. A general explanation of the mechanisms underlying this pattern therefore requires a demographic approach, that will predict the relative recruitment and mortality of sprouters and seeders respectively under the varying conditions that determine gap formation.

The prevalence of the sprouting habit among dominant shrubs in Australian heathlands could be a recent artifact of the high fire frequencies, at least in recent times (see above). However, this is unlikely because seeding shrubs have survived in areas with low foliage projective cover despite high fire frequency (Specht et al. 1958, for Dark Island heath). Specht et al. (Chapter 7, this volume) show that water availability (moisture index values) does not reach as high levels in fynbos as in Austalian heath, and thus the range of foliage projective cover values is truncated at the upper end. By this argument, there is greater potential for seeders in fynbos overall than in Australian heathland and hence their greater abundance and diversity in the former.

Sprouting hemicryptophytes and other persistent perennial herbs are more diverse and abundant in South African fynbos and Australian heath than in other mediterranean-type ecosystems (Kruger 1977a). In fynbos these plants were more abundant and diverse on frequently burnt sites than on less frequently burnt areas, and were also favoured in a measure by the opening of the canopy that occurs with long deferral of fire (Campbell and van der Meulen 1980; van Wilgen 1981). Nevertheless, this component is usually prominent irrespective of fire history.

In garrigue, perennial grasses are sparse in old, unburned stands but become abundant when vegetation is frequently burned (Trabaud 1973). In chaparral, perennials such as Stipa lepida, Paeonia californica and Marah macrocarpus persist at all stages of development, but are invariably sparse; their response to fire frequency is unknown, but their relative density post-fire is low compared with that of annuals and short-lived herbs and sub-shrubs. The vines (Convolvulus, Marah) may often have high cover after fire but populations do not increase (Horton and Kraebel 1955; Keeley et al. 1981; Westman 1981). This is a major contrast between chaparral and most other northern mediterranean-climate shrublands, and Australian heath and South African fynbos, and seems to render the post-fire environment more competitive in the latter case.

23.5 Plant demography and succession

Simple, essentially qualitative information on plant life histories and analysis of stands in such terms may be sufficient for models that predict the broader patterns of vegetation replacement, as in the "vital attributes" model of Noble and Slatyer (1981). But quantitative, demographic data are required to understand and predict the course of post-fire succession, as well as the effects of different fire regimes on the relative abundances of species.

23.5.1 PERSISTENCE AND REGENERATION IN SPROUTERS

Though most plants in a stand persist and regenerate through sprouting there are few quantitative data that relate survival and regeneration mode and rates to fire regime or to other physical and biotic environmental factors, except for the work of Horton and Kraebel (1955) and J E Keeley and colleagues (Keeley and Keeley 1981; and references therein) on the demography of seeding and sprouting in the chaparral shrubs. The prevailing dominant sprouter in chaparral, Adenostoma fasciculatum as well as certain others such as Ceanothus leucodermis are relatively plastic in their sprouting and seeding habits (Keeley and Zedler 1978), and often experience a relatively high mortality in fires (e.g. 46% - Horton and Kraebel 1955; Keeley and Zedler 1978) and high seedling recruitment, though mortality among less frequent sprouting species is low (e.g. Horton and Kraebel 1955). In Adenostoma, the proportion of the regenerating population in seedlings ranged from about 16% to 55% and increased linearly with decreasing total shrub cover at four years on different sites ($r = 0.70$, $P = 0.01$, for data in Keeley and Keeley 1981). By contrast, in Australia and South Africa, survival rate in fires of sprouters seems uniformly high. Beadle (1940) reports survival rates of 92-100% among common Australian heathland shrubs and trees. The near-zero mortality in Protea nitida (Haynes 1976) is probably indicative of the situation in fynbos. Specht et al. (1958) recorded no seedlings after fire for the dominant sprouters at Dark Island.

Aside from fire ephemerals, all mediterranean-type ecosystems have persistent herbaceous perennials of great longevity. This, with quick responses to burning, allows prolonged coexistence with dominant shrub forms and reproduction in the sometimes long-deferred post-fire environment. For example, Dixon and Pate (1978) report putative ages of up to 54 years for Drosera erythrorhiza and greater ages (>100 years) have been found in Watsonia pyramidata (Kruger 1977b). Geophytes remain semi-dormant between fires (e.g. certain Orchidaceae in fynbos - Hall 1959; Paeonia, Brodiaea and others in chaparral, Horton and Kraebel 1955). Perennial grasses also persist. In the Mediterranean region, Brachypodium ramosum is greatly reduced in abundance in well-developed stands but flourishes immediately after fire (Specht 1969; Trabaud 1973). Naveh (1974) describes other examples. Perennial grasses of a similar habit are sometimes also found in chaparral (Melica and Stipa - Horton and Kraebel 1955). In fynbos, grasses such as Merxmuellera rufa are conspicuous in the first season or two postfire, flowering profusely and having vigorous leaf rosettes. Between fires it is scarcely detectable, each tuft showing perhaps one or two brief leaf-blades (Kruger, unpublished). In fynbos, perennial herbs apparently outlive seeding shrubs (Campbell and van der Meulen 1980; van Wilgen 1981).

The demography of such species is poorly known. Data for survivorship of fire, vegetative propagation, and seedling establishment for Watsonia pryamidata show pronounced population response to fire (Kruger 1977b) but in Brodiaea, Convolvulus and other California geophytes there appears to be little change in population despite prolific growth and flowering after fire (Stone 1951; Horton and Kraebel 1955; Keeley et al. 1981).

23.5.2 SEED-BANK DYNAMICS

Westman (1979) has proposed that a large proportion (58%) at least of the herbaceous flora of chaparral depends on immigration of seed after fire from herbs in open coastal sage stands. This is unlikely, given the distances involved, the relative weakness of wind dispersal in general (Harper 1977; Cook 1980), and the very rapid appearance of herbs after fire (Keeley et al. 1981). Some species, which between fires grow on tracks and other open sites, may progressively invade the post-fire habitat, but most apparently rely on seed already present (Horton and Kraebel 1955; Vogl and

Schorr 1972; Christensen and Muller 1975; Keeley et al. 1981). Among seeding species of both herbs and shrubs the pattern appears to be one of limited dispersal and survival in situ through dormancy (Wicht 1948; Specht et al. 1958; Hanes 1977). Keeley (1980, in Keeley et al. 1981) suggests that "fire annuals" of the chaparral behave like fugitive species but since exposure of any chaparral site by fire is inevitable there is little selection for long-range dispersal. The high species turnover between the fire ephemeral floras listed for different sites by Hanes (1977) bears this out (see also Keeley et al. 1981). At least in fynbos and Australian heathlands we find that the dominant or prominent shrubs are also adapted for short- rather than long-range dispersal (Williams 1972 for Leucadendron; Berg 1981; Bond 1980).

Soil-stored seed is apparently the norm for chaparral species (Hanes 1977) but many shrubs of Australian and South African heathlands differ from their Mediterranean and New World counterparts (aside from Coniferae), in the high incidence of serotinous species that retain viable seed stored in protective organs on the parent plant (Gill and Groves 1981). Serotiny is found mainly in Proteaceae but also in Myrtaceae (Eucalyptus - Beadle 1940; Callistemon and others - Kruger, unpublished), Casuarinaceae (Beadle 1940), Ericaceae (Erica sessiliflora - Baker and Oliver 1967) and Bruniaceae (Brunia spp., Berzelia spp. - Kruger unpublished). Also, serotiny has evolved differentially and, apparently, independently in different lineages of the Proteaceae (Banksia and Hakea in Grevilloideae; Isopogon, Protea, and Leucadendron in Persoonioideae). This therefore reflects evolutionary convergence rather than phylogeny.

The seed of these shrubs are relatively large (10-760 mg), and rich in nitrogen and phosphorus (van Staden and Brown 1977; Chapter 10, this volume) with concentrations respectively 8-100 times and 30-500 times the levels found in leaves of the same plants; 65-83% of seed or embryo dry mass is in protein and oil (Kuo et al. 1982; see also van Staden and Brown 1977).

In such species seed ripens over 9-24 months (e.g. Protea repens, Jordaan 1949; Kuo et al. (1982) for Western Australian species) and is held dormant until the organ or the parent plant dies, usually from fire. Seed may be held for 3-9 years or more, (e.g. in Callistemon, Leucadendron, Protea - Kruger, unpublished) but shows no prolonged dormancy after release and imbibition (e.g. Banksia ornata, B. ericifolia, and B. serratifolia, Specht et al. 1958; Siddiqi et al. 1976; Leucadendron spp., Williams 1972) though some species require stratification to overcome seed-coat dormancy (Deall and Brown 1981). An extreme case is found in Leucadendron platyspermum, where seed germinates in the cones before falling to the pyrogenic seed bed (Williams 1972). Because seed is short-lived after release the mortality schedule and age-specific fecundities of the parent plants are critical in determining post-fire daughter populations: Bond (1980), for example, has shown how post-fire regeneration in stands of senescent Protea punctata was markedly depressed relative to that in a burnt mature stand.

Large, rich seeds are not confined to serotinous species (e.g. Williams 1972 for Leucadendron) and is a trait adaptive for seedling establishment in a competitive post-fire environment (Grundon 1972; Kuo et al. 1982) but serotiny as such would be one trait adaptive for large-seeded species in the face of predation in an otherwise resource-poor environment (Gill 1981b). Serotiny is however also a means for precisely cued release of seed onto the post-fire seed bed as well as a predator satiation mechanism (Gill 1975, 1981b).

The dynamics of canopy-stored seed have been little studied in mediterranean-climate shrubs but Gill (1981b) has indicated for Banksia ornata that annual loss is less than 2%, and that the balance of seed remain viable. By contrast, soil-stored seed of Ceanothus and Arctostaphylos in chaparral have an annual attrition rate as high as 90%, though the soil seed-bank was nevertheless 100 times or more greater than the post-fire seeding population (Keeley 1977b). Thus, serotiny is associated with a low annual production of high-quality seed with low turnover and well-cued germination, whereas soil seed-banks require high fecundities to balance annual attrition, but with the advantage that the species can persist in situ as seed for some time after death of the parent plants as well as maintaining a soil reserve after post-fire germination.

Seed banks are characteristic of successional vegetation and hence of areas subject to widespread disturbance generally (Grime 1979; Cook 1980). In mediterranean-type ecosystems they are perhaps unique in that the composition of the seed bank closely reflects the species assemblage of the current vegetation; in other situations the similarity between species as dormant seeds and those of the plant cover steadily diverges with successional change (Oosting and Humphreys 1940; Harper 1977; Cook 1980).

The cues for germination among the species of these seed banks are complex and varied, including release from allelopathic inhibition in certain California herbs and shrubs (McPherson and Muller 1969; Christensen and Muller 1975), the pyrogenic heath pulse as a predictive trigger for stimulation of dormant soil-stored seed (Quick 1935; other references in McPherson and Muller 1969), change in soil temperature regime (McPherson and Muller 1969), contact with charcoal for some chaparral herbs (Wicklow 1977), or simply release of serotinous seed on the death of the parent (see above). Persistent seed-banks are a necessary correlate of such opportunistic adaptations (Grime 1979).

The relative abundance of seed of different species in a stand will vary with the length of the fire-free interval because of differences in life span and fecundity schedules between species and differences in decay rates of seed pools. Thus the composition of the seed stock available for germination after fire will vary (Specht et al. 1958). Because of complex germination controls the responses to any given fire will vary with intensity (Gill and Groves 1981) and season of the burn (W J Bond, personal communication 1981). Therefore the interaction between the composition of the seed bank and fire regime should cause considerable variation in the relative abundance of plants. Our predictive knowledge of the effect of fire regime on mediterranean vegetation will depend greatly on understanding the dynamics of seed banks and controls in germination.

23.5.3 MORTALITY AND SENESCENCE IN SEEDING SHRUBS

Among seeders, populations each consist by and large of a single cohort dating back to the last fire. Marked differences are found among these shrubs with respect to life spans and patterns of mortality, or survivorship curves.

Among seeding shrubs the first growing season after fire is usually crucial. Specht (1981) has shown direct correlations between rate of establishment (seedling survival) and the duration of surface soil moisture exceeding 40% content, and this bears out observations by Horton and Kraebel (1955) and many others. Once plants have survived the first season, different patterns of mortality are found in the absence of fire (Fig. 3). The survivorship curve for Banksia ornata suggests typical

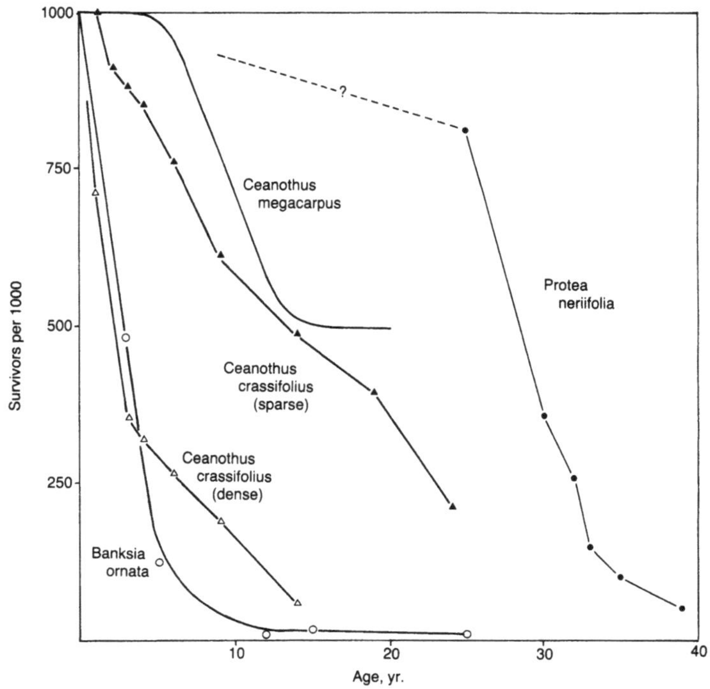

Fig. 3. Survivorship curves for a selection of mediterranean-type shrubs. Based on data in Specht et al. (1958), Horton and Kraebel (1955), Schlesinger and Gill (1978), and unpublished records of the South African Directorate of Forestry.

density-dependent mortality, with populations falling from about 74,000 plants ha^{-1} at establishment to 9000 and about 1200 at five and 15 years after fire respectively (Specht et al. 1958). In Ceanothus megacarpus rapid, apparently density-dependent thinning occurred in the first few years after fire but both dense and sparse populations then apparently remained stable at about 5000 plants ha^{-1} (Horton and Kraebel 1955). Schlesinger and Gill (1978) report a different pattern for Ceanothus megacarpus, with mortality greatest about five to 12 years after fire and populations stabilizing at about 16,000 to 30,000 plants ha^{-1}. Overall, populations thin at a relatively slow rate, so that individual shrubs are smaller at a given density than would be predicted from the two-thirds thinning rule (e.g. Schlesinger and Gill 1978).

Few data are available for fynbos but the curve for Protea neriifolia apparently represents dominant seeding shrubs. Populations have little mortality until late in life (>25 years), and then decline rapidly. This is classical population senescence in Leopold's (1961) terms.

Thus, density-dependent mortality (self-thinning) in the maturing and mature populations appears to be exceptionally low in shrub dominants of both chaparral and fynbos. Chaparral dominants maintain vigour. Even small shrubs (small diameters, low biomass) maintain height growth and their position in the canopy, despite dense crowding (Keeley and Keeley 1977; Schlesinger and Gill 1978). Growth and reproductive output rates are maintained (see Keeley and Keeley 1977, 23-year-old versus 90-year-old populations of Arctostaphylos spp.). But marked thinning from abundant post-fire populations and relatively limited life-spans (40-60 years) are evident for certain chaparral dominants, especially Ceanothus species (C. crassifolius, C. oliganthus, Hanes and Jones 1967; Hanes 1977; Keeley (1975) suggests the members of subgenus Euceanothus). Keeley (1977b) contrasted this pattern with that in Arctostaphylos glauca (low mortality, long life, few seed, few seedlings) and suggested that they represented alternative life-history strategies.

In relatively short life-spans of fynbos and Australian heathland species such as Protea neriifolia (Fig. 3) and Banksia ornata are notable. In fynbos, long-lived species like Leucadendron argenteum are those that have evolved protective mechanisms such as thick bark to ensure some survival among any cohort in most fires. They tend to occur in areas with low fuel accumulation rates because of aridity (Rourke 1972) or perhaps high fire frequency; life-history strategies are apparently entirely different.

The longest-lived seed-regenerating shrubs occur in chaparral. Davis (1973) and Keeley (1975) have described stem morphological features in long-lived seeding chaparral shrubs in the genera Arctostaphylos ("bark-striping") and Ceanothus ("stem channeling") which they suggest are adaptations for long life-spans, in that the ratio between photosynthetic tissue and supportive tissue (hence maintenance costs of carbon) is reduced (Keeley 1975). These features are absent from fynbos and Australian heath shrubs.

Reproductive output in terms of biomass in chaparral shrubs amounts to about 8-11% of total annual shoot increment (stem plus leaves plus reproductive tissues) (Mooney et al. 1977) or a maximum of about 25% in a "good" year (Keeley and Keeley 1977). Data for Banksia ornata in Specht et al (1958) indicate that reproductive tissues accounted for about 40-60% of net annual plant increment between the ages of nine and 28 years. Even if large allowance is made for leaf fall and other loss of increment it is clear that current allocation of biomass to reproduction is much greater in the Banksia than in the chaparral shrubs. This would seem to hold for fynbos seeders as well (cf. Protea repens in Kruger 1979c). Seed of these shrubs are particularly rich in nitrogen and phosphorus (see above) and the contrast in terms of mineral nutrient budgets could be more pronounced than for carbon. Thus, short lives in fynbos and Australian heathland shrubs may be owing to different resource allocation strategies that limit allocation to growth and support, relative to chaparral shrubs.

Senescence among serotinous shrubs and trees implies a high likelihood of local extinction with long intervals between fires or at least partial displacement by species with dormant seeds and by long-lived vegetative shrubs (Specht et al. 1958; Bond 1980).

All of the sclerophyllous shrublands include fast-growing perennial seeding shrubs and sub-shrubs that have intermediate life spans, which exploit the post-fire environment (Table 5). These apparently have low competitive ability and are eliminated as resources are appropriated by the longer-lived species. However, some such as Aspalathus in fynbos have very dense populations, low self-thinning, and marked senescence. It is not clear that the population dynamics in such cases are driven by competition though depletion of limiting resources such as phosphorus may limit life-spans (Chapters 12 and 13, this volume).

23.5.4 PRIMARY JUVENILE PERIODS AND FECUNDITY SCHEDULES

Information on the ages at which germinative perennials first produce seed is essential to understanding their responses to fire regime. No

Table 5. Examples of seeding shrubs with different life-spans in three mediterranean-type ecosystems. Primary juvenile periods shown in brackets[a].

Short-lived shrubs and sub-shrubs (4-10yr)	*Lotus scoparius* (Fabaceae) (2-3 yr) *Lupinus longifolius* (Fabaceae) (2-3 yr) *Nama parryi* (Asteraceae) (2-3 yr) *Senecio douglasiana* (Asteraceae) (2-3 yr)	*Aspalathus* spp. (Fabaceae) (2-4 yr) *Othonna quinquedentata*[b] (Asteraceae) (2-3 yr) *Ursinia crithmoides* (Asteraceae) (2-3 yr)	*Acacia suaveolens* (Fabaceae) *Boronia falcifolia* (Rutaceae) *Epacris pulchella* (Epacridaceae) *Persoonia shirleyae* (Proteaceae)
Intermediate shrubs (10-50 yr)	*Artemisia californica* (Asteraceae) (3 yr) *Ceanothus oliganthus* (Rhamnaceae) (< 5 yr?) *C. tomentosus* (Rhamnaceae) (< 5 yr?) *Eriogonum fasciculatum* (Polygonaceae) (< 5 yr) *Salvia mellifera* (Lamiaceae) (3 yr)	*Erica* spp. (2-4 yr) *Protea neriifolia*[b] (4-6 yr) *Protea repens* (3-4 yr) *Leucadendron floridum*, *L. macowani* (Proteaceae) (3 yr) *Leucadendron comosum* (4 yr) *Leucadendron muirii* (6 yr)	*Acrotriche serrulata* (Epacridaceae) *Adenanthos terminalis* (Proteaceae) *Banksia marginata* (Proteaceae) (c. 5 yr) *Baeckea ericaea* (Myrtaceae) *Acacia longifolia* (Fabaceae) *Gompholobium minus* (Fabaceae)
Long-lived trees and shrubs (> 50 yr)	*Arctostaphylos glauca*[b] (Ericaceae) (< 10 yr) *Ceanothus crassifolius* (Rhamnaceae) (< 10 yr)	*Protea laurifolia* (5-7 yr) *Leucadendron argenteum* (7-9 yr) *Leucospermum conocarpdendron* (Proteaceae) (7-9 yr) *L. praemorsum* (9-11 yr)	*Banksia intergrifolia* *Leptospermum laevigatum* (Myrtaceae)

a From Adamson (1935); Horton and Kraebel (1955) ; Specht et al. (1958, 1977); Van der Merwe (1969); Keeley (1975, 1977a); Hazard and Parsons (1977); Keeley and Zedler (1978); Moll et al. (1980).

b Species showing marked senescence.

studies have focused on this parameter in plants of mediterranean-type ecosystems but several report observations. The general pattern is for short-lived and intermediate shrubs to mature within 2-4 years (Horton and Kraebel 1955 for Artemisia californica; Adamson 1935 and van der Merwe 1969 for fynbos species). The relationship between age of first seeding and longevity is therefore approximately as found by Harper and White (1974), but the correlation is poor and mediterranean-type shrubs are usually precocious. Flowering and fruiting is advanced even among longer-lived dominants (Table 5). This is particularly so for fynbos shrubs. Rourke (1980), for example, reports several Protea species in cultivation as flowering within two years after germination; the median juvenile period in 42 species was four years. Observations suggest that plants in the field are not much delayed in this respect (Table 5).

Few observations are available for chaparral dominants, but they appear slower to mature, requiring 8-10 years or perhaps more (Table 5).

No data appear to be available to show patterns of age-specific seed output (fecundity schedules). Biomass accumulation graphs for Banksia ornata in Specht et al. (1958) show accelerating fruit production per plant from about nine years of age to between 20 and 25 years, with some indication of declining annual production as 25 years age is approached. Certainly these authors imply a reduced annual reproductive output between 28 and 50 years. In fynbos, it is not clear that reproductive output per plant declines as senescence approaches (Bond 1980).

Seeding chaparral dominants maintain reproductive vigour with advancing age and Keeley and Keeley (1977) found greater fruit output in a 90-year-old sample of Arctostaphylos glauca than in an adjacent 23-year-old stand.

In chaparral, sprouting shrubs also maintain their vigour and reproductive performance as they age (e.g. Arctostaphylos glandulosa in Keeley and Keeley (1977)). In fynbos, tall sprouters such as Protea nitida appear to maintain reproductive effort (Kruger, personal observation) but many shoots of the lower sprouter such as Protea cynaroides and P. speciosa appear to lose vigour within a decade or so after fire, and flowering per plant declines (Rourke 1980); the shoot systems are rejuvenated by defoliation.

23.6 Conclusion

The regions discussed here share similar regimes in that fires are of intermediate frequency and intensity; fire does not occur just about once a year as in grassland, nor at long intervals (> 100 years) as do conflagrations in coniferous or ash-type eucalyptus forests; simultaneously, fires are sufficiently intense so as to open the site effectively (more so than in grasslands, wooded savannas or eucalyptus woodlands), yet do not preclude vegetative survival of a proportion of plant individuals. These regimes may have the special characteristic of allowing more diverse modes of regeneration (various seeding and sprouting strategies) than do the regimes of grassland or forest. Comparative studies would highlight aspects of the dynamics of mediterranean-type ecosystems especially worth studying.

Although fire regimes in mediterranean-type ecosystems are predictable they are stochastic. All components of fire behaviour vary within one fire on any site, and so do intensity, frequency and seasonality in time. Interactions between the fire event and current and post-fire conditions compound this variability.

nutrient distribution, of charcoal distribution (see Wicklow 1977), post-fire erosion, and similar factors (Chapter 22, this volume) induces microsite variability and enhances pattern diversity (internal beta diversity). These effects with the temporal partitioning of resources over time that is allowed by pyric succession must play a role in maintaining plant species diversity. This role has not been studied explicitly but since managers are and will be affecting the fire factors that are relevant here, appropriate studies are urgently necessary. The Australian experience, where regular, frequent, low-intensity controlled burns markedly affected the richness and character of understorey species in eucalypt forests, to the point of modifying ecosystem nutrient cycles and forest health, is indicative of likely problems (see Gill et al. 1981).

The longer fire cycle in chaparral, with high fire intensity, would very likely account for the marked contrast between chaparral pyric succession and that in fynbos and Australian heath, while maintaining the full diversity of herbaceous species. Presumably, chaparral cannot burn so frequently that tufted grasses such as Stipa and other perennial herbs would become prominent, as grasses do in Quercus coccifera garrigue (Trabaud 1973). A regime of infrequent, intense fires in fynbos and Australian heathland would favour seeders if the fire period does not exceed the life-span of serotinous seeders, but the small annual and biennial flora would continue to be limited by nutrients. In any event, under the "normal" regime sprouting herbs and shrubs are abundant and the competitive post-fire environment not suited to annuals and short-lived perennials.

The research interests of ecologists and resource managers are largely coincident now. Alternative fire-management options must be selected on the basis of the predicted effects of different regimes on community structure and diversity. For this the effects of variations in regime must be simulated on the basis of our knowledge of the dispersal ecology, germination requirements, and demography of constituent species. Even the question of finding the "natural" regime is a matter for deduction from demography (cf. Keeley 1977a), in the absence of proxy records.

A basis for general principles must be found in geographic patterns in the structure and response to fire of vegetation. These patterns could be made tractable by approaches such as that of Specht and Morgan (1981), if linked with energy and water balance controls on vegetation dynamics (Chapters 2, 7 and 8, this volume).

Keeley (1977b), in his study of responses of chaparral shrubs to fire, proposed that each represents a different strategy, or demographic model. Gill (1981a) has outlined three different approaches to modelling vegetation-fire dynamics - species' response classification, species replacement models, and demography, each successively more complex. Presumably, convergence in plant life-history strategies will render manageable the problem of generalizing from our knowledge of a few species. Nevertheless there is considerable scope for work on the demography of representative plant species since present knowledge is simply inadequate for necessary generalizations. Studies to determine patterns in the distribution of strategy-types or fire life-form classes (Gill 1981a) are necessary to reduce apparent vegetation complexities to intelligible patterns.

References

ADAMSON RS (1935) The plant communities of Table Mountain. III. A six years' study of regeneration after burning. Journal of Ecology 23: 52-55

BAKER HA, OLIVER EGH (1967) Ericas in Southern Africa. Cape Town, Purnell, 177 pp

BANDS DP (1977) Prescribed burning in Cape fynbos catchments. In: MOONEY HA, CONRAD CE (eds) Proceedings of the symposium on the environmental consequences of fire and fuel management in mediterranean ecosystems. August 1-5, 1977, Palo Alto, California. USDA Forest Service General Technical Report WO-3, pp 245-256.

BEADLE NCW (1940) Soil temperatures during forest fires and their effect on the survival of vegetation. Journal of Ecology 28: 180-192

BELL DT, KOCH JM (1980) Post-fire succession in the northern Jarrah forest of Western Australia. Australian Journal of Ecology 5: 9-14

BERG RY (1981) The role of ants in seed dispersal in Australian lowland heathland. In: SPECHT RL (ed) Ecosystems of the world, vol 9B. Heathlands and related shrublands. Analytic studies. Elsevier, Amsterdam, pp 51-59

BOND WJ (1980) Fire and senescent fynbos in the Swartberg, southern Cape. South African Forestry Journal 114: 68-71

CAMPBELL BM, VAN DER MEULEN F (1980) Patterns of plant species diversity in fynbos vegetation, South Africa. Vegetatio 43: 43-47

CHRISTENSEN NL, MULLER CH (1975) Effects of fire on factors controlling plant growth in Adenostoma chaparral. Ecological Monographs 45: 29-55

COALDRAKE JE (1951) The climate, geology, soils and plant ecology of portion of the County of Buckingham (Ninety-Mile Plain), South Australia. CSIRO, Australia. Bulletin No 266, 81 pp

COALDRAKE JE (1961) Ecosystem of the coastal lowlands, Southern Queensland. CSIRO, Australia. Bulletin No 283, 138 pp

CODY ML, MOONEY HA (1978) Convergence versus nonconvergence in mediteranean-climate ecosystems. Annual Review of Ecology and Systematics 9: 265-322

CONNELL JH, SLATYER RO (1977) Mechanisms of succession in natural communities and their role in community stability and organization. American Naturalist 111: 1119-1144

COOK R (1980) The biology of seeds in the soil. In: SOLBRIG OT (ed) Demography and evolution in plant populations. Blackwell, Oxford, pp 107-129

COOPER WS (1922) The broad-sclerophyll vegetation of California - an ecological study of the chaparral and its related communities. Carnegie Institute, Washington, 122 p

DAVIS CB (1973) "Bark striping" in Arctostaphylos (Ericaceae). Madrono 22: 145-149

DEALL GB, BROWN NAC (1981) Seed germination in Protea magnifica Link. South African Journal of Science 77: 175-176

DEL MORAL R, WILLIS RJ, ASHTON DH (1978) Suppression of coastal heath vegetation by Eucalyptus baxteri. Australian Journal of Botany 26: 203-19

DIXON KW, PATE JS (1978) Phenology, morphology and reproductive biology of the tuberous sundew, Drosera erythrorhiza Lindl. Australian Journal of Botany 26: 441-454

DOWNING BH, ROBINSON ER, TROLLOPE WSW, MORRIS JW (1978) The influence of macchia eradication techniques in botanical composition of grasses in the Dohne sourveld of the Amatole Mountains. Proceedings of the Grassland Society of Southern Africa 13: 111-115

FLOYD AC (1976) Effects of burning on regeneration from seeds in wet sclerophyll forest. Australian Forestry 39: 210-220

GILL AM (1975) Fire and the Australian flora: a review. Australian Forestry 38: 4-25

GILL AM (1981a) Coping with fire. In: PATE JS, MCCOMB AJ (eds) The biology of Australian plants. University of Western Australia Press, Nedlands, Western Australia, pp 65-87

GILL AM (1981b) Adaptive responses of Australian vascular plant species to fires. In: GILL AM, GROVES RH, NOBLE IR (eds) Fire and the Australian biota. Australian Academy of Science, Canberra, pp 243-272

GILL AM, GROVES RH (1981) Fire regimes in heathlands and their plant-ecological effects. In: SPECHT RL (ed) Ecosystems of the world, vol 9B. Heathlands and related shrublands. Analytical studies. Elsevier, Amsterdam

GILL AM, GROVES RH, NOBLE IR (eds) (1981) Fire and the Australian biota. Australian Academy of Science, Canberra, pp 582

GRIME JP (1979) Plant strategies and vegetation processes. Wiley, New York, 222 pp

GRUNDON NJ (1972) Mineral nutrition of some Queensland heath plants. Journal of Ecology 60: 171-0181

GULMON SL (1979) Competition and coexistence: three annual grass species. American Midland Naturalist 101: 403-416

HALL AV (1959) Observation on the distribution and ecology of Orchidaceae in the Muizenberg mountains, Cape Peninsula. Journal of South African Botany 25: 265-278

HANES TL (1971) Succession after fire in the chaparral of southern California. Ecological Monographs 41: 27-52

HANES TL (1977) Chaparral. In: BARBOUR MG, MAJOR J (eds) Terrestrial vegetation of California. Wiley, New York, pp 417-469

HANES TL, JONES HW (1967) Postfire chaparral succession in southern California. Ecology 48: 259-264

HARPER JL (1977) Population biology of plants. Academic Press, London, 892 pp

HARPER JL, WHITE J (1974) The demography of plants. Annual Review of Ecology and Systematics 5: 419-463

HAYNES RA (1976) Aspects of the ecology and life-history of Protea arborea Houtt. Unpublished report, Department of Zoology, University of Rhodesia, 68 pp

HAZARD J, PARSONS RF (1977) Size-class analysis of coastal scrub and woodland, Western Port, Southern Australia. Australian Journal of Ecology 2: 187-197

HORNE IP (1981) The frequency of veld fires in the Groot Swartberg mountain catchment area, Cape Province. South African Forestry Journal 118: 56-60

HORTON JS, KRAEBEL CJ (1955) Development of vegetation after fire in the chamise chaparral of Southern California. Ecology 36: 244-262

JACKSON WD (1968) Fire, air, water and earth - an elemental ecology of Tasmania. Proceedings of the Ecological Society of Australia 3: 9-16

JORDAAN PG (1949) Aantekeninge oor die voortplanting en brandperiodes van Protea mellifera Thunb. Journal of South African Botany 15: 121-125

JORDAAN PG (1965) Die invloed van 'n winterbrand op die voortplanting van vier soorte van die Proteaceae. Tydskrif vir Natuurwetenskap 5, 27-31.

JORDAAN PG (1981) The influence of a fire in April on the reproduction of three species of Proteaceae. Journal of South African Botany 48: 1-4

KEELEY JE (1975) Longevity of nonsprouting Ceanothus. American Midland Naturalist 93: 504-507

KEELEY JE (1977a) Fire-dependent reproductive strategies in Arctostaphylos and Ceanothus. In: MOONEY HA, CONRAD CE (technical coordinators). Proceedings of the symposium on the environmental consequences of fire and fuel management in mediterranean ecosystems. USDA Forest Service General Technical Report WO-3, pp 391-396

KEELEY JE (1977b) Seed production, seed populations in the soil and seedling production after fire for two congeneric parts of sprouting and non-sprouting chaparral shrubs. Ecology 58: 820-829

KEELEY S (1977c) The relationship of precipitation to post-fire succession in southern California chaparral. In: MOONEY HA, CONRAD, CE (technical coordinators). Symposium on the environmental consequences of fire and fuel management in mediterranean ecosystems. USDA Forest Service General Technical Report WO-3, pp 387-390

KEELEY JE, KEELEY S (1977) Energy allocation patterns of sprouting and a non-sprouting species of Arctostaphylos in the California chaparral. American Midland Naturalist 98: 1-10

KEELEY JE, KEELEY S (1981) Post-fire regeneration of southern California chaparral. American Journal of Botany 68: 524-530

KEELEY JE, ZEDLER PH (1978) Reproduction of chaparral scrubs after fire: a comparison of sprouting and seeding strategies. American Midland Naturalist 99: 142-161

KEELEY S, KEELEY JE, HUTCHINSON SM, JOHNSON AW (1981) Postifre succession of the herbaceous flora in southern California. Ecology 62: 1608-1621

KRUGER FJ (1977a) A preliminary account of aerial plant biomass in fynbos communities of the mediteranean-type climate zone of the Cape Province. Bothalia 12: 301-307

KRUGER FJ (1977b) Ecology of Cape fynbos in relation to fire. In: MOONEY HA, CONRAD CE (technical coordinators) Proceedings of the symposium on the environmental consequences of fire and fuel management in mediterranean ecosystems. USDA Forest Services General Technical Report WO-3, pp 230-244

KRUGER FJ (1979a) Fire. In: DAY J, SIEGFRIED WR, LOUW GN, JARMAN ML (eds) Fynbos ecology: a preliminary synthesis. South African National Scientific Programmes Report No 40, CSIR, Pretoria, pp 43-57

KRUGER FJ (1979b) South African heathlands. In: SPECHT RL (ed) Ecosystems of the world, vol 9A. Heathlands and related shrublands. Descriptive studies. Elsevier, Amsterdam, pp 29-80

KRUGER FJ (1979c) Plant ecology. In: DAY J, SIEGFRIED NR, JARMAN ML (eds) Fynbos ecology: a preliminary synthesis. South African National Scientific Programmes Report No 40, CSIR, Pretoria, pp 88-126

KUO J, HOCKING PJ, PATE JS (1982) Nutrient reserves in seeds of selected proteaceous species from South-Western Australia. Australian Journal of Botany 30: 231-249

LEOPOLD AC (1961) Senescence in plant development. Science 134: 1727-1732

MCCUTCHEON MH (1977) Climatic features as a fire determinant. In: MOONEY HA, CONRAD CE (technical coordinators). Proceedings of the symposium on the environmental consequences of fire and fuel management in mediterranean ecosystems. USDA Forest Service General Technical Report WO-3, pp 1-11

MCPHERSON JK, MULLER CH (1969) Allelopathic effects of Adenostoma fasciculatum, "chamise", in the California chaparral. Ecological Monographs 39: 177-198

MOONEY HA, CONRAD CE (technical coordinators) (1977) Proceedings of the symposium on the environmental consequences of fire and fuel management in mediterranean ecosystems. USDA Forest Service General Technical Report WO-3, 498 pp

MOONEY HA, PARSONS DJ (1973) Structure and function of the California chaparral - an example from San Dimas. In: DI CASTRI F, MOONEY HA (eds). Mediterranean type ecosystems: origin and structure. Springer-Verlag, Berlin, pp 83-112

MOONEY HA, KUMMEROW J, JOHNSON AW, PARSONS DJ, KEELEY S, HOFFMAN A, HAYS RI, GILLBERTO J, CHU C (1977) The producers - their resources and adaptive responses. In: MOONEY HA (ed) Convergent evolution in Chile and California: mediterranean climate ecosystems. Dowden, Hutchinson and Ross, Stroudsberg, Pennsylvania, pp 85-143

NAVEH Z (1974) Effects of fire in the Mediterranean region. In: KOZLOWSKI TT, AHLGREN CE (eds) Fire and ecosystems. Academic Press, New York, pp. 401-434

NOBLE IR, SLATYER RO (1981) The use of vital attributes to predict successional changes in plant communities subject to recurrent disturbances. Vegetatio 43: 5-21

OOSTING HJ, HUMPHREYS ME (1940) Buried viable seed in a successional series of old field and forest soils. Bulletin of the Torrey Botanical Club 67: 253-273

PATRIC JH, HANES TL (1964) Chaparral succession in a San Gabriel mountain area of California. Ecology 45: 353-360

PHILPOT CW (1974) New fire control strategy developed for chaparral. Fire Management 35: 3-7

PHILPOT CW (1977) Vegetative features as determinants of fire frequency and intensity. In: MOONEY HA, CONRAD CE (technical coordinators). Proceedings of the symposium on the environmental consequences of fire and fuel management in mediterranean ecosystems. USDA Forest Service General Technical Report WO-3. pp 12-16

POSAMENTIER HG, CLARK SS, HAIN DC, RECHER HF (1981) Succession following wildfire in coastal heathland (Nadgee Nature Reserve, New South Wales). Australian Journal of Ecology 6: 165-175

QUICK CR (1935) Notes on the germination of Ceanothus seeds. Madrono 3: 23-30

ROTHERMEL RC, PHILPOT CW (1973) Predicting changes in chaparral flammability. Journal of Forestry 71: 640-643

ROURKE JP (1972) Taxonomic studies on Leucospermum R.Br. Journal of South African Botany Supplementary Volume 8: 1-194

ROURKE JP (1980) The Proteas of Southern Africa. Purnell, Cape Town, 236 pp

RUNDEL PW (1982) Fire as an ecological factor. In: LANGE OL, NOBEL PS, OSMOND CB, ZIEGLER H (eds) Encyclopedia of plant physiology: new series. Physiological plant ecology - interaction of plants with the physical environment. Springer-Verlag, Berlin. In press

RUSSELL RP, PARSONS RF (1978) The effect of time since fire on heath floristics at Wilson's Promontory, southern Australia. Australian Journal of Botany 26: 53-61

SAMPSON AW (1944) Plant succession on burned chaparral lands in northern California. University of California Bulletin 685. Berkeley, California. 144 pp

SCHLESINGER WH, GILL DS (1978) Demographic studies of the chaparral shrub, Ceanothus megacarpus, in the Santa Ynez Mountains, California. Ecology 59: 1256-1263

SIDDIQI MY, CAROLIN RC, MYERSCOUGH PJ (1976) Studies in the ecology of coastal heath in New South Wales. III. Regrowth of vegetation after fire. Proceedings of the Linnaean Society of New South Wales 101: 53-63

SPECHT RL (1969) A comparison of the sclerophyllous vegetation characteristic of mediterranean-type climates in France, California, and southern Australia. II. Dry matter, energy, and nutrient accumulation. Australian Journal of Botany 17: 293-308

SPECHT RL (1972) The vegetation of South Australia. Government Printer, Adelaide, 328 pp

SPECHT RL (1979) The sclerophyllous (heath) vegetation of Australia: the eastern and central states. In SPECHT RL (ed) Ecosystems of the world, vol 9A. Heathlands and related shrublands. Descriptive studies. Elsevier, Amsterdam, pp 125-210

SPECHT RL (1981) Responses to fires of heathlands and related shrublands. In: GILL AM, GROVES RH, NOBLE IR (eds) Fire and the Australian biota. Australian Academy of Sciences, Canberra, pp 395-414

SPECHT RL, MORGAN DG (1981) The balance between the foliage projective covers of overstorey and understorey strata in Australian vegetation. Australian Journal of Ecology 6: 193-202

SPECHT RL, RAYSON P, JACKMAN ME (1958) Dark Island heath (Ninety-Mile Plain, South Australia). VI. Pyric succession: changes in composition, coverage, dry weight, and mineral nutrient status. Australian Journal of Botany 6: 59-88

STONE EC (1951) The stimulative effect of fire on the flowering of the golden Brodiaea (Brodiaea ixioides Wats. var. Lugens Jepson Ecology 32: 534-537

TRABAUD L (1973) Experimental studies on the effects of prescribed burning on a Quercus coccifera garrigue: early results. Proceedings of the Annual Tall Timbers Fire Ecology Conference 13: 97-129

VAN DER MERWE P (1969) Die flora van Swartboskloof, Stellenbosch en die herstel van die soorte na 'n brand. Annals of the University of Stellenbosch Ser. A(14), 42: 691-736

VAN STADEN J, BROWN NAC (1977) Studies on the germination of South African Proteaceae - a review. Seed Science Technology 5: 633-643

VAN WILGEN BW (1981) Some effects of fire frequency on fynbos plant community composition and structure at Jonkershoek, Stellenbosch. South African Forestry Journal 118: 42-55

VAN WILGEN BW (1982) Some effects of post-fire age on the above-ground plant biomass of fynbos (macchia) vegetation in South Africa. Journal of Ecology 70: 217-225

VAN WILGEN BW, KRUGER FJ (1981) Recovery of fynbos plant communities at Zachariashoek, Paarl, six years after an experimental burn. Journal of South African Botany. In press

VOGL RJ, SCHORR PK (1972) Fire and manzanita chaparral in the San Jacinto mountains, California. Ecology 53: 1179-1188

WELLS PV (1968) The relation between mode of reproduction and extent of speciation in woody genera of the California chaparral. Evolution 23: 264-267

WESTMAN WE (1979) A potential role of coastal sage scrub understoreys in the recovery of chaparral after fire. Madrono 26: 64-68

WESTMAN WE (1981) Diversity relations and succession in Californian coastal sage scrub. Ecology 62: 170-184

WHELAN RJ, MAIN AR (1979) Insect grazing and post-fire plant succession in South-West Australian woodland. Australian Journal of Ecology 4: 387-398

WICHT CL (1948) A statistically designed experiment to test the effects of burning on a sclerophyll scrub community. 1. Preliminary account. Transactions of the Royal Society of South Africa 31, 479-501

WICKLOW DT (1977) Germination response in Emmenanthe penduliflora (Hydrophyllaceae). Ecology 58: 201-205

WILLIAMS IJM (1972) A revision of the genus Leucadendron (Proteaceae). Contributions from the Bolus Herbarium 3: 1-425

24. Mammal Species Diversity in Australian Heathlands: the Importance of Pyric Succession and Habitat Diversity

B. J. Fox

24.1 Introduction

Many of the ecological studies of Australian small mammals have related to the effects of fire, predominantly forest fire (Cowley et al. 1969; Christensen and Kimber 1975; Newsome et al. 1975; Recher et al. 1975, 1978; Schmidt and Mason 1975; Fox and McKay 1981). Cockburn (1978) and Fox (1980; 1982b) are the only studies dealing specifically with heathland regeneration after fire. However, the areas studied by both Newsome et al. (1975) and Recher et al. (1975) contained heath components, as did that of Catling and Newsome (1981) who recently explored the proposition that Australian vertebrate fauna has adapted to an environment subject to frequent burning over evolutionary time, as has been suggested for the flora by Mount (1964) and Gill (1975). Newsome and Catling (1979) produced a habitat complexity score, made up of components important to heathland small mammal species, and used it to analyse faunal habitat relations in lowland and highland heaths, and changes in small mammal faunas in pyric succession.

I propose to examine the hypothesis that fire is an important factor in the maintenance of heathland small mammal community structure and diversity. I use data from a five-year study on heathland regeneration after a major fire to interpret (a) the mechanisms involved in the mammalian secondary succession following fire, (b) the role this plays in facilitating coexistence and (c) the maintenance of small mammal diversity.

24.2 The study area

The study area was located in Myall Lakes National Park (32°28'S, 152°24'E), in a region with a sub-mediterranean climate. Monthly mean maximum temperatures reach 28°C and minima of coolest months, 6°C. Although there is some yearly variation, maximum precipitation usually occurs between May and June (early winter), with lowest rainfall between September and December (early summer). The annual average is 1390 mm.

The land east of Myall Lake is almost entirely sand. Thom (1965) recognizes two systems, a Pleistocene Inner Barrier system of subdued relief (a few metres), and a Holocene Outer Barrier system that rises to more than 100 m above sea level in places.

The main study area contained the full range of topography encountered in the Inner Barrier system. It is characterized by a system of parallel dune ridges and intervening swales with occasional parabolic dunes or relics of a former barrier lagoon which can hold present-day swamps. Local relief is only about 4 m and the dunes are never far above the water table. The sand is strongly podsolized and very low in nutrients.

Depending on topography the dominant vegetation alternates between wet and dry-heath and is analogous to the ground-water heaths and deep sand heaths described by Specht and Rayson (1957) and Siddiqi et al. (1976). High dunes clothed in open forest begin within 100 m of the edge of the study area. This forest does extend onto the study area, but where it does it has a dense understorey of swamp species due to the shallow water table. R C Carolin and P J Myerscough (personal communication) use the term swamp forest for this unit. Osborne and Robertson (1939) and Carolin (1970) describe a spectrum of vegetation types for the Pleistocene inner barrier system in the Myall Lakes area with swamp vegetation at one end and isolated patches of eucalypt woodland (where emergent mallee-form eucalypts are present) at the other.

24.2.1 VEGETATION TYPES

I recognized eight vegetation units in the area, based on the structural formations of Specht (1970) and attributes such as the dominant plant species and a subjective estimate of soil moisture regime (Fox and Fox 1981). A short description of each is given below. Numbers of vegetation units are as used elsewhere in the text.

1. Swamp: standing water for much of the year, with many sedges and fringed by *Gymnoschoenus sphaerocephalus*.

2. Short wet-heath (1 m high): minor flooding after heavy rains, dominated by *Persoonia lanceolata* and *Banksia aspleniifolia* with *Xanthorrhoea resinosa*. Sedges abundant in the herb layer.

2a. Hakea thicket: vegetation as for unit 2 but with the addition of patches of *Leptospermum attenuatum*, on slightly raised ground, surrounded by thickets of *Hakea teretifolia*.

3. Tall wet-heath (1 m high): bands of *Leptospermum liversidgei* which occur around swamp margins.

4. Short dry-heath (to 1.5 m high): dominated by small *Banksia aemulia* shrubs as well as *B. aspleniifolia* (open dry-heath in patches).

5. Tall dry-heath or closed scrub (1.5 to 4 m tall): dominated by mature *B. aemulia* and *Xanthorrhoea australis*, on ridges usually more than 2 m above the water table.

6. Tree-heath: on ridge tops, dense tall dry-heath understorey, with emergent mallee-form *Eucalyptus pilularis* (4-10 m high).

7. Swamp-forest: Angophora costata, E. pilularis, E. gummifera and E. robusta (to 20 m) with a dense understorey of Pteridium esculentum but including swamp species such as Blechnum cartilagineum.

24.2.2 SMALL-MAMMAL FAUNA

Ten species of small mammal were recorded from heathland in Myall Lakes National Park. Seven species were commonly caught, each comprising at least 5% of captures. These included two small dasyurid marsupials, Antechinus stuartii and Sminthopsis murina, and five murid rodents, native species Rattus fuscipes, R. lutreolus, Pseudomys novaehollandiae, and P. gracilicaudatus, and the introduced Mus musculus. Single captures were recorded for the murids Hydromys chrysogaster (native) and Rattus rattus (introduced), and the traps used were too small to catch adult bandicoots (Isoodon macrourus, a peramelid marsupial) so results for the latter three species could not be analysed.

24.2.3 WILDFIRE

A wildfire in August 1974 swept over the Myall Lakes heathland. The fire was very hot at ground level, leaving only sand and ash, and had crowned in the mallee-form eucalypts. The adjacent open forest on the high dunes was also burned and the fire crowned throughout. The area has experienced a high fire frequency for at least the last 100 years (see fire map, Fox and McKay 1981). The result has been a mosaic of patches with different regeneration ages and histories, with additional patchiness in heathland caused by different habitats as well as different fires.

24.3 Methods

24.3.1 TRAPPING

In May 1974 a 7-ha grid with a 20-m trap spacing was established in the study area. Collapsible aluminium small-mammal traps (330 x 100 x 90 mm) were used. The site was sampled at monthly or two-monthly intervals from May 1974 to May 1978 but in May only in 1979 and 1980, to give a total of 35 trapping sessions (17,050 trap-nights). For each session, trapping was carried out for four successive nights with no prebaiting. One trap was set at each of 176 trapping stations and baited with a mixture of oatmeal and peanut butter. Traps were checked in the early morning and in the evening in winter, to minimize trap mortality. On first capture each animal was marked by an individual toe-clip number and a numbered monel metal ear tag. For all captures each animal was weighed, identified to species and individual and the trap station recorded. For each trapping session the minimum number of animals known to be alive on the study plot during the session (MNA) was taken to be the simplest measure of abundance (Krebs 1966). The maximum value of MNA in any year was taken as the population peak for that year.

24.3.2 HABITAT UTILIZATION

For each species the data were expressed in captures per 100 trap nights for a given habitat, and then as a proportion of that species' captures per 100 trap nights over all habitats. The habitats were ordered along a subjective gradient of increasing dryness (average depth to water table) and increasing structural complexity, the latter determined from the number of vegetation layers present and the density of each. The median habitat on this gradient was determined for each species as the point which divided the frequency distribution of captures along the habitat gradient in two.

24.3.3 DIVERSITY

The species diversity measure used here takes into account both the number of species and the relative abundance of each, using the equivalent number of equally abundant species (Simpson 1949; Hurlbert 1971), defined as follows:

$$\Delta_3 = (\sum_i (p_i)^2)^{-1}$$

where p_i is the proportional abundance of the the ith species in a sample for a given area. Habitat diversity of a sample site was calculated using the number and areal proportions of each of the eight recognized habitat types encountered on the site. For the Victorian grids (see 24.4 below) the calculations were based on the published vegetation maps of floristic groups recognized by each author. Alpha diversity is the value calculated for a single site while regional diversity is the total number of species within a region.

24.3.4 VEGETATION REGENERATION

After the 1974 fire, permanent quadrats (1 x 2 m) were established in each vegetation type to monitor post-fire regeneration. Each quadrat was photographed about once a year from September 1974 to May 1979. Pre-fire photographs were compared with the post-fire photographs and each of the latter was scored on a 0-5 scale in terms of the percent cover and structural complexity relative to their pre-fire values with 0 indicating no recovery, 5 indicating full recovery, and other values representing intermediate conditions on a recovery scale. These scores were used as a recovery index for each vegetation type.

24.4 Results

24.4.1 VEGETATION REGENERATION

The course of regeneration of each vegetation type is shown in Fig. 1. Although the eucalypts and large shrubs (*Banksia* and *Xanthorrhoea*) were not killed by the fire there was so little green material present three weeks after the fire that all permanent quadrats were scored as zero at September 1974. Tall dry-heath and tree-heath (vegetation types 5 and 6) showed the slowest recovery, the first year's growth being due to abundant colonizers of disturbed areas such as *Actinotus helianthi* and *Pteridium esculentum*. Sprouters recover rapidly but seeders growing in intervening spaces recover more slowly (Specht 1981). Consequently the upper layers, which formed a uniform 2-3-m high closed-scrub before the fire, were still broken after five years, though most of the species present before the fire had returned by this time.

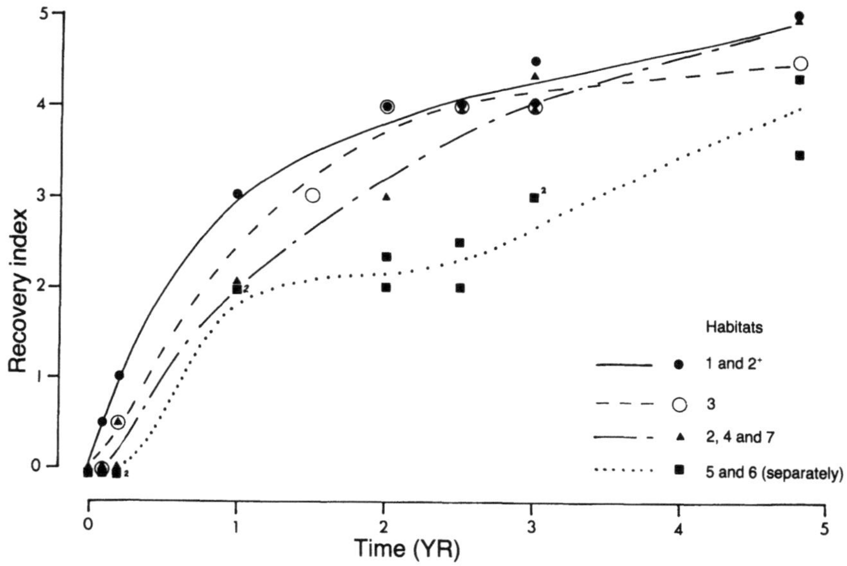

Fig. 1. Changes in recovery index with time for the vegetation in each habitat. The index estimates the increase in biomass as the vegetation regenerates following fire (see text). Curves were fitted by eye.

The vegetation of all other units regenerated much more rapidly and reached a recovery index of 4 (80% of pre-fire cover) by the end of the third year. Although recovery indices follow similar trajectories (Fig. 1) three different paths are revealed. The tall wet-heath (Type 3) was slow to recover completely, and although 100% ground cover was attained within two years the vegetation was then only 0.5 m high, and the pre-fire height of 2 m had not been reached by the end of the five-year study period.

The forms of the recovery index curves are similar to the two given for dry matter production by Groves and Specht (1965). Their wet-heath reached maximum biomass by five years while the sand-heath (Wilson's Promontary) showed two stages of recovery comparable to that seen for similar tall dry-heath here. Specht et al (1958) have also shown two stages of post-fire recovery for Dark Island heath though maximum biomass was not reached for 25 years.

24.4.2 SUCCESSIONAL CHANGES IN MAMMALIAN ABUNDANCE

Changes in the regenerating vegetation were accompanied by changes in the relative abundance of the species of small mammals (Fig. 2). The pre-fire values are tentatively plotted at six years, this being the best estimate of the age of the vegetation when it burnt. The species have been represented in taxonomic groups with Mus included with the Pseudomys group, since both genera are early colonizers. The small mammal community composition changed drastically in time. In the first few years post-fire it was a community dominated by Pseudomys. This was replaced by a community in which the importance of Pseudomys declined and the dasyurids increased. Finally, as can be seen from the pre-fire community (year 6 in Fig. 2) and from annual monitoring in May 1979 and 1980 Rattus becomes the dominant genus.

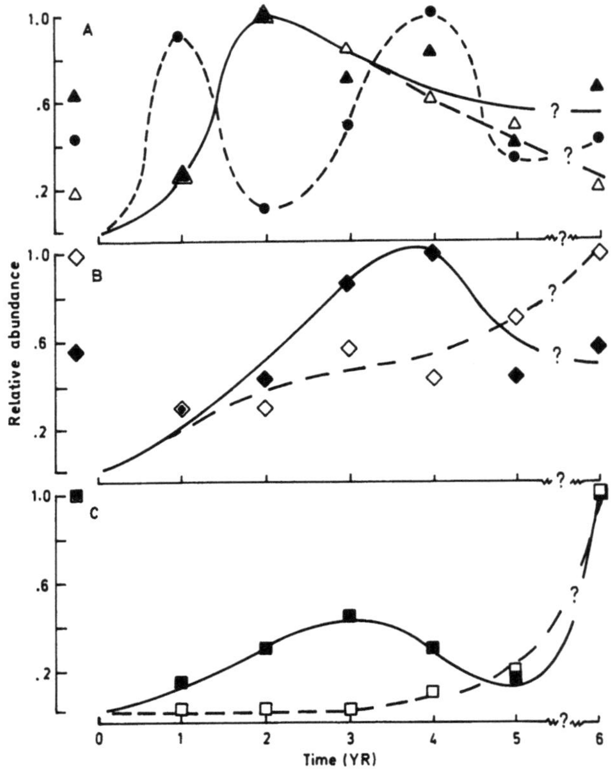

Fig. 2. Changes in abundance for each species in five successive post-fire years, shown as relative abundance i.e. observed annual population maximum relative to the maximum abundance recorded for that species (either pre- or post-fire). Pre-fire abundance is plotted to the left, and also at year six, the minimum estimated time to the previous fire. The species are shown in taxonomic groups. A. Pseudomys: P. novaehollandiae (solid triangles); P. gracilicaudatus (open triangles); and Mus musculus (dots). B. Dasyuridae; Antechinus stuartii (open diamonds); Sminthopsis murina (solid diamonds). C. Rattus: R. fuscipes (solid squares); R. lutreolus (open squares). Curves are fitted by eye.

24.4.3 FIRE-ADAPTED MAMMALS

The importance of early seral stages to Pseudomys gracilicaudatus and P. novaehollandiae at Myall Lakes is very clear (Fig. 2A). P. novaehollandiae has also been reported as abundant in the early post-fire stage by Posamentier and Recher (1974), Posamentier (1976), Kemper (1977), Fox and Fox (1978), Braithwaite and Gullan (1978) and Fox and McKay (1981). Other Pseudomys reported as fire- or disturbance-adapted species are P. shortridgei (Cockburn 1978) and P. albocinereus (C R Crouch in Cockburn 1978). Baynes (1982) has provided further convincing evidence of P. albocinereus as an opportunist species from fossil deposits in southwestern Australia and also provides similar evidence for Sminthopsis murina and Isoodon obesulus. Stoddart and Braithwaite (1979) report I. obesulus as a species reaching maximum abundance in fire-disturbed habitats, as does Fox (1980) for I. macrourus at Myall Lakes. More limited information from Ahern (1974) indicates that S. leucopus may also be an opportunist species. It appears therefore that Pseudomys, Sminthopsis and Isoodon comprise fire-adapted species. One exception to this may be Sminthopsis crassicaudata, a species of arid zones where fire is not important (Morton 1977). Baynes (1982) however, has indicated that S. crassicaudata is invading habitats in southwestern Australia in a manner consistent with an opportunist species. A final species that must be added is of course the supreme opportunist or fugitive species, the introduced Mus musculus.

24.4.4 HABITAT UTILIZATION

As well as showing preferences for different seral stages the small-mammal species had marked differential habitat preferences (Fig. 3). Within each taxonomic or ecological group the species pairs or congeners occupy distinct habitat types with one member using wetter, structurally simple habitats while the other uses the drier, more structurally complex habitats. This is clearly shown by the median habitat determined for each species.

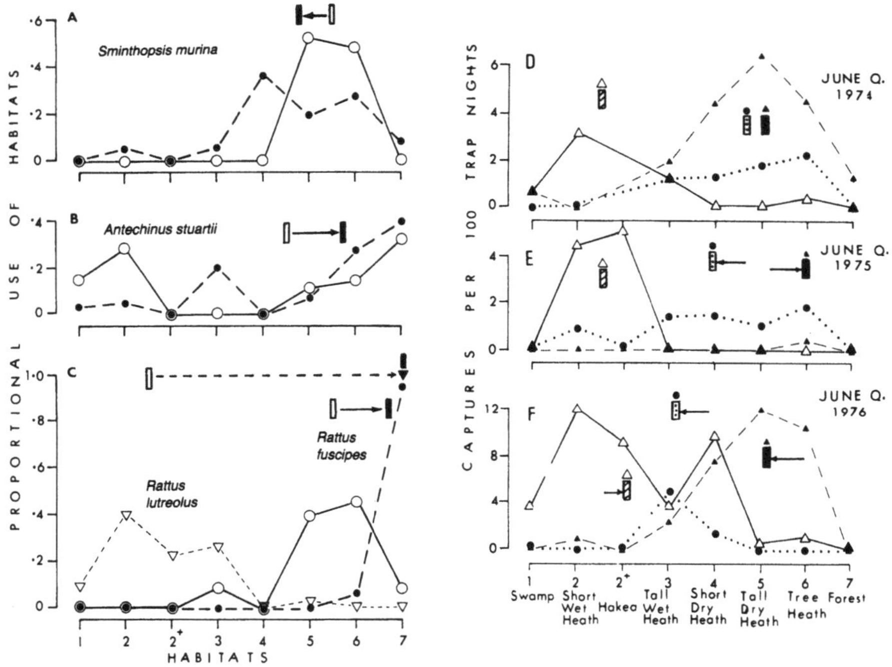

Fig. 3. Habitat utilization. The median habitat for each species shown by boxes, habitat shift by arrows. Habitats are arranged from structurally simple wet (1) to structurally complex dry (7) (see text). A-C. For individual species. Open symbols are pre-fire and solid symbols post-fire. D-F. For the species group: Pseudomys novaehollandiae (solid triangles), P. gracilicaudatus (open triangles), and Mus musculus (dots). Absolute capture rates are shown for pre-fire (D), one year post-fire (E), two years post-fire (F).

Sminthopsis murina and *Antechinus stuartii* both show a marked habitat shift when compared to the habitats occupied by them in nearby forest (Fox and McKay 1981). There is some indication that *A. stuartii* is displacing *S. murina* from the forested areas of the plot and the ecological separation between these two species has been shown by Fox (1982a). This is supported by post-fire habitat shifts as the latter's median habitat shifted away from tree habitats while the median habitat for *A. stuartii* shifted toward the forest (Fig. 3A).

Fire-induced habitat shifts were evident for other species (Fig. 3). The direction and extent of the shift of the median habitats is indicated by the length of the arrows. The most striking shifts are shown by *Rattus fuscipes* and *R. lutreolus* which for four years post-fire were only found in the swamp forest (Fig. 3C), which had the densest ground cover. Prior to the fire *R. fuscipes* occupied tall dry-heath and tree-heath and *R. lutreolus* short wet-heath. Sampling in 1979 and 1980 (five and six years post-fire) indicated that both species of *Rattus* had returned to their pre-fire habitats and that *R. lutreolus* was competing with *Pseudomys gracilicaudatus* whose numbers had increased markedly in the absence of *R. lutreolus* for four years after the fire.

The important conclusion from this section is that small-mammal species are strongly differential in their habitat selection in this community, and that this differentiation persists even when a species shifts its preference for a vegetation unit after fire.

24.5 Discussion

The species' abundance data have been used to calculate a fire response index, the logarithm of the ratio of post-fire to pre-fire abundance of each species (Fox 1980, 1982b). This measure allows species to be ranked by degree of post-fire response and reflects the extent to which each species' population is perturbed by the fire-induced habitat changes. I have equated this measure of the tendency to benefit, withstand or be detrimentally affected by such habitat change, to a "resource" that is partitioned differentially among species. Using this "resource" as a basis it is possible to apply to this community a concept similar to the "regeneration niches" proposed by Grubb (1977) for describing modes of reproduction in plants. Among the two dasyurids *Sminthopsis murina* is hypothesized to occupy an "early regeneration niche" (Fox 1982a). A similar role is proposed for *Pseudomys novaehollandiae*, *P. gracilicaudatus* and *Isoodon macrourus* (Fox 1980, in press) while at the other end of the axis species like *Rattus lutreolus* have large negative response indices reflecting the late seral stage occupied as a "late regeneration niche".

Analysis of the changes in small-mammal community structure in response to fire-induced changes in the vegetation at Myall Lakes provides insight to some of the mechanisms of coexistence of species. By attempting to generalize these findings this study can contribute to the understanding of why there are more small-mammal species on some of the east-coast heathlands than on others, as well as throwing some light on organization in heathland small-mammal communities in general.

24.6 Patterns of small-mammal diversity in Australia

Data showing the diversity and structure of small-mammal communities from heathlands in the northern (north of 30° S, southeastern Queensland

and northern New South Wales), central (30° S to 36° S, major part of New South Wales) and southern (south of 36° S, southern New South Wales and Victoria) regions of eastern Australia are given in Table 1. There are several interesting points about these communities. (a) Arranging the communities in order of increasing richness shows that additional species are successively drawn from different genera even though other species within the same genus may be present in the region. (b) The regional diversity is higher in northern and southern regions than in the central region so there is no latitudinal gradient. (c) No central site has more than one species of Antechinus while all southern sites with more than four species (total) have two species of Antechinus. (d) The richest central communities have at least two species (excluding bandicoots) that benefit from fires (F species - Pseudomys spp. and Sminthopsis murina), while the less rich central sites and all southern sites contain only one of these species. (e) Late Pleistocene fossil assemblages (Hope et al. 1977; J P Hope, personal communication) comprise richer communities with two or three species of Pseudomys.

Fox (1980) proposed for this community a mechanism for species packing in which resource utilization was such that species from each genus occupied the complete range of niche space available before sufficient resources become available to allow the addition of any congener. One species from each of the groups (Rattus, Pseudomys and Dasyuridae) had to be present in the community before a second member from any one of them could be added. Such a system appears to operate here as is shown in Table 1. M'Closkey (1978) has produced a series of assembly rules for species packing based on a minimum niche separation, and a somewhat similar set of rules may exist here, but the compatability of species seems to be important in addition to the minimum niche separation necessary in M'Closkey's community.

Further analysis shows geographical patterns in the structure of these communities. Entire communities, including introduced species, have no overall latitudinal patterns in alpha diversity; median values for the three regions are the same. But the richest communities occur in the central region. Median regional values for native species alone are again similar though the richest site once more is in the central region. However, central communities tend to comprise more F species than do those of the other regions; the number of native non-F species are significantly lower than in northern or southern communities ($\chi^2 = 4.2$, $P < 0.05$, according to the median test; Siegel (1956)).

For a more detailed analysis I have compared (in Table 2) the richer sites, i.e. sites with six or more species from the central region and sites with five or more species from the northern and southern regions. Isoodon is included in this analysis as data are available from all of the sites.

It can be seen that four of the six southern communities have an additional Antechinus drawn from the A. swainsonii-A. minimus group, and only one F species (now including Isoodon spp., Stoddart and Braithwaite, 1979). The other two sites (Syphon Road and Cranbourne) lack the extra Antechinus but each have two F species. In addition both sites lack Rattus fuscipes, while Cranbourne has no Antechinus and Syphon Road has no R. rattus.

By comparison to central communities it would appear that the southern communities could potentially contain six or seven species (in fact, Syphon Road had a single R. rattus and two A. stuartii, Braithwaite et al. (1978), but these were not considered part of the community by these authors). Two facts are important: (a) in spite of the high regional diversity

Table 1. The organization of ground-dwelling heath small mammals into taxonomic groups with sites ordered on increasing species richness. Data has been collated from the following sources: [a]Posamentier (1976); [b]Mahoney and Posamentier (1975); [c]Posamentier and Recher (1974); [d]Newsome et al. (1975); [e]L McKenzie (personal communication); [f]T Rose (personal communication); [g]J P Hope (personal communication); [h]Hope et al. (1977); [i]Braithwaite et al. (1978); [j]Braithwaite and Gullan (1978); [k]Dwyer et al. (1979) (*Isoodon* excluded on size). The three regions northern (N), central (C), southern (S) are indicated for each site.

Site		Total no. of species	Introduced spp.	Native					
				Rattus	*Antechinus*	*Pseudomys*	*Sminthopsis*	*Melomys*	Others
NO. OF EXTANT HEATH SPECIES BY SITE									
Kingscliff[a]	(N)	2	2						
Yamba[a]	(N)	2	2						
Frankston[i]	(S)	2	1	1					
Bouddi[e]	(C)	2	0	1	1				
Diggers Camp[a]	(N)	3	1	1	0	1			
Syphon Road[i]	(S)	4	1	1	1	1			
Hawks Nest[a]	(C)	4	2	1	0	1			
Cranbourne[i,j]	(S)	4	2	1	0	1			
Reeves Beach[i]	(S)	4	2	1	0	1			
Cape Liptrap[i]	(S)	4	0	2	2	0			
Beerwah[k] (*Banksia aemula*)	(N)	5	1	1	1	0	0	1	
Evans Head[a]	(N)	5	1	1	1	1		1	
Caloola Coastal[k]	(N)	5	1	2	0	0	1	1	
Kuringai Chase[f]	(C)	5	1	2	1	1			
Bonny Hills[b]	(C)	5	2	2	0	1			
Nadgee[a,d]	(S)	5	1	2	2	0			
Mt. William[i]	(S)	5	1	1	2	1			
Mirranatwa Gap[i]	(S)	5	1	1	2	1			
Kentbrook Heath[i]	(S)	5	0	2	2	1			
Bemm River[i]	(S)	5	0	2	2	1			
Beerwah Routine[k]	(N)	6	1	1	1	0	0	1	2
Brisbane Waters[b]	(C)	6	2	2	1	1			
Port Stephens[c]	(C)	6	2	2	0	1	1		
Royal N.P.[e]	(C)	6	1	2	1	1	1		
Myall Lakes N.P.	(C)	7	1	2	1	2	1		
NO. OF EXTANT HEATH SPECIES BY REGION									
Northern		11	2	3	2	1	1	1	1
Central		10	2	2	2	2	1	1	
Southern		14	2	3	4	3	2		
FOSSIL ASSEMBLAGES (LATE PLEISTOCENE)									
Wombeyan Quarry[g]	(C)	5+	-	1	1	2	0	0	1
Broom Breccia[g]	(C)	7+	-	1	1	3	1+	0	1
Kangaroo Island[h]	(S)	8+	-	2	1	3	1+	0	1

Table 2. Small-mammal species in rich communities. Bandicoot have been included. Rare species (single individuals or single captures) shown in parentheses. Sources as for Table 1. Core species and F species based on New South Wales sites (see text).

Sites	Total number of species	Exotic		Native																		
				Core species					F. Species								Others					
		Mus musculus	*Rattus rattus*	*Rattus lutreolus*	*Rattus fuscipes greyi*	*Rattus fuscipes assimilis*	*Antechinus flavipes*	*Antechinus stuartii*	*Pseudomys fumeus*	*Pseudomys novaehollandiae*	*Pseudomys shortridgei*	*Pseudomys gracilicaudatus*	*Melomys littoralis*	*Sminthopsis murina*	*Isoodon obesulus*	*Isoodon macrourus*	*Antechinus swainsonii*	*Antechinus minimus*	*Rattus tunneyi*	*Pl. maculata*	*Hydromys chrysogaster*	*Xeromys myoides*
SOUTHERN REGION																						
Mt. William	5		X	X				X	X								X					
Mirranatwa	5		X	X			X		X								X					
Bemm River	5			X		X		X	X								X					
Kentbrook Heath	5			X	X			X			X							X				
Syphon Road	5 (7)	X	(X)	X			X	(X)			X				X							
Cranbourne	5	X	X	X						X					X							
CENTRAL REGION																						
Royal Nat Park	6	X		X		X		X		X				X								
Brisbane Waters	7	X	X	X		X		X				X				X						
Port Stephens	7	X	X	X		X				X				X		X						
Myall Lakes	8 (10)	X	(X)	X		X		X		X		X		X		X					(X)	
NORTHERN REGION																						
Caloola Coastal	5	X		X									X	X					X			
Beerwah (*B. aemula*)	6	X		X			X						X			X						
Evans Head	6	X		X				X		X			X			X						
Beerwah Routine	7 (9)	X	(X)	X			X						X			X				X	X	(X)

no communities with six or seven species commonly caught have been reported in the southern region and (b) in none of the southern communities are the R. fuscipes-R. rattus and A. stuartii-M. musculus species pairs found to coexist (Braithwaite et al. 1978: 442). This would appear to indicate the number of species occupying sites in southern heaths may be limited by resource availability as the members of these pairs compete in the same food niche (Table 3).

Within the central communities there are many examples of coexistence of each, and in one case both, of the above species pairs. Such coexistence occurs in the four- and five-species communities as well as the richer six-, seven- and eight-species communities. This indicates that resources are more abundant in central communities than in their southern counterparts. However, resource abundance alone cannot account for the higher diversity in central communities as these apparently competing species pairs are present in poorer as well as rich communities.

The key is in the fact that the median number of species is similar for both states. It is individual central sites that have high diversity and richness. These sites are those with two, three or four F species (including Isoodon) in addition to the core shown in Table 2. From this I suggest that these F species are present at individual sites because those sites have had a favourable fire history. This is consistent with the response exhibited by these F species as well as that of the community as a whole.

This interpretation does not hold for the rich communities of the northern region. Here richness appears to be due simply to a greater diversity of genera in the regional biota and not because communities are structured as a core of species with added F species. Thus there are nine genera and 11 species for all northern region areas pooled, compared with seven genera and 10 species in the central region. Although the median species richness is not increased as a latitudinal effect (on the data available) the α diversity of an individual site may result from a latitudinal effect. For northern sites regional diversity is high while α diversity is similar to that of central sites.

Why should heathland in the Myall Lakes National Park have the richest small-mammal fauna in the central region? My hypothesis is that the local fire history has resulted in a mosaic of diverse habitat patches, simultaneously satisfying the divergent requirements of additional mammal species by enhancing spatial heterogeneity in vegetation type as well as through temporal, successional effects. The distribution of vegetation types and a network of tracks have favoured this pattern, and these have prevented infrequent, large, high-intensity fires that tend to create large expanses of less varied "habitat patches", and excluded too-frequent burning that leaves no late-successional patches. Detailed comparative fire histories are necessary to test this proposition.

The small-mammal species-area curve for the Myall Lakes shows the typical asymptotic form (Fig. 4A). Small-mammal species diversity (MSD) shows similar relationship with area (Fig. 4B) as does habitat diversity and area (Fig. 4C). The regression of MSD on habitat diversity (Fig 4D) is highly significant ($r = 0.95$, $P < 0.001$, 9 d.f.) and implies that it is the diversity of habitats that determines MSD in the species-area curve rather than area per se. Small-mammal species richness (Fig. 4A) behaves in much the same way as MSD and it can be seen that in general trapping grids in the Myall Lakes area need to be of the order of three to four hectares to have a reasonable chance of recording all of the species available at the site. It should also be pointed out that the same result may be obtained by a series of smaller grids in place of a single three to four-hectare grid.

Table 3. Small-mammal species occupying food niches proposed by Braithwaite et al. (1978). A generalist granivore niche has been added, and the introduced Mus musculus is shown in parenthesis.

Food niches of small mammals in Myall Lakes heathland			
	Dry habitats	Wet habitats	Total study area
Generalist granivore	P. novaehollandiae (/M. musculus)	-	P. novaehollandiae (/M. musculus)
Omnivore	R. fuscipes	(M.musculus)	R. fuscipes (/M. musculus)
Generalist herbivore	-	P. gracilicaudatus	P. gracilicaudatus
Specialist herbivore	-	R. lutreolus	R. lutreolus
Soil fossicking insectivore	S. murina /I. macrourus	I. macrourus	S. murina /I. macrourus
Scansorial insectivore	A. stuartii	A. stuartii	A. stuartii

Comparison with the four Victorian sites shows that they have much lower MSD than equal Myall Lakes areas (Fig. 4). Although the Victorian areas also have lower habitat diversity (Fig. 4C) this is not of itself sufficient explanation for the lower MSD, as the habitat diversity values show less departure from the Myall Lakes values than do the MSD values (Fig. 4D).

These comparisons support the concusions that the Myall Lakes community has a higher α diversity because historically it has been able to retain species that are elsewhere patchily distributed and has the high habitat diversity necessary for their present-day maintenance. This is consistent with Loucks's (1970) ideas on maintenance of species diversity in environments undergoing recurrent disturbance.

24.7 Conclusions

This heathland small-mammal community can be divided into associations of species, i.e. (a) Rattus lutreolus and Pseudomys gracilicaudatus; (b) Sminthopsis murina and Pseudomys novaehollandiae; (c) Rattus fuscipes and Antechinus stuartii; with Mus musculus capable of becoming a component of any of them. These associations prefer wet-heath, dry-heath and tree habitats respectively.

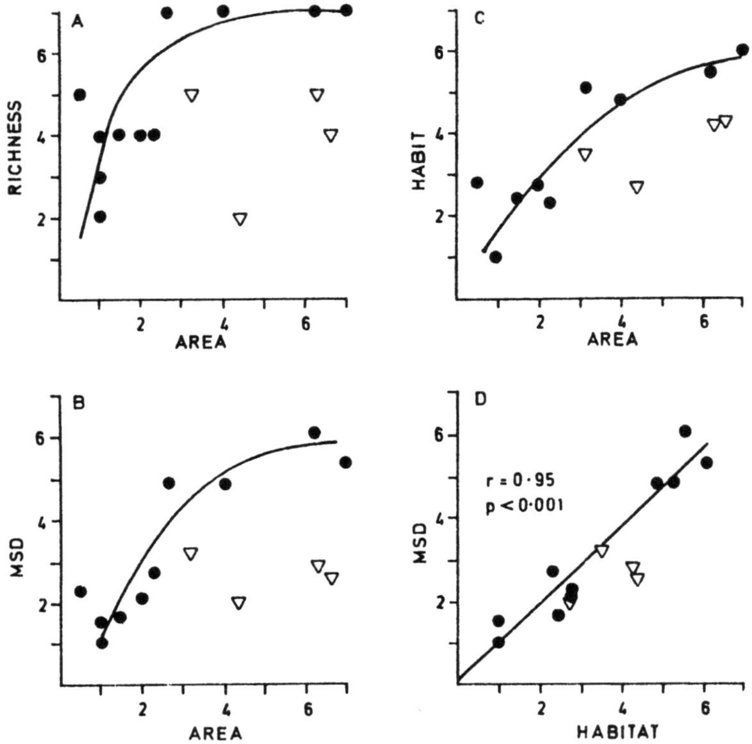

Fig. 4. Local diversity determinants for ground-dwelling small mammals (excluding bandicoots). Solid symbols are from areas in Myall Lakes National Park. Open symbols are taken from southern sites, Cranbourne (Braithwaite and Gullan 1978); Syphon Road, Frankston and Cape Liptrap (Braithwaite et al. 1978). The proportion of heath at Frankston was estimated from the vegetation maps (Braithwaite et al. 1978). Curves were fitted by eye (except for D, least squares regression). A. Small-mammal species richness as a function of area (ha). B. small-mammal species diversity (MSD) as a function of area (ha). C. habitat diversity as a function of area. D. small mammal species diversity as a function of habitat diversity.

The habitat requirements of some species are best met at particular stages of secondary succession following disturbance. This results in replacement sequences in which the ecologically dominant species are replaced in orderly shifts in relative abundance within the community. With minor exceptions the replacement sequence following regeneration after fire is as follows: M. musculus: P. novaehollandiae (or P. gracilicaudatus): S. murina: A. stuartii: R. fuscipes (or R. lutreolus) and this is similar whether the habitat is dry-heath, or wet-heath, or forest.

The members of the small-mammal community have evolved by adapting to changes in the vegetation, which provides their food and shelter, in the same way that the heath vegetation has evolved in response to specific fire regimes (Specht, 1981). This is indicated by the existence of fire specialists such as P. gracilicaudatus, and an apparently well-ordered mammalian succession.

The heathlands with the highest small-mammal diversity seem to be in areas where this regime has not been greatly altered and other man-made disturbances have been minimal. Where man's impact has altered the frequency of burning the range of different seral stages is reduced and some small-mammal species formerly present in the fossil record are lost.

A mosaic of habitat types, with many different seral stages, is needed to maintain high diversity small-mammal communities. At any given site the management policies should aim to maintain or create high habitat diversity. High regional small-mammal diversity can then follow and be maintained.

Acknowledgements

Much of the field work that I have drawn upon in this paper was completed while I was a PhD student at Macquarie University and I would like to thank the people in the School of Biological Sciences for their support. I would also like to thank Heimo Posamentier who kindly provided access to his unpublished MSc thesis which made possible many of the comparisons with other published work. Marilyn Fox critically read and commented on the manuscript as well as providing support and assistance at all stages of the project.

References

AHERN LD, (1974) A trapping study to determine habitat utilization and biology of small mammals at Sandy Point, Westernport with particular emphasis upon Sminthopsis leucopus (Gray). BSc (Hons) Thesis. Latrobe University

BAYNES A (1982) Dasyurids in Late Quaternary communities in southwestern Australia. In: ARCHER M (ed) Carnivorous marsupials. Royal Zoological Society of New South Wales, Sydney, pp 503-510

BRAITHWAITE RW, GULLAN P (1978) Habitat selection by small mammals in a Victorian heathland. Australian Journal of Ecology 3: 109-127

BRAITHWAITE RW, COCKBURN A, LEE AK (1978) Resource partitioning by small mammals in lowland heath communities of southeastern Australia. Australian Journal of Ecology 3: 423-445

CAROLIN RC (1970) Myall Lakes - an ancient and modern monument. Proceedings of the Ecological Society of Australia 5: 123-129

CATLING P, NEWSOME AE (1981) Responses of the Australian vertebrate fauna to fire: an evolutionary approach. In: GILL AM, GROVES RH, NOBLE IR (eds) Fire and the Australian biota. Australian Academy of Science, Canberra, pp 273-310

CHRISTENSEN PE, KIMBER PC (1975) Effect of prescribed burning on the flora of south-west Australian forests. Proceedings of the Ecological Society of Australia 9: 85-106

COCKBURN A (1978) The distribution of Pseudomys shortridgei (Muridae: Rodentia) and its relevance to the other heathland Pseudomys. Australian Wildlife Research 5: 213-220

COWLEY RD, HEISLERS A, EALEY EHM (1969) Effects of fire on wildlife. Victoria's Resources 11: 18-22

DWYER P, HOCKINGS M, WILLMER J (1979) Mammals of Cooloola and Beerwah. Proceedings of the Royal Society of Queensland 90: 65-84

FOX BJ (1980) The ecology of a small mammal community: secondary succession, niche dynamics, habitat partitioning, community structure and species diversity. PhD Thesis. Macquarie University

FOX BJ (1982a) Ecological separation and coexistence in Sminthopsis murina and Antechinus stuartii: a regeneration niche? In: ARCHER M (ed) Carnivorous marsupials. Royal Zoological Society of New South Wales, Sydney, pp 187-197

FOX BJ (1982b) Fire and mammal secondary succession in an Australian coastal heathland. Ecology 63. In press

FOX BJ, FOX MD (1978) Recolonization of coastal heath by Pseudomys novaehollandiae (Rodentia: Muridae) following sand mining. Australian Journal of Ecology 3: 447-465

FOX BJ, FOX MD (1981) A comparison of vegetation classifications as descriptors of small mammal habitat preference. In: GILLISON AN, ANDERSON DJ (eds) Vegetation classification in Australia. CSIRO and Australian National University Press, Canberra, Australia, pp 166-180

FOX BJ, MCKAY GM (1981) Small mammal responses to pyric successional change in a eucalypt forest. Australian Journal of Ecology 6: 29-41

GILL AM (1975) Fire and the Australian flora: a review. Australian Forestry 38: 4-25

GROVES RH, SPECHT RL (1965) Growth of heath vegetation. I. Annual growth curves of two heath ecosystems in Australia. Australian Journal of Botany 13: 261-280

GRUBB PJ (1977) The maintenance of species-richness in plant communities: the importance of the regeneration niche. Biological Reviews 52: 107-145

HOPE JP, LAMPERT RJ, EDMONDSON E, SMITH MJ, VAN TETS GF (1977) Late Pleistocene faunal remains from Seton rock shelter, Kangaroo Island, South Australia. Journal of Biogeography 4: 463-485

HURLBERT SH (1971) The non-concept of species diversity - a critique and alternative parameter. Ecology 52: 577-585

KEMPER CM (1977) Biology of the New Holland mouse. PhD Thesis. Macquarie University

KREBS CJ (1966) Demographic changes in fluctuating populations of Microtus californicus. Ecological Monographs 36: 239-273

LOUCKS OL (1970) Evolution of diversity, efficiency and community stability. American Zoologist 10: 17-23

MAHONEY JA, POSAMENTIER H (1975) The occurrence of the native rodent Pseudomys gracilicaudatus. (Rodentia: Muridae) in New South Wales. Australian Mammalogy 1: 333-346

M'CLOSKEY RT (1978) Niche separation and assembly in four species of Sonoran Desert rodents. American Naturalist 112: 683-694

MORTON SR (1977) An ecological study of Sminthopsis crassicaudata (Marsupialia: Dasyuridae) I. Distribution, study areas and methods. Australian Wildlife Research 5: 151-162

MOUNT AB (1964) The interdependence of the eucalypts and forest fires in southern Australia. Australian Forestry 28: 166-172

NEWSOME AE, CATLING P (1979) Habitat preferences of vertebrates inhabiting heathlands of coastal, montane and alpine regions of south-east Australia. In: SPECHT RL (ed) Ecosystems of the world, vol 9A. Heathlands and related shrublands. Descriptive studies. Elsevier, Amsterdam, pp 301-315

NEWSOME AE, MCILROY J, CATLING P (1975) The effects of an extensive wildfire on populations of 20 ground vertebrates in south-east Australia. Proceedings of the Ecological Society of Australia 9: 107-123

OSBORNE TG, ROBERTSON RN (1939) A reconnaissance survey of the vegetation of the Myall Lakes. Proceedings of the Linnaean Society of New South Wales 64: 279-296

POSAMENTIER HG (1976) Habital requirements of small mammals in coastal heathlands of New South Wales. MSc Thesis. University of Sydney
POSAMENTIER HG, RECHER HF (1974) Status of Pseudomys novaehollandiae. Australian Zoologist 18: 66-71
RECHER HF, LUNNEY D, POSAMENTIER HG (1975) A grand natural experiment. Australian Natural History 18: 154-163
RECHER HF, LUNNEY D, POSAMENTIER HG (1978) Effects of wildfire on small mammals at Nadgee Nature Reserve. Symposium papers, Third Fire Ecology Symposium. Forests Commission of Victoria, pp 30-36
SCHMIDT W, MASON M (1975) Fire and fauna in the northern Jarrah forest of Western Australia. Western Australian Naturalist 12: 162-164
SIDDIQI MY, CAROLIN RC, MYERSCOUGH PJ (1976) Studies in the ecology of coastal heath in New South Wales. III. Regeneration of vegetation after fire. Proceedings of the Linnean Society of New South Wales 101
SIEGEL S (1956) Nonparametric statistics for the behavioural sciences. McGraw-Hill, Tokyo, 312 pp
SIMPSON EH (1949) Measurement of diversity. Nature 163: 688
SPECHT RL (1970) Vegetation. In: LEEPER GW (ed) The Australian environment. 4th edn. (revised) CSIRO and Melbourne University Press, Melbourne, pp 44-67
SPECHT RL (1981) Responses to fires in heathlands and related shrublands. In: GILL AM, GROVES RH, NOBLE I (eds) Fire and the Australian biota. Australian Academy of Science, Canberra, Australia, pp 395-415
SPECHT RL, RAYSON P (1957) Dark Island heath (Ninety-Mile Plain, South Australia). I. Definition of the ecosystem. Australian Journal of Botany 5: 52-85
SPECHT RL, RAYSON P, JACKMAN M (1958) Dark Island heath (Ninety-Mile Plain, South Australia). VI. Pyric succession: changes in composition, coverage, dry weight and mineral nutrient status. Australian Journal of Botany 5: 52-85
STODDART DM, BRAITHWAITE RW (1979) A strategy for utilization of regenerating heathland habitat by the brown bandicoot (Isoodon obesulus; Marsupialia, Peramelidae). Journal of Animal Ecology 48: 165-179
THOM BG (1965) Late Quaternary coastal morphology of the Port Stephens-Myall Lakes area, New South Wales. Journal and Proceedings of the Royal Society of New South Wales 98: 23-36

25. Animal Demography in Relation to Fire and Shortage of Food: Some Indicative Models

A. E. NEWSOME and P. C. CATLING

25.1 Introduction

Animals of mediterranean-type ecosystems respond to the incidence per se of fire and to fire regime, just as in other ecosystem types. However, the effects of fire on quality and quantity of food and on other habitat factors can be complex and thus, faunal responses can vary strongly (Bendell 1974; Wirtz 1977). For example, numbers of black-tailed deer (Odocoileus hemionus columbianus) in California (Taber and Dasmann 1957) were not always obviously related to the fire history or to the resulting changes in food or shelter. Bendell (1974) considered that the abundance of wildlife on burnt areas was set by the amount of nutrient released as expressed in the quality of food. However, although there is good evidence for an increase in the quantity of post-fire food for deer in North America, there is scant evidence, if any, for consequent improvement in quality of browse (Bendell 1974). Levels of protein, but not of phosphorus, calcium and magnesium, may have increased but only by 1-2%.

Wirtz (1977) found a varied response in numbers of six species of small mammal in California after fire. After four to six months, trap success had declined 80% in the chaparral and 33% in grassland. However, some species (Dipodomys agilis and Peromyscus boylii) preferred the burnt habitat, whereas Perognathus californicus, Neotoma fuscipes and N. lepida preferred the unburnt habitats. Wirtz also found that predators (two raptors, Buteo jamaicensis and Bubo virginianus, and the coyote Canis latrans) had essentially the same proportions of 12 dietary items in pellets or faeces after the fire as beforehand. The success of reproduction in the raptors, however, declined after fire, indicating a shortage of food. Complexities are also inherent in many studies of habitat management for game birds (Kayll 1967; Vogl 1967; Doerr et al. 1970; Redfield et al. 1970; Scotter 1970; Chamrad and Dodd 1972). Leopold (1977) showed that increasing the ecotone between brush for shelter and grasslands for seed is what matters for the Californian quail (Lophortyx californicus).

Though there is less detailed information, similar complexities are found in the responses of the Australian fauna to fire (Christensen and

Kimber 1975; Newsome et al. 1975; Recher et al. 1975; Christensen 1977; Cockburn 1978; Lunney 1978; Recher and Christensen 1978; Kikkawa et al. 1979; Main 1980; Catling and Newsome 1981; Fox and McKay 1981). Studies have concentrated largely on defining the responses of species and communities to fire and on examining the evolutionary effects of what seems to have been a long history (more than 100 000 years) of wildfire (Singh et al. 1981). It is at least accepted that wildfire has strongly modified, if not moulded, the Australian vegetation (Gill 1975; Groves et al. 1981). Despite this, classic succession theory is thought to have had little application to the Australian vegetation with its fire-prone ecosystems (Noble and Slatyer 1977). Rather, the replacement sequence of plants after a fire depends largely on attributes vital to reproduction and survival of each species; a catalogue of these allows the sequence to be predicted for any stand.

Concepts for the responses of the vertebrate fauna to fire are not as well advanced as for the vegetation. Fire-dependence in the evolutionary sense is suspected for the fauna in general (Dwyer et al. 1979; Kikkawa et al. 1979; Catling and Newsome 1981; Fox and McKay 1981). Kikkawa et al. (1979) and Dwyer et al. (1979) concluded that heathlands in the early Tertiary occupied much of the Australian continent and that as they contracted to fragments coastally and subcoastally there was widespread extinction of forms, or radiations of others away from heathlands. They have suggested that specializations to live in the highly fire-prone heathlands arose often for the different major vertebrate taxa.

Distribution of animals in heathlands is patchy and their abundance is generally low probably due to the combined effects of fire, soils low in nutrients, low primary production and asynchronous flowering regimes (Specht 1981). Today, vertebrates found in heathland habitats in eastern Australia are basically inhabitants of the adjoining forests and woodlands (Newsome and Catling 1979). In general, heathland fauna appears to be opportunistic and largely unspecialized, except for frogs and a few birds (Dwyer et al. 1979). Main (1980) has concluded also that heathland animals in south-west Australia can tolerate fire because of the fire's usually patchy nature, and by avoiding radiant heat.

An intense fire in the tall forests of mediterranean south-west Australia caused a slight decline in the numbers of birds on the heathy understorey and on the forest floor (Kimber 1974; Christensen and Kimber 1975). Two years later, however, the number of species in these same habitats had increased beyond pre-fire levels. In eastern Australia, the seral response of birds to wildfire in a mountain habitat was non-existent or short-lived (S Wilson, personal communication 1979; Catling and Newsome 1981). In general, the data indicated that the higher a species lived in the canopy, the less the effect of the fire on its numbers (as Kimber (1974) found elsewhere). Two species of ground foragers, *Petroica phoenicia*, basically an insectivore, and *Emblema temporalis*, basically a granivore, were most affected, their numbers quadrupling in the four years after the fire. In contrast, Roberts (1970) found a sharp and persistent decline in the number of species in the understorey after a fire in a heathland in eastern Australia. Why some species should change greatly in numbers and not others, and why species diversity should change sometimes but not at other times (and places) is not understood. Perhaps the hypothesis of Huston (1979) applies, i.e. that equilibria between competing species, or suites of species, are interrupted too frequently by fire for clear succession to emerge.

According to Tyndale-Biscoe and Calaby (1976) eucalypt forests harbour the majority of the vertebrate fauna. Catling and Newsome (1981) have found evidence suggesting these fauna may be pre-adapted. Noteworthy findings as in the heathland and forest at Nadgee Nature Reserve, south-east Australia were that the seral response was absent or severely truncated (Fig. 1), and that there was a peak in abundance of small to large native mammals (rodents to kangaroos) four to five years after the fire (see also Figs 5 and 6).

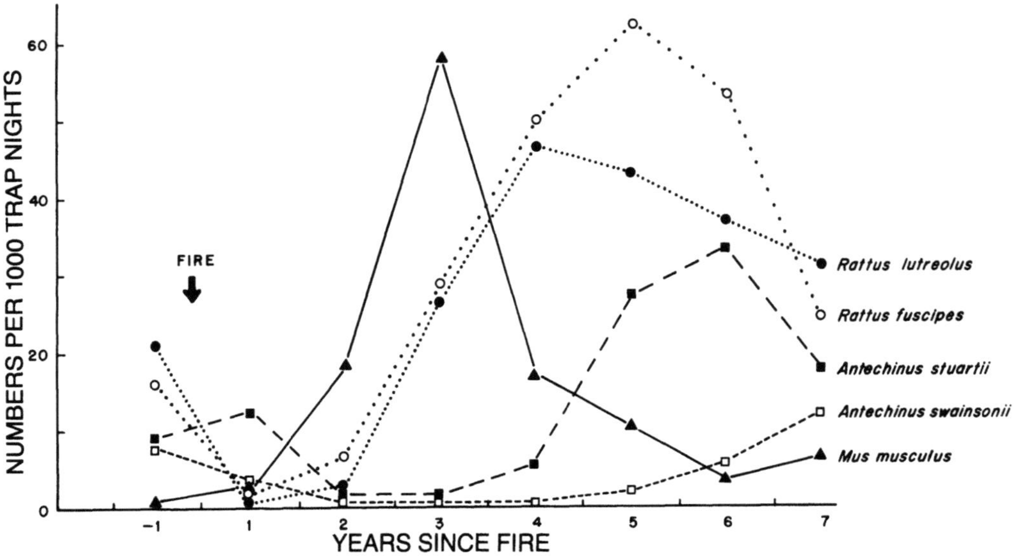

Fig. 1. Trends in annual maximum abundance of small mammals after an extensive, fierce wildfire in mixed heathland and forest in Nadgee Nature Reserve, south-east Australia (P C Catling and A E Newsome, unpublished).

By contrast, Fox and MacKay (1981) found a clear seral response on their heathland site (Fig. 2). However, two species of Pseudomys, absent from our study area for unknown reasons, were present there and largely responsible for the response observed (data presented for only one species in Fig. 2 but see Chapter 24, this volume). These species, Pseudomys gracilicaudatus and P. novaehollandiae (Keith and Calaby 1968; Fox and McKay 1981) as well as P. shortridgei (Cockburn 1978; Fig. 3) are examples of fire specialists which are otherwise rare.

There may also be adaptive reproductive responses to fire as in the possums (Catling and Newsome 1981). The common ringtail (Pseudocheirus peregrinus) nests in shrub canopy which is burned out by wild fires. It produces 1.7 young per year whose mortality rate in the first year is 0.7 with a replacement rate of 0.9 (Tyndale-Biscoe 1973). The respective values for the greater glider (Schoinobates volans), which inhabits the less fire-prone upper eucalypt canopy, are 0.6, 0.2 and 0.6-0.75. Moreover, all possums are seasonal breeders except one, the honey possum (Tarsipes spencerae). It is a highly specialized, rat-like nectar-feeder which inhabits a most highly fire-prone habitat, the Banksia heaths of southwestern Australia, and its reproduction is fittingly opportunistic (M Renfree and E Russell, personal communication 1979).

There are few data on changes to nutrient levels and nutrition due to

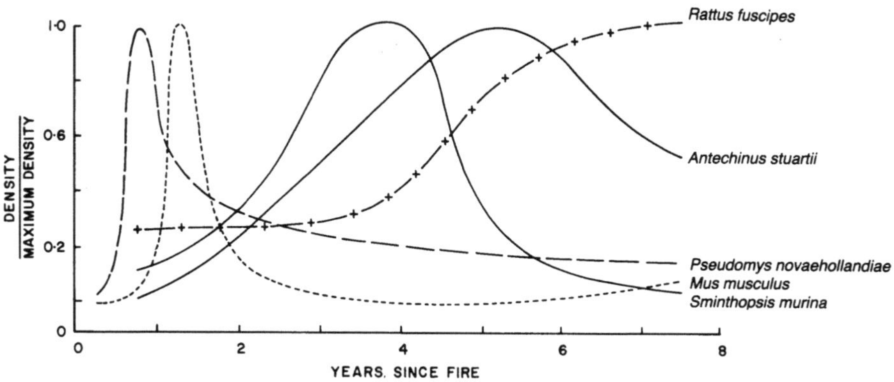

Fig. 2. Trends in abundance of small mammals after wildfire in heathland in central eastern Australia (after Fox and McKay 1981).

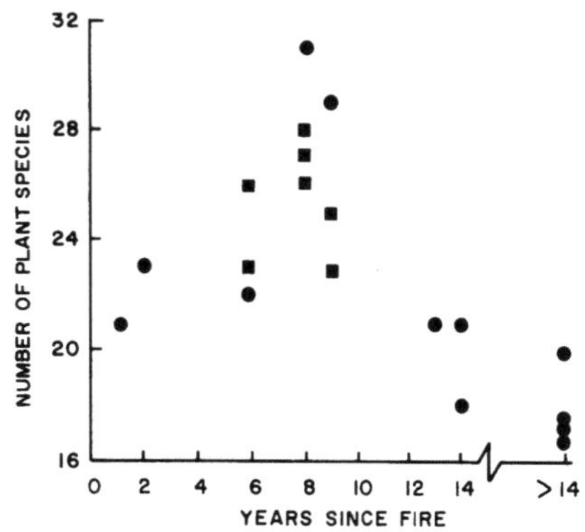

Fig. 3. Presence (squares) or absence (dots) of Pseudomys shortridgei relative to plant species diversity and time since fire in a heath in south-east Australia (after Cockburn 1978).

fires. In south-west Australia, however, litter after a severe fire (4000 kW m^{-1}) in a jarrah forest (Eucalyptus marginata) was double (515 g m^{-2}) that from an unburnt forest (264 g m^{-2}) in the first year (O'Connell et al. 1979), with concentrations of phosphorus and nitrogen which were four times higher. Levels of phosphorus were still 50% higher than in the unburnt forest in the second year. Wallabies and kangaroos congregated on the growth of new pastures after a fire there, e.g. Setonix brachyurus, Macropus irma and M. rufogriseus banksianus (Christensen and Kimber 1975). Macropus eugenii and M. giganteus respond similarly in eastern Australia (Catling and Newsome 1981). A fresh flush of green herbage may not necessarily improve breeding per se, but may improve survival of pouch-young. An example, unrelated to fire however, is Setonix which is a seasonal breeder (Sharman 1959). The wallaby mates in summer when the water and nitrogen contents of the food are lowest, 71-86% and

5-8.1%, but its young leave the pouch towards the end of the winter rains when the water and nitrogen in food are highest, 81-89% and 14.4-15.6% (Storr 1964; see summary in Main 1968).

Controlled fires are used commonly in Australia to reduce fuel in commercial forests and even National Parks. However, fuel in this instance comprises the understorey, logs, litter, and other components of the habitat which sustain populations of small mammals (Newsome and Catling 1979). The particular strata of the vegetation which members of each species of small ground mammal prefer at Nadgee Nature Reserve has been analysed recently by Catling et al. (1981).

25.2 Demographic models due to change in abundance of food, shelter and predators

From the foregoing it is clear that no useful general patterns of faunal response to fire are evident from successional data. Changes to supplies of food and shelter, and in predation are rarely measured relative to fire although their effects are likely to alter (see Lawrence 1966; Christensen 1977; Wirtz 1977). Accordingly, in an attempt to infer testable causal relationships between post-fire environments and faunal responses, demographic models have been derived from studies of Mus musculus which were unrelated to fire but in which resources were measured or provided in excess (Newsome 1967, 1969, 1970; Newsome et al. 1976). Changes in numbers, reproduction, age structure and mortality (net losses including emigration) are examined relative to sudden changes in abundance of resources and of predators. Population trends of small and large mammals in response to a fire in Nadgee Nature Reserve (Newsome et al. 1975; Recher et al. 1975; Lunney 1978; Newsome and Catling 1979; Catling and Newsome 1981) are then compared with predictions of these models.

Five models were erected on the data, and a sixth is theoretical, as follows (see Fig. 4).

Model I. Acute shortage of food (Data from Newsome (1969) from March to May, 1965). Numbers of Mus were initially high because food and burrows were both plentiful, but then the mice exhausted their food supply. Numbers faltered and crashed, reproduction fell sharply, and older mice disappeared leaving only immature ones in residence. Net losses held steady at first but then rose to 100%.

Model II. Shortage of food due to immigration (Data from Newsome (1969) from April to June, 1964). Circumstances were similar to Model I except that the surrounding habitat was ploughed causing sharply increased immigration onto the plot. Numbers and mean age initially increased before both parameters declined. Reproduction faltered and then declined as in Model I to zero. Net losses were reduced initially by immigration, but rose sharply afterwards.

Model III. Shortage of shelter, food abundant (Data from Newsome (1970) from June to August, 1967). Numbers were initially very high because mice could burrow anywhere in the moist sandy soil, and because food was supplied experimentally in excess. After a period of rapid population increase, over half the area was flooded by a rising creek. Numbers faltered and declined. The percentage of mice which were reproducing and their mean age actually rose, however, as the younger, non-breeding mice disappeared (probably dispersed). Net losses were initially high after the flood, but then declined because the remaining older residents survived.

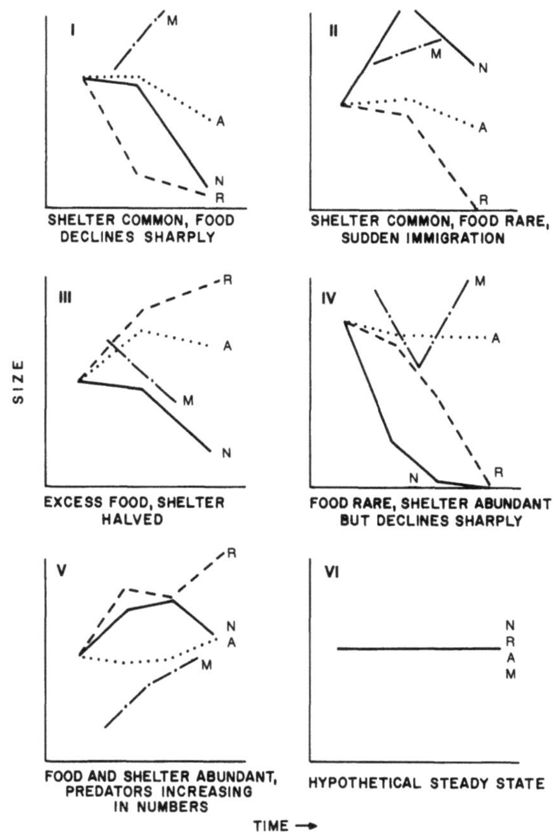

Fig. 4. Demographic models based on population studies of *Mus musculus* relative to supplies of food and shelter in the mediterranean regions of South Australia and California. The ordinate represents size for the following demographic statistics: numbers (N), mean age structure (A), rates of reproduction (R) and mortality (M), and the abscissa, time (from Newsome 1969, 1970; Newsome et al. 1976).

Model IV. Shortage of shelter, food rare (Data from Newsome (1969) from March to September combined for 1964 and 1965). Initially, food was in short supply but burrowing sites were abundant. Rains then waterlogged the soil destroying shelter and sprouting any remaining seed thereby decreasing food supply further. Numbers had increased in the preceding summer but declined sharply leading eventually to extinction. Reproduction faltered at first, and then declined steeply to zero. Again, as in Model I, the older mice disappeared leaving the young. Net losses, at first high, declined as the fall in numbers slowed, but became absolute as populations became extinct.

Model V. Increase in predators (Data from Newsome et al. (1976) from July to September 1972). Food was supplied in excess in an area of moist silty soil highly suitable for burrowing. Populations flourished, reproduction was high, age structure remained well balanced, and net losses were low. Then, numbers of several species of raptor began to increase rapidly, and mammalian predators, probably feral cats, found the study site. Mouse numbers began to decline, increasing net losses despite continued high reproduction and a balanced age structure.

Model VI. Steady state. This condition was not found in the studies of *Mus* mentioned above, but theoretically could exist.

25.3 Comparing data from fire studies with the models

Results from a fire at Nadgee Nature Reserve were examined against these models. The fire was the severest of any recent forest fire in New South Wales (P Cook, personal communication 1974). No ground vegetation remained and the forest trees were black, their canopy completely burnt or severely scorched.

Small mammals were sampled on six small and two large trapping grids located in different habitats: coastal and mountain eucalypt forest and heathland, swamp, and coastal dune-swale. Tracks of larger vertebrates were recorded from specially prepared soil plots placed every 0.3 km on roads, walking tracks and beaches. Details of methods are presented in Newsome et al. (1975). Though the models were derived from data of short-term studies on a small mammal, they were utilized to examine longer term trends from large mammals at Nadgee also. Population trends are provided in Figs 5 and 6, and summaries of changes in weight classes of the small mammals in Table 1. Indicated demographic responses and causes are presented in Table 2.

25.3.1 SMALL GROUND MAMMALS

Two native rodents (*Rattus*) and two dasyurid marsupials (*Antechinus*) are common at Nadgee.

(a) *Rattus fuscipes*. In general, these rats inhabit sclerophyllous forests. Adults weigh 100-150 g, males being slightly heavier than females. They breed from October to January, mainly in November to December (Warnecke 1971). Because the wildfire was in December, at or just after the peak in breeding, it is not possible to compare reproductive success immediately before and after the fire. (This problem holds for all the small mammals - see below.) Body masses were used as surrogate age measures, because animals could not be aged.

Numbers declined gradually in both small (0.1 ha) and large plots (2.9 ha) despite greater diversity of habitats on the latter (Newsome et al. 1975). Populations eventually became extinct on the small plots, survived on the large ones, but did not recover until well into the second year after fire. The combined mass-classes for plots of both sizes showed marked loss of larger animals post-fire (Table 1). This could mean that rats were on average either young, or small for a given age, or both. We suspect both because some individuals which survived the fire had lost weight (Newsome et al. 1975). Overall, Model IV seems most appropriate for small plots indicating an acute shortage of food and shelter after the fire: Model I and probably Model V are indicated for long-term trends on large plots with causal factors, a shortage of food, and predation.

Table 1. Population structure by size classes (live mass) of small mammals at Nadgee Nature Reserve, south-east Australia, from the April-November periods before and after the fire[a].

		Mass (g)					
		30-50	70-100	-120	-140	-160	-190
Rattus fuscipes	Before fire (1972)	0	38	54	76	40	11
	After fire (1973)	1	45	17	13	12	3
R. lutreolus	Before fire (1972)	0	27	47	57	32	3
	After fire (1973)	0	2	1	1	1	0

		Mass (g)						
		10-20	-30	-40	-50	-60	-80	-100
Antechinus stuartii	Before fire (1972)	41	72	25	2	0	0	0
	After fire (1973)	43	67	2	0	0	0	0
A. swainsonii	Before fire (1972)	0	0	1	14	19	25	6
	After fire (1973)	0	0	6	1	1	0	0

[a] The rats were both sexually mature at mass of about 120 g (Newsome et al. 1975). For the marsupials: A. stuartii mature between 20-30 g, and A. swainsonii between 40-50 g males; females between 30-40 g (Green 1973).

(b) R. lutreolus

(i) Study plots of Newsome et al. (1975) (as above): Called the swamp rat because of its usually damp habitat, this rat's demographic processes were sharply curtailed by the fire. The small study plots, which were a dense grass-sedge swamp and a moist shrubby dune-swale, were burnt bare. On the larger plot (see above) some vegetation survived near a rocky ledge, but this did not seem to help R. lutreolus. Little can be made of reproductive processes or age classes (Table 1) because rats were so rare after the fire. Because the burrows were not destroyed, a shortage of food is implicated in mortality and Model I seems to fit both plots.

(ii) Study plots of Recher et al. (1975) (from Lunney 1978): A contemporaneous study at Nadgee trapped an open grassy clearing and its forest edge which did not all burn. The numbers of R. lutreolus after fire (and of the other three species of small mammal) were higher than expected, indicating an influx. After the grassy plots were flooded by rains, immigration of R. lutreolus was recorded onto a nearby burnt plot in the forest where the species had not been caught previously. These responses best fit Model II (for other immigrant species as well).

(c) Antechinus stuartii and A. swainsonii. These marsupials (and other Antechinus) have a peculiar life history. A. swainsonii mates in July and A. stuartii in August each year whereupon all the males die (Wood 1970). Females suckle up to eight young which become independent about January and become the breeding stock in the following July/August except for an occasional old female surviving to breed twice. Reproduction immediately before and immediately after the wildfire cannot be compared because few females are caught between mating and weaning of young. Despite the wildfire, population trends and age structures indicated a successful breeding in summer 1972-73, and numbers went through an apparently normal annual cycle (Fig. 5, Table 1). However, males in samples from April to August were lighter after the fire than beforehand (23.7 ± 0.5 g vs 27.8 ± 0.6; $P < 0.01$; Newsome et al. 1975), and numbers crashed during the next summer (note the scarcity of adults in Table 1), the species disappearing on most study plots (Newsome et al. 1975; Recher et al. 1975; Lunney 1978).

Table 2. Indicated demographic models from fire studies at Nadgee Nature Reserve, south-east Australia.

Species	Nos.	Demographic factor: Age (classes by mass)	Demographic factor: Repro- duction	Demographic factor: Mort- ality	Indicated causal factors	Refs[a]
Rattus fuscipes						
Short term[b]	IV	IV	?	I	Shortage of food and shelter	1
Long term[c]	I or IV	I	I	I; V?	Acute shortage of food, plus possibly predation	1
R. lutreolus						
Short term	I	IV	?	I	Acute shortage of food	1
Long term	I	I	I	I; V?	Acute shortage of food plus predation	1
Short term	II	?	-	II	Shortage of food due to immigrants	2
Antechinus stuartii and A. swainsonii						
Short term	VI	VI	-	VI	Steady state	1
Long term	V?	V?	-	V?	Possibly predation	1
Short term	II?	?	-	II?	Possible food shortage due to immigration	2

[a] 1: Newsome et al. (1975). 2: Lunney (1978). [b] From about six months before to six months after the fire. [c] Over the two years following the fire.

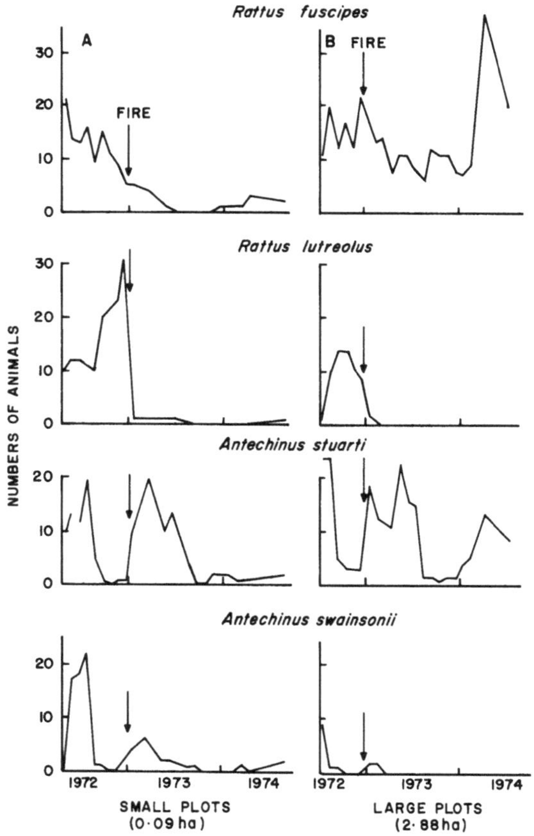

Fig. 5. Detailed fluctuations in trap records of small mammals after the fierce wildfire in Nadgee Nature Reserve, south-east Australia. A. small plots, B. large plots (from Newsome et al. 1975).

The short-term results indicate a steady state (Model VI). Why resources should have remained available for the insectivores and not for the rodents is unclear. Neither is the later decline (two years post-fire) understood. It is hard to believe that food resources could have been in short supply in that second summer, for juvenile males of A. stuartii from January to March were heavier than in the preceding year (29 ± 11 g vs 21.1 ± 0.7; $P < 0.05$). The average mass for juvenile females was also higher, but the difference was not quite significant. In the first year after the fire there had been rapid, lush regrowth of herbage, shrubs and creepers, with epicormic shoots emerging on trees over almost the entire forest, and there was a massive outbreak of insects (Fox 1978). Heavy predation may have operated on Antechinus over the long term because 20.5% of scats of European foxes (Vulpes vulpes) in the first and second year after the fire contained small mammals (Newsome and Catling, unpublished). The scant evidence is closest to Model 5.

On the study plots of Recher et al., Lunney (1978) recorded influxes of Antechinus into patches of unburnt forest soon after the fire as mentioned above, indicating Model II there (Table 1).

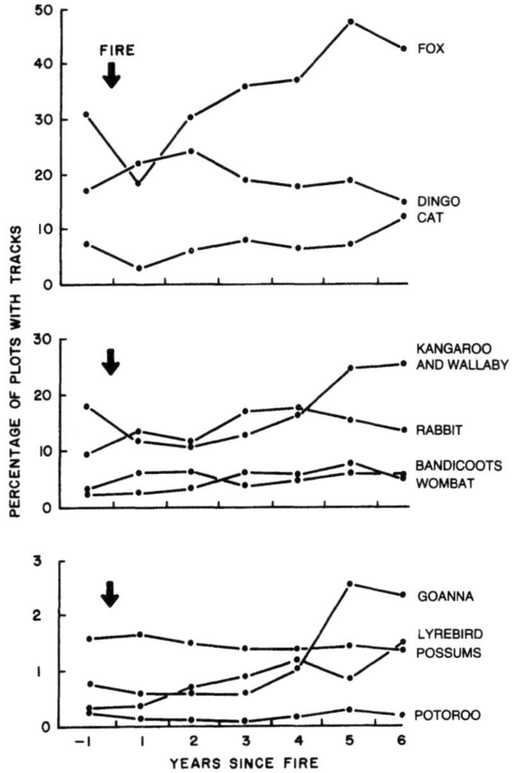

Fig. 6. Annual mean records of tracks of large vertebrates on soil plots after the fierce wildfire in Nadgee Nature Reserve, south-east Australia (Catling and Newsome, unpublished).

25.3.2 LARGE MAMMALS

Despite long life spans, the trends in abundance of larger mammals, from possums (Trichosurus vulpecula, Pseudocheirus peregrinus) to grey kangaroos (Macropus giganteus) (Fig. 6), are examined relative to the models, but without a summary table because of the speculative nature of the conclusions.

(a) Wallabies and kangaroos. Two medium-sized wallabies (Macropus rufogriseus and Wallabia bicolor) and the larger grey kangaroo (M. giganteus) declined in abundance immediately after the fire, and 41 carcasses were found (Fox 1978). The decline continued for two to three years afterwards even though the two wallabies are continuous breeders. Furthermore, good rains fell soon after the fire causing a flush of green herbage which should have promoted breeding. Medium-sized macropodid tracks (mostly wallabies but including some young kangaroos) were significantly more common than expected for the first two years post-fire ($P < 0.001$) (Newsome and Catling, unpublished). Those results indicate that, although there may have been a loss of animals in the first year, breeding success was good later on (pouch life is 8-10 months; Calaby and Poole 1971). Therefore, the model which fits is I, indicating lack of food immediately after the fire. The continued decline in numbers for a few years was probably due to severe predation from dingoes, Canis familiaris dingo (Catling and Newsome 1981), indicating Model V.

(b) Wombat (Vombatus ursinus) and European rabbit (Oryctolagus cuniculus). These two burrowing species appear to have escaped the worst of the fire. Also, within a month of the fire it rained and ground vegetation sprouted quickly. Their indices of abundance remained relatively constant probably because food and shelter did not become limiting for them. Young wombats leave the pouch in summer (McIlroy 1973) so that reproduction cannot be assessed relative to the immediate effects of the fire. These species appear to be candidates for Model VI responses both immediately after the fire, and also, unlike Antechinus, in the long term.

(c) The dingo (Canis familiaris dingo). Unlike the introduced carnivores (see below), the dingo maintained its pre-fire numbers. Evidence came not only from tracks (Fig. 6) but also from marked animals (Newsome and Catling, unpublished). We have concluded that the survival of dingoes was due to their flexibility in changing to alternative supplies of food (Catling and Newsome 1981). Six to eighteen months before the fire, small mammals comprised 18.5% of the diet, and macropodids, 44.8%. In the 6-18 months post-fire respective values were 3.7% and 68.8%. By the fourth year, small mammals were common and became important in the diet (42.4%), and the macropodids were only 23.5% (AE Newsome, PC Catling and LK Corbett, unpublished). Model VI (steady state) appears to describe the dingo's demographic responses to the fire.

(d) European fox (Vulpes vulpes) and European cat (Felis catus). After the fire the tracks of these two feral carnivores fell sharply (Newsome et al. 1975), presumably due to mortality from fire and also to the sharp decline in the numbers of small mammals (Newsome et al. 1975; see above also). Kitten tracks were absent until the second year after the fire (Newsome and Catling, unpublished), suggesting that breeding success of the feral cat was lower than usual. Though data are poor, Models I or IV may apply, indicating a collapse in the supply of food, or both food and shelter.

25.4 Discussion

There is no study known to us which traces changes in nutrient levels in soils through changes in vegetation to changes in the abundance of higher organisms. Fire, of course, recycles nutrients but resultant effects on animals must be mediated by the plants. However, in addition to food, the vegetation provides favourable microclimates, as well as home sites and refuges from predation. Besides, many heathland plants are unpalatable, or thorny, and so maybe unsuitable as food or shelter to some animals. The greatest impediment to understanding the biotic effects of fire is the lack of data on supply of resources and levels of predation, and the paucity of field experiments. Studies have been mostly historical and correlative as summarized earlier.

The data examined in this paper display the complex range of responses of fauna to intense fire. There is no one response or sequence of them. Basic complexities arise from differences in life histories of particular species, their body sizes, longevities, diets and the types of shelter they require. The most complicating factor is probably predation. Although there are many uncertainties, the six demographic models we present appear to have some relevance especially in the short-term changes after fire (less than six months). The predation model may also apply in the long term (two years) to large long-lived prey (macropodids) subject to a long-lived predator like the dingo.

Though the models present curves intended to illustrate the direction and rates of changes of demographic characteristics, the exact shapes of the curves in each model may not be as important as the trends. On that basis, Models I and IV have most in common, and may be basically the same. Model II is really a special case of I (and possibly IV) but is kept separate as an immigration model describing, essentially, a refugial circumstance. Model III may have least relevance to severe fire in the short term, unless food which is initially relatively abundant is rendered absolutely abundant by a fire which reduces shelter concomitantly. Examples of such a change to the food supply comes from a different field, herbivory. In his excellent study in the Serengeti, Bell (1970) concluded that the dicotyledonous herbs which are favoured by Thomson's gazelle (Gazella thomsoni) but often hidden by tall grasses, were exposed by the grass-eaters, zebra (Equus burchelli bohmi), wildebeest (Connochaetes taurinus albojubatus) and topi (Damaliscus korrigum jimela). Model III would require, however, a concomitant sharp decline in the supply of shelter. A possible example is the small wallaby Bettongia penicillata. In Christensen's (1977) experimental study of the effects of fire on the small rare wallaby in brushland of the mediterranean south-west region of Australia, much of the brush used for shelter was removed, but then rain produced much more herbage in the fire-opened habitat than beforehand. At the same time, however, predation became severe (see below). A combination of Models III and V may have applied in that case.

Rare species may be special cases. They may be protected from effective predation under normal circumstances simply by the denseness or some other special feature of the habitat, or, more complexly, by the abundance and availability of alternative prey. If fire decreased all populations to rarity, or exposed a rare species to predators, then its recovery time may be greatly lengthened due to local extinctions. Again, Bettongia is an example. After fire had removed its shelter, European foxes killed seven and marsupial cats (Dasyurus geoffroii) killed four of 24 survivors in the fire-opened habitat (see also Recher and Christensen 1978). The problem of survival of uncommon or rare species is heightened if, as is usually the case, the natural habitats lie inside national parks which are surrounded by cleared land. The smaller the parks the greater the risk of extinction.

The models based on Mus are likely to have many inadequacies. One failing demonstrated here is that not all mammal populations seem affected by fire. Some species (wombat and dingo) changed little in numbers, and others (Antechinus) had annual population trends which remained basically unchanged for at least the first year post-fire. Presumably the basic resources of food and shelter were not grossly altered for those species. Shelter for the wombat may have been secure due to its deep burrows dug in moist areas like creek banks (McIlroy 1973), and because rain falling one month after the fire produced herbage. The dingo appears to have coped by altering its diet. Perhaps Antechinus stuartii at least, being scansorial, did the same though its numbers crashed a year later.

We suggest that such complexities will only be unravelled through orthogonal experimentation, e.g., by providing food, shelter, water, etc., by controlling predation through exclusion fencing, removal etc., and by setting fires ranging in size and intensity over different seasons of the year.

Acknowledgements

We are grateful to Dr G Caughley for his advice on the demographic models in this paper. We are also grateful to Mr R Burt and Mr H Wakefield who assisted most with the field studies and to Mr G Richards for drawing the figures. Mrs W Guy kindly typed the manuscript. The study on fire was initiated fortuitously under Grant CS8S and CS17 from the Australian Meat Research Committee for the study of dingoes.

References

BELL RHV (1970) The use of the herb layer by grazing ungulates in the Serengeti. In: WATSON A (ed) Animal populations in relation to their food resources. Blackwell, Oxford, pp 111-123

BENDELL JF (1974) Effects of fire on birds and mammals. In KOZLOWSKI TT, AHLGREN CE (eds) Fire and ecosystems. Academic Press, New York, pp 73-138

CATLING PC, NEWSOME AE (1981) Responses of the Australian vertebrate fauna to fire: an evolutionary approach. In: GROVES RH, NOBLE IR, GILL AM (eds) Fire and the Australian biota. Australian Academy of Science, Canberra, pp 273-310

CATLING PC, NEWSOME AE, DUDZINSKI ML (1981) Small mammals, habitat components and fire in south-east Australia. In: OECHEL W, CONRAD E (eds) Proceedings of the international symposium on the dynamics and management of mediterranean-type ecosystems, San Diego, June 22-26, 1981. In press

CHAMRAD AD, DODD JD (1972) Prescribed burning and grazing for prairie chicken habitat manipulation in the Texas coastal prairie. In: Proceedings of the twelfth annual tall timbers fire ecology conference, June 8-9, 1972, Lubbock, Texas, pp 257-276

CHRISTENSEN PE (1977) The biology of _Bettongia penicillata_, Gray 1837, and _Macropus eugenii_ Desm. 1817, in relation to fire. PhD Thesis. University of Western Australia, Nedlands

CHRISTENSEN PE, KIMBER PC (1975) Effect of prescribed burning on the fauna and flora of south-west Australian forests. Proceedings of the Ecological Society of Australia 9: 85-106

COCKBURN A (1978) The distribution of _Pseudomys shortridgei_ (Muridae:Rodentia) and its relevance to that of other heathland _Pseudomys_. Australian Wildlife Research 5: 213-219

DOERR PD, KEITH LB, RUSCH DH (1970) Effects of fire on a ruffed grouse population. In: Proceedings of the tenth annual tall timbers fire ecology conference, August 20-21, 1970. Fredericton, New Brunswick, pp 25-46

DWYER PD, KIKKAWA J, INGRAM GJ (1979) Habitat relations of vertebrates in subtropical heathlands of coastal south-eastern Queensland. In: SPECHT RL (ed) Ecosystems of the world, vol 9A. Heathlands and related shrublands. Descriptive studies. Elsevier, Amsterdam, pp 281-299

FOX A (1978) The 1972 fire of Nadgee Nature Reserve. Parks and Wildlife 2: 5-24

FOX BJ, MCKAY GM (1981) Small mammal responses to pyric successional changes in eucalypt forest. Australian Journal of Ecology 6: 29-41

GILL AM (1975) Fire and the Australian flora: A review. Australian Forestry 38: 4-25

GREEN RH (1973) The mammals of Tasmania. The author, Tasmania, 74 pp

GROVES RH, NOBLE IR, GILL AM (1981) (eds) Fire and the Australian biota. Australian Academy of Science, Canberra, 582 pp

HUSTON M (1979) A general hypothesis of species diversity. American Naturalist 113: 81-101

KAYLL AJ (1967) Moor-burning in Scotland. In: Proceedings of the sixth annual tall timbers fire ecology conference, March 6-7, 1966, Tallahassee, Florida, pp 29-39

KEITH K, CALABY JH (1968) The New Holland mouse Psdeudomys novaehollandiae (Waterhouse) in the Port Stephens district, New South Wales. CSIRO Wildlife Research 13: 45-58

KIKKAWA J, INGRAM GJ, DWYER PD (1979) The vertebrate fauna of Australian heathlands - an evolutionary perspective. In: SPECHT RL (ed) Ecosystems of the world, vol 9A. Heathlands and related shrublands. Descriptive studies. Elsevier, Amsterdam, pp 231-275

KIMBER PC (1974) Some effects of prescribed burning on jarrah forest birds. Third fire ecology symposium, Monash University. Forestry Commission for Victoria, pp 49-57

LAWRENCE GE (1966) Ecology of vertebrate animals in relation to chaparral fire in the Sierra Nevada foothills. Ecology 47: 278-291

LEOPOLD AS (1977) The California quail. University of California Press, Berkeley, 281 pp

LUNNEY D (1978) Ecology of Rattus lutreolus. MSc Thesis. University of Sydney, Sydney, 146 pp

MAIN AR (1968) Physiology in the management of kangaroos and wallabies. Proceedings of the Ecological Society of Australia 3: 96-105

MAIN AR (1980) Fire tolerance of heathland animals. In: SPECHT RL (ed) Ecosystems of the world, vol 9B. Heathlands and related shrublands. Analytical studies. Elsevier, Amsterdam, pp 85-90

MCILROY JC (1973) Aspects of the ecology of the common wombat, Vombatus ursinus (Shaw 1800). PhD Thesis. Australian National University

NEWSOME AE (1967) A simple biological method of measuring the food supply of house mice. Journal of Animal Ecology 36: 645-650

NEWSOME AE (1969) A population study of house mice temporarily inhabiting a South Australian wheatfield. Journal of Animal Ecology 38: 341-359

NEWSOME AE (1970) An experimental attempt to produce a mouse plague. Journal of Animal Ecology 39: 299-311

NEWSOME AE, MCILROY JC, CATLING PC (1975) The effects of an extensive wildfire on populations of twenty ground vertebrates in south-east Australia. Proceedings of the Ecological Society of Australia 9: 107-123

NEWSOME AE, STENDELL RC, MYERS JH (1967) Free-watering a wild population of house mice - a test of an Australian hypothesis in California. Journal of Mammalogy 57(4): 677-686

NEWSOME AE, CATLING PC (1979) Habitat preferences of vertebrates inhabiting heathlands of coastal, montane and alpine regions of south-east Australia. In: SPECHT RL (ed) Ecosystems of the world, vol 9A. Heathlands and related shrublands. Descriptive studies. Elsevier, Amsterdam, pp 301-315

NOBLE IR, SLATYER RO (1977) Post-fire succession of plants in Mediterranean ecosystems. In: MOONEY HA, CONRAD CE (eds) Proceedings of the symposium on the environmental consequences of fire and fuel management in mediterranean ecosystems. USDA Forest Service General Technical Report WO-3, pp 27-36

O'CONNELL AM, GROVE TS, DIMMOCK GM (1979) The effects of a high-intensity fire on nutrient cycling in jarrah forest. Australian Journal of Ecology 4: 331-337

RECHER HF, LUNNEY D, POSAMENTIER H (1975) Effects of wildfire on small mammals at Nadgee Nature Reserve, New South Wales. Third fire ecology symposium, Monash University. Forestry Commission of Victoria, pp 30-36

RECHER HF, CHRISTENSEN PE (1978) Fire and the evolution of the Australian biota. In: KEAST A (ed) Ecological Biogeography of Australia, vol I. Junk, The Hague, pp 135-162

REDFIELD JA, ZWICKEL FC, BENDELL JF (1970) Effects of fire on numbers of blue grouse. Proceedings of the tenth annual tall timber fire ecology conference, Aug 20-21, 1970. Fredericton, New Brunswick, pp 63-83

ROBERTS PE (1970) Some effects of a bushfire on heathland bird life. Proceedings of the Royal Zoological Society of New South Wales, 1968-69: 4-43

SCOTTER GW (1970) Wildfires in relation to the habitat of barren-ground Caribou in the Taiga of Northern Canada. Proceedings of the tenth annual tall timbers fire ecology conference, Aug 20-21, 1970. Fredericton, New Brunswick, pp 85-105

SHARMAN GB (1959) Marsupial reproduction. In: BODENHEIMER FS, WEISBACH WW (eds) Biogeography and ecology in Australia. Monographiae Biologicae 8. Junk, The Hague, pp 332-368

SINGH G, KERSHAW AP, CLARK R (1981) Quarternary vegetation and fire history in Australia. In: GROVES RH, NOBLE IR, GILL AM (eds) Fire and the Australian biota. Australian Academy of Science, Camberra, pp 23-54

SPECHT RL (1981) Responses to fires by heathlands and related shrublands. In: GROVES RH, NOBLE IR, GILL AM (eds) Fire and the Australian biota. Australian Academy of Science, Canberra, pp 395-416

STORR GM (1964) Studies on marsupial nutrition. IV. Diet of the quokka Setonix brachyurus (Quoy and Gaimard) on Rottnest Island, Western Australia. Australian Journal of Biological Sciences 17: 469-481

TABER RD, DASMANN RF (1957) The dynamics of three natural populations of the deer Odocoileus hemionus hemionus. Ecology 38: 233-246

TYNDALE-BISCOE CH (1973) Life of marsupials. Edward Arnold, London

TYNDALE-BISCOE CH, CALABY JH (1976) Eucalypt forests as refuge for wildlife. Australian Forestry 38: 117-133

VOGL RJ (1967) Controlled burning for wildlife in Wisconsin. Proceedings of the sixth annual tall timbers fire ecology conference, March 6-7, 1966, Tallahassee, Florida, pp 47-96

WARNEKE RM (1971) Field study of the bush rat (Rattus fuscipes). Wildlife Contribution No. 14. Fisheries and Wildlife Department, Victoria, 115 pp

WIRTZ WO (1977) Vertebrate post-fire succession. In: MOONEY HA, CONRAD CE (eds) Proceedings of the symposium on the environmental consequences of fire and fuel management in mediterranean ecosystems. USDA Forest Service General Technical Report WO-3, pp 46-57

WOOD DH (1970) An ecological study of Antechinus stuartii (Marsupialia) in a south-east Queensland rainforest. Australian Journal of Zoology 18: 185-207

Section VI. Plant-Animal Interactions

There is growing evidence of an important role for plant-animal interactions in structuring the communities in mediterranean-type ecosystems.

This section contains two papers, both on herbivory. MORROW, while focusing on the genus *Eucalyptus*, shows considerable variation in the levels of insect herbivory found in sclerophyllous dominants of mediterranean-climate ecosystems. These levels often reach or exceed 20% of potential leaf area, despite apparently unpalatable leaf tissues owing to low nutrient and water contents and high contents of fibre and of secondary compounds. She reviews information relating levels of herbivory to various parameters of plant quality and concludes that sclerophylly and its correlates are insufficient to reduce herbivory. She then assesses explanations for the exceptionally high levels of herbivory in Australian sclerophylls, variously proposed in terms of plant phenology, special features of parasite-predator complexes, lack of host specificity in insect herbivores, and the disruptive effect of disturbance following human settlement.

FUENTES and ETCHEGARAY review the work on the effects of vertebrate and invertebrate herbivores on the structure of Chilean matorral. They show how herbivory in this environment affects plant community structure by regulating the relative abundances of species in the community, and determining spatial patterns of recruitment among dominants and the distribution of the understorey. The marked effect of introduced mammals results not only because predators and other controlling factors exert a weak influence on populations but also because patterns and levels of herbivory differ from those of native species since the exotics are not behaviourally adapted to the same controls and because of different foraging strategies. They conclude with an analysis by means of a cost-benefit model of the role of herbivory in determining community structure.

Not covered in this section is the role animals play in structuring communities through their effects as pollinators, dispersers and consumers of propagules. Interrelations of this kind have been documented in many instances recently (e.g. Rourke and Wiens 1977; Carpenter and Recher 1979; Berg 1981) and their possible relevance to intercontinental differences in community structure briefly mentioned (Cody and Mooney 1978). In a poster presentation not included in these proceedings, A V Milewski drew the contrasts between plant forms encountered in nearly identical physical environments in Western Australia and South Africa, specifically with regard to flowering and fruiting habits, and speculated that such differences were owing to co-adaptations between animal pollinators and dispersers and their host plants, that have persisted despite similar climates. Intercontinental comparative studies of this kind of phenomenon

---

This introduction was prepared by F J Kruger

should prove particularly profitable in explaining the evolution of non-convergence.

References

BERG RY (1981) The role of ants in seed dispersal in Australian lowland heathland. In: SPECHT RL (ed) Ecosystems of the world, vol 9B. Heathlands and related shrublands. Analytic studies. Elsevier, Amsterdam, pp 51-59

CARPENTER FL, RECHER HF (1979) Pollination, reproduction, and fire. American Naturalist 113: 871-879

CODY ML, MOONEY HA (1978) Convergence versus nonconvergence in mediterranean-climate ecosystems. Annual Review of Ecology and Systematics 9: 265-321

ROURKE JP, WIENS D (1977) Convergent floral evolution in South Africa and Australian Proteaceae and its possible bearing on pollination by nonflying mammals. Annals of the Missouri Botanic Garden 64: 1-17

26. The Role of Sclerophyllous Leaves in Determining Insect Grazing Damage

P. A. MORROW

26.1 Introduction

The vegetation of mediterranean-climate ecosystems is dominated by shrubs with sclerophyllous leaves. These evergreen, hard, leathery leaves, low in nutrient and water content and rich in secondary compounds are also characteristic of plants in nutrient-impoverished and/or droughty environments (Mooney and Dunn 1970; Dement and Mooney 1974; Janzen 1974). Convergence on the sclerophyllous leaf type is generally considered to be the result of an adaptation to limited availability of water (Cunningham and Strain 1969; Orians and Solbrig 1977) or a physiological response to low levels of nutrients, particularly phosphorus (Loveless 1961; Beadle 1966; Monk 1966; Small 1972; Grubb 1977). It has also been suggested that sclerophylly is an adaptation to the biotic stress imposed by herbivorous animals (Mooney and Dunn 1970; Janzen 1974; Schlesinger and Chabot 1977; Reader 1979).

In this paper I review properties of sclerophyllous leaves that are thought to render them unpalatable to herbivores. I only consider evidence for phytophagous insects although the implications are similar for herbivorous vertebrates. I then evaluate the generality of these herbivore defences with data from Australian Eucalyptus communities, reassess the hypothesis of sclerophylly as a defence and consider alternative hypotheses to explain characteristic levels of herbivore damage in these communities. The material discussed is not limited to studies done in mediterranean-climate ecosystems because there are few available data and because plants with sclerophyllous leaves are not restricted to these environments. This is especially the case in Australia where Eucalyptus and other sclerophyllous taxa are abundant over large areas including mediterranean zones.

26.2 Anti-herbivore properties of sclerophylls

It is increasingly thought that nutritive value influences the levels of insect damage that plants receive. "Nutritive value" and "food quality" are terms encompassing related parameters of leaf nitrogen and water content, kinds and concentrations of secondary compounds synthesized, and toughness of tissue produced. Numerous studies on many kinds of plants have explored the effects of each parameter on growth and reproduction of individual insects. Extrapolations from these studies have generated the assumption that decreased food quality will result in lower population densities of insects and consequently in less damage to host plants (Janzen 1974; Lawton and McNeill 1979).

Sclerophyllous leaves represent one extreme of food quality: they are tougher, lower in nitrogen and water and higher in secondary compounds than the plants from which much of the supporting data for hypotheses of nutritional effects on herbivores was gathered. Thus these plants should provide excellent examples of the relevance of this suite of characteristics to herbivorous insects, their populations dynamics and to the levels of damage plants receive. Each plant quality parameter is considered below.

26.2.1 PLANT NITROGEN

Nitrogen is the nutritional element most often limiting the growth and development of phytophagous insects (House 1965; Way and Cammell 1970; Southwood 1972). Fully expanded leaves contain, on a dry mass basis, from less than 1% nitrogen to about 6% nitrogen (6.25% to 38% protein) while insect tissue contains on the order of 6.6% to 8.6% nitrogen (44% to 54% protein) (House 1965; Southwood 1972; Fox and Macauley 1977; Slansky and Feeny 1977), much greater than the tissues on which they feed. Soo Hoo and Fraenkel (1966a,b) suggest that plants with less than 3% nitrogen are substandard food. By this criterion, most native plants of Wisconsin, particularly trees (Gerloff et al. 1964), numerous rainforest trees (Bevege and Richards 1970; Grubb 1977), trees and understorey species of Australian forests (Ashton 1975a; Fox and Macauley 1977), and evergreen perennials of mediterranean- climate scrubs (Cody and Mooney 1978) are substandard food.

Many studies have demonstrated either directly or indirectly that phytophagous insects individually benefit from increased nitrogen in their food. Benefits, expressed as increases in survivorship, individual growth rate, longevity, fecundity, and/or reproductive rate have been shown for aphids (Waring 1967; Dixon 1970; Whitham 1978), beetle larvae (Taylor and Bardner 1968; Fox and Macauley 1977; Onuf 1977), and lepidopterous larvae (Soo Hoo and Fraenkel 1966a,b; Feeney 1970; Onuf et al. 1977; Slansky and Feeny 1977). In contrast, there are many cases where fertilization of forest trees was deleterious to insects feeding on them (Stark 1965). The reasons for this are not clear.

Few studies relate increased nutrient content of plants to insect densities or to levels of damage sustained by plants. Onuf et al. (1977) found that various species of insects grow faster and are more abundant on mangroves fertilized by guano than on controls. Fertilization increased plant productivity by 37% but herbivore damage increased about four-fold so that net productivity increased by only 15% in fertilized plants. Nutrient enrichment of old fields (Hurd and Wolf 1974) and salt marshes (Valiela et al. 1975) also resulted in increased herbivore numbers and/or greater herbivore damage but Hurd et al. (1971) found no difference in herbivore numbers between enriched and control six-year-old fields.

Clearly, individual insects often benefit from increases in nitrogen content of their food and sometimes this individual benefit translates into greater damage to the host plant which may or may not be balanced by an increase in plant productivity. Increased nitrogen levels which result from trees being stressed seem to result in more insect damage to the tree than increasing nitrogen with fertilizers.

26.2.2 SECONDARY COMPOUNDS

Plants produce a diverse array of compounds that are not directly involved in primary metabolic processes. These secondary compounds vary markedly between species and since many have been shown to have deleterious effects on herbivores (Whittaker and Feeny 1971; Schoonhoven 1972; Freeland and Janzen 1974; Beck and Reese 1976) they are interpreted as being defensive chemicals.

Defensive chemicals in plant tissues have been divided into two major types, qualitative and quantitative, on the basis of their method of action against herbivores (Feeney 1976; Rhoades and Cates 1976). Qualitative defensive compounds include highly toxic chemicals present in low concentrations and typify herbaceous plants. They will not be discussed further.

Quantitative defences reduce the nutritional value of the plant tissues containing them, thus starving rather than poisoning the herbivore. Typical quantitative compounds are tannins and polyphenolic resins which may form indigestible complexes with soluble plant proteins and polysaccharides (Feeny 1969, 1970; Rhoades 1977). Other compounds such as cellulose, lignins and essential oils are indigestible or non-nutritive and reduce food value through dilution and by making tissues difficult to ingest (Feeney 1976; Swain 1979). Quantitative defensive compounds are not a very diverse group of chemicals, are present in quantities as high as 60% of leaf dry mass and act in a dosage-dependent fashion over a wide range of concentrations (Feeney 1976; Rhoades and Cates 1977). They are characteristic of woody perennials (Swain 1977).

Because the proposed mode of action of digestibility-reducing substances is so general, they are often supposed not to be important in determining which or how many species of herbivores will attack the plant. Rather, their presence should affect insect fitness and abundance in the same way as decreasing nitrogen content. Supporting data for this view are provided by studies of winter moth caterpillars feeding on oaks (Feeny 1970) and woolly aphids feeding on apple leaves (Sen Gupta 1975). Sclerophylls are particularly well endowed with presumptive quantitative defences (Janzen 1974; Fox and Macauley 1977; Reader 1979).

26.2.3 LEAF AVAILABILITY

Plant tissues vary in nutrient content not only inter- and intraspecifically, but also between different parts of a single plant. Leaf position in the crown, tissue developmental stage, age (Rhoades and Cates 1976) and the nutrient status of the soil (e.g. Onuf et al. 1977) contribute to differences in nutritive value. Nitrogen is often limiting to plant growth (Bradshaw 1969), it is one of the most mobile nutrients and is internally recycled from old tissues to meristematic regions (Loneragan et al. 1976). Thus buds and unexpanded leaves may have nitrogen levels adequate to support insect growth, while only a few centimetres away, mature tissue may contain insufficient quantities (Table 1). In addition, mature leaves are tougher, lower in water content and higher in content of quantitative chemicals (Jung et al. 1964; Feeny 1970, 1976; Dement and

Table 1. Nitrogen content of sclerophylls in relation to leaf development. Triticum included as a "standard".

Species	Plant family[a]	Growth form[b]	N, % leaf dry mass, leaf stage[c]					Authority
			1	2	3	4	5	
Paropsis velutina	Fb	DP	6.80	5.23	4.38			Cates and Rhoades 1977
Paropsis chilensis	Fb	DP	6.01	3.44	3.34			Cates and Rhoades 1977
Triticum aestivalis	P	HA			6.00		1-1.5	Milthorpe and Moorby 1974
Quercus robur	F	DP	5.0	3.2	2.0			Feeny 1970
Kalmia polifolia	E	EP	2.3	1.8	1.6	1.4	0.6	Reader 1978
Pomaderris aspera	Rh	EP	2.06		2.38			Ashton 1975b
Olearia argophylla	A	EP	2.08		1.53			Ashton 1975b
Chamaedaphne calyculata	E	EP	1.8	1.6	1.5	1.2	0.6	Reader 1978
Eucalyptus regnans	M	EP	1.74		1.52		0.99	Ashton 1975b
Eucalyptus pauciflora	M	EP	2.01		1.52	1.58	1.24	Morrow, unpublished
Eucalyptus stellulata	M	EP	1.28			1.41		Morrow, unpublished
Rhizophora mangle	Rz	EP	1.55		1.20		0.36	Onuf et al. 1977
Rhizophora mangle, fertilized			1.42		1.60		0.48	Onuf et al. 1977
Lyonia lucida	E	EP	1.30		1.10		0.74	Schlesinger and Chabot 1977

[a] Fb: Fabaceae; P: Poaceae; F: Fagaceae; Rh: Rhamnaceae; A: Asteraceae; M: Myrtaceae; Rz: Rhizophoraceae; E: Ericaceae

[b] E: evergreen; D: deciduous; H: herbaceous; A: annual; P: perennial

[c] Leaf phenological stage: 1: buds and expanding leaves; 2: fully expanded, soft leaves; 3: fully mature leaves (first season); 4: fully mature leaves (second season); 5: senescent

Mooney 1974; Rhoades and Cates 1976) which may restrict the rate at which food can be ingested and digested (Williams 1954; Southwood 1972).

Differences in nitrogen and water content and in toughness between immature and mature leaves are especially pronounced in sclerophyllous plants adapted to nutrient-deficient substrates (e.g. Rhizophora, Lyonia, Eucalyptus, Ledum, Chamaedaphne, Kalmia in Table 1). Since a large percentage of plant tissues may be only marginally adequate food, many herbivores eat only immature leaves (White 1971; Harper 1977).

Where insects are restricted to eating young tissues, insect survivorship and the resulting levels of damage to plants will be strongly influenced by the amount and timing of tissue production. For instance, successful insect attack on oak, aspen and balsam fir depends on synchrony of larval emergence and food availability, either new shoots (Feeny 1970; Whitham 1978) or new shoots plus staminate flowers (Edit and Cameron 1971). Timing of tissue production is influenced by defoliation levels (Kulman 1971; Rockwood 1974; Heichel and Turner 1976), termination of drought (Owen and Watson 1956; Hanna 1971) and associated changes in nitrate levels available to the plants (Robinson and Gacoka 1962) and fertilization (Onuf et al. 1977).

If herbivore load and/or level of grazing are negatively correlated with food quality then, by all standards above, sclerophylls ought to be free from significant herbivore damage. Sclerophyllous vegetation occupying tropical white sand soils and other extremely impoverished soils normally supports very low densities of herbivorous insects and plants are virtually free from attack (Janzen 1974). Janzen correlated this absence of herbivores to the very low productivity and high content of quantitative secondary metabolites of the plants.

The hypothesis that correlates of sclerophylly prevent substantial herbivore attack is logically satisfying but difficult to test or to distinguish from several other hypotheses to explain generally low levels of attack on plants. For instance, low levels of grazing may be the result of regulatory pressures exerted by natural enemies of herbivores (e.g. Debach 1974) or may result from herbivores being so successful in reducing the densities of their host populations that they are able to find only a small proportion of the remaining host individuals (Root 1973; Harper 1977). Supporting data for these hypotheses come primarily from the literature on biological control.

26.3 Herbivores in Australian sclerophyll communities

Sclerophyllous vegetation is widespread in Australia. In addition to its expected presence in mediterranean-climate areas, this vegetation type occupies extensive areas of nutrient-poor soils from the semi-arid interior to the moist coastal fringe (Hubble 1970; Moore and Perry 1970). Predominant in this vegetation is the genus Eucalyptus, a myrtaceous group of more than 550 species. Apparently unlike sclerophyll communities on other continents, herbivory is conspicuous in Australia. This has been well-documented for Eucalyptus where insects chronically remove 5-45% of annual leaf area in addition to damage caused by bud-feeding, sap-sucking and stem-boring insects (Readshaw and Mazanec 1969; Burdon and Chilvers 1974; Morrow 1977a,b; Morrow and LaMarche 1978; Fox and Morrow 1981, in preparation). Herbivores are also abundant in Australian sclerophyll communities dominated by other genera. For instance, Leptospermum heaths have been estimated to lose 20-50% of annual production to herbivorous insects (Jones 1968). These figures contrast to annual removal rates of

3-10% of leaf area in temperate northern hemisphere forest communities (Bray 1964; Elton 1966; Gosz et al. 1972; Mattson and Addy 1975; Nielsen and Ejlersen 1977) and to even lower levels in South African and Californian mediterranean climate scrubs (F Kruger, personal communication 1980; P A Morrow, personal observation).

Herbivorous insects in Australian sclerophyll communities are diverse and abundant. Feeding damage generally results from attack by numerous species of insects (Morrow 1977a) although there may be outbreaks where one species is primarily responsible for defoliating extensive areas of vegetation (Readshaw and Manazec 1969; Morrow 1977b). Susceptibility to insect attack can vary considerably between individuals within a population and among coexisting species. Because of this variation and because background levels of herbivore attack are consistently heavy it has been possible to investigate properties of individual plants and species which are hypothesized to influence susceptibility to attack. Studies have centred on *Eucalyptus* species.

26.3.1 SECONDARY COMPOUNDS

Eucalyptus leaves contain two important classes of secondary compounds, essential oils (Penfold and Willis 1961) and phenolics, including tannins (Hillis 1966). Properties of both groups of compounds suggest that they are quantitative defences, that is, plants with higher concentrations should be better defended than those with lower concentrations.

The concentration of total phenolics and tannins in eucalypts ranges from 4-40% and 0-27%, respectively, of leaf dry mass (Macauley and Fox 1980), as high or higher than concentrations thought to inhibit growth of insects on other groups of plants (Feeny 1970).

Noting the apparent paradox of heavy feeding on eucalypt leaves despite their high concentrations of phenolics, Fox and Macauley (1977) tested the hypothesis that feeding rate would increase with increasing tannin content provided nitrogen levels remained constant. They assumed that on plants with higher tannin contents, insects would have to consume more to compensate for nitrogen bound in tannin-protein complexes. Generally the phenolic and tannin content of leaves is very low in expanding leaves and increases throughout the growing season as leaves age (Feeny 1970; Lawton 1976). However, in eucalypts even young tissues have high concentrations of these compounds (Macauley and Fox 1980). Fox and Macauley (1977) measured growth parameters of larvae of *Paropsis atomaria*, one of a complex of over 200 species of paropsine chrysomelids that feed on *Eucalyptus*. By using tender young leaves with a wide range of phenolic and tannin contents they were able to distinguish between direct effects of these compounds and the confounding effects of toughness and increased cell wall material that accompany leaf ageing.

No index of feeding efficiency, growth or survivorship of *P. atomaria* reflected content of phenolics in their food. Although Fox and Macauley (1977) showed that eucalypt tannins are capable of binding proteins in vitro, their experiments suggest that eucalypt proteins or insect enzymes are not effectively bound in the insect gut. Since tannin binding is strongly pH dependent, they suggested that the strongly alkaline gut pH of many insect groups, including chrysomelid beetles, may render tannins relatively ineffective. A recent survey of gut pH in Lepidoptera that feed on tannin-rich and herbaceous plants supports this suggestion (Berenbaum 1980).

The second group of presumptive defensive compounds characteristic of

Eucalyptus are essential oils. Essential oils are complex mixtures of volatile terpenoids (Loomis and Croteau 1973) which, in *Eucalyptus*, are accumulated in punctate leaf glands (Carr and Carr 1970) in quantities as high as 22% of leaf dry mass. The quantity of oil produced may vary tremendously both intra- and interspecifically. For instance, adjacent individuals of *E. viminalis* Labill had oil yields of 3.7% and 21.5% of leaf dry mass and average oil yields for adjacent populations of *E. melliodora* A. Cunn ex Schau and *E. dives* Schau were 3.6% and 11.6%, respectively (Morrow and Fox 1980).

Essential oils are known to function as attractants, repellants and feeding stimulants to herbivorous insects (Schoonhoven 1972; Levin 1976). In addition, two studies suggest that increasing content of essential oils will decrease the food value of leaves even to adapted herbivores. Southwell (1974) found that koalas detoxify eucalypt oils by converting them to glucoronide esters which are then excreted in the urine, and Brattsten et al. (1977) showed that eight terpenes, all of which are commonly found in eucalypt oils, induce a dosage-dependent increase in the activity of mixed function oxidases (generalized detoxification enzymes) in the guts of *Spodoptera eridania* larvae, a polyphagous lepidopteran. These studies suggested that growth, development and reproductive rates of herbivores should decrease with an increase in oil content of their food as the resources expended on detoxification increase (Brattsten 1979).

Morrow and Fox (1980) hypothesized that essential oils decrease growth and reproductive parameters of individual insects and, in doing so, decrease herbivore population densities and grazing damage on trees with higher oil yields. The hypothesis was tested for five *Eucalyptus* species which varied in average oil content from 1.2% to 11.6% of leaf dry mass and had similar average intraspecific levels of damage. Intraspecific variability in both oil content and susceptibility to attack was high, there was no major intraspecific variation in the individual compounds comprising the essential oils. For three species with relatively low yields of oil, no correlation was found between oil yield and herbivore damage. Only for the two highest yielding species were individuals with the highest oil content less damaged than low-yielding conspecifics. However, censuses of insect eggs and adults on the trees did not show any difference in herbivore distribution that might account for the heavier attack on these low-yielding trees. There was no effect of leaf oil content on rates of insect growth and development. Analysis of leaf oils before and after passage through the guts of three *Eucalyptus*-feeding insect species showed that they did not change the composition of the oils and, presumably, did not detoxify them. Thus, if especially high essential oil yields do result in less overall damage to the plant, it does not appear to be the result of oils altering any parameters of the growth of individual insects.

Several recent studies in other systems have also failed to find a relationship between concentration of presumptive quantitative defensive compounds in plant tissues and feeding and/or growth parameters of herbivorous insects. No correlation was found between tannin content and insect herbivory in tropical forest ferns (Balick et al. 1978; Cooper-Driver 1978). Several grasshopper species that normally feed on perennial plants were unaffected by condensed or hydrolyzable tannins, but one species that feeds on grasses (which do not normally contain tannins) was adversely affected by tannins absorbed through the gut wall (Bernays 1978).

I do not suggest that terpenoid and phenolic compounds never reduce growth of herbivorous insects, only that insects are able to counter such chemicals and that their effectiveness as defences needs to be demonstrated and not assumed.

26.3.2 NITROGEN

Secondary compounds in eucalypts do not directly alter the availability of nitrogen to tested herbivores. However, the absolute levels of nitrogen in the leaves are low, generally less than 2%, even in expanding leaves. In the studies discussed above (Fox and Macauley 1977; Morrow and Fox 1980), survivorship, feeding, growth and nitrogen use in Paropsis larvae were found to be very sensitive to leaf nitrogen content.

Fox and Macauley (1977) fed larvae on young, expanding leaves with a wide range of nitrogen contents. Some leaves, from eucalypts growing on extremely impoverished soils, contained less than 1% nitrogen and failed to satisfy minimal nitrogen needs of the larvae. These plants also had little herbivore attack suggesting that below ca. 1% nitrogen, plants do not support herbivore growth. On trees whose leaves contain more than 1% nitrogen, grazing could be very heavy and Fox and Macauley (1977) suggested that a partial explanation for heavy grazing may be that each larva must consume large amounts of leaf in order to satisfy its nitrogen demands since even at the highest levels of nitrogen encountered they are still limited by this nutrient. They caution that although this may explain why individuals eat so much, it is an explanation that should not be extrapolated to community-level response of herbivores.

26.3.3 LEAF TOUGHNESS

Adults of some common beetle species feed on mature Eucalyptus leaves (Carne et al. 1974) but toughness, and the high fibre and low nitrogen and water contents correlated with toughness, restricts most species to younger tissues, particularly in early instars. Patterns of leaf production are thus of central importance to herbivorous insects.

In Australian sclerophylls, growth begins in spring and may continue in bursts throughout the summer whenever heavy showers break soil drought (Specht and Rayson 1957; Jones 1967; Specht 1969). Even in mediterranean-type climates, sclerophylls initiate new leaf tissue during the summer and so are unlike their counterparts on other continents (Specht 1969; Cody and Mooney 1978). Sclerophylls also refoliate readily following defoliation by fire or insects (Penfold and Willis 1961). Among eucalypts, production of new leaves following insect defoliation is often asynchronous within a population and depends on when and how heavily individual trees were damaged (Jacobs 1955; Cremer 1972, 1973; Morrow, personal observation; Ashton 1975a). The end result of these growth and refoliation patterns is that new tissue may be available to insects throughout much of the growing season, allowing them to complete several generations per year (L R Fox, personal communication). The readiness to refoliate may, to the extent of its uniqueness to these systems, help explain the high background levels of attack seen in Australia.

26.4 Discussion

Sclerophyllous genera growing in a variety of Australian climates represent a broad exception to the generalization that characteristics of sclerophyllous leaves render them immune to herbivore attack. For *Eucalyptus* species, presumptive quantitative defensive compounds do not affect herbivore growth parameters or explain levels of insect damage. Low levels of leaf nitrogen do decrease herbivore growth, but this does not result in low herbivore biomass or low levels of damage to trees.

The presumptive defences of sclerophylls are considered to be so generalized in their digestibility-reducing properties that herbivores cannot evolve resistance to them. This contrasts to qualitative defences such as toxins to which insects in particular may rapidly evolve resistance (Feeny 1976; Cates and Rhoades 1977; Fox 1981). That plants with ample quantitative defences support high densities of many insect species throughout Australia suggests that other or additional factors are involved in determining background levels of herbivore abundance on this and, quite likely, on other continents.

I have discussed the possible role that leaf phenology may have in contributing to high herbivore loads. Below, I briefly discuss three other factors that may contribute; I consider supporting and detracting evidence but there are so few appropriate data available that the ideas must be regarded as speculative. Some of the ideas are testable, some are historical and rely on biogeographic comparisons.

(a) Lawton and McNeill (1979) suggest that the high background levels of attack in Australia must reflect a parasite-predator complex that is ineffective in controlling insect populations. We have virtuālly no comparative data with which to evaluate this suggestion. The little that does exist suggests that invertebrate parasites and predators are no less abundant in Australia than elsewhere. For instance, parasite loads of leaf-eating chrysomelid beetles are consistently higher in Australia than in Europe and North America (L R Fox, personal communication) and predatory ants are very abundant in most Australian comunities (CSIRO 1970). However, while perhaps equally abundant, other features of invertebrate natural enemies might result in lower levels of host population control. For instance, if herbivore emergence times are not predictably timed because host plant phenology is tied to unpredictable environmental events, then parasite emergence and host finding might in turn be poorly coordinated with the availability of the host insects. By such a scenario, herbivore populations could escape parasite control more frequently than in more predictable habitats.

In contrast to the invertebrate parasite-predator complex, vertebrate predators, specifically foliage-gleaning insectivores, appear scarce compared to other communities with which I am familiar. Foliage-feeding insects, including many diurnally active species, are abundant throughout the growing season and seem sufficient to support a larger bird community. If insectivorous bird densities are substantially lower in Australia than elsewhere, this might be related to the kinds and abundance of resources available to the birds as a result of their intracontinental movement patterns (Nix 1976). Insects are scarce during the temperate zone winter months. At the same time, the majority of the Australian tropics are dry and so do not provide an alternative resource area to which birds can migrate. This shortgage of insects is not relieved by an availability of fleshy fruits since relatively few Australian species produce them. Nectar provides a winter resource for many birds, particularly for honey-eaters

(Melipagidae) in the temperate zone heathlands, although it is both patchy and unpredictable (Keast 1968). These factors may result in a low winter carrying capacity that restricts bird populations: their numbers may be too small to have much impact on insect populations building up each spring.

Data to support or refute this speculation will be difficult to obtain. Data required are estimates of bird and insect densities from similar environments on different continents, estimates of winter food resources and a better knowledge of migration and movement patterns of birds within Australia. An estimate of bird and insect densities in Australian mediterranean areas would be a useful first addition since these could be compared to counts made in the other mediterranean regions by Cody (Cody and Mooney 1978).

(b) High herbivore densities could be related to the number of alternative host species available to an insect in a particular community. Insects that have available only one potential host plant are very sensitive to changes in abundance of that plant. Species able to use more than one species of plant locally will be more buffered because they can switch to less-preferred species if the normal host becomes temporarily scarce (Fox and Morrow 1981). Alternative hosts are most likely to be closely related to the preferred host.

Australia's plant communities can be extremely rich in species, particularly those on nutrient-impoverished soils, but they tend to belong to a relatively small number of families and genera. *Eucalyptus* and *Leptospermum* in the Myrtaceae, *Banksia* and *Hakea* in the Protaceae and *Acacia* in the Fabaceae are examples. As a consequence, there are usually a number of closely related species in a community that an insect might be able to use if its preferred host is temporarily unavailable (Clark 1962; Carne 1965). In communities with fewer species-rich genera a lack of alternative hosts could result in insect population crashes whenever host species become scarce or are defoliated. Low herbivore densities could be the norm if this happended with any regularity.

(c) Heavy herbivore feeding damage may be a recent phenomenon resulting from disturbance following European settlement. Europeans have made significant changes in forests and woodlands in the ca. 200 years since settlement of Australia. Clearing for pastures has increased habitat for scarab beetles, a major defoliator of *Eucalyptus* (Carne et al. 1974) and, especially in the last decade, have massively disturbed predicted overwintering areas for southwestern and southeastern bird populations (Nix 1976).

While it is clear that disturbance has increased insect attack in some communities, other heavily attacked communities have remained largely undisturbed (Burdon and Chilvers 1974; Morrow and LaMarche 1978; Fox and Morrow, in preparation). It is important to know whether high background levels of herbivore attack are a recent or a much older feature of the Australian environment. If recent, changes that Europeans have imposed would provide clues to factors that allow or encourage insect attack. Estimates of pre-settlement levels of damage might be obtained from examination of drawings and herbarium specimens made by early explorers of Australia. Since collectors tend to take the best available plant specimens, moderate levels of damage on the specimens would indicate minimal levels of herbivore damage in undisturbed communities.

A combination of any or all of these factors as well as others I have not discussed, may affect background levels of insect attack. Data will, in time, become available to evaluate some aspects of these ideas.

I have tried to demonstrate that features of sclerophylls do not prevent herbivores from consuming them and that additional or alternative reasons for characteristic background levels of insect attack should be sought not only for the high herbivore loads in Australia but for the very low herbivore densities that are common in mediterranean-climate and other sclerophyllous vegetation outside Australia. This raises a further question that should be considered: If phenolics, terpenes and other secondary compounds produced in abundance by sclerophyllous plants are or can be ineffective against herbivorous insects, why do plants continue to allocate large amounts of carbon to their production? I shall not attempt to explore this here except to point out that these compounds may play roles in defence against pathogens and vertebrate herbivores, in litter decomposition, soil evaporation, nutrient recycling or in fire promotion.

Acknowledgements

Research summarized here has been supported by the National Science Foundation under Grant DEB 7904953, the Department of Environmental Biology, Australian National University and the Office of International Programs and the Graduate School of the University of Minnesota. I thank H Mooney for encouraging my participation in this conference, L R Fox for stimulating discussions, D Tilman for reading and D Johnson and L Carlson for typing the manuscript.

References

ASHTON DH (1975a) Seasonal growth of Eucaltyptus regnans F. Muell. Australian Journal of Botany 23: 239-252

ASHTON DH (1957b) Studies of litter in Eucalyptus regnans forests. Australian Journal of Botany 23: 413-433

BALICK MJ, FURTH DG, COOPER-DRIVER G (1978) Biochemical and evolutionary aspects of arthropod predation on ferns. Oecologia 35: 55-89

BEADLE NCW (1966) Soil phosphate and its role in molding segments of the Australian flora and vegetation, with special reference to xeromorphy and sclerophylly. Ecology 47: 992-1007

BECK SC, REESE JC (1976) Insect-plant interactions: nutrition and metabolism. Recent Advances in Phytochemistry 10: 41-92

BERENBAUM M (1980) Adaptive significance of midgut pH in larval Lepidoptera. American Naturalist 115: 138-146

BERNAYS EA (1978) Tannins: an alternative viewpoint. Entomologia Experimentalis et Applicata 24: 44-53

BEVEGE DI, RICHARDS BN (1970) Nitrogen in the growth of Araucaria cunninghamii Ait. underplanted in Pinus stands. Ecology 51: 134-142

BRADSHAW AD (1969) The ecologist's viewpoint. In: RORISON IH (ed) Ecological aspects in the mineral nutrition of plants. Blackwell, Oxford. pp 415-427

BRATTSTEN LB (1979) Biochemical defense mechanisms in herbivores against plant allelochemicals. In: ROSENTHAL GA, JANZEN DH (eds) Herbivores: their intereaction with secondary plant metabolites. Academic Press, New York, pp 200-270

BRATTSTEN LB, WILKINSON CF, EISNER T (1977) Herbivore-plant interactions: mixed-function oxidases and secondary plant substances. Science 196: 1349-1352

BRAY JR (1964) Primary consumption in three forest canopies. Ecology 45: 165-167
BURDON JJ, CHILVERS GA (1974) Fungal and insect parasites contributing to niche differentiation in mixed species stands of eucalypt saplings. Australian Journal of Botany 22: 103-114
CARNE PB (1965) Distribution of the eucalypt-defoliating sawfly, Perga affinis affinis (Hymenoptera). Australian Journal of Zoology 13: 593-612
CARNE PB, GREAVES RTG, MCINNES RS (1974) Insect damage to plantation-grown eucalypts in north coastal New South Wales, with particular reference to Christmas Beetles (Coleoptera: Scarabaeidae). Journal of the Australian Entomological Society 13: 189-206
CARR DJ, CARR SJM (1970) Oil glands and ducts in Eucalyptus L'Herit. II. Development and structure of oil glands in the embryo. Australian Journal of Botany 18: 191-212
CATES RG, RHOADES DF (1977) Prosopis leaves as a resource for insects. In: SIMPSON BB (ed) Mesquite biology in two desert ecosystems. Dowden, Hutchinson and Ross, Stroudsburg, Pennsylvania, pp 61-84
CLARK LR (1962) The general biology of Cardiaspina albitextura (Psyllidae) and its abundance in relation to weather and parasitism. Australian Journal of Zoology 10: 537-586
CODY ML, MOONEY HA (1978) Convergence versus nonconvergence in mediterranean-climate ecosystems. Annual Review of Ecology and Systematics 9: 265-321
COOPER-DRIVER GA (1978) Insect-fern associations. Entomologia Experimentalis et Applicata 24: 110-116
CREMER KW (1972) Effects of partial defoliation and disbudding on height growth of Eucalyptus regnans saplings. Australian Forest Research 6: 41-42
CREMER KW (1973) Ability of Eucalyptus regnans and associated evergreen hardwoods to recover from cutting or complete defoliation in different seasons. Australian Forest Research 6: 9-22
CSIRO, DIVISION OF ENTOMOLOGY (1970) The insects of Australia. Melbourne University Press, Carleton, 1029 pp
CUNNINGHAM GL, STRAIN BR (1969) An ecological significance of seasonal leaf variability in a desert shrub. Ecology 50: 400-408
DEBACH P (1974) Biological control by natural enemies. Cambridge University Press, London, 323 pp
DEMENT WE, MOONEY HA (1974) Seasonal variation in the production of tannins and cyanogenic glucosides in the chaparral shrub, Heteromeles arbutifolia. Oecologia 15: 65-76
DIXON AFG (1970) Quality and availability of food for a sycamore aphid population. In: WATSON A (ed) Animal populations in relation to their food resources. Blackwell, Oxford, pp 271-287
EDIT DC, CAMERON MD (1971) Delayed budbreak and spruce budworm survival. Bimonthly Research Notes 27: 28-29
ELTON CS (1966) The pattern of animal communities. Methuen, London, 181 pp
FEENY P (1969) Inhibitory effect of oak leaf tannins on the hydrolysis of proteins by trypsin. Phytochemistry 8: 2119-2126
FEENY P (1970) Seasonal changes in oak leaf tannins and nutrients as a cause of spring feeding by winter moth caterpillars. Ecology 51: 565-581
FEENY P (1976) Plant apparency and chemical defence. Recent Advances in Phytochemistry 10: 1-40

FOX LR (1981) Defense and dynamics in plant-herbivore systems. American Zoologist. In press.
FOX LR, MACAULEY BJ (1977) Insect grazing on Eucalyptus in response to variation in leaf tannins and nitrogen. Oecologia 29: 145-162
FOX LR, MORROW PA (1981) Specialization: species property or local phenomenon? Science. In press
FREELAND WJ, JANZEN DH (1974) Strategies in herbivory by mammals: the role of plant secondary compounds. American Naturalist 108: 269-289
GERLOF GC, MOORE DG, CURTIS JT (1966) Selective absorption of mineral elements by native plants of Wisconsin. Plant and Soil 25: 393-405
GOSZ FR, LIKENS GE, BORMANN FH (1972) Nutrient content of litter fall on the Hubbard Brook experimental forest, New Hampshire. Ecology 53: 769-784
GRUBB PJ (1977) Control of forest growth and distribution on wet tropical mountains. Annual Review of Ecology and Systematics 8: 83-109
HANNA LW (1971) The effects of water availability on tea yields in Uganda. Journal of Applied Ecology 8: 791-813
HARPER JL (1977) Population biology of plants. Academic Press, London, 892 pp
HEICHEL GH, TURNER NC (1976) Phenology and leaf growth of defoliated hardwood trees. In: ANDERSON J, KAYA H (eds) Perspectives of forest pest management. Academic Press, New York, pp 131-140
HILLIS WE (1966) Polyphenols in the leaves of Eucalyptus L'Herit: a chemotaxonomic survey. I. Introduction and a study of the series Globulares. Phytochemistry 5: 1075-1090
HOUSE HL (1965) Effects of low levels of the nutrient content of a food and of nutrient imbalances on the feeding and the nutrition of a phytophagous larva, Celerio euphorbiae (Linnaeus) (Lepidoptera: Sphingidae). Canadian Entomologist 97: 62-68
HUBBLE GD (1970) Soils. In: MOORE RM (ed) Australian grasslands. Australian National University Press, Canberra, pp 44-58
HURD LE, MELLINGER MV, WOLF LL, MCNAUGHTON SJ (1971) Stability and diversity at three trophic levels in terrestrial successional ecosystems. Science 173: 1134-1136
HURD LE, WOLF LL (1974) Stability in relation to nutrient enrichment in arthropod consumers of old field successional ecosystems. Ecological Monographs 44: 465-482
JACOBS MR (1955) Growth habits of the Eucalypts. Commonwealth Government Printer, Canberra, 262 pp
JANZEN DH (1974) Tropical blackwater rivers, animals, and mast fruiting by the Dipterocarpaceae. Biotropica 6: 69-103
JONES R (1967) Productivity and water use efficiency of a Victorian heathland. PhD Thesis. University of Melbourne
JONES R (1968) Estimating productivity and apparent photosynthesis from differences in consecutive measurements of total living plant parts of an Australian heathland. Australian Journal of Botany 16: 589-602
JUNG GA, LILLY B, SHIH SC, REID RL (1964) Studies with sudangrass. I. Effect of growth stage and level of nitrogen fertilizer upon yield of dry matter: estimated digestibility of energy, dry matter and protein; amino acid composition; and prussic acid potential. Agronomy Journal 56: 533-537
KEAST A (1968) Seasonal movements in the Australian honey-eaters (Meliphagidae) and their ecological significance. The Emu 67: 159-204
KULMAN HM (1971) Effects of insect defoliation on the growth and mortality of trees. Annual Review of Entomology 16: 289-324

LAWTON JH (1976) The structure of the arthropod community on bracken. Botanical Journal of the Linnean Society 73: 187-216
LAWTON JH, MCNEILL S (1979) Between the devil and the deep blue sea: on the problem of being a herbivore. Symposium of the British Ecological Society 20: 223-244
LEVIN DA (1976) The chemical defences of plants to pathogens and herbivores. Annual Review of Ecology and Systematics 6: 121-159
LONERAGAN JF, SNOWBALL K, ROBSON AD (1976) Remobilization of nutrients and its significance in plant nutrition. In: WARDLAW IF, PASSIOURA JB (eds) Transport and transfer processes in plants. Academic Press, New York, 484 pp
LOOMIS WD, CROTEAU R (1973) Biochemistry and physiology of lower terpenoids. Recent Advances in Phytochemistry 6: 147-185
LOVELESS AR (1961) A nutritional interpretation of sclerophylly based on differences in the chemical composition of sclerophyllous and mesophytic leaves. Annals of Botany 26: 168-184
MACAULEY BJ, FOX LR (1980) Variation in total phenols and condensed tannins in Eucalyptus: leaf phenology and insect grazing. Australian Journal of Ecology 5: 31-35
MATTSON WJ, ADDY ND (1975) Phytophagous insects as regulators of forest primary production. Science 190: 515-522
MILTHORPE FL, MOORBY J (1974) An introduction to crop physiology. Cambridge University Press, London, 202 pp
MONK CD (1966) An ecological significance of evergreenness. Ecology 47: 504-505
MOONEY HA, DUNN EL (1970) Convergent evolution of mediterranean-climate evergreen sclerophyll shrubs. Evolution 24: 292-303
MOORE RM, PERRY RA (1970) Vegetation. In: MOORE RM (ed) Australian grasslands. Australian National University Press, Canberra, pp 59-73
MORROW PA (1977a) Host specificity of insects in a community of three codominant Eucalyptus species. Australian Journal of Ecology 2: 89-106
MORROW PA (1977b) The significance of phytophagous insects in the Eucalyptus forests of Australia. In: MATTSON WJ (ed) The role of Arthropods in forest ecosystems. Springer-Verlag, Berlin, pp 19-30
MORROW PA, LAMARCHE VC Jr (1978) Tree ring evidence for chronic insect suppression of productivity in subalpine Eucalyptus. Science 201: 1244-1246
MORROW PA, FOX LR (1980) Effects of variation in Eucalyptus essential oil yield on insect growth and grazing damage. Oecologia 45: 209-219
NIELSEN BO, EJLERSEN A (1977) The distribution pattern of herbivory in a beech canopy. Ecological Entomology 2: 293-299
NIX HA (1976) Environmental control of breeding, post-breeding dispersal and migration of birds in the Australian region. Proceedings of the sixteenth international ornithological congress, Canberra City, Australia, August 1974, pp 272-305
ONUF CP (1977) Nutritive value as a factor in plant-insect interactions with an emphasis on field studies. In: MONTGOMERY G (ed) The ecology of arboreal folivores. Smithsonian Press, Washington DC, pp 85-96
ONUF CP, TEAL JM, VALIELA I (1977) Interactions of nutrients, plant growth and herbivory in a mangrove ecosystem. Ecology 58: 514-526
ORIANS G, SOLBRIG O (1977) A cost-income model of leaves and roots with special reference to arid and semi-arid areas. American Naturalist 111: 677-690
OWEN PC, WATSON DJ (1956) Effect on crop growth of rain after prolonged drought. Nature 177: 847
PENFOLD AR, WILLIS JL (1961) The Eucalypts. Interscience, New York, 551 pp

READER RJ (1978) Contribution of overwintering leaves to the growth of three broad-leaved, evergreen shrubs belonging to the Ericaceae family. Canadian Journal of Botany 56: 1248-1261

READER RJ (1979) Impact of leaf-feeding insects on three bog ericads. Canadian Journal of Botany 57: 2107-2112

READSHAW JL, MAZANEC Z (1969) Use of growth rings to determine past phasmatid defolations of alpine ash forest. Australian Forestry 33: 29-36

RHOADES DF (1977) The anti-herbivore defences of Larrea. in MABRY TJ, HUNZIKER J, DIFEO DR (eds) Cresote bush: biology and chemistry of Larrea in New World deserts. Dowden, Hutchinson and Ross, Stroudsburg, Pennsylvania, 368 pp

RHOADES DF, CATES RG (1976) Towards a general theory of plant antiherbivore chemistry. Recent Advances in Phytochemistry 10: 168-213

ROBINSON JBD, GACOKA P (1962) Evidence of the upward movement of nitrate during the dry season in the Kikuyu red loam coffee soil. Journal of Soil Science 13: 133-139

ROCKWOOD LL (1974) Seasonal changes in the susceptibility of Crescentia alata leaves to the flea beetle, Oedionychus sp. Ecology 55: 142-148

ROOT RB (1973) Organization of a plant-arthropod association in simple and diverse habitats: the fauna of collards (Brassica oleraceae). Ecological Monographs 43: 95-124

SCHLESINGER WH, CHABOT BF (1977) The use of water and minerals by evergreen and deciduous shrubs in Okefenokee Swamp. Botanical Gazette 138: 490-497

SCHOONHOVEN LM (1972) Secondary plant substances and insects. Recent Advances in Phytochemistry 5: 197-225

SEN GUPTA GC, MILES PW (1975) Studies on the susceptibility of varieties of apple to the feeding of two strains of woolly aphids (Homoptera) in relation to the chemical content of the tissues of the host. Australian Journal of Agricultural Research 26: 127-136

SLANSKY F Jr, FEENY P (1977) Stabilization of the rate of nitrogen accumulation by larvae of the cabbage butterfly on wild and cultivated food plants. Ecological Monographs 47: 209-228

SMALL E (1972) Photosynthetic rates in relation to nitrogen recycling as an adaptation to nutrient deficiency in peat bog plants. Canadian Journal of Botany 50: 2227-2233

SOO HOO CF, FRAENKEL G (1966a) The selection of food plants in a polyphagous isect, Prodenia eridania (Cramer). Journal of Insect Physiology 12: 693-709

SOO HOO CF, FRAENKEL G (1966b) The consumption, digestion and utilization of food plants by a polyphagous insect, Prodenia eridania (Cramer). Journal of Insect Physiology 12: 711-730

SOUTHWELL IA (1974) Novel terpenoid rearrangements in the koala. Paper delivered to the sixth international congress of essential oils. San Francisco, September 1974

SOUTHWOOD TRE (1972) The insect-plant relationship - an evolutionary perspective. In: VAN EMDEN HF (ed) Insect-plant relationships. Wiley, New York, pp 3-30

SPECHT RL (1969) A comparison of the sclerophyllous vegetation characteristic of mediterranean-type climate in France, California and southern Australia. I. Structure, morphology and succession. Australian Journal of Botany 17: 277-292

SPECHT RL, RAYSON P (1957) Dark Island heath (Ninety-Mile Plain, South Australia). Australian Journal of Botany 5: 52-85

STARK RW (1965) Recent trends in forest entomology. Annual Review of Entomology 10: 303-324

SWAIN T (1977) Secondary compounds as protective agents. Annual Review of Plant Physiology 28: 479-501

SWAIN T (1979) Tannins and lignins. In: ROSENTHAL GA, JANZEN DH (eds) Herbivores: their interaction with secondary plant metabolites. Academic Press, New York, pp 657-682

TAYLOR WE, BARDNER R (1968) Leaf injury and food consumption by larvae of Phaedon cochleariae (Coleoptera: Chrysomelidae) and Plutella maculipennis (Lepidoptera: Plutellidae) feeding on turnip and radish. Entomologia Experimentalis et Applicata 11: 177-184

VALIELA I, TEAL JM, SASS WJ (1975) Production and dynamics of salt marsh vegetation and the effects of experimental treatment with sewage sludge: biomass, production and species composition. Journal of Applied Ecology 12: 973-982

WARING CH (1967) Studies on the relations of insect and host plant. II. Effects of water stress in host plants on the fecundity of Myzus persica (Schz.) and Brevicoryne brassicae (L.) Nature 213: 1052-1053

WAY MJ, CAMMELL M (1970) Aggregation behaviour in relation to food utilization by aphids. In: WATSON A (ed) Animal Populations in relation to their food resources. Blackwell, Oxford, pp 229-247

WHITE TCR (1971) Lerp insects (Homoptera, Psyllidae) on red gum (E. camaldulensis) in South Australia. South Australian Naturalist 46: 20-23

WHITHAM TG (1978) Habitat selection by Pemphigus aphids in response to resource limitation and competition. Ecology 59: 1164-1176

WHITTAKER RH, FEENY PP (1971) Allelochemics: chemical interactions between species. Science 171: 757-770

WILLIAMS LH (1954) The feeding habits and food preferences of Acridiidae and the factors which determine them. Transactions of the Royal Entomological Society of London 105: 423-454

27. Defoliation Patterns in Matorral Ecosystems

E. R. FUENTES and J. ETCHEGARAY

27.1 Introduction

Herbivores in mediterranean shrublands face spatially and temporally varying mosaics of feeding opportunities. On the one hand, leaves of various plant species differ in their physical and chemical attributes and thus in their appeal to herbivores. On the other hand, access to food changes with season. For herbivorous insects, season, and more specifically, temperature, modifies the animal's metabolism and its capacity to move and actually ingest the leaves. In a fluctuating climate, in which winters can be too cold and wet for them, and summers too hot and dry, the length of the intermediate cool-moist season can seriously restrict the insect's impact on plants (Fuentes 1976; Atkins 1977).

Another constraint affecting both poikilothermic and homeothermic herbivores is the significant variability in leaf phenology. Grasses and herbs begin growing fairly early in the season, but also dry out sooner than shrubs (Montenegro et al. 1978, 1979). Shrubs can be roughly subdivided into drought-deciduous and evergreen types. Drought-deciduous shrubs have earlier leaf development and senescence than the latter (Montenegro et al. 1979). But even evergreen shrubs show important variations in their leaf developmental phenology. These variations impose further restrictions on the herbivores' leaf-eating activities. Young, tender leaves tend to be relatively nutritious and harmless, whereas old, dry or sclerophyllous leaves are just the opposite (Feeny 1968, 1970, 1976; Fuentes et al. 1980).

For the plant, leaf losses imply less future photosynthesis. This can be ultimately translated into a reduced seed output or diminished vegetative reproduction, and hence into reduced fitness. In addition to its implications for expected progeny, this herbivory-associated cost may affect the plant's short-term competitive abilities. Decreased photosynthesis leads to reduced root growth and a lowered ability to acquire seasonally limited water and nutrients. This would both retard recovery of the plant and increase the risk of being outgrown by competitors.

Because of the above implications of herbivory, everything else being the same, selective pressures on plants should tend to minimize the amount of foliage lost by an individual or maximize the speed of recovery after herbivory. Many anti-herbivore defence mechanisms have been described which are thought to reduce the amount of foliage loss (see Ehrlich and Raven 1965; Feeny 1976; Futuyma 1976; Rhoades and Cates 1976).

This interaction between herbivores and their food sources, in which the expected benefits for herbivores are inversely related to those for plants, generates a minimax or saddle-point solution (Maki and Thompson 1973), which we actually observe in the field. In general we neither see herbivores denude plants completely, nor herbivores completely excluded in the community; instead we observe an intermediate point of dynamic coexistence. The question therefore is to find out which factors are involved in determining the coordinates of the saddle-point solution.

Two possible factors are the abundance of herbivores and the water/nutrient availabilities in the soil. Whereas predators can reduce the effect of herbivores by either restricting their activities (see later) or reducing their population sizes, soil nutrient availability can affect the leaf's appeal by altering the content of nutrients and of defensive compounds. In both cases the pressure on the plant is modified and consequently its "strategy" changed.

In this paper we will review the evidence for the existence of the herbivore/plant conflict in the matorral by looking at:

(a) Situations in which the herbivore impact on vegetation is large (lethal or severely harmful). Here patterns emerging from the interactions between small herbivorous mammals and herbs or shrub seedlings are considered, particularly the differences which emerge from comparing co-evolved plant-herbivore-predator systems and systems in which the herbivore is exotic to both predators and plants.

(b) Situations in which the impact on the vegetation is relatively small, either because the herbivores are insects or because they are large but take relatively little at each visit. Again natural co-evolved systems will be compared with an artificial system in which goats are the herbivores.

27.2 Evidence for the conflict

27.2.1 EXTREME HERBIVORY

In mature undisturbed matorral, herbivory levels are low and plant communities diverse. However, "experimental" evidence of departures from the common matorral saddle-point and thus for the possible effects of herbivores can be seen where man has modified the landscape. Fuentes et al. (1980) have shown that in an altered shrub community at Los Dominicos, east of Santiago (33°27'S, 70°42'W), where the evergreen indigenous shrub *Muehlenbeckia hastulata* (Polygonaceae) is the only shrub species present, the number of larvae of the indigenous butterfly *Macromphalia* sp. (Lasiocampidae, Lepidoptera) is highly correlated with patch size. Where patches are large, larval density is so high that the plants are completely defoliated. After defoliation of the large patches, the larvae move to the initially less affected smaller patches and defoliate them. This eruption pattern repeats itself almost every year. By contrast, in mature matorral, *M. hastulata* never forms large patches (Parsons 1973; Mooney et al. 1977) and has low levels of defoliation (Fuentes et al. 1980).

Other artificial shrub "monocultures" also exhibit high larval densities, but have yet to be documented in detail. They conform to what seems a general pattern of complete defoliation of monocultures (Anonymous 1933; Putnam 1963; Williams 1964; Cross 1973; Ostmark 1974).

Large patches of M. hastulata are also associated with high densities of the small indigenous rodent Octodon degus (Jaksic and Yanez 1978; Jaksic et al. 1980). In these monocultures Octodon is capable of eating all shrub seedlings in the community (personal observations) and even invades neighbouring farm fields (Pefaur et al. 1979).

It is common knowledge among foresters in central Chile that planting seedlings of either introduced pine species (Pinus radiata) or native species (e.g. Quillaja saponaria and Schinus molle) cannot be done successfully without extensive poisoning of the small wild mammals in the area or protecting the plants in wire enclosures.

An even more dramatic case of the possible effect of herbivores exists in the so-called Chilean Norte Chico (29°-32°S), north of Santiago. Here, mixtures of evergreen and drought-deciduous shrublands have been devastated by mismanagement of introduced goats (UNCOD 1977; Fuentes and Hajek 1978, 1979).

It seems therefore that altering the shrub mosaic, the herbivore densities or both can result in severe defoliation usually not seen in undisturbed communities.

27.2.2 SHRUB ANTI-HERBIVORE DEFENCES

Matorral shrubs have attributes currently thought to be used in anti-herbivore defence (Futuyma 1976; Rhoades and Cates 1976; Fishbeck and Kummerow 1977; Wasserman 1979; Fuentes et al. 1980; for California chaparral see Dement and Mooney 1974; Kummerow and Fishbeck 1977 and references therein). Matorral leaf defences include among others sclerophylly, tannins, terpenes, several flavonoids, saponins, waxes, crystals and lactiferous vessels.

It is of special importance here to refer to shrub attributes that render them difficult to locate. Species with tender, presumably palatable leaves, like Muehlenbeckia hastulata, are rare in the mature matorral community (Parsons 1973) and as mentioned above, when they reach high densities at any one site, defoliation is marked. The "defence mechanism" by which plants "hide" among each other, when confronted with non-generalist herbivores, has been repeatedly documented in the literature (Pimentel 1961; Tahvanainen and Root 1972; Harris 1973; Futuyma 1976). Another way in which shrubs become less apparent (Feeny 1976), to insect defoliators at least, is associated with seasonal shifts in the mean and interspecific variation of their growth period (see later, and Fuentes et al. 1980) relative to the climatically determined peak of insect abundances.

Thus matorral shrubs exhibit a variety of anti-herbivore attributes which would not be present unless defoliation risks were important for them. Although the exact role of each one of these anti-herbivore defences is still unknown and is now being explored (G Montenegro, personal communication), they are likely to help explain the observed differences in the herbivore loads of shrubs (see Fuentes et al. 1980).

Finally, there is good experimental and observational evidence that matorral shrubs exhibit strong intra- and interspecific competition for soil resources and/or light (Hoffmann and Kummerow 1978; Gutierrez and

Fuentes 1979; Fuentes and Gutierrez 1981) and that herbs and grasses compete (Fuentes and Le Boulenge 1977). Hence one must consider herbivory not as an isolated phenomenon but as a modifier of competition between plants. Thus, everything else being the same, more severely eaten plants or plant species are more likely to be outcompeted than lightly eaten plants.

27.3 Interactive patterns with lethal consequences

It is possible to define a range of intensities of plant/herbivore interactions. At one extreme of this range are lethal effects resulting from a few plant-herbivore contacts and at the other, interactions requiring long periods of time and many contacts to kill the plant. The time of interaction is considered here in relation to the plant's generation time and the contacts referred to occur under natural conditions. A particular herbivore could shift along this gradient depending on relative plant size, its generation time, the existence of alternative food sources, its own predation risks, or management. In addition, the position along this intensity gradient would also be a function of the herbivore's energy and nutrient requirements, its foraging behaviour and so on. However, as will be seen, this scheme makes it possible to compare a wide variety of herbivores and their effects on vegetation.

One matorral example falling within the "few lethal contacts" category is that of the small diurnal rodent Octodon degus and native grasses. Fuentes and Le Boulenge (1977) (see also Le Boulenge and Fuentes 1978) showed that O. degus tends to restrict its activity to the neighbourhood of large protecting shrubs or stone walls. Outside of these refuges O. degus risks predation by raptors and foxes (Yanez and Jaksic 1978a,b and references therein). Consequently, it tends to forage intensively around the refuges and less so with increasing distances from them. Fuentes and Le Boulenge (1977) suggest that this foraging pattern accounts for the grass-free belts around M. hastulata shrubs, and in some areas, the presence of belts of the strongly aromatic herb Helenium aromaticum. In these examples the strong effect of O. degus on the grasses shifts the competitive equilibrium towards H. aromaticum. Without the mammal, the competitive equilibrium favours the grasses.

Similarly Fuentes and Jaksic (unpublished) showed that these habitat-use restrictions of O. degus result in the destruction of shrub seedlings growing in the neighbourhood of refuges. By using exclosures it was shown that seedlings within 4 m of refuges are defoliated and killed by O. degus. Recent evidence suggests that other native rodents in the matorral are also strongly associated with refuges from predators but because they are granivores and insectivores their effects are not as obvious as those of O. degus (Simonetti and Fuentes, unpublished).

Bartholomew (1970) has shown that in the California chaparral, native rabbits (Sylvilagus spp.) are among the herbivores producing the bare zone around evergreen shrub patches, in which they seek refuge (see also Orr 1940; Fitch 1947). No such effect has been described for the Mediterranean Basin's evergreen scrubland, although native rabbits there (Oryctolagus cuniculus) are also known to remain close to refuge-providing plants (Rodriguez de la Fuente 1970) and prefer areas with a high shrub cover.

There are no native rabbits in Chile but O. cuniculus has been recently introduced from Spain (Osgood 1943). In Chile this species uses mostly the open spaces between shrubs and habitats with low shrub cover, in contrast to the spatial constraints it exhibits in Spain (Jaksic et al. 1979a; Fuentes and Jaksic 1980; Jaksic and Sorigver 1981). Jaksic et al. (1979a) have shown that in spite of the existence of predators of similar size to those known to eat rabbits in California and Spain, they do not eat the Spanish rabbit in Chile. Interestingly, local foxes (Dusycion culpaeus) prey on Viscacia viscacia (Chinchillidae), the native size and food analogue of O. cuniculus, which does tend to restrict its activity to the neighbourhood of large rock outcrops (Pearson 1948; J Guzman, personal communication). This suggests then that whereas local predators are relatively ineffective in hunting the introduced rabbit, hence the rabbit's habitat release, they are effective in restricting the activity of native, co-evolved herbivores.

As a consequence of the habitat release exhibited by O. cuniculus in Chile, similar to the one exhibited by Sylvilagus in predator-free islands off California (Zoloth 1969), rabbits eat the native herbs in the open spaces between bushes, but not those under dome-shaped bushes (Fuentes and Jaksic 1980). As a result, in areas where O. degus is not present, rabbits restrict native herbs to areas under the dome-shaped bushes, and the spaces between bushes are occupied by introduced grasses (Carter 1973; Keeley and Johnson 1977). Before this effect of rabbits was known, the microhabitat restriction of native herbs, clearly differentiating the California chaparral and Chilean matorral, was thought to be a discrepancy in convergent evolution.

Spanish rabbits in Chile also affect shrubs. Fuentes and Jaksic (unpublished) show experimentally that since the introduction of rabbits, shrub seedlings are no longer "safe" if located more than 4 m away from O. degus dwellings, because rabbits eat the seedlings in the open areas. As a result, where rabbits are abundant, and they tend to be ubiquitous, very few seedlings can be seen. Recent evidence suggests that this same effect holds in mature matorral sites (Fuentes and Simonetti, unpublished). The long-term consequences of the introduction of O. cuniculus are still unknown, although the above evidence suggests that they might be interfering with the historical pattern of shrub recruitment.

These comparative effects of O. cuniculus and O. degus are an example of how predators could eventually shift a particular herbivore/plant relationship along the intensity gradient.

27.4 Low-intensity interactions

Whereas in the case of Oryctolagus cuniculus and Octodon degus relatively few contacts were sufficient to kill the plants, native herbivorous insects tend to indirectly affect the plant by way of altering the competitive dynamics of plant species. Depending on management practices, introduced goats can occupy any position along the intensity range. Below we describe some patterns in naturally occurring shrub/herbivorous insect interactions, and then compare them with those involving large natural browsers and finally with goat/shrub interactions.

27.4.1 LEAF-EATING INSECTS

Etchegaray and Fuentes (1980) have shown that leaf-eating insects of matorral shrubs show some host specificity, although they will still feed on a variety of shrub species (see Fig. 1). This dietary catholicity of the insects seems associated with the short climatically favourable season

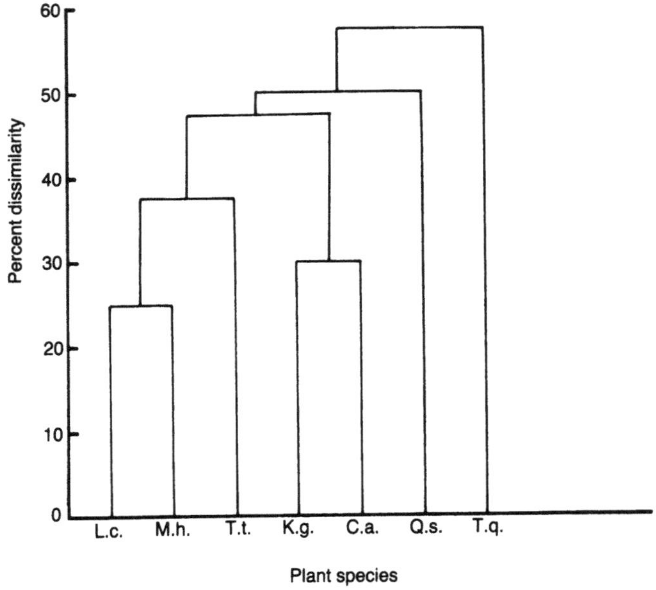

Fig. 1. Dendrogram showing degrees of dissimilarity among leaf-eating insect assemblages associated with seven common matorral shrub species. All shrubs share the same defoliators and the only differences are in the proportion of the five commonest morpho-species represented on the shrubs. Dissimilarity index as in Fuentes 1976. See the text for further details. L.c.: *Lithraea caustica*; M.h.: *Muehlenbeckia hastulata*; T.t.: *Trevoa trinervis*; K.o.: *Kageneckia oblonga*; C.a.: *Cryptocarya alba*; T.q.: *Talguenea quinquinervia*; Q.s.: *Quillaja saponaria* (from Etchegaray and Fuentes 1980).

available to them. Insect attack occurs almost simultaneously on all shrub species during the warm, wet season. There is no time spread in the attack of these insects comparable to that exhibited by shrub leaf growth (Fuentes et al. 1980). Thus whereas shrub leaf growth can start as early as July, in some drought-deciduous species, and can last until the end of February in some evergreens, herbivorous insect activity is restricted to the September-December period. Because of this spread, early- and late-growing shrub species are less exposed to herbivores than average ones.

Studies using tagged branches have shown that insects forage mostly on newly produced leaves and only eat a fraction of each one of them. From the shrub viewpoint, the important parameter is the proportion of its photosynthetic surface lost to insects. Losses here include not only what insects actually eat, but also the potential leaf area lost through eating parts of young and still growing leaves. The proportion of photosynthetic surface lost by shrub species i at any one locality can be estimated (see Fuentes et al. 1980) by:

$$E(P) = P(L)L$$

where E(P) is the expected proportion of the average leaf area of shrub species to be "lost" to herbivorous insects. P(L) is the probability of a where E(P) is the expected proportion of the average leaf area of shrub species to be "lost" to herbivorous insects. P(L) is the probability of a randomly picked leaf of shrub species exhibiting any sign of herbivore attack, and L is the average proportion of an attacked leaf that was actually eaten. We have estimated L in the field by comparing large samples of oven-dried, attacked and non-attacked leaves; and P(L) by randomly collecting at least 500 leaves from at least five different individuals at each site.

At Cuesta La Dormida (a relatively mesic site at 33°S; see Thrower and Bradbury 1977), Fuentes et al. (1980) tested the hypothesis that leaf area loss from a shrub species is correlated with its cover. A priori it was expected that if insect attack was the main factor explaining shrub abundance, the relationship between E(P) and shrub cover ought to be negative. Thus a shrub species i would have more cover than a shrub species j because it is less attacked by defoliators. On the other hand, if the effect of insect attack on the shrub was negligible compared with the role played by interspecific competition and plant resource availabilities, a positive correlation would be expected because all other things being equal, shrubs with higher cover would attract more insects simply by being more conspicuous targets as in the Macromphalia sp. larvae versus Muehlenbeckia hastulata shrubs described earlier. Therefore, the more abundant shrubs should exhibit more defoliation than less abundant ones (Fig. 2A).

Notice however, that if the herbivory were to affect competition between plants, as indicated earlier in the chapter the average value of E(P) would be uniformly small. That is, all species would minimize their E(P) values and these would therefore not be correlated with shrub abundance. Moreover, they would tend to be small for all plant species.

Results of a three-year test of this hypothesis are shown in Fig. 2B where it can be seen that E(P) exhibits between-species and between-year variation. (Analysis of variance on these data shows that, whereas between-year differences are not significant ($P > 0.90$), the between-species differences in E(P) are statistically significant ($P < 0.05$).) Nevertheless, the correlation coefficient between average E(P) and shrub cover at Cuesta La Dormida is not statistically different from zero ($P > 0.86$), in spite of the 100-fold differences in cover of the shrubs sampled.

The uniformly low average values of E(P) shown in Fig. 2B, the low scatter of the species points in the plane, and the statistically non-significant correlation between these variables, support the herbivory-competition model. There is however another hypothesis that might explain these results, namely high predation on herbivores. This will be discussed later.

Although the average between-year and between-species E(P) seem constant within a community, the actual time coupling of shrubs with their defoliating insects is quite complex. Fuentes and Etchegaray (unpublished) have shown that the average number of herbivorous insects present on different shrub species at Cuesta La Dormida is significantly correlated with the species cover at that site ($r = 0.60$) and that the coefficient of variation of insect number is also significantly correlated with average insect number ($r = 0.74$). Thus, high cover shrubs tend to have more insects and more between-year relative variation in the number of defoliating insects. Also, the between-year coefficients of variation in insect numbers and in E(P) are also highly correlated ($r = 0.79$). In other words, on the average, shrubs with high cover have more herbivorous insects, and greater yearly variation in insect numbers and in the percentage of foliage lost to herbivores.

As mentioned above E(P) varies between sites. Whereas at Cuesta La Dormida E(P) is on the average 7% (SE = 2%), at Salamanca, 109 km to the

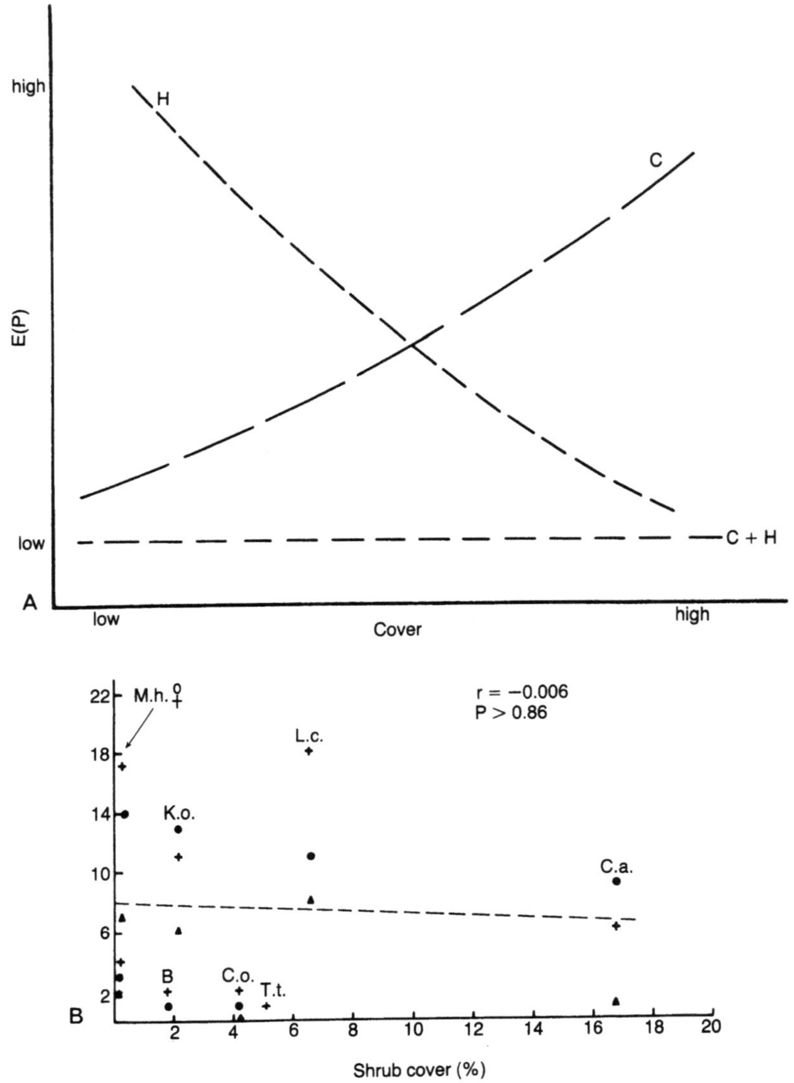

Fig. 2. Shrub cover and defoliation. A. The three basic a priori hypotheses on possible relations between defoliation (E(P)) and shrub species cover at one site are shown. The curve H represents the expected relationship given the hypothesis that herbivory governs plant abundance, and C that where interplant competition is most important. The C + H curve is that expected under the combined defoliation-competition hypothesis. See the text for further discussion. B. The actual relationship found between E(P) and cover. Results correspond to a three-year sampling programme at Cuesta La Dormida. Data for 1977 (+) 1978 (dots) and 1980 (triangles) are shown for seven shrub species encompassing a broad spectrum of percentage covers. Cover remained constant through the study period. M.h.: *Muehlenbeckia hastulata*; Q.s.: *Quillaja saponaria*; B.r.: *Baccharis rosmarinifolia*; C.o.: *Colliguaya odorifera*; L.c.: *Lithraea caustica*; C.a.: *Cryptocaria alba*; Ko: *Kageneckia oblonga*. See the text for a discussion of the biological significance.

north, where climate is drier and considerably more variable between years (Di Castri and Hajek 1976; Fuentes and Hajek 1978), average community E(P) is only 2% (SE = 1%). Lower insect abundance at Salamanca than at Cuesta La Dormida could partly explain this difference in average E(P), but there may be other factors involved. Current research will help clarify this problem.

Just as average between-communities E(P) values are not constant, the E(P) for one shrub species at different localities is not constant either. For example in the centre of its range (on the south-facing slopes of the dry Andean ranges east of Santiago) Kageneckia oblonga forms relatively dense stands and shows average E(P) values of 7% (Fuentes et al. 1980). At the same latitude in the more mesic coastal ranges at Cuesta La Dormida, where K. oblonga is relatively scarce, it has higher average E(P) values of 11% and in moist ravines at Cuesta La Dormida, where the species is at the limits of its distribution and even rarer, the E(P) can be as high as 70%. These observations conform well with our herbivory-competition explanation suggested above.

Geographical variation in E(P) is found also in Colliguaya spp. and Lithraea caustica. Average values of E(P) increase significantly with increasing altitude or latitude (see Figs 3 and 4). But since other species (i.e. Quillaja saponaria) show weak and even negative correlations along the same axes, further research is needed to explain the geographical variation in E(P).

The between-communities variation in E(P) and the somewhat independent variation in E(P) for the species sampled further suggest that the uniformly low E(P) values within the community are not due to random effects. Moreover, the existence of patterns in the values of E(P) and the fact that these patterns are not the same for all species, even when measured in the same collection of stands, further suggests that shrub species are not equally susceptible to herbivore attack. That is, the E(P) value at any one site would be at least partially dependent on the characteristics of the plant. However, to completely explain the actual E(P) value the role of predators must be clarified. They may play an important role in depressing the herbivore populations, thus partially accounting for the E(P) value. We are presently examining this in a study of insectivorous birds.

Our present lack of understanding of the geographical patterns in E(P) is partly due to lack of data on herbivores and insectivores and on geographical variations of plant defences, particularly those, which while acting as anti-herbivore defences, might also be responses to other environmental challenges. Two such "defences" as examples of the complexity involved in unravelling causal factors are described below.

There is a significant correlation ($r = 0.72$, $P < 0.05$) between average leaf size and E(P), when leaves of different shrub species are considered (Fig. 5). This correlation is in part due to an intraspecific correlation effect (i.e. K. oblonga, $r = 0.91$, $P < 0.001$, but not L. caustica, $r = 0.10$, $P > 0.70$; or Quillaja saponaria, $r = 0.15$, $P > 0.60$), but other still unknown effects related to leaf size may also be involved. This effect has been substantiated experimentally. Larval growth rates, presumably correlated with their fitness in the fluctuating matorral environment, clearly indicate that they have different efficiencies in their use of circular leaf samples of different sizes. Larvae of Macromphalia sp., 30-40 mm long, were collected from M.hastulata and fed ad libitum on 7-and 2-mm-radius samples of C. odorifera, L. caustica and M. hastulata in the laboratory. Growth rates were significantly higher with larger samples (t-test, $P < 0.05$) than small, irrespective of plant species.

These results suggest that, everything else being equal, insects would be selected to prefer large leaves. However, because small leaves are also optimal in dry climates, and leaf size covaries with toughness and perhaps

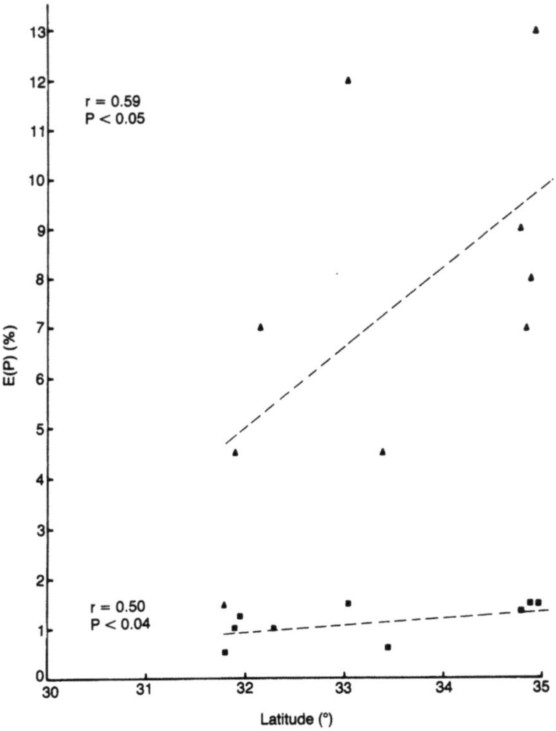

<u>Fig. 3.</u> Relationship between defoliation and latitude. Values of E(P) for <u>Lithraea caustica</u> (triangles) and <u>Colliguaya</u> spp. (squares) within the area of mediterranean climate in central Chile.

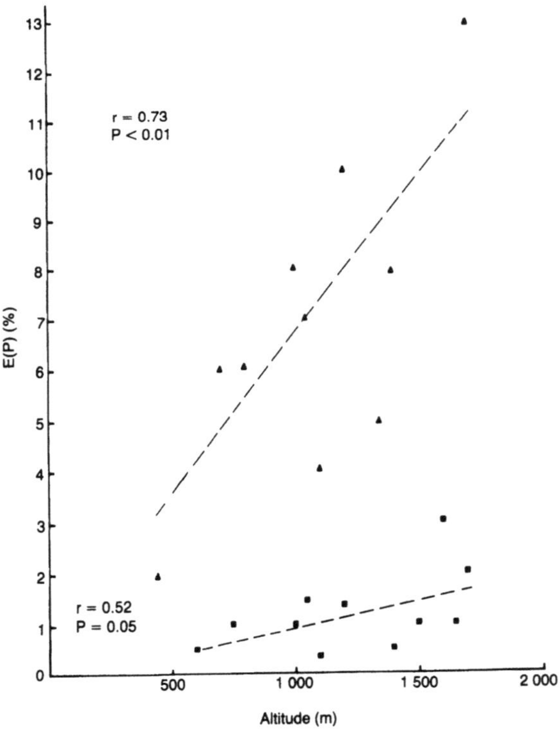

<u>Fig. 4.</u> Defoliation and altitude. As in Fig. 4 but for Andean altitudes within the area of mediterranean climate in central Chile. See the text for discussion.

other variables, the role of E(P) as a selective factor is difficult to pinpoint.

Obviously more research is needed to clarify the relationships between leaf phenotypes, anti-herbivore defences and other environmental attributes showing geographical variation.

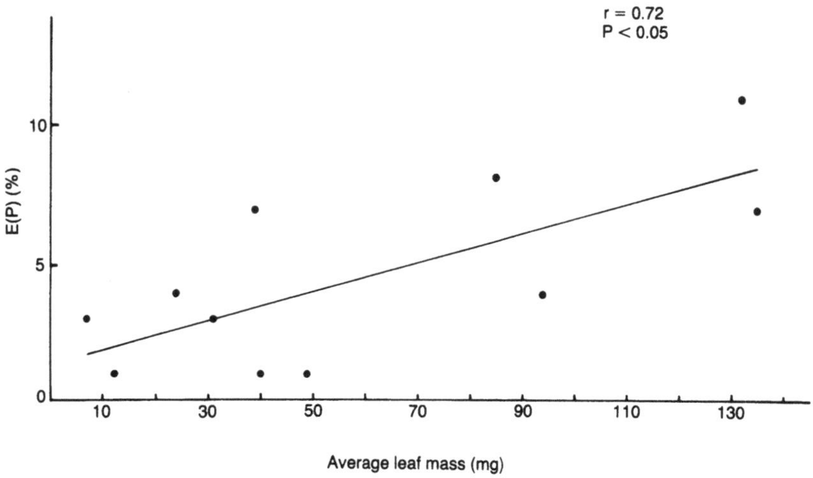

Fig. 5. Relationship between average leaf mass and E(P) for 11 species of common shrubs in central Chile, including the seven species indicated in Fig. 2B. See the text for further discussion.

27.4.2 MAMMALIAN HERBIVORES

Goats and other large browsers. "Cafeteria tests" performed with individual goats presented with pairs of equal-size artificial "shrubs" made out of freshly cut branches (Fuentes and Etchegaray, 1981) suggest:

(a) Juvenile goats, up to six months old, do not exhibit statistically significant preferences for either *Lithraea caustica*, *Kageneckia oblonga*, *Colliguaya odorifera* or *Quillaja saponaria*.

(b) Adult females, older than two years of age, eat significantly less *L. caustica* than of the other three shrub species, but do not discriminate between the remaining three.

This catholic foraging behaviour pattern contrasts with that shown by insects as a whole: matorral insects, although capable of feeding on most shrub species, exhibit clear preferences in the field. This is also reflected by the differences in E(P) values between shrub species.

The importance of the difference in the diet of introduced goats and naturally occurring insects lies in their potential impact on their food plants.

Torres et al. (1980) have shown that matorral shrubs with high E(P) values have compensation mechanisms to balance normal levels of foliage loss. They artificially defoliated shrubs of *L. caustica* (with high E(P)) and C. odorifera (with low E(P)) and found that at 25% defoliation levels only L. caustica showed a strong vegetative response to compensate for leaf surface losses. At 100% leaf loss both show a strong vegetative response. Plants with intermediate E(P) values seem to have intermediate responses to defoliation (G Montenegro, personal communication).

It follows that if goats had strongly preferred plants with high E(P) values, then at moderate levels of browsing their overall impact on the vegetation would have been less than was actually found. It has been found that in areas where there are few goats their foraging activities are capable of changing the shrub species equilibrium to favour species like L. caustica, the cacti Trichocereus spp., and the toxic weedy shrub Cestrum parqui. Where goat numbers are high, they completely remove the vegetation

Thus it can be seen that matorral shrubs may behave in qualitatively different ways when grazed on by unspecialized, introduced goats, or when attacked by various densities of naturally occurring insects.

Unfortunately, the camelid (Lama guanicoe), the only natural large browser of comparable size, and perhaps potential effect, to goats is now locally extinct. Neither their natural densities nor their feeding behaviour is known.

In California there is the relatively well studied deer (Odocoileus hemionus), in addition to introduced sheep and goats. However, the patterns of insect shrub defoliation are unknown there and a complete comparison is not possible at this time. Nevertheless, pertinent evidence from California suggests that differences between natural deer and introduced goats or sheep are likely to be particularly important.

Biswell et al. (1952) reported that in northern California O. hemionus, when feeding on shrubs, eat mostly Adenostoma fasciculatum, Quercus wislizenii, Q. dumosa and to lesser extent Ceanothus cuneatus. In these same habitats they report as most abundant A. fasciculatum, Ceanothus cuneatus, C. foliosus on the south-facing slopes, and Q. wislizenii and Q. dumosa on the north-facing slopes. Thus the deer are eating the most abundant shrubs (see also Longhurst et al. 1979). However, "cafeteria tests" show that these shrubs are not their preferred ones (Longhurst et al. 1979), suggesting a compromise between cost and preference in their diet selection in the field.

It is interesting to note that shrub preferences in introduced sheep and native deer correlate in rank by species despite overall differences in habitat and food (grass versus shrub) preferences (Longhurst et al. 1979). The combined effect of both these animals, however, is strong enough to reduce seedling survival (oaks), and thus produce local changes in the composition of the northern California chaparral.

In southern California, in a habitat where the dominant shrubs are mainly Arctostaphylos glandulosa and Ceanothus greggii, with somewhat lesser covers of Adenostoma fasciculatum and Quercus dumosa, Ahmed et al. (1979) showed experimentally that goats preferred Q. dumosa and A. fasciculatum. However, in the field they mainly eat the most abundant shrubs.

The California studies suggest that these large ungulates eat grasses and forbs when available, and otherwise browse especially on young tissues. As with goats in the Chilean matorral they seem to show relatively weak preferences, and on the whole they are relatively "fine-grained" herbivores. Their effect on the shrub community would therefore be strongly dependent upon their density and on management practices.

The ways in which goats have changed the composition of the shrub cover have been documented for shrublands in Australia and the Mediterranean region, but since species-specific shrub responses or effects of natural browsers are unknown to us, we will not discuss them further here (see McKell et al. 1972).

27.5 Final comments and conclusions

High- and low-intensity herbivory can affect the composition of plant communities. High-intensity herbivory kills the plant before it reproduces and can thus result in local extinctions and generate shrub-free belts around the refuge-providing plants. Thus they can be one factor affecting the spatial distribution of shrubs. However, as the intensity of herbivory decreases, the effect through altering the competitive success of plants is likely to become progressively more important, and progressively more time will be required for the complete expression of important differences resulting from herbivory load. Nevertheless, even at low defoliation levels, herbivores at any one site would sift from the potential shrub flora a subset of species capable of competitive long-term coexistence. That is, the natural herbivory patterns we see in the mediterranean areas today are likely to be the result not only of selection by climatic and soil patterns but also of competitive co-existence under herbivore pressure.

If the original herbivores change we would expect to see changes in composition of the plant community. In Chile we are today observing the transition from one equilibrium domain to another, unknown one, under the influence of goats and rabbits; the most important factor being that neither of the new herbivores is held at a low density by predation. In the case of rabbits, local predators have not yet learned to hunt them whereas potential predators on goats, like foxes and pumas (Felis concolor), are being exterminated by man.

These two cases represent departures from a natural saddle-point solution, in which there has not been enough time for natural selection to operate on both the components. Natural selection might be expected to favour plant phenotypes with low herbivore loads, because herbivores seem capable of reducing the plants' capacity to obtain useful resources in a competitive environment. Herbivory would affect not only how strongly individuals of a plant species "perceive" individuals of another competing species, but also the actual competitive pressure on the plant, since competitors may have more resources available to them for growth, if the first species is weakened. However, the optimum solution, even from the shrub's point of view, might not be zero herbivory. There are at least two factors involved which will be examined in some detail.

Let C be the energetic cost associated with producing and depositing anti-herbivore defences on the average leaf. The C value has an opportunity cost function m (in calories) related to unrealized photosynthesis. This is the potential energy lost to the plant by making fewer defended leaves (which may also have decreased photosynthetic efficiency) instead of many cheap leaves with high photosynthetic efficiency. Because of the positive feedbacks between present photosynthesis and future photosynthesis, m functions should be concave and increase concomitantly with C (Fig. 6A).

C has a second opportunity cost associated with it, namely the cost in future photosynthesis of having invested only C additional calories per leaf in anti-herbivore defences, and therefore of incurring some herbivore-related loss of photosynthetic surfaces. This second cost function (p) should also be concave, for the same reasons as m, but decrease with C (Fig. 6B).

Both opportunity costs are present if between-shrub competition and leaf losses due to herbivores are important structuring factors. Plant performance should therefore be dependent on the reduction in the photosynthetic expectancy due to both costs. That is, natural selection would be expected to minimize the sum of the two opportunity costs, associated with defoliation damages, and thus optimize the amount of net photosynthesis through their lifespan (see Fig. 6C, D).

Since environments differ in the importance of competition and herbivore pressure, different optimal combinations are possible. The higher the importance of competition the lower the C investment in defences (Fig. 6C). On the other side, environments with high herbivore pressure (i.e. more mesic habitats) should select for higher investments in defences than lower ones (Fig. 6B). Consequently from this model, the following ranking in cost of defence systems is expected: high herbivory/low competition, low competition/high herbivory, low competition/low herbivory, and high competition/low herbivory (Fig. 6E). This hypothesis should be testable within an area of mediterranean climate.

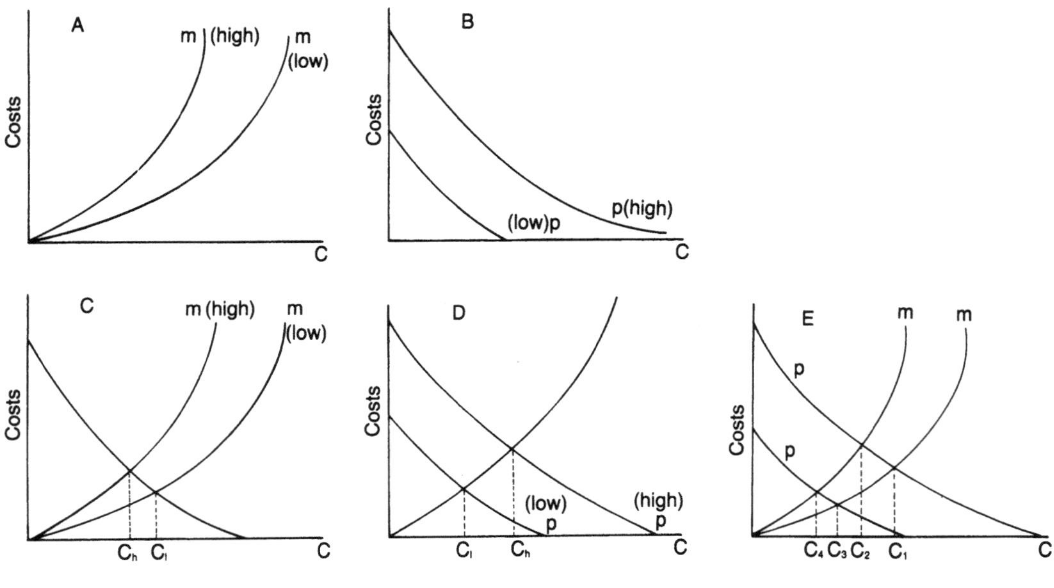

Fig. 6. Shrub defences model. Ordinates are opportunity costs associated with the addition of various C costs (abscissas) as anti-herbivore defences to the average leaf. A. Cost functions (m) associated with not having made more "cheap" leaves with C = 0. The m functions of a shrub in highly and weakly competitive environments are indicated. B. Cost functions (p) associated with defoliation because only an amount C was invested in defending the average leaf. The p functions with high and low herbivore densities are shown. C. Minimum total costs solution in high and low competition situations for a given herbivore pressure. D. Minimum total costs solution in high and low herbivore-pressure environments. E. Comparison of the solutions in the four situations described. See the text for further explanation.

Notice that in the model, predators, by reducing herbivore pressure, would favour a decrease in the amount of defence costs, and thus reduce photosynthetic losses to the plant. A similar result would hold if low soil nutrients led to the production of leaves low in nutrients, and thus reduced the number of defoliating insects (this is equivalent to lowering of p function). In either case the low-C category would be favoured.

In all of the four cases though, the optimum solution from the plant's side is not zero herbivory, but a C value and herbivory levels that are greater than zero. (In passing it should be noticed that because of analogous compromises the optimum from the herbivore's point of view might not be to completely detoxify its food. In this case the compromise would be between opportunity costs associated with expected progeny associated with (a) having a "heavy" detoxifying machinery and (b) the actual physiological performance when being somewhat poisoned.)

In the above discussion we have assumed that herbivore pressure is exactly the same for all shrubs. But, as we discussed before, when herbivores are not completely specialized, a certain amount of herbivore-sharing and transfer is expected. Thus, if shrub species i at a particular point in time were relatively palatable, because of a low investment in defences, and therefore increased its competitive ability and cover, natural selection on the herbivores would be expected to increase the herbivore load on this species. This increased number of actual herbivores would again select shrubs with higher C and lower E(P) values. In general, the lower the degree of specialization of defoliating insects and the higher the importance of shrub cover as a cue for herbivores, the narrower should be the belt of E(P) values in the community.

In conclusion, our results suggest that herbivores in evolutionary equilibrium with their food might tend towards a compromise in which although they would not be very important in the actual nutrient flow of matorral ecosystem, they have a role in selecting the phenotypes of plant species involved in that flow.

Acknowledgements

Research described in this contribution is part of the Chilean contribution to UNESCO-MAB 6 programme. The manuscript itself benefited from the comments of an anonymous reviewer and the editors of this volume. Their help is greatly appreciated.

References

AHMED ES, MORRIS JG, RADOSEVITCH SR (1979) Relationships between the botanical and the chemical constituents of the diet selected by Spanish goats grazing chaparral. WSAS Proceedings 30: 166-169

ANONYMOUS (1933) Locusts in the cane fields. International Sugar Journal 54: 161

ATKINS M (1977) Insect biomass and diversity. In: THROWER NJW, BRADBURY DE (eds) Chile-California mediterranean scrub atlas. A comparative analysis. Dowden, Hutchinson and Ross, Stroudsburg, Pennsylvania, pp 130-183

BARTHOLOMEW B (1970) Bare zone between California shrub and grassland communities: The role of animals. Science 170: 1210-1212

BISWELL HH, TABER RD, HEDRICK DW, SCHULTZ AM (1952) Management of chamise brushlands for game in the north coast region of California. California Fish and Game 38(4): 453-484

CARTER SI (1973) A comparison of the pattern of herb and shrub growth in comparable sites in Chile and California. MSc Thesis. California State University, San Diego
CROSS WH (1973) Biology, control and eradication of the boll weevil. In: SMITH RF, MITTLER TE, SMITH CN (eds) Annual Review of Entomology 18: 17-46
DEMENT WA, MOONEY HA (1974) Seasonal variation in the production of tannins and cyanogenic glucosides in the chaparral shrub Heteromeles arbutifolia. Oecologia 25: 65-76
DI CASTRI F, HAJEK ER (1976) Bioclimatologia de Chile. Vicerrectoria Academica de la Universidad Catolica de Chile, 128 pp
EHRLICH PR, RAVEN PH (1965) Butterflies and plants: a study in coevolution. Evolution 18: 586-608
ETCHEGARAY J, FUENTES ER (1980) Insectos desfoliadores asociados a siete especies arbustivas del matorral. Anales Museo Historia Natural (Valpso) 13: 159-166
FEENY P (1968) Seasonal changes in the tannin content of oak leaves. Phytochemistry 7: 871-880
FEENY P (1970) Seasonal changes in oak leaf tannins and nutrients as a cause of spring feeding by winter moth caterpillars. Ecology 51: 565-581
FEENY P (1976) Plant apparency and chemical defense. Recent Advances in Phytochemistry 10: 1-40
FISHBECK K, KUMMEROW J (1977) Comparative wood and leaf anatomy. In: THROWER NJW, BRADBURY DE (eds) Chile-California mediterranean scrub atlas. A comparative analysis. Dowden, Hutchinson and Ross, Stroudsburg, Pennsylvania, pp 148-161
FITCH HS (1947) Ecology of a cottontail rabbit (Sylvilagus auduboni) population in central California. California Fish and Game 33: 159-184
FUENTES ER (1976) Ecological convergence of lizard communities in Chile and California. Ecology 17: 3-17
FUENTES ER, LE BOULENGE P (1977) Predation et competition dans la structure d'une communaute herbacee secondaire du Chile central. La Terre et la Vie 31: 313-326
FUENTES ER, ETCHEGARAY J (1981) Preferencias de cabras por arbustos del Matorral. Ensayos Con 4 especies. Ciencia e Investigacion Agraria 8: 53-56
FUENTES ER, GUTIERREZ (1981) Intra and interspecific competition between matorral shrubs. Oecologia Plantarum 2: 283-289
FUENTES ER, HAJEK ER (1978) Interaccion hombre-clima en la desertificacion del Norte Chico chileno. Ciencia e Investigacion Agraria 5: 137-142.
FUENTES ER, HAJEK ER (1979) Patterns of landscape modification in relation to agricultural practice in central Chile. Environmental Conservation 6: 265-271
FUENTES ER, JAKSIC FM (1979) Lizards and rodents: An explanation for their relative species diversity in Chile. Archivos de Biologia y Medicina Experimentales 12: 179-190
FUENTES ER, JAKSIC FM (1980) Consideraciones teoricas para el manejo del conejo Oryctolagus cuniculus en Chile central. Medio Ambiente 8: 45-49
FUENTES ER, ETCHEGARAY J, ALJARO ME, MONTENEGRO G (1981) Shrub defoliation by matorral insects. In: DI CASTRI F, GOODALL DW, SPECHT R (eds) Ecosystems of the world, vol 11. Mediterranean-type shrublands. Elsevier, Amsterdam, pp 345-359
FUTUYMA DJ (1976) Food plant specialization and environmental predictability in Lepidoptera. The American Naturalist 110(972): 285-292

FUTUYMA DJ, GOULD F (1979) Associations of plants and insects in a deciduous forest. Ecological Monographs 49(1): 33-50

GUTIERREZ JR, FUENTES ER (1979) Evidence for intraspecific competition in the Acacia caven (Leguminosae) savanna of Chile. Oecologia Plantarum 14: 151-158

HARRIS P (1973) Insects in the population dynamics of plants. In: VAN EMDEN HF (ed) Insect-plant relationships. Blackwell, Oxford, pp 201-210

HOFFMANN A, KUMMEROW J (1978) Root studies in the Chilean matorral. Oecologia 32: 57-69

JAKSIC FM, YANEZ JL (1978) Variacion anual de la composicion comunitaria de micromamiferos en Los Dominicos, Santiago. Noticiario Mensual. Museo Nacional de Historia Natural 267: 10-11

JAKSIC FM, FUENTES ER (1980) Why are herbs in the Chilean matorral more abundant beneath bushes: microclimate or grazing? Journal of Ecology 68: 665-669

JAKSIC FM, SORIGUER R (1981) Predation upon the European rabbit (Oryctolagus cuniculus L.) in Mediterranean habitats of Chile and Spain: a comparative analysis. Journal of Animal Ecology 50: 269-281

JAKSIC FM, FUENTES ER, YANEZ JL (1979a) Effective predation and spatial distribution of the rabbit (Oryctolagus cuniculus) in central Chile. Journal of Mammalogy 60: 207-209

JAKSIC FM, FUENTES ER, YANEZ JL (1979b) Two types of adaptation of vertebrate predators to their prey. Archivos de Biologia y Medicina Experimentales 12: 143-152

KEELEY SC, JOHNSON AW (1977) A comparison of the pattern of herb and shrub growth in comparable sites in Chile and California. The American Midland Nauralist 97: 120-132

LE BOULENGE E, FUENTES ER (1978) Quelques donnees sur la dynamique de population chez Octodon degus. La Terre et la Vie 32: 325-342

LONGHURST WM, CONNOLLY GE, BROWNING BM, GARTON EO (1979) Food interrelationships of deer and sheep in parts of Mendocino and Lake Counties, California. Hilgardia 47: 191-247

MAKI DP, THOMPSON M (1973) Mathematical models and applications. Prentice-Hall, Englewood Cliffs, New Jersey

MCKELL CM, BLAISDELL JP, GOODIN JR (1972) Wildland shrubs. Their biology and utilization. An International symposium. Utah State University, Logan, Utah. USDA Forest Service General Technical Report INT-1

MONTENEGRO G, RIVERA O, BAS F (1978) Herbaceous vegetation in the Chilean matorral. Oecologia 36: 237-244

MONTENEGRO G, ALJARO ME, KUMMEROW J (1979) Growth dynamics of Chilean matorral shrubs. Botanical Gazette 140: 114-119

MOONEY HA, JOHNSON A, PARSONS D, KEELEY S, HOFFMANN A, HAYS R, GILIBERTO J, CHU C (1977) The producers - their resources and adaptive response. In: MOONEY HA (ed) Convergent evolution in Chile and California. Mediterranean climate ecosystems. Dowden, Hutchinson and Ross, Stroudsberg, Pennsylvania, 224 pp

ORR RT (1940) The rabbits of California. California Academy of Sciences. Occasional Papers 19, 227 pp

OSGOOD W (1943) The mammals of Chile. Field Museum Natural History Publications, Zoological Series, Chicago. 30: 1-268

OSTMARK HE (1974) Economic insect pests of bananas. In: SMITH RF, MITTLER TE, SMITH CN (eds) Annual Review of Entomology 19: 161-176

PARSONS DJ (1973) A comparative study of vegetation structure along analogous climatic gradients in California and Chile. PhD Thesis. Stanford University, California

PEARSON O (1948) Life history of mountain vischacha in Peru. Journal of Mammalogy 4(29): 345-374

PEFAUR JE, YANEZ JL, JAKSIC FM (1979) Biological and environmental aspects of a mouse outbreak in the semi-arid region of Chile. Mammalia 43(3): 313-322

PIMENTEL D (1961) Species diversity and insect population outbreaks. Annals of the Entomological Society of America 54: 76-86

PUTNAM WL (1963) The codling moth, *Carpocapsa* (*Cydia*) *pomonellas* (L.): A review of its bionomics, ecology and control on apple with special reference to Ontario. Proceedings of the Entomological Society of Ontario 93: 22-60

RHOADES DF, CATES RG (1976) Toward a general theory of plant antiherbivore chemistry. Recent Advances in Phytochemistry 10: 168-205

RODRIGUEZ DE LA FUENTE F (1970) Enciclopedia Salvat de la fauna. Salvat, Pamplona

TAHVANAINEN JO, ROOT RB (1972) The influence of vegetational diversity on the population ecology of a specialized herbivore, *Phyllotreta cruciferae*. Oecologia 10: 321-346

TAHVANAINEN JO, ROOT RB (1972) The influence of vegetational diversity on the population ecology of a specialized herbivore, *Phyllotreta cruciferae*. Oecologia 10: 321-346

THROWER NJW, BRADBURY DE (1977) Chile-California mediterranean scrub atlas. A comparative analysis. Dowden, Hutchinson and Ross, Stroudsburg, Pennsylvania, 237 pp

TORRES JC, GUTIERREZ JR, FUENTES ER (1980) Vegetative responses to defoliation of two Chilean matorral shrubs. Oecologia 46: 161-163

UNCOD (United Nations Conference on Desertification) (1977) Case-study on desertification, region of Combarbala, Chile. Annnual Conference 74/9, Nairobi, Kenya, 162 pp (mimeographed)

WASSERMAN S (1979) Allelochemic diversity and plant apparency: Evidence from the detoxificaiton systems of caterpillars. American Midland Naturalist 102: 401-403

WILLIAMS JR (1964) Cane pests. Mauritius Sugar Industrial Research Institution Annual Report: 91-93

YANEZ JL, JAKSIC FM (1978a) Rol ecologico de los zorros (*Dusicyon*) en Chile central. Anales del Museo de Historia Natural (Valparaiso) 11: 105-112

YANEZ JL, JAKSIC FM (1978b) Presas de *Parabuteo unicinctus* en los alrededores de Santiago (Falconiformes: Accipitridae). Noticiario Mensual. Museo Nacional de Historia Natural 264: 8-9

ZOLOTH SR (1969) Observations of the populations of brush rabbits on Ano Nuevo Island, California. Wassman Journal of Biology 27: 149-161

Subject Index

Ecological Studies

Analysis and Synthesis

Editors: W.D. Billings, F. Golley, O.L. Lange, J.S. Olson, H. Remmert

Volume 36
C.B. Osmond, O. Björkman, D.J. Anderson

Physiological Processes in Plant Ecology

Toward a Synthesis with *Atriplex*

1980. 194 figures, 76 tables. XI, 468 pages
ISBN 3-540-10060-1

Contents: Physiological Processes in Plant Ecology: the Structure for a Synthesis. - Systematic and Geographical State of *Atriplex*. - Genecological Differentiation. - Genetic and Evolutionary Relationships in *Atriplex*. - *Atriplex* Communities: Regional Environments and Their Ecological Analysis. - Germination and Seedling Establishment. - Absorption of Ions and Nutrients. - Water Movement and Plant Response to Water Stress. - Photosynthesis. - Productivity and Environment. - Epilog. - References. - Taxonomic Index. - Subject Index.

Volume 37
H. Jenny

The Soil Resource

Origin and Behavior

1980. 191 figures. XX, 377 pages
ISBN 3-540-90543-X

Contents: Ecosystems and Soils. - Processes of Soil Genesis: Water Regimes of Soils and Vegetation. Behavior of Ions in Soils and Plant Responses. Origin, Transformation, and Stability of Clay Particles. Biomass and Humus. Soil Colloidal Interactions and Hierarchy of Structures. Pedogenesis of Horizons and Profiles. - Soil and Ecosystem Sequences: State Factor Analysis. The Time Factor of System Genesis. State Factor Parent Material. State Factor Topography. State Factor Climate. Biotic Factor of System Genesis. Integration of Factors and Overview of Book. - Appendix I: Names of Vascular Species Cited. - Index.

Volume 38

The Ecology of a Salt Marsh

Editors: **L.R. Pomeroy, R.G. Wiegert**
1981. 57 figures. XIV, 271 pages
ISBN 3-540-90555-3

Contents: Ecosystem Structure and Function: Ecology of Salt Marshes: An Introduction. The Physical and Chemical Environment. - Salt Marsh Populations: Primary Production. Aquatic Macroconsumers. Grazers on *Spartina* and Their Predators. Aerobic Microbes and Meiofauna. Anaerobic Respiration and Fermentation. - The Salt Marsh Ecosystem: The Cycles of Nitrogen and Phosphorus. A Model View of the Marsh. The Salt-Marsh Ecosystem: A Synthesis. - References. - Index.

Springer-Verlag
Berlin
Heidelberg
New York
Tokyo

Ecological Studies

Analysis and Synthesis

Editors: W. D. Billings, F. Golley, O. L. Lange, J. S. Olson, H. Remmert

Volume 39

Resource Use by Chaparral and Matorral

A Comparison of Vegetation Function in Two Mediterranean Type Ecosystems

Editor: **P.C. Miller**
1981. 118 figures. XVIII, 455 pages.
ISBN 3-540-90556-1

Contents: Conceptual Basis and Organization of Research. - Resource Availability and Environmental Characteristics of Mediterranean Type Ecosystems. - The Plant Communities and Their Environments. - Biomass, Phenology, and Growth. - Microclimate and Energy Exchange. - Water Utilization. - Energy and Carbon Acquisition. - Carbon Allocation and Utilization. - Mineral Nutrient and Nonstructural Carbon Utilization. - Nutrient Cycling in Mediterranean Type Ecosystems. Models of Plant and Soil Processes. - Similarities and Limitations of Resource Utilization in Mediterranean Type Ecosystems. - Literature Cited. - Resumen. - Index.

Volume 40
B. Nievergelt

Ibexes in an African Environment

Ecology and Social System of the Walia Ibex in the Simen Mountain, Ethiopia

1981. 40 figures. IX, 189 pages. ISBN 3-540-10592-1

Contents: Introduction. - A Reader's Guide. - The Simen, a Unique Afroalpine Community. - Methods, Techniques and Population Estimates. - The Niche and Habitat of the Walia Ibex, the Klippspringer and the Gelada Baboon. - The Social System of the Walia Ibex. - Conservational Outlook. - References. - Subject Index.

Volume 41

Forest Island Dynamics in Man-Dominated Landscapes

Editor: **R. L. Burgess, D.M. Sharpe**
With contributions by numerous experts
1981. 61 figures. XVII, 310 pages.
ISBN 3-540-90584-7

Contents: Introduction. - The Minimum Critical Size of Ecosystems. - Woodlots as Biogeographic Islands in Southeastern Wisconsin. - The Groundlayer Vegetation of Forest Islands in an Urban-Suburban Matrix. - Mammals in Forest Islands in Southeastern Wisconsin. - The Importance of Edge in the Structure and Dynamcis of Forest Islands. - Biogeography of Forest Plants in the Prairie-Forest Ecotone of Western Minnesota. - Effects of Forest Fragmentation on Avifauna of the Eastern Deciduous Forest. Modeling Recolonization by Neotropical Migrants in Habitats with Changing Patch Structure, with Notes on the Age Structure of Populations. - Modeling Seed Dispersal and Forest Island Dynamcis. - Optimization of Forest Island Spatial Patterns: Methodology for Analysis of Landscape Pattern. - Artifical Succession - A Feeding Strategy for the Megazoo. - Summary and Conclusions. - References. - Index.

Volume 42

Ecology of Tropical Savannas

Editors: **B.J. Huntley, B.H. Walker**

1982. 262 figures. XI, 669 pages
ISBN 3-540-11885-3

Contents: Introduction. - Structure. - Determinants. - Function. - Nylsvley, a South African Savanna. - Dynamics and Management. - Conclusion: Characteristic Features of Tropical Savannas. - Subject Index.

Springer-Verlag
Berlin
Heidelberg
New York
Tokyo

Ecological Studies 43
Mediterranean-Type Ecosystems

ERRATUM

The following is the beginning of the first paragraph of page 467, which was mistakenly omitted in the text:

These factors affect the competitive relations among species in two ways. First, fire-to-fire variation induces fluctuational change in the vegetation like that reported by Gulmon (1979) for annual grasses though on a longer time scale, but also serving to preclude competitive exclusion of any member of the community. Second, the patterning of fire intensities of. . . .

Springer-Verlag
Berlin Heidelberg New York Tokyo 1983

MIX
Papier aus verantwortungsvollen Quellen
Paper from responsible sources
FSC® C105338

If you have any concerns about our products,
you can contact us on
ProductSafety@springernature.com

In case Publisher is established outside the EU,
the EU authorized representative is:
Springer Nature Customer Service Center GmbH
Europaplatz 3, 69115 Heidelberg, Germany

Printed by Libri Plureos GmbH
in Hamburg, Germany